Water for the Environment

Water for the Environment
From Policy and Science to Implementation and Management

Edited by

Avril C. Horne

J. Angus Webb

Michael J. Stewardson

Brian Richter

Mike Acreman

ACADEMIC PRESS

An imprint of Elsevier

Academic Press is an imprint of Elsevier
125 London Wall, London EC2Y 5AS, United Kingdom
525 B Street, Suite 1800, San Diego, CA 92101-4495, United States
50 Hampshire Street, 5th Floor, Cambridge, MA 02139, United States
The Boulevard, Langford Lane, Kidlington, Oxford OX5 1GB, United Kingdom

Notices

Knowledge and best practice in this field are constantly changing. As new research and experience broaden our understanding, changes in research methods, professional practices, or medical treatment may become necessary.

Practitioners and researchers must always rely on their own experience and knowledge in evaluating and using any information, methods, compounds, or experiments described herein. In using such information or methods they should be mindful of their own safety and the safety of others, including parties for whom they have a professional responsibility.

To the fullest extent of the law, neither the Publisher nor the authors, contributors, or editors, assume any liability for any injury and/or damage to persons or property as a matter of products liability, negligence or otherwise, or from any use or operation of any methods, products, instructions, or ideas contained in the material herein.

British Library Cataloguing-in-Publication Data
A catalogue record for this book is available from the British Library

Library of Congress Cataloging-in-Publication Data
A catalog record for this book is available from the Library of Congress

ISBN: 978-0-12-803907-6

For Information on all Academic Press publications
visit our website at https://www.elsevier.com/books-and-journals

Working together
to grow libraries in
developing countries

www.elsevier.com • www.bookaid.org

Publisher: Candice Janco
Acquisition Editor: Louisa Hutchins
Editorial Project Manager: Hilary Carr
Production Project Manager: Punithavathy Govindaradjane
Cover Designer: Greg Harris and Emma Jennings
Cover Art: Emma Jennings

Typeset by MPS Limited, Chennai, India

Contents

SECTION I INTRODUCTION

SECTION II HISTORY AND CONTEXT OF ENVIRONMENTAL WATER MANAGEMENT

SECTION III VISION AND OBJECTIVES FOR THE RIVER SYSTEM

SECTION IV HOW MUCH WATER IS NEEDED: TOOLS FOR ENVIRONMENTAL FLOWS ASSESSMENT

SECTION V ENVIRONMENTAL WATER WITHIN WATER RESOURCE PLANNING

SECTION VII REMAINING CHALLENGES AND WAY FORWARD

About the Editors

Avril C. Horne is an environmental water policy specialist, with 15 years' experience across a range of interdisciplinary projects, and worked extensively across consulting, government, and academia. This unique interdisciplinary perspective means that her work in all three areas transcends traditional academic, government, and industry silos. She was the project manager for the Australian national assessment of stressed river catchments and aquifers. As an assistant director in the water group at the Australian Competition and Consumer Commission, she was heavily involved in the development of the water trading rules for the Murray–Darling Basin Plan. She returned to academia in 2014 and is currently a Research Fellow working in the Environmental Hydrology and Water Resources Group at the University of Melbourne. Her research focuses on developing tools and systems to assist in the active management of environmental water, ensuring the greatest value to consumers and the environment from this scarce resource. She was part of the organizing committee for the 11th International Symposium on Ecohydraulics, held in Melbourne, 2016, and has also previously sat on the Project Advisory Group for the development of assurance standards for water accounts in Australia. *Water for the Environment* represents the fulfilment of an idea first hatched in 2012—that there was no single book that covered the full diversity of environmental water policy, science, and management.

J. Angus Webb is a Senior Lecturer in Environmental Hydrology and Water Resources at the University of Melbourne, Australia. He originally trained as a marine ecologist before moving into the study and restoration of large-scale environmental problems in freshwater systems. Much of his research centers on improving the use of the existing knowledge and data for such problems. To this end he has developed innovative approaches to synthesizing information from the literature, eliciting knowledge from experts, and analyzing large-scale data sets. His teaching at the University of Melbourne is focused on monitoring and evaluation in aquatic systems. Angus is heavily involved in the monitoring and evaluation of ecological outcomes from environmental water delivered under the Australian government's Murray–Darling Basin Plan, leading the program for the Goulburn River, Victoria, and advising on data analysis at the basin scale. He has authored over 100 publications in the international literature, including 58 journal papers. In addition to this book, he is currently editing two journal special issues on different aspects of environmental water science and management, and is an associate editor for the journal *Environmental Management*. He was awarded the 2013 prize for Building Knowledge in Waterway Management by the River Basin Management Society in Australia, and the 2012 Australian Society for Limnology Early Career Achievement Award.

Michael J. Stewardson's research, over the last 24 years, has focused on interactions between hydrology, geomorphology, and ecology in rivers (http://www.findanexpert.unimelb.edu.au/display/person14829). This has included physical habitat modeling, flow-ecology science, and innovation in environmental water practice. He has participated in Australia's water reforms through advisory roles at all levels of government. More recently, his research has focused on the physical, chemical, and biological processes in streambed sediments and their close interactions in regulating stream ecosystem services. He leads the Environmental Hydrology and Water Resources Group in Infrastructure Engineering at The University of Melbourne (http://www.ie.unimelb.edu.au/research/water/).

Brian Richter has been a global leader in water science and conservation for more than 30 years. After leading The Nature Conservancy's global water program for two decades, he now serves as President of Sustainable Waters, a global water education organization. In this role, Brian also promotes sustainable water use and management with governments and local communities, and serves as a water advisor to some of the world's largest corporations, investment banks, and the United Nations, and has testified before the US Congress on multiple occasions. Brian has consulted on more than 150 water projects worldwide. He also teaches a course on Water Sustainability at the University of Virginia.

Brian has developed numerous scientific tools and methods to support river protection and restoration efforts, including the Indicators of Hydrologic Alteration software that is being used by water managers and scientists worldwide. Brian was featured in a BBC documentary with David Attenborough on "How Many People Can Live on Planet Earth?" He has published many scientific papers on the importance of ecologically sustainable water management in international science journals, and co-authored a book with Sandra Postel entitled *Rivers for Life: Managing Water for People and Nature* (Island Press, 2003). His new book, *Chasing Water: A Guide for Moving from Scarcity to Sustainability*, has now been published in six languages.

Mike Acreman is Science Area Lead on Natural Capital at the Centre for Ecology and Hydrology, Wallingford, UK and visiting Professor of Eco-hydrology at University College London. He has over 30 years' research experience at the interface of hydrology and freshwater ecology. His PhD was on flood risk estimation at the University of St Andrews. He worked for the Institute of Hydrology as a flood modeler. In the 1990s, he was freshwater management advisor to the IUCN-The World Conservation Union. He has worked in numerous countries worldwide for DFID, The World Bank, European Commission, IUCN, Ramsar Convention, Biodiversity Convention, and national governments. At CEH he leads a team of 40 scientists studying catchment processes, river ecology, and wetland hydrology. His interests include hydro-ecological processes in wetlands and definition of ecological flow requirements of rivers, particularly at extremes of floods and droughts. He led the World Bank program on environmental flows and was hydrological advisor to DFID with major input to the World Commission on Dams. He is a member of the WWF-UK board, the Ramsar Convention science panel, and the Natural England Science Advisory Board. He is a co-editor of *Hydrological Sciences Journal*, has edited Special Issues on Ecosystem Services of Wetlands and Environmental Flows, and has published over 170 scientific papers.

About the Contributors

Catherine Allan is an associate professor of Environmental Sociology and Planning at the Institute of Land, Water and Society at Charles Sturt University, Australia. Her research focuses on adaptive management, in particular the social learning, evaluation, and institutional aspects of natural resource management. Adaptive management and social learning cut across many disciplinary areas, including water governance, which has been one of Catherine's focal areas for many years, with particular interest in the Murray–Darling Basin.

Meenakshi Arora is a senior lecturer in the Department of Infrastructure Engineering at the University of Melbourne. The focus of her research is on understanding the influence of anthropogenic activities on urban stream health, contaminant fate and transport in subsurface and integrated water cycle management.

Angela H. Arthington is an adjunct emeritus professor in the Australian Rivers Institute at Griffith University, Brisbane. Her research on the science of environmental water has focused on coastal and arid zone floodplain rivers, producing the book, *Freshwater Fishes of North-Eastern Australia* (2004, CSIRO Publishing), a special issue of "Freshwater Biology on Environmental Flows: Science and Management" (2010), and a book, *Environmental Flows: Saving Rivers in the Third Millennium* (2012, University of California Press). She has an international reputation for pioneering work on ecosystem level environmental flows assessment methods (e.g., DRIFT, ELOHA) and has offered advice to many agencies and programs in Australia, the United States, South Africa, and New Zealand.

Beth Ashworth has over 10 years' experience in Victoria's water industry. She has been the executive officer of the Victorian Environmental Water Holder since its creation in 2011. Prior to this, she gained significant water resource management policy experience at the then Department of Environment and Primary Industries (and previously Department of Sustainability and Environment). She has particular experience in water resource allocation and reallocation issues, including through the development of sustainable water strategies. Beth has an honors degree in Environmental Science from Monash University, Melbourne, Australia. She has been significantly involved in the development of policy frameworks and practices for the active management of environmental water rights in Australia.

Lindsay C. Beevers is a civil engineer and has a PhD in morphological modeling of fluvial impoundments. Since then she has been involved in a wide range of projects using numerical modeling to solve environmental problems in fluvial settings, specifically relating to hydrological extremes (floods and droughts) and their associated impacts on the hydromorphological environment and freshwater ecosystems. Her research investigates these relationships and explores the interactions between flow, river form, society vulnerability, and resilience as well as ecosystem services and the issue of climate-related uncertainty.

Nick R. Bond is Professor of Freshwater Ecology at La Trobe University and Director of the Murray–Darling Freshwater Research. He has more than 20 years of experience working on applied research questions related to river hydro-ecology. His primary interests are focused on understanding the effects of hydro-climatic variability on river ecosystems and combined empirical field-based sampling with a wide range of quantitative modeling approaches. He has worked extensively on assessing the ecological impacts of flow regulation, and he has been involved in both developing and applying holistic methodologies for determining environmental flow requirements for a number of rivers in Australia, including work to support the development of the Murray–Darling Basin Plan. From 2007 to 2012 he also worked extensively on river health and environmental flow projects in China as part of the Australia–China Environment Development Program. He has authored or co-authored more than 50 peer-reviewed journal papers and numerous technical reports, and has developed several software tools for the analysis of hydrologic time-series data.

Roser Casas-Mulet is a postdoctoral researcher in the Department of Infrastructure Engineering at the University of Melbourne. Her research experience lies in the analysis of environmental changes occurring in rivers, with a focus on physical and ecological interactions in regulated systems, where more often difficult decisions on how best to balance water demands and environmental protection need to be made.

John C. Conallin is a visiting scientific researcher at IHE Delft, Delft, The Netherlands, working specifically on various water management projects internationally. His main interests include water, food and energy security, social-ecological systems thinking, adaptive management of environmental water, and sustainable hydropower.

Justin F. Costelloe is a senior researcher in the Department of Infrastructure Engineering at the University of Melbourne. He has 12 years of experience in investigating surface water–groundwater connectivity and salinity across a range of environments from arid to humid climates, and has published a number of journal articles in this area.

Katherine A. Daniell is a senior lecturer in the Fenner School of Environment and Society at the Australian National University. Katherine's current research work focuses on the challenges of implementing collaborative approaches to policy and action for sustainable development. In this field, she has recently worked in Europe and the Asia-Pacific on projects related to water governance, risk management, sustainable urban development, climate change adaptation, and international science and technology cooperation. Katherine's recent works include *Co-engineering and Participatory Water Management: Organisational Challenges for Water Governance* (Cambridge University Press 2012, monograph), *River Basin Management in the 21st Century: Understanding People and Place* (CRC Press, 2014, co-editor), *Understanding and Managing Urban Water in Transition* (Springer, 2015, co-editor), and over 80 other academic publications. Katherine has received many awards and honors for her work including a John Monash Scholarship and being elected as a Fellow of the Peter Cullen Water and Environment Trust. Katherine currently serves as a member of the National Committee on Water Engineering (Engineers Australia), a member of the Initiatives of the Future of Great Rivers, and is the editor of the *Australasian Journal of Water Resources*.

Chris Dickens is the head of office and principal researcher for the International Water Management Institute, Southern Africa. He is an aquatic scientist with an international profile and 27 years of experience working with aquatic ecosystems and the management of water resources. His main interests are aquatic ecosystems health, environmental water requirements, water resources management, and governance.

Benjamin B. Docker has worked with the Australian Government Department of the Environment and Energy since 2007, helping to establish and implement the Commonwealth Environmental Water Holder function, a statutory position responsible for managing the environmental water entitlements owned by the federal government in the Murray–Darling Basin. He has provided policy advice on governance, risk management, planning, and decision making for environmental flows and overseen their implementation and monitoring and evaluation of outcomes across all catchments of the Murray–Darling Basin. He was part of the interjurisdictional working group negotiating key aspects of the Basin Plan with state governments and the Murray–Darling Basin Authority. Ben is currently working on climate change mitigation in agriculture within the Department of the Environment and Energy and has previously worked for the United Nations in Italy, the University of Sydney, and as an Endeavour Executive Fellow and consultant with the Mekong River Commission in Lao PDR. He has a PhD in geomorphology from the University of Sydney and an MSc in environmental economics from the University of London.

Jane M. Doolan is currently a professorial fellow in Natural Resource Governance at the University of Canberra, Australia. She has extensive experience in sustainable water resource management, providing policy advice at senior levels to governments on issues such as urban and rural water supply and security, national water reform, water allocation, river and catchment management, and water sector governance. She has led key policy initiatives in river health policy, environmental water allocation, institutional arrangements for catchment management, and the management of water during drought and climate change. She has been significantly involved in the development of policy frameworks and practices for the active management of environmental water rights in Australia.

Brian L. Finlayson is a physical geographer with specialist expertise in geomorphology and environmental hydrology. He is an honorary principal fellow in the School of Geography at the University of Melbourne and guest professor at East China Normal University, Shanghai. He has over 40 years' experience as a university academic in undergraduate and graduate teaching programs, graduate research supervision, research, and consulting. His current research is focused on the downstream impacts of the Three Gorges Dam on the Yangtze River, China.

Tim D. Fletcher has written over 300 peer-reviewed publications on urban stormwater hydrology, quality, effects, and management. He is Professor of Urban Ecohydrology at the University of Melbourne (Waterway Ecosystem Research Group) and was formerly an Australian Research Council Future Fellow. With colleague Chris Walsh, he leads the Little Stringybark Creek project, a world-first attempt at catchment-scale retrofit of stormwater control measures, with the aim of simultaneously restoring the ecological health of the creek and delivering an entirely new urban water supply. He is co-chair of the prestigious Novatech Urban Water Management Conference, held every 3 years in France.

Keirnan J.A. Fowler is an Australian hydrologist with experience across industry and academia. He previously worked as a consulting hydrologist specializing in modeling catchment hydrology, including farm dams as discussed in Chapter 20. Currently, Keirnan is a PhD candidate at the University of Melbourne investigating methods for predicting the amount of runoff that occurs during long-term droughts and/or climate change.

Dustin E. Garrick is currently Lecturer in Environmental Management at the Smith School of Enterprise and the Environment at the University of Oxford, United Kingdom. Prior to this, Dustin was the assistant professor and Philomathia Chair of Water Policy at McMaster University in Ontario, Canada. His comparative water policy research examines water allocation institutions and climate change adaptation in transboundary rivers of western North America and southeast Australia, where he completed a Fulbright Fellowship in 2011. He currently serves on the Organisation for Economic Cooperation and Development (OECD)/Global Water Partnership task force on water security and sustainable growth, and as Colorado River Basin Chair of the Food, Energy, Environment, and Water (FE2W) network.

David J. Gilvear undertook a PhD on sediment transport in regulated rivers and continues to be interested in the fluvial impacts of flow regulation. Over the next three decades he has also been involved in a range of research projects at the interface of hydrology, geomorphology, and ecology, and has embraced the interdisciplinary concept of river science—he is president of the International Society of River Science. Most of his research has been applied in nature and directed at environmentally sound river restoration and management, and in recent years this has included exploring the utility of the ecosystem service concept and river resilience to 21st-century river management.

Barry T. Hart is the director of the environmental consulting company—Water Science Pty Ltd. He is also emeritus professor at Monash University, where previously he was director of the Water Studies Centre. Prof. Hart has established an international reputation in the fields of ecological risk assessment, environmental flow decision making, water quality and catchment management, and environmental chemistry. He is well known for his sustained efforts in developing knowledge-based decision-making processes in natural resource management in Australia and Southeast Asia. Prof. Hart is currently a board member of the Murray–Darling Basin Authority and a nonexecutive director of Alluvium Consulting Australia Pty Ltd. He is also deputy chair of the Scientific Inquiry into Hydraulic Fracturing of Onshore Unconventional Reservoirs in the Northern Territory, which commenced in December 2016.

Christoph Hauer is a civil engineer with an educational background in landscape ecology and management. His PhD thesis was dealing with morphodynamic aspects in habitat modeling and environmental water assessments. He is a senior scientist at BOKU University teaching ecohydraulics and natural-orientated hydraulic engineering. His main interests are numerical habitat modeling, physical laboratory studies, and management issues in terms of hydropower use or sediment dynamics.

Declan Hearne works for the International Water Centre, Brisbane, Australia, as a program manager with a strong community development focus through the implementation of Natural Resources Management (NRM) and Integrated Water Resource Management (IWRM) programs.

Sue Jackson is a geographer with over 20 years' experience in researching the social dimensions of natural resource management in Australia, particularly community-based conservation initiatives and institutions. She has research interests in systems of resource governance, including customary indigenous resource rights, and indigenous capacity building for improved participation in natural resource management and planning, as well as the social and cultural values associated with water. She has conducted research with communities throughout many regions of Australia. Her recent research has improved the capacity of indigenous communities to advance their claims for water rights.

She holds a Future Fellowship awarded by the Australian Research Council to undertake research on indigenous water management and rights in the Murray–Darling Basin.

Sue is co-convenor of a new working group of the Sustainable Water Future Program, an initiative of FutureEarth. With the title *Rivers, flows and people: Connecting ecosystems with human communities, cultures, and livelihoods*, this group aims to address the pressing need of ensuring that all the world's rivers have in place environmental flows that sustain ecosystems and human livelihoods and well-being. It will synthesize experience and knowledge, share lessons, and facilitate project-based research to advance transdisciplinary approaches in this new field of socio-ecological science.

Sue has published over 50 journal articles in international geography, hydrology, ecology, law, and planning journals and 30 book chapters.

Sharad K. Jain has research, development, and teaching experience of nearly 30 years in the field of water resources. He is a scientist at the National Institute of Hydrology, Roorkee, India. He was a postdoctoral fellow at the National Research Institute for Earth Science and Disaster Prevention (Japan) and served as Visiting Professor at the Louisiana State University, United States, for 1 year. During April 2009 to April 2012, he was NEEPCO Chair Professor at IIT Roorkee. His research interests include surface water hydrology, water resources planning and management, hydropower, and impact of climate change. He has also been involved in many research and consultancy projects.

Hilary L. Johnson is a director within the Commonwealth Environmental Water Office, a division of the Australian Government Department of the Environment and Energy. The Office supports the Commonwealth Environmental Water Holder, a statutory position responsible for managing the environmental water entitlements owned by the Australian government in the Murray–Darling Basin. Hilary has worked for over 8 years in environmental water management and policy development for Australian Government agencies (including at the Commonwealth Environmental Water Office and the Murray–Darling Basin Authority). In this time, he has participated in and led the development of environmental water governance frameworks, legislative reforms, decision-support tools, and communication and engagement activities. His current role sees him responsible for advising and implementing decisions on the use of commonwealth environmental water in the southern Murray–Darling Basin. Hilary has a Bachelor of Environmental Science (Hons.) from the University of Wollongong.

Giri R. Kattel is currently a CAS-PIFI professorial fellow at the Chinese Academy of Sciences (Nanjing). He is an honorary fellow at the University of Melbourne, and has more than 15 years' experience in ecology, paleoecology, and water resource management in lakes and rivers of both hemispheres.

Christopher Konrad is a research hydrologist with the US Geological Survey in Tacoma, Washington. His research focuses on hydrologic analyses of stream, river, and floodplain ecosystems, spanning a range of topics including hydraulics and sediment transport, groundwater and surface water interactions, fluvial geomorphology, streamflow regimes, and hydrologic effects of land use. Dr. Konrad served as the National River Science Coordinator for the Nature Conservancy and USGS from 2007 to 2011 working on the development and evaluation of ecological flow requirements for rivers. He earned a BS degree in Biological Sciences from Stanford University and MS and PhD degrees in Civil and Environmental Engineering from the University of Washington.

Nicolas Lamouroux is a civil engineer who jumped into stream ecology during his PhD at the University in Lyon, France. He leads a team of 20 people at Irstea targeting the development, testing, and transfer of ecohydrological models. He is particularly interested in the development of statistical models of the hydraulic-habitat of organisms, and field tests of their ecological predictions.

Gaisheng Liu is an associate scientist at Kansas Geological Survey, University of Kansas. He has 15 years of research experience on issues related to groundwater resources development and management with a particular emphasis on different aquifer systems across Kansas.

Lisa Lowe is a senior project officer within the Hydrological Risks and Planning Team of the Department of Environment, Land, Water, and Planning in Victoria, Australia. In her current role, she draws upon current scientific knowledge to provide guidance to water corporations on the impact that climate change and changes in land use will have on water supplies. Previously, she spent 10 years working as a water resource engineer in consulting. During this time, she was involved in numerous environmental flow projects, including making scientifically based environmental water recommendations, developing sustainable diversion limits for rivers across Victoria, and creating indicators of environmental flow stress. She holds a PhD in Hydrology that focused on understanding the uncertainty in the information commonly used by water resource managers.

Matthew P. McCartney is a principal researcher specializing in water resources, and wetland and hydro-ecological studies. He is currently theme leader of the International Water Management's (IWMI) theme on ecosystem services. His experience stems from participation in a wide range of research and applied projects, conducted in both Africa and Asia.

Michael E. McClain is Chair Professor in Ecohydrology and head of the Hydrology and Water Resource Chair Group at IHE Delft, Delft, The Netherlands. He has over 25 years of experience in catchment hydrology and water quality, environmental flows, and land–water interactions.

Alexander H. McCluskey is a researcher in the Department of Infrastructure Engineering at the University of Melbourne and commenced a Technical University of Munich Foundation Fellowship in November 2016. Alexander's recent PhD examined steady and unsteady mechanisms driving hyporheic exchange, and his research interests span hydrological and geomorphic processes in ecohydraulics.

Carlo R. Morris is a student and tutor at the University of Melbourne, Australia. He recently submitted his PhD thesis titled "Management of the environment using a combination of policy instruments: A case study of farm dams in Victoria."

Erin L. O'Donnell is an environmental water policy and governance specialist. She has worked in environmental management and water governance since 2003, in the private and public sectors. Most recently, Erin was responsible for developing the Victorian Environmental Water Holder, as well as policy on environmental water trading and water market rules enabling environmental water managers to access the water market. Erin is a senior fellow at the University of Melbourne Law School, Australia, where she is currently completing her PhD. She is investigating the opportunities created by using a corporation to hold and manage environmental water, and the governance challenges of delivering both independence and accountability in managing such a precious public resource. Erin has a BSc (first-class honors in ecology) from Adelaide University, and an LLB (first-class honors) from Deakin University.

Jay O'Keeffe is a river ecologist. He gained a PhD in Applied Biology from Imperial College of Science and Technology, University of London, in 1983. Since the mid-1980s, based at Rhodes University in South Africa, and UNESCO-IHE in the Netherlands, his research and educational focus has been on environmental flows and their assessment (EFA). He has facilitated more than 40 EFAs on rivers in over 20 countries, and was involved in the development of the concept of the ecological reserve, included in the South African Water Act of 1998. In 2004, he was awarded the Gold Medal of the Southern African Society of Aquatic Scientists.

Julian D. Olden is the H. Mason Keeler Endowed Professor in the School of Aquatic and Fishery Sciences, University of Washington, and Adjunct Research Fellow at the Australian Rivers Institute, Griffith University. His research focuses on the ecology and conservation of freshwater ecosystems.

Stuart Orr is currently the water practice lead at World Wildlife Fund (WWF) and has a background in business and academic research. He has worked on a broad spectrum of water issues helping to support WWF in devising and testing new approaches to conservation through engaging business and finance, the water–food–energy nexus, economic incentives, and water-related risk. He has published widely on these and other topics while also acting on various advisory panels and boards ranging from the World Economic Forum's Water Security Council to the IFC Infrastructure & Natural Resources (INR) Advisory Steering Committee. He oversees a growing and talented team within WWF dedicated to proving that there are better ways to develop and grow economies, protect the environment, and work with business. He holds an MSc in environment and development from the School of International Development at the University of East Anglia. Recent projects include linking water ecosystems to economic and finance concerns of water users, supporting WWF offices in Kenya, Suriname, Bhutan/Nepal, Zambia, Mekong, and Turkey.

Ian Overton is an expert advisor in integrated water resource management and environmental systems. He has worked extensively within the Murray–Darling Basin in Australia but also in the United Kingdom, France, Germany, United States, Nepal, China, and Chile. He has a Bachelor of Science Degree with Honours in environmental science and spatial information systems, a graduate certificate in management, and a PhD in river science. Ian is also a graduate of the Australian Institute of Company Directors. After a period in academia, Ian established a spatial information company and won awards for research, development, and commercialization. He is an innovative thought leader with expertise in managing complex multidisciplinary projects in the private and public sectors. He spent 15 years at CSIRO in Australia researching solutions in environmental flows, ecohydrology, floodplain inundation modeling, and predictive modeling of vegetation health. He has developed innovative frameworks for linking natural capital to human well-being and adaptive management systems for water resource management. Ian is now the Principal at Natural Economy, a research and consulting business in Australia, specializing in systems understanding and sustainability, including environmental, social, and economic decision making and planning. He has expertise in business management and corporate governance across entrepreneurship and strategic public sector management. He has authored over 150 publications and is an associate editor of the Hydrological Sciences journal.

Murray C. Peel is a senior researcher in the Department of Infrastructure Engineering at the University of Melbourne. He has 17 years of experience investigating interannual variability of annual runoff around the world, the hydrologic impacts of climate change and land use change, and improving techniques for hydrologic prediction under changing conditions.

Tim J. Peterson is a researcher in the Department of Infrastructure Engineering at the University of Melbourne. His research focuses on understanding catchment hydrological resilience through deterministic and statistical modeling and data-driven approaches for hydrological insights such as the drivers of groundwater trends or surface–groundwater interactions.

N. LeRoy Poff is a professor in the Department of Biology at Colorado State University, and he holds a partial appointment as distinguished Professorial Chair in Riverine Science and Environmental Flows at the University of Canberra. He is an internationally recognized pioneer in the research area of hydroecology and its application to sustainable river management. His research orients on how hydrologic variability and its modification by water infrastructure and climate change regulate the interactions among species and the structure and function of aquatic and riparian ecosystems, using species-based and traits-based approaches. He is a former president of the Society of Freshwater Science (SFS), a fellow of the Ecological Society of America (ESA), and an elected fellow of the American Association for the Advancement of Science (AAAS).

David E. Rheinheimer is a postdoctoral researcher at the Department of Civil & Environmental Engineering at the University of Massachusetts Amherst, where he is developing innovative modeling tools for collaborative decision making in water system planning and management. His primary research interest is in integration of environmental water needs into water planning and management models, for which a detailed understanding of a wide range of river processes is critical. His love of rivers and interest in environmental water management began while an intern with the freshwater science team at the World Wildlife Fund, United States. Since then, his experience has focused primarily on environmental aspects of hydropower, particularly in the Sierra Nevada mountains of California while at the University of California, Davis, and the Yangtze River and the Three Gorges Reservoir while at Wuhan University, China. More recently, he investigated the geologic, hydrologic, political, social, and economic threats to and of hydropower development in the Himalayas while a Fulbright-Nehru fellow at the Indian Institute of Technology, Roorkee, India.

Robert J. Rolls is a research fellow with the Institute for Applied Ecology at the University of Canberra, Australia. He is a freshwater ecologist with a background in research underpinning the management and delivery of environmental flows in temperate, subtropical, and dryland river systems. His recent research focused on the effects of habitat fragmentation and connectivity on riverine fish at landscape scales, disturbance–biodiversity relationships across spatial scales, experimental assessment of the ecological consequences of environmental flows, and identifying ecological mechanisms underpinning effects of hydrological events and regimes on freshwater ecosystems. Dr. Rolls's current research is centered on conceptualizing and quantifying aquatic foodweb consequences of climate change and species invasions, multiscale biodiversity responses to catchment disturbances, and the role of hydrological regimes in trophic dynamics of river-floodplain ecosystems. Each of Rob's past and current research themes has been developed with clear applications to the conservation and management of freshwaters.

Claire Settre is a PhD student with the Centre for Global Food and Resources at the University of Adelaide and an Australian Endeavor Fellowship holder. Her thesis investigates how water markets can be used and adapted to provide water to the environment in ways that increase ecological benefit and reduce costs to farming communities. Claire holds a Bachelor of Civil and Environmental Engineering (honors) and a Bachelor of Arts with a major in international politics.

Wenxiu Shang is a PhD student in the School of Civil Engineering, Tsinghua University. Her research mainly focuses on water policies and river health.

Jody Swirepik is currently an executive general manager at the Clean Energy Regulator that administers climate change legislation aimed at accelerating carbon abatement. Prior to 2015, Jody spent 25 years working in roles within the water industry in different jurisdictions and levels of government. Most recently she spent 14 years at the Murray–Darling Basin Authority (and its predecessor) with the latter period dominated by Australia's largest water reform—the development of the Murray–Darling Basin Plan. She has worked extensively in water planning/security, environmental watering and river operations, hydrologic and inundation modeling, river health monitoring, water trade, and on the management of water quality, salinity, native fish, algal blooms, and droughts. She has overseen the planning of major infrastructure projects to deliver environmental water and has worked as a regulator controlling point and nonpoint source pollution. She has been significantly involved in the development of policy frameworks and practices for the active management of environmental water rights in Australia.

Joanna Szemis is a research fellow in the Infrastructure Engineering Department at the Melbourne School of Engineering, University of Melbourne, Australia. Her PhD research focused on developing a framework for the scheduling of environmental water management alternatives in the South Australian Murray River using ant colony optimization algorithms. Currently, she is involved in an Australian Research Council Linkage Project, which is developing a decision support tool using optimization to help environmental water managers make transparent and informed seasonal environmental watering decisions in the Yarra, Goulburn, and Murrumbidgee river system in Australia.

Rebecca E. Tharme is the director of Riverfutures, focused on international research for development in river basins, and an adjunct principal research fellow of the Australian Rivers Institute. She has considerable experience working with partners in developing countries across Africa, Latin America, and Asia to build capacity and provide policy-appropriate solutions to challenges in environmental water management. She has served on the Ramsar Convention on Wetlands' Scientific and Technical Review Panel and contributed to several high-profile global water initiatives.

Rebecca is a coauthor of a wide range of publications on environmental flows, agriculture—wetlands interactions, and water security.

Gregory A. Thomas has practiced natural resources law, and planning for over 35 years as an advocate, professor, and project manager. For the past 26 years, he has served as the founder and president of the Natural Heritage Institute (NHI). Founded in 1989 by a multidisciplinary group of experienced environmental professionals, NHI specializes in rehabilitating heavily engineered river systems to restore their natural functions and protect the natural functions that support water-dependent ecosystems and the services they provide to sustain and enrich human life. NHI designs and then demonstrates restoration tools and techniques in local settings around the world, usually at a river basin-wide scale, often in a transboundary context. As the CEO and project manager, Mr. Thomas has led NHI's work on hydropower systems throughout the United States, in irrigation and flood management systems in the Central Valley of California, in the binational river system that defines the US—Mexican border, in the Okavango River system in southern Africa, and throughout continental Africa for the World Bank. For one of his current projects, Mr. Thomas is partnering with the four national governments of the Lower Mekong River, with funding from USAID and the MacArthur Foundation to transform the official dam development plans throughout the entire Mekong River system to maintain the flows of water, sediments, nutrients, and migratory fish to sustain its exceptional biological productivity.

Mr. Thomas' areas of expertise include water resources management, preserving biological diversity, hydropower reoperation, energy policy, control of pollution and hazardous substances, public lands, marine resources, and international environmental law and conservation. His professional skills encompass nonprofit management, the development and implementation of complex projects, administrative trials, legislative advocacy, policy analysis, strategic planning, and negotiations and consensus processes.

Geoff J. Vietz is a fluvial geomorphologist and hydraulic modeler. He is a senior research fellow with the Waterway Ecosystem Research Group at the University of Melbourne, and the director and principal scientist of Streamology Pty Ltd. Geoff's research and consulting is focused on understanding the physical mechanisms that link hydrology to waterway ecosystem health and the application of scientifically based solutions to the management of urban and rural waterways.

Andrew T. Warner is an Oak Ridge Institute for Science and Education (ORISE) Senior Fellow at the US Army Corps of Engineers' Institute for Water Resources. He was formerly the Deputy Director of Water Infrastructure and Senior Advisor for Water Management at The Nature Conservancy, where he was the architect and national coordinator of the Sustainable Rivers Project. Andrew has 30 years of experience in environmental and conservation projects and policy relating to water, water infrastructure, and floodplain management, and has worked on rivers in the United States, Asia, Latin America, and Africa.

Robyn J. Watts is a professor of Environmental Science in the Institute of Land, Water and Society at Charles Sturt University, Australia. She teaches undergraduate and postgraduate students and undertakes research on the ecology, biodiversity, management, and restoration of aquatic ecosystems. Specific areas of expertise include flow-ecology relationships, river rehabilitation through the modification of the operation of dams, ecosystem responses to environmental flows, and adaptive management of environmental flows. She has considerable experience in leading interdisciplinary research projects on ecosystem responses to environmental flows, working with ecologists, hydrologists, geomorphologists, social scientists, water management agencies and the community. She is particularly interested in bringing together expert and local knowledge to contribute to the adaptive management of water. Robyn has supervised 18 PhD students and 14 Honours students and has written over 90 publications, including journal papers, book chapters, conference papers and technical reports.

Sarah A. Wheeler is an Australian Research Council Future Fellow and Associate Director Research with the Centre for Global Food and Resources at the University of Adelaide. She graduated with her PhD in 2007, and has over 100 peer-reviewed publications in the research areas of irrigated farming, organic farming, water markets, water scarcity, crime, and gambling. She is an associate editor of the Australian Journal of Agricultural and Resource Economics and Water Resources and Economics, and editor of Water Conservation Science and Engineering. She has been a guest editor for a special issue of Agricultural Water Management and is currently on the editorial boards of Economics, Australasian Journal of Water Resources, Water Resources and Economics, and Agricultural Science.

Sarah M. Yarnell is an associate project scientist at the Center for Watershed Sciences at the University of California, Davis. Her studies focus on integrating the traditional fields of hydrology, ecology, and geomorphology in the river environment. She is currently conducting research that applies understanding of river ecosystem processes to managed systems in the Sierra Nevada range of California, with a focus on the development and maintenance of riverine habitat. She is a recognized expert in the ecology of the Foothill yellow-legged frog (*Rana boylii*), a Californian species of special concern, and she is the first researcher to apply sediment transport and two-dimensional hydrodynamic modeling techniques to the evaluation of instream amphibian habitat. Her experience includes consultation as a technical expert for various hydroelectric power relicensing projects, where she has work closely with government resource agencies and the private sector to assess the impacts of environmental flows on aquatic biota and provide recommendations for developing flows that improve the functioning of river ecosystems. More recently, her experience has expanded to include evaluation and restoration of headwater systems, particularly montane meadows in the Sierra Nevada and Cascade ranges of California.

Acknowledgments

Thanks to the large number of chapter authors who have contributed to the book, engaged in the process, and made this an enjoyable experience.

We would like to thank a number of people who provided independent reviews of book chapters: Angela Arthington, Andrew Western, William Chen, Kate Rowntree, Ruth Beilin, Amy Mannix, Kira Russo, Peter Davies, Michael McClain, Nick Bond, Geoff Vietz, Piotr Parasiewicz, Rory Nathan, David Tickner, Stephen Hodgson, Dustin Garrick, Daniel Connell, Denis Hughes, Sharad Jain, Tony Ladson, Barry Hart, and Michael Dunbar.

Suzie Porter produced all the graphics for the book and was patient and helpful through this process. Similarly, thanks to Julie Cantrill for her copyediting and ensuring consistency between chapters.

Thanks also to the various staff at Elsevier who have worked on the book, and in particular Hilary Carr.

And a final thanks—which cannot be done justice in a single line here—to our families for their ongoing support and tolerance of projects such as these!

Dr. Avril Horne was employed through ARC LP130100174 during the development of this book.

INTRODUCTION

THE ENVIRONMENTAL WATER MANAGEMENT CYCLE

Avril C. Horne[1], Erin L. O'Donnell[1], J. Angus Webb[1], Michael J. Stewardson[1], Mike Acreman[2], and Brian Richter[3]

[1]*The University of Melbourne, Parkville, VIC, Australia* [2]*Centre for Ecology and Hydrology, Wallingford, United Kingdom* [3]*Sustainable Waters, Crozet, VA, United States*

1.1 WHAT IS THIS BOOK ABOUT?

"The tragedy of the commons" (Hardin, 1968) describes how a common-access sheep-grazing pasture will degrade as each individual herder seeks to improve their own outcome by increasing their own sheep numbers (an individual gain), while the common pasture becomes overgrazed (a collective cost). As Hardin eloquently states:

> *Therein is the tragedy. Each man is locked into a system that compels him to increase his herd without limit—in a world that is limited. Ruin is the destination toward which all men rush, each pursuing his own best interest in a society that believes in the freedom of the commons.*

We see many parallels in attitudes to freshwater resources. Their management is a complex and enduring policy problem, involving multiple participants at the individual and jurisdictional level, each with their own interests and purposes for water use. We can conceptualize this by considering freshwater resource management as a *commons* resource challenge.

Humans have modified most aspects of the terrestrial phase of the world's hydrological cycle. Consumptive water use, hydroelectric power generation, flood engineering, deforestation, agriculture, and urbanization all contribute to altered water regimes in river channels, groundwater aquifers, wetlands, and estuaries (Vorosmarty et al., 2010). These hydrological pressures are often accompanied by disturbances to water quality and riparian vegetation, barriers to movement of aquatic biota within rivers (and between rivers and their floodplains), and channel engineering. Aquatic ecosystems and freshwater biodiversity are in decline worldwide (Dudgeon et al., 2006), and the pressures and impacts are expected to increase with shifting climates and environmental change (Meyer et al., 1999; Poff et al., 2002).

For Hardin, the tragedy is the result of an inevitable decline that is futile to oppose, as it is driven by the mathematics of individual gains set against collective losses. However, Ostrom (1990) shows that the *tragedy* can be avoided if communities develop reasonable governance arrangements for common resources when faced with scarcity. So there is hope, but avoiding the tragedy requires effective management of limited resources. For freshwater resources, effective management requires altering our attitudes and historical patterns of water use.

Water for the Environment. DOI: http://dx.doi.org/10.1016/B978-0-12-803907-6.00001-2

This book addresses the commons resource challenge of how to maintain ecologically sustainable rivers in a world where humans rely on rivers for so much. The book focuses on *environmental water*—the growing management trend of restoring or maintaining water regimes to protect aquatic and riparian ecosystems (Box 1.1). In practice, this often requires imposing a limit on consumptive water use, but it can also include limits to hydropower generation capacity and on flood control infrastructure, and constraints on the development of floodplains. There is now wide recognition that future human use of rivers depends on maintaining healthy aquatic ecosystems, which requires a new balance between the requirements for human water and river use, and conservation of freshwater ecosystems (Richter, 2014).

This book provides a synthesis of environmental water practice drawn from many parallel efforts to develop context- and discipline-specific knowledge, tools, governance, and policy. We highlight common elements and principles of environmental water practice without prescribing a one-size-fits-all approach. Local context, and particularly the existing water resource policy, is critical in the design and implementation of environmental water management practices. Where environmental water rights are already legally protected around the world, the technical, legal, and institutional approaches vary widely, and we expect this diversity of solutions to continue. We hope this book can strengthen the capacity of local practitioners and communities to shape and implement an effective environmental water program, building on the best practices around the world.

BOX 1.1 KEY DEFINITIONS—WHAT IS ENVIRONMENTAL WATER?

The term *environmental water* and related phrases, such as instream flow, are familiar to most water scientists and managers. Broadly, we use these terms to define the provision of water for environmental outcomes, and to maintain or improve the ecological condition of rivers, wetlands, and estuaries. However, water policy is heavily path-dependent and context-specific, so most jurisdictions and disciplines have developed their own definitions. As a result, the meaning and precise use of these terms can vary among different users and settings. For example, consider these two different definitions of environmental flows:

> *Environmental flows are the quantity and timing of water flows required to maintain the components, functions, processes and resilience of aquatic ecosystems and the goods and services they provide to people.*
>
> **(TNC, 2016)**

> *The water regime provided within a river, wetland or coastal zone to maintain ecosystems and their benefits where there are competing water uses and where flows are regulated.*
>
> **(Dyson et al., 2003)**

The first definition takes a scientific view and refers to the water *required*. The second takes a water resource management view as to what was *actually provided*—a potentially major difference depending on the processes for implementation in a given system. Clarity of discussion throughout the book, and particularly in Sections IV and V, relies on clearly defining terms. For transparency and consistency in this book, we have provided an extensive glossary to which all chapters conform. At the outset, however, we provide the definitions below for especially important terms. We appreciate that these definitions (plus those in the glossary) may conflict with some local terminologies and do not expect universal agreement on them, but they provide a consistent language for use throughout this book.

Environmental flows assessment is the process used to determine the environmental water requirement (see below) for targeted ecological endpoints (Tharme, 2003). This may be based on a combination of hydrological, hydraulic, ecological, and social knowledge; possibly in combination with the use of expert knowledge and opinion.

By convention, we retain the use of *flow* although we define this term to include water volume requirements for wetlands and other ponded water bodies.

(Continued)

BOX 1.1 (CONTINUED)

Environmental water requirement is the water regime required to sustain a targeted ecological endpoint within freshwater and estuarine ecosystems (based on an environmental flows assessment).

Environmental water regime is the quantity, timing, and quality of water required to sustain freshwater and estuarine ecosystems and the human livelihoods and well-being that depend on these ecosystems (Brisbane Declaration, 2007). The key aspect is that the word *regime* captures the time-varying nature of flow. The regime needed to attain multiple ecological endpoints may vary from day-to-day, seasonally, and interannually, presenting challenges in both expressing the conclusion of an environmental flows assessment, as well as in providing the intended variability through water management. This builds on environmental water requirements to include consideration of combined ecosystem objectives for the river. Broadly, we are favoring the term *environmental water* (a water volume) instead of environmental flow (a discharge) as this concept is applicable across both ponded bodies such as wetlands, and flowing water bodies such as rivers and estuaries.

Environmental allocation mechanisms are the policy mechanisms available to provide environmental water. There are two general approaches: (1) those that impose regulations on the behaviors of other users (i.e., caps, conditions on storage operators, or conditions on license holders) and (2) those that provide the environment a direct right to water (environmental reserve or environmental water rights).

Environmental water encompasses all water legally available to the environment through the array of possible allocation and legislative mechanisms. Each year, the precise volume of environmental water actually allocated or remaining under these legal mechanisms may vary depending on overall water availability, demands, and priorities.

Environmental water release is a release from storage made specifically for the purposes of meeting a downstream environmental objective. The environmental water regime can be delivered through a combination of environmental flow releases and *exogenous flows*, including unregulated inflows and releases for other water uses (i.e., agriculture or hydropower).

Environmental water management includes the process of determining, allocating, implementing, and managing environmental water. This management activity is located on a spectrum from passive to active water management. *Passive* management is associated with establishing long-term plans and rules that do not require further action to provide environmental water. *Active* environmental water management refers to those allocation mechanisms that require ongoing decision making concerning when and how to use environmental water to achieve the desired outcomes.

1.2 ENVIRONMENTAL WATER MANAGEMENT

Typically, the evolution of environmental water protection in any particular river basin includes several phases that are covered sequentially in this book. The first step is often advocacy to protect the river ecosystem and biodiversity, usually led by nongovernment organizations and local communities. The case for protection is built on arguments concerning the impacts of water use and river developments on riverine ecosystems (Section II). Once there is broader societal and political support for environmental water protection, the second phase is to establish the environmental objectives or vision to inform a balance between human water use and protection of the river ecosystem (Section III). The third phase is to determine the environmental water needed to achieve this vision, which typically includes a volume of environmental water, but also a specified environmental water regime in terms of the timing, frequency, and duration of flow events (Section IV). Environmental water planning and management must then be integrated within the broader water-planning framework (Section V) to ensure that environmental water is offered sufficient protection, and that impacts on other water and river users are considered in the planning process. This balance typically involves assessing trade-offs among different water allocation options and competing uses. In some cases, the delivery of an environmental water entitlement must be actively managed to optimize its use for environmental benefits (Section VI). All phases of water management must be situated within a framework of environmental change in which visions, objectives, and water allocations need to be adaptable to meet new scenarios of resource availability and shifting social attitudes.

The book is structured around these key elements of environmental water management (Fig. 1.1). The sequence in which these elements are presented reflects the sequence in which they would normally be addressed in practice, recognizing that iteration through some or all of these stages is common over years to decades, and linear processes are seldom realized.

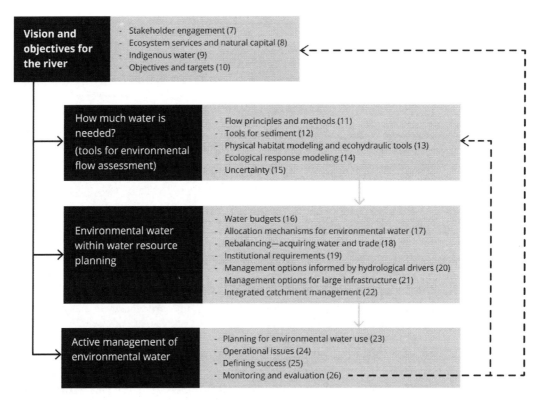

FIGURE 1.1

Overview of the key elements of environmental water management. The different boxes refer to the different sections of the book (with Sections I and II not shown), and the bracketed numbers refer to the individual chapter numbers pertaining to the specific topics. The reverse arrows imply adaptation of vision and objectives, and/or adjustment of environmental flows assessment.

1.2.1 **VISION AND OBJECTIVES FOR THE RIVER**

Human uses of rivers are extremely diverse, with a wide range of people, agencies, and communities relying directly or indirectly on rivers for their livelihoods and well-being (Fig. 1.2). Successful environmental water management is built on successful engagement with these stakeholders and balancing their priorities and competing demands (Chapter 7). The various stakeholder views and priorities for the river must be translated into a clear shared vision and set of formal objectives for environmental water management, and indeed water resource management more broadly.

In setting the vision and objectives, all stakeholders need to gain an understanding of the direct and indirect benefits provided by the river system and its associated floodplains. The direct utilitarian benefits of river systems for human consumptive uses (i.e., agriculture or power generation) are generally better understood and quantified than the indirect but essential services provided by the environment (i.e., soil fertility, water self-purification, cultural values, and aesthetics). Environmental flows assessments that evaluate ecosystem services provide a clearer perspective on these direct and indirect benefits, and enable a more balanced consideration of the benefits derived from consumptive use of water versus water allocated to the environment (Chapter 8).

One important class of benefits that has been insufficiently considered in past environmental water planning is the role that healthy rivers play in the cultural heritage of indigenous communities. Indigenous communities have much to offer in sharing their understanding of how our rivers work and in their management, but this knowledge has often not been recognized. Clearly, there is the potential for environmental water to help protect and restore culturally important river habitats and species. However, there are also occasions that require distinction between environmental water and indigenous water rights (Chapter 9).

Once these benefits are understood and articulated, objectives for environmental water management should be specific, measurable, and time limited, and expressed at an appropriate level of precision (Chapter 10). There are considerable challenges in establishing objectives that address the spatial and temporal complexities of environmental responses. Objectives must also lend themselves to monitoring and adaptive management (Chapter 25); it must be possible to use monitoring to assess whether an objective has been met, and use adaptive management to improve performance against the objective (Fig. 1.1).

1.2.2 **HOW MUCH WATER IS NEEDED: TOOLS FOR ENVIRONMENTAL FLOWS ASSESSMENT**

The underlying premise for the provision of environmental water is that there is a causal relationship between the water regime and environmental condition—that is, there is an ecological response to flow. There is a significant research effort aimed at better understanding and representing such flow−environment relationships and synthesizing this knowledge to support management decisions (Chapters 11−15). Fig. 1.3 illustrates some of the challenges associated with defining these relationships. Some systems and environmental endpoints will be highly sensitive to flow, with some hydrological alterations leading to a significant change in conditions. Other systems and ecological conditions will be more resistant, and tolerate much larger hydrological

FIGURE 1.2

The multiple users of a river system all have an interest in a sustainable river ecosystem.

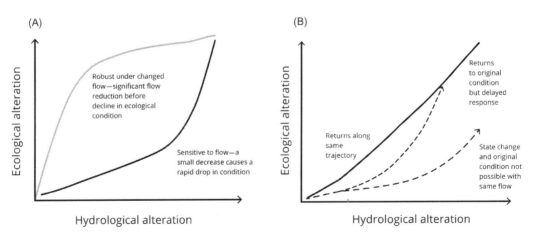

FIGURE 1.3

(A) Some systems will be robust under changed flow conditions, whereas others will be sensitive to flow reduction. (B) When reinstating a water regime, some systems will rebound quickly, others slowly, and some may never return to the same state.

alteration before environmental impacts are apparent (Fig. 1.3A). These responses may occur quickly or take years to appear. Once the environmental condition of a system is altered, returning additional flows will not necessarily return the system to its original state and the recovery may take a long time (known as hysteresis; Fig. 1.3B). A further challenge exists in taking individual environmental outcomes or endpoints, and understanding how these pieces fit together to provide an overall outcome of river or floodplain condition.

This book provides a high-level overview of environmental flows assessment approaches (Chapter 11) and tools for modeling responses to flow alteration of geomorphology and sediment (Chapter 12), physical habitat (Chapter 13), and ecology (Chapter 14). It also considers how uncertainty impacts on the environmental flows assessment process (Chapter 15). There already exists extensive literature and a number of books targeted specifically at these fields (Arthington, 2012). We do not aim to replicate or replace these texts, but rather highlight the information from the existing literature that is relevant for interdisciplinary management of environmental water.

Although the discipline of environmental water has traditionally been driven from the perspectives of ecology and hydrology, it has transformed over time into a truly multi-disciplinary field. Ecological and hydrological sciences continue to be central elements of environmental water management; however, management also needs to consider how this information links to the wider factors that influence the drivers for, and success of, environmental water provisions. Often these elements do not fit neatly together (Fig. 1.4), but all remain essential to successful environmental water programs. The evolution of environmental flows assessment methodologies reflects this transition (Chapter 11); moving from techniques originally rooted in hydrology and hydraulics, through to those that considered ecological outcomes directly, and to present-day approaches that consider the socio-ecological system holistically. Explicitly engaging with these multiple components at the outset of environmental water management programs will be beneficial for assessing the success of the overall program (Chapters 25 and 26 in Section VI).

1.2.3 ENVIRONMENTAL WATER WITHIN WATER RESOURCE PLANNING

Environmental water management is but one component of integrated water resource management (Section V). Rather than being an alternative or competing focus, managing rivers for environmental outcomes should be a responsibility that sits within holistic water resource planning (Fig. 1.5). A crucial first step in this process is understanding the water available in the catchment and those uses that depend on it (Chapter 16).

Within this context, there are a number of options for providing environmental water. The appropriate approach will depend on the existing policy framework. Five broad allocation mechanisms can be used to provide environmental water: a cap on abstractions, a defined environmental reserve, operating conditions for abstractive uses, conditions on storage operators, and environmental water rights (Chapter 17). Each of these approaches requires different levels of ongoing engagement and management. Therefore, effective environmental water management institutions and organizations are an essential component of successful implementation (Chapter 19).

FIGURE 1.4

Multidisciplinary nature of environmental water management.

Specific approaches can be used to allocate environmental water, depending on the system and regulatory context. Where a system is closed to new allocations, institutions such as water markets may allow entry of new users by redressing the imbalance in the system and provide a transition to environmental water provision (Chapter 18). In systems with large onstream dams, an alternative to allocating additional volumes of environmental water is to re-engineer the system, changing operational rules and delivery paths to better achieve conjunctive benefits across the environmental and other water users (Chapter 21). This option highlights the potential for conjunctive use of water and the limitations of regarding environmental water allocation as a distinctive and competing process to consumptive management. Indeed, even in systems where formal environmental water rights exist, targeted environmental releases often aim to enhance a flow event partially provided by exogenous flows. Although a formal legal allocation of water to the environment is of significant value for establishing the environmental water regime, this potential to achieve multiple benefits

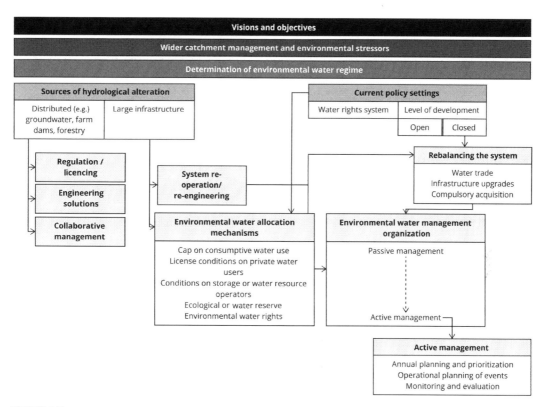

FIGURE 1.5

Environmental water management within water resource planning.

from the same *parcel* of water means that environmental water management is not necessarily a zero-sum game.

Although onstream dams are the most widely known infrastructure with direct effects on river flow regimes and floodplains, a range of other drivers of hydrological alteration are distributed through the catchment such as urban and agricultural development, groundwater abstraction, forestry, and farm dams (Chapter 3). These diffuse drivers of hydrological alteration require alternative management approaches that address the challenge of dispersed pathways and impacts, often limited data and monitoring, and varying levels of recognition within existing policy frameworks (Chapter 20). Similarly, although the water regime is a key driver of river ecosystem condition, many other impacts on river ecosystems are driven by changing land use and catchment condition leading to altered sediment supply, degraded water quality, removal of riparian vegetation, and introduction of exotic species. Integrated catchment management promotes the concept of considering the totality of catchment links between human activity and nature, and is a widely promoted approach (Chapter 22). Despite its prominence, there continues to be limited implementation of truly integrated land and water management, with ongoing fragmentation across policy and management domains of different catchment elements, and different degrading and restorative processes. In addition, ecosystems are naturally dynamic and may include natural changes to species distribution and community interactions, food web structures, and flows of energy and nutrients. Thus, ecosystems may change even if external drivers (i.e., river flow regimes) remain unchanged.

1.2.4 ACTIVE MANAGEMENT OF ENVIRONMENTAL WATER

Chapter 19 introduces the concepts of active and passive environmental water management, where *active* environmental water management refers to those allocation mechanisms that require ongoing decision making concerning when and how to use environmental water to achieve the desired outcomes. Whereas a cap on consumptive water use can be set as part of a long-term plan and then monitored (*passive* management), environmental water rights, and some forms of reserves, require ongoing decisions around when and how to best use the water available, and the temporal sequencing (both within and between years) of environmental water releases. This is most apparent when using environmental water held in storage. However, a similar case can also be made for temporary leases of rights to provide environmental water in rivers that lack instream storages: without a decision to continue the leasing arrangement, the environmental water will not be provided. Ultimately, active management is the corollary of flexibility. The more flexible the environmental water allocation mechanism, the more it will need an active decision maker to decide when and how to use (acquire, trade, or deliver) the water.

With active water management, there comes a suite of planning requirements and operational challenges. A planning structure is required that determines which environmental endpoints or locations, and which elements of the water regime to target in a particular season or year. An initial step in the planning process is priority setting across assets and watering actions; weighing the value of individual assets (e.g., conservation status of a species or the international significance of a wetland), the likelihood and significance of a successful watering outcome, the long-term capacity to provide ongoing benefit at a location, and the cost-effectiveness of a watering action. The planning process must also contemplate how priorities and watering actions may change within a variable climate. The annual planning process can then be nested within the longer-term prioritization, to consider the sequencing of flow delivery among years, the antecedent condition of assets, and the likely climate for the year ahead (Chapter 23).

When environmental water is held in storage, there can also be significant challenges associated with delivering water (Chapter 24). River basin infrastructure has often not been designed to support delivery of environmental water needs to specific areas. For example, flooding risks for private landholders may be associated with delivering larger environmental watering events. In addition, there are often complex administrative processes associated with the management of an individual environmental water release. These include permits to authorize release, permission to inundate wetlands (which can be on private or public land), and coordination among agencies, particularly where a flow release crosses jurisdictional boundaries.

Finally, adaptive management is an essential element of both active and passive environmental water management, as managers are making long- and short-term decisions amid significant uncertainty (Chapter 25). The decisions that environmental water managers are making regarding prioritization, sequencing of flows, and trade-offs often require extrapolation of our current scientific knowledge. The central concept of adaptive management is the iterative learning process—a cycle of *plan*, *do*, *monitor*, and *learn*. Monitoring plays two critical roles in active management of environmental water: it can be used to demonstrate the *return-on-investment* of environmental water; and it can be used to build our knowledge base of environmental water requirements and ecological responses. To successfully implement adaptive management, a significant commitment is required from managers and stakeholders, along with the patience to see the process through. This extends to the evaluation framework for environmental management institutions, which must consider not only the efficiency and effectiveness of environmental watering programs, but also the longer-term factors that underpin success such as legitimacy, organizational capacity, and partnerships (Chapter 26).

1.3 WHAT REMAINS AHEAD?

In this book, we have attempted to capture the complete environmental water management cycle and demonstrate the significant progress that has been made in this rapidly maturing field. However, along with identifying the factors contributing to success of environmental water management, there are also lessons from past mistakes and many remaining challenges. Key among these is the challenge of implementation. Environmental water is now formally recognized through the policies and laws of many countries (Le Quesne et al., 2010). Although there have been significant advances toward sustainable water resource management, the problem remains of how to *implement* a well-managed commons that fulfills human needs, but also leaves enough water to protect the core ecosystem functions and values of a river system. The final section of the book (Chapter 27) outlines the challenges—under the broader heading of implementation—that face environmental water management in the coming decade.

Critical among the challenges identified are:

1. *How much water do rivers need?* Richter et al. (1997) asked this question 20 years ago, and in Chapter 9, Jackson recast this question as "How much water does a culture need?" The challenge of defining an environmental water regime is twofold; understanding the shared objectives and values of the river, and understanding the scientific links between flow and ecosystem health. The answer to the question of how much water a river needs will indeed change over time as part of an ongoing discussion about what sort of river the community wants. Importantly, cultures value and experience the natural environment differently and environmental water programs need the flexibility to align with and incorporate these different perspectives. Many of the world's rivers are managed using large infrastructure such as dams, where it may be more appropriate to design a flow regime that meets the multiple objectives (consumptive and environmental) of the system rather than attempt to emulate the natural flow regime (Acreman et al., 2014). There remains a significant challenge as to how to design and manage a flow regime to ensure that the complex needs of the environment are supported in the longer term (Acreman et al., 2014; Arthington, 2012; Arthington et al., 2006). The emergence of allocation mechanisms for environmental water that require active management has led to more probing questions from managers around the sequencing of water allocations, the marginal improvement in biological condition from incremental changes in streamflow, and the recovery

rate postwatering. These management questions test the boundaries of our current scientific understanding (Horne et al., in press). A core element of scientific knowledge required in the future will be aimed at understanding the implications of climate change for environmental water management (Poff et al., 2002). For example, are organisms adapting to altered hydrological and geomorphological regimes? How resilient are particular ecosystems to change, and are there thresholds beyond which a state change in ecosystem function would occur?

2. *How do we increase the number of rivers where environmental water is provided?* Implementing environmental water regimes in more locations will require better representations of the socio-economic costs and benefits, political will, stakeholder engagement, and financial resources. Representing the value of ecosystem services relative to other consumptive and commercial uses is important for understanding trade-offs where they are required. This will be particularly important in developing countries where there is an immediate need to support basic human needs (Christie et al., 2012). Sustainable financing will be an essential element. There may be a role for the rapid roll out of precautionary methods that are cheap to implement, with later prioritization of more robust detailed environmental flows assessments (a triage-type approach).

3. *How can we embed environmental water management as a core element of water resource planning?* This book has clearly demonstrated that successful environmental water management requires integration with broader water resource policy, and ideally links to broader catchment management. Climate change will continue to highlight the importance of an inclusive water resource planning process, where more explicit discussion will be required around how climate change impacts are distributed among water users and what changes may be required for environmental objectives (Acreman et al., 2014) and allocation frameworks to support sustainable river systems for the future. Reduced water availability may require a change in the objectives for environmental water, potentially managing for adaptation and resilience rather than restoration of a particular species or ecosystem state. Integrating environmental water within broader water resource management will allow more nimble and informed responses to change (Dalal-Clayton and Bass, 2009). It also allows for more novel approaches that support conjunctive water uses.

4. *How can knowledge and experience be transferred and scaled?* Although there are numerous ad hoc literature reviews, there is limited systematic global review around implementation and effectiveness of environmental watering under different levels of development, administrative settings, and political systems (Pahl-Wostl et al., 2013). Many countries are grappling with challenges of poverty alleviation, livelihoods, and economic development. There may be opportunities to demonstrate success in similar climatic and ecological systems, or in more achievable circumstances (rather than necessarily in the most stressed rivers) through the use of demonstration sites. Similarly, there remains a challenging scientific question around how, or under what circumstances, ecological knowledge from one location can be validly transferred to another (Arthington et al., 2006; Poff et al., 2010). Living labs provide one avenue to support the transfer of knowledge and management tools across regions.

5. *How do we enhance the legitimacy of environmental water programs?* This requires a partnership across all stakeholders, which takes vision, time, effort, and humility to achieve. Real gains can be made in acknowledging both the diversity of values and the local and indigenous knowledge that can inform environmental water management. Although stakeholder engagement is commonly noted as an element of environmental water management, there are limited examples of true

partnerships. Introducing legitimacy as a key performance measure for environmental water programs may help ensure that stakeholder engagement is a more central element of the environmental water management process. Establishing demonstration catchments and a review of the institutional structures that encourage partnerships would be useful initial steps in this regard.

6. *How can we support the inclusion of adaptive management as standard practice?* Adaptive management is a crucial concept in environmental water management, but better documentation of decisions involving partnerships between managers, scientists, and the public could improve local learning and knowledge transfer (Poff et al., 2003). This approach needs to be underpinned by long-term monitoring or modeling of processes to support both the inner and outer loops of adaptive management (Fig. 1.1). The design, sustainable funding, and administration of such monitoring programs need to be identified as early as possible in the environmental water management cycle. Although widely advocated, the adaptive management approach is rarely well implemented, and will become even more essential as climates, environmental conditions, and societal aspirations change over time (Acreman et al., 2014).

We believe that these are the critical tipping points for the most significant improvements in the socio-ecological health of the world's rivers. The final contribution of this book is to frame the future research efforts toward answering these challenging questions.

REFERENCES

Acreman, M.C., Arthington, A.H., Colloff, M.J., Couch, C., Crossman, N., Dyer, F., et al., 2014. Environmental flows — natural, hybrid and novel riverine ecosystems. Front. Ecol. Environ. 8, 466—473.

Arthington, A., 2012. Environmental Flows — Saving Rivers in the Third Millennium. University of California Press.

Arthington, A.H., Bunn, S.E., Poff, N.L., Naiman, R.J., 2006. The challenge of providing environmental flow rules to sustain river ecosystems. Ecol. Appl. 16, 1311—1318.

Brisbane Declaration, 2007. Environmental flows are essential for freshwater ecosystem health and human well-being. Brisbane, Australia. 10th International River Symposium and International Environmental Flows Conference, 3—6 September 2007.

Christie, M., Cooper, R., Hyde, T., Fazey, I., 2012. An evaluation of economic and non-economic techniques for assessing the importance of biodiversity and ecosystem services to people in developing countries. Ecol. Econ. 83, 67—78.

Dalal-Clayton, B., Bass, S., 2009. The challenges of environmental mainstreaming — experience of integrating environment into development institutions and decisions. International Institute for Environment and Development, London.

Dudgeon, D., Arthington, A.H., Gessner, M.O., Kawabata, Z., Knowler, D.J., Leveque, C., et al., 2006. Freshwater biodiversity: importance, threats, status and conservation challenges. Biol. Rev. 81, 163—182.

Dyson, M., Bergkamp, G., Scanlon, J., 2003. Flow. The Essentials of Environmental Flows. IUCN, Gland, Switzerland and Cambridge, UK.

Hardin, G., 1968. The tragedy of the commons. Science 162, 1243—1248.

Horne, A., Szemis, J., Webb, J.A., Kaur, S., Stewardson, M., Bond, N., et al. (in press). Informing environmental water management decisions: using conditional probability networks to address the information needs of planning and implementation cycles. Environ. Manage.

Le Quesne, T., Kendy, E., Weston, D., 2010. The implementation challenge. Taking stock of governmental policies to protect and restore environmental flows. The Nature Conservancy and WWF.

Meyer, J.L., Sale, M.J., Mulholland, P.J., Poff, N.L., 1999. Impacts of climate change on aquatic ecosystem functioning and health. J. Am. Water Resour. Assoc. 35, 1373−1386.

Ostrom, E., 1990. Governing the Commons: The Evolution of Institutions for Collective Action. Cambridge University Press, Cambridge.

Pahl-Wostl, C., Arthington, A., Bogardi, J., Bunn, S.E., Hoff, H., Lebel, L., et al., 2013. Environmental flows and water governance: managing sustainable water uses. Curr. Opin. Environ. Sustain. 5, 341−351.

Poff, N.L., Brinson, M.M., Day, J.W.J., 2002. Aquatic ecosystems and gobal climate change − potential impacts on inland freshwater and coastal wetland ecosystems in the United States. Prepared for the Pew Center on Global Climate Change.

Poff, N.L., Allan, J.D., Palmer, M.A., Hart, D.D., Richter, B.D., Arthington, A.H., et al., 2003. River flows and water wars: emerging science for environmental decision making. Front. Ecol. Environ. 1, 298−306.

Poff, N.L., Richter, B.D., Arthington, A.H., Bunn, S.E., Naiman, R.J., Kendy, E., et al., 2010. The ecological limits of hydrologic alteration (ELOHA): a new framework for developing regional environmental flow standards. Freshw. Biol. 55, 147−170.

Richter, B., Baumgartner, J., Wigington, D.B., 1997. How much water does a river need? Freshw. Biol. 37, 1.

Richter, B.D., 2014. Chasing Water: A Guide for Moving from Scarcity to Sustainability. Island Press, Washington.

Tharme, R.E., 2003. A global perspective on environmental flow assessment: emerging trends in the development and application of environmental flow methodologies for rivers. River Res. Appl. 19, 397−441.

TNC. 2016. Environmental flow concepts website [Online]. The Nature Conservancy. Available from: http://www.conservationgateway.org/ConservationPractices/Freshwater/EnvironmentalFlows/Concepts/Pages/environmental-flows-conce.aspx (accessed 01.02.16).

Vorosmarty, C.J., Mcintyre, P.B., Gessner, M.O., Dudgeon, D., Prusevich, A., Grenn, P., et al., 2010. Global threats to human water security and river biodiversity. Nature 467, 555−561.

HISTORY AND CONTEXT OF ENVIRONMENTAL WATER MANAGEMENT

DRIVERS AND SOCIAL CONTEXT

Mike Acreman[1], Sharad K. Jain[2], Matthew P. McCartney[3], and Ian Overton[4]

[1]Centre for Ecology and Hydrology, Wallingford, United Kingdom [2]National Institute of Hydrology, Roorkee, UK, India [3]IWMI, Vientiane, Laos [4]Natural Economy, Adelaide, SA, Australia

2.1 WATER USE AND HUMAN DEVELOPMENT

Water is important for many aspects of our lives. We all recognize the importance of the direct use of water for drinking, washing, growing food, generating power, and supporting industry. Increasingly, we also understand the values associated with the indirect use of water such as providing water to rivers, lakes, wetlands, and estuarine ecosystems to provide natural benefits including fisheries, fertile floodplain land, timber, wild fruits, and medicines (Acreman, 1998).

Early civilizations flourished along the floodplains of major rivers such as the Tigris-Euphrates, Indus, Ganges, and Nile (Solomon, 2010), where people could easily benefit from both direct and indirect water use. Hydrological modifications using simple water control structures were employed to store and divert water to enhance irrigation and reduce vulnerability to the naturally varying water cycle (Maltby and Acreman, 2011). Although populations were small, impacts on the natural benefits of the river were minor and there were few conflicts in meeting direct and indirect water use objectives. However, as populations expanded, there was a requirement for more structural control, including large dams and barrages, to meet rapidly increasing direct needs for water such as to generate electricity to operate factory machinery during the Industrial Revolution. Artificial lakes dating back to the 5th-century BC have been found in ancient Greece (Wilson, 2009). The aim was to store water during wet periods for use during dry periods, thus evening out natural hydrological variations. Grey and Sadoff (2007) argue that there is a direct relationship between our ability to control the water cycle and economic growth, with countries lacking storage "remaining hostage to hydrology" and "typically among the world's poorest." For example, existing surface storage capacity in India is just 11% of the annual river flow, whereas in the *Murray–Darling Basin (MDB)* of Australia, storage is 150% of the annual flow (CWC, 2007).

Some dam operations included releases of water to the downstream river, but historically this was to allow further exploitation of the available water for public drinking water supply, irrigation, or power generation, using the river as a convenient water conduit; dam releases were seldom made for environmental purposes. As more hydrological control was exerted, the trade-off between direct and indirect water use increased and more environmental degradation resulted. A most striking example of this economic driver for direct water use is the Aral Sea, formerly one of the four largest lakes in the world with an area of 68,000 km^2. It started shrinking in the 1960s after the rivers that fed the Sea were diverted for irrigation (direct water use); by 2007 it was 10% of its original size (Zavialov, 2005), devastating the fisheries (indirect water use) and creating major local health problems due to windblown dust from dried lakeshores. A similar trade-off is anticipated in other river basins such as the Mekong, where upstream dams, being constructed for hydropower generation, are likely to put fisheries resources supporting millions of people at risk downstream (Ziv et al., 2012). Different direct water uses are also coming into conflict such as irrigation in upstream reaches of the Rufiji River in Tanzania with hydropower generation downstream; both these direct uses are increasingly in conflict with indirect use for wildlife conservation (Acreman et al., 2006).

2.2 ENVIRONMENT AND WATER MANAGEMENT

The idea that the lives of people and the environment are fundamentally interrelated has been reported for more than 150 years (e.g., Marsh, 1864) and, in response to major environmental degradation, since the 1960s (e.g., Carson, 1962). By the time of the United Nations Conference on Environment and Development in Rio de Janeiro in 1992, there was full global recognition of the importance of the environment to human well-being.

Ecological processes maintain the planet's capacity to deliver goods and services such as water, food, and medicines and much of what we call *quality of life* (Acreman, 2001). The concept of ecosystem services (Barbier, 2009; Dugan, 1992; Fischer et al., 2009) brought to prominence in the Millennium Ecosystem Assessment (MEA, 2005) demonstrated that healthy freshwater ecosystems provide economic security, for example, fish, medicines, and timber (Cowx and Portocarrero, 2011; Emerton and Bos, 2004); social security, for example, protection from natural hazards such as floods; and ethical security, for example, upholding the rights of people and other species to water (Acreman, 2001). Thus, water allocated for the environment is a sound objective as it also supports people by maintaining the ecosystem services on which we depend (Acreman, 1998; MEA, 2005; Chapter 8).

This connection between ecosystem health and human well-being has been further institutionalized in intergovernmental agreements. The Millennium Development Goals, adopted by 189 countries in 2000, included the need for environmental sustainability such as reducing the rate of loss of species threatened with extinction. Subsequently, the Rio + 20 meeting in 2012 (http://www.uncsd2012.org/) called for action to protect and sustainably manage ecosystems (including maintaining water quantity and quality), and recognized that the global loss of biodiversity and the degradation of ecosystems undermines global development (Costanza and Daly, 1992), affecting food security and nutrition, the provision of and access to water, and the health of the rural poor. Rio + 20 also launched a process to develop a set of *sustainable development goals (SDGs)*, which will build upon the Millennium Development Goals and converge with the post-2015 development agenda. Goal 6 of the SDGs calls for sustainable water withdrawals and protection and restoration of ecosystems, including forests, wetlands, rivers, aquifers, and lakes.

2.3 DEVELOPMENT OF NATIONAL AND INTERNATIONAL POLICIES

The UK *Water Resources Act 1963* was one of the world's first pieces of environmental water legislation; it required minimum acceptable flows to maintain natural beauty and fisheries of UK rivers. This was followed by the US *Clean Water Act 1972* that set the objective of restoring and maintaining the chemical, physical, and biological integrity of surface and ground waters. In recent times, environmental water protection has been integrated into water management in many countries. For example, the *1998 National Water Act* of South Africa decreed that water for the maintenance of the environment is accorded the highest priority along with that for basic human needs (King and Pienaar, 2011; Rowlston and Palmer, 2002). This concept was followed by other countries such as Tanzania (Acreman et al., 2009). Within the European Union, the *Water Framework Directive* (*WFD*; European Commission and Parliament, 2000) required member states to achieve good status in all water bodies that includes the need to maintain river hydromorphology.

Environmental water has been incorporated into the agreed activities of the 169 national signatories to the International Convention on Wetlands (initiated in Ramsar, Iran, in 1971). Environmental water has also become a central part of the policies of major institutions, including the World Bank (Hirji and Davis, 2009) and the International Union for the Conservation of Nature (Dyson et al., 2003). Although not explicitly environmental water, the concept of maintaining aquatic ecosystems is now a key element in many international policies such as the Ecosystem Approach (Maltby et al., 1999) adopted by the Convention on Biological Diversity, which opened in 1992 and has 196 parties. Furthermore, water allocation for ecosystem maintenance is a central part of *Integrated Water Resources Management* (*IWRM*; Falkenmark, 2003) promoted by the Global Water Partnership (Brachet et al., 2015), and environmental impact assessment (Wathern, 1998).

2.4 SETTING OBJECTIVES

Environmental water regimes encompass the quantity, timing, and quality of water flows required to sustain freshwater and estuarine ecosystems and the human livelihoods and well-being that depend on these ecosystems (Arthington, 2012). This definition does not define exactly what state the ecosystem should be in or what ecosystem services should be delivered, as the same ecosystem can have many forms or conditions that can deliver many different sets of benefits. Objectives for rivers vary around the world and may be set at international, national, or river basin level.

There are generally two ways in which environmental water objectives are set (Acreman and Dunbar, 2004). First, specific objectives for many rivers are set by legislation. For example, the WFD provides a clear set of objectives for European rivers. Member states are required to achieve at least *good ecological status (GES)* in all water bodies (Acreman and Ferguson, 2010). High ecological status is the target for rivers of greatest conservation interest, whereas *good ecological potential (GEP)* is the target for highly modified rivers (i.e., those with dams in place). In South Africa, rather than employing a single set of objectives for all rivers, the Department of Water Affairs and Forestry defines objectives according to different ecological management targets

Table 2.1 *Environmental Management Classes (EMC)* **and Management Perspective**

EMC	Description	Management Perspective
A	Natural rivers with minor modification of instream and riparian habitat.	Protected rivers and basins. Reserves and national parks. No new water projects (dams, diversions, etc.) allowed.
B	Slightly modified and/or ecologically important rivers with largely intact biodiversity and habitats despite water resources development and/or basin modifications.	Water supply schemes or irrigation development present and/or allowed.
C	The habitats and dynamics of the biota have been disturbed, but basic ecosystem functions are still intact. Some sensitive species are lost and/or reduced in extent. Alien species present.	Multiple disturbances associated with the need for socio-economic development, e.g., dams, diversions, habitat modification, and reduced water quality.
D	Large changes in natural habitat, biota, and basic ecosystem functions have occurred. A clearly lower than expected species richness. Much lowered presence of intolerant species. Alien species prevail.	Significant and clearly visible disturbances associated with basin and water resources development, including dams, diversions, transfers, habitat modification, and water quality degradation.
E	Habitat diversity and availability have declined. A strikingly lower than expected species richness. Only tolerant species remain. Indigenous species can no longer breed. Alien species have invaded the ecosystem.	High human population density and extensive water resources exploitation. Generally, this status should not be acceptable as a management goal. Management interventions are necessary to restore flow pattern and to *move* a river to a higher management category.
F	Modifications have reached a critical level and ecosystem has been completely modified with almost total loss of natural habitat and biota. In the worst case, the basic ecosystem functions have been destroyed and the changes are irreversible.	This status is not acceptable from the management perspective. Management interventions are necessary to restore flow pattern, river habitats (if still possible/feasible) etc.—to move a river to a higher class.

(Rogers and Bestbier, 1997). There are four target classes, A–D (Table 2.1). Two additional classes, E and F, may describe the present ecological status, but all rivers must have a target class of D or above. Some rivers may be designated under national law or international conventions such as the Ramsar Convention on Wetlands of International Importance and this is accompanied by an explicit statement of desired status of the ecosystem for which appropriate flows need to be defined. Some rivers are associated with an icon species such as the Yangtze River dolphin (*Lipotes vexillifer*) in Southeast Asia or the River Murray cod (*Maccullochella peelii*) in the River Murray. Environmental water releases of 7500 million m^3 of water were made from the Manatali dam in Mali to ensure inundation of 50,000 ha of the Senegal River floodplain that supported specific ecosystem services, for instance flood recession agriculture and fisheries for local communities in Senegal and Mauritania (Acreman, 2009). In India, environmental water regimes have been set to achieve particular water levels for religious festivals along the Ganges River.

A second way to define objectives is where they are not fixed by legislation. Instead stakeholders such as water users and local community representatives are invited to define their expectations for a river ecosystem and water uses, and to negotiate an agreed decision. This is particularly relevant where there are competing demands for water, including for public supply, for irrigated agriculture, and for industry, which cannot all be fully met. The negotiation considers a set of scenarios within which there are various trade-offs with different amounts of water for each sector. A final decision may be reached by consensus or decided by a judge or arbitrator. The process of negotiation can be socially inclusive, but is often nonspecific and subjective. People may want the river to be natural, or they may have a golden age in mind (i.e., a view of the landscape in a painting from 1850), or memories of how nice the river was when they were young, which can influence their vision. Desires are often driven by a cultural or spiritual connection with the river. Given the high demand for water in many river basins it is often impossible to meet everyone's needs, and compromises are required. Reaching agreement can be very difficult if expectations are unrealistic, for example, if the river has been heavily managed and will continue to be so for local or national economic prosperity. Setting objectives for environmental water through stakeholder engagement is thus a socio-political process rather than a solely scientific procedure.

2.5 PATHWAYS TO ENVIRONMENTAL WATER POLICY

What leads to the implementation of environmental water policy in a river basin and how do these policies progress over time? Although the impetus for environmental water policy is now well recognized, the pathway to implementation varies significantly across the globe. Moore (2004) surveyed 272 individuals involved in environmental water (through a range of organizations) and asked respondents how the concept of environmental water regimes was initially established in their various different river basins and countries. The results show that public awareness and recognition of the importance to local livelihoods were both important factors. Interestingly, the introduction of *environmental flow assessment* projects and expertise was also seen as a major driver. Environmental flow assessments, particularly when there is stakeholder engagement, can lead to significant changes in community attitudes around sustainable water resource management (King et al., 2003; Moore, 2004).

The following sections detail the development of environmental water policy in four places: the United Kingdom, India, Southeast Asia, and Australia. These overviews aim to show the differences and commonalities in the pathways that led to concern and implementation of environmental water policy.

2.5.1 ENVIRONMENTAL WATER IN THE UNITED KINGDOM

UK rivers are utilized for a wide range of purposes, including water supply and recreation such as fishing, boating, and bird watching. Water is abstracted (either from the river itself or adjacent aquifers) primarily for public supply, industry, and power station cooling (although irrigation is important locally in dry areas) and returned after cleaning at sewage treatment works. Thus water is recycled along many rivers. All abstractors must obtain a license and plans are in place to allow trading of licenses between abstractors (OfWat, 2015). Increasingly, households are being metered and charged by volume for their water use as a demand management measure. Storage reservoirs have long been a feature of the United Kingdom landscape with masonry dams dating back to the 17th century, particularly of upland areas, although they also exist in the lowlands such as around London.

Water releases from reservoirs have been made in the United Kingdom since the 1800s, but these were to permit abstraction downstream (using the river as a conduit) or to provide water for downstream riparian users and were termed *compensation flows* (Gustard et al., 1987). These releases cannot be considered environmental water as we understand them today, even though they may have unintentionally supported the river ecosystem to some extent. The first flow management for ecosystems per se focused on the concept of a minimum flow for diluting polluted discharges, based on the notion that, as long as the flow is maintained at or above a critical minimum, water quality will be protected. This was manifested in the UK *Water Resources Act 1963* that required minimum acceptable flows to maintain natural beauty and fisheries. Although conceptually groundbreaking, the Act was never fully implemented in terms of linking it to licenses to abstract from or discharge water to rivers.

The United Kingdom's Environment Agency was established in 1996 and had a duty to define Catchment Abstraction Management Strategies for all river basins in England and Wales. To implement the strategies, the Agency developed the *Resource Assessment and Management (RAM)* Framework tool to determine environmental water requirements (Environment Agency, 2002). A key feature of the RAM Framework was the concept of *hands-off flows*; threshold flows below which abstraction should cease or be very limited (Barker and Kirmond, 1998) to protect the river ecosystem.

In the 1990s, environmental water objectives in the United Kingdom were largely set by local stakeholder negotiations, facilitated by the Environment Agency. For example, in 1998, a Public Inquiry was held to define the amount of water that could be abstracted for public supply from the River Kennet in Southern England, and hence the amount needed to be left in the river to provide environmental water regimes. The inquiry inspector assessed evidence from many parties, but debate focused on the economic value of water for different purposes using data provided by the private water supply company, angling associations, local nature organizations, and the Environment Agency. The outcome was that water abstraction should be limited to that which resulted in a maximum 10% loss in river habitat, requiring the use of hydraulic-habitat river models such as *Physical Habitat Simulation* (*PHABSIM*; Elliott et al., 1999).

To harmonize legislation across the European Union, the European Water Framework Directive (European Commission and Parliament, 2000) was established, which required member states to achieve at least good status in all its water bodies. This effectively gave river ecosystems the highest priority for water and replaced objectives negotiated by stakeholders with those fixed by law. Good status was a combination of good chemical status and GES, with GES defined as slight deviation from the reference conditions; for instance a river should contain the assemblages of component species (fish, macro-invertebrates, macrophytes and phytobenthos, and phytoplankton) that would be found in an *undisturbed* (natural) river of that type. Rivers of great conservation value had a target of high ecological status. If rivers were not at GES, restoration measures were required. It was recognized that supporting elements such as channel form and river flow (termed *hydromorphology*) also affect the biology of rivers. Hence, setting appropriate environmental water regimes was a key step in achieving good status. To implement the Directive, maximum abstraction allowances were set for different river types by a panel of experts (Acreman et al., 2008). This is quite problematic in the United Kingdom as most rivers are managed (by impoundments, diversions, discharges, and channel alterations) to some extent and for some river types there are no reference conditions remaining. Aiming for natural rivers is thus unrealistic and often not desired. River Itchen, for example, is protected for its current highly cherished ecosystem even though this has resulted from significant management for many centuries and the river is quite unlike its natural condition (Acreman et al., 2014).

Under the WFD, rivers affected by embankments and dams were considered *heavily modified* and were required to meet a different objective—that of GEP. To achieve this objective, environmental water releases need to be made from dams in the form of specific flow elements (e.g., summer low flows and spring floods) that deliver particular objectives such as fish migration and spawning, or wetting of channel backwaters to provide fish-rearing habitat (Fig. 2.1). These elements combine to form a flow regime that meets, as far as possible, all specified objectives and desirable ecological/social outcomes such as recreation or cultural indigenous perspectives (Acreman et al., 2009). This method was based on the Building Block Methodology developed in South Africa (King et al., 2000).

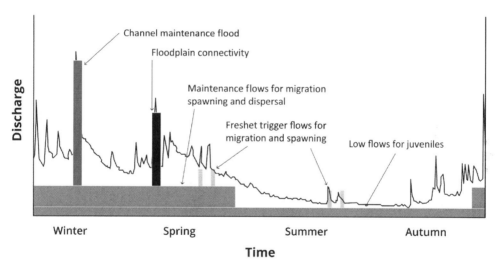

FIGURE 2.1

Environmental water releases from a reservoir designed to achieve specific river ecosystem responses.

Source: Acreman et al. (2009)

2.5.2 ENVIRONMENTAL WATER IN INDIA

Since earliest times, agriculture has been the main occupation of Indian society. As rainfall is highly erratic, uncertain, and mainly confined to the monsoon season, the objective for historical water resources development was to divert river flows for irrigation through canals. For example, the Upper

Ganga Canal system in Western Uttar Pradesh was completed in 1854. It was one of the largest canal systems in the world at that time. For this canal, water is diverted from the Ganga River at Haridwar. Subsequently, many such canals were built in the 19th and 20th centuries. Agriculture now accounts for about 85%–90% of all the consumptive water abstractions in India (FAO, 2012). The other major uses are domestic and industrial supply and hydropower generation (nonconsumptive).

With the increase in water demands, construction of multipurpose water storage projects such as Teri Dam on the Ganga River were implemented in the 1900s. As progressively more and more water was diverted from the rivers and withdrawn from aquifers to support expanding agricultural land, river flows continued to decrease, particularly during the lean (dry) season (Jain et al., 2015). In extreme cases, some rivers have become dry for short reaches during parts of the lean season leading to local habitat loss and loss of longitudinal connectivity in the river system. High dams also cause loss of connection between the river and groundwater (vertical fragmentation) and between the river and its floodplains (horizontal fragmentation). These types of fragmentation are damaging for river ecosystems and their benefits to people such as fisheries. This demonstrates the incompatibility of objectives and the trade-off of direct water use for agriculture and indirect water use for ecosystem services.

In addition to catering for society's direct water needs, cultural, social, religious, and ecological services are clearly important objectives for Indian rivers, notably the Ganga. Most rivers are worshipped as a mother and many Indian customs and festivals are linked to them. River waters are used in many social rituals such as marriage, worshipping, and cremation. Most people believe that rivers such as Ganga have immense cleaning power—physical and spiritual (Fig. 2.2). Hence, on auspicious occasions such as the full moon and no-moon days, melas (fairs, e.g., Kumbh) are held where millions of pilgrims gather on river banks and take baths for spiritual reasons. Bathing requires clean flowing water with sufficient depth. Such events have a transient impact on river water quality as a large amount of human waste is generated in small areas over short periods. Consequently, these events put enormous pressure on water and sanitation and the carrying capacity of the area is exceeded manifold during a brief period.

People believe that due to immense self-cleansing power, the waste joining the river cannot harm it—a belief that is slowly changing. However, more efforts are needed to change the mindset so that people refrain from polluting rivers. Of late, the thinking has emerged that healthy rivers can be realized best and most sustainably by integrated environmental management or overall cleaning and to that end, the Government of India has launched the *Swatchh Bharat* or Clean India Mission (MDWS, 2014).

Nongovernmental organizations, such as the *Worldwide Fund for Nature (WWF)*, have identified conservation of icon species such as the Ganga freshwater dolphin (*Platanista minor*) as an objective for Indian rivers in addition to spiritual needs.

India is a union of states and is governed by a central government and many state governments. Under the Indian constitution, each state government has the power to make laws with respect to their water resources. The Parliament has the power to legislate on regulation and development of interstate rivers. Thus, the state governments exercise authority over water management in respective states, subject to certain limitations that may be imposed by the Parliament. Clearly, any

FIGURE 2.2

The Ganga River has immense cultural religious and cultural significance and is worshipped as mother. This photo shows Ganga Aarti (prayer to Ganga), which is performed at Har Ki Paidi, Haridwar, every day at sunset.

Source: Photo M. Acreman

large-scale water management effort involving multiple states requires close cooperation between the central and concerned state governments. The National Water Policy of India (Government of India, 2012) recognizes that "Water is essential for sustenance of ecosystem, and therefore, minimum ecological needs should be given due consideration."

Research and development work on environmental flows assessments in India started only recently. It was initiated from a realization that certain minimum flows should be always left in the rivers for maintaining aquatic ecosystems. These minimum flows were typically prescribed as percentages of lean season flows. Motivated by the need for healthy rivers and based on research and practices in other countries, Indian practitioners began applying hydrological methods for environment water determination in the past two decades. Some recent studies have also incorporated habitat requirements of aquatic species in assessments. In the meantime, environmental water regimes as a percentage of uninterrupted seasonal flows have been suggested for water resources development projects until better estimates are made. Jain and Kumar (2014) have reviewed environmental water case studies in India.

WWF-India initiated a program of environmental flows assessments in the Ganga basin (WWF, 2012). Environmental water requirements were estimated at representative sites in four zones of the river and included religious and cultural needs as well as for biodiversity.

Currently, a large number of organizations and individuals are involved in research and development for environmental flows assessments in India and the number is growing. The database of properties and requirements of aquatic ecosystems is inadequate and many efforts are underway to collect the requisite data and improve the understanding of flow—ecology relationships. Attempts are being made to involve stakeholders in environmental water determination and implementation through public hearings, conferences, social media, and so on, but more efforts are needed.

2.5.3 ENVIRONMENTAL WATER IN SOUTHEAST ASIA

The *Greater Mekong Subregion (GMS)* is very diverse physically, socially, culturally, politically, and economically, and is rapidly changing. In recent decades, the countries within the subregion have focused on economic growth that has contributed positively to the alleviation of poverty and improvement in living conditions. Increases in gross domestic product across the region have broadly translated into increases in the Human Development Index, which captures life expectancy, knowledge, and standard of living (UNDP, 2012) and declines in infant mortality and chronic poverty. However, this development has been both dependent on, and come at significant cost to, the natural capital of the region (Ziv et al., 2012). Across the region water security is deemed low and worsening, in large part because of environmental degradation (ADB, 2013).

During recent decades, investment in infrastructure is set to increase significantly. For example, the implementation plan of the Regional Investment Framework of the GMS has an investment objective of $30.1 billion, primarily in river transport, but also some water resource infrastructure such as dams, between 2014 and 2018 (ADB, 2015). It is recognized that while creating economic opportunities, this level of investment is likely to be associated with high environmental and social costs that have yet to be fully understood. The potential impact on natural capital—including aquatic ecosystems—could be very high and ultimately might undermine the sustainability of much of the investment (ADB, 2015). The region faces important decisions on its objectives for direct and indirect water use.

Many millions of people are dependent on the natural resources of the Mekong River along its 4350 km. The Lower Mekong Basin supports the world's largest inland freshwater fishery, with an annual turnover of $1.4—3.9 billion (ADB, 2015). This includes the wild fishery of the Tonle Sap Great Lake in Cambodia, which is critically dependent on the annual flood pulse. By blocking migratory fish passage and by altering flow regimes, dam construction—primarily for hydropower—on both the mainstream of the Mekong River (where 12 dams are planned south of China) and the tributaries (where 78 dams are planned) could significantly undermine these fisheries as well as other ecosystem services (e.g., flood recession gardens), which are vital for the livelihoods of approximately 60 million people (ICEM, 2010). This demonstrates a significant trade-off between direct use of water facilitated by dam construction and indirect use through provision of fisheries (Ziv et al., 2012).

In recent years, awareness of the need for *green* economic growth has increased in the GMS, but translation into practical measures has been limited. To date, there has been very little attention given to environmental flow assessments and only very limited incorporation of the concept into national legislation and water resources planning. One exception is Vietnam where the law on Water Resources (1998) and the National Water Resources Strategy recognize that environmental water is required to protect aquatic ecosystems (Lazarus et al., 2012).

The *Mekong River Commission's (MRC's)* Environment Programme works to support cooperation among MRC member countries to secure a balance between economic development, environmental protection, and social sustainability. However, beyond articulating the urgent need, limited consideration has yet been given to implementing environmental water programs, and in particular, getting the cross-sector and transboundary commitment required for practical implementation in the GMS. Factors that are limiting the adoption of environmental water regimes in the region include: (1) lack of awareness and understanding of the concepts associated with environmental water regimes; (2) lack of political will; and (3) lack of readiness to accept and face up to complexity and uncertainty when making decisions about rivers and water resources (Lazarus et al., 2012). One key constraint is that most dams in the region are constructed by private companies using loans from private banks with no safeguard policies and currently no requirements to ensure environmental water regimes are provided.

Against this background, very few environmental flow assessments have been conducted to date. One study conducted in the Nam Songkhram River Basin, a tributary of the Mekong located in northeast Thailand, tested environmental water approaches incorporating economic, ecological, social, and transboundary dimensions (Lazarus et al., 2012). Another has been conducted in the Huong River Basin in Vietnam. Both of these illustrated not only the potential of environmental flow assessment approaches in the context of IWRM, but also the challenges of implementation and specifically the difficulties of translating internationally derived tools and the application of interdisciplinary processes into local contexts (IUCN, 2005). More recently, a study has illustrated how climate change is also likely to affect ecologically important aspects of the Mekong flow regime (Thompson et al., 2014).

Things are changing. For example, in Laos, the National Water Law is presently being revised and the National Water Resources Strategy 2020 is under preparation. Both these documents emphasize the need to integrate social and environmental concerns into water resource planning. They also promote stronger environmental management and protection for sustainability, including broader issues for transboundary water allocations to downstream Cambodia and Vietnam. If such policies can be translated into meaningful action, the relevant expertise can be built, and political will can be expanded further, then an environmental water approach could contribute significantly to water governance and sustainable water resource management throughout the region.

2.5.4 ENVIRONMENTAL WATER IN AUSTRALIA

Australia is often referred to as the driest continent. Furthermore, climate change is likely to increase the intensity of extreme weather events and alter the amounts and patterns of precipitation with up to 20% more months of drought by 2030. The natural flow variability in Australian rivers is the highest in the world. For example, the Darling River has flood peak flows, which are as much as 1000 times the flow in dry periods. Coupled with this variability is the fact that most of Australia's surface water supplies (northern Australia) are not located where the majority of people live (east and southeast). South and east Australia, like much of the world, has a high number of dams and high water abstraction in many of its rivers to support direct water. This has put considerable strain on these rivers in supplying water for irrigated agriculture, stock and domestic water supply, and hydropower generation.

Abstraction from rivers to support large irrigated agricultural regions has led to a significant decline in river flows. This impact of extraction is magnified during drought periods and has led to severe environmental decline along many Australian rivers. Environmental water allocations only came into use in Australia in the 1980s (Tharme, 2003) with the protection of low flows from dam releases. Minimum water releases from dams are still a major means of environmental water provision such as in the Mersey River in Tasmania and the Cotter River in the Australian Capital Territory (Overton and Acreman, 2012). The setting and release of environmental water regimes have been strongly linked to water-sharing plans, defined as water that is provided to meet agreed environmental and other public benefit objectives (*National Water Initiative [NWI]*). More recently, provision of environmental water has changed to recognize that all elements of the flow regime have a role in influencing the freshwater ecosystem, including floods, average and low flows (Bunn and Arthington, 2002).

With increasing concerns over environmental decline the *Council of Australian Governments (CoAG)* introduced significant reforms in the 1990s that are assisting environmental restoration of rivers and water-dependent ecosystems across Australia. These included a limit on water diversions called the Cap. These initial water reforms happened at the same time as the National Competition Policy, with a strong push for economic principles. This led to water policies addressing water pricing through a water market to allow continued economic growth with a capped resource. One of the most important was disentangling water ownership from land ownership. This has made water tradable between users and, in the southern MDB, tradable between catchment regions. The environment has been recognized as a legitimate user of water since the 1994 CoAG Water Reform Framework.

The National Water Initiative (2004) further established the environment as a key objective of water management and set a national blueprint for water reform in Australia. The NWI is a shared commitment by the state governments in Australia to increase the efficiency of Australia's water use, leading to greater certainty for investment and productivity, for rural and urban communities, and for the environment. The NWI requires state governments to: prepare comprehensive water plans; achieve sustainable water use in overallocated or stressed water systems; introduce registers of water rights and standards for water accounting; expand trade in water rights; improve pricing for water storage and delivery; and better manage urban water demands.

Although northern Australia is gaining increasing attention for its water resource potential, the historic focus of Australian water policy reforms has been the MDB. The MDB is Australia's largest river basin, occupying one-seventh of Australia and containing 65% of Australia's irrigated lands. Abstraction for irrigation began in the MDB in the 1880s and by the 1980s reached up to 56% of the average available water (CSIRO, 2008). A cap was placed on water abstraction from the MDB in 1995. The Millennium Drought of 1998–2009 had the greatest impact on record for river flows in the MDB. Higher than normal temperatures increased evapotranspiration, which compounded the reduced runoff. This reduced runoff caused significant environmental degradation and was the impetus for major water reform.

The *Water Act 2007* was introduced to implement a range of policies including establishment of the Murray–Darling Basin Authority, which has planning and management responsibility for water in the Basin, in the interests of the whole Basin. The *Water Act 2007* represents a very large investment by the federal government in Australia's water resources and this investment was required to get state government agreement. Previously water was managed independently by the four states and one territory within the Basin and the federal government could only intervene in matters of water resources on the basis of their international commitments to Ramsar and other international agreements. The *Water Act 2007* also outlined two new mechanisms for providing environmental water in the Basin: the establishment of a sustainable diversion limit (a revision to the Cap), and the creation of the *Commonwealth Environmental Water Holder (CEWH)* with a mandate to manage and trade environmental water rights across Australia. The CEWH manages their environmental water holdings by strategic releases to protect key environmental assets in the MDB, usually in conjunction with other environmental water releases made by the Murray–Darling Basin Authority or one of the Australian state governments. The environmental water rights are held in reservoirs and then used to manage in-channel habitats, wetlands connected to the river and broad floodplains. Examples include the Barmah Forest in Victoria, the Chowilla Floodplain in South Australia, and the Macquarie Marshes in New South Wales. Environmental water rights are also being used to mitigate depleting groundwater resources (e.g., the Gnangara Mound in Western Australia), improve water quality issues (i.e., the Fitzroy River in Queensland), and protect unregulated rivers (i.e., the Daly River in the Northern Territory). Management of environmental water rights remains a relatively new field and is explored in detail in Section IV.

A significant investment in science has been made to inform environmental flow assessments in the MDB. This includes detailed water quantity and quality modeling (CSIRO, 2008), environmental condition monitoring and evaluation, predictive modeling (Saintilan and Overton, 2010), and adaptive trials. Recently, investigations have included *cultural flows* for indigenous benefits (Weir, 2010). The Murray–Darling Basin Authority released the Basin Plan in 2012 that guides governments, regional authorities, and communities to sustainably manage and use the waters in the MDB.

There is considerable debate in Australia concerning what is trying to be achieved environmentally, given the importance of food security, economic growth, the rights of the Aboriginal people and community identity. Many people identify key environmental assets that should not be compromised. However, most are realistic that there is less water in the environment than would be the case naturally and that it may not be possible to conserve *living museums*. Environmental water in Australia is often targeted at specific ecological outcomes such as maintaining floodplain forests for stimulating a wetland bird breeding event. The Australian government through the Murray–Darling Basin Authority and CEWH have developed an environmental water management decision framework that identifies the water availability and defines a range of ecological goals from maintaining refugia during low water availability to promoting regeneration and building resilience during high water availability times. Australia is now considering a move toward assessing ecosystem services outcomes from environmental water as these are explicitly referred to in the *Water Act 2007*.

2.6 CONCLUSION

Despite the growing knowledge base for active environmental water management in some locations, most of Australia is highly unknown in terms of ecological response to changing flow regimes. It is recognized that each location is a complex socio-ecological system with substantial knowledge needs (eWater, 2009). Some water plans in Australia include groundwater as part of environmental flows assessments, although this is not undertaken as comprehensive or as widespread as the common surface water–groundwater interactions in most parts would indicate is necessary. Groundwater extraction limits are determined in most high-use areas and do in some cases take into consideration base flow to rivers.

The political situation in Australia allows for large investment and control by state governments and the federal government. Australia is accepting of this scale of government intervention that does not occur in most other countries. The broad-reaching water reforms and the water resource developments in irrigation infrastructure, particularly in the MDB, have been essentially state- and federally funded.

Australia has developed a highly sophisticated approach to environmental water and is considered one of the leading countries in IWRM. Water allocation is now a complex balance between environmental, social, and economic benefits. It is also starting to recognize the benefits of natural capital and ecosystem-based approaches to water management (Overton et al., 2014) and the multidisciplinary science and other forms of knowledge required (Acreman et al., 2014).

2.6.1 COMMON THREADS AND CONTRASTS

There has been a general global move toward recognition of the importance of the environment to people and that freshwater ecosystems need water to deliver ecosystem services. However, converting this to practical action is not straightforward due to historical water allocations and differences in approach that reflect different regional priorities. Objectives for water use vary significantly. The Industrial Revolution in the United Kingdom dominated water management in the 18th century. As populations grew and public water supply expanded, this became the priority. In the last 50 years, the water management challenge has shifted to balancing river ecosystem and public supply demands. In India and Australia, historical objectives have been to expand agriculture. Recognition of the major religious importance of rivers in India has been a significant driver to consider restoring environmental water to rivers, whereas in Australia loss of floodplain forests and birds has led to changes in the objectives toward ecosystem restoration. Reallocating water from one sector to another, to deliver environmental water, cannot be achieved quickly. Historical allocations have led to long-established practices such as in agriculture. Reallocation thus requires technological changes such as more efficient water use for crops, establishment of new systems such as water markets, repeal of old and passing of new water laws, and changes in attitudes of people toward water. In Australia, this has already been a 30-year journey and there are still many steps left to be implemented. As water demands and aspirations change over time, systems need adaptation, so addressing environmental water needs will require continued attention, rather than final closure.

There remains a sharp contrast between developed and developing regions in terms of objectives. In the United Kingdom, the objective for environment water is largely supporting recreation, aesthetics, and ethical needs; the trade-off is primarily with public supply, though locally with agriculture and industry. In the Mekong, there is a more immediate connection between environmental water and subsistence of poor people, with trade-off principally with hydropower expansion.

All regions of the world human populations are expanding and putting ever more pressure on water resources. It is becoming increasingly impossible to meet all objectives for direct and indirect water use. Thus, we are all facing critical decisions concerning trade-offs that have important economic, cultural, social, and ecological implications. The imperative of sustainable development requires that these decisions are made wisely with full understanding of what is gained and lost.

REFERENCES

Acreman, M.C., 1998. Principles of water management for people and the environment. In: de Shirbinin, A., Dompka, V. (Eds.), *Water and Population Dynamics.* American Association for the Advancement of Science. pp. 25−48.

Acreman, M.C., 2001. Ethical aspects of water and ecosystems. Water Policy J. 3 (3), 257−265.

Acreman, M.C., 2009. Senegal river basin. In: Ferrier, R., Jenkins, A. (Eds.), Handbook of Catchment Management. Blackwell, Oxford.

Acreman, M.C., Dunbar, M.J., 2004. Methods for defining environmental river flow requirements − a review. Hydrol. Earth Syst. Sci. 8 (5), 861−876.

Acreman, M.C., Ferguson, A., 2010. Environmental flows and European Water Framework Directive. Freshw. Biol. 55, 32−48.

Acreman, M.C., King, J., Hirji, R., Sarunday, W., Mutayoba, W., 2006. Capacity building to undertake environmental flow assessments in Tanzania. Proceedings of the International Conference on River Basin Management. Sokoine University, Morogorro, Tanzania, March 2005.

Acreman, M.C., Dunbar, M.J., Hannaford, J., Wood, P.J., Holmes, N.J., Cowx, I., et al., 2008. Developing environmental standards for abstractions from UK rivers to implement the Water Framework Directive. Hydrol. Sci. J. 53 (6), 1105−1120.

Acreman, M.C., Aldrick, J., Binnie, C., Black, A.R., Cowx, I., Dawson, F.H., et al., 2009. Environmental flows from dams; the Water Framework Directive. Eng. Sustain. 162, 13−22.

Acreman, M.C., Overton, I.C., King, J., Wood, P., Cowx, I.G., Dunbar, M.J., et al., 2014. The changing role of ecohydrological science in guiding environmental flows. Hydrol. Sci. J. 59 (3−4), 433−450.

ADB, 2013. Asian Water Development Outlook 2013: Measuring Water Scarcity in Asia and the Pacific. Asian Development Bank, Mandaluyong City, Philippines.

ADB, 2015. Investing in Natural Capital for a Sustainable Future in the Greater Mekong Subregion. Asian Development Bank, Mandaluyong City, Philippines.

Arthington, A., 2012. Environmental flows. Saving rivers in the Third Millennium. University of California Press, Berkeley, California.

Barbier, E.B., 2009. Ecosystems as natural assets. Found. Trends Microecon. 4 (8), 611−681.

Barker, I., Kirmond, A., 1998. Managing surface water abstraction. In: Wheater, H., Kirby, C. (Eds.), Hydrology in a Changing Environment, Vol. 1. British Hydrological Society, p. 249.

Brachet, C., Thalmeinerova, D., Magnier, J., 2015. The Handbook for Management and Restoration of Aquatic Ecosystems in River and Lake Basins. INBO, GWP, IOW. Available from: http://www.gwp.org.

Bunn, S.E., Arthington, A.H., 2002. Basic principles and ecological consequences of altered flow regimes for aquatic biodiversity. Environ. Manage. 30, 492−507.

Carson, R., 1962. Silent Spring. Houghton Mifflin, New York.

CEWH, 2011. A Framework for Determining Commonwealth Environmental Water Use. Australian Department of Sustainability, Environment, Water, Population and Communities, Canberra, Australia.

Costanza, R., Daly, H.E., 1992. Natural capital and sustainable development. Conserv. Biol. 6, 37−46.

Cowx, I.G., Portocarrero-Aya, M., 2011. Paradigm shifts in fish conservation: moving to the ecosystem services concept. J. Fish Biol. 79, 1663−1680.

CSIRO, 2008. Water availability in the Murray-Darling Basin. A Report to the Australian Government from the CSIRO Murray-Darling Basin Sustainable Yields Project. CSIRO, Collingwood, Australia.

CWC, 2007. Report of Working Group to Advise WQAA on the Minimum Flows in the Rivers. Central Water Commission, Ministry of Water Resources, Government of India, July 2007.

Dugan, P., 1992. Wetland Conservation. IUCN, Gland, Switzerland.

Dyson, M., Bergkamp, G., Scanlon, J. (Eds.), 2003. Flow. The Essentials of Environmental Flows. IUCN, Gland, Switzerland.

Elliott, C.R.N., Dunbar, M.J., Gowing, I., Acreman, M.C., 1999. A habitat assessment approach to the management of groundwater dominated rivers. Hydrol. Process. 13, 459–475.

Emerton, L., Bos, E., 2004. Value. Counting Ecosystems as an Economic Part of Water. IUCN, Gland, Switzerland and Cambridge, UK.

European Commission and Parliament. Directive establishing a framework for community action in the field of water policy. (2000/60/EC). Off. J., 2000.

eWater, 2009. Emerging practice in active environmental water management in Australia. eWater Cooperative Research Centre and the Australian Government National Water Commission, Canberra, Australia.

Falkenmark, M. 2003. Water Management and Ecosystems: Living With Change. Global Water Partnership/Swedish International Development Agency, Stockholm, Sweden TEC Background Papers, no. 9.

Fisher, B., Turner, R.K., Morling, P., 2009. Defining and classifying ecosystem services for decision making. Ecol. Econ. 68, 643–653.

Food and Agriculture Organization, 2012. Irrigation in Southern and Eastern Asia in Figures. FAO Water Report 37. Food and Agriculture Organization, Rome.

Grey, D., Sadoff, C.W., 2007. Sink or swim? Water security for growth and development. Water Policy J. 9, 545–571.

Gustard, A., Cole, G., Marshall, D., Bayliss, A., 1987. A Study of Compensation Flows in the UK. Report 99. Institute of Hydrology, Wallingford, UK.

Hirji, R., Davis, R., 2009. Environmental Flows in Water Resources Policies, Plans, and Projects. World Bank, Washington, DC.

ICEM, 2010. MRC Strategic Environmental Assessment (SEA) of hydropower on the Mekong mainstream: summary of the final report, Hanoi, Vietnam. International Centre for Environmental Management.

IUCN, 2005. Environmental flows – ecosystems and livelihoods – the impossible dream? Report of 2nd Southeast Asia Water Forum, 31 August 2005. International Union for Nature Conservation.

Jain, S.K., Kumar, P., 2014. Environmental flows in India: towards sustainable water management. Hydrol. Sci. J. 59 (3–4), 1–19.

Jain, S.K., Kumar, P., Mishra, P.K., Agarwal, Y.S., Gaur, S., Qazi, N., 2015. Assessment of environmental flows for Himalayan rivers. Project Report Prepared for the Ministry of Earth Science (MoES), New Delhi, India.

King, J.M., Brown, C.A., Sabet, H., 2003. A scenario-based holistic approach for environmental flow assessments. River Res. Appl. 19 (5–6), 619–639.

King, J., Pienaar, H. (Eds.), 2011. Sustainable use of South Africa's inland waters: a situation assessment of Resource Directed Measures 12 years after the 1998 National Water Act. Report No. TT 491/11. Water Research Commission, Pretoria, South Africa.

Lazarus, K., Blake, D.J.H., Dore, J., Sukrarort, W., Hall, D.S., 2012. Negotiating flows in the Mekong. In: Öjendal, J., Hansson, S., Hellberg, S. (Eds.), Politics and Development in a Transboundary Watershed: the Case of the Lower Mekong Basin. Springer, New York, pp. 127–153.

Maltby, E., Acreman, M.C., 2011. Ecosystem services of wetlands: pathfinder for a new paradigm. Hydrol. Sci. J. 56 (8), 1–19.

Maltby, E., Holdgate, M., Acreman, M.C. Weir, A. (Eds.), 1999. *Ecosystem Management; Questions for Science and Society*. Sibthorp Trust.

Marsh, G.P., 1864. Man and Nature. Physical Geography as Modified by Human Action. Charles Scribner & Sons, New York.

Millennium Ecosystem Assessment, 2005. Ecosystems and Human Well-being. Island Press, Washington, DC.

Ministry of Drinking Water and Sanitation, 2014. Swachh-Bharat Mission. Central Water Commission, New Delhi.

Moore, M. 2004. Perceptions and Interpretation of Environmental Flows and Implications for Future Water Resources Management: A Survey Study (M.Sc. thesis). Department of Water and Environment Studies, Linköping University, Sweden.

National Water Initiative, 2004. Intergovernmental Agreement on a National Water Initiative. Council of Australian Governments, Canberra.

Ofwat, 2015. Towards Water 2020 — Meeting the Challenges for Water and Wastewater Services in England and Wales. Ofwat, London.

Overton, I.C., Acreman, M.C., 2012. Environmental Flow Methods in Australia. CSIRO, Collingwood, Australia.

Overton, I.C., Smith, D.M., Dalton, J., Barchiesi, S., Acreman, M.C., Stromberg, J.C., et al., 2014. Implementing environmental flows in integrated water resources management and the ecosystem approach. Hydrol. Sci. J. 59 (3−4), 860−877. Available from: http://dx.doi.org/10.1080/02626667.2014.897408.

Rogers, K.H., Bestbier, R.X., 1997. Development of a Protocol for the Definition of the Desired State of Riverine Systems in South Africa. Department of Environmental Affairs and Tourism, Pretoria, South Africa.

Rowlston, W.S., Palmer, C.G., 2002. Processes in the development of resource protection provisions on South African Water Law. Proceedings of the International Conference on Environmental Flows for River Systems, Cape Town, South Africa, March 2002.

Saintilan, N., Overton, I.C. (Eds.), 2010. Ecosystem Response Modelling in the Murray-Darling Basin. CSIRO Publishing, Canberra, Australia.

Solomon, S., 2010. Water: the Epic Struggle for Wealth, Power, and Civilization. Harper Collins Publishers, New York.

Tharme, R.E., 2003. A global perspective on environmental flow assessment: emerging trends in the development and application of environmental flow methodologies for rivers. River Res. Appl. 19, 397−441. Available from: http://dx.doi.org/10.1002/rra.736.

Thompson, J.R., Laizé, C.L.R., Green, A.J., Acreman, M.C., Kingston, D.G., 2014. Climate change uncertainty in environmental flows for the Mekong River. Hydrol. Sci. J. 59 (3−4), 935−954.

UNDP, 2012. The Human Development Concept. United Nations Development Programme, Nairobi, Kenya.

Weir, J.K., 2010. Cultural flows in the Murray River country. Aust. Human. Rev. 2010, 48.

Wilson, N., 2009. Encyclopedia of Ancient Greece. Routledge, New York, London.

WWF, 2012. Summary report. Assessment of environment flows for Upper Ganga Basin. WWF, New Delhi, India.

Zavialov, P.O., 2005. Physical Oceanography of the Dying Aral Sea. Springer, Chichester, UK.

Ziv, G., Baran, E., Rodriguez-Iturbe, I., Levin, S.A., 2012. Trading-off fish biodiversity, food security, and hydropower in the Mekong River Basin. Proc. Natl. Acad. Sci. 109, 15.

UNDERSTANDING HYDROLOGICAL ALTERATION

3

Michael J. Stewardson[1], Mike Acreman[2], Justin F. Costelloe[1], Tim D. Fletcher[1], Keirnan J.A. Fowler[1], Avril C. Horne[1], Gaisheng Liu[3], Michael E. McClain[4], and Murray C. Peel[1]

[1]*The University of Melbourne, Parkville, VIC, Australia* [2]*Centre for Ecology and Hydrology, Wallingford, United Kingdom* [3]*The University of Kansas, Lawrence, KS, United States* [4]*IHE Delft, Delft, Netherlands*

3.1 INTRODUCTION

Environmental variability is a dominant feature of freshwater habitats. Much of this variability is produced by dynamics in the Earth's water cycle at scales of individual storm events, seasons, and multiyear sequences of wet and dry years. In particular, hydrological variability reflects fluctuating precipitation and *evapotranspiration (ET)* associated with climate, vegetation, and other hydrological controls on the water balance at the land surface. Precipitation in excess of ET enters the terrestrial phase of the water cycle and drains through a catchment along multiple surface and subsurface flow paths (Fig. 3.1). These flow paths may include particularly long travel times produced by periods of transient storage in soil, snowpack, natural lakes, wetlands, and aquifers. Aquatic ecosystems experience a characteristic water regime as a net result of these upstream runoff generation, drainage, and storage processes, including interactions with the groundwater store.

Over the last half century, a vast body of literature has established hydrological variability as the dominant abiotic control on the structure, function, and *integrity* of freshwater ecosystems (Poff et al., 1997). Importantly, hydrological variability largely determines the extent, depth, and duration of surface inundation; flow velocities and hence advective transport of nutrients and contaminants; hydrodynamic forces on objects within the water body; and both mass exchange and connectivity between aquatic units including floodplains, the hyporheic zone, and river reaches.

Human use of land and water has altered the terrestrial phase of the global water cycle, leading to widespread impacts on freshwater ecosystems. Such freshwater ecosystems are described as flow stressed. Many environmental water managers have sought to minimize these ecological impacts with a particular focus on rivers affected by large dams. However, large dams are not the only cause of hydrological change. Managers may also need to consider environmental water requirements in freshwater ecosystems affected by farm dams, surface water diversion, groundwater extraction, and modification to natural drainage networks for flood protection. There is also emerging interest in environmental water management for rivers affected by urbanization and land clearance, both of which alter catchment runoff generation and hence streamflow volumes. These are all discussed in this chapter.

Water for the Environment. DOI: http://dx.doi.org/10.1016/B978-0-12-803907-6.00003-6

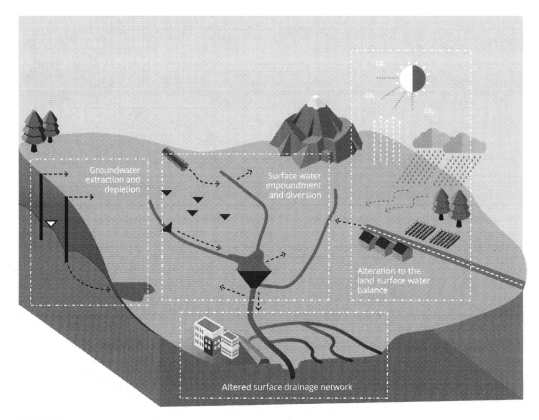

FIGURE 3.1

The hydrologic cycle and hydrological alteration.

We organize hydrological human disturbance into four groups based on the component of the terrestrial water cycle affected (Fig. 3.1):

1. First, we consider human activities that alter the land surface water balance. These include: climate change, effects of increased CO_2 concentration on plant water use, changes in vegetation cover and particularly tree clearing, and the construction of impervious surfaces as part of urbanization.
2. The second type is the artificial impoundment and diversion of surface water. This includes operation of dams of all sizes from the many small dams constructed for farm water supply to large dams constructed mostly for water supply, hydropower generation, and flood control.
3. The third type is groundwater withdrawal.
4. The fourth type is the modification of the natural surface water drainage system including channelization and wetland drainage.

In combination, these human impacts have had a dramatic and widespread effect on the water regime of aquatic habitats worldwide. The chapter begins with a discussion of methods for assessing hydrological alteration. This is followed by a discussion of these four types of hydrological disturbance.

3.2 ASSESSING THE LEVEL AND SIGNIFICANCE OF HYDROLOGICAL ALTERATION

Water regime refers to characteristics of the temporal and spatial variability in water volume and its movement within an aquatic habitat, including wetlands, aquifers, and estuaries; *streamflow regime* is more specific to the water regime of river channels. Much of the literature on hydrological variability focuses on streamflow regimes, characterized using widely available time series observations of river discharge. The streamflow regime of rivers can be considered a combination of flow freshes and baseflow(s). Flow freshes (or pulses, events, and spates) are a direct response to precipitation. The temporal characteristics of a flow fresh are similar to the rainfall event that produced it, allowing for the time lag for surface flow to points more distant downstream from the source catchment and dispersion of the flow wave. In contrast, baseflow(s) are produced by the drainage of natural catchment water storages including the soil and groundwater stores. In the absence of any replenishment from precipitation, baseflow(s) can be expected to decline gradually as the stores drain. The rate at which stores decline can vary widely up to many decades for large confined continental-scale groundwater systems. Seasonal variation in water stored within a catchment produces seasonal variation in baseflow(s).

Stream discharge is an obvious variable for environmental flows assessments because of widespread availability of modeled or observed discharge time-series. However, discharge may not have direct relevance for studies assessing environmental water requirements. The limitations of *discharge* for characterizing environmental water requirement are discussed in Chapter 13, along with a discussion of alternate ecohydraulic measures that can have more direct relevance to aquatic processes and organisms. Measures such as water depth, extent, volume, and velocity may provide alternate measures for use in environmental flows assessment studies, and in particular, specifying environmental water requirements. Fortunately, these alternate measures can be closely related to streamflow, with water depth, volume, extent, and velocity generally increasing with river discharge. Furthermore, many of the statistical and modeling methods widely used to assess streamflow regimes can be adapted or extended for analysis of these other closely related water regime variables.

Evaluating the level and significance of hydrological alteration is a core task in environmental water planning. The focus of such evaluations has evolved over the last four decades with advances in understanding of the role of water regime in river ecosystems (Stewardson and Gippel, 2003). The *Instream Flow Incremental Methodology (IFIM)* (Bovee, 1982), which applied the widely used *Physical Habitat Simulation (PHABSIM)* method (Bovee and Milhous, 1978), also included recommendations for evaluating hydrological variability. However, these aspects of IFIM were frequently overlooked in PHABSIM studies that focused on identifying optimum or minimum flows to protect habitat for individual species. Perhaps in response, the natural flow paradigm emphasized the importance of all aspects of flow variability for the integrity of river ecosystems including "magnitude, frequency, duration, timing and rate of change of hydrologic conditions" (Poff et al., 1997) and has been widely adopted in environmental water management (Lytle and Poff, 2004; Richter et al., 1997). These different aspects of flow regimes have been characterized using a very wide range of hydrological statistics (Olden and Poff, 2003). It has become common practice to evaluate the human impact on hydrological variability by comparing flow regimes subject to water management with a flow regime without human alterations (including Clausen and Biggs, 2000; Olden and

Poff, 2003; Puckridge et al., 1998; Richards, 1989, 1990; Richter et al., 1996). Many of these studies use generic hydrological statistics with no specific link to ecosystem function. Approaches such as the Flow Events Method (Stewardson and Gippel, 2003) or Environmental Flow Components analysis (Mathews and Richter, 2007) incorporate ecological knowledge into the characterization of flow variability.

Measures of hydrological alteration attempt to represent changes in the magnitude, seasonality, and variability in the water regime relative to a reference state, usually conditions in the absence of human impacts. This reference has been described as the *natural flow regime* or *predevelopment flow regime* where the reference state might retain some anthropogenic hydrological alteration such as land clearance. These measures include metrics that describe the frequency, timing, and duration of high- and low-flow events. Olden and Poff (2003) highlight the importance of careful selection of the indices to ensure redundant indicators are minimized and suggest that a subset of the available indices can be used to represent the predominant features of the hydrological regime. Careful specification of methods is required to ensure reproducibility. In particular, analysis of high- or low-flow events can be particularly sensitive to criteria used to define start and end of events. Gordon et al. (2004) provides a thorough description of the various standard statistical approaches to analyzing streamflow series.

More recently, it has been suggested that the natural or predevelopment flow regime may not be a useful reference or benchmark for designing environmental water regimes, particularly in rivers with a long history of flow alteration (Acreman et al., 2014). In such rivers, it may be more appropriate to design environmental water components to achieve specific ecosystem functions regardless of the natural flow regime. With this approach, we must choose measures of flow variability with some knowledge of their relationship to ecosystem function (Box 3.1).

BOX 3.1 DERIVING A NATURAL FLOW SERIES

An assessment of hydrological alteration in rivers usually requires a streamflow series representing impacted conditions and a streamflow series representing a reference or target state. Derivation of these two streamflow series can itself pose a challenge. There can be significant nonstationarity in gauged streamflow as a result of changes in water- or land-use management and long-term climate variability. There are a number of different approaches to deriving a target streamflow series with each approach including different assumptions and uncertainties. These challenges are discussed further in Chapter 15.

Despite the challenges and assumptions required in deriving measures of hydrological alteration, the clear benefit is the ability to assess regional hydrological alteration in a way that allows comparison both within and across catchments. This makes it a valuable planning tool to highlight potentially flow-stressed systems that require more thorough investigation into environmental water needs.

In catchments with significant levels of human activity during the time of gauged records, modeling techniques are required to derive an estimated target flow series (Blöschl et al., 2013). Three approaches are described here.

1. *Rainfall Runoff Modeling:* A rainfall runoff model may be calibrated to gauged data from the preimpact time period. This model can then be used to estimate reference streamflows for the postimpact period. Abstractions/returns may be added to create any scenario of target regime. This approach requires that streamflow records commence sometime before the onset of hydrological alteration.

(Continued)

BOX 3.1 (CONTINUED)

2. *Streamflow Transposition:* If there is no record available for a site prior to the onset of hydrological alteration, then streamflow records from a nearby unimpacted catchment may be *transposed* to estimate the unimpacted streamflows at the site of interest using a transposition factor (Lowe and Nathan, 2006). There is a significant body of work around the transposition of information from paired or experimental catchments. This is particularly relevant when assessing the alterations due to land use changes (Brown et al., 2005). This method requires that there are streamflow records available for a nearby *hydrologically similar* catchment, which is unimpacted by human disturbances.

3. *Reversing Water Use and Regulation:* Where there is information available about the level of human impact (including the timing and volume of water extractions and catchment changes such as farm dam development), these effects can be removed from the gauged record for the catchment. In a river system managed for consumptive water supply, there may be a water resource model that represents the hydrological impacts of the water supply system. These water resource models can often be used to model streamflows that would have occurred in the absence of water resource developments.

3.3 ALTERATION TO THE LAND SURFACE WATER BALANCE

The land surface water balance refers to the balance of precipitation and ET, with the residual providing a contribution to the terrestrial phase of the water cycle. This balance can be altered through changes in meteorological drivers produced by anthropogenic climate change. Current understanding of hydrological impacts of climate change is synthesized by Jiménez-Cisneros et al. (2014) and will only be briefly described here. Climate change is understood to alter precipitation and ET producing reductions in surface water yields and groundwater recharge in most dry subtropical regions of the world and increasing yields at high latitudes. It also generally increases streamflow variability with more frequent extreme floods (Arnell and Gosling, 2016) and droughts. Further, rising temperatures are leading to a decline in snow in various parts of the world and hence the seasonal distribution of runoff. Also, the enhanced atmospheric CO_2 concentrations can affect the land surface water balance by altering vegetation physiology (i.e., stomatal response) and foliage area (Piao et al., 2007), although the net effect is unclear (Ukkola and Prentice, 2013).

In addition to changes in climate drivers, anthropogenic changes produced by land management can alter the land surface water balance including vegetation changes and urbanization. Climate change impacts are only briefly described above but are nonetheless potentially significant sources of hydrological alteration, which should be considered in environmental water planning. The following sections detail effects of vegetation change and urbanization.

3.3.1 VEGETATION CHANGE

The most widespread hydrological impact of humans is the removal of almost half of the world's trees (Crowther et al., 2015). Vegetation plays a key role in the ET component of the surface water balance of a catchment via transpiration of water through leaf stomata and evaporation of intercepted precipitation from vegetated surfaces. Direct measurement of transpiration and interception over large regions is difficult, leading to large uncertainty in the relative contributions to total ET over the global land surface of transpiration (38%−77%) and interception (10%−27%) (Blyth and Harding, 2011; Crockford and Richardson, 2000; Jasechko et al., 2013; Miralles et al., 2011; Schlesinger and Jasechko, 2014; Wang et al., 2007, 2014). Transpiration and interception varies between vegetation types (forest, shrubs, grasses; Schlesinger and Jasechko, 2014; Wang et al., 2007) and anthropogenic modification of catchment vegetation through deforestation has been known to increase runoff since antiquity (see reference to Pliny the Elder in Andréassian, 2004). In modern times, paired catchment studies show higher actual ET from forested catchments relative to nonforested catchments and hence reduced streamflow volumes and in particular reductions in baseflow(s) (Andréassian, 2004; Brown et al., 2005, 2013; Peel, 2009), a result confirmed in a global comparative hydrology study of nonpaired catchments (Peel et al., 2010). Sterling et al. (2013), investigating the impact of global anthropogenic land-use change on actual ET, noted that approximately 41% of the Earth's land surface has been altered. Conversion of wetlands and forests to nonirrigated agriculture and grazing has decreased ET, whereas conversion of barren land to irrigated cropland or submersion of land under reservoirs has increased ET (Sterling et al., 2013).

In addition to direct modification of vegetation, indirect impacts on vegetation are expected through anthropogenic climate change. An enriched CO_2 atmosphere is expected to modify vegetation in three ways: (1) reducing stomatal opening, thus reducing transpiration and increasing water use efficiency; (2) increased growth; and (3) changes in species composition (Field et al., 1995). Although forest water use efficiency has increased (Keenan et al., 2013; Peñuelas et al., 2011), forest growth has not increased as expected possibly due to increased temperature and water and nutrient limitation (Peñuelas et al., 2011). ET rates can also be enhanced in regions with widespread irrigation, intensifying the hydrological cycle and leading to increased rainfall and runoff (Lo and Famiglietti, 2013).

3.3.2 URBANIZATION

The hydrological impact of land-use change is most obvious in urban areas and specifically the effects of impervious areas and constructed drainage systems (Walsh et al., 2012). Urbanization has profound impacts on the water balance, through alterations to the land surface (by creation of impervious areas, soil modification, and compaction), vegetation (typically a reduction in vegetation cover, but not universally), and critically, through the modification of drainage pathways (by construction of hydraulically efficient drains, pipes, and channels).

Creation of impervious areas (roads, roofs, footpaths, etc.) is perhaps the most obvious impact of urbanization. On natural pervious surfaces, with the exception of very large or long precipitation events that will result in surface runoff (Hill et al., 1998), much of the precipitation is intercepted by vegetation or infiltrates into porous soils where it will then either be evapotranspired or make its way to groundwater and potentially contribute to stream base flow (Hamel et al., 2013; Price, 2011). Typical runoff coefficients (the annual proportion of precipitation that becomes streamflow) from forested and grassland catchments range from 0.05 to 0.4 (Fig. 3.2), depending on precipitation (Zhang et al., 2001). In contrast, approximately 90% of precipitation landing on impervious surfaces becomes runoff (Boyd et al., 1993).

Changes to the properties of remaining pervious areas in the urban landscape can also influence the water balance. Soil stripping, a loss of soil organic matter, and soil compaction all result in decreased infiltration and increased surface runoff (Konrad and Booth, 2005). Changes to vegetation will also have an impact (as discussed above). In most cases there will be a significant reduction in overall ET, as a result of reduced vegetation cover (Grimmond and Oke, 1999), but in some cases, for example in arid areas where urbanization results in water imports and planting of higher water-using vegetation, the net ET may actually increase (Bhaskar et al., 2016).

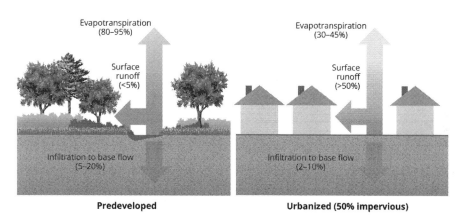

FIGURE 3.2

Changes to the urban water balance as a function of urbanization.

Although the creation of impervious surfaces is often cited as the principal cause of increases to the runoff coefficient in urban catchments, several authors have demonstrated that the creation of hydraulically efficient drainage networks (pipes, drains, and channels) is at least as important. For example, Leopold (1968) identified the critical role that the drainage connection has in determining the hydrologic consequences of a given impervious area. More recently, Walsh et al. (2012) showed that the flow regime in two catchments with similar impervious percentage but with different drainage configuration (one with a stormwater pipe network discharging to the stream, the other with many informal drainage lines, and pipes that discharge into forest several hundred meters from the stream) differed substantially. Such a distinction is important, because it points to possible alternative stormwater management approaches to avoid disturbance to flow regimes.

In addition to the physical changes to catchment properties, urbanization commonly involves the import (water supply) or export (wastewater disposal) of water, with the potential to change the catchment water balance and streamflow regime. For example, leakage from water and wastewater networks can contribute to groundwater and streamflows, whereas groundwater or surface water extraction may deplete streamflows (Hamel et al., 2013; Price, 2011).

Changes to water balance and streamflow regime as a result of urbanization are well-documented and include an overall increase in the streamflow volume (Fig. 3.3) and an increase in the frequency and magnitude of peak flows. As an example, Wong et al. (2000) showed that the

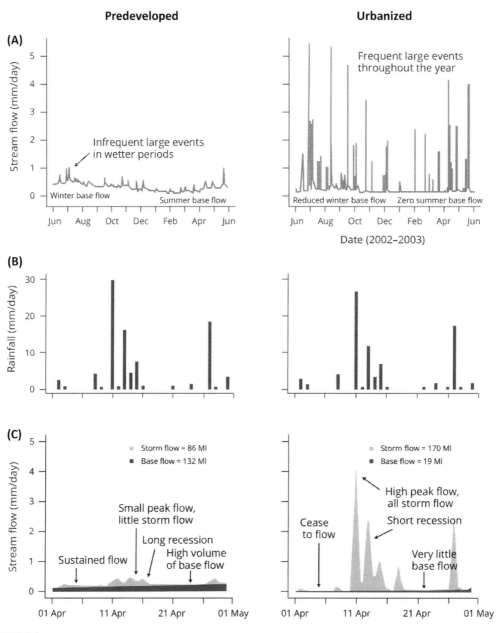

FIGURE 3.3

Typical changes to flow regimes as a consequence of urbanization including: (A) an annual hydrograph; and a 1-month sequence of (B) rainfall and (C) runoff separated into base flow and storm flow.

magnitude of the flow that occurs on average once per year would increase by an order of magnitude, even at moderate levels of imperviousness, whereas Burns et al. (2013) showed that the number of days with catchment-wide surface runoff discharging to the stream would increase from less than 5 days per year on average to around 100 days, in a Mediterranean climate. Such an increase can be explained by the loss of the *sponge effect* offered by vegetated pervious soils; Hill et al. (1998) demonstrated that, in southeastern Australia, around 25 mm is needed to generate runoff from grassland or forest areas, whereas runoff from impervious areas is generated after around 1 mm of rainfall (Boyd et al., 1993).

Urbanization commonly produces a decrease in base flow (Hamel et al., 2013) as a result of the reduced infiltration and thus contributions to groundwater, caused by the creation of formally drained impervious areas. However, this response is not universally observed, with factors such as anthropogenic water sources and loss of ET (through loss of vegetation) being observed to maintain or even increase groundwater and base flow in some situations (Bhaskar et al., 2016). Urbanization also decreases lag time (time between rainfall commencement and the delivery of flow to the receiving water) and recession time (time for the hydrograph to recede to its prerainfall event magnitude). This can have unexpected downstream impacts if tributaries drain urban and nonurban catchments. As a result, flood peaks may coincide or become separated (i.e., flood peaks may increase or decrease) with the result depending on the position of the rural and urban land within the catchment.

Urbanization also results in major changes to water quality in receiving waters, in part through the generation of pollutants associated with urban land use and activities, but also in part due to the increased capacity of impervious surfaces and hydraulically efficient urban drainage systems to mobilize and transport pollutants, respectively (for a full review of these issues, see Fletcher et al., 2013; Novotny and Olem, 1994). Pollutants may include sediments, nutrients, toxicants, organics, as well as emerging pollutants such as micro-pollutants, herbicides, pesticides, and endocrine disruptors (Fletcher et al., 2013). The mobilization and subsequent transport of pollutants involves the deposition on urban surfaces during both wet and dry weather and the subsequent washoff from these surfaces during rainfall, followed by transport in the stormwater pipe network. Not surprisingly, therefore, Hatt et al. (2004) demonstrated that *effective imperviousness*, the proportion of a catchment made up of impervious surfaces directly connected to receiving waters, was a better predictor of instream water quality than other factors such as the density of septic tanks, unsealed roads, or the total impervious area (ignoring the nature of the drainage connection).

3.4 SURFACE WATER IMPOUNDMENT AND DIVERSION

Humans have constructed artificial water storages (also referred to as reservoirs or dams) across the world to retain surface water for a range of purposes including consumptive water use, hydropower generation, and to reduce flood peaks. The total cumulative storage capacity of large dams is currently 20% (ICOLD, 2007) of the natural annual river discharge to the world's oceans, estimated to be around 40,000,000 Gl/year (Müller Schmied et al., 2014) although estimates vary considerably (Wada and Bierkens, 2014). However, human water withdrawals are less than the total storage volume at around 10% of the global discharge (Wada and Bierkens, 2014; Wada et al., 2011). These withdrawals combined with operation of dams have substantial effects on at least half of the world's rivers frequently characterized by reduced discharge (Müller Schmied et al., 2014), reduced amplitude of seasonal flow variations (Biemans et al., 2011), increasing frequency of low annual flows (Wada et al., 2013; Wanders and Wada, 2015), and river fragmentation (Grill et al., 2015). A global overview of dam-based impacts shows that over half of the world's largest river basins (172 out of 292) are fragmented by dams, including the eight most biogeographically diverse basins (Nilsson et al., 2005). A review of US dams showed that dams altered river hydrology on a national scale (Magilligan and Nislow, 2005b).

The hydrological effects of these storages depend primarily on their purpose and size relative to inflow, but details of the design of the dam including water release structures, and operational policies also produce a wide variety of impacts even for storages managed for the same purpose. The following sections describe the effects of four important types of storage: large water supply reservoirs, hydropower dams, flood alleviation dams, and small farm dams.

3.4.1 LARGE WATER SUPPLY DAMS

Water supply dams (or reservoirs) are constructed to hold water during periods when streamflows exceed volumes needed for human use, to make it available during dry periods when demands exceed streamflows. The effect of water supply dams on downstream flow regimes depends on their size, point of water withdrawal (at the dam or downstream), patterns in water demand, and operating policy. Dams with storage capacity that is large relative to inflow volumes can capture all inflows except the exceptionally large floods when the dam fills completely and then overflows through a spillway.

Water may be diverted directly from the reservoir. However, very large dams are frequently constructed in the river headwaters and water is released into the river and diverted from the river some distance downstream closer to where it is required. In this case, streamflow immediately downstream of the dam is typically increased during the drier months when consumptive demands are at their peak and decreased at wetter times with lower consumptive needs. This effect is most severe with irrigation release storages where irrigation demands occur during the dry season. In some cases, this results in a complete reversal of the seasonal pattern of streamflow (Cottingham et al., 2010). Downstream of the water withdrawal point, streamflows are typically reduced year-round except when the dam is spilling. Unregulated tributary inflows can mitigate the severity of hydrological alteration with distance downstream of the dam and water diversion.

3.4.2 HYDROPOWER DAMS

Roughly 16% of the 58,000 dams in the World Register of Dams are currently fitted for hydropower generation, contributing more than 1000 GW of installed potential (ICOLD, 2015; IHA, 2015). Since 2004, the global pace of hydropower development has been increasing rapidly, with a 27% increase in installed capacity, mainly in Asia, Latin America, and to a lesser extent Africa (WEC, 2015). Growth is slower in the already developed parts of North America and Europe, but there too more growth is expected. Projections are that over the next three decades the global installed capacity will double to more than 2000 GW (WEC, 2015). Some of this will come by refitting existing dams with hydropower, but most will come from the construction of new dams. Those already under construction or planned are also concentrated in Asia (e.g., the Mekong), South America (e.g., the Amazon), and Africa (e.g., the Nile; Zarfl et al., 2015).

The theoretical relationship describing the power (P in watts) that can be generated from falling water is $P = \rho g Q H$, where ρ and g are density (approximately 1000 kg/m^3) and acceleration of gravity (9.81 m/s^2), respectively, Q is the water flow (m^3/s), and H is the falling height (head, m). As ρ and g are constants, the main design and operation parameters of hydropower generation are water flow and head. A maximum flow capacity and minimum head will be built into the design, and each of these variables may also be regulated depending on the storage capacity and other design elements built into the hydropower project. With these considerations in mind, there are three main types of hydropower: storage, run-of-river, and pumped storage, which differ in their effects on river flow regimes.

Storage hydropower uses water released, as needed, from a reservoir that often holds water for multiple uses. The sum effect of the reservoir on river flows will therefore include the effects already described above for large water supply reservoirs, but releases to meet seasonal, daily, and subdaily electricity demands will also impart a unique hydroelectric signature on downstream flows (Magilligan and Nislow, 2005a). Seasonal changes in demand are mainly related to differing electricity use to warm indoor spaces in winter and cool them in summer. Weekly patterns often include higher electricity use during weekdays and reduced use on weekends. In addition, subdaily patterns involve maximizing electricity generation during high-demand hours, which are typically during waking hours (e.g., 07:00−21:00 daily), with peak demand in the evening.

Run-of-river hydropower involves minimal storage and diverts water from the river at a more constant rate. Minimal storage may allow for some flexibility on a subdaily basis (e.g., hydropeaking), but the main objective of these schemes is usually to contribute base load power to the electrical grid. The effect of run-of-river projects on river flow regimes is therefore concentrated at lower flow levels, having the potential for full control when river flows are less than the intake capacity of diversion channels.

Conversely, pumped storage hydropower is intended for use during periods (usually daily) of peak power demands. In these schemes, excess base load power during low-demand periods (e.g., late night) is used to pump water to upper reservoirs, which may or may not lie on the same river. The effect on river flows, especially in conjunction with run-of-river schemes, is to withdraw an additional fixed volume of water from the river daily, only to be returned as an added flow during peak electricity demand periods.

Although hydropower is a largely nonconsumptive use of water (disregarding evaporation from hydropower reservoirs and water losses in transmission structures), river reaches between the water intake point and eventual water return point will experience flow reductions equal to what has been diverted. For very high dams, where head can be fully regulated locally, water may be released directly from the dam to downstream reaches, but in many schemes additional head is gained by locating power plants at some point downstream of the intake point. In these schemes the dewatered reaches may extend over many kilometers, and in the case of interbasin transfers water may never be returned to the source river.

The magnitudes of flow diversions vary depending on the scale of the hydropower scheme. The lower extremes of hydropower generation lie at the household level using pico hydropower systems (up to 5 KW). At the other extreme, the Three Gorges Hydroelectric Plant on the Yangtze River in China has the potential to generate 22.5 GW of power using nearly 1000 m^3/s. What is more important, however, is the volume of flow diverted (design discharge) of the scheme relative to the natural discharge in the river. Many considerations go into the determination of the design discharge, including physical features of the site, river hydrology, environmental water requirements, other water demands, electricity demand, and other environmental, social, and economic factors. In a run-of-river scheme without other large competing water demands, the standard design discharge will lie near the 30th percentile value of the annual flow duration curve (i.e., annual flow exceeded in 30% of years) or in the range of 100%−150% of the mean annual flow (IFC, 2015). From a design perspective, these flow levels would be considered sufficiently reliable to warrant alignment of other infrastructure design elements.

Flow alterations linked to hydropower affect aquatic ecosystems on multiple spatial and temporal scales, but the most distinctive impact of hydropower likely comes from the subdaily variations related to hydropeaking (Cushman, 1985). This operational practice causes frequent and rapid changes in downstream flow regimes, which may remain within normal ranges of flow magnitude in the reach, but are decidedly unnatural in their frequency, rate of change, and number of reversals (Bevelhimer et al., 2015). This causes extreme variations in hydraulic, thermal, and even water quality conditions. Of particular concern is the effect of hydropeaking in creating cycles of repeated wetting and drying on the margins of the river (Fisher and LaVoy, 1972; Geist et al., 2005; Korman and Campana, 2009; Strayer and Findlay, 2010).

3.4.3 FLOOD ALLEVIATION DAMS

Floods can be a major cause of human fatalities, infrastructure damage, and disruption (Jonkman and Kelman, 2005). One option for managing flood hazards is using dams or detention basins to reduce flood peak magnitudes (Green et al., 2000). Many dams worldwide are operated to provide some degree of flood alleviation, although most are multipurpose and serve other functions such as hydropower generation (Bakis, 2007). Very recently, dams have also been used to generate floods where they are required for environmental purposes (Acreman, 2009), although the capacity of dam release valves is an upper limit on controlled releases for this purpose (McMahon and Finlayson, 1995). Detention basins are small storages commonly used to store floodwaters in urban catchments (Guo, 2001). In contrast to large dams, detention basins are typically offstream in lowland areas, filling only with the rise in floodwaters (Green et al., 2000).

Dams operated for flood alleviation will generally reduce the frequency of floods for a range of flood peak magnitudes. For example, flood alleviation dams were reported to have reduced flood flows by half in the 1997 California floods (Green et al., 2000). Flood volumes may be less affected depending on the operation of the reservoir during the flood. Typically, dams and detention basins achieve flood alleviation by storing water during the flood peak. The capacity to attenuate flood peaks diminishes as flood volumes increase relative to the size of the dam. Once full, there is only a minor attenuation of flood peak as a result of flood wave attenuation through the reservoir. As stored water in flood alleviation dams needs to be released (to make room to store water during future floods), the downstream river often experiences extended periods of high flow when the river would normally be low, such as on the Green

River in Kentucky, United States, resulting in further ecological alteration (Richter and Thomas, 2007).

3.4.4 SMALL FARM DAMS

Private landowners, and particularly farmers, often build smaller dams to impound water for consumptive use onsite. These may be built on a watercourse or offstream and filled via surface runoff (including constructed drainage diverting runoff to the dam) or pumping water from a nearby watercourse. Although the resulting water bodies are typically small (less than 100 ML) (Nathan and Lowe, 2012), the cumulative impact of many such water bodies across a landscape may be significant both on hydrology (Fowler et al., 2016) and ecology (Mantel et al., 2010a, 2010b). These privately owned water bodies are known as *farm ponds* in the United States (Arnold and Stockle, 1991), *tanks* in India, and *farm dams* in Australia and Africa (Hughes and Mantel, 2010; Schreider et al., 2002). We use the term *farm dams* here.

Farm dams cause reductions in river flows because, in addition to capturing precipitation or sometimes storing water pumped from rivers, they intercept local surface runoff that may otherwise flow into streams and rivers (Schreider et al., 2002). Understanding hydrological alterations caused by farm dams requires consideration of farm dam size (i.e., maximum volumetric capacity), location relative to the stream network, and key water fluxes into and out of the water bodies such as precipitation, evaporation, inflows due to surface runoff, consumptive use, seepage, and downstream spills or releases (Arnold and Stockle, 1991; Nathan et al., 2005). There may be hundreds of farm dams distributed across a catchment, so generalizations are sometimes necessary to develop catchment-scale insights (Lowe et al., 2005). However, in hotter climates evaporation is likely to increase markedly due to creation of open water surfaces.

Various studies have estimated reductions in flow caused by farm dams. In most cases, the reported interception is less than 5% of mean annual flow (Chaney et al., 2012; Neal et al., 2001; Schreider et al., 2002), but cases in excess of 10% have also been reported (Fowler and Morden 2009; Hughes and Mantel, 2010; Lett et al., 2009). Evidence from satellite imagery (Liebe et al., 2009), hydrologic simulation (Nathan et al., 2005), and field experience suggest that farm dams generally go through seasonal cycles of filling and emptying, and thus it is important to also consider impacts on a season-by-season basis. Although the rate of water interception may be greatest in absolute terms during times of high flow, the proportion of flows intercepted may be highest during times of low flow, as shown in Fig. 3.4 (Fowler et al., 2016; Mantel et al., 2010a). Depending on the number of dams, their size, and position in the landscape, flow reductions during seasonal dry periods can be relatively high compared to the annual averages stated above. For example, Fowler and Morden (2009) found that farm dams intercepted as much as 80% of flow during summer in irrigated horticultural regions of southwestern Australia. In response to potentially high seasonal impacts, some regulatory authorities have considered compulsory *low-flow bypasses*, devices that allow surface runoff to pass unimpeded around a dam during times of low flow (Cetin et al., 2013).

In addition to these changes in flow volume, farm dams may also cause a shift in the timing of the onset of seasonal high flows (Fowler and Morden, 2009), as the *first flush* of the wet season is absorbed to replenish storages rather than allowed to pass further downstream. Furthermore, farm dams may reduce the connectivity of the stream network regardless of season or flow magnitude (Malveira et al., 2011).

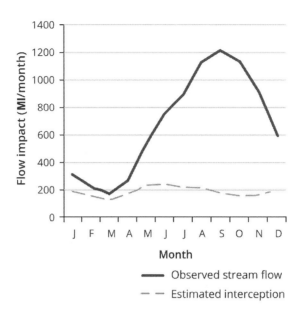

FIGURE 3.4

Estimated seasonal interception (i.e., diversion of water) by farm dams for a catchment in southeast Australia. The average flow at the catchment outlet (i.e., downstream of the dams, shown with a solid line), over a 47-year period, is compared on a month-by-month basis with the farm dam interception (shown with a dashed line) over the same period. The latter was estimated using a water balance model. The flow that would have occurred at the outlet in the absence of farm dams is the sum of the two lines (not shown). This case study was based on a catchment with relatively high farm dam development, with an estimated combined farm dam volume of 3150 ML and average annual flow of 8500 ML/year (116 mm/year).

Source: Adapted from Fowler et al. (2016)

Farm dam construction may occur in response to changes in land use, particularly to provide semi-permanent water sources for livestock. For example, in the 1930s, government initiatives in the United States constructed thousands of farm dams to assist transitions from degraded cropland to pasture (Walter and Merritts, 2008). Farm dams can supply or supplement water for irrigation (Wisser et al., 2010) and be used in periurban areas for domestic supply or as ornamental lakes (Ignatius and Jones, 2014). Severe droughts may lead to accelerated development of farm dams as landholders seek alternative water supplies, a trend observed in Australia (GA, 2008) and Africa (Hatibu et al., 2000). Thus, an assessment of hydrological alteration due to farm dams in a given catchment needs to consider the currency of data, for example the age of aerial photography used to count dams and characterize their size.

3.5 GROUNDWATER EXTRACTION AND DEPLETION

Groundwater is the world's largest source of freshwater in liquid form. It provides drinking water to nearly half of the world's population and supplies 20%−40% of total water uses in agricultural, industrial, and domestic activities (Margat and Van Der Gun, 2013; Wada and Bierkens, 2014; Zektser and Everett, 2004). Groundwater withdrawal in 2010 was estimated to be 1200 km^3/year, of which almost 60% was for irrigated agriculture (Pokhrel et al., 2015), and 60% of the total withdrawal was not replenished, resulting in a net loss of groundwater storage. Satellite-based observations and modeling of groundwater storage change also indicate high rates of groundwater depletion at 113 km^3/year (Doll et al., 2014). Falling water levels affect surface water systems that are dependent on groundwater, including spring-fed wetlands, and streams with base flow from groundwater discharge, which is critical to sustaining flow during dry spells (Nevill et al., 2010).

Worldwide, groundwater extraction has been contributing to an increasing volume for irrigation and other water uses (Wada et al., 2014). Pumping from a groundwater well creates a cone of depression centered on the well that draws in groundwater from other parts of an aquifer. Spatially, broad-scale declines in water levels often result when pumping occurs from wells distributed over a large area of the aquifer. Groundwater extraction in hydraulically connected systems can result in decreases in river flow via decreasing or reversing the hydraulic gradient between the aquifer and the river so that the flow between them will decrease or reverse (in the latter case the river loses water to the aquifer). It is also common for the increased uptake of groundwater for irrigation to be linked to policy changes affecting some part of the water cycle or energy supply. In Australia, a 1994 cap placed on surface water extraction of the rivers of the Murray—Darling Basin without complementary groundwater reform led to increased groundwater extraction in many parts of the Basin (Nevill, 2009). India has the largest groundwater use of any country with the uptake significantly driven by subsidized energy policy (Scott and Shah, 2004), which has resulted in streamflow depletion (Nune et al., 2014) and water table lowering.

The disturbance to streamflow by groundwater pumping is typically concentrated on low-flow periods when groundwater discharge makes the greatest percentage contribution to streamflow. This can even lead to the cessation of streamflow for some periods. Groundwater level reductions can also have an effect on other components of the flow regime, such as bank storage. In streams with hydraulic connection between streamflow and groundwater, decreasing groundwater levels can lead to increases in bank storage capacity, that is, volume of alluvial aquifer available to store water during a high-flow event (Lamontagne et al., 2014). Bank storage can be an important component of transmission losses (Costa et al., 2012) and thus affect the delivery of environmental water to floodplains through natural flood mechanisms.

Groundwater level rises from anthropogenic activities can also result in hydrologic alteration. Recharge from excess irrigation water has been observed to lead to groundwater mounding and increased base flow into rivers in Montana (Kendy and Bredehoeft, 2006). Similarly, increased diffuse precipitation recharge can occur in response to land clearing of deep-rooted vegetation and lead to increased fluxes of groundwater to streams and rivers across a range of scales. This has been observed in small paired catchment studies (Brown et al., 2005) and is also a key driver of increased saline regional groundwater flow toward large river systems (Knight et al., 2005). In the middle and lower reaches of the Murray River, this process, in combination with river levels kept artificially high by locks and summer flow releases that limit the discharge of regional saline groundwater into the river, has led to the salinization of floodplain environments. This salinization has produced die-back and poor health of riparian tree assemblages, which become a key focus for the delivery of environmental water (Box 3.2).

BOX 3.2 GROUNDWATER DEPLETION IN THE HIGH PLAINS AQUIFER, CENTRAL UNITED STATES

The High Plains aquifer of the central United States is one of the well-known examples of groundwater depletion due to excessive irrigation pumping (Kustu et al., 2010; Luckey and Becker, 1999; Scanlon et al., 2012). Lying below portions of eight states with a total area of 454,000 km^2, it is one of the world's most productive aquifers and provides most of the water for a region of irrigated agriculture referred to as the *grain basket* of the United States (Fig. 3.5). The aquifer

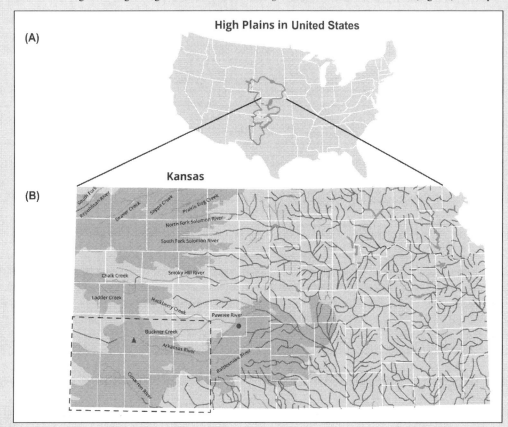

FIGURE 3.5

(A) Location of the High Plains aquifer in the United States (marked with polygons); and (B) the High Plains aquifer distribution in Kansas (shaded area). Major streams and rivers are shown in (B), where the lighter stream segments are perennial prior to intensive groundwater development and now become dry most of the year primarily due to pumping-induced groundwater level declines in the High Plains aquifer. The dashed box on the southwestern corner of the Kansas map indicates the area for which groundwater pumping from the High Plains aquifer is plotted in Fig. 3.6. The triangle and circle on the Arkansas River are the locations for which the measured river flow and a photo of dry riverbed are shown in Fig. 3.7.

Source: Fross et al. (2012)

(Continued)

BOX 3.2 (CONTINUED)

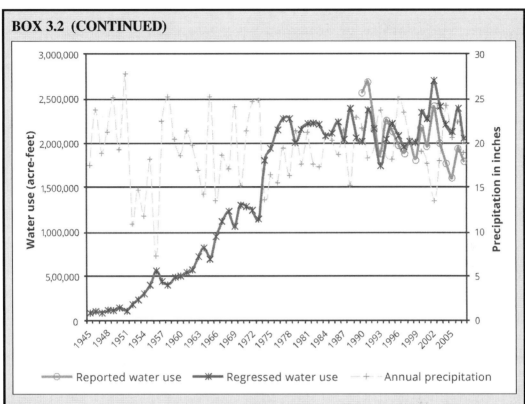

FIGURE 3.6

Groundwater pumping from the High Plains aquifer of southwestern Kansas (see Fig. 3.5 for the calculated area). The lighter solid (open circles) line is the reported water use, the darker (crosses) line is estimated water use based on regression analysis between climatic parameters and the reported water use.

is mainly composed of sands, gravels, silts, and clays, all of which were deposited from sediments washed off the Rocky Mountains to the west, as well as local sources, over the last several million years (Buchanan et al., 2015).

Much of the water in the aquifer is generally considered to originate from recharge during the wetter climate of the last ice age. The natural recharge rate is highly variable spatially, with high rates in the northern portion of the aquifer (25–210 mm/year) and lower rates in the central and southern portions (2–25 mm/year) due to the differences in climate and shallow soil conditions (higher precipitation and coarser soils in northern portion; Scanlon et al., 2012). As irrigation pumping has greatly exceeded recharge over the last few decades, groundwater storage has been quickly depleted in the central and southern portions of the aquifer, posing a significant threat to the long-term economic and ecological viability of the region.

Groundwater pumping for irrigation began to intensify in the High Plains aquifer of Kansas in the 1960s and reached the peak level by the late 1970s (Liu et al., 2010; Fig. 3.6). Owing to groundwater depletion, the pumping rate has gradually dropped over the past 30 years and portions of the aquifer have already been exhausted for large-volume water uses. The pumping-induced water level declines have also significantly decreased groundwater discharge to the

(*Continued*)

BOX 3.2 (CONTINUED)

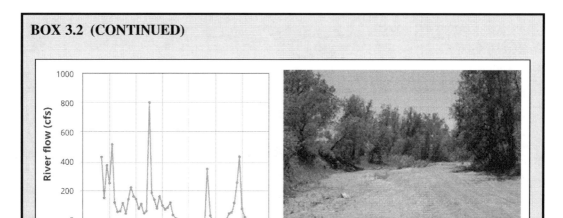

FIGURE 3.7

Measured Arkansas River annual flow at a US Geological Survey gage (left) and a photo of the dry riverbed (right). The locations of gauge and photo are shown in Fig. 3.5.

Source: Photo taken by Kansas Geological Survey. www.kgs.ku.edu/HighPlains/complete/mid_ark_model.shtml

rivers in the area (Fig. 3.7). The reduced groundwater discharge, and in many areas a reversed hydraulic gradient that causes water loss from rivers to the aquifer, have led to many formerly perennial rivers of the area becoming dry during most times of the year (Fig. 3.5). In addition to the damaging impacts of reducing or periods of no-flow on river ecology, the declining water table is found to have influenced the characteristics of riparian zone vegetation (Butler et al., 2007). Between 2003 and 2004, the water table was declining at a faster rate than the root network of the cottonwood trees could adapt to, causing the trees to become stressed. The inability of cottonwood to draw water from the rapidly declining water table was corroborated by the lack of diurnal water level fluctuation that were otherwise clearly noted in 2003 when the water table was within the reach of the cottonwoods' roots (Butler et al., 2007).

3.6 ALTERED SURFACE DRAINAGE NETWORK

The use of floodplains for agricultural and urban developments has intensified throughout the world. As a result, flood embankments and enlarged river channels with higher flow capacity have been constructed along many of the larger lowland rivers in populated areas. Flood embankments (also referred to as levees, dikes, and bunds) are typically earthen constructions that prevent flood-water spilling onto part or all of the floodplain (Green et al., 2000). Described as *river improvement*, channel capacity is sometimes increased by widening, deepening, straightening, or reducing the roughness of river channels. Unfortunately, such schemes have often increased flood risk such as on the Mississippi River (Criss and Shock, 2001). Although reducing flooding locally, loss of floodwater storage can mean significant increase in downstream flood risk. For example, projected economic costs of enhanced flooding from disconnecting floodplains along the Charles River, United States, prompted the purchase and reconnection of the floodplain to protect Boston rather than building structural defenses (Ogawa and Male, 1986). Fig. 3.8 shows similar results for the River Charwell (Acreman et al., 2003).

Engineering works to increase flood capacity of the river channel are frequently accompanied by wetland drainage schemes (Biebighauser, 2007). In rural area wetlands, these schemes reduce waterlogging of farmland and similarly in urban areas they allow for urban development. They reduce surface and subsurface water retention in floodplains and can have significant ecological impacts.

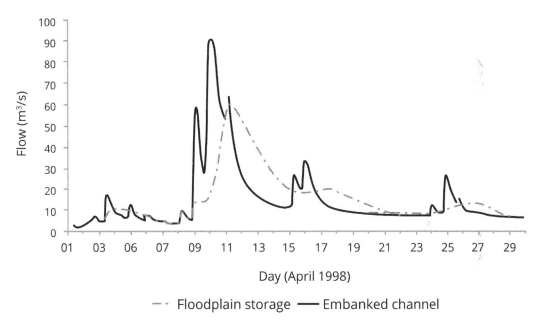

FIGURE 3.8

Observed flows on the River Cherwell (dashed line) compared with modeled flows (solid line) if floodplain storage is removed by embanking the river.

Source: Acreman et al. (2003)

Flood alleviation and wetland drainage works have isolated rivers from their floodplains, altered the hydrological regime, and significantly changed riverine ecosystems (Krause et al., 2011). By concentrating flow in the main channel of many rivers, flood works have reduced the surface area of floodplains and wetlands worldwide. In just 15 years, a series of global satellite observations shows a 6% decline in freshwater area, with greatest declines in regions experiencing rapid population increases in South America and South Asia (Prigent et al., 2012). In England and Wales, embankments and channel modifications have separated over 2/5 (42% by area) of all flood-plains (defined by the 100-year flood envelope) from their rivers (UK National Ecosystem Assessment, 2014). These hydrological impacts can entirely convert riverine habitats to terrestrial habitats, arguably the greatest impact on riverine ecosystems of all the hydrological alterations discussed in this chapter.

3.7 SYNTHESIS OF HYDROLOGICAL ALTERATIONS

Land and water management activities combined with climate change can alter baseflow(s), flood regimes, and flow variability. Table 3.1 summarizes the conditions leading to each of these hydrological impacts. Although these generalizations provide a useful guide to the impacts that can be expected, there can be great variation in the magnitude of these impacts and even the direction of change. These variations, in part, result from variation in the hydrological setting. For example, rivers with high streamflow variability at the scale of days to weeks may respond differently to construction of farm dams than rivers with a very predictable seasonal hydrograph. The type and size of water management infrastructure, along with operational policies can affect patterns of water demand and also play a significant part in shaping hydrological impacts. Indeed, these aspects often respond to the hydrological setting so that it is difficult to separate the relative effects of catchment hydrology and management. For example, larger water supply dams are generally required in catchments with great interannual variability in streamflow.

3.8 FUTURE OUTLOOK

It is clear that there has been global-scale impact of humans on the terrestrial component of the global water cycle including surface and groundwater and this has modified the water regimes of flowing and standing water ecosystems. There is little analysis of the distribution and severity of these impacts at the global scale. However, it is clear that in developed basins, land and water developments have resulted in dramatic hydrological change. It is of great concern that these hydrological pressures are all expected to increase in the future with ongoing deforestation (Crowther et al., 2015), urban expansion (Seto et al., 2011), increases in atmospheric CO_2 concentrations (IPCC, 2014), construction of new reservoirs (Grill et al., 2015), including many hydropower dams under construction or planned (Zarfl et al., 2015) and expansion of irrigation (Siebert et al., 2015). Significantly, the global volume of water withdrawals in excess of sustainable levels has increased since the late 1990s, despite a wetter climate during this period, and this trend is projected to continue over the coming century (Wada and Bierkens, 2014). This increasing trend in hydrological alteration needs to be met with a redoubled effort to implement environmental water management regimes to protect freshwater ecosystems.

Table 3.1 Synthesis of Hydrological Impacts Produced by Anthropogenic Disturbances

	Alteration to the Land Surface Water Balance	Surface Water Impoundment and Diversion	Groundwater Extraction and Depletion	Altered Surface Drainage Network
Base flows	**Reductions** Urbanization commonly reduces base flow. **Increases:** Deforestation can increase baseflow(s) by reducing transpiration. Climate change can alter seasonal patterns of base flow in rivers receiving snowmelt with reductions in snowmelt and increased runoff during winter. Climate change may also reduce baseflow(s) in dry subtropical regions and increase baseflow(s) in high latitudes.	**Reductions:** Water supply reservoirs can reduce downstream, baseflow(s) either during the wetter season when water is being stored for delivery in the drier season or year-round downstream of the location of water withdrawal. Farm dams produce a reduction in streamflows, generally year-round, which is more apparent in the drier periods. **Increases:** Dry season releases can increase baseflow(s) downstream of large water supply dams. Release of water from flood alleviation dams is necessary to maintain capacity to store floodwaters and this can lead to extended periods of increased baseflow(s).	**Reductions:** Groundwater extractions reduce baseflow(s) where groundwater would normally discharge to the stream and in extreme cases, perennial streams can become intermittent or ephemeral.	
Flood regime and floodplain connectivity	**Increases:** Urbanization increases magnitude of runoff during precipitation events. Climate change is increasing flood magnitude and frequency in many regions of the world.	**Reductions:** Water supply reservoirs and flood alleviation dams reduce the frequency and magnitude of floods with effect persisting with larger (less frequent) floods where the dam size is large relative to the mean annual streamflow. Reduced flood frequency and duration downstream of large dams will reduce floodplain inundation.		**Increases:** Flood alleviation works that eliminate flood storage can lead to increased downstream flood peaks. **Reductions:** Flood alleviation works isolate floodplains from the river channel reducing flood extents. Wetland drainage.
Discharge variability	**Increases:** Typically, urbanization leads to faster travel times through catchments and hence enhanced rates of streamflow rise and fall in response to storms. Climate change is expected to increase extremes in streamflows.	**Reductions:** Operation of water supply reservoirs generally reduces variability in streamflows. **Increases:** Hydropower dams can increase flow variations as flow releases respond to fluctuations in power demand.		**Increases:** Flood alleviation works that eliminate flood storage can lead to more rapid rise and fall in flood levels.

REFERENCES

Acreman, M.C., 2009. Senegal river basin. In: Ferrier, R., Jenkins, A. (Eds.), Handbook of Catchment Management. Blackwell, Oxford.

Acreman, M.C., Riddington, R., Booker, D.J., 2003. Hydrological impacts of floodplain restoration: a case study of the River Cherwell. UK Hydrol. Earth Syst. Sci. Discuss. 7 (1), 75−85.

Acreman, M., Arthington, A.H., Colloff, M.J., Couch, C., Crossman, N.D., Dyer, F.G., et al., 2014. Environmental flows for natural, hybrid, and novel riverine ecosystems in a changing world. Front. Ecol. Environ. 12 (8), 466−473.

Andréassian, V., 2004. Waters and forests: from historical controversy to scientific debate. J. Hydrol. 291, 1−27.

Arnell, N.W., Gosling, S.N., 2016. The impacts of climate change on river flood risk at the global scale. Clim. Change 134 (3), 387−401.

Arnold, J.G., Stockle, C.O., 1991. Simulation of supplemental irrigation from on-farm ponds. J. Irrigat. Drain. Eng. 117 (3), 408−424.

Bakis, R., 2007. Electricity production opportunities from multipurpose dams. Renew. Energy 32 (10), 1723−1738.

Bevelhimer, M.S., McManamay, R.A., O'Connor, B., 2015. Characterizing sub-daily flow regimes: implications of hydrologic resolution on ecohydrology studies. River Res. Appl. 31, 867−879.

Bhaskar, A.S., Beesley, L., Burns, M.J., Fletcher, T.D., Hamel, P., Oldham, C.E., et al., 2016a. Will it rise or will it fall? Managing the complex effects of urbanization on base flow. Freshw. Sci 35 (1), 293−310.

Biebighauser, T.R., 2007. Wetland Drainage, Restoration, and Repair. University Press of Kentucky, Lexington, Kentucky.

Biemans, H., Haddeland, I., Kabat, P., Ludwig, F., Hutjes, R.W.A., Heinke, J., et al., 2011. Impact of reservoirs on river discharge and irrigation water supply during the 20th century. Water Resour. Res. 47 (3).

Blöschl, G., Sivapalan, M., Wegener, T., Viglione, A., Savenije, H., 2013. Runoff prediction in ungauged basins: synthesis across processes, places and scales. Cambridge University Press, Cambridge.

Blyth, E., Harding, R.J., 2011. Methods to separate observed global evapotranspiration into the interception, transpiration and soil surface evaporation components. Hydrol. Process. 25 (26), 4063−4068.

Bovee, K.D., 1982. A guide to stream habitat assessment using the Instream Flow Incremental Methodology. Instream Flow Information Paper 12, FWS/OBS-82/26. US Fish and Wildlife Service, Fort Collins, Colorado.

Bovee, K.D., Milhous, R., 1978. Hydraulic simulation in instream flow studies: theory and techniques. Instream Flow Information Paper 5, FWS/OBS-78/33. US Fish and Wildlife Service, Fort Collins, Colorado.

Boyd, M.J., Bufill, M.C., Kness, R.M., 1993. Pervious and impervious runoff in urban catchments. Hydrol. Sci. J. 38 (6), 463−478.

Brown, A.E., Western, A.W., McMahon, T.A., Zhang, L., 2013. Impact of forest cover changes on annual streamflow and flow duration curves. J. Hydrol. 483, 39−50.

Brown, A.E., Zhang, L., McMahon, T.A., Western, A.W., Vertessy, R.A., 2005. A review of paired catchment studies for determining changes in water yield resulting from alterations in vegetation. J. Hydrol. 310, 28−61.

Buchanan, R.C., Wilson, B.B., Buddemeier, R.R., Butler Jr., J.J., 2015. The High Plains Aquifer. Kansas Geological Survey.

Burns, M.J., Fletcher, T.D., Walsh, C.J., Ladson, A.R., Hatt, B., 2013. Setting objectives for hydrologic restoration: from site-scale to catchment-scale (Objectifs de restauration hydrologique: de l'échelle de la parcelle à celle du bassin versant). In: Bertrand-Krajewski, J.-L., Fletcher, T. (Eds.), Novatech. GRAIE, Lyon, France.

Butler Jr., J.J., Kluitenberg, G.J., Whittemore, D.O., Loheide II, S.P., Jin, W., et al., 2007. A field investigation of phreatophyte-induced fluctuations in the water table. Water Resour. Res. 43, W02404.

Cetin, L.T., Alcorn, M.R., Rahmanc, J., Savadamuthu, K., 2013. Exploring variability in environmental flow metrics for assessing options for farm dam low flow releases. 20th International Congress on Modelling and Simulation, Adelaide, Australia. pp. 2430–2436.

Chaney, P.L., Boyd, C.E., Polioudakis, E., 2012. Number, size, distribution, and hydrologic role of small impoundments in Alabama. J. Soil Water Conserv. 67 (2), 111–121.

Clausen, B., Biggs, B., 2000. Flow indices for ecological studies in temperate streams: groupings based on covariance. J. Hydrol. 237, 184–197.

Costa, A.C., Bronstert, A., De Araújo, J.C., 2012. A channel transmission losses model for different dryland rivers. Hydrol. Earth Syst. Sci. 16, 1111–1135.

Cottingham, P., Stewardson, M.J., Roberts, J., Oliver, R., Crook, D., Hillman, T., et al., 2010. Ecosystem response modeling in the Goulburn River: how much water is too much. In: Saintilan N., Overton I. (Eds.), Ecosystem Response Modelling in the Murray-Darling Basin, pp. 391–409.

Criss, R.E., Shock, E.L., 2001. Flood enhancement through flood control. Geology 29, 875–878.

Crockford, R.H., Richardson, D.P., 2000. Partitioning of rainfall into throughfall, stemflow and interception: effect of forest type, ground cover and climate. Hydrol. Process. 14, 2903–2920.

Crowther, T.W., Glick, H.B., Covey, K.R., Bettigole, C., Maynard, D.S., Thomas, S.M., et al., 2015. Mapping tree density at a global scale. Nature 525, 201–205.

Cushman, R.M., 1985. Review of ecological effects of rapidly varying flows downstream from hydroelectric facilities. N. Am. J. Fish. Manage. 5 (3A), 330–339.

Doll, P., Muller Schmied, H., Schuh, C., Portmann, F.T., Eicker, A., 2014. Global-scale assessment of ground-water depletion and related groundwater abstractions: combining hydrological modeling with information from well observations and GRACE satellites. Water Resour. Res. 50, 5698–5720.

Field, C.B., Jackson, R.B., Mooney, H.A., 1995. Stomatal responses to increased CO_2: implications from the plant to the global scale. Plant Cell Environ. 18, 1214–1225.

Fisher, S.G., LaVoy, A., 1972. Differences in littoral fauna due to fluctuating water levels below a hydroelectric dam. J. Fish. Res. Bd. Canada 29, 1472–1476.

Fletcher, T.D., Andrieu, H., Hamel, P., 2013. Understanding, management and modelling of urban hydrology and its consequences for receiving waters; a state of the art. Adv. Water Resour. 51, 261–279.

Fowler, K.J.A., Morden, R.A., 2009a. Investigation of strategies for targeting dams for low flow bypasses. Hydrology and Water Resources Symposium, pp. 1185–1193.

Fowler, K., Morden, R., Lowe, L., Nathan, R., 2016. Advances in assessing the impact of hillside farm dams on streamflow. Aust. J. Water Resour.

Fross, D., Sophocleous, M., Wilson, B.B., Butler Jr., J.J., 2012. Kansas High Plains Aquifer Atlas. Kansas Geological Survey.

GA, 2008. Mapping the growth, location, surface area and age of man made water bodies, including farm dams, in the Murray-Darling Basin. Geoscience Australia Report for the Murray Darling Basin Commission, Canberra, Australia.

Geist, D.R., Brown, R.S., Cullinan, V., Brink, S.R., Lepla, K., Bates, P., et al., 2005. Movement, swimming speed, and oxygen consumption of juvenile white sturgeon in response to changing flow, water temperature, and light level in the Snake River, Idaho. Trans. Am. Fish. Soc. 134 (4), 803–816.

Gordon, N.D., McMahon, T.A., Finlayson, B.L., Gippel, C.J., Nathan, R.J., 2004. Stream Hydrology: an Introduction for Ecologists. Wiley, West Sussex, UK.

Green, C.H., Parker, D.J., Tunstall, S.M., 2000. Assessment of flood control and management options, thematic review IV.4 prepared as an input to the World Commission on Dams, Cape Town. Available from: http://www.dams.org.

Grill, G., Lehner, B., Lumsdon, A.E., MacDonald, G.K., Zarfl, C., Reidy Liermann, C., 2015. An index-based framework for assessing patterns and trends in river fragmentation and flow regulation by global dams at multiple scales. Environ. Res. Lett. 10 (1), 015001.

Grimmond, G.S.B., Oke, T.R., 1999. Evapotranspiration rates in urban areas. Impacts of Urban Growth on Surface Water and Groundwater Quality. Proceedings of IUGG 99 Symposium HS5. IAHS, Birmingham, UK (Publication No. 259).

Guo, Y., 2001. Hydrologic design of urban flood control detention ponds. J. Hydrol. Eng. 6 (6), 472–479.

Hamel, P., Daly, E., Fletcher, T.D., 2013. Source-control stormwater management for mitigating the effects of urbanisation on baseflow: a review. J. Hydrol. 485, 201–213.

Hatibu, N., Mahoo, H.F., Kajiru, G.J., 2000. The role of RWH in agriculture and natural resources management: from mitigating droughts to preventing floods. In: Hatibu, N., Mahoo, H.F. (Eds.), Rainwater Harvesting for Natural Resources Management: a Planning Guide for Tanzania, Regional Land Management Unit (RELMA). Swedish International Development Cooperation Agency (Sida), Nariobi, Kenya.

Hatt, B.E., Fletcher, T.D., Walsh, C.J., Taylor, S.L., 2004. The influence of urban density and drainage infrastructure on the concentrations and loads of pollutants in small streams. Environ. Manage. 34 (1), 112–124.

Hill, P., Mein, R., Siriwardena, L., 1998. How much rainfall becomes runoff?: Loss modelling for flood estimation. Cooperative Research Centre for Catchment Hydrology (Report 98/5), Melbourne, Australia.

Hughes, D.A., Mantel, S.K., 2010. Estimating the uncertainty in simulating the impacts of small farm dams on streamflow regimes in South Africa. Hydrol. Sci. J. 55 (4), 578–592.

ICOLD, 2007. World Register of Dams. International Commission on Large Dams, Paris, France.

ICOLD, 2015. World Register of Dams. International Commission on Large Dams, Paris, France.

IFC, 2015. Hydroelectric Power: a Guide for Developers and Investors. International Finance Corporation.

Ignatius, A.R., Jones, J.W., 2014. Small reservoir distribution, rate of construction, and uses in the upper and middle Chattahoochee Basins of the Georgia Piedmont, USA, 1950–2010. ISPRS Int. J. Geo-Inform. 3 (2), 460–480.

IHA, 2015. Hydropower Status Report. International Hydropower Association.

IPCC, 2014. In: Core Writing Team, Pachauri, R.K., Meyer, L.A. (Eds.), Climate Change 2014: Synthesis Report. Contribution of Working Groups I, II and III to the Fifth Assessment Report of the Intergovernmental Panel on Climate Change. IPCC, Geneva, Switzerland.

Jasechko, S., Sharp, Z.D., Gibson, J.J., Birks, S.J., Yi, Y., Fawcett, P.J., 2013. Terrestrial water fluxes dominated by transpiration. Nature 496 (7445), 347–350.

Jiménez-Cisneros, B.E., Oki, T., Arnell, N.W., Benito, G., Cogley, J.G., Döll, P. (Eds.), Climate Change, 2014. Impacts, Adaptation, and Vulnerability. Part A: Global and Sectoral Aspects. Contribution of Working Group II to the Fifth Assessment Report of the Intergovernmental Panel on Climate Change. Cambridge University Press, Cambridge.

Jonkman, S.N., Kelman, I., 2005. An analysis of the causes and circumstances of flood disaster deaths. Disasters 29 (1), 75–97.

Keenan, T.F., Hollinger, D.Y., Bohrer, G., Dragoni, D., Munger, J.W., Schmid, H.P., et al., 2013. Increase in forest water-use efficiency as atmospheric carbon dioxide concentrations rise. Nature 499, 324–327.

Kendy, E., Bredehoeft, J.D., 2006. Transient effects of groundwater pumping and surface-water-irrigation returns on streamflow. Water Resour. Res. 42, W08415.

Knight, J.H., Gilfedder, M., Walker, G.R., 2005. Impacts of irrigation and dryland development on groundwater discharge to rivers—a unit response approach to cumulative impacts analysis. J. Hydrol. 303, 79–91.

Konrad, C.P., Booth, D.B., 2005. Hydrologic changes in urban streams and their ecological significance. Amercian Fisheries Society Symposium 47. Amercian Fisheries Society, pp. 157–177.

Korman, J., Campana, S.E., 2009. Effects of hydropeaking on nearshore habitat use and growth of age-0 rainbow trout in a large regulated river. Trans. Am. Fish. Soc. 138 (1), 76–87.

Krause, B., Culmsee, H., Wesche, K., Bergmeier, E., Leuschner, C., 2011. Habitat loss of floodplain meadows in north Germany since the 1950s. Biodivers. Conserv. 20, 2347–2364.

Kustu, M.D., Fan, Y., Robock, A., 2010. Large-scale water cycle perturbation due to irrigation pumping in the US High Plains: a synthesis of observed streamflow changes. J. Hydrol. 390, 222–244.

Lamontagne, S., Taylor, A.R., Cook, P.G., Crosbie, R.S., Brownbill, R., Williams, R.M., et al., 2014. Field assessment of surface water–groundwater connectivity in a semi-arid river basin (Murray-Darling, Australia). Hydrol. Process. 28, 1561–1572.

Leopold, L.B., 1968. Hydrology for Urban Land Planning: a Guidebook on the Hydrological Effects of Urban Land Use. 554, U.S. Geological Survey, Washington, DC.

Lett, R.A., Morden, R., McKay, C., Sheedy, T., Burns, M., Brown, D., 2009. Farm dam interception in the Campaspe Basin under climate change. Hydrology and Water Resources Symposium, pp. 1194–1204.

Liebe, J.R., Van De Giesen, N., Andreini, M., Walter, M.T., Steenhuis, T.S., 2009. Determining watershed response in data poor environments with remotely sensed small reservoirs as runoff gauges. Water Resour. Res. 45 (7).

Liu, G., Wilson, B.B., Whittemore, D.O., Jin, W., Butler Jr., J.J., 2010. Ground-water model for Southwest Kansas Groundwater Management District No. 3: Kansas Geological Survey.

Lo, M.-H., Famiglietti, J.S., 2013. Irrigation in California's Central Valley strengthens the southwestern U.S. water cycle. Geophys. Res. Lett. 40 (1–6).

Lowe, L., Nathan, R.J., 2006. Use of similarity criteria for transposing gauged streamflows to ungauged locations. Aust. J. Water Resour. 10 (2), 161–170.

Lowe, L., Nathan, R.J., Morden, R., 2005. Assessing the impact of farm dams on streamflows, Part II: Regional characterisation. Aust. J. Water Resour. 9 (1), 13–26.

Luckey, R.L., Becker, M.F., 1999. Hydrogeology, water use, and simulation of flow in the High Plains aquifer in northwestern Oklahoma, southeastern Colorado, southwestern Kansas, northeastern New Mexico, and northwestern Texas, U.S. Geological Survey.

Lytle, D.A., Poff, N.L., 2004. Adaptation to natural flow regimes. Trends Ecol. Evol. 19 (2).

Magilligan, F.J., Nislow, K.H., 2005a. Changes in hydrologic regime by dams. Geomorphology 71 (1), 81–178.

Magilligan, F.J., Nislow, K.H., 2005b. Changes in hydrologic regime by dams. Geomorphology 71 (1–2), 61–78.

Malveira, V.T.C., Araújo, J.C.D., Güntner, A., 2011. Hydrological impact of a high-density reservoir network in semiarid Northeastern Brazil. J. Hydrol. Eng. 17 (1), 109–117.

Mantel, S.K., Hughes, D.A., Muller, N.W., 2010a. Ecological impacts of small dams on South African rivers Part 1: Drivers of change-water quantity and quality. Water SA 36 (3), 351–360.

Mantel, S.K., Muller, N.W., Hughes, D.A., 2010b. Ecological impacts of small dams on South African rivers Part 2: Biotic response-abundance and composition of macroinvertebrate communities. Water SA 36 (3), 361–370.

Margat, J., Van Der Gun, J., 2013. Groundwater Around the World. CRC Press, London.

Mathews, R., Richter, B.D., 2007. Application of the indicators of hydrologic alteration software in environmental flow setting. J. Am. Water Resour. Assoc. 43 (6), 1400–1413.

McMahon, T.A., Finlayson, B.L., 1995. Reservoir system management and environmental flows. Lakes Reserv. Res. Manage. 1 (1), 65–76.

Miralles, D.G., De Jeu, R.A.M., Gash, J.H., Holmes, T.R.H., Dolman, A.J., 2011. Magnitude and variability of land evaporation and its components at the global scale. Hydrol. Earth Syst. Sci. 15 (3), 967–981.

Müller Schmied, H., Eisner, S., Franz, D., Wattenbach, M., Portmann, F.T., Flörke, M., et al., 2014. Sensitivity of simulated global-scale freshwater fluxes and storages to input data, hydrological model structure, human water use and calibration. Hydrol. Earth Syst. Sci. 18 (9), 3511−3538.

Nathan, R., Lowe, L., 2012. The hydrologic impacts of farm dams. Aust. J. Water Resour. 16 (1), 75−83.

Nathan, R.J., Jordan, P., Morden, R., 2005. Assessing the impact of farm dams on streamflows, Part I: Development of simulation tools. Aust. J. Water Resour. 9 (1), 1−12.

Neal, B.P., Nathan, R.J., Schreider, S., Jakeman, A.J., 2001. Identifying the separate impact of farm dams and land use changes on catchment yield. Aust. J. Water Resour. 5 (2), 165.

Nevill, C.J., 2009. Managing cumulative impacts: groundwater reform in the Murray-Darling Basin, Australia. Water Resour. Manage. 23, 2605−2631.

Nilsson, C., Reidy, C.A., Dynesius, M., Revenga, C., 2005. Fragmentation and flow regulation of the world's large river systems. Science 308, 405−408.

Novotny, V., Olem, H., 1994. Water Quality: Prevention, Identification and Management of Diffuse Pollution. Van Nostrand Reinhold, New York.

Nune, R., George, B.A., Teluguntla, P., Western, A.W., 2014. Relating trends in streamflow to anthropogenic influences: a case study of Himayat Sagar catchment, India. Water Resour. Manage. 28, 1579−1595.

Ogawa, H., Male, J.W., 1986. Simulating the flood mitigation role of wetlands. J. Water Resour. Plan. Manage. 112, 114−127.

Olden, J.D., Poff, L.N., 2003. Redundancy and the choice of hydrologic indices for characterizing streamflow regimes. River Res. Appl. 19, 101−121.

Peel, M.C., 2009. Hydrology: catchment vegetation and runoff. Prog. Phys. Geogr. 33 (6), 837−844.

Peel, M.C., McMahon, T.A., Finlayson, B.L., 2010. Vegetation impact on mean annual evapotranspiration at a global catchment scale. Water Resour. Res. 46 (9), W09508.

Peñuelas, J., Canadell, J.G., Ogaya, R., 2011. Increased water-use efficiency during the 20th century did not translate into enhanced tree growth. Global Ecol. Biogeogr. 20 (4), 597−608.

Piao, S., Friedlingstein, P., Ciais, P., De Noblet-Ducoudre, N., Labat, D., Zaehle, S., 2007. Changes in climate and land use have a larger direct impact than rising CO_2 on global river runoff trends. Proceedings of the National Academy of Sciences of the United States of America 104 (39), 15242−15247.

Poff, N.L., Allan, J.D., Bain, M.B., Karr, J.R., Prestegaard, K.L., Richter, B.D., et al., 1997. The natural flow regime, a paradigm for river conservation and restoration. BioScience 47, 769−784.

Pokhrel, Y.N., Koirala, S., Yeh, P.J.F., Hanasaki, N., Longuevergne, L., Kanae, S., et al., 2015. Incorporation of groundwater pumping in a global Land Surface Model with the representation of human impacts. Water Resour. Res. 51 (1), 78−96.

Price, K., 2011. Effects of watershed topography, soils, land use, and climate on baseflow hydrology in humid regions: a review. Prog. Phys. Geogr. 1−28. Available from: http://dx.doi.org/10.1177/0309133311402714.

Prigent, C., Papa, F., Aires, F., Jimenez, C., Rossow, W.B., Matthews, E., 2012. Changes in land surface water dynamics since the 1990s and relation to population pressure. Geophys. Res. Lett. 39, 8.

Puckridge, J., Sheldon, F., Walker, K., Boulton, A., 1998. Flow variability and the ecology of large rivers. Mar. Freshw. Res. 49, 55−72.

Richards, R., 1989. Measures of flow variability for Great Lakes tributaries. Environ. Monitor. Assess. 12, 361−377.

Richards, R., 1990. Measures of flow variability and a new flow-based classification of Great Lakes tributaries. J. Great Lakes Res. 16, 53−70.

Richter, B.D., Baumgartner, J.V., Powell, J., Braun, D.P., 1996. A method for assessing hydrologic alteration within ecosystems. Conserv. Biol. 10 (4), 1163−1174.

Richter, B.D., Baumgartner, J.V., Wigington, R., Braun, D.P., 1997. How much water does a river need? Freshw. Biol. 37, 231–249.

Richter, B.D., Thomas, G.A., 2007. Restoring environmental flows by modifying dam operations. Ecol. Soc. 12 (1).

Scanlon, B.R., Faunt, C.C., Longuevergne, L., Reedy, R.C., Alley, W.M., McGuire, V.L., et al., 2012. Groundwater depletion and sustainability of irrigation in the US High Plains and Central Valley. Proc. Natl. Acad. Sci. USA 109 (24), 9320–9325.

Schlesinger, W.H., Jasechko, S., 2014. Transpiration in the global water cycle. Agricult. Forest Meteorol. 189–190, 115–117.

Schreider, S.Y., Jakeman, A.J., Letcher, R.A., Nathan, R.J., Neal, B.P., Beavis, S.G., 2002. Detecting changes in streamflow response to changes in non-climatic catchment conditions: farm dam development in the Murray-Darling basin, Australia. J. Hydrol. in press.

Scott, C.A., Shah, T., 2004. Groundwater overdraft reduction through agricultural energy policy: insights from India and Mexico. Water Resour. Dev. 20, 149–164.

Seto, K.C., Fragkias, M., Güneralp, B., Reilly, M.K., 2011. A meta-analysis of global urban land expansion. PloS One 6 (7), e23777.

Siebert, S., Kummu, M., Porkka, M., Döll, P., Ramankutty, N., Scanlon, B.R., 2015. A global data set of the extent of irrigated land from 1900 to 2005. Hydrol. Earth Syst. Sci. 19 (3), 1521–1545.

Sterling, S.M., Ducharne, A., Polcher, J., 2013. The impact of global land-cover change on the terrestrial water cycle. Nat. Clim. Change 3 (4), 385–390.

Stewardson, M.J., Gippel, C.J., 2003. Incorporating flow variability into environmental flow regimes using the Flow Events Method. River Res. Appl. 19, 1–14.

Strayer, D.L., Findlay, S.E., 2010. Ecology of freshwater shore zones. Aquat. Sci. 72 (2), 127–163.

UK National Ecosystem Assessment, 2014. The UK National Ecosystem Assessment: synthesis of the key findings.

Ukkola, A.M., Prentice, I.C., 2013. A worldwide analysis of trends in water-balance evapotranspiration. Hydrol. Earth Syst. Sci. 17 (10), 4177–4187.

Wada, Y., Bierkens, M.F.P., 2014. Sustainability of global water use: past reconstruction and future projections. Environ. Res. Lett. 9 (10), 104003.

Wada, Y., Van Beek, L.P.H., Viviroli, D., Dürr, H.H., Weingartner, R., Bierkens, M.F.P., 2011. Global monthly water stress: 2. Water demand and severity of water stress. Water Resour. Res. 47 (7).

Wada, Y., Van Beek, L.P.H., Wanders, N., Bierkens, M.F.P., 2013. Human water consumption intensifies hydrological drought worldwide. Environ. Res. Lett. 8 (3), 034036.

Wada, Y., Wisser, D., Bierkens, M.F.P., 2014. Global modeling of withdrawal, allocation and consumptive use of surface water and groundwater resources. Earth Syst. Dynam. 5, 15–40.

Walsh, C.J., Fletcher, T.D., Burns, M.J., 2012. Urban stormwater runoff: a new class of environmental flow problem. PloS One 7 (8), e45814.

Walter, R.C., Merritts, D.J., 2008. Natural streams and the legacy of water-powered mills. Science 318 (5861), 299–304.

Wanders, N., Wada, Y., 2015. Human and climate impacts on the 21st century hydrological drought. J. Hydrol. 526, 208–220.

Wang, D., Wang, G., Anagnostou, E.N., 2007. Evaluation of canopy interception schemes in land surface models. J. Hydrol. 347, 308–318.

Wang, L., Good, S.P., Caylor, K.K., 2014. Global synthesis of vegetation control on evapotranspiration partitioning. Geophys. Res. Lett. 41, 6753–6757.

WEC, 2015. World Energy Resources: Charting the Upsurge in Hydropower Development. World Energy Council.

Wisser, D., Frolking, S., Douglas, E.M., Fekete, B.M., Schumann, A.H., Vörösmarty, C.J., 2010. The significance of local water resources captured in small reservoirs for crop production — a global-scale analysis. J. Hydrol. 384 (3), 264–275.

Wong, T.H.F., Lloyd, S.D., Breen, P.F., 2000. Water sensitive road design — design options for improving stormwater quality of road runoff. Cooperative Research Centre for Catchment Hydrology, Melbourne, Australia.

Zarfl, C., Lumsdon, A.E., Berlekamp, J., Tydecks, L., Tockner, K., 2015. A global boom in hydropower dam construction. Aquat. Sci. 77 (1), 161–170.

Zektser, I.S., Everett, L.G. (Eds.), 2004. Groundwater Resources of the World and Their Use, IHP-VI Ser. Groundwater, vol. 6,. U. N. Education. Science and Cultural Organization, Paris.

Zhang, L., Dawes, W.R., Walker, G.R., 2001. Response of mean annual evapotranspiration to vegetation changes at catchment scale. Water Resour. Res. 37 (3), 701–708.

ENVIRONMENTAL AND ECOLOGICAL EFFECTS OF FLOW ALTERATION IN SURFACE WATER ECOSYSTEMS

Robert J. Rolls[1] and Nick R. Bond[2]

[1]University of Canberra, Canberra, ACT, Australia [2]La Trobe University, Wodonga, VIC, Australia

4.1 INTRODUCTION

Flow regime change caused by human alterations to the water cycle is evident in all permanently inhabited continents of the Earth (Chapter 3) and these impacts to hydrological regimes have affected freshwater and estuarine ecosystems worldwide. The hydrological regimes of 172 out of 292 of the world's largest rivers are altered by large dams (Nilsson et al., 2005). Analyses based on data at regional and global spatial extents have identified that 77% and 65%, respectively, of the total volume of water discharged in rivers is impacted by dams, dam operations, and water diversion (Dynesius and Nilsson, 1994; Vörösmarty et al., 2010). Freshwater ecosystems support a disproportionately high contribution to global biodiversity and ecosystem services (e.g., food and water), given the small proportion of the Earth covered by freshwater (Carrete Vega and Wiens, 2012). Freshwater biodiversity and ecosystem services are under multiple threats (stressors), with the alteration of hydrological regimes by dams, weirs, and water extraction being the most globally prevalent (Collares-Pereira and Cowx, 2004; Dudgeon et al., 2006; Strayer and Dudgeon, 2010).

Flow alteration resulting from these various forms of disturbances has impacted the ecological functioning and biodiversity of rivers, wetlands, floodplains, and estuaries worldwide (Poff and Zimmerman, 2010). These collective impacts arise from a variety of direct and indirect causes such as habitat loss and fragmentation, altered water quality and thermal regimes, the loss of important life-history cues, changes to food webs and patterns of energy production, and modification of environments in ways that promote ecological invasion (Bunn and Arthington, 2002). Ecological effects of flow regime alteration have been the topic of research and regular synthesis in aquatic ecology since the 1970s, with this research spanning a broad range of taxa, including algae and biofilms, aquatic and riparian plants, benthic invertebrates, fishes, amphibians, and waterbirds. Such studies have further considered multiple levels of ecological organization (individual, population, community, and ecosystem; Box 4.1). Research has also long considered the effects of flow regime change on physical, chemical, and ecosystem processes such as sediment dynamics, nutrient cycling, and energy fluxes, although ecosystem perspectives have perhaps been less well studied.

Water for the Environment. DOI: http://dx.doi.org/10.1016/B978-0-12-803907-6.00004-8

BOX 4.1 PROCESSES AND PRINCIPLES OF RIVER ECOLOGY RELEVANT TO UNDERSTANDING ECOLOGICAL CONSEQUENCES OF FLOW REGIME CHANGE

Ecology is defined as the study of how the physical, chemical, and biological features of the environment interact to determine the distribution and abundance of living organisms (Begon et al., 2006). Freshwater ecology applies the study of ecology to freshwater environments such as rivers, lakes, and wetlands. Conceptualizing the fundamental and unique features of river ecosystems is essential to understanding how human activities such as flow regime change modify the ecological processes that influence the distribution and abundance of species (Boulton et al., 2014).

Spatial scaling: By combining physical geography and biology, ecology requires a clear understanding of spatial scaling (Ward, 1989; Wiens, 1989). Spatial scaling is critical in ecology to understand how ecological processes vary in space across global, continental, regional, and local spatial extents. Freshwater environments are viewed as being organized in a nested hierarchy, whereby microhabitats (e.g., the spaces in between rocks in the river bed) are nested within slow-flowing pools and fast-flowing turbulent riffles. In turn, pool−riffle sequences are nested within reaches, which form separate segments (tributaries) of an entire river system (Fausch et al., 2002; Frissell et al., 1986).

Linkages across scales: A hierarchical view emphasizes the dimensions relevant to spatial scaling in freshwater ecology (Ward, 1989): the lateral dimension (the linkages between stream channels, the riparian zone, and floodplains), the longitudinal dimension (linking between headwater, lowland, and terminal zones), and the vertical dimension (depth of habitats, linkages between surface waters and groundwaters). The temporal dimension describes how interactions along these three spatial dimensions vary in time (Ward, 1989). The presence and movement of water is the strongest determinant of the connections and linkages across spatial dimensions.

Mechanistic roles of water: Within freshwater environments, it is the presence and movement of water that mediates the extent to which physical, chemical, and biological processes determine the distribution and abundance of organisms. Sponseller et al. (2013) classify the ecological roles of water into three distinct mechanisms whereby water acts as: (1) a resource or habitat for organisms; (2) a vector for the exchange and movement of organisms, nutrients, and material by controlling hydrological connectivity (Fullerton et al., 2010); and (3) a driver of disturbance and geomorphology.

Levels of ecological organization: Finally, ecological processes influence different levels of ecological organization. Organisms live as individuals, independently sourcing energy and occupying space. All individuals within a habitat are a population, which vary in abundance, biomass, and structure (size, sex, age). Communities are formed by the co-occurrence and interaction of multiple species populations. These interactions are best known as predation, competition, and parasitism (Begon et al., 2006). At the ecosystem level, communities and their nonliving environment are linked by the exchange of material between living and nonliving components. Separate groups of organism (e.g., birds, viruses, bacteria, phytoplankton, macroinvertebrates, fish, amphibians, mammals, plants, and reptiles) all occupy or interact with freshwater environments.

Bringing these concepts together: Ecological impacts of flow regime change are mediated by the effect of anthropogenic drivers (see Chapter 3) on hydrological components of the flow regime, which have consequences across different levels of ecological organization via distinct ecohydrological mechanisms.

Increasingly, the restoration of both the structure and function of aquatic ecosystems degraded by flow alteration is being sought through the delivery of environmental water (Naiman et al., 2012). Although there are presently few published studies evaluating the success of such efforts (Olden et al., 2014), it is clear that flow restoration will be an essential component in restoring many degraded river systems worldwide affected by flow alteration.

As well as there being a strong conceptual basis for why flow alteration can impact aquatic ecosystems (see e.g., Bunn and Arthington, 2002), numerous more systematic reviews have found strong evidence of the effects of flow regulation on ecosystems (Dewson et al., 2007; Gillespie et al., 2015b; Poff and Zimmerman, 2010). For example, 92% of the studies examined by Poff and Zimmerman (2010) reported ecological effects associated with flow regulation. There is thus a compelling evidence base from which to consider the need for environmental water in hydrologically altered river systems, and to minimize the degree of harmful hydrological change in rivers with low levels of water demand and/or water infrastructure.

Ecological effects of flow regime change are not caused by the driver of hydrological alteration per se (e.g., dams, weirs, and urbanization), but rather by the way in which these drivers alter specific hydrological attributes. The purpose of this chapter is to introduce and illustrate the broad range of ecological responses to changes in specific hydrological components that are often altered by anthropogenic flow regime change (Chapter 3). A clear understanding of the impacts of change in specific hydrological attributes is both essential and beneficial to predicting the ecological consequences of flow regime restoration using environmental water regimes (Box 4.1). We first identify how the drivers of flow regime change (sensu Chapter 3) alter specific components of the flow regime, and then summarize the subsequent ecological responses, building on existing reviews and conceptual papers (e.g., Bunn and Arthington, 2002; Dewson et al., 2007; Poff and Zimmerman, 2010). We also give consideration to the local factors that might influence the degree to which hydrological impacts translate into ecological changes such as channel size and morphology and local species traits, and the implications this has for efforts to synthesize impacts across broad geographic settings. We recognize and discuss briefly the capacity for dams and weirs to alter rivers' temperature regimes (Olden and Naiman, 2010) and disrupt habitat and hydrological connectivity (Fullerton et al., 2010). However, this chapter primarily focuses on the effects of changes to the hydrological regime. Our goal is to emphasize the variety of ecological consequences that arise from different types of flow alteration in different contexts, which in turn contribute to decisions and solutions for effective management of environmental water.

4.2 HYDROLOGICAL COMPONENTS: LINKING DRIVERS OF CHANGE WITH ECOLOGICAL RESPONSES

The most pervasive and widespread cause of flow regime change is the storage and regulation of water by dams for the reliable supply of water for human use (Nilsson et al., 2005; see Chapter 3). Storages can be managed to meet a range of human needs, including flood mitigation, hydroelectric power generation, irrigation, and domestic water supply. Many reservoirs are managed in a way that spans multiple uses, often varying across different seasons (Ahmad et al., 2014; see Chapter 3). For example, many storages used for water supply during dry periods, or for hydropower generation, are also used for flood control following high rainfall. Each of these human needs tends to be associated with a distinctive hydrological fingerprint in terms of downstream river flow alterations, for example by reducing flood magnitude and frequency, reducing winter baseflow(s), and increasing summer flows. Describing these patterns of hydrological change has been an active area of research in its own right. There are a large number of hydrological indices used to describe these different elements of hydrological regimes, and much work has gone into trying to identify small sets of indices that characterize the regime as a whole as efficiently as possible (Olden and Poff, 2003). In assessing hydrological change, Richter et al. (1996) proposed a set of 64 indicators of hydrologic alteration statistics that could be used to quantify the degree of flow regime change on the basis of ecologically meaningful metrics. A number of other similar approaches have been proposed. Common to each of these methods is the focus on distinctive *components* of the flow regime (defined in terms of magnitude, timing, duration, frequency, and rate of change). Although the important and distinctive flow components within any given river will likely vary with geography and climate, common elements include the characteristics of in-channel and overbank floods, seasonal low and high flows, and cease-to-flow periods (Bunn and Arthington, 2002). By understanding the natural patterns of these distinctive hydrological components of the natural flow regime (i.e., the flow regime occurring in the absence of water extraction and/or storage), one can begin to consider the various biophysical processes that might be linked to each flow component, and hence to conceive of which components might need restoring as part of an environmental water regime.

Here we begin by summarizing some of the distinctive impacts of different forms of river regulation on major flow components (baseflow(s), high flows, etc.), before moving on to discuss the consequences arising when each of those flow components is altered in the context of our conceptual understanding of freshwater ecosystems (Box 4.1). It is worth noting as an aside that although recognizing distinctive flow components has proven useful in assessing hydrological impacts and environmental water requirements, rarely are flow components impacted in complete isolation from one another, nor do ecosystems respond in that way. Such complexities present distinctive challenges for scientists seeking to understand flow—ecology linkages and isolate particular events from the broader flow regime (Konrad et al., 2011; Stewart-Koster et al., 2014). Nevertheless, we regard the concept of flow components as a useful construct for organizing research into the biophysical impacts of hydrological change.

4.2.1 REDUCED BASEFLOW(S)

Water extraction for human consumption is a major cause of reduced baseflow(s) (i.e., flows occurring in the absence of overland runoff). Water extraction can occur directly from dams, in river reaches downstream of dams, or from groundwater storages that are connected to surface water ecosystems. Many older storages have no capacity to deliver water downstream except from water overtopping a dam wall or a spillway, and thus it is not uncommon for rivers to receive no base flow immediately downstream of large dams, with some baseflow(s) only being restored by tributary inflows. When water is extracted directly either from dams or unregulated rivers, flow volume is reduced and can lead to much lower flows than expected under natural conditions (Brown and Bauer, 2010; White et al., 2012), or even the complete cessation of flow in what would normally be perennially flowing streams (Ibanez et al., 1996; Martinez et al., 2013). Where river channels are used as conduits to transport water between storages and downstream users, baseflow(s) may increase and/or become more stable, although this may progressively decline downstream via extraction and tributary inflows (e.g., Humphries et al., 2008; Reich et al., 2010). Removal of water from groundwater also reduces streamflow in connected surface water systems (Falke et al., 2011; Kustu et al., 2010). Therefore, the process and location of water extraction has variable impacts on flow magnitude, which in turn determines aquatic ecosystem size and driving ecological effects.

4.2.2 **REDUCED FLOODS**

Perhaps one of the most pervasive impacts of dam construction is the storage and capture of flood waters, which results in increased constancy (reduced variability) of flow from daily to inter-annual time scales. Across the continental United States, flow regulation has reduced the size of maximum floods (Magilligan and Nislow, 2005; Mims and Olden, 2013). In an analysis of river regulation impacts across Canada, Assani et al. (2006) found that reservoirs altered all aspects of the flooding regime, including the size and frequency of spring flooding peaks, and a loss of most flow events with a recurrence interval of more than 10 years. It is also common for dams to remove moderately sized floods from the regime (e.g., Aristi et al., 2014; Sammut and Erskine, 1995), as these are insufficient to fill storages (Table 4.1). Another important but sometimes overlooked impact of river regulation is a reduction in the duration and size of floods and other high-flow events such as in-channel freshes (Maheshwari et al., 1995; Fig. 4.1). As discussed in the next section, this has important implications for river—floodplain connections and a range of ecological processes.

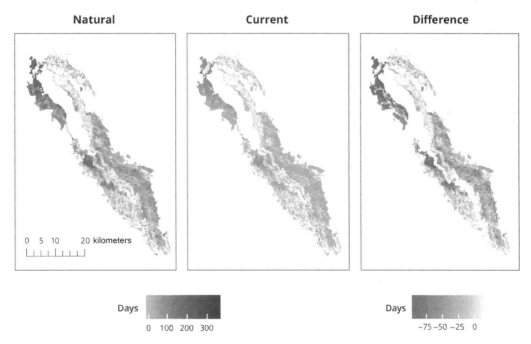

FIGURE 4.1

Plot showing the natural, current, and change in the number of days of floodplain inundation at Koondrook—Perricoota Forest on the Murray River in southeastern Australia. Changes in flood duration as well as magnitude is an important driver of river—floodplain ecology.

Source: Data from CSIRO, Overton et al. (2006)

Table 4.1 Summary of the Effects of Flow Regime Change Caused by Alteration to Specific Components of the Hydrological Regime and Underlying Ecological Mechanism

Hydrological Component	Ecological Mechanism	Ecological Responses
Reduced base flow magnitude	Reduced or complete loss of fast-flowing, turbulent habitats (riffles); reduced wetted area	Increased density and richness of organisms during initial habitat loss (crowding; Dewson et al., 2007) and decline in population size or local extinction of species (Kupferberg et al., 2012) and reduced species richness (Benejam et al., 2010; Boulton, 2003) and encroachment of terrestrial organisms into the riparian and channel zone (Stromberg et al., 2007)
Reduced flood frequency and size	Decline in life-history cues stimulating reproduction (Bunn and Arthington, 2002); reduced disturbance frequency and severity; loss of nutrient inputs and productivity from floodplains	Reduced species richness due to the loss of organisms requiring flowing habitats for feeding or reproduction (Alexandre et al., 2013; Growns and Growns, 2001; Meador and Carlisle, 2012) and reduced cover of bryophytes and macroalgae due to desiccation and invasion of nonnative aquatic and terrestrial species into the riparian and channel zone (Catford et al., 2011, 2014; Dolores Bejarano et al., 2011; Greet et al., 2011; Reynolds et al., 2014; Stromberg et al., 2007) and decline in species richness and functional diversity (Kingsford et al., 2004; Nielsen et al., 2013) and lower ecosystem productivity such as fisheries (Bonvechio and Allen, 2005; Gillson et al., 2009)
Reduced floodplain inundation	Decline in wetland area used for feeding, reproduction, and rearing	Reduced size and density of populations of floodplain-dependent organisms, lower species richness (Kingsford and Thomas, 1995)
Increased base flow (artificial *perennialization*)	Absence of nonflowing habitats and loss of drying events	Altered composition of species to reflect permanently flowing reaches or rivers (Alexandre et al., 2013; Chessman et al., 2010; Reich et al., 2010; Rolls and Arthington, 2014) and increased metabolic activity and productivity of chlorophyll a (Ponsatí et al., 2015) and increased biomass of consumers (Miserendino, 2009)
Increased discharge variability	Frequent drying and flooding of stream channel as disturbances	Population declines (Schmutz et al., 2015) due to increased mortality of eggs, larvae, and adults (Bishop and Bell, 1978; Casas-Mulet et al., 2015; Freeman et al., 2001; Miller and Judson, 2014; Nagrodski et al., 2012) and reduced species richness and altered species composition (Paetzold et al., 2008; Perkin and Bonner, 2011)

4.2.3 INCREASED BASEFLOW(S) (ANTIDROUGHT)

Storage and use of water for dry-season irrigation typically leads to increased discharge down-stream of dams during the normal low-flow season (termed *antidrought*; McMahon and Finlayson, 2003). These heightened dry-season flows are often also associated with cold-water pollution (Lugg and Copeland, 2014; Table 4.1). Seasonal flow reversal and reduced flooding is now evident across many of the world's climate regions, including humid tropical, humid continental, and Mediterranean temperate regions (e.g., Assani et al., 2006; Batalla et al., 2004; Li et al., 2015). These changes often mean that rivers that normally cease to flow (termed *intermittent*) are managed so that flows are now permanent (e.g., Bunn et al., 2006; Reich et al., 2010). As low flows and cease-to-flow periods play an important role in the ecology of these systems (Rolls et al., 2012), antidrought can have major ecological consequences, and is an important aspect of environmental flow restoration.

4.2.4 INCREASED SHORT-TERM VARIABILITY (HYDROPEAKING)

Another pervasive impact of hydroelectric dams is increased short-term (diel) flow variability due to hydropeaking, in which flows are increased to coincide with daily periods of heightened electricity demand (Meile et al., 2011; Table 4.1). Hydropeaking can be avoided by operating generation facilities as a run-of-river, but this is typically associated with reduced economic value (due to the short-term spikes in electricity price) and decreased peak-production capacity, which are two of the primary drivers for hydroelectric generation (see Chapter 3). Even when managed as run-of-river (i.e., with minimal impacts on the overall hydrograph), dams disrupt longitudinal connectivity and can convert large sections of river from lotic to lentic habitats, both of which have strong negative ecological effects (Bunn and Arthington, 2002). Although not considered further here, these impacts must be taken into account when setting objectives and assessing the feasibility of using environmental water releases to offset the impacts of dam construction.

4.3 ECOLOGICAL EFFECTS OF FLOW ALTERATION

A fundamental tenet of river ecology is that the flow regime is a *master variable* (Power et al., 1995) and that organisms have adapted to the natural flow regime (Lytle and Poff, 2004). Therefore, anthropogenic changes to the natural flow regime by dams, weirs, diversion, and extraction of water can be expected to cause ecological change (Bunn and Arthington, 2002; Dewson et al., 2007; Poff et al., 1997). These impacts occur via a number of potential mechanisms, however, are all driven by the fundamental ecological roles of water within freshwaters; its role as a habitat and resource, a vector for connectivity and transport of material and organisms, and an agent for disturbance and geomorphic change (Sponseller et al., 2013).

4.3.1 REDUCED BASEFLOW(S) AND INCREASED INTERMITTENCY

Low-flow (base-flow) periods are an important component of the hydrological regime for freshwater ecosystems because the area and depth of aquatic habitats is often compared with higher discharge periods, and both physical and chemical conditions within those habitats can change rapidly as flow declines (Rolls et al., 2012). Reduced discharge can thus have substantial ecological consequences. Declines in discharge volume disproportionately affect habitats that exhibit large elevational gradients and therefore experience fast flows such as riffles, specifically due to changes in area, depth, and velocity (Boulton, 2003). By reducing base-flow discharge due to flow regulation, loss of flowing habitats causes impacts to population size and extinctions of organisms that require flowing habitats such as the foothill yellow-legged frog (*Rana boylii*) in western North American rivers (Kupferberg et al., 2012). Increased densities of organisms occur during flow recession, leading to initial crowding and competition for resources (Dewson et al., 2007). Differences in the composition of communities between regulated and unregulated rivers are reported worldwide (e.g., Grubbs and Taylor, 2004), and are driven by the loss or decline of organisms that have a strong preference to specific hydraulic habitats that are altered by flow regulation (e.g., Gillespie et al., 2015a). For terrestrial organisms, effects of flow regulation on reduced aquatic habitat size contribute to encroachment of riparian plant species such as the invasive *Tamarix* spp. (Stromberg et al., 2007), emphasizing that effects can occur in both the aquatic and riparian terrestrial zone.

Reduced baseflow(s) due to flow regulation also increases flow intermittency that is the frequency and duration of zero flow. Intermittency can be increased in rivers downstream of locations where water is extracted or diverted and also causes marked ecological changes. Such effects are most apparent in *perennial* rivers that naturally flow permanently. For example, water abstraction converted naturally perennial-flowing rivers to intermittently flowing rivers in Spain, leading to a decline in fish species richness by 35% (Benejam et al., 2010). Therefore, ecological effects of flow regime change on altered discharge occur via changes in discharge magnitude and the duration and frequency of flow intermittency, and depend highly on the geomorphologic characteristics of the river channel.

4.3.2 REDUCED FLOOD MAGNITUDE AND FREQUENCY

Alterations to the magnitude and frequency of flood events are widely reported effects of flow regulation (Assani et al., 2006), and are a primary driver of ecological effects of altered flow regimes. Floods influence aquatic ecosystems as a disturbance (where organisms and material are removed and transported from locations), determine habitat structure, and facilitate connectivity and therefore movement (Sponseller et al., 2013). Reduced flood frequency and magnitude by flow regulation causes reduced abundance and richness of flowing specialist taxa (Alexandre et al., 2013) or those that spawn during flood events (Perkin and Bonner, 2011). For example, the decline of flow variability by flooding in regulated rivers of the eastern United States is attributed to have caused the overall decline in 35% of native fish species, and a loss of over 50% of riffle-dwelling taxa (Meador and Carlisle, 2012). Species richness of macroinvertebrates, periphyton, and macrophytes has also been found to be lower in regulated rivers with reduced flood magnitude (e.g., Growns and Growns, 2001).

Loss of temporal variation in flow and reduced flood disturbance frequency with regulation both influence the persistence and establishment of organisms in aquatic and riparian habitats. Percentage cover of bryophytes and macroalgae on rocks was found to be lower in regulated upland rivers when compared with unregulated reaches, and these differences were attributed to reduced temporal variability in flow over fine temporal scales (daily, weekly; Downes et al., 2003).

Reduced flow variability also facilitates the invasion of (often) nonnative species. By reducing flow velocity and flooding disturbances, flow regulation promotes the invasion of nonnative fish in the regulated Ebre River (Spain; Caiola et al., 2014). In the riparian zone, reduced scouring frequency increases richness of terrestrial vegetation due to the increased prevalence of *dry* species intolerant of flooding (Dolores Bejarano et al., 2011; Greet et al., 2011; Reynolds et al., 2014; Stromberg et al., 2007), particularly when flooding seasonality is altered and synchronized with the release of seeds by invasive species (Mortenson et al., 2012). In floodplain wetlands, reduced flooding frequency increases the invasion and dominance of nonnative vegetation (e.g., Catford et al., 2011, 2014). Additionally, increased water level stability in floodplain wetlands in regulated river systems reduces taxonomic and functional diversity of birds, aquatic plants, and microfauna (Kingsford et al., 2004; Nielsen et al., 2013), emphasizing that effects of altered flood disturbance regimes affect ecosystems across aquatic, bankside, and floodplain habitats.

In lowland rivers and their estuaries, floods act less as a disturbance and more as a driver of ecological productivity by linking resources between riverine and floodplain habitats. Termed the *flood pulse advantage* (Bayley, 1991), the productivity of rivers is driven by the size of floods. In turn, reduced flooding size and frequency by regulation drives declines in productivity, mostly reported in estuarine fishery productivity. For example, catch per unit effort of commercial and recreational fish species is positively linked with freshwater discharge to estuaries (e.g., Gillson et al., 2009), or flow volume at the time of spawning (Bonvechio and Allen, 2005; Growns and James, 2005). On the Barmah—Milewa floodplain of the Murray River in Australia, Robertson et al. (2001) found controlled flooding during spring—summer or summer to result in higher net primary productivity of woody trees when compared with spring flooding only or no flooding, yet spring flooding promoted the highest primary production of algae and macrophytes. These findings suggest that both the magnitude *and* seasonal timing of flooding are strong determinants of ecosystem productivity, and alterations to floodplain inundation dynamics by flow regulation could influence levels of primary production for lowland river ecosystems.

4.3.3 REDUCED OVERBANK FLOODING

The decline in area inundated during flood events is a clear consequence of reduced flood magnitude and frequency imposed by flow regulation. In the Murrumbidgee River, Southern Australia, flow regulation has permanently inundated some low-level floodplain wetlands (particularly those associated with weir pools), yet has reduced the duration and frequency of inundation by 40% overall (Frazier and Page, 2006). Floodplain wetland size (spatial extent or area) is often reduced in terminal floodplain wetlands due to upstream extraction of water for irrigation (Kingsford and Thomas, 1995) or because of altered channel morphology due to constant regulated discharge (Gorski et al., 2012). This loss of inundated habitat has substantial impacts for organisms that rely on floodplain wetlands for nesting, recruitment, or feeding (e.g., Kingsford and Thomas, 1995).

4.3.4 INCREASED BASEFLOW(S) (ANTIDROUGHT)

In contrast to flooding impacts, the storage and release of water for dry-season irrigation leads to increased discharge downstream of dams. Unregulated flows can either decline to very low levels or cease completely during dry seasons, but constant elevated discharge under regulated conditions prevents natural low flows from occurring (antidroughts; McMahon and Finlayson, 2003). Flow regulation can consequently cause streams that naturally cease to flow (intermittent rivers; Leigh et al., 2016) permanently (Bunn et al., 2006; Reich et al., 2010). Constant flow releases from large dams in the Mekong River have increased inundation of lowland floodplains in Cambodia (Arias et al., 2014). As low flows play an important ecological role in all rivers (Rolls et al., 2012), anti-drought can have significant ecological consequences and is a novel aspect of environmental water regimes (Lake et al., in prep.).

Ecological effects of increased baseflow(s) due to flow regulation are best understood by changes in community composition to reflect those typical of permanently flowing rivers. Comparisons between regulated and unregulated rivers in southeast Queensland, Australia, identi-fied that the largest shifts in fish assemblage composition occurred in streams that were originally intermittent but had become perennial under current regulated conditions (Rolls and Arthington, 2014). Artificially perennial flow in rivers in Portugal caused an increase in the abundance of taxa with broad environmental tolerances, including nonnative taxa (Alexandre et al., 2013). Across multiple dryland rivers of Australia, changes from intermittent to permanent flow regimes have caused biotic communities to change substantially in composition (e.g., Chessman et al., 2010; Reich et al., 2010). In addition, artificial flow permanence can increase metabolic activity of bio-films and productivity of chlorophyll a (Ponsatí et al., 2015), which in turn can increase density and biomass of consumers such as macroinvertebrates (Miserendino, 2009).

4.3.5 INCREASED SHORT-TERM FLOW VARIABILITY (HYDROPEAKING)

Hydropeaking, pulsing releases of water downstream of hydroelectric dams, causes rapid changes in flow that impact freshwater ecosystems by frequent partial or entire drying of the stream channel and unstable, persistent habitats. Rapid dewatering regimes by hydropeaking causes fish population declines (Schmutz et al., 2015), resulting in changes in community composition (Perkin and Bonner, 2011). These declines occur due to the rapid and frequent dewatering of riffles that severely increase egg mortality of riffle-spawning fish such as Atlantic salmon (*Salmo salar*; Casas-Mulet et al., 2015), stranding, and death of adult individuals (e.g., Bishop and Bell, 1978; Miller and Judson 2014; Nagrodski et al., 2012), or lack of persistent shallow habitats for juvenile fish recruitment (e.g., Freeman et al., 2001). Increased flow variability due to hydropeaking also alters species richness, in both the stream channel and riparian zone. Reduced species richness of riparian arthropod assemblages has been linked with flow regime change due to intradaily hydropeaking (Paetzold et al., 2008). These examples suggest that the combination of habitat loss and disturbance frequency determine the extent to which species richness is altered by flow regime change.

4.3.6 NONHYDROLOGICAL IMPACTS OF FLOW REGULATION REQUIRING CONSIDERATION FOR HOLISTIC MANAGEMENT OF ENVIRONMENTAL WATER

The impoundment and storage of water in dams, weirs, and reservoirs alters aquatic ecosystems due to the process of habitat loss and fragmentation (sensu Fahrig, 2003). Such effects of habitat loss and fragmentation by barriers are not due to changes in the flow regime itself (and therefore unable to be readily addressed by environmental water releases). However, these effects are an important component of the broader consequences of water resource development and therefore necessary to consider simultaneously with flow regime management. Lentic-adapted organisms are better suited to regulated, stable environments, and respond positively to water impoundment (e.g., Taylor et al., 2008). Conversion of lotic to lentic habitats by dams and weirs also alters ecosystem functioning such as reduced wood decomposition due to reduced physical movement and abrasion of detritus in regulated Mediterranean climate streams (Abril et al., 2015). Here, differences in decomposition between river and impoundment habitats were most apparent during winter when hydrological differences between regulated and unregulated rivers were largest (Abril et al., 2015).

Rivers have constrained (narrow) connections linking channel habitats, and are therefore vulnerable to habitat fragmentation by barriers such as dams (Beger et al., 2010). However, depending on catchment topography, dam characteristics (e.g., size), and position in the stream network, the ecological effects of dams can be contradictory. A widely observed effect of large dams is the occurrence of distinct communities of species, primarily fish, between reaches fragmented by dams. For example, fish communities have become fragmented by the effects of the Tallowa Dam (Shoalhaven River, Australia) on restricting movement, particularly for diadromous species (Gehrke et al., 2002). The construction of 1356 dams between 1634 and 1860 reduced connectivity to almost the entire stream network in coastal Maine, United States (Hall et al., 2011). However, natural barriers such as waterfalls also cause significant discontinuities in freshwater fish communities such as in the Madeira River, Brazil (Torrente-Vilara et al., 2011). Inundation and subsequent removal of natural barriers by reservoir impoundments can promote the dispersal of species among previously fragmented reaches (Vitule et al., 2012).

In addition to hydrological effects of flow regime change, the design and operation of dams alters the physicochemical characteristics of water downstream of dams. Altered temperature regimes (the daily and seasonal fluctuations in water temperature) are a global consequence of large dams (Olden and Naiman, 2010). Selective removal and transport of unnatural or unseasonal cold or warm water from large, stratified reservoirs can be the cause of ecological responses to flow regime change (Olden and Naiman, 2010). For example, hypolimnetic releases of water from large dams can reduce water temperatures by up to 15°C for hundreds of kilometers downstream (Casado et al., 2013; Dickson et al., 2012; Lugg and Copeland, 2014; Olden and Naiman, 2010; Preece and Jones, 2002). Additionally, under regulated conditions, stratification of lotic habitats (e.g., temperature and salinity) can be more pronounced when compared with unregulated systems due to reduced turbulence contributing to vertical mixing of the water column (Frota et al., 2012).

4.4 THE IMPORTANCE OF LOCAL FACTORS

The preceding sections of this chapter summarize some of the more consistent impacts that can arise from modifying particular aspects of the flow regime. Yet, perhaps surprisingly, efforts to distill these patterns into simple statistical relationships that would support rules of thumb around the limits to hydrological alteration (Poff et al., 2010) have proven difficult. For example, Poff and

Zimmerman (2010) found that, despite consistent reporting of impacts on the indictors they examined, there was little consistency in response when examined in terms of the relative degree of hydrological alteration (as measured, e.g., by a relative change in high- or low-flow magnitudes). Although this reflects a range of factors, including differences in measurement and analytical approaches, it is likely that *real* differences also arise in different rivers. This has important implications for practitioners involved in providing environmental water recommendations.

To provide several examples, in the first instance, because environmental water regimes will often have to be defined in quite specific terms (e.g., a flow of 1500 per day for 10 days during spring) to a river operator, there is a need to go beyond hydrology alone to consider the hydraulic relationships that link hydrology with biophysical processes. This could be in terms of whether high flows are sufficiently high to dislodge organisms from the streambed or provide access to the floodplain, or whether low flows are causing silt to smother cobbled habitats. Ultimately, it is these interactions between runoff variability and the slope and morphology of the river channel, and the size and shape of the substrate that determines the environment experienced by the biota. Such interactions cannot be discerned from hydrological data alone, which is why most environmental flows assessment depend heavily on hydraulic models. In fact we contend that some information on hydraulics is critical in interpreting or predicting the effects of hydrological alteration.

A second issue is that both life-history characteristics and ecological characteristics (*traits*) play a significant role in determining how individual taxa respond to flow regime change in rivers and streams. The ways in which individual taxa respond to flow regime change culminates in how patterns of biodiversity are linked with flow alteration. For example, fish species that have *opportunistic* life-history traits (e.g., small body size and early maturation) are better adapted to rivers with high interannual runoff variability. Conversely, those with *equilibrium* life histories (e.g., intermediate maturation and high juvenile survivorship) appear to be better adapted to rivers with more stable flow regimes. As a result, the general seasonal stabilization of runoff by river regulation tends to favor equilibrium fauna (Mims and Olden, 2013). However, species and ecological traits also play a large part in explaining why flow regime change facilitates invasion and establishment of nonnative species, and simultaneous decline of native species adapted to historical conditions (Gido et al., 2013). Understanding the broader evolutionary context and biology of organisms within the regional species pool (both of which vary spatially) may be helpful in trying to understand likely impacts arising from particular patterns of hydrological change.

4.5 CONCLUSION

Understanding the effects of flow regime change is essential for the planning, implementation, delivery, and evaluation of the outcomes of environmental water regimes. As shown in this synthesis, flow regulation alters multiple components of the hydrological regime in both consistent but sometimes idiosyncratic ways. Idiosyncrasies may arise because dams and weirs are rarely operated in the same way, and local variation in topography, geomorphology, and evolutionary history of the biota may lead to differences in the way that changes in hydrology manifest themselves both physically and biologically. However, in spite of such variation, there are numerous consistencies in impact that occur as particular flow components are modified in particular ways. Many of these are predictable if one considers biotic adaptations to the natural flow regime. By understanding the effects of flow regime change on the desirable structural and functional attributes of ecosystems, the relevant specific hydrological components can be manipulated and restored. For example, recruitment of some riverine fish is facilitated by the access to the resources available in prolonged inundated floodplain habitats—providing the combination of energy-rich and slow-flowing, warm habitats beneficial to growth of newly hatched individuals (King et al., 2003). In this example, fundamental knowledge of the importance of floodplain inundation timing, extent, and duration has helped inform the delivery and evaluation of artificially extended floodplain inundation using environmental water allocations (King et al., 2009). Combining evidence of the effects of flow regime change and the responses of environmental water releases will only improve further development in prediction, monitoring, and evaluation of environmental water requirements (Konrad et al., 2011; Olden et al., 2014). Such evidence is especially central to the application of environmental water regimes as a tool for climate change adaptation (Poff and Matthews, 2013).

The provision of environmental water will typically have large financial costs, either in terms of the need to purchase water and manage it specifically for environmental benefits or the lost opportunities for economic revenue arising particularly from industrial and agricultural water use (see Chapters 18 and 23). It is therefore essential the environmental water is delivered in ways that either maximize ecological benefits while minimizing economic costs, or provide simultaneous benefits for both environmental and economic outcomes. Such outcomes will likely be improved if there is clear knowledge of the context-specific effects of flow regulation on ecosystems, so that the limits of extrapolation and transferability of evidence across local, regional, continental, and global scales are sufficiently understood.

ACKNOWLEDGMENTS

We thank Will Chen, Avril Horne, and Angus Webb for constructive comments that improved the logic and structure of this chapter.

REFERENCES

Abril, M., Munoz, I., Casas-Ruiz, J.P., Gómez-Gener, L., Barcelo, M., Oliva, F., et al., 2015. Effects of water flow regulation on ecosystem functioning in a Mediterranean river network assessed by wood decomposition. Sci. Total Environ. 517, 57−65.

Ahmad, A., El-Shafie, A., Razali, S.F.M., Mohamad, Z.S., 2014. Reservoir optimization in water resources: a review. Water Resour. Manage. 28, 3391−3405.

Alexandre, C.M., Ferreira, T.F., Almeida, P.R., 2013. Fish assemblages in non-regulated and regulated rivers from permanent and temporary Iberian systems. River Res. Appl. 29, 1042−1058.

Arias, M.E., Piman, T., Lauri, H., Cochrane, T.A., Kummu, M., 2014. Dams on Mekong tributaries as significant contributors of hydrological alterations to the Tonle Sap Floodplain in Cambodia. Hydrol. Earth Syst. Sci. 18, 5303−5315.

Aristi, I., Arroita, M., Larrañaga, A., Ponsatí, L., Sabater, S., Von Schiller, D., et al., 2014. Flow regulation by dams affects ecosystem metabolism in Mediterranean rivers. Freshw. Biol. 59, 1816−1829.

Assani, A.A., Stichelbout, E., Roy, A.G., Petit, F., 2006. Comparison of impacts of dams on the annual maximum flow characteristics in three regulated hydrologic regimes in Quebec (Canada). Hydrol. Process. 20, 3485–3501.

Batalla, R.J., Gómez, C.M., Kondolf, G.M., 2004. Reservoir-induced hydrological changes in the Ebro River basin (NE Spain). J. Hydrol. 290, 117–136.

Bayley, P.B., 1991. The flood pulse advantage and the restoration of river-floodplain systems. Regul. Rivers. 6, 75–86.

Beger, M., Grantham, H.S., Pressey, R.L., Wilson, K.A., Peterson, E.L., Dorfman, D., et al., 2010. Conservation planning for connectivity across marine, freshwater, and terrestrial realms. Biol. Conserv. 143, 565–575.

Begon, M., Townsend, C.R., Harper, J.L., 2006. Ecology: from Individuals to Ecosystems. Blackwell Publishing Ltd, Malden, MA.

Benejam, L., Angermeier, P.L., Munné, A., García-Berthou, E., 2010. Assessing effects of water abstraction on fish assemblages in Mediterranean streams. Freshw. Biol. 55, 628–642.

Bishop, K.A., Bell, J.D., 1978. Observations of the fish fauna below Tallowa Dam (Shoalhaven River, New South Wales) during river flow stoppages. Aust. J. Mar. Fresh. Res. 29, 543–549.

Bonvechio, T.F., Allen, M.S., 2005. Relations between hydrological variables and year-class strength of sportfish in eight Florida waterbodies. Hydrobiologia 532, 193–207.

Boulton, A.J., 2003. Parallels and contrasts in the effect of drought on stream macroinvertebrate assemblages. Freshw. Biol. 48, 1173–1185.

Boulton, A.J., Brock, M.A., Robson, B.J., Ryder, D.S., Chambers, J.M., Davis, J.A., 2014. Australian Freshwater Ecology: Processes and Management. Wiley-Blackwell, Chichester, United Kingdom.

Brown, L.R., Bauer, M.L., 2010. Effects of hydrologic infrastructure on flow regimes of California's Central Valley rivers: implications for fish populations. River Res. Appl. 26, 751–765.

Bunn, S.E., Arthington, A.H., 2002. Basic principles and ecological consequences of altered flow regimes for aquatic biodiversity. Environ. Manage. 30, 492–507.

Bunn, S.E., Thoms, M.C., Hamilton, S.K., Capon, S.J., 2006. Flow variability in dryland rivers: boom, bust and the bits in between. River Res. Appl. 22, 179–186.

Caiola, N., Ibanez, C., Verdu, J., Munné, A., 2014. Effects of flow regulation on the establishment of alien fish species: a community structure approach to biological validation of environmental flows. Ecol. Indic. 45, 598–604.

Carrete Vega, G., Wiens, J.J., 2012. Why are there so few fish in the sea? Proc. R. Soc. B. Biol. Sci. 279, 2323–2329.

Casado, A., Hannah, D.M., Peiry, J.-L., Campo, A.M., 2013. Influence of dam-induced hydrological regulation on summer water temperature: Sauce Grande River, Argentina. Ecohydrology 6, 523–535.

Casas-Mulet, R., Saltveit, S.J., Alfredsen, K., 2015. The survival of Atlantic salmon (*Salmo salar*) eggs during dewatering in a river subjected to hydropeaking. River Res. Appl. 31, 433–446.

Catford, J.A., Downes, B.J., Gippel, C.J., Vesk, P.A., 2011. Flow regulation reduces native plant cover and facilitates exotic invasion in riparian wetlands. J. Appl. Ecol. 48, 432–442.

Catford, J.A., Morris, W.K., Vesk, P.A., Gippel, C.J., Downes, B.J., 2014. Species and environmental characteristics point to flow regulation and drought as drivers of riparian plant invasion. Divers. Distrib. 20, 1084–1096.

Chessman, B.C., Jones, H.A., Searle, N.K., Growns, I.O., Pearson, M.R., 2010. Assessing effects of flow alteration on macroinvertebrate assemblages in Australian dryland rivers. Freshw. Biol. 55, 1780–1800.

Collares-Pereira, M.J., Cowx, I.G., 2004. The role of catchment scale environmental management in freshwater fish conservation. Fisheries Manage. Ecol. 11, 303–312.

Dewson, Z.S., James, A.B.W., Death, R.G., 2007. A review of the consequences of decreased flow for instream habitat and macroinvertebrates. J. N. Am. Benthol. Soc. 26, 401–415.

Dickson, N.E., Carrivick, J.L., Brown, L.E., 2012. Flow regulation alters alpine river thermal regimes. J. Hydrol. 464–465, 505–516.

Dolores Bejarano, M., Nilsson, C., Gonzalez Del Tanago, M., Marchamalo, M., 2011. Responses of riparian trees and shrubs to flow regulation along a boreal stream in northern Sweden. Freshw. Biol. 56, 853–866.

Downes, B.J., Entwisle, T.J., Reich, P., 2003. Effects of flow regulation on disturbance frequencies and in-channel bryophytes and macroalgae in some upland streams. River Res. Appl. 19, 27–42.

Dudgeon, D., Arthington, A.H., Gessner, M.O., Kawabata, Z.-I., Knowler, D.J., Lévêque, C., et al., 2006. Freshwater biodiversity: importance, threats, status and conservation challenges. Biol. Rev. 81, 163–182.

Dynesius, M., Nilsson, C., 1994. Fragmentation and flow regulation of river systems in the northern third of the world. Science 266, 753–762.

Fahrig, L., 2003. Effects of habitat fragmentation on biodiversity. Ann. Rev. Ecol. Evol. Syst. 34, 487–515.

Falke, J.A., Fausch, K.D., Magelky, R., Aldred, A., Durnford, D.S., Riley, L.K., et al., 2011. The role of groundwater pumping and drought in shaping ecological futures for stream fishes in a dryland river basin of the western Great Plains, USA. Ecohydrology 4, 682–697.

Fausch, K.D., Torgersen, C.E., Baxter, C.V., Li, H.W., 2002. Landscapes to riverscapes: bridging the gap between research and conservation of stream fishes. BioScience 52, 483–498.

Frazier, P., Page, K., 2006. The effect of river regulation of floodplain wetland inundation, Murrumbidgee River, Australia. Marine Freshw. Res. 57, 133–141.

Freeman, M.C., Bowen, Z.H., Bovee, K.D., Irwin, E.R., 2001. Flow and habitat effects on juvenile fish abundance in natural and altered flow regimes. Ecol. Appl. 11, 179–190.

Frissell, C.A., Liss, W.J., Warren, C.E., Hurley, M.D., 1986. A hierarchical framework for stream habitat classification: viewing streams in a watershed context. Environ. Manage. 10, 199–214.

Frota, F.F., Paiva, B.P., Franca Schettini, C.A., 2012. Intra-tidal variation of stratification in a semi-arid estuary under the impact of flow regulation. Braz. J. Oceanogr. 61, 23–33.

Fullerton, A.H., Burnett, K.M., Steel, E.A., Flitcroft, R.L., Pess, G.R., Feist, B.E., et al., 2010. Hydrological connectivity for riverine fish: measurement challenges and research opportunities. Freshw. Biol. 55, 2215–2237.

Gehrke, P.C., Gilligan, D.M., Barwick, M., 2002. Changes in fish communities of the Shoalhaven River 20 years after construction of the Tallowa Dam, Australia. River Res. Appl. 18, 265–286.

Gido, K.B., Propst, D.L., Olden, J.D., Bestgen, K.R., 2013. Multidecadal responses of native and introduced fishes to natural and altered flow regimes in the American Southwest. Can. J. Fish. Aquat. Sci. 70, 554–564.

Gillespie, B.R., Brown, L.E., Kay, P., 2015a. Effects of impoundment on macroinvertebrate community assemblages in upland streams. River Res. Appl. 31, 953–963.

Gillespie, B.R., Desmet, S., Kay, P., Tillotson, M.R., Brown, L.E., 2015b. A critical analysis of regulated river ecosystem responses to managed environmental flows from reservoirs. Freshw. Biol. 60, 410–425.

Gillson, J., Scandol, J., Suthers, I., 2009. Estuarine gillnet fishery catch rates decline during drought in eastern Australia. Fish. Res. 99, 26–37.

Gorski, K., Van Den Bosch, L.V., Van De Wolfshaar, K.E., Middelkoop, H., Nagelkerke, L.A.J., Filippov, O. V., et al., 2012. Post-damming flow regime development in a large lowland river (Volga, Russian Federation): implications for floodplain inundation and fisheries. River Res. Appl. 28, 1121–1134.

Greet, J., Webb, J.A., Downes, B.J., 2011. Flow variability maintains the structure and composition of in-channel riparian vegetation. Freshw. Biol. 56, 2514–2528.

Growns, I.O., Growns, J.E., 2001. Ecological effects of flow regulation on macroinvertebrate and periphytic diatom assemblages in the Hawkesbury-Nepean River, Australia. Regul. Rivers. 17, 275–293.

Growns, I., James, M., 2005. Relationships between river flows and recreational catches of Australian bass. J. Fish Biol. 66, 404–416.

Grubbs, S.A., Taylor, J.M., 2004. The influence of flow impoundment and river regulation on the distribution of riverine macroinvertebrates at Mammoth Cave National Park, Kentucky, USA. Hydrobiologia 520, 19—28.

Hall, C., Jordaan, A., Frisk, M., 2011. The historic influence of dams on diadromous fish habitat with a focus on river herring and hydrologic longitudinal connectivity. Landscape Ecol. 26, 95—107.

Humphries, P., Brown, P., Douglas, J., Pickworth, A., Strongman, R., Hall, K., et al., 2008. Flow-related patterns in abundance and composition of the fish fauna of a degraded Australian lowland river. Freshw. Biol. 53, 789—813.

Ibanez, C., Prat, N., Canicio, A., 1996. Changes in the hydrology and sediment transport produced by large dams on the lower Ebro river and its estuary. Regul. Rivers. 12, 51—62.

King, A.J., Humphries, P., Lake, P.S., 2003. Fish recruitment on floodplains: the roles of patterns of flooding and life history characteristics. Can. J. Fish. Aquat. Sci. 60, 773—786.

King, A.J., Tonkin, Z., Mahoney, J., 2009. Environmental flow enhances native fish spawning and recruitment in the Murray River, Australia. River Res. Appl. 25, 1205—1218.

Kingsford, R.T., Thomas, R.F., 1995. The Macquarie Marshes in arid Australia and their waterbirds: a 50-year history of decline. Environ. Manage. 19, 867—878.

Kingsford, R.T., Jenkins, K.M., Porter, J.L., 2004. Imposed hydrological stability on lakes in arid Australia and effects on waterbirds. Ecology 85, 2478—2492.

Konrad, C.P., Olden, J.D., Lytle, D.A., Melis, T.S., Schmidt, J.C., Bray, E.N., et al., 2011. Large-scale flow experiments for managing river systems. BioScience 61, 948—959.

Kupferberg, S.J., Palen, W.J., Lind, A.J., Bobzien, S., Catenazzi, A., Drennan, J., et al., 2012. Effects of flow regimes altered by dams on survival, population declines, and range-wide losses of California river-breeding frogs. Conserv. Biol. 26, 513—524.

Kustu, M.D., Fan, Y., Robock, A., 2010. Large-scale water cycle perturbation due to irrigation pumping in the US High Plains: a synthesis of observed streamflow changes. J. Hydrol. 390, 222—244.

Leigh, C., Boulton, A.J., Courtwright, J.L., Fritz, K., May, C.L., Walker, R.H., et al., 2016. Ecological research and management of intermittent rivers: an historical review and future directions. Freshw. Biol. 61, 1181—1199.

Li, R., Chen, Q., Tonina, D., Cai, D., 2015. Effects of upstream reservoir regulation on the hydrological regime and fish habitats of the Lijiang River, China. Ecol. Eng. 76, 75—83.

Lugg, A., Copeland, C., 2014. Review of cold water pollution in the Murray—Darling Basin and the impacts on fish communities. Ecol. Manage. Restor. 15, 71—79.

Lytle, D.A., Poff, N.L., 2004. Adaptation to natural flow regimes. Trends Ecol. Evol. 19, 94—100.

Magilligan, F.J., Nislow, K.H., 2005. Changes in hydrologic regime by dams. Geomorphology 71, 61—78.

Maheshwari, B.L., Walker, K.F., McMahon, T.A., 1995. Effects of regulation on the flow regime of the River Murray, Australia. Regul. Rivers. 10, 15—38.

Martinez, A., Larranaga, A., Basaguren, A., Perez, J., Mendoza-Lera, C., Pozo, J., 2013. Stream regulation by small dams affects benthic macroinvertebrate communities: from structural changes to functional implications. Hydrobiologia 711, 31—42.

McMahon, T.A., Finlayson, B.L., 2003. Droughts and anti-droughts: the low flow hydrology of Australian rivers. Freshw. Biol. 48, 1147—1160.

Meador, M.R., Carlisle, D.M., 2012. Relations between altered streamflow variability and fish assemblages in eastern USA streams. River Res. Appl. 28, 1359—1368.

Meile, T., Boillat, J.L., Schleiss, A.J., 2011. Hydropeaking indicators for characterization of the Upper-Rhone River in Switzerland. Aquat. Sci. 73, 171—182.

Miller, S.W., Judson, S., 2014. Responses of macroinvertebrate drift, benthic assemblages, and trout foraging to hydropeaking. Can. J. Fish. Aquat. Sci. 71, 675—687.

Mims, M.C., Olden, J.D., 2013. Fish assemblages respond to altered flow regimes via ecological filtering of life history strategies. Freshw. Biol. 58, 50−62.

Miserendino, M.L., 2009. Effects of flow regulation, basin characteristics and land-use on macroinvertebrate communities in a large arid Patagonian river. Biodivers. Conserv. 18, 1921−1943.

Mortenson, S., Weisberg, P., Stevens, L., 2012. The influence of floods and precipitation on Tamarix establishment in Grand Canyon, Arizona: consequences for flow regime restoration. Biol. Invasions 14, 1061−1076.

Nagrodski, A., Raby, G.D., Hasler, C.T., Taylor, M.K., Cooke, S.J., 2012. Fish stranding in freshwater systems: sources, consequences, and mitigation. J. Environ. Manage. 103, 133−141.

Naiman, R.J., Alldredge, J.R., Beauchamp, D.A., Bisson, P.A., Congleton, J., Henny, C.J., et al., 2012. Developing a broader scientific foundation for river restoration: Columbia River food webs. Proc. Natl. Acad. Sci. 109, 21201−21207.

Nielsen, D., Podnar, K., Watts, R.J., Wilson, A.L., 2013. Empirical evidence linking increased hydrologic stability with decreased biotic diversity within wetlands. Hydrobiologia 708, 81−96.

Nilsson, C., Reidy, C.A., Dynesius, M., Revenga, C., 2005. Fragmentation and flow regulation of the world's large river systems. Science 308, 405−408.

Olden, J.D., Naiman, R.J., 2010. Incorporating thermal regimes into environmental flows assessments: modifying dam operations to restore freshwater ecosystem integrity. Freshw. Biol. 55, 86−107.

Olden, J.D., Poff, N.L., 2003. Redundancy and the choice of hydrologic indices for characterizing streamflow regimes. River Res. Appl. 19, 101−121.

Olden, J.D., Konrad, C.P., Melis, T.S., Kennard, M.J., Freeman, M.C., Mims, M.C., et al., 2014. Are large-scale flow experiments informing the science and management of freshwater ecosystems? Front. Ecol. Environ. 12, 176−185.

Overton, I.C., McEwan, K., Gabrovsek, C., Sherrah, J.R., 2006. The River Murray Floodplain Inundation Model (RiM-FIM) Hume Dam to Wellington. CSIRO Water for a Healthy Country Flagship Technical Report 2006, Adelaide.

Paetzold, A., Yoshimura, C., Tockner, K., 2008. Riparian arthropod responses to flow regulation and river channelization. J. Appl. Ecol. 45, 894−903.

Perkin, J.S., Bonner, T.H., 2011. Long-term changes in flow regime and fish assemblage composition in the Guadalupe and San Marcos rivers of Texas. River Res. Appl. 27, 566−579.

Poff, N.L., Matthews, J.H., 2013. Environmental flows in the Anthropocene: past progress and future prospects. Curr. Opin. Environ. Sustain. 5, 667−675.

Poff, N.L., Zimmerman, J.K.H., 2010. Ecological responses to altered flow regimes: a literature review to inform the science and management of environmental flows. Freshw. Biol. 55, 194−205.

Poff, N.L., Allan, J.D., Bain, M.B., Karr, J.R., Prestegaard, K.L., Richter, B.D., et al., 1997. The natural flow regime. BioScience 47, 769−784.

Poff, N.L., Richter, B.D., Arthington, A.H., Bunn, S.E., Naiman, R.J., Kendy, E., et al., 2010. The ecological limits of hydrologic alteration (ELOHA): a new framework for developing regional environmental flow standards. Freshw. Biol. 55, 147−170.

Ponsatí, L., Acuña, V., Aristi, I., Arroita, M., García-Berthou, E., Von Schiller, D., et al., 2015. Biofilm responses to flow regulation by dams in Mediterranean rivers. River Res. Appl. 31, 1003−1016.

Power, M.E., Sun, A., Parker, G., Dietrich, W.E., Wootton, J.T., 1995. Hydraulic food-chain models: an approach to the study of food-web dynamics in large rivers. BioScience 45, 159−167.

Preece, R.M., Jones, H.A., 2002. The effect of Keepit Dam on the temperature regime of the Namoi River, Australia. River Res. Appl. 18, 397−414.

Reich, P., McMaster, D., Bond, N., Metzeling, L., Lake, P.S., 2010. Examining the ecological consequences of restoring flow intermittency to artificially perennial lowland streams: patterns and predictions from the Broken−Boosey creek system in northern Victoria, Australia. River Res. Appl. 26, 529−545.

Reynolds, L.V., Shafroth, P.B., House, P.K., 2014. Abandoned floodplain plant communities along a regulated dryland river. River Res. Appl. 30, 1084−1098.

Richter, B.D., Baumgartner, J.V., Powell, J., Braun, D.P., 1996. A method for assessing hydrological alteration within ecosystems. Conserv. Biol. 10, 1163−1174.

Robertson, A., Bacon, P., Heagney, G., 2001. The responses of floodplain primary production to flood frequency and timing. J. Appl. Ecol. 38, 126−136.

Rolls, R.J., Arthington, A.H., 2014. How do low magnitudes of hydrologic alteration impact riverine fish populations and assemblage characteristics? Ecol. Indic. 39, 179−188.

Rolls, R.J., Leigh, C., Sheldon, F., 2012. Mechanistic effects of low-flow hydrology on riverine ecosystems: ecological principles and consequences of alteration. Freshw. Sci. 31, 1163−1186.

Sammut, J., Erskine, W.D., 1995. Hydrological impacts of flow regulation associated with the Upper Nepean water supply scheme, NSW. Aust. Geogr. 26, 71−86.

Schmutz, S., Bakken, T.H., Friedrich, T., Greimel, F., Harby, A., Jungwirth, M., et al., 2015. Response of fish communities to hydrological and morphological alterations in hydropeaking rivers of Austria. River Res. Appl. 31, 919−930.

Sponseller, R.A., Heffernan, J.B., Fisher, S.G., 2013. On the multiple ecological roles of water in river networks. Ecosphere 4, 17.

Stewart-Koster, B., Olden, J.D., Gido, K.B., 2014. Quantifying flow−ecology relationships with functional linear models. Hydrol. Sci. J. 59, 629−644.

Strayer, D.L., Dudgeon, D., 2010. Freshwater biodiversity conservation: recent progress and future challenges. J. N. Am. Benthol. Soc. 29, 344−358.

Stromberg, J.C., Lite, S.J., Marler, R., Paradzick, C., Shafroth, P.B., Shorrock, D., et al., 2007. Altered streamflow regimes and invasive plant species: the Tamarix case. Glob. Ecol. Biogeogr. 16, 381−393.

Taylor, C.M., Millican, D.S., Roberts, M.E., Slack, W.T., 2008. Long-term change to fish assemblages and the flow regime in a southeastern U.S. river system after extensive aquatic ecosystem fragmentation. Ecography 31, 787−797.

Torrente-Vilara, G., Zuanon, J., Leprieur, F., Oberdorff, T., Tedesco, P.A., 2011. Effects of natural rapids and waterfalls on fish assemblage structure in the Madeira River (Amazon Basin). Ecol. Freshw. Fish 20, 588−597.

Vitule, J.R.S., Skóra, F., Abilhoa, V., 2012. Homogenization of freshwater fish faunas after the elimination of a natural barrier by a dam in Neotropics. Divers. Distrib. 18, 111−120.

Vörösmarty, C.J., McIntyre, P.B., Gessner, M.O., Dudgeon, D., Prusevich, A., Green, P., et al., 2010. Global threats to human water security and river biodiversity. Nature 467, 555−561.

Ward, J.V., 1989. The four-dimensional nature of lotic ecosystems. J. N. Am. Benthol. Soc. 8, 2−8.

White, H.L., Nichols, S.J., Robinson, W.A., Norris, R.H., 2012. More for less: a study of environmental flows during drought in two Australian rivers. Freshw. Biol. 57, 858−873.

Wiens, J.A., 1989. Spatial scaling in ecology. Funct. Ecol. 3, 385−397.

GEOMORPHOLOGICAL EFFECTS OF FLOW ALTERATION ON RIVERS

5

Geoff J. Vietz and Brian L. Finlayson

The University of Melbourne, Parkville, VIC, Australia

5.1 INTRODUCTION

The geomorphic characteristics of stream channels and their floodplains provide the physical template through which flow is translated into physical and hydraulic conditions. These conditions and their interactions with, and effects upon flow, are what drives the distribution and functioning of biota and ecosystem processes. Lignon et al. (1995) stated that "if the stream's physical foundation is pulled out from under the biota, even the most insightful biological research program will fail to preserve ecosystem integrity" (p. 183).

Not only is flow translated into the hydraulic conditions experienced by biota, but the geomorphic characteristics of rivers and their floodplains are a function of flow. Alterations to the flow regime directly modify river morphology and geomorphic functions such as channel shape and sediment transport and distribution (Harvey, 1969). There is a feedback mechanism that the altered channel morphology has on the interaction of flow with the channel. So the primary response of altered flow and the secondary response of altered channel morphology both have implications for aquatic ecosystems.

The geomorphic effects of flow alteration are profound. Reviews of the scientific literature on the effects of flow alteration on river geomorphology provide strong evidence of these effects over time scales that range from millennia, such as climatically induced channel size reductions (e.g., Dury, 1970), to depositional features resulting from daily flow fluctuations of a single or sequence of flow events (e.g., Vietz et al., 2012). In fact, long-term climatic changes may be recognized as one of the most influential geomorphic agents, and it is only because of a focus on management time scales that this has not been included in this chapter.

Geomorphic changes to channel form and processes, such as bed and bank erosion, can be related to specific characteristics of flow regime alteration. Important flow characteristics can include: (1) increased frequency of high-magnitude flows or larger overall flow volume (i.e., in urban rivers) that can lead to greater channel scour; (2) changes in the rate of flow recession such as rapid flow drawdown below a hydropower dam, leading to increased likelihood of mass failure where saturated riverbanks overcome the strength of the bank material retaining them; and (3) prolonged low flows, such as below an irrigation storage, leading to subaerial preparation of the bank through drying and desiccation, facilitating removal of dislodged sediments during subsequent flow events. Changes to deposition are equally driven by flow. Floodplain formation through deposition can be reduced below a regulated river through decreases in both the frequency of overbank flows and the lower sediment concentrations (Marren et al., 2014).

Water for the Environment. DOI: http://dx.doi.org/10.1016/B978-0-12-803907-6.00005-X

In understanding the geomorphic effects of flow alteration, we must remain aware of the complexities and multiple drivers of physical form and process. There are numerous attributes that alter the sensitivity of stream morphology to flow alteration including: the presence of bedrock, which influences the rate of migration and the ability of the stream to influence channel form (e.g., Cenderelli and Cluer, 1998); vegetation cover that can provide resistance to flow such as the role of dense vegetation (e.g., Osterkamp and Hupp, 2010); or the role of woody debris in increasing channel resistance to flow (e.g., Gippel et al., 1992). Furthermore, another fluvial input that alters channel geomorphology is sediment. Sediment transport is a dominant driver of river morphology (and one that is commonly co-related to flow alteration). For example, dams often retain more than 75% of the river's sediment load and this can lead to changes in channel dimensions and substrate character (Petts and Gurnell, 2005). Nevertheless, one way or another, flow is most commonly the dominant driver of changes to morphology and in-channel processes.

In this chapter, geomorphic changes to river systems under common flow alteration settings are described such as those detailed in Chapter 2. We focus on three common and overwhelming flow alteration scenarios that have a significant impact on the geomorphology and hydraulics of rivers: watershed vegetation changes, watershed urbanization, and the construction of dams on rivers. The significant influence of clearing of watershed vegetation can partly be attributed to it commonly being the first significant human-induced perturbation on the fluvial system. Urban development is another significant diffuse source flow alteration and the effects can be dramatic, with, for example, channels enlarging by 2−4 times (MacRae, 1996), initiated by changes to less than 5% of the watershed (Vietz et al., 2014). The geomorphic influence of dams is well-known and highly pervasive. In some cases, more than 100 km of river have been degraded below dams (Petts and Gurnell, 2005), and flow reductions have led to channel narrowing of up to 85% (Poff et al., 1997).

Geomorphic changes are considered to the physical form (morphology) and functions (processes). Some settings, such as groundwater influences, are not covered here because their relative geomorphic role (e.g., compared to dam impacts) is minor. Although other geomorphic agents such as sediment load and direct modification to channels are not the focus of this chapter, these drivers are discussed within each flow alteration setting. Importantly, geomorphic change (i.e., erosion) is not necessarily a negative attribute. It is rather the extent and rate of change, influenced by anthropogenic changes to the flow and sediment regime, which we seek to understand and manage. First, however, we place the role of geomorphology in an ecological context by highlighting its role and the influence of flow alteration.

5.2 THE ROLE OF GEOMORPHOLOGY IN AQUATIC ECOSYSTEMS

Flow is a major determinant of physical habitat in streams, which in turn is a major determinant of biotic composition.

First principle on flow alteration and ecosystems, Bunn and Arthington (2002, p. 492)

The form and function, as well as the hydraulics, of river systems are altered by changes to the flow regime, and these in turn impact physical habitat and aquatic ecosystems (Fig. 5.1). Examples of physical and hydraulic attributes that may be ecologically important include: the size and diversity of bed sediments and how they influence hyporheic exchange, hydraulic diversity at the reach and patch scale, pools and slow water refuges, sediment deposits such as bars and benches and the role they play in vegetation diversity, erosional niches such as undercut banks and the distribution of large wood (O'Connor, 1991; Steiger et al., 2001; Vietz et al., 2013). For any given discharge, the more diverse the channel morphology, the greater the range of flow velocities and water depth, and hence the higher the diversity of physical habitats and resulting opportunities for biological diversity (Gippel, 2001).

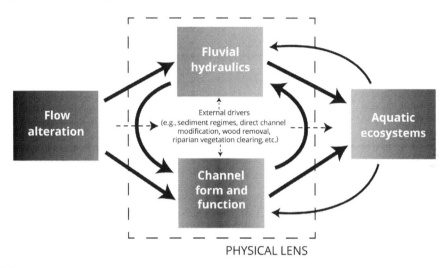

FIGURE 5.1

Interactions between flow and ecology require consideration of the physical lens that translates flow into the hydraulic and geomorphic changes to river systems.

Importantly, although managed flow alteration produces changes in channel morphology and geomorphic processes that often have negative ecological impacts, it is important to recognize that rivers are naturally dynamic. Poff et al. (1997) state that the ecological integrity of river ecosystems depends on their natural dynamic character. For example, some level of bank erosion is a natural part of river behavior and is integral to the functioning of most river ecosystems (Florsheim et al., 2008). It is most commonly excessive extents or rates of change that we aim to understand and manage. Instream biota are adaptable to change within natural rates, but may not be able to quickly adapt to sudden changes in channel characteristics such as morphology or sediment movement.

The role of feedbacks in ecogeomorphic systems is well recognized (Stoffel et al., 2013). Flow alters geomorphic conditions, facilitating ecological responses that in turn further alter geomorphic conditions. Examples of this include: the development of benches, islands, or floodplains in aggrading systems that encourage vegetation colonization, encouraging further deposition, with the process repeating itself (Corenblit et al., 2009; Steiger et al., 2001); or large wood liberated from floodplains through bank erosion, increasing roughness, encouraging deposition within the channel and floodplain, and providing niches for vegetation establishment as a future source of wood (Osterkamp and Hupp, 2010). Furthermore, flow alteration alters channel morphology, which influences hydraulics and can exacerbate flow alteration. An incised channel resulting from urban stormwater flows, for example, will contain larger flow events that can increase physical disturbance (Vietz et al., 2016a).

There is a range of flows that can influence channel form or geomorphic processes. Bankfull and overbank flows are most commonly considered to be the geomorphically *effective* or *dominant* discharges based predominantly on their magnitude and the energy expended on the channel (Pickup and Warner, 1976). However, these flows occur less frequently than smaller flows. Medium-sized events such as half-bankfull may be more geomorphically effective. Not only are they more frequent, but they are often capable of reaching required sediment transport thresholds such as by mobilizing bed sediments and initiating channel incision (Hawley and Vietz, 2016; see also Chapter 12). Smaller flows, such as baseflow(s), are rarely considered but can also have geomorphic influence through actions such as infilling pools in sand bed streams, and driving the distribution of aquatic vegetation. The change in these flows is dependent on the type of flow alteration and these settings are the focus of the next section.

5.3 COMMON SETTINGS THAT INFLUENCE RIVER GEOMORPHOLOGY

There are numerous influences that alter river flows, including atmospheric, anthropogenic, and physiographic changes to watersheds such as changes to land use. These are described in Chapter 3. Some of the common flow alteration scenarios and their effect on river geomorphology are described in Table 5.1. The three most geomorphically effective flow scenarios (vegetation changes, urbanization, and dams) are described in more detail in the following sections.

5.3.1 VEGETATION CHANGE

Most geomorphic studies on rivers and vegetation focus on the role of riparian and channel vegetation (e.g., Osterkamp and Hupp, 2010) or large wood (e.g., Tonkin et al., 2016). A common observation in this regard is that stream channels with intact riparian vegetation cover and more wood are wider, less incised, have increased bed roughness and lower average velocity than streams with deforested riparian zones. Sweeney et al. (2004) found that the high-quality geomorphic characteristics of rivers in forested riparian zones led to improved ecosystem attributes such as increased mass of organic matter, improved nutrient dynamics, and a greater abundance of macroinvertebrates. The clearing of vegetation from the watershed or floodplain has also long been linked to an increased likelihood of catastrophic events. Examples of this include river avulsion (Brizga and Finlayson, 1990), floodplain stripping (Nanson, 1986), or gully development (Prosser and Soufi, 1998). Nevertheless, the benefits of vegetation in the riparian or floodplain zone are not necessarily flow-related effects and are more attributable to shading, hydraulic roughness, soil binding, and the supply and retention of organic matter.

Where watershed-scale clearing is the subject of a study, it is commonly focused on runoff and sediment yield (e.g., Ngo et al., 2015), rather than channel change. The clearing of vegetation from a watershed, although more diffuse and less obvious than vegetation change at the stream, influences stream geomorphology through changes in flow and sediment delivered to the stream. The case of the lower Bega River, NSW, Australia, provides a good example where clearing of vegetation from a watershed within a few decades following European settlement increased the geomorphic effectiveness of rainfall.

Table 5.1 Common Flow Alteration Scenarios That Impact River Geomorphology

Flow Alteration Scenario	Common Geomorphic Response of Receiving Channel	Example Citations
Vegetation clearing	– Channel incision and enlargement – Variable sediment loads	Brookes and Brierley, 1997; Garcia-Ruiz et al., 2010
Urbanization	– Channel incision and enlargement – Channel simplification – Sediment increases then decreases	Chin, 2006; Vietz et al., 2014
Large water supply reservoirs	– Channel reduction – Channel simplification – Variable sediment loads – Bed sediment armoring	Erskine, 1996; ICOLD, 2009; Kondolf et al., 1997
Hydropower	– Channel enlargement (run-of-river) – Reduction in channel sinuosity – Episodic erosion – Bed sediment armoring	Assani and Petit, 2004
Flood alleviation dams	– Enlarged low-flow channel – Potentially reduced high-flow channel – Reduced sediment supply – Reduced floodplain formation	Petts and Gurnell, 2005
Farm dams	– Channel reductions and connectivity interruptions – Potential for clearwater erosion under sediment starvation	Callow and Smettem, 2009
Groundwater extraction and depletion	– Minor influence through vegetation or changed salinity	Boulton and Hancock, 2006; Brunke and Gonser, 1997
Altered surface drainage network	– Loss of small-order streams – Channelization – Concentrated flow leading to channel enlargement	Pavelis, 1987

Scenarios in bold are discussed in more detail in the following section (with the term Dams *used to categorize scenarios involving reservoirs or dams). Note that there is considerable variability in the geomorphic responses that cannot always be generalized.*

With more than 70% of the granitic watershed cleared of vegetation, and much of the riparian zone, the river was transformed from a narrow, relatively deep, mixed-load system with pools and a fine-grained floodplain, to a shallow, wide, channel with sandy bed-load and floodplain (Brookes and Brierley, 1997).

Similarly, in the Spanish Pyrenees, watershed clearing and intensive agricultural activity over recent centuries has led to unstable, eroding channels (Garcia-Ruiz et al., 2010). Subsequent depopulation, commencing at the beginning of the 20th century and particularly in the 1960s, resulted in farmland abandonment and expansion of the shrub and forest cover. These changes led to a secondary fluvial response including channel narrowing, vegetation colonization of mid-channel islands, and less erosional activity, attributed to declining streamflow and reduced sediment transport. In contrast, Davies-Colley (1997) found stream channels in forested watersheds to be wider than those with pasture. They hypothesized that pasture grasses gradually colonize and stabilize marginal sediment deposits that creep streamward. The strength of grasses in stabilizing channel banks and riparian zones has been shown by Blackham (2006).

It is well-known that suspended sediment yield from watersheds commonly increases following deforestation. These changes, however, can also be exacerbated by land-use practices such as forest operations, mining, and construction. For example, forest roads are known to cause net increases in sediment production (Wemple et al., 2001), and Prosser and Abernethy (1999) demonstrated gully erosion to be 1.5 times more likely where wood piles (log-rows) confined flow.

5.3.2 URBANIZATION

Urbanization of a watershed is arguably the most effective geomorphic agent of change for receiving rivers. First, small streams (low-order streams) are often buried underground and placed in pipes or conduits (Elmore and Kaushal, 2008). Second, urban development and associated drainage infrastructure significantly modifies stormwater runoff from a watershed (see Chapter 2—urban impacts on hydrology) and leads, at least initially, to increased channel erosion and sediment transport (Wolman, 1967). Third, channel erosion commonly elicits a management response whereby hard-lining, such as rock and concrete, is used to stabilize rivers in the often constrained urban environment (Stein et al., 2013).

The term *urban stream* commonly conjures images of a channel that has been straightened and hard-lined, such as with rock and concrete. These changes, however, are a management response to flow and sediment alteration, rather than a direct consequence of urbanization of the watershed. Before stream channels subject to watershed urbanization are stabilized or directly modified they undergo numerous morphologic and physical changes, including:

1. *Increased channel instability*: Although erosion is a natural process, it is commonly accelerated in urban streams. Vietz et al. (2014) demonstrated that bank erosion is more extensive and active in urban streams, with erosion commonly occurring on both banks. Channel instability has been associated with low-quality habitat (Booth and Henshaw, 2001).
2. *Channel enlargement*: commonly a 2.5-times increase but often more (Chin, 2006). The process of channel incision involves deepening of the channel through bed erosion, commonly followed by channel widening (Hawley et al., 2012).
3. Trimble (1997) found that for streams in California, United States, increases in channel capacity contributed more than two-thirds of the sediment supply from the watershed over a 10-year period.
4. *Increased rates of lateral migration*: where channels are not constrained (Nelson et al., 2006).
5. *Loss of riffles*: Hawley et al. (2013) found that for streams in Kentucky, United States, urbanization led to streams with shorter riffles and longer and deeper pools.
6. *Altered substrate*: the increased stream power associated with urban flows can coarsen the size of bed sediments by preferentially removing finer material (Hawley et al., 2013), and can lead to reduced depth of mobile streambed sediments (Vietz et al., 2014).
7. *Changes in bars and benches*: Sediment deposits such as bars and benches have been found to be reduced in number in urban streams where there is an increase in stream energy but a

reduction in available sediment (Vietz et al., 2014), although in the early phases of urbanization there is commonly an increase in sediment load (Chin, 2006).

The driver of these geomorphic changes has increasingly been recognized as not just an increase in the *total imperviousness (TI)* of a watershed, but the way in which stormwater is conveyed from impervious surfaces to the stream by efficient drainage systems (Booth, 1991). *Effective imperviousness (EI)*, the proportion of TI connected to the stream, has been found to be a considerably stronger predictor of geomorphic response than TI (Vietz, 2013; Fig. 5.2).

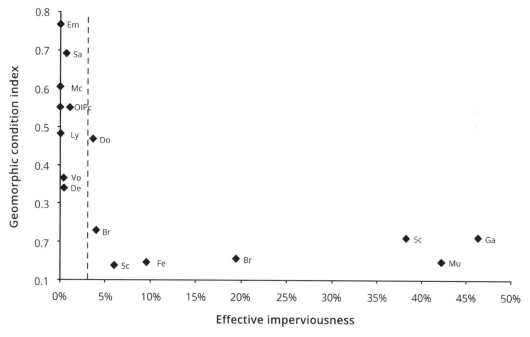

FIGURE 5.2

Geomorphic condition score for urban streams relative to the impervious surfaces directly connected to the stream (effective imperviousness). The dashed line represents 3% EI, the threshold beyond which geomorphic condition is consistently low.

Source: Based on data subsequently published in Vietz et al. (2014)

It is not, however, just the flow to the stream that is affecting such geomorphic changes. It is well established that streams are also a product of the sediment supply provided to them (Schumm, 1977; Fig. 4.3). The way in which urbanization increases flow volume, while decreasing sediment loads to streams, has been considered a "double-edged sword" greatly accelerating geomorphic degradation (Vietz et al., 2016b). Wolman (1967) conceptualized changes to suspended sediment from an urban watershed that included: (1) an initial increase in sediment yield from the watershed during the construction phase and (2) stabilization of sediment yield once the watershed is dominated by urban land cover and streams are stabilized or buried as pipes. The lag times in streams responding to these phases of urbanization can often be decadal (Chin, 2006) and recognition of these phases and response times is critical to the aim to develop an understanding of a *common* response of streams to urbanization (Vietz et al., 2016a). The sources and changes of coarse-grained sediments (as opposed to suspended sediment) are much less well understood (Russell et al., 2016). The management challenges of protection, particularly recovery, of streams in urban watersheds through flow and sediment regime management is made most challenging by these multiple impacts and the variability in response.

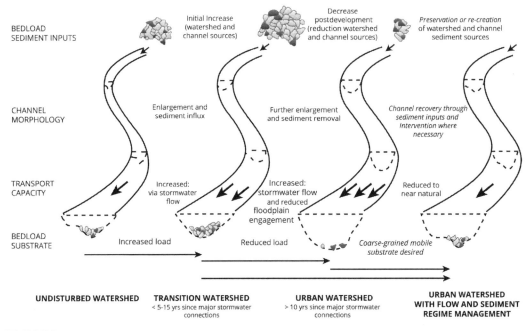

FIGURE 5.3

Conceptual model of the impacts of urbanization on stream bed-load sediment and flow inputs, sediment transport capacity, and channel morphology, including the potential for future management.

Source: Adapted from Vietz et al. (2016a)

Considerable efforts are being directed toward opportunities for reducing the impacts of urban-induced flow alteration on stream geomorphology. This is principally being done through addressing stormwater flows that are so effective in increasing erosion and transport capacity (Fig. 5.3). Hawley and Vietz (2016) describe an approach for protection of stream geomorphology where the shear stress that mobilizes the streambed sediments is calculated to target reductions for flows leading to channel incision. Flow reductions are achieved through watershed-scale activities, including detention basins and distributed stormwater control measures (Fletcher et al., 2014; Vietz et al., 2016b). Urban-induced changes to the sediment regime must be considered and addressed where possible, consistent with flow regime management, by maintaining appropriate supply of coarse-grained sediments to the stream (Vietz et al., 2016a). All stages of urbanization require consideration with regard to the impacts on receiving rivers. For example, rapid conversion of agricultural land to housing can see significant increases in flow as well as initial sediment loads such as described by Ebisemiju (1989) for a town in Nigeria.

Adequately addressing the twin causes (flow and sediment regimes) of geomorphic degradation of streams in urban watersheds is a significant future challenge (Vietz et al., 2016b). Whether or not urban stream channels can be managed in a relatively stable, quasi-equilibrium state appears to be still an open question.

5.3.3 **DAMS**

Hydrologic influences on rivers can be highly localized, and dams are likely to represent the greatest point source of flow alteration (Petts, 1984). More than half of the world's large rivers are regulated by dams (Nilsson et al., 2005; see Chapter 3). Although dams are found on streams of all sizes, down to farm dams on small first-order streams, it is the large dams that attract most attention in the literature because of the scale and significance of their effects. Dams can affect the geomorphological characteristics of streams in three ways: (1) dams interrupt the continuity of sediment movement down rivers by trapping all but the finest particles; (2) dams can change the magnitude, frequency, duration, and regime of flows released downstream, and hence the way in which stream energy is exerted on the channel; and (3) dams can divert flow out of the river system and into other streams or to consumptive uses, reducing (or sometimes increasing) the overall volume of flow. ICOLD (2009) lists the factors that determine the magnitude of the geomorphological impact of dams as: the storage capacity of the impoundment in relation to mean annual runoff, operational procedures used on the dam, bed materials in the downstream channel, water outlet structures on the dam, and the sediment load of the river prior to dam construction.

Although all dams interrupt the transport of sediment to some degree, it is the size of the dam storage relative to the mean annual flow that determines the magnitude of downstream flow alteration that is possible. Many large dams, particularly in areas with high interannual flow variability, are designed to hold 100% or more of the mean annual flow, and can therefore make substantial changes to the flow regime, even changing the season in which most flow occurs (e.g., Gippel and Finlayson, 1993). Similarly, there are large dams that may only store a relatively small proportion of the mean annual flow and therefore have only minor impacts on the pattern of flows. The most notable example of such a dam is the Three Gorges Dam on the Yangtze River in China. Although its dam wall is the largest in the world, it can store only 4% of the mean annual flow and is therefore not capable of making significant changes to the downstream flow pattern (Chen et al., 2016). The resulting downstream channel changes are dominated by incision resulting from the sediment-trapping effects of the dam (Yuan et al., 2012). As pointed out below, adjustment times can be very long and it is not yet clear just how the Yangtze will respond to these changes or how long it will take (Yuan et al., 2012).

Here we consider dams built for water supply (irrigation and urban water), hydropower dams, and dams for flood control, noting, however, that it is often the case that a dam has more than one of these uses. It should also be noted that the persistence of the changes downstream depends on the nature of the watershed below the dam. In dry climates where the downstream river is an exotic stream, the changes will persist to the mouth of the stream (e.g., the Nile River in Egypt) or to the next major tributary (e.g., the Murray River in Australia). In contrast, where the climate downstream of the dam is humid, the impacts of the dam are ameliorated by runoff and sediment generated below the dam (e.g., the Yangtze River in China).

The geomorphic impacts of sediment storage in dams (and extraction of sediment from river channels by mining) have been described in detail by Kondolf (1997), who describes this as *"hungry water"* because of the interruption to sediment continuity. Channel incision occurs downstream of dams, and the lowering of the bed level in turn destabilizes the channel banks. Continued erosion without an upstream sediment supply causes the bed sediment to become coarser, and the bed can become armored by the coarser fraction overlying the finer sediments, often halting river-bed incision as described by Erskine (1996) for the Goulburn River in Victoria.

The ability of a dam to store sediment being delivered from upstream is commonly assessed using the capacity/inflow ratio (Brune, 1953). Using this approach, Vörösmarty et al. (2003) estimate that more than 30% of global sediment flux is intercepted by dams. Walling (2006), however, cautions that the sediment transport data are generally poor, and that there have also been widespread increases in sediment load that date back much earlier than the period during which large dams have been built. This means that there are both many examples of the effects of increased sediment supply to coasts (Syvitski et al., 2005), and also of coastal erosion resulting from the storage of sediment in reservoirs. For example, Yang et al. (2011) report the erosive effect on the Yangtze delta of the >50,000 dams in the catchment. In the 1960s and 1970s, the delta accumulated $\sim 125 \times 10^6 \mathrm{m}^3$/a but by the first decade of the 21st century this changed to $\sim 100 \times 10^6 \mathrm{m}^3$/a of erosion. Major erosion has been reported for the Nile delta, driven by the storage of sediment in the Aswan Dams and the diversion of water for irrigation (Inman and Jenkins, 1984; Stanley, 1996). Similarly, Milliman and Farnsworth (2011) report that dam construction in the Danube River system has reduced the river's sediment load by 35%, causing coastal erosion at the mouth of the river.

The net effect of dams built for water supply is that water will be diverted from the river system and most commonly not returned directly to the river. This process can lead to a reduction in flow (in some cases a complete loss of flow) in the downstream channel. It can also result in an increase in flow where water is diverted into another system for distribution. In both cases there will be an impact on the channel size and form. River channels tend to be in a state of quasi-equilibrium determined by the range of flows that do the most work, referred to as the *dominant discharge* or the *effective discharge*, though this is usually a range rather than a single discharge. Adjustment of the channel is achieved by alteration to the channel depth, width, and slope in relation to flow and sediment load. ICOLD (2009) provide a detailed description of the adjustment of channels as a result of the changes imposed by dams. They also give a detailed discussion of the regime equations available to predict the response of the channel to a change in flow regime. Here, we illustrate the process using an example of rivers from the Snowy River system in Australia.

Tilleard (2001), using effective discharge defined in terms of excess stream power, pointed out that the actual rate of erosion downstream of a dam depends on the power available to move material and the resistance of that material to being moved. In some cases, there is little erosion as the material is too coarse to be moved by the available flow. Using this approach, properties of both the flow and the material forming the bed and banks are taken into consideration in determining channel response to changes in flow. The Snowy Mountains Hydro-electric Scheme involved the diversion of flow from the Snowy River to rivers flowing inland into the Murray River system in southeastern Australia. The Snowy River experienced a loss of 99% of its discharge with the completion of the Jindabyne Dam in 1967, with much of the flow being diverted into the Tumut River that flows into the Murray–Darling system.

The effect on the morphology of the Snowy River downstream of the Jindabyne Dam was a dramatic reduction in channel size (Fig. 5.4). The change occurred over a period of nearly 15 years, indicating that lack of sediment caused by the trapping effect of the dam has slowed the rate of channel response.

In the case of the receiving river, the Tumut River experienced increased flow volume and excess stream energy (Fig. 5.5). In this case, the effective flow based on critical shear stress for the bed is the same in both the pre- and postdiversion flows because of the coarse bed material. However, there is a significant change in effective flow for the banks. The regulated flows are concentrated at close to bankfull stage, while at the same time, the material forming the banks is finer than the bed material. The increases in flow have thus resulted in significant channel widening (Gippel et al., 1992; Tilleard, 2001).

In addition, because the channel is filled to near bankfull level throughout most of the summer, this inhibits the growth of vegetation on the banks and limits the bank-stabilizing effects of vegetation. This exacerbates the erosive effects of the increases in flow. In its unregulated state, summer was the low-flow season for the Tumut River, which allowed vegetation to grow on the banks. As shown in Fig. 5.6, adjustment of the channel to changed flow conditions has been a long-term process; channel widening continued for 30 years in the period covered in Fig. 5.6 and may have continued further beyond that.

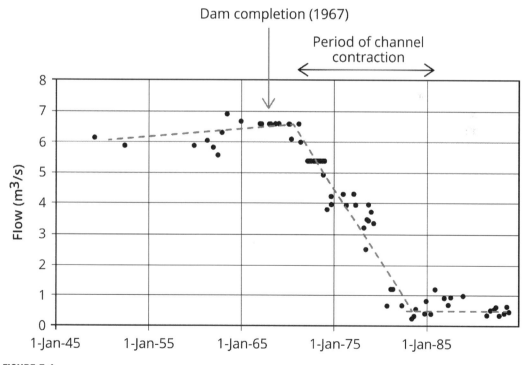

FIGURE 5.4

Snowy River at Dalgety. Time-series of the flow corresponding to a stage that is representative of the contemporaneous bankfull stage. Dashed line graphically fitted.

Source: Tilleard (2001)

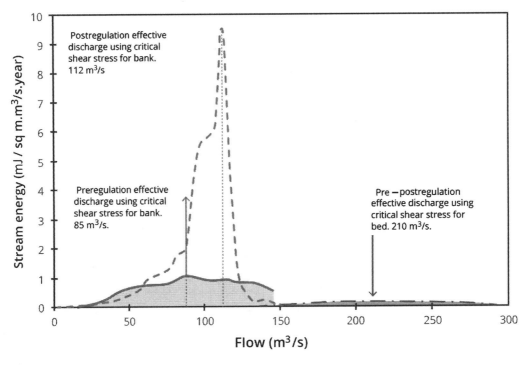

FIGURE 5.5

Distribution of excess energy with flow for pre- and postregulation periods showing estimates of effective discharge for bed and bank materials.

Source: Tilleard (2001)

One of the commercial advantages of hydroelectric power generation is that the system can be started and stopped almost instantaneously, and so hydroelectric dams have commonly been used only in peak load periods. These dams cause significant changes to the flow patterns downstream. Flow events are constrained within a narrow range of discharges that are the most efficient for electricity generation (McMahon and Finlayson, 1995). The impact on the channel form and sediment properties depends on the relationship between this range of flows and the shear resistance of the channel bed and banks, as well as both the amount and grain size distribution of the sediment available to be transported. However, there are now increasing numbers of large dams generating base load power and therefore running virtually full time and these obviously have very different effects on flows. One example of this is the Three Gorges Dam on the Yangtze River in China, discussed earlier.

Assani and Petit (2004) monitored the impact of the Butgenbach peak load hydropower dam on the downstream bed morphology and sediment properties of the Warche River in Belgium. The hydropower dam began its operation in 1932 and the most efficient flow for electricity generation is approximately 0.6 of bankfull discharge. Flows at this level are therefore common and the

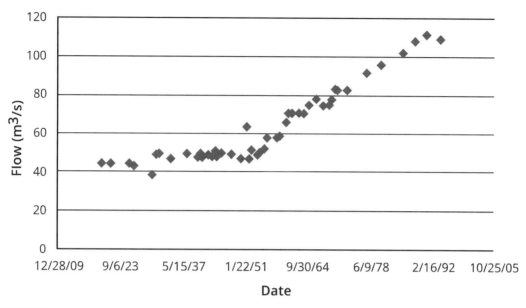

FIGURE 5.6

Time-series of flows at bankfull. From rating tables applying at the Tumut River pump station gauge from 1919 to 1993.

Source: Tilleard (2001)

frequency and duration of larger flows significantly reduced. Floods larger than half the mean annual flood have been reduced in frequency and duration, and floods with a return period >10 years no longer occur. Bed erosion below the dam occurred quickly after the hydropower dam began operation, but was then halted by the exposure of bedrock in the channel bed. Channel width continued to increase for 60 years, and there was a reduction in channel sinuosity. Pool—riffle sequences have virtually disappeared from the channel to be replaced by gravel bars, islets, and bedrock outcrops. River bed sediments below the dam have become coarser, with D_{50} now three times that above the dam. Sediment below the dam is now better sorted following the selective removal of the fine fraction; the difference in particle size between the surface and subsurface layers has been effectively removed.

In the examples we have discussed here, the morphological response has been slow. This slow rate of response to regulation has also been noted by Petts and Gurnell (2005), who point out that in arid and semi-arid regions channel changes may be rapid, but in more humid climate zones the time scale of response may extend to millennia. The examples we report here range from 15 to 60 years, though in the cases of the Tumut and Yangtze Rivers, the channels are still adjusting. The length of time that river channels can continue to adjust their morphology presumably has adverse consequences for biological responses that have a different time cycle. These examples also show that although there are some general principles that govern the geomorphic effects of dams, in each case there are local characteristics that determine the detail of effects.

McMahon and Finlayson (1995), in reviewing the provision of environmental water in regulated rivers, pointed out that the outlet structures of dams have invariably been constructed for the operational requirements of the dam and are often not capable of releasing the range of flows necessary for environmental restoration. Constraints on flow, however, are often operational and are determined by the loss of other functions such as hydropower generation (Pennisi, 2004). Thus the management and design arrangements for outlet structures are another factor controlling the geomorphological response of the channel. The case of the Jindabyne Dam on the Snowy River, which we have mentioned above, is an example of this limit imposed by the outlet structure (see Fig. 5.4). The dam was constructed with a fixed outlet pipe with a capacity of only 50 ML/day ($\sim$1% of natural flow) and incapable of releasing the range of flows necessary for channel maintenance and environmental sustainability. When eventually public pressure forced the release of environmental water into the river downstream, a complete new outlet structure had to be built on the dam to increase the release to 400 ML/day (Snowy Hydro, 2015) at a cost of US$216 million (Pigram, 2000). Opportunities for modifying dam outlet structure management to minimize geomorphic impacts require further consideration, including: (1) increased outlet capacities (up to bankfull and floodplain flows); (2) sediment mobilization from the base of storages; and (3) operational strategies that minimize impacts (i.e., reduced rates of drawdown) or maximize benefits (i.e., "piggy backing" on downstream tributary inflows).

5.4 CONCLUSION

Flow alteration to river systems has increased significantly over the last half century, and is set to increase further under climate change (Vörösmarty et al., 2010). By better understanding the relationship between flow alteration and geomorphic change, and the resulting influence on aquatic ecosystems, we can elucidate the implications for managing river systems. In particular, simply linking flow alteration and ecological response reduces our mechanistic understanding and limits our opportunities for addressing the underlying processes of degradation. Seeing flow alteration through the *physical lens* of changes to river morphology and function will assist in identifying required actions to restore environments, whether they are in the watershed or within the river channel.

When considering the role of geomorphology in understanding the implications of landscape change to ecosystems, Gilvear (1999) proposed four tenets. The first of these is to recognize the importance of lateral, vertical, and downstream connectivity in the fluvial system. Whether aiming to minimize point or diffuse sources of flow alteration, consideration of the receiving channel and the links this has with the floodplain or groundwater, requires watershed-scale strategies. Second, Gilvear (1999) stresses the importance of understanding fluvial history and chronology over a range of time scales. Whether it is the immediate geomorphic response of a channel to dam closure, or the adjustment of a channel to vegetation clearing a century later, the time scales over which these changes can occur need to be central to any proactive strategy for river management. Third, recognizing the sensitivity of geomorphic systems to environmental change and the dynamics of the natural systems are important to distinguish from the excessive extents or rates of change induced by flow alteration. Finally, Gilvear (1999) highlights the importance of landforms and processes in controlling and defining the hydraulic environment for biota and ecosystems. In this respect, the understanding and addressing of flow regime changes that negatively impact the physical form and processes of rivers should be central to river restoration and protection. These tenets apply equally to any study or management of the implications of flow alteration on the geomorphic form and function of river systems.

ACKNOWLEDGMENTS

We would like to thank Prof. Kate Rowntree and Dr. Angus Webb for reviews that greatly improved this chapter, and Dr. Avril Horne for making it a pleasant process.

REFERENCES

Assani, A.A., Petit, F., 2004. Impact of hydroelectric power releases on the morphology and sedimentology of the bed of the Warche River (Belgium). Earth Surf. Process. Landf. 29 (2), 133−143.

Blackham, D.M., 2006. The resistance of herbaceous vegetation to erosion: implications for stream form. Unpublished PhD thesis, School of Geography, University of Melbourne, Melbourne, Australia.

Booth, D.B., 1991. Urbanization and the natural drainage system − impacts, solutions, and prognoses. Northwest Environ. J. 7, 93−118.

Booth, D.B., Henshaw, P.C., 2001. Rates of channel erosion in small urban streams. In: Wigmosta, M., Burges, S. (Eds.), Land Use and Watersheds: Human Influence on Hydrology and Geomorphology in Urban and Forest Areas. AGU Monograph Series, Water Science and Application. Washington, DC, pp. 17−38.

Boulton, A.J., Hancock, P.J., 2006. Rivers as groundwater-dependent ecosystems: a review of degrees of dependency, riverine processes and management implications. Aust. J. Botany 54, 133−144.

Brizga, S.O., Finlayson, B.L., 1990. Channel avulsion and river metamorphosis: the case of the Thomson River, Victoria, Australia. Earth Surf. Process. Landf. 15 (5), 391−404.

Brookes, A., Brierley, G., 1997. Geomorphic responses of lower Bega River to catchment disturbance, 1851−1926. Geomorphology 18, 291−304.

Brune, G.M., 1953. Trap efficiency of reservoirs. EOS 34 (3), 407−418.

Brunke, M., Gonser, T., 1997. The ecological significance of exchange processes between rivers and groundwater. Freshw. Biol. 37, 1−33.

Bunn, S.E., Arthington, A.H., 2002. Basic principles and ecological consequences of altered flow regimes for aquatic biodiversity. Environ. Manage. 30 (4), 492−507.

Callow, J.N., Smettem, K.R.J., 2009. The effect of farm dams and constructed banks on hydrologic connectivity and runoff estimation in agricultural landscapes. Environ. Model. Softw. 24 (8), 959−968.

Cenderelli, D.A., Cluer, B.L., 1998. Depositional processes and sediment supply in resistant-boundary channels: examples from two case studies. Fluvial Processes in Bedrock Channels, Rivers Over Rock.

Chen, J., Finlayson, B.L., Wei, T., Sun, Q., Webber, M., Li, M., et al., 2016. Changes in monthly flows in the Yangtze River, China − with special reference to the Three Gorges Dam. J. Hydrol. 536, 293−301.

Chin, A., 2006. Urban transformation of river landscapes in a global context. Geomorphology 79 (3), 460−487.

Corenblit, D., Steiger, J., Gurnell, A., Tabacchi, E., Roques, L., 2009. Control of sediment dynamics by vegetation as a key function driving biogeomorphic succession within fluvial corridors. Earth Surf. Process. Landf. 34, 1790−1810.

Davies-Colley, R.J., 1997. Stream channels are narrower in pasture than in forest. N. Z. J. Mar. Freshw. Res. 31, 599–608.

Dury, G.H., 1970. General theory of meandering valleys and underfit streams. Rivers and River Terraces. The Geographical Readings Series, Macmillan 264–275.

Ebisemiju, F.S., 1989. Patterns of stream channel response to urbanization in the humid tropics and their implications for urban land use planning: a case study from southwestern Nigeria. Appl. Geogr. 9 (4), 273–286.

Elmore, A.J., Kaushal, S.S., 2008. Disappearing streams: patterns of stream burial due to urbanization. Front. Ecol. Environ. 6, 308–312.

Erskine, W.D., 1996. Downstream hydrogeomorphic impacts of Eildon Reservoir on the Mid-Gouldburn River. Victoria. Proc. Royal Soc. Vic. 108 (1), 1–15.

Fletcher, T.D., Vietz, G.J., Walsh, C.J., 2014. Protection of stream ecosystems from urban stormwater runoff; the multiple benefits of an ecohydrological approach. Prog. Phys. Geogr. 38 (5), 543–555.

Florsheim, J.L., Mount, J.F., Chin, A., 2008. Bank erosion as a desirable attribute of rivers. BioScience 58 (6), 519–529.

Garcia-Ruiz, J.M., Lana-Renault, N., Beguería, S., Teodoro, L., Regues, D., Nadal-Remoero, E., et al., 2010. From plot to regional scales: interactions of slope and catchment hydrological and geomorphic processes in the Spanish Pyrenees. Geomorphology 120, 248–257.

Gilvear, D.J., 1999. Fluvial geomorphology and river engineering: future roles utilizing a fluvial hydrosystems framework. Geomorphology 31, 229–245.

Gippel, C., 2001. Geomorphic issues associated with environmental flow assessment in alluvial non-tidal rivers. Aust. J. Water Resour. 5 (1).

Gippel, C.J., Finlayson, B.L., 1993. Downstream environmental impacts of regulation of the Goulburn River, Victoria. Institution of Engineers Australia, pp. 33–38.

Gippel, C.J., O'Neill, I.C., Finlayson, B.L., 1992. The Hydraulic Basis of Snag Management. Centre for Environmental Applied Hydrology, Department of Civil and Agricultural Engineering and Department of Geography. University of Melbourne, Melbourne, Australia, p. 116.

Harvey, A.M., 1969. Channel capacity and the adjustment of streams to hydrologic regime. J. Hydrol. 8, 82–98.

Hawley, R.J., Bledsoe, B.P., Stein, E.D., Haines, B.E., 2012. Channel evolution model of semiarid stream response to urban-induced hydromodification. J. Am. Water Resour. Assoc. 48 (4), 722–744.

Hawley, R.J., MacMannis, K.R., Wooten, M.S., 2013. Bed coarsening, riffle shortening, and channel enlargement in urbanizing watersheds, northern Kentucky, USA. Geomorphology 201, 111–126.

Hawley, R.J., Vietz, G.J., 2016. Addressing the urban stream disturbance regime. Freshw. Sci. 35 (1), 278–292.

ICOLD, 2009. Sedimentation and sustainable use of reservoirs and river systems. International Committee on Large Dams, Committee on Reservoir Sedimentation Draft Bulletin. Available from: http://www.icold-cigb.org/userfiles/files/CIRCULAR/CL1793Annex.pdf.

Inman, D.L., Jenkins, S.A., 1984. The Nile littoral cell and man's impact on the coastal zone of the South East Mediterranean. 19th Coastal Engineering Conference Proceedings, Houston, Texas, pp. 1600–1617.

Kondolf, G., 1997. Hungry water: effects of dams and gravel mining on river channels. Springer-Verlag New York Inc.

Lignon, F.K., Dietrich, W.E., Trush, W.J., 1995. Downstream ecological effects of dams. BioScience 45 (3), 183–192.

MacRae, C., 1996. Experience from morphological research of Canadian streams: is control of the two-year frequency event the best basis for stream channel protection? In: Roesner, L.A. (Ed.), Effects of Watershed Development and Management on Aquatic Ecosystems. American Society of Civil Engineers, New York.

Marren, P.M., Grove, J.R., Webb, J.A., Stewardson, M.J., 2014. The potential for dams to impact lowland meandering river floodplain geomorphology. Sci. World J. Article 309673.

McMahon, T.A., Finlayson, B.L., 1995. Reservoir system management and environmental flows. Lake Reserv. Res. Manage. 1, 65−76.

Milliman, J.D., Farnsworth, K.L., 2011. River discharge to the coastal ocean − a global synthesis. Cambridge University Press, Cambridge, UK.

Nanson, G.C., 1986. Episodes of vertical accretion and catastrophic stripping: a model of disequilibrium floodplain development. Geol. Soc. Am. Bull. 97, 1467−1475.

Nelson, P.A., Smith, J.A., Miller, A.J., 2006. Evolution of channel morphology and hydrologic response in an urbanizing drainage basin. Earth Surf. Process. Landf. 31, 1063−1079.

Ngo, T.S., Nguyen, D.B., Rajendra, P.S., 2015. Effect of land use change on runoff and sediment yield in Da River Basin of Hoa Binh province, Northwest Vietnam. J. Mountain Sci. 12 (4), 1051−1064.

Nilsson, C., Reidy, C.A., Dynesius, M., Revenga, C., 2005. Fragmentation and flow regulation of the world's large river systems. Am. Assoc. Advance. Sci. 405. Available from: https://ezp.lib.unimelb.edu.au/login?url = https://search.ebscohost.com/login.aspx?direct = true&db = edsjsr&AN = edsjsr.3841248&site = eds-live&scope = site.

O'Connor, N.A., 1991. The effects of habitat complexity on the macroinvertebrates colonising wood substrates in a lowland stream. Oecologia 85, 504−512.

Osterkamp, W.R., Hupp, C.R., 2010. Fluvial processes and vegetation — Glimpses of the past, the present, and perhaps the future. Geomorphology 116 (3−4), 274−285.

Pavelis, G., 1987. Farm drainage in the United States: history, status and prospects. US Department of Agriculture, Economic Research Service, Washington, DC.

Pennisi, E., 2004. The grand (Canyon) experiment. Science 306, 1884−1886.

Petts, G., 1984. Impounded Rivers: Perspectives for Ecological Management. John Wiley & Sons, New York.

Petts, G.E., Gurnell, A.M., 2005. Dams and geomorphology: research progress and future directions. Geomorphology 71 (1), 27−47.

Pickup, G., Warner, R.F., 1976. Effects of hydrologic regime on magnitude and frequency of dominant discharge. J. Hydrol. 29 (1−2), 51−75.

Pigram, J.J., 2000. Viewpoint − options for rehabilitation of Australia's Snowy River: an economic perspective. Regul. Rivers 16, 363−373.

Poff, L.N., Allan, J.D., Bain, M.B., Karr, J.R., Prestegaard, K.L., Richter, B.D., et al., 1997. The natural flow regime, a paradigm for river conservation and restoration. BioScience 47 (11), 769−784.

Prosser, I.P., Abernethy, B., 1999. Increased erosion hazard resulting from log-row construction during conversion to plantation forest. Forest Ecol. Manage. 123 (2−3), 145−155.

Prosser, I.P., Soufi, M., 1998. Controls on gully formation following forest clearing in a humid temperate environment. Water Resour. Res. 34 (12), 3666−3671.

Russell, K., Vietz, G., Fletcher, T.D., 2016. Not just a flow problem: how does urbanization impact on the sediment regime of streams? 8th Australian Stream Management Conference, Blue Mountains, NSW, Australia.

Schumm, S.A., 1977. The Fluvial System. John Wiley & Sons, New York.

Snowy Hydro Limited, 2015. Snowy Hydro Water Report 2014−2015. Snowy Hydro Limited, Cooma, NSW, Australia.

Stanley, D.J., 1996. Letter section: Nile delta: extreme case of sediment entrapment on a delta plain and consequent coastal land loss. Mar. Geol. 129, 189−195.

Steiger, J., Gurnell, A., Petts, G., 2001. Sediment deposition along the channel margins of a reach of the Middle Severn, UK, Regul. Rivers Res. Manage. 17 (4−5), 443−460.

Stein, E.D., Cover, M.R., Elizabeth Fetscher, A., O'Reilly, C., Guardado, R., Solek, C.W., 2013. Reach-scale geomorphic and biological effects of localized streambank armoring. J. Am. Water Resour. Assoc. 49 (4), 780−792.

Stoffel, M., Rice, S., Turowski, J.M., 2013. Process geomorphology and ecosystems: disturbance regimes and interactions. Geomorphology 202, 1–3.

Sweeney, B.W., Bott, T.L., Jackson, J.K., Kaplan, L.A., Newbold, J.D., Standley, L.J., et al., 2004. Riparian deforestation, stream narrowing, and loss of stream ecosystem services. Proc. Natl. Acad. Sci. 101 (39), 14132–14137.

Syvitski, J.P.M., Vorosmarty, C.J., Kettner, A.J., Green, P., 2005. Impact of humans on the flux of terrestrial sediment to global coastal ocean. Science 308, 376–380.

Tilleard, J., 2001. River channel adjustment to hydrologic change. PhD thesis, Department of Civil and Environmental Engineering, The University of Melbourne.

Tonkin, Z., Kitchingman, A., Ayres, R.M., Lyon, J., Rutherfurd, I.D., Stout, J.C., et al., 2016. Assessing the distribution and changes of instream woody habitat in south-eastern Australian rivers. River Res. Appl. 32, 1576–1586.

Trimble, S.W., 1997. Contribution of stream channel erosion to sediment yield from an urbanizing watershed. Science 278 (5342), 1442–1444.

Vietz, G.J., 2013. Water(way) sensitive urban design: addressing the causes of channel degradation through catchment-scale management of water and sediment. Proceedings of the 8th International Water Sensitive Urban Design Conference (Institution of Engineers Australia, ed.), Gold Coast, Australia, 25–29 November 2013, pp. 219–225.

Vietz, G.J., Stewardson, M.J., Rutherfurd, I.D., Finlayson, B.L., 2012. Hydrodynamics and sedimentation of concave benches in a lowland river. Geomorphology 147–148, 86–101.

Vietz, G.J., Sammonds, M.J., Stewardson, M.J., 2013. Impacts of flow regulation on slackwaters in river channels. Water Resour. Res. 49 (4), 1797.

Vietz, G.J., Sammonds, M.J., Walsh, C.J., Fletcher, T.D., Rutherfurd, I.D., Stewardson, M.J., 2014. Ecologically relevant geomorphic attributes of streams are impaired by even low levels of watershed effective imperviousness. Geomorphology 206, 67–78.

Vietz, G.J., Rutherfurd, I.D., Fletcher, T.D., Walsh, C.J., 2016a. Thinking outside the channel: challenges and opportunities for stream morphology protection and restoration in urbanizing catchments. Landsc. Urban Plan. 145, 34–44. Available from: http://dx.doi.org/10.1016/j.landurbplan.2015.09.004.

Vietz, G.J., Walsh, C.J., Fletcher, T.D., 2016b. Urban hydrogeomorphology and the urban stream syndrome: treating the symptoms and causes of geomorphic change. Prog. Phys. Geogr. 40 (3), 480–492.

Vörösmarty, C.J., Meybeck, M., Fekete, B., Sharma, K., Green, P., Syvitski, J.P.M., 2003. Anthropogenic sediment retention: major global impact from registered river impoundments. Glob. Planet. Change 39, 169–190.

Vörösmarty, C.J., McIntyre, P.B., Gessner, M.O., Dudgeon, D., Prusevich, A., Green, P., et al., 2010. Global threats to human water security and river biodiversity. Nature 467 (7315), 555–561.

Walling, D.E., 2006. Human impact on land–ocean sediment transfer by the world's rivers. Geomorphology 79, 192–216.

Wemple, B.C., Swanson, F.J., Jones, J.A., 2001. Forest roads and geomorphic process interactions, Cascade Range, Oregon. Earth Surf. Process. Landf. 26 (2), 191–204.

Wolman, M.G., 1967. A cycle of sedimentation and erosion in urban river channels. Geogr. Ann. Phys. Geogr. 49 (2/4), 385–395.

Yang, S.L., Milliman, J.D., Li, P., Xu, K., 2011. 50,000 dams later: erosion of the Yangtze River and its delta. Glob. Planet. Change 75, 14–20.

Yuan, W., Yin, D., Finlayson, B., Chen, Z., 2012. Assessing the potential for change in the middle Yangtze River channel following impoundment of the Three Gorges Dam. Geomorphology 147–148, 27–34.

IMPACTS OF HYDROLOGICAL ALTERATIONS ON WATER QUALITY

6

Meenakshi Arora, Roser Casas-Mulet, Justin F. Costelloe, Tim J. Peterson, Alexander H. McCluskey, and Michael J. Stewardson

The University of Melbourne, Parkville, VIC, Australia

6.1 NATURAL AND ANTHROPOGENIC DRIVERS OF WATER QUALITY

Water quality is a fundamental aspect of aquatic ecosystems, defining the conditions (e.g., salinity, temperature, nutrients, and oxygen) for aquatic flora and fauna. Whether water quality elements support or degrade ecosystems depend on concentrations, for example nutrients such as phosphorus and nitrogen are essential, but at high levels can become contaminants. Changes in water quality may result in immediate changes in the structure and function of ecosystems, including the numbers and types of organisms that can survive in the altered environment (ANZECC and ARMCANZ, 2000; Boulton et al., 2014). Such changes can also have significant effects on other water values such as drinking water quality, industrial water-use, and recreational, aesthetic, cultural, and spiritual values. Many natural and anthropogenic factors influence water quality in freshwater ecosystems. The influences of water quantity on quality are highly complex, and depend on the characteristics of individual catchments and the overlay of local anthropogenic influences. This chapter is primarily concerned with the dominant role of water flow and volume on water quality. It addresses specifically the impacts of anthropogenic hydrological alterations and the use of environmental water management to address water quality problems.

Many of the physicochemical characteristics of freshwater ecosystems are determined by eroded sediments, nutrients, salts, toxins, pathogens, and other contaminants delivered from the surrounding catchment into rivers after rainfall (Hoven et al., 2008; Thomas, 2014; Zarnetske et al., 2012), or from groundwater discharge to rivers. Changes in water quality can be directly attributed to source of water, hydrological variability, and in stream processes (SKM, 2013). The source of water may vary due to altitude, topography, soil type, and vegetation cover (Baker, 2003; Byers et al., 2005). Whether from snowmelt, rainfall/surface runoff, groundwater, tidal inflows, irrigation transfers, point and diffuse sources, the type of source is also a major determinant of the water quality. It can define water quality differences in terms of key parameters such as salinity, temperature, nutrients, and dissolved oxygen. Water flowing through catchments has a central role in the transport of contaminants to and within freshwater ecosystems and hydrological alterations in aquatic ecosystems can have significant impacts on water quality.

Human modifications to catchments have altered the quantity, quality, and balance of these natural sources, and introduced new water flow paths, for example irrigation transfers, flow releases from dams, and point and diffuse sources (Brabec et al., 2002; Walsh et al., 2005; Young and Huryn, 1999). Anthropogenic activities, such as river regulation, catchment land-use, and water extraction alter the natural flow and associated water quality characteristics. Some water quality indicators respond almost immediately to environmental change, particularly in response to changes in flow or water volume, but in other cases (e.g., salinity responses to land clearing) responses can have decadal response times. Therefore, a thorough understanding of the relationships between water quantity and quality is essential to guide management responses in a timely manner.

For this chapter, we have considered four representative measures of water quality: salinity, temperature, nitrogen concentration, and dissolved oxygen, to illustrate the effects of hydrological alteration and how they respond to environmental water deployment. Salinity comprises the dissolved load in streamflow and is affected by both hydrological and catchment land-use alteration. Water temperature is a physical water quality parameter and is affected by flow regulation and riparian changes. Nitrogen provides the essential nutrient for the growth of flora and fauna and is affected by land-use changes as well as hydrological flow variations. Dissolved oxygen is required by aquatic fauna for respiration and fluctuates with flow and temperature variation.

6.2 SALINITY

Stream salinization occurs as a result of complex interactions between the climate, hydrogeology, land-use, and catchment relief, and can arise naturally (henceforth, *primary salinity*) or as a result of land-use or water-use change (henceforth, *secondary salinity* or *salinization*). Salinity principally occurs in subhumid to arid regions where evaporation during dry periods is sufficient to concentrate salts within the soil, water table, or in zero-flow refuge ponds. Within these climate regions, the occurrence of secondary salinity is dependent upon local land-use and topography and is a major issue within Central and South America, Africa, the Middle East, central Asia, and southern mainland Australia (Williams, 1999).

The significance of stream salinity for environmental water is predominantly during periods of low flow, when saline groundwater in flow to streams may be a major contributor to total discharge, or when evaporative concentration may increase salinity in refuge ponds (Simpson and Herczeg, 1991). At times of low flow, saline marine water may also ingress inland into estuaries and low reaches of rivers. However, more complex processes can arise from altered peak flows, whereby reduced flood recharge can cause increased floodplain aquifer salinity causing significant stress to riparian vegetation (Jolly et al., 2008); and vegetation stress can be a major focus of environmental water releases. Whatever aspect of the flow regime is of concern, the interaction of the stream with the aquifer is often critical, and consequently the slow response times of aquifers can cause the changes to be semi-permanent within a management time frame (Cheng et al., 2014) and potentially irreversible (Peterson et al., 2014a, 2014b). In the following sections, the mechanisms are further discussed, followed by an overview of the anthropogenic influences and management options.

6.2.1 STREAM SALINITY PROCESSES

Stream salinity is the culmination of spatial and temporal processes that span from near-stream to intercatchment, and from subdaily to multiple decades, with short-term processes predominately occurring at fine spatial scales. Long-term processes are dominated by groundwater flow. Groundwater predominantly flows from fresher recharge zones in the upper slopes of a catchment to discharge into lower elevation streams (Fig. 6.1). Along the flow path, solutes within the aquifer are mobilized, causing groundwater salinity to increase. The accumulation of solutes and salinization of groundwater is facilitated by drier climatic conditions where mean annual potential evapotranspiration exceeds mean annual rainfall (Jackson et al., 2009). If there is a hydraulic gradient from the groundwater elevation to the stream then saline groundwater can flow into the stream.

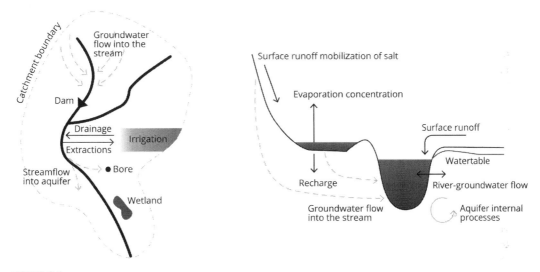

FIGURE 6.1

Major processes controlling stream salinity. The plane view of the catchment illustrates that groundwater flow in the upper catchment is often from the recharge locations into the stream. In the lower catchment, groundwater can flow from the stream to the aquifer. The cross-section demonstrates the former, and illustrates that surface runoff can also mobilize salt on the soil into the surface waters while standing water bodies can also experience high salinity due to evaporative concentration.

The influence of saline groundwater on the stream system is strongly affected by reach-scale hydraulic gradients between groundwater and surface water (Lamontagne et al., 2014). Consistently gaining gradients between groundwater and stream tend to occur higher in the catchment where groundwater is fresher (Fig. 6.1). Consistently losing gradients between stream and groundwater can occur in various parts of the catchment (Fig. 6.1), but result in little groundwater discharge to the stream and can minimize the effects of saline groundwater on riparian vegetation (Banks et al., 2011). Saline groundwater discharge would typically have most influence in reaches with saline unconfined groundwater and where the hydraulic gradient varies with stream stage, and these generally occur in the middle to lower reaches of catchments (Braaten and Gates, 2003; Ivkovic, 2009). At high-flow levels, recharge occurs into the stream bank storage zone and freshens the unconfined groundwater (Cartwright et al., 2010). At low-flow levels, the saline groundwater can discharge into the stream and result in orders-of-magnitude changes in stream salinity (Costelloe and Russell, 2014).

In stream evaporative concentration can have a significant influence on the salinity of standing pools or during low flows, particularly where saline groundwater is already influencing stream salinity. In semi-arid to arid catchments with high potential evapotranspiration rates, the reduction in volume by evaporation leaves behind the solutes and can result in significant increases in salinity during periods of no flow and also exacerbate salinity rises driven by groundwater discharge (Costelloe et al., 2005).

6.2.2 ANTHROPOGENIC INFLUENCES ON STREAM SALINITY

Anthropogenic influences on stream salinity occur both through river regulation and catchment land-use change. Both influences can drive diffuse or concentrated spatial effects on stream salinity and also have profound influences on the health of the riparian and floodplain zone of the river.

Catchment changes that occur far from the stream can drive an increase in the regional unconfined groundwater elevation, resulting in increased stream salinity. Such actions include land clearing and irrigation, both of which increase groundwater recharge. Land clearing affects groundwater because clearing of deep-rooted vegetation for shallow-rooted crops decreases evapotranspiration from the soil profile, and allows more recharge (Zhang et al., 2001). Irrigation can affect groundwater through over-irrigation, but efficient irrigation also requires periodic recharge to occur to flush solutes from the root zone. This irrigation drainage can cause groundwater mounding beneath irrigation areas (Sarwar et al., 2001). The increased infiltration and recharge from land clearing and irrigation can mobilize high solute levels in the vadose (unsaturated) zone and lead to increases in groundwater levels without any reduction in groundwater salinity (or even increased salinity; Salama et al., 1999). These increased groundwater levels provide an increased pressure gradient from the aquifer to the stream and subsequent higher salt loads being discharged by groundwater to streamflow. In addition, shallow saline groundwater in the riparian–floodplain zone can cause solutes to rise into the unsaturated zone through capillary action, and even be deposited at the surface, causing significant osmotic stress for riparian plant communities (Jolly et al., 2008).

Changes to the flood regime through river regulation can also impact stream salinity. For example, large in-stream storages can reduce the frequency of a range of discharges, from overbank events to subbankfull flows (see Chapter 3). In reaches experiencing saline groundwater discharge, small flow releases may not be sufficient to dilute stream salinity levels to desired levels or to effectively flush dense saline water from deep pools (Turner and Erskine, 2005; Western et al., 1996). Reducing the frequency of large flood events will reduce floodplain recharge and can cause a reduction in the extent of fresh bank storage lenses around the river or allow the buildup of solutes in the floodplain unsaturated zone (Cartwright et al., 2010). Reaches with a fresh bank storage zone are buffered against regional saline groundwater discharge. They are also vulnerable to near-stream groundwater pumping driving groundwater drawdown and increased saline regional groundwater flow into the riparian zone.

Other forms of river regulation, such as through locks and weirs that are in place to aid river navigation or water delivery, can maintain high river water levels throughout the natural low-flow season. As a result, the discharge of saline regional groundwater to the stream is limited and this can further increase the unconfined groundwater levels in the floodplain. As a result, saline groundwater can cause significant stress, or even die-back, of riparian vegetation, particularly large trees (Jolly et al., 2008).

6.2.3 MANAGEMENT OPTIONS TO CONTROL STREAM SALINITY

Given that stream salinity problems can be driven by both river regulation and catchment land-use changes, typically, integrated catchment management actions are required rather than solely relying on environmental water approaches (see Chapter 22). In addition, quantifying the contribution of most sources of stream salinity is very challenging. With the exception of irrigation drainage, the fluxes are determined by spatially heterogeneous processes acting at subdaily to multidecadal time scales. Furthermore, saline subsurface flow into the stream is determined by both the salinity in the aquifer and stream, and their respective water levels.

With these challenges in mind, options to control catchment land-use drivers of stream salinity include catchment revegetation with perennial pastures or woody vegetation, but the time scale of response is often multidecadal, or groundwater interception schemes.

Environmental water can be used to control stream salinity in a variety of ways. Flow releases can be used to dilute streamflow salinity during low-flow phases or else to flush saline water from standing pools. The latter option requires consideration of the stream discharge required to cause turbulent flow and full mixing in deeper pools, as density differences between saline water in pools and fresh streamflow can otherwise limit pool flushing (Western et al., 1996). Environmental water flows at bankfull to overbank levels are often aimed at providing water to riparian and floodplain vegetation experiencing osmotic stress due to groundwater and soil water salinity. The aim of these flows is to promote bank storage and floodplain recharge to lower the salinity of the soil water and upper phreatic zone (Holland et al., 2009). However, the *freshening* by environmental water releases may only bring short- to medium-term relief if unconfined groundwater levels in the floodplain are kept high by increased catchment recharge and/or flow-regulating structures. In some cases, pumping of saline groundwater into surface disposal basins has been explored to lower groundwater levels and facilitate the effectiveness of environmental water releases (George et al., 2005).

6.3 WATER TEMPERATURE

Water temperature is a key variable in freshwater systems. The diversity of thermal habitats can directly or indirectly influence the structure and dynamics of aquatic ecosystems (Dent et al., 2000; Frissell et al., 1986; Townsend et al., 1997). Temperature provides fundamental links between physical, chemical, and ecological processes, and plays an important role in the overall river ecosystem health and functioning (Poole and Berman, 2001). Water temperature conditions affect physical and biochemical processes in rivers such as suspended sediment concentration, organic matter decomposition, and dissolved oxygen (Johnson, 2004; Poole and Berman, 2001; Webb et al., 2008). It also affects growth, distribution, and survival of stream organisms (Boulton et al., 2014; Caissie, 2006; Ebersole et al., 2003b; McCullough et al., 2009). In addition, water temperature has economic importance, for example in industry (e.g., electricity and drinking water production), agriculture, aquaculture, and recreation activities (Van Vliet et al., 2011; Webb et al., 2008).

The thermal mass in rivers increases with water volume and stream temperature changes are inversely proportional to discharge (Sullivan et al., 1990). Therefore, hydrological alterations can have a significant effect on stream water temperature, particularly during periods of low flows, but also as a result of flow releases from thermally stratified reservoirs. Moreover, climate change may further affect river hydrology and temperatures globally (Beechie et al., 2013; Schneider et al.,

2013; Van Vliet et al., 2011), with associated consequences for aquatic ecosystems (Caissie, 2006; Poole and Berman, 2001; Webb et al., 2008; Wilby et al., 2010). Predicted future climatic change scenarios have been a significant driver of increasing interest in stream water temperature within the scientific community, environment managers, and regulators (Hannah et al., 2008), and is particularly challenging in regulated rivers where multiple water demands must be met.

6.3.1 THERMAL PROCESSES AND CONTROLLING MECHANISMS IN RIVERS

Heat exchange occurs at the air—water boundary and at the water—streambed boundary (Fig. 6.2A) in addition to heat transport within surface- and groundwater. At the air—water interface, the dominant source of heat is solar radiation or net short-wave radiation; this is followed by long-wave radiation and to a lesser extent evaporation and convective heat flux (Fig. 6.2A). Heat fluxes at the water—streambed interface are mainly a function of streambed conduction (from geothermal heating), and advective heat transport with groundwater and hypo-rheic exchange (Fig. 6.2A). Fluid friction can also influence heat exchange, but its contribution is usually small compared to the aforementioned components (Caissie, 2006; Constantz et al., 1994; Olden and Naiman, 2010).

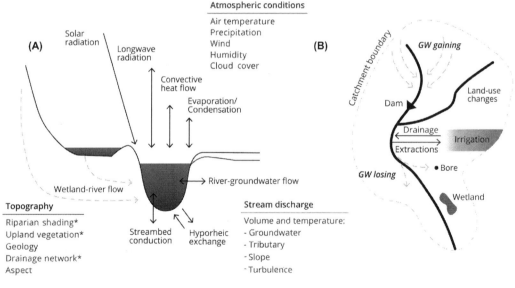

FIGURE 6.2

River heat exchange processes, drivers, and impacts, including (A) heat exchange processes in rivers and streams (arrows) and the physical drivers that control the rate of heat and water delivery to stream and river ecosystems (lists) at the transect level, and (B) impacts to the physical drivers (*) controlling rate of heat at the catchment scale.

Source: Based on Caissie (2006) and Olden and Naiman (2010)

Mechanisms influencing the rate of heat and water delivery to streams can be classified into four groups: atmospheric conditions, topography, stream discharge, and streambed (Caissie, 2006; Fig. 6.2B). Atmospheric conditions include air temperature, precipitation, wind speed, humidity, and cloud cover. They are at the same time largely influenced by topography (geology, aspect, latitude/altitude, and shading by riparian and aquatic vegetation), and hence both factors have a major effect on water temperature (Olden and Naiman, 2010). Although stream discharge and streambed are secondary drivers, they influence heating capacity through (1) mixing of water volumes from different sources and (2) sediment conduction and facilitation of hyporheic exchange and groundwater influx. Hence, the extent of natural or artificial hydrological alterations will dictate the extent of thermal modification in river systems.

Stream temperature generally increases longitudinally in the downstream direction or with increasing stream order (Allan and Castillo, 2007; Gordon et al., 2004). However, water temperatures in rivers are not spatially uniform and can vary significantly among nearby habitat patches (Hauer and Hill, 1996). Lateral (off-channel) and vertical (within substrate) patterns of temperature are increasingly recognized as an important aspect of habitat heterogeneity in streams (Boulton et al., 2014; Ebersole et al., 2003a; Poole and Berman, 2001).

Vertical temperature stratification is uncommon in flowing streams due to strong vertical mixing by turbulent flow conditions, but it can be found in river stretches where deep pools with stagnant water are present (Gordon et al., 2004). Overall, temperature stratification usually occurs in natural lakes or reservoirs. Warmer water has a lower density and sits above the cooler water forming a clear vertical temperature profile with depth (Boulton et al., 2014; Jorgensen et al., 2012; Moss, 2009; Fig. 6.3). Such natural or artificially driven phenomena may have a significant thermal influence on downstream reaches (see Section 6.2.2 on anthropogenic impacts).

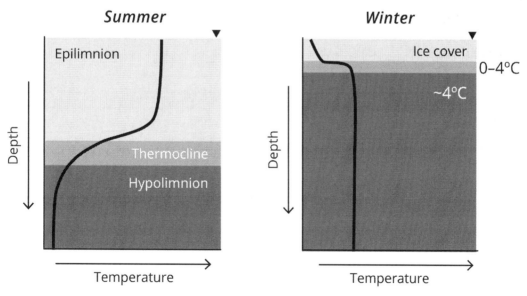

FIGURE 6.3

Seasonal stratification of reservoirs and lakes during summer (left) and winter (right).

Important temporal changes in stream temperature occur seasonally and daily, given that solar radiation is the dominant source of heat in rivers (Constantz et al., 1994). In temperate or cold climates, the major seasonal changes in stream temperature are driven by snowmelt in spring and evapotranspiration in summer. Although ice cover during winter may actually provide shelter to organisms from low air temperatures to the underlying stream water (Huusko et al., 2007), stream temperatures may be significantly reduced downstream by snowmelt events. Evapotranspiration losses may reduce discharges, and reduce in-flow from cool groundwater during the day, leading to warming stream temperatures (Constantz et al., 1994). Large differences in temperature can also occur between shaded headwater reaches and open lowland reaches, where water temperature will be minimally influenced by shading (Hauer and Hill, 1996; Stanford et al., 1988).

Depending on regional geology, the relative surface versus groundwater discharge, and the degree of connectivity between the channel and the groundwater, upwelling areas may provide cooler groundwater sources in summer or warmer in winter (Allan and Castillo, 2007; Hauer and Hill, 1996), creating a wide range of thermal habitats that in turn may influence organisms' life cycles and reproduction (Jorgensen et al., 2012).

The influence of fluvial geomorphology via hyporheic flows to stream temperature has been considered in alluvial streams (Burkholder et al., 2008). For example, more porous substrates will promote hyporheic flows (Johnson, 2004) that can potentially cool, buffer, or lag water temperature (Arrigoni et al., 2008). Blackwater deposition areas in rivers may show different thermal patterns from the main channel (Hauer and Hill, 1996). An integrated consideration of how fluvial geomorphology alters the spatial variability in stream temperature has only been recently established as a result of novel thermal imagery advances via remote sensing. This has partly been driven by an increasing interest in thermal refugia for aquatic organisms (Dugdale et al., 2015; Kurylyk et al., 2015; Torgersen et al., 2001).

6.3.2 ANTHROPOGENIC IMPACTS TO STREAM THERMAL PROCESSES

Anthropogenic impacts on catchments can drive significant changes in the natural thermal regimes. The combined effect of climate change and other human-driven modification of the environment indicate a rising trend in water temperature (Van Vliet et al., 2011; Wilby et al., 2010), but the consequences of such a rise may have differing thermal effects on rivers depending on the climatic region and season (Caissie, 2006). Here, we will focus on the effects of hydrological alterations due to river regulation in the natural thermal regime. Such effects are significant (Acaba et al., 2000; Casado et al., 2013; Horne et al., 2004; McCullough, 1999; Ryan et al., 2001; Zolezzi et al., 2011) and can have major impacts on aquatic biodiversity (Boulton et al., 2014; Bunn and Arthington, 2002; McCullough et al., 2009; Rutherford et al., 2009; Vatland et al., 2015).

Impoundments usually increase the residence time of water and, for example, increase stream temperature in cold climates (Allan and Castillo, 2007). In thermally well-stratified reservoirs, thermal changes in the downstream water course should be considered. Releases from the deeper reservoir water layer (hypolimnion) may be significantly cooler in summer and warmer in winter. This can cause important negative impacts on the life cycle of aquatic organisms (Chapter 4).

They include advances or delays in spawning, hatching, and development (McCullough, 1999; Preece and Jones, 2002). Particularly in Australia, the unseasonably cold water release from the bottom of stratified reservoirs to satisfy irrigation and water supply demands has a recognized impact on native fish. It affects their reproduction, feeding, growth, and survival, as well as slowing down all biological metabolic processes in the aquatic systems. Moreover, cold-water reservoir releases may put native fish species at a disadvantage as introduced colder-water species adapt better to such thermal changes (Lugg and Copeland, 2014; Rutherford et al., 2009). The persistence of water temperature alterations downstream will be dependent on the volume and temperature of the released water from the reservoir, and the tributary and groundwater in flow downstream. More importantly, the recovery distances will depend on the heat exchange processes at the air–water interface (Webb, 1995).

Hydrological alterations due to river regulation can also mean a reduced amount of water in the stream (see Chapter 3). This is likely to increase the importance of groundwater influx in the system and affect the overall river thermal regime, being most critical during summer months. In the context of climate change, this may potentially lead to temperatures above the living threshold of certain aquatic organisms, with significant effects to overall community structure and functional shifts in trophic levels (Woodward et al., 2010).

6.3.3 MANAGEMENT OPTIONS

Potential measures to mitigate increased water temperature have been developed, in order to buy time for aquatic ecosystems to adapt (Boulton et al., 2014; Hansen et al., 2003; Nõges et al., 2010; Wilby et al., 2010). Although several management options to mitigate human-induced thermal impacts include riparian replanting to control temperature in situ and downstream targeted reaches (Capon et al., 2013; Wilby et al., 2014), here we focus on options to mitigate thermal impacts from hydrological alterations.

Options to mitigate thermal impacts from cold-water releases or dam stratification have been summarized in Sherman (2000) and later in Olden and Naiman (2010). Such options are divided into two main categories, including the selective withdrawal of water of the desired temperature along the stratified profile, or artificially breaking up the stratification before releasing the water.

The first category includes multilevel intake structures with several openings located at fixed levels above the bottom of the reservoir. Selective withdrawal removes water from different thermal strata within the water column, by replacing one or more bulkheads in the multilevel outlet structure with trash racks to screen out large debris while allowing water to flow through. Variations of the above include floating intakes or trunnions, which intake water from the selected strata through two pipes hinged at the dam wall.

Artificial destratification aims to reduce the temperature difference between the bottom and the surface of the reservoir, and it is achieved by inducing large-scale circulation. The effectiveness of such a category largely depends on the local climate and the size of the storage to be destratified. Specific measures include: (1) bubble plume systems pushing cold water up to the surface, typically applied in eutrophic reservoirs to improve overall water quality; (2) surface pumps sending surface water downward into the intake structure through floating platforms; (3) draft tube mixers, which are a variation of the former using telescopic tubes with floats to propel surface water to targeted depths; (4) submerged weirs or curtains that can be used to provide a barrier and force warm or cold water above or below the curtain; and (5) stilling basins that are used to delay the downstream release of water, so that water temperature reaches an equilibrium with air temperature.

Sherman (2000) recognized that because of the inverse relationship between discharge and water temperature, such options could not possibly mimic natural daily temperature variations unless the hydrological regime was also mimicked. Taking the irrigation flows experienced during summer in Australia as an example, while such options may modify the mean water temperature of the release, they are unlikely to fully restore the natural temperature range. More recently, Rheinheimer et al. (2014) presented an exploratory approach for optimizing selective withdrawal from reservoirs to mitigate climate change effects. Specifically, they proposed selecting the release from different thermal layers in reservoirs to minimize deviation from target downstream temperatures.

Accurately predicting stream thermal regimes is challenging given their spatial and temporal heterogeneity (Webb et al., 2008). Recent technological developments, however, have facilitated long-term river temperature monitoring at multiple sites, including low-cost accurate, programmable, and reliable miniature loggers (Johnson, 2003; Johnson et al., 2005). Moreover, the more recent use of remotely sensed thermal infrared imagery for documenting spatiotemporal variability in river temperatures has addressed many uncertainties and problems associated with interpretation, resolution, calibration, and geo-referencing (Dugdale et al., 2013, 2015). These, together with the development of new data analysis techniques, are providing numerous tools to assess, understand, and mitigate anthropogenic impacts on thermal regimes in rivers (Webb et al., 2008).

6.4 NUTRIENTS

Nitrogen is the most common nutrient found in river systems and is essential at low levels for ecosystem functioning. However, at elevated levels this nutrient can pose detrimental impacts on stream health and becomes a pollutant (Sheibley et al., 2003). High nitrogen in waterways has been reported worldwide. Excessive application of nitrogen-based fertilizer on pastures and agriculture fields, improper management of landfill leachate, and exhaust from cars and other internal combustion engines have all resulted in increased nitrogen in groundwater and surface water sources (Hoven et al., 2008; Thomas, 2014). Nitrogen exists in various forms in aquatic systems. In particular, ammonia/ammonium, nitrite, and nitrate are water-soluble species affecting the water quality of rivers and streams (Marzadri et al., 2011). The most significant impacts of increased nitrogen levels in streams are eutrophication due to excessive growth of blue−green algae. This leads to reduced dissolved oxygen and loss of flora and fauna. Higher levels of nitrate in drinking water can lead to a disease called methemoglobinemia or blue baby syndrome in infants (Super et al., 1981).

6.4.1 NITROGEN TRANSFORMATION PROCESSES AND CONTROLLING MECHANISMS IN RIVERS

Nitrogen from the catchment enters streams via surface runoff after rainfall events (nonpoint source) as well as point source discharge from wastewater treatment plants or industrial discharges. A large number of chemical and biological reactions transform different forms of nitrogen within the water column. Most of these reactions happen near the sediment—water interface, known as hyporheic exchange (Arrigoni et al., 2008; Santschi et al., 1989; Storey et al., 2004; Zhou et al., 2014). This zone provides the longer residence times for biogeochemical reactions to occur through a series of biogeochemical processes (Boano et al., 2014; Grimm and Fisher, 1984; Sheibley et al., 2003) as presented in Fig. 6.4. The two major processes that remove nitrogen from aquatic system processes are nitrification and denitrification. Nitrification converts ammonium into nitrate and requires aerobic conditions to convert ammonia to nitrates. Denitrification requires anaerobic conditions suitable for denitrifying bacteria to convert nitrates to nitrogen gas, thus removing it from the ecosystem (Zarnetske et al., 2015).

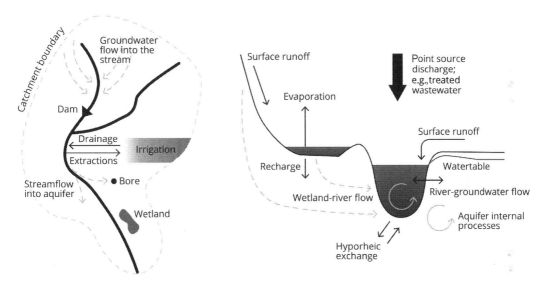

FIGURE 6.4

Major processes controlling stream nutrient levels. The plane view of the catchment illustrates that groundwater flow in the upper catchment is often from the recharge locations into the stream. In the lower catchment, groundwater can flow from the stream to the aquifer. The cross-section demonstrates the former, and illustrates that surface runoff mobilizes nutrients on the soil into the surface waters, point sources discharge a significant amount of nutrients in streams, and standing water bodies can also experience high levels of nutrients due to evaporative concentration.

Shorter residence time results in nitrification only, whereas longer residence time can lead to nitrification followed by denitrification, and thus complete attenuation of the nitrogen. These in-stream transformation processes can significantly affect surface water quality (Thouvenat et al., 2006). The interplay between these processes and net rates of these reactions determines the nitrate concentration in surface water or sediments. Therefore, the hyporheic zone is a potential source or a sink of nitrate depending on the demand for biological nitrogen and the availability of required substrates and abiotic factors (Moore and Schroeder, 1970; Rivett et al., 2008; Thomas, 2014; Zarnetske et al., 2012; Zhou et al., 2014).

Sediment permeability and water velocity also play a vital role in affecting biogeochemical reactions in the hyporheic zone (Bardini et al., 2012). Both these factors directly affect the nutrient supply and residence time in the sediments. The reaction rates for nitrogen transformation processes are sensitive to dissolved oxygen levels. Oxygen levels in water can change with a change in temperature (Veraart et al., 2011). Additionally, the supply of the initial substrate, for instance ammonium (for nitrification), and nitrate and organic carbon (for denitrification) are dependent on hydraulic conductivity. The attenuation of nitrogen in the water–sediment interface is transport limited and highly dependent upon the streamflows. It has also been observed that the stream water temperature, bed morphology, permeability of sediments, and sediment size affect the nitrogen exchange, all of which are dependent on the environmental water (Marzadri et al., 2013; Ronan et al., 1998; Storey et al., 2003).

In temperate rivers, low flows, increased light penetration, and warm temperatures in summer increase rates of biological processes such as primary production, and carbon and nutrient cycling, and may reduce in-stream nutrient levels. Conversely, high flows in the winter months lead to increased dissolved nutrients and carbon inputs to the river. The lower temperatures and water clarity reduce rates of biological processes. However, in tropical rivers, low flows and cease-to-flow events occur during the winter dry season and high flows occur during the summer wet season. The water quality of urban streams is highly variable. Urban stormwater runoff transports a variety of chemicals including nutrients and biological materials to the receiving water body (Walsh et al., 2005).

6.4.2 ANTHROPOGENIC INFLUENCES ON STREAM NITROGEN LEVELS

Anthropogenic influences on stream nitrogen levels occur both through river regulation and catchment land-use change. Nitrogen can enter waterways by point or nonpoint (diffuse) pathways. Point sources are typically continuous and from a specific location through a pipe or drain such as stormwater, sewage treatment plant, or industrial discharges. Nonpoint sources of nitrogen include runoff from agricultural land, roads, or lawns, and leaking wastewater infrastructure.

River regulation leading to altered flow regimes also significantly impacts nitrogen levels in streams (Feldman et al., 2015). River regulation can cause lower variability in flow, overall lower flow magnitudes, and, in some cases, seasonal reversal of the flow regime in rivers downstream of storages (see Chapter 3). In-stream storages (i.e., reservoirs) can lead to reduced flows downstream, concentrating the nitrogen. It is most strongly felt in the reached receiving point source discharge of excess nitrogen, for example, sewage treatment plant during the low-flow conditions, which due to reduced buffering capacity of the streamflow dilutes the discharged nitrogen from effluent. Increases in flow during such dry spells generally enhance nitrogen attenuation. Flooding, however, causes erosion and greater concentrations of nitrogen to arrive at waterways from agricultural lands. Onstream dams lead to more variable or more consistent flows as well as reduce connectivity along the river length that affects nutrient and sediment transport, and therefore downstream trophic structure and function (Bunn and Arthington, 2002).

6.4.3 MANAGEMENT OPTIONS TO CONTROL STREAM NITROGEN LOADS

As stream nitrogen load problems are generally driven by both river regulation and catchment land-use changes, a multipronged approach providing the whole of catchment management is most effective (see Chapter 22). Some of the suggested management options include treatment at-source for stormwater runoff from roads, gardens, and other urban areas. This can significantly reduce the nutrients and sediments reaching the waterways. Various Water Sensitive Urban Design strategies including swales, bio-filters, sedimentation tanks, and wetlands have been utilized for such purposes (Walsh, 2000). Better planning for irrigation and fertilizer applications in farmlands, along with improved river bank management to reduce the cattle invasion, can significantly reduce the nitrogen loads in streams. Use of environmental water to achieve optimal levels of nitrification/denitrification, in combination with engineered bedforms and other channel geometries, can also be used to mitigate high nitrogen levels in streams.

6.5 DISSOLVED OXYGEN

6.5.1 PROCESSES AND CONTROLLING MECHANISMS IN RIVERS

Dissolved oxygen concentrations in streams reflect the balance of oxygen added by both reaeration at the water surface and photosynthesis, and lost through ecosystem respiration and oxidation of contaminants in the water column and boundary sediments. When oxygen consumption exceeds supply, oxygen concentrations decline (Diaz, 2001). Importantly, rates of both reaeration and consumption of oxygen within bed sediments can be transport limited, and hence dependent on flow conditions. Specifically, these processes require the exchange of oxygen across the air–water interface and the water–sediment interface, respectively.

Oxygen exchange at the air–water interface

The reaeration rate, defined as the net mass flux of oxygen from air to water, can be larger than either oxygen production by photosynthesis or consumption by respiration/oxidation (Aristegi et al., 2009), making it a critical control on trends in dissolved oxygen concentration. For reaeration to occur, oxygen gas must be transported across the air and water-side boundary layers (Gulliver, 2011), as illustrated in Fig. 6.5. For a poorly soluble gas such as oxygen, the processes regulating exchange on the water-side limit reaeration rates (Gualtieri and Doria, 2008). The two models commonly applied to represent water-side gas exchange emphasize the role of water-side near-surface turbulence (Gualtieri and Doria, 2008). In the classical two-film model, turbulence reduces the thickness of the concentration boundary layer, thereby increasing oxygen concentration gradients within the diffusive sublayer and hence mass flux for a given molecular diffusion coefficient (Gulliver, 2011). In the surface renewal model, turbulent eddies deliver oxygen-rich water close to the interface, displacing oxygen-poor water. Empirical observations indicative of the importance of turbulence include the high correlations between field measurements of near-surface turbulent energy dissipation and interfacial gas exchange rates in natural stream over a range of sizes and flow conditions (Vachon et al., 2010).

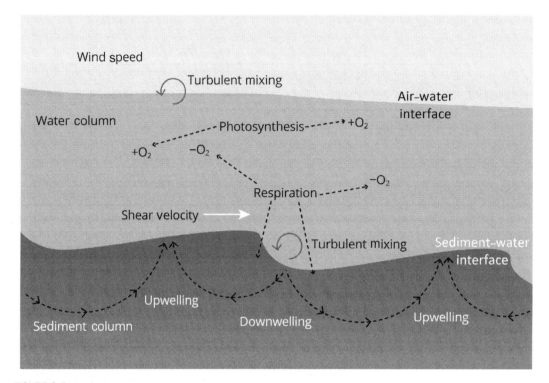

FIGURE 6.5

Schematic side view showing major processes controlling stream and benthic dissolved oxygen concentrations. Turbulent diffusions of oxygen at the air–water interface make oxygen available for respiration in the water column, whereas photosynthesis in the water column adds to dissolved oxygen. Downwelling of streamborne water into the sediment bed provides benthic supply of oxygen, which is respired and depleted, the longer it is resident in the hyporheic zone.

Reaeration rates are usually modeled as the product of a reaeration coefficient and the oxygen deficit given by the difference between saturated oxygen concentration, which occurs at the water surface, and oxygen concentration below the water-side concentration boundary layer (Gualtieri et al., 2002). Estimation of the reaeration coefficient is a major challenge in modeling stream oxygen balance (Grace et al., 2015). Large variations are produced by complex interactions of flow with boundary roughness and the effect of wind shear on the water surface, including formation of surface waves, all of which contribute to water-side near-surface turbulence. Direct measurement of the reaeration coefficient is possible using a gas tracer added to water, but such measurements need to be repeated for any change in flow or wind conditions. Empirical equations based on repeat measurements at multiple sites have related reaeration coefficient to wind speed, flow velocity, water depth, boundary roughness, and streams slope as proxies for flow turbulence (Aristegi et al., 2009; Vachon et al., 2010). However, such models perform poorly when transferred to different sites (Gualtieri and Gualtieri, 1999; Melching and Flores, 1999; Parker and DeSimone, 1992; Riley and Dodds, 2013; Wilson and Macleod, 1974).

Grace et al. (2015) favor an alternate method where reaeration coefficients are estimated indirectly by fitting an oxygen balance model to a time series of dissolved oxygen concentration measurements. This is based on a method first proposed by Odum (1956) to infer rates of gross primary production and respiration of streams from diel variations in observed oxygen concentrations. Respiration rates are assumed constant with diel variations in oxygen assumed to be the result of photosynthesis decreasing to zero during the night.

Oxygen exchange at the sediment–water interface

Benthic mass flux (Bardini et al., 2012; O'Connor and Harvey, 2008) and, in particular, dissolved oxygen (Marzadri et al., 2013) are significant for numerous stream and ecosystem functions (Boulton et al., 1998; Krause et al., 2011; Marion et al., 2014; Stanford and Ward, 1988). Mixing between streams and sediments, often referred to as hyporheic exchange, occurs in three stages: (1) the influx of streamborne water into the sediment bed in downwelling zones; (2) the residence and biogeochemical reaction within the sediment; and (3) the release of pore water back into the water column. As illustrated in Fig. 6.5, while resident in the streambed, dissolved oxygen is depleted in response to hyporheic respiration (Findlay et al., 1993; Naegeli and Uehlinger, 1997) and associated biogeochemical feedback processes. For instance, biofilms developing in downwelling zones, where oxygen and nutrient fluxes are high, act to restrict the downwelling fluxes into the hyporheic zone (Hendricks, 1993). However, biofilms provide a food supply for bioturbators that may increase sediment permeability and hyporheic exchange (Boulton, 2007). As a result of the periodic nature of the flushing and reaction of streamborne solutes, the hyporheic zone has been described as the *liver* of a stream (Fischer et al., 2005).

Fluxes across the sediment–water interface undergo a transition from free-surface turbulent flows to laminar pore water flow in the sediment bed. The transitional region, known as the brinkman layer, penetrates 4–5 times the sediment grain size by which 90% of water column turbulence has diminished (Vollmer et al., 2002). Therefore, hyporheic exchange depends on both water and sediment column conditions. In the water column, hyporheic exchange depends on flow pressure variation at the sediment–water interface and flow turbulence penetrating into the streambed. Sediment oxygen demand has been related to flow velocity in the water column (Nakamura and Stefan, 1994), and the permeability and microbial activity of the sediment (Higashino and Stefan, 2005). Streambed features such as bedforms (Elliott and Brooks, 1997), heterogeneity of sediment at the sediment–water interface (Aubeneau et al., 2014; Cardenas et al., 2004), and periodic changes in stream velocity (Shum, 1993) lead to the formation of upwelling–downwelling cells (Boano et al., 2011; Grant and Marusic, 2011) promoting mass flux across the sediment–water interface.

Advective fluxes of dissolved oxygen into the sediment bed can be quantified by the mass transfer coefficient, which is related to the mass flux normalized by the difference between downwelling and upwelling concentrations (McCluskey et al., 2016). If all oxygen is consumed by benthic respiration, then the system is said to be mass transfer-limited (Grant et al., 2014). Collectively, the mass transfer coefficient and residence time distribution determine both the influx of oxygen and reaction time within the sediment. Therefore, the mass transfer coefficient describes the limit to which benthic respiration can occur.

6.5.2 ANTHROPOGENIC INFLUENCES

Elevated dissolved or particulate organic matter loading often reduces oxygen concentrations by supporting heterotrophic microbial activity and hence oxygen consumption. The additional organic matter may be contributed by allochthonous sources including blackwater draining of floodplains, where dissolved organic matter is leached from leaf litter and sewage spills (Hladyz et al., 2011; McCarthy et al., 2014), or from autochthonous processes stimulated by eutrophication such as algal blooms (Mallin et al., 2006). Consequences of low dissolved oxygen concentrations, or hypoxia, in streams include: altered fish behavior (Dwyer et al., 2014); increased mortality of fish (Small et al., 2014) and freshwater crustaceans (McCarthy et al., 2014); and reduced richness and abundance of zooplankton (Ning ct al., 2015).

Reaeration is highly dependent on flow-mediated mixing across the air—water interface. Reduced flow velocities, and hence turbulence, will reduce reaeration rates and can lead to declining dissolved oxygen concentrations within the water column. These conditions are likely in ponded water (e.g., upstream of weirs), particularly if this occurs in combination with reduced flows.

Density stratification of water bodies can occur as a result of either surface water heating (thermal stratification) or inputs of salt from seawater, groundwater, or other sources (salinity stratification; Williams, 2006). A vertical temperature or salinity gradient (thermocline or halocline) produces density differences, with the more buoyant (higher temperature or lower salinity) water *floating* at the surface above the denser (colder or higher salinity) water. Under these conditions, reaeration of the bottom layer requires the downward exchange of dissolved oxygen across the interface between these two layers. Alternatively, reaeration can occur by mixing of the water column and elimination of the stratified conditions. Both these processes are flow-dependent. Increases in flow generally enhance exchange of dissolved oxygen and mixing of the two layers by producing greater shear and hence turbulence at the interface. As with the oxygen exchange at the air—water interface, reduced flows in regulated rivers or impoundment of rivers by weirs will increase the risk of establishing stratified conditions and reduce gas exchange to the lower layer, resulting in hypoxic conditions lower in the water column.

Reservoirs are particularly prone to the formation of temperature stratification as surface waters are heated during the warmer season. The colder hypolimnion often becomes anoxic. If reservoir releases draw water from the hypolimnion, then the downstream river will experience low dissolved oxygen. However, this effect will be attenuated rapidly downstream, with high reaeration rates produced by a large gradient in oxygen concentrations at the water surface.

Anthropogenic influences on morphology of river channels often reduce channel irregularities including removal of meanders and bedforms (Hancock, 2002). These features are critical to hyporheic exchange and their removal changes the distribution of hydraulic controls in river systems, generally reducing the supply of dissolved oxygen to streambed sediments (Hanrahan, 2008). In addition, elevated suspended sediment loads in response to catchment disturbances may clog bed sediments and also inhibit dissolved oxygen exchange across the sediment—water interface.

6.5.3 MANAGEMENT OPTIONS

Various management strategies have been proposed to mitigate hypoxia in streams including delivery of dilution flows; mechanical reaeration using pumps, paddle wheels, and overflow at structure;

and diversion of blackwater (with high dissolved organic matter concentrations) into shallow off-channel storages (Whitworth et al., 2013). It is also possible to use environmental water to enhance the natural process of stream reaeration. This last approach is well suited to adaptive management (see Chapter 25), where an initial environmental water release can be used to calibrate an oxygen balance model (e.g., using a modified approach to Grace et al., 2015), which can subsequently be used for optimizing environmental water releases to mitigate hypoxia through enhanced reaeration. This approach takes advantage of the sensitivity of reaeration coefficients to flow turbulence, which is dependent on flow velocity and hence stream discharge.

6.6 OTHER CONTAMINANTS

There are various other contaminants of concern such as heavy metals, pesticides, pharmaceutical compounds, and contaminants of emerging concerns. Most of these contaminants are transported to the waterways along with runoff, point source discharge from industries, and/or groundwater discharge into streams. The fate and transport of each of these contaminants depends on the complex array of factors, both physical and biogeochemical, including the stream advection.

6.7 CONCLUSION

Four key variables of water quality in freshwater ecosystems: salinity, temperature, nitrogen, and dissolved oxygen have been discussed in this chapter. Atmospheric and topographic conditions are the major drivers of water quality in streams; however, stream discharge can significantly influence the water quality. As a consequence, the effects of hydrological alterations due to river regulation on stream water salinity, natural thermal regimes, nitrogen loads, and dissolved oxygen can be significant and have major impacts on aquatic life (Table 6.1).

In general, water quality problems (i.e., excessive concentrations of contaminants, high water temperatures, or low dissolved oxygen) can be mitigated by increasing environmental water releases. Increased flows will dilute contaminants and remove them from the river more quickly through advective transport. Increases in flow volume will increase thermal mass that will dampen diurnal temperature extremes. Environmental water releases can also promote reaeration and the breakdown of density stratification within the water column. However, there may be some situations where delivery of environmental water will be detrimental. Increased flows can reduce residence times and hence dissolved nitrogen transformations and removal through denitrification, poor water quality captured near the streambed in density stratified flows may be mobilized leading to impacts downstream, and there may be short-term consequences of blackwater events if carbon stores in the river bank and floodplain are mobilized by rising water levels. In addition, water quality is highly sensitive to catchment land-use change and so environmental water in isolation can rarely address a number of water quality issues (e.g., high salinity and nutrient loads). Fundamental to understanding the use of environmental water to manage water quality is the ability to represent flow-dependent transport processes in freshwater ecosystems.

Table 6.1 Summary of Impacts of Hydrological Variations on Salinity, Temperature, Nitrogen, and *Dissolved Oxygen (DO)*

Hydrological Processes	Anthropogenic Change	Water Quality Response			
		Salinity	Temperature	Nutrients	DO
Surface water impoundment and diversion	Large water supply reservoirs (hydropower, flood alleviation)	Less flushing of saline pools and buildup of saline groundwater in floodplain	Increase of water residence time and thermal stratification resulting in cold or warm water releases	Increased nitrogen levels downstream due to reduced dilution	De-oxygenation through reduced mixing and consumption
	Farm dams/small storages	Decreased groundwater discharge to streams	Minimal	As above	Reduced turbulent mixing and oxygenation of water column
Groundwater extraction and depletion	Groundwater pumping and depletion	Decreased groundwater discharge to streams	Decreased potential of using groundwater upwelling to mitigate temperature extremes	Reduced groundwater discharge to streams, lesser dilution of nutrients	Reduced pore water flushing across the sediment–water interface
	Groundwater mounding from irrigation	Increased saline groundwater discharge to streams	As above	Greater dilution of nutrients in streams due to increased groundwater discharge	As above
Altered surface drainage network	Dam water transfers: reduced low flows	May increase floodplain salinity	Extreme water temperatures (e.g., warm water temperatures in summer)	Critical high levels of nitrogen downstream of point discharge	Minimal
	Dam water transfers: increased flows		Alteration of natural thermal regime	Increased nitrogen loads in surface runoff from agricultural fields	
Alteration to land surface water balance	Vegetation change	Increased saline discharge to river and floodplain environment			Reduced photosynthesis and respiration in water column
	Urbanization	Minimal			Minimal

REFERENCES

Acaba, Z., Jones, H., Preece, R., Rish, S., Ross, D., Daly, H., 2000. The Effects of Large Reservoirs on Water Temperature in Three NSW Rivers Based on the Analysis of Historical Data. Centre for Natural Resources, NSW Department of Land and Water Conservation, Sydney, Australia.

Allan, J.D., Castillo, M.M., 2007. Stream Ecology: Structure and Function of Running Waters. Springer Science & Business Media.

ANZECC, ARMCANZ, 2000. Australian and New Zealand Guidelines for Fresh and Marine Water Quality. Australian and New Zealand Environment and Conservation Council and Agricultural and Resource Management Council of Australia and New Zealand.

Aristegi, L., Izagirre, O., Elosegi, A., 2009. Comparison of several methods to calculate reaeration in streams, and their effects on estimation of metabolism. Hydrobiologia 635, 113−124. Available from: http://dx.doi.org/10.1007/s10750-009-9904-8.

Arrigoni, A.S., Poole, G.C., Mertes, L.A., O'Daniel, S.J., Woessner, W.W., Thomas, S.A., 2008. Buffered, lagged, or cooled? Disentangling hyporheic influences on temperature cycles in stream channels. Water Resour. Res. 44, 1−13.

Aubeneau, A.F., Hanrahan, B., Bolster, D., Tank, J.L., 2014. Substrate size and heterogeneity control anomalous transport in small streams. Geophys. Res. Lett. 41, 8335−8341. Available from: http://dx.doi.org/10.1002/2014GL061838.

Baker, A., 2003. Land use and water quality. Hydrol. Process. 17, 2499−2501.

Banks, E.W., Simmons, C.T., Love, A.J., Shand, P., 2011. Assessing spatial and temporal connectivity between surface water and groundwater in a regional catchment: implications for regional scale water quantity and quality. J. Hydrol. 404, 30−49.

Bardini, L., Boano, F., Cardenas, M.B., Revelli, R., Ridolfi, L., 2012. Nutrient cycling in bedform induced hyporheic zones. Geoch. Cosmoch. 84, 47−61. Available from: http://dx.doi.org/10.1016/j.gca.2012.01.025.

Beechie, T., Imaki, H., Greene, J., Wade, A., Wu, H., Pess, G., et al., 2013. Restoring salmon habitat for a changing climate. River Res. Appl. 29, 939−960.

Boano, F., Revelli, R., Ridolfi, L., 2011. Water and solute exchange through flat streambeds induced by large turbulent eddies. J. Hydrol. 402, 290−296. Available from: http://dx.doi.org/10.1016/j.jhydrol.2011.03.023.

Boano, F., Harvey, J.W., Marion, A., Packman, A.I., Revelli, R., Ridolfi, L., et al., 2014. Hyporheic flow and transport processes: mechanisms, models, and biogeochemical implications. Rev. Geophys. 52, 603−679. Available from: http://dx.doi.org/10.1002/2012RG000417.

Boulton, A, 2007. Hyporheic rehabilitation in rivers: restoring vertical connectivity. Freshw. Biol. 52 (4), 632−650.

Boulton, A.J., Marmonier, P., Davis, J.A., 1998. Hydrological exchange and subsurface water chemistry in streams varying in salinity in southwestern Australia. Int. J. Salt Lake Res. 8, 361−382.

Boulton, A., Brock, M., Robson, B., Ryder, D., Chambers, J., Davis, J., 2014. Australian Freshwater Ecology: Processes and Management. John Wiley & Sons, NY.

Braaten, R., Gates, G., 2003. Groundwater-surface water interaction in inland New South Wales: a scoping study. Water Sci. Technol. 48 (7), 215−224.

Brabec, E., Schulte, S., Richards, P.L., 2002. Impervious surfaces and water quality: a review of current literature and its implications for watershed planning. J. Plan. Lit. 16, 499−514.

Bunn, S.E., Arthington, A.H., 2002. Basic principles and ecological consequences of altered flow regimes for aquatic biodiversity. Environ. Manage. 30, 492−507.

Burkholder, B.K., Grant, G.E., Haggerty, R., Khangaonkar, T., Wampler, P.J., 2008. Influence of hyporheic flow and geomorphology on temperature of a large, gravel-bed river, Clackamas River, Oregon, USA. Hydrol. Process. 22, 941−953.

Byers, H.L., Cabrera, M.L., Matthews, M.K., Franklin, D.H., Andrae, J.G., Radcliffe, D.E., et al., 2005. Phosphorus, sediment, and *Escherichia coli* loads in unfenced streams of the Georgia Piedmont, USA. J. Environ. Qual. 34, 2293−2300.

Caissie, D., 2006. The thermal regime of rivers: a review. Freshw. Biol. 51, 1389−1406.

Capon, S.J., Chambers, L.E., Mac Nally, R., Naiman, R.J., Davies, P., Marshall, N., et al., 2013. Riparian ecosystems in the 21st century: hotspots for climate change adaptation? Ecosystems 16, 359−381.

Cardenas, M.B., Wilson, J.L., Zlotnik, V.A., 2004. Impact of heterogeneity, bed forms, and stream curvature on subchannel hyporheic exchange. Water Resour. Res. 40, W08307. Available from: http://dx.doi.org/10.1029/2004wr003008.

Cartwright, I., Weaver, T.R., Simmons, C.T., Fifield, L.K., Lawrence, C.R., Chisari, R., et al., 2010. Physical hydrogeology and environmental isotopes to constrain the age, origins, and stability of a low-salinity groundwater lens formed by periodic river recharge: Murray River, Australia. J. Hydrol. 380, 203−221.

Casado, A., Hannah, D.M., Peiry, J.L., Campo, A.M., 2013. Influence of dam-induced hydrological regulation on summer water temperature: Sauce Grande River, Argentina. Ecohydrology 6, 523−535.

Cheng, X., Benke, K.K., Beverly, C., Christy, B., Weeks, A., Barlow, K., et al., 2014. Balancing trade-off issues in land use change and the impact on streamflow and salinity management. Hydrol. Process. 28, 1641−1662. Available from: http://dx.doi.org/10.1002/hyp.9698.

Constantz, J., Thomas, C.L., Zellweger, G., 1994. Influence of diurnal variations in stream temperature on streamflow loss and groundwater recharge. Water Resour. Res. 30, 3253−3264.

Costelloe, J.F., Grayson, R.B., McMahon, T.A., Argent, R.M., 2005. Spatial and temporal variability of water salinity in an ephemeral arid-zone river, central Australia. Hydrol. Process. 19, 3147−3166. Available from: http://dx.doi.org/10.1002/hyp.5837.

Costelloe, J.F., Russell, K.L., 2014. Identifying conservation priorities for aquatic refugia in an arid zone, ephemeral catchment: a hydrological approach. Ecohydrology 7, 1534−1544.

Dent, C., Schade, J., Grimm, N., Fisher, S., 2000. Subsurface influences on surface biology. In: Jones, J.B., Mulholland, P.J. (Eds.), Streams and Ground Waters. Academic Press, San Diego, California, pp. 381−402.

Diaz, J.D., 2001. Overview of hypoxia around the world. J. Environ. Qual. 30, 275−281.

Dugdale, S.J., Bergeron, N.E., St-Hilaire, A., 2013. Temporal variability of thermal refuges and water temperature patterns in an Atlantic salmon river. Remote Sens. Environ. 136, 358−373.

Dugdale, S.J., Bergeron, N.E., St-Hilaire, A., 2015. Spatial distribution of thermal refuges analysed in relation to riverscape hydromorphology using airborne thermal infrared imagery. Remote Sens. Environ. 160, 43−55.

Dwyer, G.K., Stoffels, R.J., Pridmore, P.A., 2014. Morphology, metabolism and behaviour: responses of three fishes with different lifestyles to acute hypoxia. Freshw. Biol. 59, 819−831. Available from: http://dx.doi.org/10.1111/fwb.12306.

Ebersole, J.L., Liss, W.J., Frissell, C.A., 2003a. Cold water patches in warm streams: physicochemical characteristics and the influence of shading. J. Am. Water Resour. Assoc. 39, 355−368.

Ebersole, J.L., Liss, W.J., Frissell, C.A., 2003b. Thermal heterogeneity, stream channel morphology, and salmonid abundance in northeastern Oregon streams. Can. J. Fish. Aquat. Sci. 60, 1266−1280.

Elliott, A.H., Brooks, N.H., 1997. Transfer of nonsorbing solutes to a streambed with bed forms: theory. Water Resour. Res. 33, 123−136. Available from: http://dx.doi.org/10.1029/96wr02784.

Feldman, D.A., Sengupta, A., Pettigrove, V., Arora, M., 2015. Governance issues in developing and implementing offsets for water management benefits: can preliminary evaluation guide implementation effectiveness. WIRE-Water, 2 (2), 121−130.

Findlay, S., Strayer, D., Goumbala, C., Gould, K., 1993. Metabolism of streamwater dissolved organic carbon in the shallow hyporheic zone. Limnol. Oceanogr. 38, 1493−1499.

Fischer, H., Kloep, F., Wilzcek, S., Pusch, M.T., 2005. A river's liver — microbial processes within the hyporheic zone of a large lowland river. Biogeochemistry 76, 349—371.

Frissell, C.A., Liss, W.J., Warren, C.E., Hurley, M.D., 1986. A hierarchical framework for stream habitat classification: viewing streams in a watershed context. Environ. Manage. 10, 199—214.

George, R., Dogramaci, S., Wyland, J., Lacey, P., 2005. Protecting stranded biodiversity using groundwater pumps and surface water engineering at Lake Toolibin, Western Australia. Aust. J. Water Resour. 9, 119—128.

Gordon, N.D., Finlayson, B.L., McMahon, T.A., 2004. Stream Hydrology: An Introduction for Ecologists. John Wiley and Sons, NY.

Grace, M.R., Giling, D.P., Hladyz, S., Caron, V., Thompson, R., 2015. Fast processing of diel oxygen curves: estimating stream metabolism with BASE (Bayesian Single-station Estimation). Limnol. Oceanogr. Method. 13, 103—114.

Grant, S.B., Marusic, I., 2011. Crossing turbulent boundaries: interfacial flux in environmental flows. Environ. Sci. Technol. 45, 7107—7113. Available from: http://dx.doi.org/10.1021/es201778s.

Grant, S.B., Stolzenbach, K., Azizian, M., Stewardson, M.J., Boano, F., Bardini, L., 2014. First-order contaminant removal in the hyporheic zone of streams: physical insights from a simple analytical model. Environ. Sci. Technol. 48, 11369—11378. Available from: http://dx.doi.org/10.1021/es501694k.

Grimm, N.B., Fisher, S.G., 1984. Exchange between interstitial and surface water: implications for stream metabolism and nutrient cycling. Hydrobiologia 111 (3), 219—228.

Gualtieri, C., Doria, G.P., 2008. Gas-transfer at unsheared free-surfaces. In: Gualtieri, C., Mihailovic, D.T. (Eds.), Fluid Mechanics of Environmental Interfaces. CRC Press, Taylor and Francis Group, NW.

Gualtieri, C., Gualtieri, P., 1999. Statistical analysis of reaeration rate in streams. International Agricultural Engineering Conference (ICAE) '99, Washington DC.

Gualtieri, C., Gualtieri, P., Doria, P.G., 2002. Dimensional analysis of reaeration rate in streams. J. Environ. Eng. 128, 12—18.

Gulliver, J.S., 2011. Air-water mass transfer coefficients. In: Thibodeaux, L.J., Mackay, D. (Eds.), Chemical Mass Transfer in the Environment. CRC Press, Taylor and Francis Group, NW.

Hancock, P.J., 2002. Human impacts on the stream—groundwater exchange zone. Environ. Manage. 29, 763—781. Available from: http://dx.doi.org/10.1007/s00267-001-0064-5.

Hannah, D.M., Malcolm, I.A., Soulsby, C., Youngson, A.F., 2008. A comparison of forest and moorland stream microclimate, heat exchanges and thermal dynamics. Hydrol. Process. 22, 919—940.

Hanrahan, T.P., 2008. Effects of river discharge on hyporheic exchange flows in salmon spawning areas of a large gravel-bed river. Hydrol. Process. 22, 127—141.

Hansen, L.J., Biringer, J.L., Hoffman, J., 2003. Buying time: a user's manual for building resistance and resilience to climate change in natural systems. World Wildlife Fund, Washington, DC.

Hauer, F., Hill, W., 1996. Temperature, light and oxygen. In: Hauer, F., Hill, W. (Eds.), Methods in Stream Ecology. Academic Press, New York.

Hendricks, S.P., 1993. Microbial ecology of the hyporheic zone: a perspective integrating hydrology and biology. J. N. Am. Benthol. Soc. 12, 70—78. Available from: http://dx.doi.org/10.2307/1467687.

Higashino, M., Stefan, H.G., 2005. Oxygen demand by a sediment bed of finite length. J. Environ. Eng. 131, 350—358.

Hladyz, S., Watkins, S.C., Whitworth, K.L., Baldwin, D.S., 2011. Flows and hypoxic blackwater events in managed ephemeral river channels. J. Hydrol. 401, 117—125.

Holland, K.L., Charles, A.H., Jolly, I.D., Overton, I.C., Gehrig, S., Simmons, C.T., 2009. Effectiveness of artificial watering of a semi-arid saline wetland for managing riparian vegetation health. Hydrol. Process. 23, 3474—3484.

Horne, B.D., Rutherford, E., Wehrly, K.E., 2004. Simulating effects of hydro-dam alteration on thermal regime and wild steelhead recruitment in a stable-flow Lake Michigan tributary. River Res. Appl. 20, 185—203.

Hoven, S.J.V., Fromm, N.J., Peterson, E.W., 2008. Quantifying nitrogen cycling beneath a meander of a low gradient, N-impacted, agricultural stream using tracers and numerical modelling. Hydrol. Process. 22, 1206−1215. Available from: http://dx.doi.org/10.1002/hyp.6691.

Huusko, A., Greenberg, L., Stickler, M., Linnansaari, T., Nykänen, M., Vehanen, T., et al., 2007. Life in the ice lane: the winter ecology of stream salmonids. River Res. Appl. 23, 469−491.

Ivkovic, K.M., 2009. A top-down approach to characterise aquifer-river interaction processes. J. Hydrol. 365, 145−155.

Jackson, R.B., Jobbágy, E.G., Nosetto, M.D., 2009. Ecohydrology in a human-dominated landscape. Ecohydrology 2, 383−389.

Johnson, S.L., 2003. Stream temperature: scaling of observations and issues for modelling. Hydrol. Process. 17, 497−499.

Johnson, S.L., 2004. Factors influencing stream temperatures in small streams: substrate effects and a shading experiment. Can. J. Fish. Aquat. Sci. 61, 913−923.

Johnson, A.N., Boer, B.R., Woessner, W.W., Stanford, J.A., Poole, G.C., Thomas, S.A., et al., 2005. Evaluation of an inexpensive small-diameter temperature logger for documenting ground water−river interactions. Ground Water Monit. Remediat. 25, 68−74.

Jolly, I.D., McEwan, K.L., Holland, K.L., 2008. A review of groundwater-surface water interactions in arid/semi-arid wetlands and the consequences of salinity for wetland ecology. Ecohydrology 1, 43−58.

Jorgensen, S.E., Tundisi, J.G., Tundisi, T.M., 2012. Handbook of Inland Aquatic Ecosystem Management. CRC Press, Taylor and Francis group, NW.

Krause, S., Hannah, D.M., Fleckenstein, J.H., Heppell, C.M., Kaeser, D., Pickup, R., et al., 2011. Inter-disciplinary perspectives on processes in the hyporheic zone. Ecohydrology 4, 481−499. Available from: http://dx.doi.org/10.1002/eco.176.

Kurylyk, B.L., MacQuarrie, K.T., Linnansaari, T., Cunjak, R.A., Curry, R.A., 2015. Preserving, augmenting, and creating cold-water thermal refugia in rivers: concepts derived from research on the Miramichi River, New Brunswick (Canada). Ecohydrology 8, 1095−1108.

Lamontagne, S., Taylor, A.R., Cook, P.G., Crosbie, R.S., Brownbill, R., Williams, R.M., et al., 2014. Field assessment of surface water−groundwater connectivity in a semi-arid river basin (Murray-Darling, Australia). Hydrol. Process. 28, 1561−1572.

Lugg, A., Copeland, C., 2014. Review of cold water pollution in the Murray−Darling Basin and the impacts on fish communities. Ecol. Manage. Restor. 15, 71−79. Available from: http://dx.doi.org/10.1111/emr.12074.

Mallin, M.A., Johnson, V.L., Ensign, S.H., MacPherson, T.A., 2006. Factors contributing to hypoxia in rivers, lakes, and streams. Limnol. Oceanogr. 51, 690−701.

Marion, A., Nikora, V., Puijalon, S., Bouma, T., Koll, K., Ballio, F., et al., 2014. Aquatic interfaces: a hydro-dynamic and ecological perspective. J. Hydraul. Res. 52, 744−758. Available from: http://dx.doi.org/10.1080/00221686.2014.968887.

Marzadri, A., Tonina, D., Bellin, A., 2011. A semianalytical three-dimensional process-based model for hyporheic nitrogen dynamics in gravel bed rivers. Water Resour. Res. 47, 1−14.

Marzadri, A., Tonina, D., Bellin, A., 2013. Quantifying the importance of daily stream water temperature fluctuations on the hyporheic thermal regime: implication for dissolved oxygen dynamics. J. Hydrol. 507, 241−248. Available from: http://dx.doi.org/10.1016/j.jhydrol.2013.10.030.

McCarthy, B., Zukowski, S., Whiterod, N., Vilizzi, L., Beesley, L., King, A., 2014. Hypoxic blackwater event severely impacts Murray crayfish (*Euastacus armatus*) populations in the Murray River, Australia. Austral Ecol. 39, 491−500. Available from: http://dx.doi.org/10.1111/aec.12109.

McCluskey, A.H., Grant, S.B., Stewardson, M.J. 2016. Flipping the thin film model: mass transfer by hypor-heic exchange in gaining and losing streams. Water Resour. Res. 52 (10) 7806−7818.

McCullough, D.A., 1999. A review and synthesis of effects of alterations to the water temperature regime on freshwater life stages of salmonids, with special reference to Chinook salmon, Seattle, Washington, U.S. Environmental Protection Agency, Region 10.

McCullough, D.A., Bartholow, J.M., Jager, H.I., Beschta, R.L., Cheslak, E.F., Deas, M.L., et al., 2009. Research in thermal biology: burning questions for coldwater stream fishes. Rev. Fish. Sci. 17, 90–115.

Melching, C.S., Flores, H.E., 1999. Reaeration equations derived from U.S. Geological Survey Database. J. Environ. Eng. 125, 407–414.

Moore, S.F., Schroeder, E.D., 1970. An investigation of the effects of residence time on anaerobic bacterial denitrification. Water Res. 4, 685–694.

Moss, B.R., 2009. Ecology of fresh waters: man and medium, past to future. John Wiley & Sons, NY.

Naegeli, M.W., Uehlinger, U., 1997. Contribution of the hyporheic zone to ecosystem metabolism in a prealpine gravel-bed-river. J. N. Am. Benthol. Soc. 16, 794–804.

Nakamura, Y., Stefan, H.G., 1994. Effect of flow velocity on sediment oxygen demand: theory. J. Environ. Eng. 120, 996–1016.

Ning, N.S.P., Petrie, R., Gawne, B., Nielsen, D.L., Rees, G.N., 2015. Hypoxic blackwater events suppress the emergence of zooplankton from wetland sediments. Aquat. Sci. 77, 221–230. Available from: http://dx. doi.org/10.1007/s00027-014-0382-3.

Nõges, T., Nõges, P., Cardoso, A.C., 2010. Review of published climate change adaptation and mitigation measures related with water. Scientific and Technical Research Series EUR, p. 24682.

O'Connor, B.L., Harvey, J.W., 2008. Scaling hyporheic exchange and its influence on biogeochemical reactions in aquatic ecosystems. Water Resour. Res. 44, W12423. Available from: http://dx.doi.org/10.1029/2008wr007160.

Odum, H.T., 1956. Primary production in flowing waters. Limnol. Oceanogr. 1, 102–117.

Olden, J.D., Naiman, R.J., 2010. Incorporating thermal regimes into environmental flows assessments: modifying dam operations to restore freshwater ecosystem integrity. Freshw. Biol. 55, 86–107.

Parker, G.W., DeSimone, L.A., 1992. Estimating reaeration coefficients for low-slope streams in Massachusetts and New York, 1985–1988, Water-Resources Investigation Report, U.S. Geological Survey.

Peterson, T.J., Western, A.W., 2014a. Multiple hydrological attractors under stochastic daily forcing: 1. Can multiple attractors exist? Water Resour. Res. 50, 2993–3009. Available from: http://dx.doi.org/10.1002/2012WR013003.

Peterson, T.J., Western, A.W., Argent, R.M., 2014b. Multiple hydrological attractors under stochastic daily forcing: 2. Can multiple attractors emerge? Water Resour. Res. 50, 3010–3029. Available from: http://dx. doi.org/10.1002/2012WR013004.

Poole, G.C., Berman, C.H., 2001. An ecological perspective on in-stream temperature: natural heat dynamics and mechanisms of human-caused thermal degradation. Environ. Manage. 27, 787–802.

Preece, R.M., Jones, H.A., 2002. The effect of Keepit Dam on the temperature regime of the Namoi River, Australia. River Res. Appl. 18, 397–414.

Rheinheimer, D.E., Null, S.E., Lund, J.R., 2014. Optimizing selective withdrawal from reservoirs to manage downstream temperatures with climate warming. J. Water Resour. Plan. Manage. 141, 04014063.

Riley, A.J., Dodds, W.K., 2013. Whole-stream metabolism: strategies for measuring and modeling diel trends of dissolved oxygen. Freshw. Sci. 32, 56–69. Available from: http://dx.doi.org/10.1899/12-058.1.

Rivett, M.O., Buss, S.R., Morgan, P., Smith, J.W., Bemment, C.D., 2008. Nitrate attenuation in groundwater: a review of biogeochemical controlling processes. Water Res. 42, 4215–4232. Available from: http://dx.doi.org/10.1016/j.watres.2008.07.020.

Ronan, A.D., Prudic, D.E., Thodal, C.E., Constantz, J., 1998. Field study and simulation of diurnal temperature effects on infiltration and variably saturated flow beneath an ephemeral stream. Water Resour. Res. 34 (9), 2137–2153.

Rutherford, J.C., Lintermans, M., Groves, J., Liston, P., Sellens, C., Chester, H., 2009. Effects of cold water releases in an upland stream. eWater Technical. eWater Cooperative Research Centre, Canberra, Australia.

Ryan, T., Webb, A., Lennie, R., Lyon, J., 2001. Status of cold water releases from Victorian dams. Report produced for Department of Natural Resources and Environment, Melbourne, Australia.

Salama, R.B., Otto, C.J., Fitzpatrick, R.W., 1999. Contributions of groundwater conditions to soil and water salinization. Hydrogeol. J. 7, 46−64.

Santschi, P., Höhener, P., Benoit, G., Buchholtz-ten Brink, M., 1989. Chemical processes at the sediment-water interface. Mar. Chem. 30, 269−315.

Sarwar, A., Bastiaanssen, W.G.M., Feddes, R.A., 2001. Irrigation water distribution and long-term effects on crop and environment. Agricult. Water Manage. 50, 125−140.

Schneider, C., Laizé, C., Acreman, M., Florke, M., 2013. How will climate change modify river flow regimes in Europe? Hydrol. Earth Syst. Sci. 17, 325−339.

Sheibley, R.W., Jackman, A.P., Duff, J.H., Triska, F.J., 2003. Numerical modeling of coupled nitrification− denitrification in sediment perfusion cores from the hyporheic zone of the Shingobee River, MN. Adv. Water Resour. 26, 977−987. Available from: http://dx.doi.org/10.1016/s0309-1708(03)00088-5.

Sherman, B., 2000. Scoping options for mitigating cold water discharges from dams. CSIRO Land and Water, Canberra, Australia.

Shum, K.T., 1993. The effects of wave-induced pore water circulation on the transport of reactive solutes below a rippled sediment bed. J. Geophys. Res. Oceans 98, 10289−10301. Available from: http://dx.doi.org/10.1029/93JC00787.

Simpson, H.J., Herczeg, A.L., 1991. Salinity and evaporation in the River Murray Basin, Australia. J. Hydrol. 124, 1−27.

SKM, 2013. Characterising the relationship between water quality and water quantity. A report prepared by Sinclair Knight Merz for Melbourne Water, Melbourne.

Small, K., Kopf, R.K., Watts, R.J., Howitt, J., 2014. Hypoxia, blackwater and fish kills: experimental lethal oxygen thresholds in juvenile predatory lowland river fishes. PloS One 9, e94524. Available from: http://dx.doi.org/10.1371/journal.pone.0094524.

Stanford, J.A., Hauer, F.R., Ward, J.V., 1988. Serial discontinuity in a large river system. Verhandlugen Internationale Vereingen Theoretische Angewa. Limnologie 23, 1114−1118.

Stanford, J.A., Ward, J.V., 1988. The hyporheic habitat of river ecosystems. Nature 335, 64−66.

Storey, R.G., Howard, K.W., Williams, D.D., 2003. Factors controlling riffle-scale hyporheic exchange flows and their seasonal changes in a gaining stream: a three-dimensional groundwater flow model. Water Resour. Res. 39 (2).

Storey, R.G., Williams, D.D., Fulthorpe, R.R., 2004. Nitrogen processing in the hyporheic zone of a pastoral stream. Biogeochemistry 69, 285−313.

Sullivan, K., Tooley, J., Doughty, K., Caldwell, J., Knudsen, P., 1990. Evaluation of prediction models and characterization of stream temperature regimes in Washington. Washington Department of Natural Resources Timber/FishRep., Wildlife Report TFW-WQ3-90-006.

Super, M., Heese, H., MacKenzie, D., Dempster, W.S., Plessis, J., et al., 1981. An epidemiological study of well water nitrates in a group of South West African/Namibian infants. Water Res. 15, 1265−1270.

Thomas, L., 2014. The stream subsurface: nitrogen cycling and the cleansing function of hyporheic zones. Sci. Find. 166, 1−6.

Thouvenot, M., Billen, G., Garnier, J., 2006. Modelling nutrient exchange at the sediment−water interface of river systems. J. Hydrol. 341 (1−2), 55−78.

Torgersen, C.E., Faux, R.N., McIntosh, B.A., Poage, N.J., Norton, D.J., 2001. Airborne thermal remote sensing for water temperature assessment in rivers and streams. Remote Sens. Environ. 76, 386−398.

Townsend, C., Doledec, S., Scarsbrook, M., 1997. Species traits in relation to temporal and spatial heterogeneity in streams: a test of habitat templet theory. Freshw. Biol. 37, 367–387.

Turner, L., Erskine, W.D., 2005. Variability in the development, persistence and breakdown of thermal, oxygen and salt stratification on regulated rivers of southeastern Australia. River Res. Appl. 21, 151–168.

Vachon, D., Prairie, Y.T., Cole, J.J., 2010. The relationship between near-surface turbulence and gas transfer velocity in freshwater systems and its implications for floating chamber measurements of gas exchange. Limnol. Oceanogr. 55, 1723–1732. Available from: http://dx.doi.org/10.4319/lo.2010.55.4.1723.

Van Vliet, M., Ludwig, F., Zwolsman, J., Weedon, G., Kabat, P., 2011. Global river temperatures and sensitivity to atmospheric warming and changes in river flow. Water Resour. Res. 47 (2), 1–19.

Vatland, S.J., Gresswell, R.E., Poole, G.C., 2015. Quantifying stream thermal regimes at multiple scales: combining thermal infrared imagery and stationary stream temperature data in a novel modeling framework. Water Resour. Res. 51, 31–46.

Veraart, A.J., de Klein, J.J.M., Scheffer, M., 2011. Warming can boost denitrification disproportionately due to altered oxygen dynamics. PLoS One 6 (3), e18508.

Vollmer, S., de los Santos Ramos, F., Daebel, H., Kühn, G., 2002. Micro scale exchange processes between surface and subsurface water. J. Hydrol. 269, 3–10. Available from: http://dx.doi.org/10.1016/S0022-1694 (02)00190-7

Walsh, C.J., 2000. Urban impacts on the ecology of receiving waters: a framework for assessment, conservation and restoration. Hydrobiologia. 431. Available from: http://findanexpert.unimelb.edu.au/individual/publicationS1031012.

Walsh, C.J., Roy, A.H., Feminella, J.W., Cottingham, P.D., Groffman, P.M., Morgan, R.P., 2005. The urban stream syndrome: current knowledge and the search for a cure. J. N. Am. Benthol. Soc. 24, 706–723.

Webb, B., 1995. Regulation and thermal regime in a Devon river system. In: Foster, I., Gurnell, A.M., Webb, B.W. (Eds.), Sediment and Water Quality in River Catchments. John Wiley & Sons, Chichester, UK.

Webb, B.W., Hannah, D.M., Moore, R.D., Brown, L.E., Nobilis, F., 2008. Recent advances in stream and river temperature research. Hydrol. Process. 22, 902–918.

Western, A.W., O'Neill, I.C., Hughes, R.L., Nolan, J.B., 1996. The behaviour of stratified pools in the Wimmera River, Australia. Water Resour. Res. 32, 3197–3206.

Whitworth, K.L., Kerr, J.L., Mosley, L.M., Conallin, J., Hardwick, L., Baldwin, D.S., 2013. Options for managing hypoxic blackwater in river systems: case studies and framework. Environ. Manage. 52, 837–850. Available from: http://dx.doi.org/10.1007/s00267-013-0130-9.

Wilby, R., Orr, H., Watts, G., Battarbee, R., Berry, P., Chadd, R., et al., 2010. Evidence needed to manage freshwater ecosystems in a changing climate: turning adaptation principles into practice. Sci. Total Environ. 408, 4150–4164.

Wilby, R.L., Johnson, M.F., Toone, J., 2014. Nocturnal river water temperatures: spatial and temporal variations. Sci. Total Environ. 482, 157–173.

Williams, W.D., 1999. Salinisation: a major threat to water resources in the arid and semi-arid regions of the world. Lake Reserv. Res. Manage. 4, 85–91.

Williams, B.J., 2006. Hydrobiological Modelling. University of Newcastle, NSW, Australia. Available from: http://www.lulu.com.

Wilson, G.T., Macleod, N., 1974. A critical appraisal of empirical equations and models for the prediction of the coefficient of reaeration of deoxygenated water. Water Res. 8, 341–366.

Woodward, G., Perkins, D.M., Brown, L.E., 2010. Climate change and freshwater ecosystems: impacts across multiple levels of organization. Phil. Trans. R. Soc. B Biol. Sci. 365, 2093–2106. Available from: http://dx.doi.org/10.1098/rstb.2010.0055.

Young, R.G., Huryn, A.D., 1999. Effects of land use on stream metabolism and organic matter turnover. Ecol. Appl. 9, 1359–1376.

Zarnetske, J.P., Haggerty, R., Wondzell, S.M., Bokil, V.A., González-Pinzón, R., 2012. Coupled transport and reaction kinetics control the nitrate source-sink function of hyporheic zones. Water Resour. Res. 48, 1−15. Available from: http://dx.doi.org/10.1029/2012wr011894.

Zarnetske, J.P., Haggerty, R., Wondzell, S.M., 2015. Coupling multiscale observations to evaluate hyporheic nitrate removal at the reach scale. Freshw. Sci. 34, 172−186. Available from: http://dx.doi.org/10.1086/680011.

Zhang, L., Dawes, W.R., Walker, G.R., 2001. Response of mean annual evapotranspiration to vegetation changes at catchment scale. Water Resour. Res. 37, 701−708.

Zhou, N., Zhao, S., Shen, X., 2014. Nitrogen cycle in the hyporheic zone of natural wetlands. Chinese Sci. Bull. 59, 2945−2956. Available from: http://dx.doi.org/10.1007/s11434-014-0224-7.

Zolezzi, G., Siviglia, A., Toffolon, M., Maiolini, B., 2011. Thermopeaking in alpine streams: event characterization and time scales. Ecohydrology 4, 564−576.

VISION AND OBJECTIVES FOR THE RIVER SYSTEM

STAKEHOLDER ENGAGEMENT IN ENVIRONMENTAL WATER MANAGEMENT

7

John C. Conallin[1], Chris Dickens[2], Declan Hearne[3], and Catherine Allan[4]

[1]IHE Delft, Delft, Netherlands [2]International Water Management Institute, Pretoria, South Africa
[3]International Water Centre, Brisbane, QLD, Australia [4]Charles Sturt University, Albury, NSW, Australia

7.1 INTRODUCTION

Effective conservation planning is a social process informed by science, not a scientific process which engages society.

Knight et al. (2011)

Water policy makers, water managers, and local communities have sought both the meaning and means of engagement for the past two or three decades, with stakeholder engagement becoming increasingly included as a key component in policy and planning (Reed, 2008). However, stakeholder engagement has often been conducted on an ad hoc basis, and primarily tokenistic, jeopardizing the long-term sustainability of many programs, and has rarely been institutionalized (Reed et al., 2009). The concept of *Integrated Water Resource Management (IWRM)* is an attempt to co-consider and coordinate a range of technical and social elements of water use and management, and necessarily involves some form(s) of stakeholder engagement (Saravanan et al., 2009). IWRM and related practices have created a *new paradigm* for water management that has stakeholder engagement and adaptive management as central characteristics; however, implementation remains a challenge (Pahl-Wostl et al., 2011). The increase in participatory approaches has not achieved the reforms in water management as expected and emphasizes the continued need to ensure the stakeholder engagement is further institutionalized with a focus on greater trust and ownership of both the processes and outcomes (Cook et al., 2013a,b). Trust is a fundamental component of relationships, and influences an individual's or group's willingness to collaborate or believe the information provided (including scientifically defensible information). Trust processes are yet to be effectively incorporated within many participatory approaches (Flitcroft et al., 2010). Jacobs et al. (2016) cites that "the costs of capacity building to allow meaningful stakeholder engagement in water-management decision processes was not widely recognized, and that a failure to appreciate the associated costs and complexities may contribute to the lack of successful engagement."

Decisions about water allocation have important social-ecological impacts, meaning that societal needs, values, and aspirations need to be integrated into planning, implementation, and evaluation of water resource management (Richter et al., 2003). Identifying, understanding, and valuing the

Water for the Environment. DOI: http://dx.doi.org/10.1016/B978-0-12-803907-6.00007-3

views of stakeholders on how water should be allocated are key factors to improving water management (Rogers, 2006). Environmental watering has impacts beyond individual wetlands or river reaches, influencing the wider social-ecological system, and prioritizing stakeholder engagement strategies within environmental water programs is needed if they are to be accepted and implemented (Pahl-Wostl et al., 2013; Poff and Matthews, 2013). Indeed, Richter (2014) suggests that the first step in attaining sustainability within water management is centered on the people involved, and building a shared vision for the water resource. This is rarely straightforward, with uncertainty in management of water resources compounding the challenges of bringing together stakeholders with different interests. Successful implementation of environmental water programs requires the balancing of priorities of competing demands (perceived and real) between different stakeholder groups. Uncertainty coupled with the complexity of diverse stakeholder groups means that environmental water programs become not such a matter of determining outcomes, but rather finding common ground and negotiating a *shared* way forward between many competing interests (Mott Lacroix et al., 2016; see Box 7.1).

BOX 7.1 SHARED VISION PLANNING—A COLLABORATIVE APPROACH TO WATER RESOURCE MANAGEMENT

Shared vision planning (SVP) provides a collaborative approach to water resource management, concentrating on three main areas: (1) traditional water resources planning; (2) structured public participation; and (3) collaborative computer systems modeling. Stakeholders form the basis of the approach as opposed to technical components of the planning process. The goal is to improve the social, environmental, and economic outcomes of water management decisions by focusing on creating a common understanding of a system, and providing a consensus-based forum for stakeholders to identify trade-offs and new management options. Created and housed at the US Army Engineers Institute for Water Resources background information, case studies and a toolkit are provided for users (US Army Engineers Institute for Water Resources, 2012).

Insufficient progress in the effective implementation of environmental water programs at a global level is evident, despite the creation of laws and management plans (Le Quesne et al., 2010). Effective (or ineffective) stakeholder engagement is viewed as a major contributor to this lack of integration/implementation of environmental water programs (Moore, 2004). An increasing focus on stakeholder engagement also represents part of a shift away from top-down technocratic centralized governance in favor of bottom-up collaborative approaches to environmental water management, and has become a driver for alternative governance models. Complexity and complications notwithstanding, some level of stakeholder engagement is needed in planning, implementation, and evaluation, for environmental water programs to be effective and to meet anticipated outcomes (Acreman et al., 2014). How that is done is individual to the situation, but appropriate resourcing is essential and principles, guiding steps, and techniques are available, which form the basis of this chapter.

This chapter explores the concept of stakeholder engagement, puts it in the context of wider resource management and outlines 10 key principles, five key steps, and some techniques that can guide stakeholder engagement approaches in environmental water management, illustrated by reference to three short case studies.

KEY POINTS

1. Stakeholder engagement is a central component of IWRM as traditional top-down governance structures are challenged.
2. Programs involving environmental water sit within wider social-ecological systems, and therefore stakeholders are central to the programs.
3. Environmental water legislation and planning have rarely materialized into implementation; and inadequate stakeholder engagement is a contributing factor of failure to implement.

7.2 STAKEHOLDER ENGAGEMENT WITHIN ENVIRONMENTAL WATER MANAGEMENT

Stakeholder engagement is a central component to the governance/participatory turn that has been occurring within water resource management for at least the past three decades. Engagement of a range of people in decision making and implementation is used to reduce conflict, enhance fairness, build capacity for decision making, and reduce the transaction costs of traditional top-down, technocratic approaches (Conley and Moote, 2003). Programs can only really be considered successful when the key stakeholders acknowledge that they are a success, and continue support long term (Cook et al., 2013a). Expectations change and effective stakeholder engagement can aid in both managing expectations, and dealing with entrenched win–lose perceptions. Stakeholder engagement needs to both understand and manage these expectations, and for managers to then deliver on those expectations or be able to communicate why they cannot be met, and provide alternatives (Mott Lacroix et al., 2016).

7.2.1 STAKEHOLDERS

Simply put, stakeholders are individuals or groups who have some sort of *stake* or *interest* in the issue or object under consideration. What a stake represents, and who can be a stakeholder, are thus important questions, and ones that are not straightforward (Harrington et al., 2008). Note the multiple scales of natural resource management communities, proposing that interest may be locality-based, but may also include interest generated by being affected by the issue or proposed action. Thus, stakeholders in the case of an environmental watering event may be the local people who live near the target water body, but may also be people situated elsewhere who are legislatively required to participate, or other people/groups who assign values to, and have aspirations for, the water body in question (e.g., local or international nongovernmental organization, tourist, or recreational users from other regions). Any attempt at stakeholder engagement must: (1) identify spatial and temporal boundaries of the proposed action(s); (2) consider how stakeholders are to be defined and delineated in the context of the action; and (3) acknowledge that stakeholder interest and roles are dynamic and can change over time (Steyaert and Jiggins, 2007).

7.3 ENGAGEMENT

The second part of *stakeholder engagement* is a broad term that could stand for any type of communication. In the field of IWRM *engagement* is often equated with *participation*, that is with individuals, groups, and organizations choosing to take an active role in making decisions that affect them (Reed, 2008). Therefore, engagement and participation are used interchangeably in this chapter. The purpose of the engagement can vary greatly, for example from seeking social acceptance of an action (Jacobs et al., 2016), collaboration for social and/or knowledge outcomes, or it may be full participation in some form of co-management-based governance arrangement (Margerum and Robinson, 2015). Stakeholder engagement may be sought in any or all of the three main management phases of planning, implementation, and evaluation phases, and/or with research or policy development.

Yee (2010) provides a simple definition that can be summarized as:

A framework of policies, principles, and techniques which ensure that citizens and communities, individuals, groups, and organizations have the opportunity to be engaged in a meaningful way in the process of decision-making that will affect them, or in which they have an interest.

Stakeholder engagement (including fostering trust) has been recognized by the *Organization for Economic Cooperation and Development (OECD)* as integral to good water management (OECD, 2015). In addition, stakeholder engagement can increase the legitimacy of environmental water programs (see Chapter 26); however, stakeholder engagement remains constrained by a number of factors. One frequently raised constraint is that of the transaction costs of including multiple parties (e.g., Crase et al., 2013). These (sometimes perceived) costs can lead to proponents (e.g., government, industry) choosing to make decisions on proposed actions with limited engagement, or only engaging with stakeholders on a needs or must-do basis (often set through legislation), and such engagement may be mostly tokenistic. This tokenism often creates conflict, which invariably slows down effective long-term management, and invariably increases transaction costs (IFC, 2007). Positive management performance and effectiveness of sustainable development goals have been directly linked to participation by local inhabitants (Schultz, 2011). The costs of engagement must therefore be weighed against the benefits; the first is up front, whereas the latter are often only apparent in the medium to long term. Thus, exclusion of stakeholder engagement may hasten the initial planning phase, but in the end may lead to significant delays moving into the implementation phase as nonengaged stakeholders react negatively (IFC, 2007).

KEY POINTS

1. The terms *stakeholder engagement* and *participation* are often used interchangeably within IWRM.
2. Stakeholder engagement is a key aspect of all management phases, including planning, implementation, and evaluation.
3. Transaction costs of engagement must be weighed against the benefits; the first is up front, whereas the latter are often only apparent in the medium to long term.

7.4 INTEGRATING STAKEHOLDER ENGAGEMENT INTO THE MANAGEMENT FRAMEWORK

Stakeholder engagement is not separate from other management processes, but needs to be captured within the overarching management framework, and throughout the planning, implementation, and evaluation phases. One such management framework is adaptive management, which is increasingly being used in water management (Rist et al., 2013). Adaptive management is a cyclical process of using the results of prior actions to inform future actions; that is, it is about learning from practice (Argent, 2009). Adaptive management is suited to situations of great complexity and uncertainty, and requires the participation of a range of stakeholders to enable learning and comanagement to occur (see Chapter 25). An adaptive management operational framework can be a valuable tool for stakeholder engagement, helping with defining boundaries, identifying stakeholders within those boundaries, defining their roles and responsibilities, and situating the stakeholder engagement strategy within an overarching management framework. Strategic adaptive management is a refinement of adaptive management that overtly recognizes the multistakeholder nature of natural resources. It notes that there are two fundamental conditions necessary for effective natural resource management to (1) learn and adapt and (2) do so purposefully with relevant stakeholders (Roux and Foxcroft, 2011; see Box 7.2).

BOX 7.2 STRATEGIC ADAPTIVE MANAGEMENT AS A STAKEHOLDER ENGAGEMENT PLATFORM

Strategic adaptive management (SAM) recognizes the importance of stakeholders as a central component of any initiative. Integral within SAM is development of an objectives hierarchy, which commences with an overall stakeholder agreed vision (future desired state), using the criteria of *Values, Social, Technical, Environmental, Economic, Political (VSTEEP)*. The vision is then broken down into objectives and subobjectives with increasing focus and rigor, culminating in explicit and measureable scientific endpoints. The adaptive planning phase involves collaboration between different stakeholders with different levels of participation, and this sets up the implementation and evaluation phases of SAM.

KEY POINTS

1. An adaptive management framework provides the opportunity to embed stakeholder engagement strategies within an overarching management framework.
2. Adaptive management requires the participation of a range of stakeholders to enable learning and comanagement to occur.
3. SAM has an explicit focus on stakeholder engagement, colearning, and adapting as new information is generated or values change.

7.5 PRINCIPLES FOR ENGAGEMENT

When seeking to negotiate trade-offs, there are various factors that will influence the ability and willingness of stakeholders to engage, trust being central to a willingness to listen and participate (Leahy and Anderson, 2008). Principle-based engagement follows sets of principles that should occur throughout engagement processes (Cornwall, 2008). In addition, it allows flexibility and responsiveness in design/implementation to unique conditions (Kilvington et al., 2011), and can form the basis for inclusion in adaptive management frameworks and SVP processes. A range of principles considered important for guiding genuine and effective stakeholder participation have been listed by various authors (Acland, 2008; Irvine and O'Brien, 2009; Jackson et al., 2012; Mostert, 2015). In attempting to clarify these, it emerges that 10 principles exist including three foundation principles: *inclusiveness*, *transparency*, and *commitment* are the pillars to building *trust* and *ownership* (see also Box 7.3 for developing trust and ownership). These 10 principles form the basis for successful engagement and in turn successful implementation of any initiative. Table 7.1 outlines the 10 principles needed for effective stakeholder engagement and the goal ascribed to each principle.

BOX 7.3 CONDITIONS FOR TRUST IN RIO CHALMER VALLEY OF NORTHERN NEW MEXICO

In discussions between government, environmentalists, and local farmers in the Rio Chalmer Valley of northern New Mexico, after many setbacks the group developed a list for "Future Conditions for Trust":

1. Expect to make mistakes; acknowledge them, learn from them, try to forgive them.
2. Reveal relevant information, records, relationships.
3. Communicate immediately when there is a problem; don't let it fester.
4. Believe what I am saying; don't assume I am lying.
5. Don't set me up.
6. Don't bash me in the media.
7. Keep the door open.
8. Be able to differ with me in a respectful way, to disengage without malice.

Source: Moore (2013)

The ability to build trust and create ownership needed for functional relationships between different stakeholder groups is central to stakeholder engagement (Hamm et al., 2016). Social capital (i.e., connections/networks between people) shows how trust, moral obligations and norms, values, and social networks are fundamental for understanding relationships and aiding in the decision-making process. Where social capital is high, there is confidence that groups and individuals will demonstrate greater reciprocity, where they are more willing to take higher-risk decisions and invest in collective actions. This confidence is also associated with a reduced transaction cost in decision making. Conversely, where there is low social capital between groups, collaboration is unlikely without some form of coercion resulting in higher transaction costs (Tan et al., 2008).

Three main types of social capitals exist: *bonding*—single group bonded together over a shared interest, *bridging*—multiple groups brought together by a common goal, and *linking*—multiple groups connected at different hierarchical levels (Hearne and Powell, 2014; Fig. 7.1). An important point within social capital is the ability of *champions* within stakeholder groups to enable the different levels of social capital to occur, and to build trust (Diedrich et al., 2016). This is not always an individual, as a particular group perceived as neutral can also play a fundamental role in

Table 7.1 Principle-based Engagement. Three Foundation Principles Within 10 Core Principles That Are Needed for Effective Stakeholder Engagement

	Principle	Goal (Overall Goal Centered on Building Trust and Ownership)
1	Inclusiveness[1]	Ensuring the participation of all stakeholders who have an interest in or who would be affected by a specific decision, including *hard-to-reach* groups. Valuing stakeholders input and different forms of knowledge.
2	Transparency	Ensuring that stakeholders have the information they need and in a way they can understand, that they are told where information is lacking or uncertain, and what they can or cannot influence.
3	Commitment	Showing respect for stakeholders by giving engagement the appropriate priority, and by demonstrating that it is a genuine attempt to commit, understand, and incorporate other opinions, even when these conflict with the existing point of view.
4	Resourcing	Providing adequate time, monetary, and expertise resources needed to facilitate engagement and build decision-making capability among different stakeholders.
5	Accountability	Ensuring that, as soon as possible, during stages of the engagement process participants receive an unambiguous account of how and why their contributions have, or have not, influenced the outcome, and ensuring that there are routes for follow-up, including reporting on final decisions, strategies, and/or implementation plans.
6	Adaptiveness	Ensuring that those involved in the consulting must be open to the notion that their existing ideas can be improved (or are wrong), that they change as new information is presented, and that the ideas will, if necessary, be amended.
7	Willingness to learn	Encouraging both the engagers and the engaged to learn from each other, with a style of process that is as interactive and as incremental as possible to build increasing layers of mutual understanding and respect.
8	Productivity	Establishing from the outset how the engagement process will make something better, and that resources will clearly lead to benefits. Being able to show clear outcomes.
9	Accessibility	Providing different ways for people to be engaged and ensuring that people are not excluded through barriers of language, culture, or opportunity, and that stakeholders have access to talk to the right people.
10	Responsiveness	Responding to and timely turnaround times on questions, submissions, or meetings, so that stakeholders are not left wondering if they are listened to, and what the next steps will be.

[1]*Fundamental principles are underlined.*
Source: Adapted from Acland (2008) and Irvine and O'Brien (2009)

ensuring that bridging and linking occurs within the stakeholder engagement process (Fig. 7.1). Engagement without thoughtful facilitation and strategic attention to understanding the relationships between different stakeholder groups inhibits trust and shared ownership of decisions, which may result in perceptions of win–lose outcomes (Mountjoy et al., 2014). Understanding social capital and encouraging champions makes it possible to move away from win–lose perceptions to more shared decision making and hopefully win–win solutions.

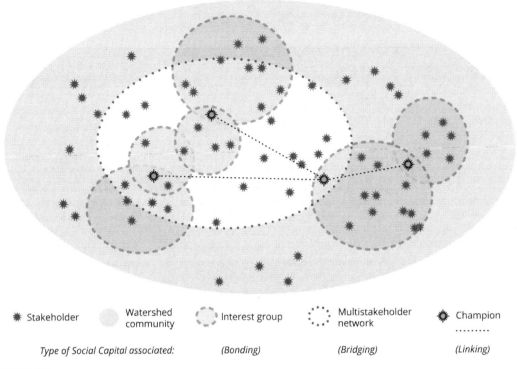

FIGURE 7.1

Principle stakeholder groups are associated with three main types of social capital: bonding, bridging, and linking. Champions within the different groups are needed to facilitate bridging and linking stages to occur and be maintained.

Source: Hearne and Powell (2014)

Conflict and conflict resolution are often associated with stakeholder engagement processes. Conflict can occur between the proponent and stakeholder group(s), between different stakeholder groups themselves, or a combination of both, and at many different levels (Butler et al., 2015). It is very important in stakeholder engagement to address and continually readdress conflict, and to be aware that it can arise in any stage of a stakeholder engagement. Paradoxically, conflict and cooperation can coexist (Zeitoun and Mirumachi, 2008), and stakeholder engagement is not so much about eliminating conflict, simply managing it so that processes can move forward. Conflict identification, understanding, and resolution form an important component of any initial stakeholder analysis, and throughout the whole engagement process. Ensuring that stakeholder engagement is principle-based helps reduce conflict, and there are established methods for conflict resolution discussed (Mayer, 2012).

KEY POINTS

1. Building trust and ownership is key to successful engagement and in turn successful implementation of any initiative.
2. High social capital reduces transaction costs, whereas champions enable different levels of social capital to occur.
3. Conflict resolution is an important component of stakeholder engagement, its inclusion imperative, and conflict and cooperation can coexist.

7.6 ESTABLISHED TOOLS FOR STAKEHOLDER ENGAGEMENT INCLUDING CONFLICT RESOLUTION

A central purpose of engagement in water resource management should be to enable redistribution of power associated with decision making between the proponent (e.g., government, industry) and the other stakeholders (e.g., community, nongovernmental organization). According to Arnstein's ladder (Arnstein, 1969), which seeks to understand participation through a lens of power, eight participation levels within three main categories occur (Arnstein, 1969; Fig. 7.2). Manipulation and therapy levels are bottom rungs of the ladder and considered nonparticipatory, and only the top rungs—partnership, delegated power, and citizen control—can result in appropriate power redistribution and production of shared outputs. Typologies for engagement and role of stakeholders have multiplied since the publication of Arnstein's ladder. Similar to Arnstein's ladder, Callon (1999) distinguished three modes of participation between the proponent and other stakeholders; the Public Education Model where specialists educate the public and public opinion is not valued; the Public Debate Model where debate can occur between specialists and nonspecialists; and the Coproduction of Knowledge Model where both specialist and nonspecialist knowledge is valued, and used in decision making (Callon, 1999; Fig. 7.2). In Callon's case he was referring to the relationships between specialists and lay people, but it is broadly applicable to any proponent and other stakeholders. Building on these former models, the "Spectrum of Public Participation" (IAP2, 2014) developed by the International Association for Public Participation provides targeted goals for the level of participation, sets out the commitment required from the proponent, and techniques for delivery of different levels of participation (IAP2, 2014; Fig. 7.2, Table 7.2). Although centered on the public, it is broadly applicable to all stakeholders within any initiative.

The combination of these three models provides a sound basis for understanding the levels of engagement and that higher levels are associated with more effective engagement, although this is particular to the initiative and where the stakeholders themselves want to be positioned. They form useful planning tools for defining levels of engagement for different stakeholder groups. It is important to highlight that stakeholder engagement is not a linear process, and that different stakeholder groups will require different levels of engagement through the process. These changing levels should be clearly agreed upon in the engagement strategy devised during the planning phase. This highlights the need for stakeholder engagement strategies to be housed within the broader

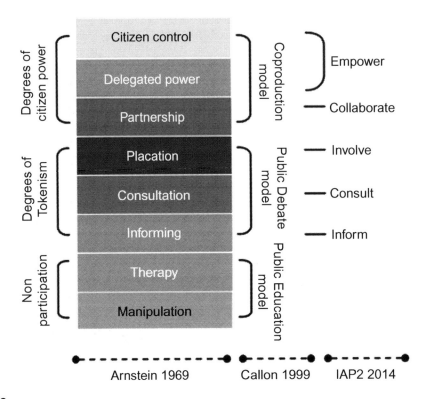

FIGURE 7.2

Arnstein's ladder forms the foundation for defining levels of participation and successive models have refined this. The complementary nature of these three models provides a good basis for understanding the different levels, and all agree that higher levels of participation are desirable for more effective stakeholder engagement.

Source: Adapted from Arnstein (1969), Callon (1999), and IAP2 (2014)

management framework, and ideally an adaptive management framework where stakeholder engagement is explicit. A conflict assessment of all stakeholder groups is essential as part of any stakeholder engagement strategy, and should be conducted during the initial stakeholder mapping/ analysis. This entails the systematic collection of information about the dynamics of conflict within the process. The Wehr Conflict Mapping Guide (Wehr, 1979) and the Hocker–Wilmot Conflict Assessment Guide (Hocker and Wilmot, 1985) are two examples of conflict mapping methods. Tan et al. (2008) highlight other tools such as participatory river models that can assist in development of a shared vision, and a focus on managing trade-offs and win–win situations between different interest groups. Serious gaming is also becoming a popular tool for supporting participatory decision making by allowing joint exploration of the different possible outcomes from different scenarios (Van der Wal et al., 2016). SimBasin and Wat-A-Game toolkit are examples of serious games currently being used.

Table 7.2 Spectrum of Public Participation Building on Arnstein's Ladder Providing Goals, Promises, and Techniques for Effective Stakeholder Engagement

	Inform	**Consult**	**Involve**	**Collaborate**	**Empower**
Goal	To provide the public with balanced and objective information to assist them in understanding the problems, alternatives, opportunities, and/or solutions.	To obtain public feedback on analysis, alternatives, and/or decisions.	To work directly with the public throughout the process to ensure that public concerns and aspirations are consistently understood and considered.	To partner with the public in each aspect of the decision including the development of alternatives and the identification of the preferred solution.	To place final decision making in the hands of the public.
Promise	We will keep you informed.	We will keep you informed but we like to hear your view. We listen to and acknowledge your concerns and provide feedback on how your inputs influenced the final decision.	We will work with you to ensure your concerns and aspirations are directly reflected in the alternatives developed and provide feedback on how your input influenced the decision.	We will look to you for direct advice and innovation in formulating solutions and will incorporate your advice and recommendations into the decisions to the maximum extent possible.	We will implement what you decide.
Techniques	Fact sheets Web sites Open houses.	Public comment Focus groups Surveys Public meetings.	Workshops Deliberate polling.	Citizen Advisory Committees Consensus building Participatory decision making.	Citizen juries Ballots Delegated decisions.

KEY POINTS

1. Typologies for stakeholder engagement have grown, but all are centered around different engagement levels, and expectations and deliverables at those levels.
2. Current practice recognizes the need that stakeholders need to be directly involved in deciding the levels of their participation in a given initiative.
3. Conflict resolution is a key component of stakeholder engagement, should be explicit in engagement strategies, and is expected to occur and reoccur.
4. Stakeholder engagement is not a linear process, and stakeholder engagement levels can change throughout time.

7.7 FIVE STEPS FOR DEVISING AN EFFECTIVE STAKEHOLDER ENGAGEMENT STRATEGY

Appropriate resources need to be given to the process by which stakeholders are engaged within environmental water programs (Mott Lacroix et al., 2016). An objective of stakeholder engagement is to reach decisions, or create approaches that are accepted by stakeholders. This may mean reaching consensus, or perhaps compromise, or, ideally, an emergent shared view that was not apparent before the engagement (e.g., Collins and Ison, 2010). It should also be recognized that expectations of stakeholders are not fixed, and can change as the situation changes. Monitoring and evaluation are important to provide information for adjustments, and for adapting the program in relation to new information. Communication is needed to relay and take in information, manage expectations, but it can also help with negative perceptions and/or unrealistic expectations so that they become more achievable within the context of the intended outcomes. Conversely, ineffective communications can create the perception of failure in the mind of a stakeholder even when the project is delivered on time, on budget, and within the specified scope (Box 7.4).

> **BOX 7.4 SEVEN-POINT CHECK LIST FOR STAKEHOLDER ENGAGEMENT**
>
> 1. Show results and communicate to them.
> 2. Design multiple channels of interaction.
> 3. Provide multitiered levels of engagement.
> 4. Get personal, target, and customize.
> 5. Reinforce sense of civic duty and collectiveness.
> 6. Get precommitment from citizens.
> 7. Learn to experiment and experiment to learn.
>
> Source: Coursera (2016)

The remainder of this chapter builds on the 10 key stakeholder engagement principles and focuses on outlining five key steps for developing and implementing a stakeholder engagement strategy within environmental water programs, including examples from case studies from different areas in the world. The five key steps are explained below in detail.

7.8 PREENGAGEMENT PREPARATION PHASE: BE PREPARED!

There are a series of steps that need to occur by the proponent before formalizing engagement strategies with stakeholder groups, followed by a series of steps completed during implementation and evaluation/adaption phases of the overall program to ensure acceptable outcomes.

7.8.1 STEP 1: INTERNAL ENGAGEMENT STRATEGY AND INTEGRATION WITHIN OVERALL MANAGEMENT FRAMEWORK

1.1. *Overarching management framework*: The stakeholder engagement strategy needs to be embedded within the overarching environmental water initiative framework at all three project phases: planning, implementation, and evaluation. It should also act as a tool for guiding the broader process in relation to the social components.

1.2. *Purpose of engagement*: Before developing an engagement strategy, the proponent must first understand what stakeholder engagement means to their initiative, and understand and agree internally on the motivation and purpose of their intended engagement. A clear purpose and scope of the overall environmental water program should be established by the proponent, with the intention that this will be adapted once stakeholders are involved. If the initiative is to build trust and ownership, the purpose of stakeholder engagement should be focused on creating trust and ownership of the overall environmental water initiative. The focus should be around setting a common vision and objectives, and creating shared outcomes and learning. Although the proponent must have a clear understanding of what they are seeking to achieve, setting objectives before engagement can create power imbalances and the principal of codesign suggests that stakeholders need to be engaged right from the conception phase. As this is an internal stage it is not expected or advised to engage all stakeholders at this stage but engaging target representatives can enhance the legitimacy of this first step, provide valuable information, and also help to facilitate further steps.

1.3. *Choosing an engagement lead*: Prioritization and neutrality are important components of stakeholder engagement, and in large complex initiatives such as environmental water programs where many real and perceived win–lose scenarios exist, sufficient resources should be committed to a position or group responsible for the whole stakeholder engagement process. Internally, champions and owners of the engagement strategy should be identified and resourced. This could be a position within the organization, but also could take the form of a steering committee (both internal and external members), which takes the role of an *honest broker* to have the mandate of facilitating the process toward an agreed set of goals or outcomes. Initially, this may appear cost-prohibitive, but in the long term should reduce transaction costs as delays and conflicts are reduced.

1.4. *Know your boundaries*: Before moving further the proponent must have a clear understanding of what they are seeking to achieve within the initiative and be able to clearly articulate what their needs are to others. The proponent should have a clear understanding of the spatial and temporal boundaries, and capacity and resourcing constraints. These boundaries may also change as further engagement and implementation occurs, but initially they must be understood for identifying the range of stakeholder that needs to be engaged.

7.8.2 STEP 2: WHO TO ENGAGE: STAKEHOLDER MAPPING/ANALYSIS

2.1. *Stakeholder mapping/analysis*: Effective engagement must ensure inclusive opportunities for all stakeholders to participate at some level in decision making (highlighting the first foundation principle; inclusiveness). Identifying and engaging stakeholders should be viewed as a reiterative process, and stakeholders added as the initiative develops. Stakeholder mapping/

analysis is an established process for considering who may be interested/impacted and the appropriate levels of engagement for each. Stakeholder analyses allow the proponent to identify different stakeholder groups and key players (including champions or saboteurs). Stakeholders can be analyzed by different categories such as interest, impact, and influence, helping to identify potential conflict, and so on. The goal should be to identify and devise how best to approach each group, and how, and when/if to bring different groups together. The choice of stakeholder analysis is program/focus/expertise-specific, but many generic simple approaches such as tables and matrices exist that easily facilitate this task; see Reed et al. (2009) for a list of examples. Conflict should also be assessed at this step, and should consider conflict not only between both the proponent and relevant stakeholders, but also between stakeholder groups themselves. Using methods such as the Wehr Conflict Mapping Guide or the Hocker–Wilmot Conflict Assessment Guide would be sufficient for identifying potential sources of conflict and devising ways to deal with them within the initiative (see Box 7.5).

BOX 7.5 COMMUNITY ENGAGEMENT/CONFLICT RESOLUTION—THE KEY TO WATER REFORM IN AUSTRALIA'S FOODBOWL, THE MURRAY–DARLING BASIN

In the heart of Australia's foodbowl, the Murray–Darling Basin, major water reform through the Murray–Darling Basin Plan is taking place where water once used for consumptive purposes is being transferred to the environment either through the government entering the water market (through water buybacks) or funding infrastructure-based water-saving projects. While the Murray–Darling Basin Authority (MDBA) invested significant effort in public meetings throughout the Basin Plan development, there were significant limitations in the way decisions were made and information was communicated, with many stakeholders believing their views were either inappropriately considered or simply dismissed (Commonwealth of Australia, 2013). In the initial planning, although a formal stakeholder engagement process was followed (MDBA, 2009), the three fundamental principles (inclusiveness, transparency, and commitment) were not successfully implemented, perhaps in part due to the speed at which the draft plan had to be formulated. The MDBA has responded to criticism surrounding its early stakeholder engagement processes and invested considerable effort in further stakeholder engagement during implementation. A key element of this has been the recognition of the need for localism— "This includes reflecting the knowledge of the local people, and local land and catchment managers in how to best manage environmental flows and recognizing the work done to date by local communities in developing state water sharing plans" (House of Representatives Standing Committee on Regional Australia, 2011). There remains a challenge as to how to effectively balance centralized decision-making roles with the need for localism. Perhaps a major step forward for the MDBA is the recent announcement that they will begin a process to decentralize operations throughout the basins, allowing local communities and stakeholders more opportunity to engage directly with the MDBA (Long, 2016).

Note: It is essential that an initial analysis needs to be conducted before engaging with stakeholders; however, it must be also revisited directly after engaging with identified stakeholder groups, updated, and ideally agreed upon by each stakeholder group (Fundamental Principle 2: Transparency). If there are large unknowns (e.g., in influence or attitude), then these should be the focus of initial engagement. The stakeholder analysis should be regularly revisited and updated as the program is implemented, and moves through the planning, implementation, evaluation phases of the overall program (Fundamental Principle 3: Commitment).

2.2. *Levels of engagement*: Deciding on the levels of engagement is a critical component and Arnstein's ladder is still considered the standard approach for considering the different levels of engagement, and a good starting point when discussing with stakeholders where they feel they should be positioned. In addition, the pragmatism of the International Association for Public Participation 2 (IPA2) framework, showing the goals, promises, and techniques for engagement at the different levels builds from Arnstein's ladder and provides a basis for use within the initial stakeholder analysis on what techniques may be used for engagement at the different levels (IPA2, 2014; Fig. 7.2, Table 7.2). The proponent should assess and identify a clear perspective of what levels of engagement are sought from different stakeholders. This should be confirmed or adjusted once full engagement has begun and emphasize the different components of principle-based engagement. The proponent should also allow the opportunity, even encourage stakeholders to move between levels. If resources are lacking to allow this to occur, but this is seen as essential, then resources should be allocated (highlighting Principles 3, 6, 9, and 10). For example, a group may want to be at the IPA2 *Involve* level during the Planning Phase, at the *Inform* level during the Implementation Phase, and at the *Collaborate* level during the Evaluation and Adaption Phase. In addition, a group may also start at the *Inform* level but want to be at the *Partnership* level by a certain stage of the initiative. They may lack adequate training or financial skills to perform at different levels so resources should be available to accommodate this (showing commitment and inclusiveness). These changing levels should be clearly agreed upon in the engagement strategy devised during the planning phase, but be flexible enough to change throughout the initiative (Principles 6, 7, and 10). The different levels of engagement should be agreed upon by both parties, and ideally agreed upon by all stakeholders in a collaborative workshop. The proponent's commitment or promise to stakeholders is key for ensuring transparency and accountability (primarily Principles 1−7) (see Box 7.6).

BOX 7.6 LESOTHO HIGHLANDS ENVIRONMENTAL FLOWS CASE STUDY—STAKEHOLDER ENGAGEMENT AS A CONDUIT FOR RESOURCE SHARING, AND FINDING A SUSTAINABLE COMPROMISE

The Senqu (upper Orange) River originates high in the Maluti Mountains in Lesotho and is home to many small-scale and subsistence farmers who are reliant on the river for water and other ecosystem services. Major dams and future development bring both opportunity and risk to the communities reliant on the river. Through three levels of engagement (central government, local government, and local communities), the Lesotho Highlands Water Project has been able to produce a number of possible dam development scenarios and associated trade-offs, aimed at finding a sustainable compromise between the different uses. A particular strength was the engagement with the rural communities, the ones most affected by changes to the river flow. Historically, there have been challenges in trust and ownership between the different stakeholders, so by ensuring appropriate levels of engagement to those most affected, and adherence to the foundation principles, agreements are being reached.

Note: Although higher levels of engagement are desirable, external processes already in place and outside the control of the proponent may also inhibit which level becomes achievable within the proposed process. For example, if the broader political base is driven by a top-down process, stakeholder engagement at levels above *Involvement* may never be realistically achievable, as opposed to the political base where bottom-up approaches are more common and *Empowerment* levels of stakeholders are achievable (Table 7.2). Although this is less desirable from a trust and ownership perspective, clearly articulating the levels attainable to stakeholders at the very beginning of engagement is critical.

7.9 FULL ENGAGEMENT COMMENCEMENT STEPS
7.9.1 STEP 3: WHEN AND HOW TO FULLY ENGAGE

3.1. *When*: Early engagement is essential, but full structured engagement (e.g., bringing different stakeholder groups together) should not commence before the steps above have been carried out. Having an initial understanding of the needs, interests, and potential conflicts (with the proponent or between other groups) is essential to guide full engagement. Having a good background understanding highlights that you have prepared, and the foundation principles to building trust and ownership within the initiative should be followed. By completing steps 1−2, the proponent will be able to properly articulate to the stakeholders what their ambitions are for both initiative and stakeholders, what they expect of themselves, and what they expect of the stakeholders, and have information presented in an appropriate format (Principle 4). Three main questions form the basis of comparing the different stakeholder groups: (1) what would they like to see (desired future state); (2) what can they live with; and (3) what is completely unacceptable? This provides the information needed to look at what is achievable and what shared management outcomes with win−win situations can be achieved. These should be clarified at the initial stages of engagement. This then forms the basis for further developing the stakeholder engagement strategy to accompany the final planning stages of the overarching framework and successive implementation-evaluation phases.

3.2. *How*: Understanding the different means by which stakeholders are engaged is critical to inform the likely awareness and understanding of complex issues, which in turn can influence the likely level of engagement of stakeholders on different issues. The IAP2 Spectrum (Table 7.2) ascribes different techniques for the different levels of engagement. Although these approaches are generic tools for engagement they are appropriate for shifting the focus away from unilateral technical fixes toward values-based negotiations. This is also an important step in recognizing social capital and to show support and commitment to bridging and linking groups, to identify potential champions to aid in taking the process forward, and helping in finding win−win outcomes.

3.3. *Honesty and fairness in engaging*: When discussing what is and what is not possible, honesty and fairness are essential, as it is better to not commit to something that later on cannot be done. In addition, by conducting the earlier steps the proponent will have an understanding of group dynamics and potential conflicts between groups. Collaborative workshops are a good tool for working on shared outcomes, but only if groups are comfortable enough to work together. Conducting the steps above should gauge if this can happen, and similarities between values as opposed to differences should be the focus of the engagement. However, there are always surprises, and the proponent should be aware of this. An advantage of having an expert facilitator at such meetings/workshops is that they have techniques available to deal with conflict, but the overall person(s)/group from the proponent responsible for stakeholder engagement should always be present and take an active and lead role. A facilitator is only there to facilitate. Scenario forecasting and providing choice of different outcomes through scenarios is critical for moving engagement forward. Scenarios provide a range of options, and allow stakeholders' input in finding shared outcomes.

3.4. *Forms of knowledge*: One of the common mistakes in stakeholder engagement is the ability of the proponents to value and utilize different forms of knowledge. This avoidance or inability to use all forms of knowledge not only leads to power imbalances but also creates conflict. Often stakeholder groups have information in various forms (e.g., indigenous and local knowledge) that is useful to decision making. The proponent needs to provide a procedure for utilizing all forms of knowledge in the planning, implementation, and evaluation stages of the initiative (Principle 1). This not only enhances trust and ownership from stakeholders, but also adds information to what often are data-poor situations, helping to form a multiple lines of evidence approach.

Note: Silence from stakeholders does not necessarily mean that the stakeholder engagement process is working, indeed usually it means the opposite, as stakeholders may feel undervalued and so disengaged that they feel completely powerless and unable to contribute to the process in a meaningful way. This may seem an ideal situation for actions to be completed faster but it usually means that an *attack* on the process will occur at the first opportunity, and inevitably leads to substantial delays.

7.9.2 STEP 4: DEVELOP AND IMPLEMENT STAKEHOLDER ENGAGEMENT PLAN

4.1. *The plan*: The stakeholder engagement plan needs to become a document that captures the strategy, and provides a plan that informs both the proponent and stakeholders what their obligations are, and a document all stakeholders can refer to when negotiating certain aspects of the initiative itself.

4.2. *Providing security*: An engagement strategy document provides security for stakeholders that their values and needs have been captured, engagement priorities and steps set, and everything properly documented. The document should always remain a *living* document, and adapted as new information becomes available or the situation changes. However, this should not be used as an excuse to keep it in continually draft form, and a final product should be produced, then revisited at key evaluation stages (Step 5). This final product should not be finalized by the proponent until input from the stakeholders has been included, they have had an opportunity to comment on the final product, and signed off by key stakeholder groups operating above the inform level of engagement. All key principles are played out in this stage, especially inclusiveness, transparency accessibility, and accountability. The final product must be shared among the different stakeholder groups.

7.9.3 STEP 5: IMPLEMENTATION AND EVALUATION OF ENGAGEMENT STRATEGY

5.1. *Enacting plan*: Implementation of the strategy should be integrated into the overarching initiative implementation phase. What was agreed upon in the strategy should now be enacted as the overall initiative is implemented. During implementation of specific environmental water releases, some stakeholders may be involved at a monthly, weekly, or even daily level depending on the action, and how this will occur needs to be captured and understood before implementation.

5.2. *Monitor*: The situation then needs to be monitored from a stakeholder engagement perspective. This information should allow for adjustments to be made as the relationship between the proponent and stakeholder group develops, or if the relationship between other stakeholder groups changes. Communication of progress is important, and should be in line with what was agreed upon in the strategy document.

5.3. *Evaluate*: Monitoring is a continual process at all steps and for timely decision making and corrective action (if necessary). Evaluation is different as it is a selectively staged exercise that attempts to systematically and objectively assess progress toward achievement of the outcomes. At the most basic level, the proponent has to look back on the objectives they set for stakeholder engagement, assess if they are being met, and also assess if stakeholder engagement processes are helping or hindering in other objectives of the overall initiative. Evaluation should be at a time step that is particular to the initiative, but should be timely enough that stakeholders see progress and have the ability to comment and suggest changes. For example, in a 5-year initiative an engagement strategy may need evaluating at a minimum yearly, with half-yearly stakeholder engagement workshops to discuss challenges and recent developments.

5.4. *Communication*: Providing opportunity for continual communication to and between stakeholders is essential during all phases of an initiative. Champions are essential in this. This should be clearly articulated in the plan as to how and when this will occur and the procedure followed. There should be feedback to and from stakeholders, especially highlighting any changes or how their inputs have helped decision making. Lack of communication and any inability for stakeholders to provide feedback will be very destructive to the relationship.

5.5. *Learning and adaption*: Learning and adapting are critical at all steps to make adjustments, and usually operating at small scales within different steps throughout the entire process. For example, a stakeholder group may have a particular issue that can be resolved in a day, through negotiation of a small change to the process. However, a full evaluation step allows for the overall engagement strategy to be scrutinized, and changes to be made (if necessary). These changes need to be agreed upon by the respective stakeholders as done in Step 4, and then integrated into the overall initiative in what is a repetitive cyclic process.

Communication is an essential component throughout all steps, but especially important in adapting the strategy to clearly articulate with stakeholders what has been learned and what next steps are being proposed.

7.10 CONCLUSION

Trust is fundamental to relationships... the "power factor" that essentially describes the willingness of individuals to believe the source of information, and their willingness to participate.

Flitkroft (2010)

When climate change expert Bill McKibben was asked where he would move to have the best chances of surviving climate change, his answer was not a geographical one. He said, "I would look for a place to go where there is a community that can make decisions together. That is the kind of place where survival is possible" (Moore, 2013). In reality, stakeholder engagement is the same as any human relation; it takes time and understanding to build the trust and ownership needed to meet comanagement-based outcomes associated with effective stakeholder engagement strategies. Stakeholders need to be at the center of the process due to the social importance of environmental water initiatives, and the implications at a wider social-ecological system perspective. There is no panacea for stakeholder engagement with every case specific; however, principle-based engagement sets a foundation for going forward, and there are many pragmatic tools available to help with the process.

REFERENCES

Acland, A., 2008. A handbook of public and stakeholder engagement. Dialogue by Design, Surrey, U.K.

Acreman, M.C., Overton, I.C., King, J., Wood, P.J., Cowx, I.G., Dunbar, M.J., et al., 2014. The changing role of ecohydrological science in guiding environmental flows. Hydrol. Sci. J. 59, 433–450.

Argent, R.M., 2009. Components of adaptive management. In: Allan, C., Stankey, G. (Eds.), Adaptive environmental management: a practitioner's guide. Springer, Dordrecht, Netherlands.

Arnstein, S.R., 1969. A ladder of citizen participation. J. Am. Inst. Plan. 35, 216–224.

Butler, J.R., Young, J.C., McMyn, I.A., Leyshon, B., Graham, I.M., Walker, I., et al., 2015. Evaluating adaptive co-management as conservation conflict resolution: learning from seals and salmon. J. Environ. Manage. 160, 212–225.

Callon, M., 1999. The role of lay people in the production and dissemination of scientific knowledge. Sci. Technol. Soc. 4, 81–94.

Collins, K.B., Ison, R.L., 2010. Trusting emergence: some experiences of learning about integrated catchment science with the Environment Agency of England and Wales. Water Resour. Manage. 24, 669–688.

Commonwealth of Australia, 2013. Impact of the Basin Plan on Rural Communities, Localism and Stakeholder Engagement (Chapter 7) in The Management of the Murray-Darling Basin, Parlimentary Inquiry report, 13 March 2013.

Conflict assessment, 1985. In: Hocker, J., Wilmot, W. (Eds.), Interpersonal Conflict. 2nd ed. Wm. C. Brown Publishers, Dubuque, Iowa, Chapter 6.

Conley, A., Moote, M.A., 2003. Evaluating collaborative natural resource management. Soc. Nat. Resour. 16, 371–386.

Cook, B., Kesby, M., Fazey, L., Spray, C., 2013a. The persistence of 'normal' catchment management despite the participatory turn: exploring the power effects of competing frames of reference. Soc. Stud. Sci. 43.

Cook, B.R., Atkinson, M., Chalmers, H., Comins, L., Cooksley, S., Deans, N., et al., 2013b. Interrogating participatory catchment organisations: cases from Canada, New Zealand, Scotland and the Scottish-English Borderlands. Geogr. J. 179, 234–247.

Cornwall, A., 2008. Democratising engagement. What the UK can Learn from International Experience. Demos, London.

Coursera, 2016. World Bank Group Course 2016 — Engaging citizens: a game changer for development, https://www.coursera.org/course/engagecitizen.

Crase, L., O'Keefe, S., Dollery, B., 2013. Talk is cheap, or is it? The cost of consulting about uncertain reallocation of water in the Murray—Darling Basin, Australia. Ecol. Econ. 88, 206—213.

Diedrich, A., Stoeckl, N., Gurney, G.G., Esparon, M., Pollnac, R., 2016. Social capital as a key determinant of perceived benefits of community-based marine protected areas. Conserv. Biol. 31 (2), 311—321.

Flitcroft, R., Dedrick, D.C., Smith, C.L., Thieman, C.A., Bolte, J.P., 2010. Trust: the critical element for successful watershed management. Ecol. Soc. 15 (3), r3.

Hamm, J.A., Hoffman, L., Tomkins, A.J., Bornstein, B.H., 2016. On the influence of trust in predicting rural land owner cooperation with natural resource management institutions. J. Trust Res. 6, 37—62.

Harrington, C., Curtis, A., Black, R., 2008. Locating communities in natural resource management. J. Environ. Pol. Plann. 10, 199—215.

Hearne, D., Powell, B., 2014. Too much of a good thing? Building social capital through knowledge transfer and collaborative networks in the southern Philippines. Int. J. Water Resour. Dev. 30, 495—514.

House of Representatives Standing Committee on Regional Australia, 2011. Of drought and flooding rains: Inquiry into the impact of the Guide to the Murray-Darling Basin Plan, Canberra, May 2011, p. 73. http://www.aph.gov.au/parliamentary_business/committees/house_of_representatives_committees?url=ra/murraydarling/report.htm

IAP2, 2014. Spectrum of Public Participation. Available from: www.iap2.org.au (accessed 01.02.17).

IFC, 2007. Stakeholder Engagement: A Good Practice Handbook for Companies Doing Business in Emerging Markets. International Finance Corporation, Washington, DC.

Irvine, K., O'Brien, S., 2009. Progress on stakeholder participation in the implementation of the water framework directive in the Republic of Ireland. Biol. Environ. Proc. Royal Irish Acad. 109B, 365—376.

Jackson, S., Tan, P.-L., Mooney, C., Hoverman, S., White, I., 2012. Principles and guidelines for good practice in indigenous engagement in water planning. J. Hydrol. 474, 57—65.

Jacobs, K., Lebel, L., Buizer, J., Addams, L., Matson, P., McCullough, E., et al., 2016. Linking knowledge with action in the pursuit of sustainable water-resources management. Proc. Natl. Acad. Sci. 113, 4591—4596.

Kilvington, M., Atkinson, M., Fenemor, A., 2011. Creative platforms for social learning in ICM: the Watershed Talk project. N. Z. J. Mar. Freshw. Res. 45, 557—571.

Knight, A.T., Cowling, R.M., Boshoff, A.F., Wilson, S.L., Pierce, S.M., 2011. Walking in STEP: lessons for linking spatial prioritisations to implementation strategies. Biol. Conserv. 144, 202—211.

Leahy, J.E., Anderson, D.H., 2008. Trust factors in community—water resource management agency relationships. Landsc. Urban Plan. 87, 100—107.

Le Quesne, T., Kendy, E., Weston, D., 2010. The Implementation Challenge: taking stock of government policies to protect and restore environmental flows. WWF Report.

Long, W., 2016. Murray-Darling Basin Authority begins decentralising jobs, ABC rural News, 15 August, 2016. Available from: http://www.abc.net.au/news/2016-08-12/mdba-moves-jobs-to-regional-areas/7726272.

Margerum, R.D., Robinson, C.J., 2015. Collaborative partnerships and the challenges for sustainable water management. Curr. Opin. Environ. Sustain. 12, 53—58.

Mayer, B., 2012. The dynamics of conflict: a guide to engagement and intervention. 2nd ed. Jossey-Bass, San Francisco.

MDBA, 2009. Stakeholder Engagement Strategy. Involving Australia in the development of the Murray—Darling Basin Plan.

Moore, M., 2004. Perceptions and interpretations of environmental flows and implications for future water resource management; A survey study. Department of Water and Environmental Studies, Linkoping University, Sweden, 1—67.

Moore, L., 2013. Common ground on hostile turf: stories from an environmental mediator. Island Press, Washington, DC.

Mostert, E., 2015. Who should do what in environmental management? Twelve principles for allocating responsibilities. Environ. Sci. Pol. 45, 123−131.

Mott Lacroix, K.E., Xiu, B.C., Megdal, S.B., 2016. Building common ground for environmental flows using traditional techniques and novel engagement approaches. Environ. Manage. 57, 912−928.

Mountjoy, N.J., Seekamp, E., Davenport, M.A., Whiles, M.R., 2014. Identifying capacity indicators for community-based natural resource management initiatives: focus group results from conservation practitioners across Illinois. J. Environ. Plan. Manage. 57, 329−348.

OECD, 2015. OECD Principles on Water Governance: welcomed by Ministers at the OECD Ministerial Council Meeting on 4 June 2015. Organisation for Economic Co-operation and Development. Online: Directorate for Public Governance and Territorial Development.

Pahl-Wostl, C., Jeffrey, P., Isendahl, N., Brugnach, M., 2011. Maturing the new water management paradigm: progressing from aspiration to practice. Water Resour. Manage. 25, 837−856.

Pahl-Wostl, C., Arthington, A., Bogardi, J., Bunn, S.E., Hoff, H., Lebel, L., et al., 2013. Environmental flows and water governance: managing sustainable water uses. Curr. Opin. Environ. Sustain. 5, 341−351.

Poff, N.L., Matthews, J.H., 2013. Environmental flows in the Anthropocene: past progress and future prospects. Curr. Opin. Environ. Sustain. 5, 667−675.

Reed, M.S., 2008. Stakeholder participation for environmental management: a literature review. Biol. Conserv. 141, 2417−2431.

Reed, M.S., Graves, A., Dandy, N., Posthumus, H., Hubacek, K., Morris, J., et al., 2009. Who's in and why? A typology of stakeholder analysis methods for natural resource management. J. Environ. Manage. 90, 1933−1949.

Richter, B., 2014. Chasing water: A guide for moving from scarcity to sustainability. Island press/Centre for Economics, Washington DC, ISBN 9781597264624.

Richter, B.D., Mathews, R., Harrison, D.L., Wigington, R., 2003. Ecologically sustainable water management: managing river flows for ecological integrity. Ecol. Appl. 13, 206−224.

Rist, L., Campbell, B.M., Frost, P., 2013. Adaptive management: where are we now? Environ. Conserv. 40, 5−18.

Rogers, K.H., 2006. The real river management challenge: integrating scientists, stakeholders and service agencies. River Res. Appl. 22, 269−280.

Roux, D.J., Foxcroft, L.C., 2011. The development and application of strategic adaptive management within South African National Parks. Koedoe 53 (2), 1−5.

Saravanan, V.S., McDonald, G.T., Mollinga, P.P., 2009. Critical review of Integrated Water Resources Management: moving beyond polarised discourse. Nat. Resour. Forum 33, 76−86.

Schultz, P.W., 2011. Conservation means behavior. Conserv. Biol 25, 1080−1083. Available from: http://dx.doi.org/10.1111/j.1523-1739.2011.01766.x.

Steyaert, P., Jiggins, J., 2007. Governance of complex environmental situations through social learning: a synthesis of SLIM's lessons for research, policy and practice. Environ. Sci. Pol. 10, 575−586.

Tan, P.-L., Jackson, S., MacKenzie, J., Proctor, W., Ayre, M., 2008. Collaborative water planning: context and practice literature review. Vol. 1. Report to the Tropical Rivers and Coastal Knowledge (TRaCK) program. Land and Water Australia, Canberra, Australia.

US Army Engineer Institute for Water Resources, 2012. Shared Vision Planning website. Available from: www.sharedvisionplanning.us (accessed 10.04.16).

Van Der Wal, M.M., De Kraker, J., Kroeze, C., Kirschner, P.A., Valkering, P., 2016. Can computer models be used for social learning? A serious game in water management. Environ. Model. Softw. 75, 119−132.

Wehr, P., 1979. Conflict regulation. Westview Press, Boulder, CO.

Yee, S., 2010. Stakeholder engagement and public participation in environmental flows and river health assessment. Australia-China Environment Development Partnership. Project code P0018, May.

Zeitoun, M., Mirumachi, N., 2008. Transboundary water interaction I: reconsidering conflict and cooperation. Int. Environ. Agree. Pol. Law Econ. 8, 297–316.

ENVIRONMENTAL WATER REGIMES AND NATURAL CAPITAL: FREE-FLOWING ECOSYSTEM SERVICES

David J. Gilvear[1], Lindsay C. Beevers[2], Jay O'Keeffe[3], and Mike Acreman[4]

[1]*Plymouth University, Plymouth, United Kingdom* [2]*Heriot-Watt University, Edinburgh, Scotland*
[3]*Rhodes University, Grahamstown, South Africa* [4]*Centre for Ecology and Hydrology, Wallingford, United Kingdom*

8.1 INTRODUCTION

The idea that people and the environment are inextricably linked is not new (Acreman, 2001); only the language has changed. The great philosophers, Socrates, Laozi, and the Buddha, talked of the harmony of heaven, Earth, and humans as far back as 500 BC. In 1864, G. P. Marsh published *Man and Nature* arguing among others things that deforestation in the Mediterranean could lead to droughts. In the 1990s, the concept of *ecosystem services (ESs)* came into use and developed largely in wetlands (Maltby and Acreman, 2011) to describe the way nature provides benefits to people, and the ecosystem approach was adopted by the Convention on Biological Diversity as a framework for managing natural resources. The suffix *eco* is primarily the language of biologists, who see biodiversity at the center (Mace et al., 2012), while in other fields similar holistic approaches have developed (e.g., *integrated water resources management* in water resources). Many species, communities, and ecosystems, particularly in aquatic systems, are in decline (Vörösmarty et al., 2010). Part of the problem may be that the biologically based language used to influence social values has not been successful. If this is the case, what language should we use that could build improved awareness of the link between people and nature?

Many decision makers in governments and business have been trained in politics, finance, and economics, so perhaps using the language of economics would be more successful. Indeed everyone is familiar with purchasing a good in the marketplace using money, and in that sense the terminology of economics is ubiquitous. Economics reframes decisions in terms of values, value maximization, and exploring trade-offs using a common metric with a utilitarian philosophical basis. The concept of *capital* is very familiar to economists; businesses have staff (human capital), buildings, offices, machinery (manufactured capital), money in the bank (financial capital), and ways of working together (social capital). We can add to this list the natural capital that provides benefits to people resulting from the interaction and spatial configuration of the elements of nature, for instance soils and rocks, animals, plants, and water. This concept recognizes the dependencies between different forms of capital (Fig. 8.1). The hydrological cycle is a good example of the interaction and spatial configuration of the elements of nature that produce stores of water in rivers, lakes, and aquifers. However, for people located at a distance from the water resource to benefit

Water for the Environment. DOI: http://dx.doi.org/10.1016/B978-0-12-803907-6.00008-5

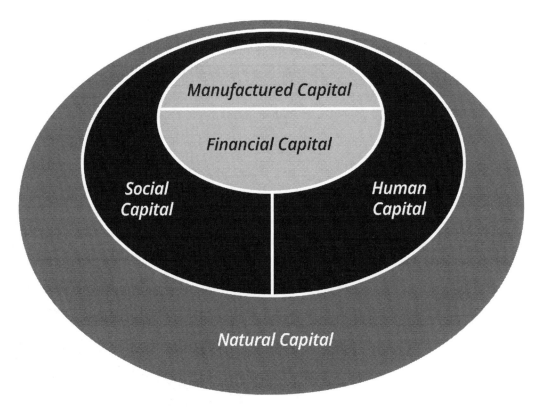

FIGURE 8.1

The five capitals model showing the relationship of natural capital to the other forms of capital.

Source: Forum for the Future (2015)

from a delivered water supply a pipeline (manufactured capital) is required, with money to fund it (financial capital), people to manage it (human capital), and a process of interaction with customers (social capital). Thus, while important and essential to life, people cannot live in today's society by natural capital alone and the concept of *natural capital* is emerging as a new driver for ecosystem management. The UK government, for example, created the Natural Capital Committee that has developed a 25-year plan for restoration of natural capital assets as it recognizes that they underpin economic growth and human welfare (Natural Capital Committee, 2015).

Environmental water is a good example of how natural capital assets provide multiple benefits to people including fisheries, recreation, and spiritual importance. The capital framework lends itself to assessing trade-off in decisions (Acreman et al., 2011). For example, the waters of the Mekong River are a key natural asset that delivers many benefits, both already realized and potential, to people. The Tonle Sap is one of the world's most important fisheries providing around 230,000 tons per year (Baran et al., 2001), providing essential protein for many hundreds of thousands of local people. Another use of the water is to dam the rivers and generate hydropower that can be used locally and exported for important international revenue. However, this alters the flow regime of the river and will reduce fisheries yields in the Tonle Sap (Ziv et al., 2012). Similarly, trade-offs in exploitation of benefits resulting from different water management options has been investigated for many international rivers (e.g., Mekong, Baran et al., 2001; the Nile, Goor et al., 2010; and Zambezi, Tilmant et al., 2010). Although the use of economic language in natural capital is particularly suited to monetary valuation (e.g., Emerton, 2004), some natural capital assets are important for their mere existence rather than being directly used (such as the landscapes of Antarctica) and not readily valued in such terms, which poses significant challenges.

This chapter introduces the concepts of natural capital and ES as a way of incorporating environmental value into water resource management. Here, we use the definition of natural capital, where the biodiversity associated with land and water can be seen as a capital stock (Ehrlich et al., 2012) and from this capital stock different ESs can be derived. We use the following definition for ESs: "Ecosystem services are the contributions of ecosystem structure and function—in combination with other inputs—to human well-being" (Burkhard et al., 2012).

The seminal work by Ehrlich and Ehrlich (1981) coined the term *ecosystem services*, and in subsequent decades this concept was developed. Publications by Dugan (1990), Costanza et al. (1997), and Daily (1997) raised the profile of the idea, and made links to the notion of natural capital (Fig. 8.2). The global recognition of the concept of ESs led to the Millennium Ecosystem Assessment (MEA, 2005), which aimed "to assess the consequences of ecosystem change for human well-being and the scientific basis for action needed to enhance the conservation and sustainable use of those systems and their contribution to human well-being" across the world. The assessment used four categories of services; namely provisioning (e.g., fresh water, timber, and food), supporting (e.g., soil formation, nutrient cycling, and primary production), regulating (e.g., flood regulation, air quality, climate regulation, and water purification), and cultural (e.g., spiritual, recreational, aesthetic, health, and well-being).

The concept of ES (Costanza et al., 1997; Daily, 1997) deserves attention in that it links ecosystem structure and functions to anthropogenic interests. The linking of natural and socio-economic systems should be integral to 21st-century river management (Gilvear et al., 2016) because the ESs provided by rivers are many and varied. These range from the easily valued services such as hydropower production, fisheries, and irrigation water supply, to less tangible benefits such as recreational and cultural values (Table 8.1). Adopting an ES framework to assess rivers can be used to examine the trade-offs between ESs under freely flowing conditions (Auerbach et al., 2014) and *dam-controlled* conditions. Despite this, only a few studies have actually utilized ES concepts and terminology in relation to environmental flows assessment (Fig. 8.2) such as in East Africa (Notter et al., 2012) and the Mekong (Ziv et al., 2012). Wide adoption would encourage a more complete assessment of the full range of ESs rather than a piecemeal approach often used in the past.

Here, we examine how natural capital and ES concepts relate to environmental water and river environments, and the nature of the linkage between natural capital, ESs, and river flows. The chapter discusses how ESs are identified, measured, and assessed, and the associated challenges. River ESs can be seen as part of a cascade, as introduced by Potschin and Haines-Young (2011), and this chapter explores this from the biophysical realm. The focus of this chapter is on biophysical structure and process, and the functions of rivers that provide ESs and benefits. It concentrates less on the socio-economic and cultural processes, while recognizing the importance of this part of the cascade within the ES concept (Fig. 8.3). Economic valuation methods of ESs are not explicitly covered, and we argue that some ES components cannot be valued economically.

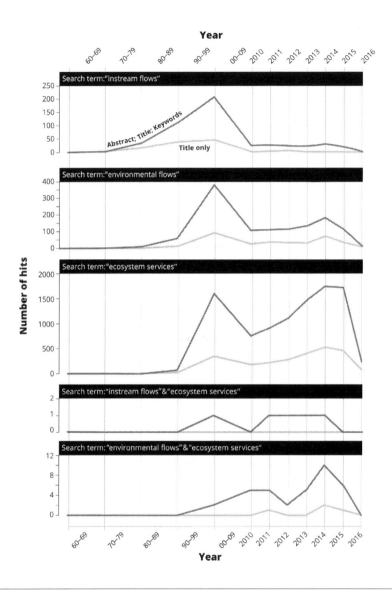

FIGURE 8.2

A chronology of the evolution of environmental flows assessment and ecosystem services assessment and implementation as indexed by a Scopus search of papers with keywords in the scientific literature.

The chapter concludes with a number of international case studies to demonstrate and highlight the key themes. Natural capital and ESs provide a sound conceptual framework to better understand the pros and cons of differing environmental water regimes with the identification of the full range of cause–effect factors. These terms also provide a vehicle for communication of environmental water regimes to stakeholders, particularly those with an economics/finance focus, so that they can better understand trade-offs between differing river flow scenarios.

Table 8.1 Ecosystem Services Delivered by Freshwater River Systems and Potential Indicators

Ecosystem Services Type	Ecosystem Services	Examples of Potential Indicators
Provisioning	Fresh water supply for human consumption and energy production	Flow indices (e.g., Q95), chemical properties (e.g., compared to drinking water guidelines)
	Fertile moist lands for food, timber, and firewood	Crop-production figures
	Flows and habitat to support fishing	Fish catch data
Supporting and maintenance	Alluvial soil formation,	Floodplain sedimentation types and rates
	Nutrient input and cycling	Nitrogen and phosphorus levels; occurrence of algal blooms
	Primary production	Diatom assessment
	Habitat	Instream/riparian vegetation; river bed and floodplain sediment characteristics, geomorphological features
	Seed dispersal	Seedling germination numbers
	Hydro-hazard regulation	Flood flow indices (mean annual flood); floodplain inundation frequency
	Erosion regulation	Water turbidity values and suspended solids concentrations; bank erosion rates
	Climate regulation	Water temperature regimes
	Water purification/quality	Biological indicators and hydrochemical indicators (e.g., macroinvertebrate-based measures of water quality; dissolved oxygen concentrations)
Cultural	Spiritual	Customs and festivals (e.g., numbers of people visiting)
	Recreational	Presence of boating, swimming, and walking (numbers; commercial income); recreational fishing (license/permit numbers)
	Aesthetic	Landscape designations (e.g., areas of outstanding natural beauty in England and Wales); heat maps of photographs on social media
	Health and well-being	Quality-of-life indicators

8.2 ENVIRONMENTAL WATER, NATURAL CAPITAL, AND ECOSYSTEM SERVICES

Environmental water regimes support a range of ESs that otherwise would be reduced or lost due to river flow regulation or catchment modifications (e.g., afforestation) while recognizing that in most instances, provision of these flows will lead to a reduction in the level of service to direct or consumptive water users (Arthington et al., 2006; Arthington, 2012).

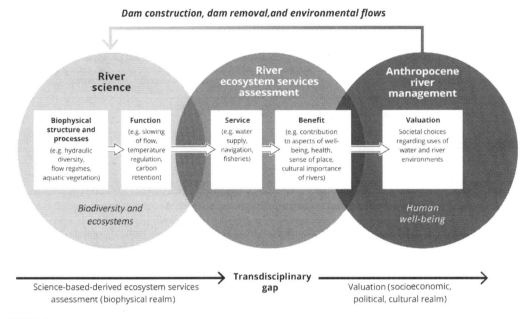

FIGURE 8.3

Adaptation of the Potschin and Haines-Young (2011) ecosystem services cascade model showing its relevance as a conceptual framework for environmental water and ecosystem services assessment.

Environmental flows assessment methodologies have evolved over time (Chapter 11). They were originally developed as hydrological models linked to some measures of hydraulic-habitat availability, either for single target species (e.g., *Physical Habitat Simulation [PHABSIM]*; Milhous et al., 1989), or for overall ecosystem health (e.g., Tennant, 1976). Many of these methodologies are still in widespread use, but gradually socio-cultural and economic components were added on to the assessments (e.g., *Downstream Response to Imposed Flow Transformations [DRIFT]*; King et al., 2003). The instream flow paradigm or what is now more commonly termed the *environmental flow paradigm* started in the Pacific Northwest of the United States in the 1970s and 1980s in relation to concerns over dwindling stocks of salmonids in the face of flow regulation by impoundments, and also the perceived water rights (involving a range of regulating, supporting, provisioning, and cultural ESs) of Native Americans where rivers flowed through their reservations. This latter fact has been partially forgotten in the intervening time period with the science of environmental flow assessments and determination becoming the main focus. However, there is an increasing awareness once again that the whole environmental flows assessment process needs to be conducted within a socio-cultural framework, within which stakeholders at different levels are educated as to the meaning and worth of environmental water regimes, and are engaged through interviews and meetings, to provide information on the priority uses of a river, the most valued goods and services, and to participate in setting the environmental objectives that will guide the assessment of flow requirements for the river (Chapter 7).

Over the next decade, environmental flows assessment may enter a new phase that recognizes this larger context, whereby they are conducted within an integrated biophysical and socio-cultural framework. Such a framework will enable evaluation of how changing river flows impacts the complex array of values of rivers to society and better integrates what stakeholders value on their particular river. Within a "socio-cultural framework" is the influence of politics and governance arrangements, which may need adaptation to fully utilize ES studies. Bennett et al. (2009) highlight the need for an integrated socio-ecological approach to enable identification and assessment of multiple ESs. The need for a better understanding of the relationship between multiple ESs and trade-offs between competing ESs has also been identified by a number of other authors (Baran et al., 2001; Fisher et al., 2009; Palmer and Bernhardt, 2006; Pascual et al., 2010). The approach will also facilitate understanding and particularly awareness of short-term benefits that can often be quantified economically versus less economically tangible long-term environmental and social/cultural impacts. Undertaking environmental flows assessments under the umbrella of natural capital and ESs will provide a more balanced consideration of the economic gains from water usage and that of water allocated for less tangible environmental benefits such as aesthetics and water self-purification.

One of the problems in deciding on the required environmental water regime is that the effects of gradually reducing flows are incremental, rather than catastrophic, at least until extreme thresholds (i.e., complete cessation of flow or dry river beds) are reached. It is therefore seldom possible to identify sudden thresholds, or state changes at flow levels, before irreversible crises will occur. One solution is to identify what Biggs and Rogers (2003) term *thresholds of potential concern (TPC)*. These are measurable monitoring endpoints beyond which there is likely to be an unacceptable ecological condition in the river, but before any irreversible changes have occurred. This may be a reduction in densities of a particularly flow-dependent fish species, the absence of seedlings of a riparian tree, a rise in salinity above 95th percentile levels, and/or the seasonal disappearance of net-spinning *Trichoptera* larvae, and so on. TPCs are set in terms of presently available knowledge, and are then monitored. If any TPC for ESs of importance is exceeded, there must be a management response, which may be to institute research into the possible cause, or simply to adjust the TPC in the light of increased knowledge over time.

8.3 IDENTIFYING AND ASSESSING ECOSYSTEM SERVICES

Different methods of identifying and assessing ES exist and depend on which kind of service is available. Techniques vary; simple methods are ideal for such easily valued services as hydropower production, fisheries, and irrigation water supply, whereas for the less tangible benefits such as recreational and cultural values more complicated methods are needed (Table 8.1).

Standard techniques can be employed in order to identify relevant ESs delivered by rivers, which rely on the experience of specialists (e.g., freshwater ecologists, hydrologists, geomorphologists, water quality and hydraulic experts, etc.). In this manner, there is significant overlap between ES identification and environmental flows assessment. Methods to identify services can include:

1. Desktop assessments;
2. Field surveys;
3. Interviews and community engagement;
4. Detailed modeling exercises;
5. Experimental water releases.

Desktop assessment will afford the identification of the most obvious services, for example large-scale hydropower, irrigation, and water supply, and perhaps large-scale flood regulation works. However, for the more detailed and nuanced services, a mixture of field-based methods and interviews of local people and stakeholders may be necessary. Those services identified through examination of the physical environment (e.g., habitat provision, water purification, and nutrient cycling) will require field-based surveys of river reaches and sectors, specifically mapping physical features that contribute to service provision (e.g., geomorphological features for habitat provision or nutrient levels for nutrient cycling). For the *silent* or hidden services (e.g., cultural, recreational, and health and well-being), a different approach may be necessary. Here community engagement through targeted interviews, workshops, and discussion forums is likely to be necessary in order to appreciate the importance of the river and tap into local knowledge. The cultural value of the river will be dependent upon the beliefs and values of society, and these will not be the same in different parts of the world and may be influenced significantly by the level of economic development.

Once the services are identified, understanding the spatial and temporal variability is central to understanding river ecosystems, and thus ES delivery (Burkhard et al., 2012). Imperative within the spatiotemporal framework, and within the three dimensions of connectivity (longitudinal, horizontal, and vertical), is to identify *hotspots* and *hot moments* of ES delivery. Across the suite of ESs delivered by rivers, a critical component is the identification of reaches on the river network that are ES-*rich* (e.g., Large and Gilvear, 2014) and moments in time that supply abundant ESs. Understanding exactly where and when ESs are most abundant and determination of the correct flow to try to optimize their delivery is highly important. For example, the network dynamics hypothesis (Benda et al., 2004) highlights tributary junctions as being important for biodiversity. Similarly, reaches with backwaters and wetlands (e.g., billabongs on the Murray—Darling River, Australia) are also likely to be important for carbon sequestration. Consideration also needs to be given to how ESs may diminish or be enhanced with the passage of water downstream from the site of regulation to the area of demand.

For each identified ES, there is then the challenge of assessing or measuring the level of provision provided. This can be extremely challenging. One approach to quantifying this can be measuring the delivery through indicators or proxies. For example, fish production can simply be estimated through fish catch, or more indirectly through their associated habitat (e.g., instream vegetation, geomorphological features) and food sources (e.g., macroinvertebrates; Garbe et al., 2016). Table 8.1 suggests some indicators that may be used in freshwater ES delivery monitoring and assessment; these should, however, be tailored for specific rivers and particular service provision. Once established, the ecosystem indicators can be used to establish the influence different flows have on the indicator response, and the TPC can be established. Once more this has significant symmetry to the environment flows assessment methodologies.

To use the established relationships within trade-off assessments may, but not always, require economic valuation of the services. Different techniques exist to quantify the services, for example contingent value methods or willingness to pay. These techniques have significant drawbacks, however, and are known to underestimate the economic value of ecological assets. This can put nature conservation at a disadvantage if the failure to capture the true values means it cannot compete economically with other uses of water. This will be discussed later in relation to the Zambezi River and impacts of flow regulation on wildlife. There is also the need to evaluate both the change in capital value of the ecosystem supplying the service in addition to the *annual return*. Finally, using these relationships, trade-offs can be explored.

As an example, the highly regulated Zambezi River Basin has been subject to environmental flows assessments (e.g., Beilfuss and Brown, 2010) and significant ecological research (e.g., Gope et al., 2014; Ncube et al., 2012; Timberlake, 2000). This has sought to identify key areas of ecosystem importance, and understand their flow dependence. Within a river basin the size of the Zambezi it is clear that several important hot spots exist, which are rich in terms of the ESs they provide, but that have been impacted to differing degrees as a result of hydropower development (a provisioning service). The system is highly regulated with four main dams: Kariba Dam (built 1959) and Cahorra Bassa (built 1974) on the main stem, and Kafue Gorge Dam (built 1973) and Itezhitezhi Dam (built 1977) on the Kafue River (Fig. 8.4). Although there are many areas of importance along the river, perhaps the most critical ecological hot spots can be categorized as the Barotse Plains in the Upper Zambezi (upstream of significant dam-controlled conditions); the Mana floodplains in the Middle Zambezi (downstream of Kariba Dam); the Kafue flats on the Kafue tributary and the delta area. Each of these areas provides a wealth of ESs; for example the flood regulation services of the Barotse Plain are key to the functioning of the river and timing of the water delivery through the basin (regulating service). The Mana floodplains form part of the Mana Pools National Park (Ramsar Site; World Heritage Site), and support a significant large mammal population that is closely linked to the tourism industry in the region (cultural/supporting services). Similarly, the delta supports a wealth of users, dependent on water for its provisioning service (e.g., small-scale subsistence agriculture to coastal and freshwater fisheries), its regulation service (e.g., flow regulation), and its supporting services (e.g., habitat, biodiversity). These hot spots require targeted flows at specific times of the year to maintain the desired ESs.

An interesting example of the concept of hot moments is evidenced by recent environmental flows assessment work on the Ganges River, India (O'Keeffe, 2012), which supports a population of nearly 500 million people and is a critical natural resource. Over several thousand years the river has been important for providing recreational, livelihood, and deep spiritual/cultural ESs, as well as a multitude of domestic, agricultural, industrial, and power generation uses. Climate change and a more unpredictable monsoon in terms of timing and volume along with decreased glacial melt contributions will put further pressures on water resources and the environmental water regime/ES issue. The hot moment in the delivery of ESs for the Ganges was the elevated flow release in 2013 for the 6 weeks of Kumbh, a religious festival occurring once every 12 years, at the confluence of the Ganges and Yamuna Rivers. Kumbh attracts between 80 and 100 million pilgrims to perform ritual bathing in the Ganges (cultural service; Fig. 8.5A and B). The state government agreed to provide the flows (between 200 and 300 m^3/s) by reducing the diversion of water into upstream irrigation canals (provisioning service). Interestingly, farmers who risked having their irrigation flows substantially reduced were proud to be able to contribute to the religious festival. An ecological benefit was that the 6-week period matched very well with the habitat requirements for the endemic river dolphins (Lokgariwar et al., 2013).

The Zambezi River Basin

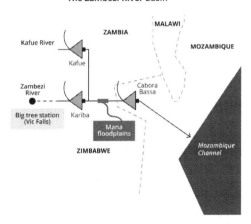

The Middle Zambezi Basin schematic layout

FIGURE 8.4

The geography of the Zambezi River, Africa, and a schematic of dams and regulated flows.

8.4 CHALLENGES OF MEASURING AND VALUING FRESHWATER FLUVIAL ECOSYSTEM SERVICES

ESs are critical to our survival (Costanza et al. 1997; Daily, 1997). However, measuring, valuing, and considering these services in trade-off assessments is fraught with difficulties (Heal, 2000; Momblanch et al., 2016; Seifert-Dahnn et al., 2015), and if the concept is to be applied to environmental flows assessments these must be understood. Several difficulties arise, and have been reported in literature specifically: economic valuation of ES (Grizzetti et al., 2016; Heal, 2000), modeling and methodological approaches (Portman, 2013; Seifert-Dahnn et al., 2015), and handling the uncertainty associated with both biophysical and economic variables (Momblanch et al. 2016; Seifert-Dahnn et al., 2015).

(A)   (B)

FIGURE 8.5

Photographs of the Ganges and Kumbh festival demonstrating (A) the cultural value of the river and (B) the human pressure on the river.

Perhaps, first and foremost, challenges lie with the valuation of ESs, both the usefulness of the concept as well as the method selection can be disputed (Seifert-Dahnn et al., 2015). Valuation can have different connotations; specifically economic valuation versus importance (Heal, 2000), although often stakeholders feel most comfortable with the former (Seifert-Dahnn et al., 2015). Different methods exist to assign value to services, for example direct consumptive uses of water tend to be reasonably easily valued, while the *hidden* or *silent* services (e.g., hidden or silent ESs—not generally appreciated by society, e.g., carbon sequestration, cultural, and recreation) are much harder (Stocker, 2015). Standard typologies for valuation exist in the literature (e.g., Costanza et al., 2011; Grizzetti et al., 2016; TEEB, 2010) and these are summarized by Momblanch et al. (2016) as:

1. *Market value*: Used when goods are delivered by a specific ES (e.g., fish catch);
2. *Production-based*: Applied to services that are used in the production of a good (e.g., water for cooling in energy production);
3. *Cost-based*: Here replacement costs are estimated (e.g., if the ES can be used to prevent damages);
4. *Revealed preference*: Methods in this category expose the added value for some of the hidden services (e.g., recreational value of ES can be tested using travel cost methods that estimate the cost associated with the expense incurred to enjoy it);
5. *Stated preference*: Methods in this category often use willingness-to-pay surveys to expose the value of a particular service to the respondent and then economic methods such as contingent valuation or choice experiments can be used to estimate value;
6. *Benefit transfer*: This is a type of meta-analysis and can be used to apply estimates of ES delivery from one site to another.

Each group of methods introduces uncertainty into the valuation process and the literature reports skepticism on the robustness of methods (Seifert-Dahnn et al., 2015). For example, the values that are captured using the ES concept depend heavily on how this is operationalized and implemented, which is dictated by the method and approach used. Laurans et al. (2013) highlighted the lack of scientific evidence for ES valuation and its role in decision making, as both monetary and nonmonetary valuation of services are important (Grizzetti et al., 2016).

In ES assessment it is important to consider political, social, or ecological issues, which may be weighed alongside economic valuation when a decision is being made regarding the use of water. Political considerations may include the obligations of a state under international conventions. Some states also have formal agreements to ensure that certain quantities of water flow downstream to their neighbors along international rivers. Social considerations may include the decision to maintain traditional river-focused *ways of life*, which depend on river flow such as fishing and flood recession farming, which maintain the social fabric of the indigenous societies, thus effectively giving them a high value. Some say that such considerations are "priceless" and putting monetary values on such issues seems unethical. As such, there are strong arguments that willingness to pay (or to accept payment) should not be the only criterion used to make decisions about riverine ESs and water usage. Also, some believe that we should have moral obligations to species beyond any economic value. The World Conservation Strategy advises against the extinction of species and promotes species diversity to maintain biological stability (and by implication the stability of economic production dependent on biological resources) and to keep options (option value) open for the future. Freshwater environments have experienced the most rapid species decline over recent decades (WWF, 2016). Here, the precautionary principle comes to the fore. The issue of endangered species is developed further in Section 8.6.2. The challenge is therefore how to combine these wider issues that evade monetary valuation with those that can be more readily valued; various methods of multicriteria analysis have been develop for this purpose (e.g., Saarikoski et al., 2016).

Wider challenges also exist. No standard methods exist for ES assessment and this can introduce uncertainty into the assessment process. A wide range of methods currently exist, from the very simple (oversimplified) to the complex. Often these methods do not incorporate the linked nature of ES delivery, for example the trade-offs across spatial and temporal scales (highlighted by this chapter). Methods that address and optimize these trade-offs are thus a useful way forward (Volk, 2013). The lack of consistent methods is considered to be one of the main challenges to the development of ES into wider policy areas (e.g., Volk, 2013). Addressing the uncertainties in both physical process representation and valuation thus remain significant challenges going forward.

8.5 OPPORTUNITIES AND CHALLENGES FOR ENVIRONMENTAL WATER DELIVERY OF ECOSYSTEM SERVICES

The ES concept applied to environmental water regimes provides an opportunity for consideration of the many values that society and scientists assign to rivers. Only by consideration of the full suite of benefits that are accrued from rivers can the inevitable compromises that have to be made be fully appreciated. Decision making to date has been too focused on the economic value of consumptive water use or conservation of individual species, with cultural values, for instance, rarely featuring in the process. Multiple use of rivers needs to be given greater emphasis in future assessments with a move away from simple economic assessment. Such an approach will provide stakeholders better representation in the process and allow the views of citizens of the river to be fully incorporated into decision making. The cultural value of rivers together with the hidden ESs of rivers, however, is still not well developed (Tengburg et al., 2012). Improved methods of taking into account stakeholder views across the suite of supporting, regulating, provisioning, and cultural ESs will also be required within an ecosystems service framework where difficult trade-off decisions are going to be made.

Environment water regimes and ES interactions present enormous challenges for the science of understanding how flow modifications impact whole ecosystem functioning, and identification of hot spots and hot moments. The uncertainty in flow assessments with metaphorically large error bars on the flow recommendation lends itself to the idea of TPC and adaptive flow management strategies. In such circumstances, monitoring of indicators of ES values will become important. Climate change impacts on flow will add another dimension of uncertainty highlighting further that an adaptive approach will be required.

8.6 AFRICAN CASE STUDIES IN ECOSYSTEM SERVICE AND ENVIRONMENTAL FLOWS ASSESSMENT RESEARCH AND IMPLEMENTATION

8.6.1 ZAMBEZI RIVER

As described in Section 8.3, the Zambezi River is a highly regulated system, with many competing sectoral demands for water, the most dominant currently being hydropower. The river rises in Zambia and flows through nine riparian countries along its 2750 km length. The basin covers 1.39 Mkm^2 and outfalls into the Indian Ocean in Mozambique, where the river is characterized by a wide, flat, marshy area with extensive floodplains (Tilmant et al., 2010). There are three distinct geographical stretches: the Upper Zambezi from its source to Victoria Falls (including the Barotse Plains), the Middle Zambezi from Victoria Falls to Cahora Bassa (including the main tributary the Kafue River, the Kafue Flats, and the Mana Pools), and the Lower Zambezi from Cahora Bassa to the delta (Fig. 8.4).

Many studies have been undertaken on the ecology and associated ESs of the Zambezi River Basin and, according to Timberlake (2000), the existing dams (Kariba, Cahora Bassa, Kafue, and Itezhitezhi) pose a risk to particular areas in the basin including the Kafue Flats (upstream of Kafue Gorge Dam), Middle Zambezi floodplains (downstream of Kariba), and the Lower Zambezi floodplains and delta (downstream of Cahora Bassa). Several studies have been conducted on the environmental water requirements of the delta (e.g., Beilfuss and Brown, 2010), which identify that the current flows to the delta have become near-constant during the year with a mean annual post-regulation discharge of 3800 m^3/s. Reported bankfull discharges are of the order of 4500 m^3/s. The environment flows assessment suggests several different target pulse flows for the peak flow season in February and March, for example up to 7000 m^3/s. Providing this pulse flow during the wet season aims to restore part of the seasonal component of the flow regime, which is important to the ecosystem. During predamming this seasonal flow was achieved 50% of the time and resulted in the total inundation of mid-channel islands and the floodplain (Tilmant et al., 2010).

Similarly, studies undertaken on the Mana Pools have investigated the link between certain ESs and the flow regime (Gope et al., 2014; Ncube et al., 2012). The main tourist attraction of the Mana floodplains is associated with the dry season high wildlife densities on these floodplains and the Zambezi River, which creates a unique wilderness (Du Toit, 1983). During the dry season (August—October), the wildlife moves toward these floodplains where there are extensive grasslands and *Faidherbia albida* (acacia) foliage (Jarman, 1972). *F. albida* becomes an important source of foliage in the dry season as it is in full fruit and leaves (Dunham, 1989), and this is often a critical time for wildlife survival as there will be limited foliage elsewhere (Barnes and Fagg, 2003). *F. albida*'s reverse phenology, in which it is leafless during the summer wet season (November—March) and in leaf in the winter dry season (May—October), distinguishes it from other acacia species. Studies have investigated the evolution of the *F. albida* stand in the Mana Pools park to understand its links to the influence of the regulated flow regime (supporting and provisioning services), and the subsequent impacts to large mammal populations (supporting services), and hence tourism as a cultural service (Gope et al., 2014; Ncube et al., 2012). These studies have concluded that the flows (both high and low) regulated by the Kariba Dam have adversely impacted the downstream ecology, changing the population dynamics of the *F. albida* stand structure, hence influencing the large mammal population and by extension the tourist potential of the region.

In further studies of the river basin, other authors have tried to tie these competing intersectoral requirements (e.g., irrigation—a growing requirement [provisioning service], hydropower [provisioning service], and ecology [providing a range of services from supporting and regulating through to cultural as detailed in Section 8.2]) for water together to understand the trade-offs (both temporal and spatial) and marginal water values in the context of the water, food, energy nexus (Tilmant et al., 2010, 2011). These studies have used approaches, which apply neoclassical economic theory to the allocation of water at the basin scale, that require economic estimates of each water use. This is reasonably straightforward in the case of irrigation and hydropower demands, but much more complex for ecosystem requirements (Tilmant et al., 2011). The studies show that for the ecosystem requirements, allocation of water to support some of the hot spots is more likely than others (i.e., further upstream, and at particular times). When economic estimates are attached to the ecosystem requirements, these estimates may not be significant enough to force trade-offs toward the environment and away from hydropower. If we bring together this work on the Zambezi, we can suggest a few conclusions for the consideration of environmental flows assessments:

1. Current methods attaching economic value to freshwater systems are complex and their results uncertain. This may lead to the undervaluation of freshwater systems, and thus artificially reduce their importance in allocation decisions.
2. Linking the required flows for the desired ecosystem function and the subsequent ES upon which people depend may be a useful tool. Understanding the linked ESs can provide context to stakeholders and increase understanding and therefore value.
3. Spatial and temporal considerations are crucial at the basin scale. In the context of ESs, areas and times that are rich in ES delivery (hot spots and hot moments) are inevitable and understanding these can inform true trade-off discussions. Methods that allow the unbiased trade-offs to be quantified require some development.

8.6.2 BALANCING WATER RESOURCE USE AND ECOSYSTEM SERVICES WITHIN AREAS OF NATURE CONSERVATION VALUE

As part of *Worldwide Fund for Nature (WWF)* river basin projects environmental flows assessments (LVBC and WWF-ESARPO, 2010; O'Keeffe, 2012; WWF-TCO, 2010) were carried out on the Mara River (Kenya and Tanzania from 2006 to 2008) and the Kihansi and the Great Ruaha River (Tanzania from 2007 to 2009). All three are East African rivers, both the Mara River and Great Ruaha flow into national parks, and each demonstrates contrasting lessons.

The Mara River

The Mara River flows from the Mau Escarpment in central Kenya, through natural forest that has been cleared extensively in the past 20 years, through irrigated farmland, the grazing lands of the Masai, and then into the Masai Mara game reserve, Kenya, and the Serengeti National Park, Tanzania. The river provides the only perennial water in the conservation areas, from which it flows in a westerly direction, through extensive wetlands, into Lake Victoria. The river flows have been modified historically, both inadvertently and deliberately by catchment deforestation, irrigation abstraction, and are now threatened by plans for further abstraction and interbasin transfer from the upper Kenyan parts of the basin. Water quality has also deteriorated, as a result of the increasing number of tourist camps and hotels, which use water from the river, and dispose of their wastewater back into it. The flows, however, are not impounded substantially by dams, and the overall flow regime remains substantially natural.

The conservation areas of the Masai Mara and Serengeti provide in the region of 30% of the foreign exchange earnings for each country, through tourism revenue (provisioning, supporting, and cultural services). However, the bulk of the river flows are generated upstream of the conserved areas, mainly in the Mau Escarpment, a naturally forested upland area that has been mostly cleared and settled for agriculture during the past three decades (provisioning service). In addition, the main commercial irrigated agriculture (provisioning), and planned future developments are also in the basin upstream of the conserved areas. There is therefore a mismatch between the major economically important ecological goods and services (in the conserved areas) and the source of the flow changes and reductions—flashier flows from the cleared uplands and irrigation abstractions from the farmland.

A process of economic compensation for ecological services has been suggested, in which the tourist areas compensate upstream settlers and farmers for reduced development, reafforestation, off-channel storage, and so on. Currently, the Masai communities, who lost traditional grazing land to the Masai Mara, receive 18% of the gate entrance fees in compensation. Nevertheless, there has been resistance to further proposals for compensation from tourism interests, and it would probably require government intervention to promote it.

The Great Ruaha

The Great Ruaha River rises in central Tanzania, provides irrigation water to an area of intensive rice farming (provisioning service), then flows into the Usangu wetlands, which have recently been included in the Ruaha National Park (supporting). A natural rocksill across the river backs water into the wetlands, varying in area from $200\,km^2$ during the dry season to $600\,km^2$ in the wet season. During the dry season, the wetlands act as a refuge area for wildlife. From the wetlands, the river flows down into the Ruaha National Park, into the Mtera Dam, used to generate hydropower for Dar es Salaam (provisioning). The river flow regime is monsoonal and highly seasonal, with wet season flows replenishing the impoundment. The dry season flows, reduced naturally by evapo-transpiration from the wetlands, have been drying up more frequently and for longer over the last two decades. This is due primarily to extensive irrigation upstream for rice but possibly other land-use changes. The river within the national park has been reduced to an average of one pool per kilometer by the end of the dry season. The result, apart from a serious reduction in fish and inver-tebrates, has been the concentration of hippos and crocodiles in the remnant pools, resulting in increased aggressive animal interactions, as well as a concentration of large mammals around the pools. This congregation of animals in the pools and overgrazing surrounding the pools has resulted in water quality deterioration and channel erosion and long term is likely to reduce animal num-bers. This could have severe repercussions in terms of the tourist revenues brought in to the local community.

The major lessons that emerged from the work completed on the Ruaha suggested that it was essential to employ a proactive integrative approach, identifying all the services and involving the stakeholders who relate to these, in order to manage the river in a sustainable manner (O'Keeffe, 2012). When a river is already over-allocated in terms of flow, compromise short-term solutions may be required to restore some flow and minimize long-term irreversible damage. However, these still require negotiation involving the lost opportunities for certain services (e.g., rice cultivation in this scenario), the benefits for the downstream river and the Ruaha National Park (supporting ser-vices), and the reduction of the wetland area during the dry season (supporting services). In such cases, a time-consuming environmental flows assessment is not the priority. Where there is no flow in a formally perennial river, it is obvious that some flow needs to be restored. The precise flow requirements and potentially the economic trade-offs can be adjusted in the future.

Eventually, the success of environmental water regimes, as with all other water management issues, becomes a societal choice, dependent on political, economic, and social factors. The require-ment is to balance the losses in revenue, job opportunities, and livelihoods from reduced rice pro-duction, against the preservation of national and international heritage values, and tourism revenue from the national park, and against the loss of ecological goods and services from reduced wetland area during the dry season. In the longer term, investment in more efficient irrigation could provide for increased rice production, improved downstream river flow, and the maintenance of wetland area, but this would require initial development costs or government incentives for farmers to improve irrigation.

Even with the best scientific information available, environmental water regimes will only be implemented where there is a general understanding of, and support for, the importance of sustain-ing flows for ESs (i.e., wildlife for nature conservation), where there is the political will to use riv-ers within sustainable limits, and where there is a recognition of the long-term economic benefits of protecting water resources, rather than the short-term maximization of consumptive uses.

The Kihansi River

The Kihansi River in central Tanzania has similar issues relating to ES delivery challenges. In July 1995, the Government of Tanzania began construction of the 180-MW Lower Kihansi Hydropower Project (provisioning service) in order to meet the growing electricity demands of its mining and tourism industries (provisioning, supporting, and cultural services). A 25-m high dam situated upstream of the waterfall collects water, diverts it into a series of tunnels running into and out of the power plant, and returns the water to the river at the bottom of the gorge, 6 km downstream. There are reduced flows in the intervening 6 km reach impacting habitat provision. The Kihansi spray toad, *Nectophrynoides asperginis*, is endemic to a 2-ha area at the base of a waterfall (Channing et al., 2006) within the Kihansi Gorge reach of the river. Despite attempts to maintain spray habitat through the installation of artificial spray, the spray toad, first discovered in 1996 (Poynton et al., 1999), disappeared from its natural habitat by 2003. This was probably due to a number of factors: reduction of the spray habitat due to flow reduction; an introduced chytrid fungus (known to have caused extinctions in amphibians worldwide and detected in the gorge), and high concentrations of endosulfan-based pesticides (Hirji and Davis, 2009). In the meantime, captive breeding programs had been undertaken at the Bronx and Toledo zoos in the United States, and the toads were returned to the Kihansi Gorge in 2012. After negotiations, a final water right was granted by the river basin water board. This included an environmental requirement of $1.5-2.0$ m^3/s (Hirji and Davis, 2009).

The spray toad is an interesting and important endemic species, but it could be argued that basing the environmental water regime entirely on the survival of this species in this African setting is unjustified, when first it is threatened by other pressures and when, during dry periods, there may be a conflict between electricity supply to major centers. This trade-off for the continued provision of artificial habitat for a reintroduced toad species may not really reflect the priorities of the local communities or of national policy. If we refer back to the argument of placing flow requirements into the larger socio-political landscape, the position of the International Union for Conservation of Nature regarding endangered species (i.e., moral obligation) can help in this case. For example, an integrated river ecosystem-ESs and trade-off analysis framework may have led to a different outcome from that of conservation. A recent study of flow requirements for the entire Kilombera floodplain (McClain et al., in prep) has included studies of many biophysical components and processes, and an extensive investigation of livelihoods and important ecological goods and services, through meetings with institutional and direct use stakeholders, and more than 700 interviews. This will provide a much more convincing motivational basis for continued implementation of sustainable flows for the Kihansi and the rest of the basin.

8.7 CONCLUSION AND WAY FORWARD

This chapter has examined environmental water within a natural capital and ES framework and more generally sustainable water resource management in the 21st century.

It has demonstrated that this framework is useful in that it can provide a convincing basis for consideration of the compromises that need to be made between increasing consumptive water use and loss of other benefits that rivers bring to society through ESs. It suggests that the consideration of environmental water regimes as important natural capital enables river ecosystems to be considered within wider economic frameworks that are frequently used by businesses and government decision makers. It has also emphasized that implementation of environmental water regimes has to be undertaken in a socio-economic framework because ultimately people who work and live on rivers will be the people who control whether environmental water regimes are implemented.

Important future developments for environmental flows assessments should include a concentration on holistic methodologies for the quantitative valuation of flow-related ecosystem goods and services, perhaps using the processes suggested by Costanza et al. (2014). At present, the flow environment frequently loses out to consumptive water uses (generally provisioning services), whose short-term financial profits can easily be quantified, although the long-term nonfinancial benefits of maintaining flows in the river are difficult to quantify and compare. In this regard, traditional economic models discount benefits into the future, so that long-term sustainable benefits are less valued than immediate profits, regardless of consequent degradation.

The Indian and East African examples used in this chapter highlight two valuation issues that are fundamental to quantifying the ESs of rivers and the role of flow regulation: in the Ganges the spiritual value of the river as the holy mother river of India is inestimable, but nevertheless the river is usually reduced to a polluted trickle during the dry season. Traditional ES assessments would unlikely prevent this. The Kumbh festival, for example, demonstrates that, given the motivation, the government and most people (including the farmers who bore the costs) value healthy flows in the river very highly, but it would be extremely difficult to quantify that value realistically. The Kihansi case study in Africa emphasizes that even environmental and conservation values are extremely variable, so that quantifying or comparing the value of a rare endemic in the European, American, or Australian context will be quite different from that in a developing country, where the priorities are more basic. Societal values put on healthy rivers vary geographically and decisions made on environmental water need to be careful to appreciate this fact.

As long ago as the 4th-century BC, Plato recognized the value of healthy flows, but we do not seem to have progressed much since then in our ability to conserve them: "What now remains of the formerly rich land is like the skeleton of a sick man with all the fat and soft earth having wasted away and only the bare framework remaining. Once the land was enriched by yearly rains, which were not lost, as they are now, by flowing from the bare land into the sea. The soil was deep, it absorbed and kept the water …. and the water that soaked into the hills fed springs and running streams everywhere!"

REFERENCES

Acreman, M.C., 2001. Ethical aspects of water and ecosystems. Water Policy J. 3 (3), 257–265.

Acreman, M.C., Harding, R.J., Lloyd, C., McNamara, N.P., Mountford, J.O., Mould, D.J., et al., 2011. Trade-off in ecosystem services of the Somerset Levels and Moors wetlands. Hydrol. Sci. J. 56 (8), 1543–1565.

Arthington, A.H., 2012. Environmental Flows: Saving Rivers in the Third Millennium. University of California Press, Berkeley, California.

Arthington, A.H., Bunn, S.E., Poff, N.L., Naiman, R.J., 2006. The challenge of providing environmental flow rules to sustain river ecosystems. Ecol. Appl. 16, 1311–1318.

Auerbach, D.A., Deisenroth, D.B., McShane, R.R., McClunet, K.E., Poff, L.N., 2014. Beyond the concrete: accounting for ecosystem services from free-flowing rivers. Ecosyst. Serv. 10, 1–5.

Baran, E., van Zalinge, N., Peng Bun, N., Baird, I., Coates, D., 2001. Fish resource and hydrobiological modelling approaches in the Mekong Basin. ICLARM, Penang, Malaysia and the Mekong River Commission Secretariat, Phnom Penh, Cambodia.

Barnes, R.D., Fagg, C.W., 2003. *Faidherbia albida*: monograph and annotated bibliography. Tropical Forestry Papers 41. Oxford Forestry Institute, Oxford, UK.

Beilfuss, R., Brown, C., 2010. Assessing environmental flow requirements and trade-offs for the Lower Zambezi River and Delta, Mozambique. Int. J. River Basin Manage. 8 (2), 127–138. Available from: http://dx.doi.org/10.1080/15715121003714837.

Benda, L., Poff, N.L., Miller, D., Dunne, T., Reeves, G., Pess, G., et al., 2004. The network dynamics hypothesis: how channel networks structure riverine habitats. BioScience 54 (5), 413–427.

Bennett, E.M., Peterson, G.D., Gordon, L.J., 2009. Understanding relationships among multiple ecosystem services. Ecol. Lett. 12 (12), 1394–1404. Available from: http://dx.doi.org/10.1111/j.1461-0248.2009.01387.

Biggs, H.C., Rogers, K.H., 2003. An adaptive system to link science, monitoring and management in practice. In: Du Toit, J.T., Rogers, K.H., Biggs, H.C. (Eds.), The Kruger Experience: Ecology and Management of Savanna Heterogeneity. Island Press, Washington, DC, pp. 59–80.

Burkhard, B., Kroll, F., Nedkov, S., Muller, F., 2012. Mapping ecosystem service supply, demand and budgets. Ecol. Indic. 21, 17–19.

Channing, A., Finlow-Bates, S., Haarklaum, S.E., Hawkes, P.G., 2006. The biology and recent history of the critically endangered Kihansi Spray Toad *Nectophrynoides asperginis* in Tanzania. J. East African Nat. Hist. 95 (2), 117–138.

Costanza, R., D'Arge, R., Groot, R.D., Farber, S., Grasso, M., Hannon, B., et al., 1997. The value of the world's ecosystem services and natural capital. Nature 387, 253–260.

Costanza, R., Kubiszewski, I., Ervin, D., Bluffstone, R., Boyd, J., Brown, D., et al., 2011. Valuing ecological systems and services. F1000 Biol. Rep. 3, 14.

Costanza, R., Kubiszewski, I., Giovannini, E., Lovins, H., McGlade, J., Pickett, K.E., et al., 2014. Time to leave GDP behind. Comm. Nat. 505, 283–285.

Daily, G., 1997. Nature's Services: Societal Dependence on Natural Ecosystems. Island Press, Washington, DC.

Du Toit, R.F., 1983. Hydrological changes in the Middle Zambezi System. Zimbabwe Sci. News 17, 121–126.

Dugan, P. (Ed.), 1990. Wetland Conservation: a Review of Current Issues and Required Action. IUCN, Gland, Switzerland.

Dunham, K.M., 1989. Long-term changes in Zambezi riparian woodlands, as revealed by photopanoramas. African J. Ecol. 27, 263–275.

Ehrlich, P., Kareiva, P., Daily, G., 2012. Securing natural capital and expanding equity to rescale civilization. Nature 486, 68–73. Available from: http://dx.doi.org/10.1038/nature11157.

Ehrlich, P.R., Ehrlich, A.H., 1981. Extinction: the Causes and Consequences of the Disappearance of Species. Random House, New York.

Emerton, L., Bos, E., 2004. Value. Counting Ecosystems as an Economic Part of Water Infrastructure. IUCN, Gland, Switzerland and Cambridge, UK.

Fisher, B., Turner, R.K., Morling, P., 2009. Defining and classifying ecosystem services for decision making. Ecol. Econ. 68, 643–653.

Forum for the Future, 2015. The Five Capitals Model—A Framework for Sustainability. Forum for the Future, London.

Garbe, J., Beevers, L., Pender, G., 2016. The interaction of low flow conditions and spawning brown trout (*Salmo trutta*) habitat availability. Ecol. Eng. 88, 53–63.

Gilvear, D.J., Greenwood, M., Thoms, M.C., Wood, P., 2016. River Science; Research and Application for the 21st Century. Wiley.

Goor, Q., Halleux, C., Mohamed, Y., Tilmant, A., 2010. Optimal operation of a multipurpose multireservoir system in the Eastern Nile River Basin. Hydrol. Earth Syst. Sci. 14, 1895–1908. Available from: http://dx.doi.org/10.5194/hess-14-1895-2010.

Gope, E., Sass-Klasson, U., Irvine, K., Beevers, L., Hes, E., 2014. Effects of flow alteration on Apple-ring Acacia (*Faidherbia albida*) stands, Middle Zambezi floodplains, Zimbabwe. Ecohydrology 201. Available from: http://dx.doi.org/10.1002/eco.1541.

Grizzetti, B., Lanzanova, D., Liquete, C., Reynaud, A., Cardoso, A., 2016. Assessing water ecosystem services for water resource management. Environ. Sci. Pol. 61, 194–203.

Heal, G., 2000. Valuing ecosystem services. Ecosystems 3, 24–30.

Hirji, R., Davis, R., 2009. Environmental Flows in Water Resources Policies, Plans, and Projects. Case Studies. Paper Number 117, Environment Department Papers, Natural Resource Management Series. The World Bank Environment Department.

Jarman, P.J., 1972. Seasonal distribution of large mammal populations in the unflooded middle Zambezi Valley. Appl. Ecol. 9, 283–299.

King, J., Brown, C., Sabet, H., 2003. A scenario-based holistic approach to environmental flow assessments for rivers. River Res. Appl. 19, 619–639.

Large, A.R.G., Gilvear, D.J., 2014. Using GoogleEarth, a virtual-globe imaging platform for ecosystem services-based river assessment. River Res. Appl. Available from: http://dx.doi.org/10.1002/rra.2798.

Laurans, Y., Rankovic, A., Billé, R., Pirard, R., Mermet, L., 2013. Use of ecosystem services economic valuation for decision making: questioning a literature blindspot. J. Environ. Manage. 119, 208–219.

Lokgariwar, C., Chopra, R., Smakhtin, V., Bharati, L., O'Keeffe, J., 2013. Including cultural water requirements in environmental flow assessment: an example from the upper Ganga River, India. Water Int. Available from: http://dx.doi.org/10.1080/02508060.2013.863684.

LVBC and WWF-ESARPO, 2010. Assessing reserve flows for the Mara River, Nairobi and Kisumu, Kenya. Eastern and Southern Africa Regional Programme Office Lake Victoria Basin Commission. Unpublished report of WWF.

Mace, G., Norris, K., Fitter, A.H., 2012. Biodiversity and ecosystem services: a multilayered relationship. Trend. Ecol. Evol. 27 (1), 19–26.

Maltby, E., Acreman, M.C., 2011. Ecosystem services of wetlands: pathfinder for a new paradigm. Hydrol. Sci. J. 56 (8), 1–19.

McClain, M., Tharme, R., O'Keeffe, J., et al., (in prep). Environmental flows in the Rufiji River Basin, assessed from the perspective of planned development in the Kilombera and Lower Rufiji sub-basins. Report to USAID.

MEA, 2005. Ecosystems and Human Well-being: Synthesis. Millennium Ecosystem. Assessment. Island Press, Washington, DC.

Milhous, R.T., Updike, M.A., Schneider, D.M., 1989. Physical habitat simulation system reference manual—Version 2. Instream flow information paper 26. USDI Fish and Wildlife Services. Biol. Rep. 89, 16.

Momblanch, A., Connor, J., Crossman, N., Paredes-Arquiola, J., Andreu, J., 2016. Using ecosystem services to represent the environment in hydro-economic models. J. Hydrol. 538, 293–303.

Natural Capital Committee, 2015. The state of natural capital protecting and improving natural capital for prosperity and wellbeing. Third report to the Economic Affairs Committee. Natural Capital Committee, London.

Ncube, S., Beevers, L., Hes, E., 2012. The interactions of the flow regime and the terrestrial ecology of the Mana floodplains in the middle Zambezi River Basin. Ecohydrology. Available from: http://dx.doi.org/10.1002/eco.1335.

Notter, B., Hurni, H., Wiesmann, U., Abbaspour, K.C., 2012. Modelling water provision as an ecosystem service in a large East African river basin. Hydrol. Earth Syst. Sci. 16 (1), 69–86.

O'Keeffe, J.H., 2012. Environmental flow allocation as a practical aspect of IWRM. In: Boon, P.J., Raven, P. J. (Eds.), River Conservation and Management. Wiley-Blackwell, pp. 43–55.

Palmer, M.A., Bernhardt, E.S., 2006. Hydroecology and river restoration: ripe for research and synthesis. Water Resour. Res. 42, W03S07.

Pascual, U., Muradian, R., Brander, L., Gómez-Baggethun, E., Martín-López, M., Verman, M., et al., 2010. The economics of valuing ecosystem services and biodiversity. In: Kumar, P. (Ed.), The Economics of Ecosystems and Biodiversity Ecological and Economic Foundations. Earthscan, London and Washington, DC.

Portman, M., 2013. Ecosystem services in practice: challenges to real world implementation of ecosystem services across multiple landscapes — a critical review. Appl. Geogr. 45, 185–192.

Potschin, M., Haines-Young, R., 2011. Ecosystem Services: Exploring a geographical perspective. Prog. in Phys. Geogr. 35 (5), 575–594.

Poynton, J.C., Howell, K.M., Clarke, B.T., Lovett, J.C., 1999. A critically endangered new species of Nectophrynoides (Anura, Bufonidae) from the Kihansi Gorge, Udzungwa Mountains, Tanzania. African J. Herpetol. 47, 59–67.

Saarikoski, H., Barton, D.N., Mustajoki, J., Keune, H., Gomez-Baggethun, E., Langemeyer, J., 2016. Multi-criteria decision analysis (MCDA) in ecosystem service valuation. In: Potschin, M., Jax, K. (Eds.): OpenNESS Ecosystem Services Reference Book. EC FP7 Grant Agreement no. 308428. Available from: www.openness-project.eu/library/reference-book.

Seifert-Dahnn, I., Barkved, L., Interwies, E., 2015. Implementation of the ecosystem service concept in water management — challenges and ways forward. Sustain. Water Qual. Ecol. 5, 3–8.

Stocker, T.F., 2015. The silent services of the world's oceans. Science 350, 764–765.

TEEB, 2010. In: Kumar, P. (Ed.), The Economics of Ecosystems and Biodiversity Ecological and Economic Foundations. Earthscan, London and Washington, DC.

Tengberg, A., Fredholm, S., Eliasson, I., Knez, I., Saltzman, K., Wetterberg, O., 2012. Cultural ecosystem services provided by landscapes: assessment of heritage values and identity. Ecosyst. Serv. 2, 14–26.

Tennant, D.L., 1976. Instream flow regimes for fish, wildlife, recreation and related environmental resources. In: Orsborne, J., Allman, C. (Eds.), Instream Flow Needs, volume 2. American Fisheries Society, Western Division, Bethesda, Maryland, pp. 359–373.

Tilmant, A., Beevers, L., Muyunda, B., 2010. Restoring a flow regime through the coordinated operation of a multireservoir system — The case of the Zambezi River Basin. Water Resour. Res. 46, W07533. Available from: http://dx.doi.org/10.1029/2009WR008897.

Tilmant, A., Kinzelbach, W., Juizo, D., Beevers, L., Senn, D., Cassarotto, C., 2011. Economic valuation of benefits and costs associated with the coordinated development and management of the Zambezi river basin. Water Policy. Available from: http://dx.doi.org/10.2166/wp.2011.189.

Timberlake, J.R., 2000. Biodiversity of the Zambezi Basin Wetlands. IUCN ROSA, Harare, Zimbabwe.

Volk, M., 2013. Modelling ecosystem services — challenges and promising future directions. Sustain. Water Qual. Ecol. 1–2, 3–9.

Vörösmarty, C.J., McIntyre, P.B., Gessner, M.O., Dudgeon, D., Prusevich, A., Green, P., et al., 2010. Global threats to human water security and river biodiversity. Nature 467 (7315), 555.

WWF, 2016. The Living Planet Report. Available from: http://assets.wwf.org.uk/custom/lpr2016/.

WWF-TCO, 2010. Assessing environmental flows for the Great Ruaha River, and Usangu Wetland, Tanzania. World Wildlife Fund-Tanzania Country Office. Report of WWF.

Ziv, G., Baran, E., Nam, S., Rodriguez-Iturbe, I., Levin, S.A., 2012. Trading-off fish biodiversity, food security and hydropower in the Mekong River basin. Proc. Natl. Acad. Sci. 109 (15), 5609–5614.

HOW MUCH WATER DOES A CULTURE NEED? ENVIRONMENTAL WATER MANAGEMENT'S CULTURAL CHALLENGE AND INDIGENOUS RESPONSES

Sue Jackson

Griffith University, Nathan, QLD, Australia

9.1 INTRODUCTION

Twenty years ago, aquatic ecologists Richter et al. (1997) authored an article titled *How much water does a river need?* to provoke the scientific community to better appreciate the complexity of aquatic ecosystem processes, functions, and interactions in its efforts to give priority of water to river ecosystems. This chapter recasts this question to assist the environmental water management sector—its scientists, policymakers, managers, and supporting nongovernmental organizations—to better appreciate the social and cultural complexity of people's relationships with water and, in particular, the multifaceted dependence of indigenous peoples on river ecosystems. A further aim is to critique the way in which indigenous values and interests relating to water are treated in Australian policy, discourse, and scientific practice.

The article by Richter et al. (1997) was motivated by an awareness that reductionist approaches to river management and restoration were on the rise in response to escalating rates of river regulation and flow regime alteration, and that these were at risk of failing to protect hydrological variation and integrity in water negotiations and regulatory proceedings. The rhetorical question was posed to deliberate on the "chasm between applied river management and current theories of aquatic ecology," one that they hoped their techniques could help to bridge. It was subsequently taken up as a sort of catch-cry for a new paradigm of river management in which instream or environmental flows assessment tools proliferated (Postel and Richter, 2012). New approaches challenged the utility of such a simplistic question in defining and scoping scientific inquiry into the relationships between the physical and biotic health of riverine systems, fisheries, human health, and the security of livelihoods (see Arthington, 2012).

In many years of social research conducted with indigenous communities in Australia, I too, have observed theoretical weaknesses in the management and governance of water, but the weaknesses observed pertain to understandings of the social and cultural significance of water and rivers—how people relate to and through water—and the nature of cultural differences in those

Water for the Environment. DOI: http://dx.doi.org/10.1016/B978-0-12-803907-6.00009-7

ways of relating. An impulse toward reductionism among those involved in the environmental management sector is one consequence of popular usage and understandings of *culture*. This is manifest in the tendency to reduce the complex challenge of meeting indigenous water requirements, including access to water and participation in decision making, to the question: *How much water does a culture need*? The ways in which indigenous water values, rights, and interests are framed within environmental water management has a number of other consequences, which will be described. In considering the implications, the responses, and counterstrategies being articulated and deployed by indigenous groups and their supporters will also be analyzed.

In order to support the argument, this chapter will draw heavily on recent Australian debates and policy development relating to indigenous water rights and values in the context of an historic reallocation of water to the environment, the existence of complex water marketing institutions, and a flourishing environmental water management sector. In their water rights struggles, indigenous groups face what is referred to here as environmental water management's cultural imperative or challenge. In this context, the cultural challenge has three distinguishable dimensions. First, water management processes typically emphasize reified and objectified aspects of indigenous relationships with river systems and thereby create a dichotomy between the material (e.g., environmental and economic) and a separate symbolic sphere of meaning (e.g., belief and value), with the latter understood to be *cultural*. Second, policymakers, practitioners, and researchers tend to reduce complex forms of indigenous attachment to and uses of water to a simplistic notion of those deemed *traditional*, or supportive of precolonial *essences* of indigenous identity. This *essentializing* tendency has been observed elsewhere and in other water resource development contexts (e.g., Babidge, 2015; Perreault, 2008; Prieto, 2014). Third, the values and perspectives of the nonindigenous society or culture, which have a significant bearing on allocation frameworks, assessment methodologies, and resulting water-sharing decisions, tend to be hidden from view in the parochial *Western* paradigm of resource management. Culture is thought of as an attribute of ethnic or indigenous minorities, not the *mainstream* society (Head et al., 2005). Allied to this is the idea that culture occupies a separate sphere; that it does not pervade all our lives and institutions, including scientific ones. In the Australian context, framing indigenous water issues or concerns in the ways described here relegates indigenous interests to a realm of negligible significance to the political economy of regional agricultural development (Jackson, 2006, 2008, 2017, in press) and marginalizes them within environmental evaluations and technical processes, such as environmental flows assessments (see also Finn and Jackson, 2011; Jackson and Langton, 2012). Indigenous peoples must respond to these ways of defining and deploying culture in their water rights struggles.

This chapter is structured as follows. In the next section, water's role in social and cultural life is briefly introduced, with a focus on the meaning of water to indigenous societies. This section also includes a discussion about research that links hydrological flow to elements in indigenous religious life and economic systems. In the section to follow, the major developments in Australian water governance reform are provided by way of context for the analysis to follow. Attention is given in this contextual section to the *Murray–Darling Basin (MDB)*, which is the country's agricultural heartland. The chapter then turns to the institutions and negotiating arenas through which indigenous peoples seek to establish the legitimacy of their water-related values, rights, ethics, and practices as well as to define, increase, and control their access to water. This penultimate section examines in more detail the cultural challenge facing environmental water management. The final section provides concluding comments.

9.2 **WATER CULTURES AND INDIGENOUS WATERSCAPES**

Human societies developed around water and, in doing so, forged meanings that continue to be ascribed to water and its flow. Numerous religious systems and poetic and musical forms show the human capacity to revere rivers and celebrate symbols or rituals relating to water. In the deified Ganges River, Lokgariwar et al. (2014, p. 82) show that the "rituals of and philosophies of riparian communities are synchronic with and reflective of the natural rhythms" of their river. The material connections and affective attachments between people and rivers are eternal; indeed, the source of life itself is attributed to water in many cultures (Barber and Jackson, 2014; Klaver, 2012; Palmer, 2015; Russo and Smith, 2013). Throughout time, learned patterns of behavior have transformed water, shaping water access and use, as well as the institutional principles that mediate human—water interactions, generating both conflict and consensus (Johnston et al., 2012).

At this point it is worthwhile reflecting more closely on the meaning of culture, which is a widely used term that even in technical academic analysis has a range of meanings (Barber and Jackson, 2011). In natural resource management discourse a term such as cultural values is rarely defined, despite being an object of growing interest. Culture is typically defined as a "'collective' phenomenon that is approximately 'shared' among members of a culture" (Fischer, 2009, p. 29). Social researchers characterize culture as a shared meaning system that is learned, not transmitted genetically (Fischer, 2009). Social researchers tend to focus on the social processes by which humans generate and transmit shared meanings and understandings of the world through "communication of key symbols, ideas, knowledge, and values between individuals from one generation to the next" (Fischer, 2009, p. 29). In the context of this article, "values" is an important word too. Strang (1997, p. 178), an anthropologist who has researched the meaning of water to many different social groups, offers an explanation of how values are formed:

> Beliefs and values are received, inculcated and passed on through a process of socialisation that creates a culturally specific relationship with the environment. This process consists of several elements: the creation of categories, the learning of language, and the acquisition and dissemination of cultural knowledge. Each involves an interaction with the physical, social and cultural environment and contributes to the formation of individual and cultural identity.

In light of these theoretical insights, it is clear that all human groups have culture, or create cultural forms and processes, and are socialized to think about land, water, and nature in particular ways (Barber and Jackson, 2011; Head et al., 2005).

Although considerable literature construes water as an object of value to human communities, scholars from geography and anthropology conceive of water as an integral part of social and political relationships, emphasizing the fact that "rather than being imposed, water's meanings are emergent from these relationships" (Krause and Strang, 2016, p. 634; see also Budds and Hinojosa, 2012; Jackson and Barber, in press; Palmer, 2015). Such a relational view of water is now of interest to scholars who see its potential to bring to light a much deeper understanding of the role of water in human lives; to reveal how deeply water has "permeated social and cultural life" (Krause and Strang, 2016, p. 634).

The significance of water for indigenous communities is a strong focus of studies of water and its socio-cultural dimensions (Berry, 1997; Boelens et al., 2012; Langton, 2002; Lansing et al., 1998; Perreault, 2008; Perreault et al., 2012; Strang, 2005; Weir, 2009). Indigenous peoples use land and water resources in a variety of interrelated ways, including for subsistence use of water-dependent resources, recreation, commercial activities, and in cultural practices. For indigenous communities, water is described in much literature as central to defining complex attachments to place and numerous studies have demonstrated the wider role of water as creative and sustaining, as well as socially unifying (see Johnston et al., 2012). Water emphasizes the interconnectedness of places and associates the material and the economic with notions of sociality, sacredness, identity, and life-giving (Jackson, 2006).

Australian researchers report that water is described by indigenous people as an element that embodies a life force, as *living water* (Rose, 2004; Toussaint et al., 2005; Weir, 2009). In other parts of the world, free-flowing rivers hold special significance (Anderson and Veilleux, 2016; Tipa, 1999). In the Amazon for instance, the Shawi bathe in rivers, gathering strength from water carried down from mountains and ancestors (Anderson and Veilleux, 2016). Impoundment for dams is identified as a threat to a valued cultural principle held by the numerous indigenous groups of that region and many others (Jackson et al., 2009; Toussaint et al., 2005). For example, the Daly River in Australia has been described by indigenous landowners as a "significant ceremonial track" (Jackson, 2006, p. 23). In the neighboring Roper River catchment, the *Big River Country*, as it is known locally, is enlivened by the Dreaming creators and the spirits of ancestors and governed by principles of traditional ownership and management (Jackson and Barber, 2013). In the Pilbara region of Western Australia, Barber and Jackson (2011) found that those indigenous people directly associated with water places have responsibilities to sustain and protect them; contemporary guardians feel that they have obligations to ancestors, subsequent generations, and living people with kin and historical connections to those places. Care of water sites also implies a responsibility to those living downstream—water is not a static resource associated with a particular country alone, but part of an interconnected series of surface and subterranean flows.

For indigenous peoples, the contamination, diversion, and depletion of water bodies represent an attack on collective identities and survival as peoples as well as their direct health and well-being. Environmental change such as river regulation and upstream diversion can endanger both biological and cultural diversity of the river basins in which indigenous territories lie, threatening distinct *river cultures* (Wantzen et al., 2016) and undermining their survival and stability, or their *biocultural health* (Johnston and Fiske, 2014). Disruption and regulation of river flows in the United States has affected many wildlife communities intensively utilized by Indian tribes, for example, the carefully managed salmon runs that sustained the indigenous economies of the Pacific Northwest (Perreault et al., 2012). In many Andean regions, overallocation of water by governments is increasingly at odds with aquatic habitat protection, environmental water requirements, and the ability of indigenous peoples to exercise their rights for consumptive uses (Boelens et al., 2012).

In response to these pressures on indigenous communities and others reliant on aquatic resources, scientists have included social assessments in their approaches to environmental flows assessments. Attempts have been made in a number of flow studies to quantitatively define the water requirements of local communities that are dependent on aquatic resources for subsistence, not all of them being indigenous (see Esselman and Opperman, 2010; Finn and Jackson, 2011; Jackson et al., 2015; King and Brown, 2010; Lokgariwar et al., 2014; Meijer and Van Beek, 2011; Tipa, 1999; Wantzen et al., 2016). For example, researchers in Honduras drew on ecological knowledge collected from local and indigenous communities to form hypotheses about flow-dependent ecological and social factors that may be vulnerable to disruption from dam-modified river flows (Esselman and Opperman, 2010). There are also examples of studies that relate intangible cultural values and environmental flows assessment, specifically the relationship between flow requirements of: (1) sacred places and those required to maintain ecological conditions in South Africa; (2) ghats used in ritual ablutions in India (Lokgariwar et al., 2014); and (3) culturally significant aquatic biota such as river dolphins, or qualities such as the presence of life (Tipa, 1999).

This kind of research is needed urgently because in many places modernist reforms have empowered the formal systems of the state to dominate or marginalize parallel, extant (and often-times communal) customary systems of water governance maintained by indigenous peoples (Boelens, 2012; Jackson and Palmer, 2012). Indigenous peoples' fundamental interest in water management has been recognized in international sustainable development forums (e.g., World Water Forums) and human rights instruments (e.g., the UN Declaration on the Rights of Indigenous Peoples), and in some national water policy frameworks, although the degree to which indigenous claims to property in water and a determinative stake in development decisions and wider water management remains contested (Behrendt and Thompson, 2004; Ramazotti, 2008; Ruru, 2009).

As we will see in the following section, in nations such as Australia and South Africa, wide-ranging governance reforms have resulted in a "radical shift in the manner in which water is conceived in legal terms and the regulatory frameworks that control its allocation and distribution" (Godden, 2005, p. 181). In the growing number of studies concentrating on restructuring of water governance, including new environmental water management institutions, too little attention is given to the critical role of cultural processes and the wider social impacts of water regulation on diverse human societies.

9.3 TRENDS IN THE RECOGNITION OF INDIGENOUS WATER INTERESTS IN AUSTRALIA

Over the past 20 years, policy objectives have shifted in Australia from a preoccupation with developing inland water resources to conserving and reallocating water to the environment. Under a new environmental management paradigm (Hillman and Brierley, 2012), the provision of environmental water is seen as a promising strategy for integrating freshwater management into the broader scope of ecological sustainability (Arthington, 2012; Arthington et al., 2010). This historic shift in emphasis was accompanied by neoliberal policy instruments that altered property regimes (e.g., separating land and water titles), established markets in water, and decentralized water allocation decisions.

No formal consideration was given to the effect of these reforms on indigenous peoples and their rights and norms (Tan and Jackson, 2013). A major legal decision in Australia's High Court in 1992 (the *Mabo* case) had recognized native title and national legislation was promulgated for its protection. The *Native Title Act 1993* included water bodies within its definition of native title and litigation has led to the recognition of hunting, gathering, and fishing rights for the purposes of satisfying indigenous personal, domestic, and noncommercial needs. However, this historic decision made negligible impression on the water sector during this first phase of reform and subsequent case law did little to advance recognition of water rights (McAvoy, 2006; Tan and Jackson, 2013). It took a further decade for national water policy (the *National Water Initiative [NWI]*) to seek improvements in indigenous participation in water planning and access to water. The NWI now requires jurisdictions to include indigenous customary, social, and spiritual objectives in water plans (see Tan and Jackson, 2013). In this way, Australian governments have deployed the concept of customary practice to signal culturally distinctive forms of water rights and practices capable of being recognized by Australian law. Under Australian legal frameworks for water, indigenous rights are narrowly prescribed, discretionary, and contain no substantive restitutionary measures to redress the historical pattern of exclusion from the water economy. Tan and Jackson (2013, p. 13) highlight that commercial benefits are not likely to flow to indigenous people from this influential national policy framework.

> *In light of Australian Government commitments to overcome Indigenous disadvantage, it is significant that no explicit obligation is placed on parties to utilise the market-based water policy framework to advance Indigenous peoples' economic standing.*

For example, appeals to state governments to finance the purchase of water entitlements on behalf of indigenous peoples and to direct water to indigenous purposes have not been successful.

As a result, the consequent policy prescriptions have failed to address the marked inequity in water distribution or to significantly empower indigenous people within the structures that influence water-use decisions (Jackson and Langton, 2012). Indigenous specific water entitlements are at present estimated at less than 0.01% of Australian water diversions (Jackson and Langton, 2012).

Water represents a means of empowering and mobilizing people, and indigenous groups from many Australian regions are now organizing at the regional scale to address the implications of water governance reform for their communities. In Australia's southeast for example, where irrigated agriculture is concentrated, some 40 indigenous groups have experienced loss of access to country, inability to holistically manage land and water, exercise custodial authority, and prevent further ecological degradation. They have also been excluded from the water economy by virtue of their low rates of land ownership. Devastating environmental consequences of water regulation and excessive extraction, combined with the lack of legal recognition of indigenous rights and interests, has instigated the establishment of indigenous water advocacy groups (Weir, 2009). Numerous groups are pursuing strategies that will allow traditional owners to exercise custodial rights, fulfill cultural responsibilities, pursue social and economic interests, and protect culturally sensitive sites and burial grounds from alterations to water levels.

In the MDB, the recent multijurisdictional water-sharing plan that aims to reduce agricultural diversions (referred to as the Basin Plan) includes only modest provision for indigenous water management (Jackson, 2011). It places requirements on state and regional water planners to identify and provide for indigenous customary (noncommercial) uses and values and to consult over environmental water management. The submissions from a number of indigenous groups to a recent review of the *Water Act 2007* argue that indigenous rights and interests are treated in a tokenistic way by the legislation and its architecture (Department of Environment, 2015; Federation of Victorian Traditional Owner Corporations, 2014).

Although there is now a greater appreciation that subjective social objectives are critical to both the design and implementation of socially acceptable environmental water management targets and strategies, recent moves to recognize the water rights and interests of indigenous people show a tendency toward technocratic and reductionist reasoning of the kind that underpins the belief in an uncontested, standardized, and simple answer to the question: *How much water does a river need?*

For the reasons outlined below, I argue that this formulation of the problem of how to meet indigenous people's multifaceted water requirements (including a need to participate in decision making) is most unhelpful.

9.4 INDIGENOUS PEOPLES AND ENVIRONMENTAL WATER MANAGEMENT

Australia has seen the emergence of a relatively large environmental water governance system with institutional arrangements to acquire and manage substantial volumes of water. Commonwealth Government purchase of entitlements is the favored policy approach in the MDB. When the purchasing program is complete, the Commonwealth Environmental Water Holder will hold more than one-fourth of the MDB's extractive water rights and, in some regions, the figure could be 50% (Young, 2012). Management of a water reserve of this magnitude could present indigenous people with an opportunity to access water and restore some environments, as well as reaffirm and rebuild socio-ecological relationships and water-dependent livelihoods.

Whether such an opportunity can be realized will depend on a number of factors, not least, the capacity of the environmental water sector, with its cultures, techniques, practices, and regulatory regimes, to critically reflect on its ontological premises, epistemological foundations and norms, and to make room for other cultures and their ways of knowing the world. All groups, including scientists, bring to any such process assumptions, values, and particular forms of knowledge about how the world was formed (ontology) and theories about the nature, methods, and limits of human knowledge (epistemology).

Studies suggest the existence of numerous significant impediments to more inclusive environmental water policy and practices (Jackson and Barber, 2013; Jackson et al., 2012; McAvoy, 2006; Tan and Jackson, 2013; Weir, 2009). In the MDB and other Australian regions, flow assessments undertaken by water resource agencies have made little attempt to understand the pattern and significance of indigenous resource use and its role in flow ecology, nor indeed the wider socio-cultural context, which informs the development of values, beliefs, and ideas about the environment and gives rise to differences in environmental philosophy across cultures (Finn and Jackson, 2011).

Although the Basin Plan contains an obligation to consult with indigenous peoples, the degree to which consultation will make a material difference is conditional upon any identified indigenous values aligning with or enhancing *environmental outcomes*. Environmental objectives and outcomes are generated by biocentric assessments that prioritize water features according to universal conservation principles such as rarity or international significance. The new Commonwealth Environmental Water Holder, who has the role of buying, holding, and managing water, is similarly tightly prescribed in purpose and activities, reflecting the prioritization of ecological over socio-ecological objectives.

The policies and practices of environmental water management are grounded in the biophysical sciences and those practices can be quite exclusionary. Based on my observations, scientists tend to divorce aquatic ecological components from social relationships, cultural practices, belief systems, and social context (see also Brugnach and Ingram, 2014). For instance, presently the determination of environmental water regimes takes place in situations characterized by competition for and conflict over water distributions and, in Australia at least, the introduction of water markets has placed considerable pressure on those involved to nominate definitive volumes of water to particular environmental objectives. The intangible values that indigenous people regard as critical to their sense of identity, cultural practices, spiritual beliefs, customary management practices, and livelihoods, are consistently raised as a challenge to the quantitative and competitive methods of resource allocation currently favored by market-based reform programs.

A technical preoccupation with a scientifically determined river flow regime comes at the risk of neglecting the critical relational water values characteristic of indigenous water cultures (Jackson and Palmer, 2012; Strang, 2005). Brugnach and Ingram (2012) argue that indigenous frames of reference and their underpinning ontologies are not taken seriously in current water management models and so these groups are often excluded or marginalized from technical approaches in which the environment is "subservient to the humans who control it and descriptions and understandings of the natural system are externalized and independent from human experience" (p. 50).

Utilitarian values are prioritized over relational ones and consideration of indigenous interests have been accorded an even lower priority, if they are recognized at all in Australian environmental water management policies and programs (Jackson, 2006; Weir, 2009). For example, although the volume and timing of river flows are of great concern to many indigenous people (Jackson et al., 2011; Weir, 2009), numerous indigenous groups report that environmental water has not been directed to the features that they consider to be of the greatest significance or value or at the appropriate time of year. These concerns exist alongside wider aspirations to maintain and reaffirm relationships with country (customary land and waterscapes), fulfill intergenerational responsibilities, and apply and teach traditional knowledge and skills to younger generations, as well as pursue livelihoods that may rely on access to water and/or bountiful aquatic ecosystems use (e.g., fishing, hunting and gathering, tourism). With these objectives in mind, indigenous people are stressing a relational rather than commodity value of water; in interacting with water and the places it creates, traditional owners produce and create relationships between each other and nonhuman nature (see Babidge, 2015).

Indigenous objectives are currently only one of many Basin Plan objectives that sit alongside requirements to implement international agreements (e.g., Ramsar Treaty) and meet the water needs of ecological assets. In discussions about priorities for environmental water, indigenous people do not subscribe to these universalizing approaches to conservation or restoration, prioritizing instead local connections and localized measures of significance (e.g., sacred sites, conception sites). Furthermore, existing water governance arrangements do not acknowledge place-based responsibilities, as a member of the Ngemba community of the Barwon—Darling region told researchers (Maclean et al., 2012). Failure to understand how landscapes and water places are connected can lead to a lack of social responsibility to look after the country and to environmental degradation:

> You need to keep company with those places, otherwise it is just a story. So to understand those places, [and those] stories you need to keep company [with the landscape]. If you don't know about a place then there is less responsibility, nothing to do to heed to the repercussions. But those repercussions are now coming to have a hold on us, as is (sic) the implications of our actions.
>
> **Maclean et al. (2012, p. 43)**

Given the intensity of competition for environmental water, the low priority given to indigenous needs, and the way in which their needs have been framed, indigenous groups now have little confidence that the sites, places, ecological functions, social and environmental relationships they regard as important will receive water or other management attention (Jackson et al., 2012; Weir, 2009). In the following section, I will describe some of the strategies developed by indigenous groups that respond to, and at times reflect, the ways that the dominant society is framing their needs.

9.5 INDIGENOUS RESPONSES AND THEIR WATER RIGHTS STRATEGIES

Some indigenous advocates are closely examining the various policy options developed to acquire water to improve environmental water regimes for their social justice potential. In the pursuit of opportunities to secure water for indigenous use, instruments that deliver water to the environment could serve as model institutions through which to redress the historical neglect of indigenous water rights and interests and the transparently inequitable distribution of water. Deploying a political strategy that is seen by some indigenous groups as analogous to the campaign for recognition of the rights of aquatic ecosystems to water, considerable effort is going to the development of water entitlements to protect culture. Representative organizations such as the Murray Lower–Darling Rivers Indigenous Nations are calling for *cultural flows*, which would entitle them to water under a separate category of use. Cultural flows are defined by the MDB's indigenous water alliances in the following terms:

> *Cultural Flows are water entitlements that [would be] legally and beneficially owned by the Indigenous Nations of a sufficient and adequate quantity and quality to improve the spiritual, cultural, environmental, social and economic conditions of those Indigenous Nations.*
>
> **Weir (2009)**

Matt Rigney, one of the indigenous leaders who promulgated the concept, explained why the indigenous leadership developed the concept and the shape it took:

> *Maybe the paradigm needs to be changed a little, so that religious and spiritual aspects are included within cultural values. We have to talk more about our spiritual and religious values. We as Aboriginal people don't have policies on land and water—we are following those of the governments. So we said "we are going to develop our own charter, policies and programs and see where the government can then fit in with us" .... It's all Western world values. They talk about ethnocentrism, but they don't take our world view as the basis. We need to develop structures and processes to get people to see culture as a living thing.*
>
> **Rigney (2006)**

Advocates of cultural flows seek to leverage water allocations off the success of the *environmental flow* concept. The strategy has had some immediate traction. Research funds have been allocated by the lead MDB management agency to explore new institutional configurations to define and apply cultural flows, to determine the volumes or flow regimes required, and to measure the social, economic, and health benefits that may ensue. During consultations for the Basin Plan in 2012–2013, over 400 individual indigenous people and a further 21 indigenous organizations made submissions. A frequent recommendation was for specific cultural flow entitlements and allocations to be managed by indigenous people; another was for the establishment of an indigenous water holder to manage that water. Submissions also called for cultural flows to be protected by legislation. Beyond laying out a discursive claim for water for *culture*, a figure has been put on the share of the MDB's water that indigenous groups are claiming. The Northern Basin Aboriginal Nations, for instance, has called for an allocation of 5% of the entitlements of each water resource plan to be directed to indigenous people as a cultural flow (Northern Basin Aboriginal Nations, 2014). Thus far, the legally binding Basin Plan has not committed to allocating any water to indigenous needs.

In this maneuver, indigenous advocates are presenting a symbol of "traditional, place-specific, and often culturally distinctive resource use practices" (Perreault, 2008, p. 840), in a manner somewhat similar to the water rights strategies of the indigenous and campesino peoples in Bolivia. Jackson and Langton (2012) have recognized the symbolic potency of the cultural flows concept but caution against use of the term, arguing that it may be counterproductive to local water rights struggles. They argue that the concept has not been adequately and precisely defined: the water rights underpinning the notion are unclear and the vague terms pertaining to cultural value concepts from heritage management discourse do not readily translate into present water policy frameworks. It is the economic use within the ambit of a cultural flow that would appear particularly hard for the environmental water management framework in Australia to accommodate because of the history of overallocation for industrial (mainly agricultural) uses. Indigenous advocates are more recently discussing the merits of this terminology and some see the need to change the term to indigenous flows, or to indigenous environmental water (see Northern Basin Aboriginal Nations, 2014), and to explicitly address their economic exclusion from current water allocation regimes.

In their critique of the initial cultural flow concept, Jackson and Langton (2012) presented an argument that is central to environmental water management's cultural challenge. Framing indigenous water rights claims around ethnic or cultural differences gives rise to a paradox that operates on two levels: (1) by obscuring the cultural attitudes and assumptions of the dominant society, and (2) by assuming indigenous water requirements will be minimal. Justifying this assertion will require that I elaborate in some detail on the two distinct levels of this paradox.

On the first level, the role of culture in constructing the waterscapes and features of conservation value deserving of environmental water allocations is rarely, if ever, understood or recognized by environmental advocates, aquatic ecologists, or water planners who exclude indigenous epistemologies and ontologies from their resource assessments. We can think here of the attention that is given to efforts to ascertain the water requirements of a fish or bird species favored by the nonindigenous population. Where hydrological regimes have been altered, the concept of environmental water allocation, including which features should be protected and how much water should be applied, entails consideration of a set of choices and preferences driven by human objectives that are influenced by cultural processes.

At this point, it is worth recalling a number of important details in the history of river science, especially the influence of social objectives and political–economic context on techniques, methods, and decision making. Emerging from the American West during the post-World War II era of rampant dam construction, instream flow studies exhibited a strong bias toward objectives that favored socially valuable fisheries at risk from river regulation (Postel and Richter, 2012; Matthews et al., 2014). In the United States, during the 1970s and 1980s, minimum flow standards were heavily oriented toward meeting the needs of the recreational fishing sector (even though these same species were critical to economies of Indian tribes and were in many cases protected by treaties (Colombi, 2005), it was not until much later that instream flow policy and practice engaged with tribes).

In criticizing these methods for stressing the needs of a single or small number of species over those of a variable and complex whole, Postel and Richter (2012) single out the work of Australian and South African aquatic ecologists. These ecologists made significant advances in methods for prescribing ecological flows for rivers. Holistic approaches that reflected the natural flow regime served as an alternative to those predominantly fish-oriented methods and standards. According to Postel and Richter, because the antipodeans were removed from the entrenched and constrained debates about how much water could be spared from human use, they brought a "fresh and objective perspective to the challenge of prescribing ecological flows" (p. 52). For instance, Arthington turned the minimalist defensive position that had developed in the United States on its head, asking not how much water does a river need, but how much alteration to the natural flow regime was too much? (Postel and Richter, 2012, p. 57).

An intention, indeed requirement, to define the flows that meet *desired* features of the ecosystem is common to both the simplest and more comprehensive, or holistic, objective-based environmental flows assessment methods. Acreman and Dunbar (2004) argue that "although environmental flow setting is a practical river management tool, there remains an element of expert or political judgement" (p. 863). Thus, the choice of objectives and their ecological requirements are socially determined.

In Australia, as a consequence of existing power relations, indigenous people have little or no say over decisions affecting those processes that select water management objectives, and ultimately, over environmental quality. These decisions are currently made by experts and powerful social groups on narrow economic and ecological criteria that tend to exclude indigenous interests as merely "cultural." In doing so, individuals and groups appear unable to recognize that this categorization is itself a product of a culturally distinct way of seeing and distinguishing human uses of water. Greater thought therefore needs to be given to reflecting on the question: Whose knowledge and values hold sway in appraising the environment, prioritizing features of importance or desired conditions, and directing water to those features, species, or conditions?

In their article, Jackson and Langton (2012) questioned whether these *lenses* on our environmental relationships and problems can be challenged if indigenous people pursue cultural flows in isolation from environmental water allocations and management. There is the risk, that in doing so, other (nonindigenous) groups going about their research, deliberation, and decision making may assume that their preferences are objective, neutral, or natural; that they are devoid of values, biases, and political consequences, and are inherently more meritorious for those attributes.

In relation to the second level of the paradox, by foregrounding traditional cultural uses or values in water allocation decisions, indigenous requirements and needs are relegated to a reified, essentialist category of use that counterintuitively tends to require negligible amounts of water, for example water for ceremonies. If precolonial, *traditional use* was set as the standard for allocating water to indigenous needs, it would require that we take seriously the dependence of indigenous societies and their economies on the unregulated river systems that, at the time of British colonization, were high in water quality, had low rates of sedimentation, flowed through diverse vegetation communities, and were abundant in fish and wildlife (see Humphries, 2007). Doing so would require allocations to the environment of a magnitude far larger than is contemplated in current water reform debates in developed regions of Australia, and perhaps other choices, such as dam releases, would likely be unacceptable to entrenched agricultural interests.

Some indigenous activists, practitioners, and scholars are especially wary of invoking mass mediated portraits of traditional indigenous culture in their water strategies. Indigenous lawyer, Tony McAvoy (2006), for example, argues that: "There is no place in modern river management systems for the protection of indigenous spiritual values" and that "real impact on the commercial market in water and therefore river management will only occur when indigenous people are water owners themselves" (p. 97). To that end, suggestions have been made for the establishment of an independent Indigenous Water Fund or Trust to allow indigenous peoples to participate in the water market and allocate water to meet self-determined objectives. Those funds could be used to purchase licences and would be managed to provide for necessary infrastructure and other costs associated with accessing water entitlements. Purchases and use costs could be funded on an ongoing basis from a small levy on water trades (McAvoy, 2006). Under such an arrangement, indigenous peoples might choose to direct water to the environment.

In addition to reflecting on these explicitly cultural politics, there is valuable work to be done adapting the scientific procedures developed within the field of environmental water management to more effectively integrate indigenous engagement practices and knowledge systems. Recent efforts to identify indigenous water management objectives for features of social and environmental significance in New South Wales, Australia (including socio-ecological relationships), and the flow requirements that might meet these objectives (Jackson et al., 2015), serves as a pointer toward more inclusive water planning frameworks. Such planning frameworks will need to: (1) improve methods of eliciting indigenous water values and specification of water requirements; (2) explore opportunities to realize multiple benefits from environmental allocations; (3) consider equity of access (both across the indigenous and nonindigenous communities and within the indigenous community; (4) address indigenous expectations for direct participation in environmental water governance; and (5) reflect indigenous water rights recognition from an indigenous viewpoint.

9.6 CONCLUSION

The struggle for water rights for indigenous people has involved disputes over access to water and contest over the "contents of water rules and rights, the recognition of legitimate authority, and the discourses that are mobilized to sustain water governance structures and rights orders" (Boelens, 2008, p. 48). Mobilization against state reform of the water sector is yet to fundamentally alter the institutional basis of rural water management and water rights, yet progress can be discerned in the formulation and articulation of indigenous critiques of water policy and institutions. In their focus on water acquisition mechanisms, significant effort is being directed by indigenous people toward "the right to culturally define, politically organize and discursively shape" water-use systems (Boelens, 2008, p. 48). Such efforts to negotiate the relationships of inclusion and exclusion are integral to water rights struggles worldwide.

Throughout the period of rapid water sector reform described above, Australia's water crisis has been framed as an allocation problem where scarce water is unsustainably distributed between three broad categories of use: agriculture, cities, and the environment. This specification of the water governance problem contains a critical blind spot; it overlooks the interests, perspectives, knowledge, and rights of indigenous Australians who have unmet water needs, unresolved claims for political, economic, and cultural recognition, and a body of knowledge to contribute to the resolution of water conflicts.

In the 1990s, the environment was introduced as a new *water user* with legal instruments to meet environmental imperatives. Protection and formal legal status for environmental water renders it equivalent to consumptive rights in the system (Foerster, 2011). It appears that it is this equivalence that indigenous people are seeking from a regime in which competition for water is intense and reallocation to the environment is an ascendant objective. However, having been omitted from water allocation frameworks for more than two centuries and doubtful that environmental gains will be shared equitably across society, indigenous groups are mobilizing an explicitly *cultural* politics of water to gain access to public sympathy and state resources. In their campaigns, indigenous people will talk about culture as a highly distinct and separate domain, often one associated with the past, as part of a strategic claim for recognition by the nation-state (Merlan, 1989). *Culture* as a category of water use also seems to appeal to the settler society because it accords with a preconception that indigenous uses are premodern and therefore do not compete with so-called productive and highly water-intensive uses. Perhaps it is this appeal that underpins the desire to answer the question: How much water does a culture need? It is a strategy that nonetheless carries some risks and for this reason indigenous groups are also seeking to use the market mechanisms at hand to acquire rights to water (Barber and Jackson, 2014; McAvoy, 2006).

Having found little satisfaction in the native title adjudications and determinations that should have confirmed their status as prior water users, indigenous people instead come to Australian water debates and allocation processes with little negotiating power and perceived as *latecomers*. Assuming state provision of funds, they may find a strategy of buying water from willing sellers and engaging in trade more promising than recovering water through political processes, especially in regions experiencing water scarcity. It is too early to tell whether the multiple strategies described above will result in a fairer distribution of water and more effective indigenous representation in water governance and, moreover, whether they will do justice to indigenous ontologies of reciprocal, ethical relations with land and water.

The counterstrategies currently being promoted and enacted nonetheless represent key sites for experimentation in alternate forms of water resource governance. They also represent a challenge for professionals in the field of environmental water management to develop more inclusive processes and give greater priority to indigenous water management objectives in their water purchases and management programs. Environmental researchers can assist by partnering with indigenous communities to explore methods and means to derive water and other requirements to sustain the valued associations and customary relationships indigenous people maintain with their rivers and waters. Unless social and cultural dimensions are integral considerations in water management governance, water resource developments will continue to significantly threaten local livelihoods and ways of life.

REFERENCES

Acreman, M.C., Dunbar, M.J., 2004. Defining environmental river flow requirements? A review. Hydrol. Earth Syst. Sci. 8 (5), 861–876.

Anderson, E., Veilleux, J., 2016. Cultural costs of tropical dams. Science 352 (6282), 159.

Arthington, A., 2012. Environmental Flows. Saving Rivers in the Third Millennium. University of California Press, Berkeley, CA.

Arthington, A., Naiman, R., McClain, M., Nillson, C., 2010. Preserving the biodiversity and ecological services of rivers: new challenges and research opportunities. Freshw. Biol. 55, 1–16.

Babidge, S., 2015. Contested value and an ethics of resources: water, mining and indigenous people in the Atacama Desert, Chile. Aust. J. Anthropol. 27 (1), 84–103.

Barber, M., Jackson, S., 2011. Indigenous people, water values and resource development pressures in the Pilbara region of north-west Australia. Aust. Aborig. Stud. 2, 32–49.

Barber, M., Jackson, S., 2014. Autonomy and the intercultural: historical interpretations of Australian Aboriginal water management in the Roper River catchment, Northern Territory. J. Roy. Anthropol. Inst. 20, 670–693.

Behrendt, J., Thompson, P., 2004. The recognition and protection of Aboriginal interests in New South Wales rivers. J. Indig. Policy 3, 37–140.

Berry, K., 1997. Of blood and water. J. Southwest 39, 79–111.

Boelens, R., 2008. Water rights arenas in the Andes: upscaling the defence networks to localize water control. Water Altern. 1, 48–65.

Boelens, R., 2012. The politics of disciplining water rights. Dev. Change 40, 307–331.

Boelens, R., Duarte, B., Manosalvas, R., Mena, P., Roa Avendaño, T., Vera, J., 2012. Contested territories: water rights and the struggles over Indigenous livelihoods. Int. Indig. Policy J. 3, Available from: http://ir.lib.uwo.ca/iipj/vol3/iss3/5 (accessed 13.02.16).

Brugnach, M., Ingram, H., 2014. Rethinking the role of humans in water management: toward a new model of decision-making. In: Johnston, B.R., Hiwasaki, L., Klaver, I., Castillo, A.R., Strang, V. (Eds.), Water, Cultural Diversity, and Global Environmental Change: Emerging Trends, Sustainable Futures? Springer, Dordrecht, The Netherlands, pp. 49–64.

Budds, J., Hinojosa, L., 2012. Restructuring and rescaling water governance in mining contexts: the co-production of waterscapes in Peru. Water Altern. 5, 119–137.

Colombi, B., 2005. Dammed in Region Six. The Nez Perce tribe, agricultural development, and the inequality of scale. Am. Indian Quart. 29, 560–744.

Department of Environment, 2015. Independent review of the *Water Act 2007*. Available from: http://www.agriculture.gov.au/water/policy/legislation/water-act-review#submissions (accessed 02.09.16).

Esselman, P., Opperman, J. 2010. Overcoming information limitations for the prescription of an environmental flow regime for a Central American river. Ecol. Soc. 15, 6. [online] Available from: http://www.ecologyandsociety.org/vol15/iss1/art6/ (accessed 24.02.16).

Federation of Victorian Traditional Owner Corporations, 2014. Submission to the Review of the *Water Act 2007*. Available from: http://www.environment.gov.au/water/legislation/water-act-review (accessed 14.02.16).

Finn, M., Jackson, S., 2011. Protecting Indigenous values in water management: a challenge to conventional environmental flow assessments. Ecosystems 14 (8), 1232–1248.

Fischer, R., 2009. Where is culture in cross cultural research? An outline of a multilevel research process for measuring culture as a shared meaning system. Int. J. Cross Cult. Manage. 9, 25–49.

Foerster, A., 2011. Developing purposeful and adaptive institutions for effective environmental water governance. Water Resour. Manage. 25, 4005–4018.

Godden, L., 2005. Water law reform in Australia and South Africa. J. Environ. Law 17, 181–205.

Head, L.M., Trigger, D., Mulcock, J., 2005. Culture as concept and influence in environmental research and management. Conserv. Soc. 3 (2), 251–264.

Hillman, M., Brierley, G., 2002. Information needs for environmental-flow allocation: A case study from the Lachlan River, New South Wales, Australia. Ann. Assoc. Am. Geogr. 92 (4), 617–630.

Humphries, P., 2007. Historical Indigenous use of aquatic resources in Australia's Murray-Darling Basin, and its implications for river management. Ecol. Manage. Restor. 8, 106–113.

Jackson, S., 2017. Indigenous peoples and water justice in a globalizing world. In: Concu, K., Weinthal, E. (Eds.), Oxford Handbook on Water Politics and Policy. Oxford University Press, Oxford. Available from: http://doi: 10.1093/oxfordhb/9780199335084.013.5.

Jackson, S., (in press). Enduring injustices in Australian water governance. In: Lukasiewicz, A. (Ed.), Resources, Environment and Justice: the Australian Experience. CSIRO Publishing, Melbourne, Australia.

Jackson, S., 2006. Compartmentalising culture: the articulation and consideration of Indigenous values in water resource management. Aust. Geogr. 37, 19–32.

Jackson, S., 2008. Recognition of Indigenous interests in Australian water resource management, with particular reference to environmental flow assessment. Geogr. Compass 2, 874–898.

Jackson, S., 2011. Indigenous water management in the Murray-Darling Basin: priorities for the next five years. In: Grafton, Q., Connell, D. (Eds.), Basin Futures: Water Reform in the Murray-Darling Basin. ANU E-Press, Canberra, Australia, pp. 163–178.

Jackson, S., Altman, J., 2009. Indigenous rights and water policy: perspectives from tropical northern Australia. Aust. Indig. Law Rev. 13, 27–48.

Jackson, S., Barber, M., (in press). Historical and contemporary waterscapes of north Australia – Indigenous attitudes to dams and water diversions. Water Hist.

Jackson, S., Barber, M., 2013. Indigenous water values and resource governance in Australia's Northern Territory: current progress and ongoing challenges for social justice in water planning. Plann. Theory Pract. 14, 435–454.

Jackson, S., Langton, M., 2012. Trends in the recognition of indigenous water needs in Australian water reform: the limitations of 'cultural' entitlements in achieving water equity. J. Water Law 22, 109–123.

Jackson, S., Palmer, L., 2012. Modernising water: articulating custom in water governance in Australia and Timor Leste. Int. Indig. Policy J. 3. Available from: http://ir.lib.uwo.ca/iipj/vol3/iss3/7 (accessed 15.02.16).

Jackson, S., Pollino, C., Maclean, B., Bark, R., Moggridge, B., 2015. Meeting Indigenous people's objectives in environmental flow assessments: case studies from an Australian multi-jurisdictional water sharing initiative. J. Hydrol. 52, 141–151.

Jackson, S., Tan, P., Mooney, C., Hoverman, S., White, I., 2012. Principles and guidelines for good practice in Indigenous engagement in water planning. J. Hydrol. 474, 57–65.

Johnston, B., Fiske, S., 2014. The precarious state of the hydrosphere: why biocultural health matters. WIREs Water 1, 1–9.

Johnston, B., Hiwasaki, L., Klaver, I., Castillo, A.R., Strang, V. (Eds.), 2012. Water, Cultural Diversity, and Global Environmental Change: Emerging Trends, Sustainable Futures? Springer, Dordrecht, The Netherlands.

King, J., Brown, C., 2010. Integrated basin flow assessments: concepts and method development in Africa and South-east Asia. Freshw. Biol. 55, 127–146.

Klaver, I., 2012. Placing water and culture. In: Johnston, B., Hiwasaki, L., Klaver, I., Castillo, A.R., Strang, V. (Eds.), Water, Cultural Diversity, and Global Environmental Change: Emerging Trends, Sustainable Futures? Springer, Dordrecht, The Netherlands, pp. 9–20.

Krause, F., Strang, V., 2016. Thinking relationships through water. Soc. Nat. Resour. 29, 633–638.

Langton, M., 2002. Freshwater. Background Briefing Papers: Indigenous Rights to Waters. Lingiari Foundation, Broome, Australia, pp. 43–64.

Lansing, J., Lansing, P., Erazo, J., 1998. The value of a river. J. Pol. Ecol. 5, 1–22.

Lokgariwar, C., Chopra, R., Smakhtin, V., Bharati, L., O'Keeffe, J., 2014. Including cultural water requirements in environmental flow assessment: an example from the upper Ganga River, India. Water Int. 39, 81–96.

Maclean, K., Bark, R., Moggridge, B., Jackson, S., Pollino, C., 2012. Ngemba Water Values and Interests: Ngemba Old Mission Billabong and Brewarrina Aboriginal Fish Traps (Baiame's Nguunhu). CSIRO, Australia.

Matthews, J., Forslund, J., McClain, M., Tharme, R., 2014. More than the fish: environmental flows for good policy and governance, poverty alleviation and climate adaptation. Aquat. Procedia 2, 16−23.

McAvoy, T., 2006. Water − fluid perceptions. Transform. Cultures eJ. 1, 97−103.

Meijer, K., Van Beek, E., 2011. A framework for the quantification of the importance of environmental flows for human well-being. Soc. Nat. Resour. 24, 1252−1269.

Merlan, F., 1989. The objectification of 'culture': an aspect of current political process in Aboriginal affairs. Anthropol. Forum 6, 105−116.

Northern Basin Aboriginal Nations, 2014. Submission to the review of the *Water Act 2007* (Cth).

Palmer, L., 2015. Water Politics and Spiritual Ecology: Custom, Environmental Governance and Development. Routledge, London.

Perreault, T., 2008. Custom and contradiction: rural water governance and the politics of usos y costumbres in Bolivia's irrigators' movement. Ann. Assoc. Am. Geogr. 98, 834−854.

Perreault, T., Wraight, S., Perreault, M., 2012. Environmental injustice in the Onondaga Lake waterscape, New York State, USA. Water Altern. 5, 485−506.

Postel, S., Richter, B., 2012. Nature. second ed. Island Press, Washington, DC.

Prieto, M., 2014. Privatizing water and articulating indigeneity: the Chilean water reforms and the Atacameño people (Likan Antai). PhD thesis. University of Arizona, Tucson, Arizona.

Ramazotti, M., 2008. Customary water rights and contemporary water legislation: Mapping out the interface. FAO Legal Papers Online No. 76. FAO, Rome, Italy (accessed 14.02.16).

Richter, B.D., Baumgartner, J.V., Wigington, R., Braun, D.P., 1997. How much water does a river need? Freshw. Biol. 37, 231−249.

Rigney, M., 2006. The role of the Murray Lower Darling Rivers Indigenous Nations (MLDRIN) in protecting cultural values in the Murray Darling rivers. In: Jackson, S. (Ed.), Recognising and Protecting Indigenous Values in Water Management: a Report from a Workshop. CSIRO, Darwin, Australia, pp. 5−6, April.

Rose, D.B., 2004. Freshwater rights and biophilia: Indigenous Australian perspectives. Dialogue 23, 35−43.

Ruru, J., 2009. Undefined and unresolved: exploring Indigenous rights in Aotearoa New Zealand's freshwater legal regime. Water Law 20, 236−242.

Russo, K., Smith, Z., 2013. What Water is Worth: Overlooked Non-Economic Value in Water Resources. Palgrave McMillan, New York.

Strang, V., 1997. Uncommon Ground: Cultural Landscapes and Environmental Values. Berg, Oxford.

Strang, V., 2005. Water works: agency and creativity in the Mitchell River catchment. Aust. J. Anthropol. 16, 366−381.

Tan, P., Jackson, S., 2013. Impossible Dreaming − does Australia's water law and policy fulfil Indigenous aspirations? Environ. Plann. Law J. 30, 132−149.

Tipa, G., 1999. Environmental Performance Indicators: Taieri River Case Study. Ministry for the Environment, Wellington, New Zealand.

Toussaint, S., Sullivan, P., Yu, S., 2005. Water ways in Aboriginal Australia: an interconnected analysis. Anthropol. Forum 15, 61−74.

Wantzen, K.M., Ballouche, A., Longuet, I., Bao, I., Bocoum, H., Cissé, L., et al., 2016. River culture: an eco-social approach to mitigate the biological and cultural diversity crisis in riverscapes. Ecohydrol. Hydrobiol. 16, 7−18.

Weir, J., 2009. Murray River Country: an Ecological Dialogue with Traditional Owners. Aboriginal Studies Press, Canberra, Australia.

Young, M., 2012. Chewing on the CEWH: improving environmental water management. In: Quiggan, J., Mallawaarachchi, T., Chambers, S. (Eds.), Water Policy Reform: Lessons in Sustainability from the Murray-Darling Basin. Edward Elgar, Cheltenham, UK, pp. 129−135.

VISIONS, OBJECTIVES, TARGETS, AND GOALS

10

Avril C. Horne[1], Christopher Konrad[2], J. Angus Webb[1], and Mike Acreman[3]

[1]*The University of Melbourne, Parkville, VIC, Australia* [2]*U.S. Geological Survey, Washington, WA, United States*
[3]*Centre for Ecology and Hydrology, Wallingford, United Kingdom*

10.1 INTRODUCTION

Without an operational definition of the desired endpoint, effective management is unlikely; and without broad consensus on that definition, acceptance within wider societal value systems is equally unlikely.

Rogers and Biggs (1999)

Rivers and associated wetlands are used and appreciated for a wide range of social, economic, and ethical purposes (Acreman, 2001) that hinge on the environmental condition of the system. Communities and cultures interact and value these river systems in complex ways (Chapters 7 and 9). Establishing a shared and defined vision is a critical initial step in managing river resources and implementing environmental water regimes. The vision represents the social narrative statement, which may contain specific numerical parameters, of the desired state of the river: What do we want the river to *look* like? What do we want to achieve from environmental water? It is generally considered that setting a vision is part of the social processes to balance competing needs among stakeholders (Poff et al., 2010; Richter et al., 2006; U.S. Department of the Interior, 2012). Stakeholder endorsement of this vision stimulates broad ownership and responsibility, which is an essential element for ongoing community and political support for the changes in water management required to achieve this vision. The vision will sometimes be constrained by overriding legislation or government policy. For example, the European Water Framework Directive dictates that all rivers should achieve Good Ecological Status (with some exceptions); here stakeholder participation focuses on the measures to achieve the Directive (Acreman and Ferguson, 2010).

However it is defined, the vision provides a long-term and overarching endpoint for environmental water programs. Progression toward this ultimate endpoint will be achieved with many contributing activities over decadal timescales. To support progressive implementation of such a vision for a water resource system it should be elaborated in subordinate *objectives*, *targets*, and *goals*. These are critical for ensuring alignment of component environmental water management efforts, including complementary works. They provided for targeted monitoring and assessment of environmental watering program outcomes to evaluate progress toward the vision (Fig. 10.1).

The focus of this chapter is on the approach to defining objectives and translating these into measurable targets and goals. These objectives, targets, and goals frame all future stages of

Water for the Environment. DOI: http://dx.doi.org/10.1016/B978-0-12-803907-6.00010-3

environmental water management including the selection of environmental assessment techniques, the derivation of flow recommendations, and the implementation of adaptive management and monitoring. Best results are achieved when targets and goals are SMART—Specific, Measurable, Achievable, Realistic, and Time-bound. Clear statements about desired ecological conditions contribute to the broader recognition of benefits from environmental water management (Harnish et al., 2014). This is a key element in the management of water resources for ecological purposes (Olden et al., 2014; Richter et al., 2006).

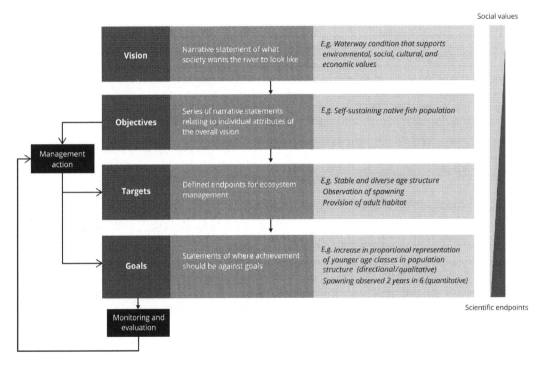

FIGURE 10.1

Hierarchy of objectives for an environmental water management program.

It is important to recognize that there are no universally accepted definitions or terminology to describe the hierarchy of objectives, with the terms *objective*, *vision*, *goal*, *outcome*, and *target* used differently across studies or jurisdictions. Fig. 10.1 defines the terminology that will be adopted for this chapter. It also shows the transition of social values through to scientific endpoints that support and inform management (Rogers and Biggs, 1999).

10.2 ESTABLISHING A VISION FOR A RIVER

A *vision* guides the overall management and purpose of an environmental water regime with value-based statements of strategic intent that have been informed by stakeholder opinion (Rogers and

Biggs, 1999). Disagreements over allocating water for ecological purposes can reflect fundamental conflicts among social, economic, utilitarian, and environmental values. Stakeholders may support environmental water allocations and improved resource condition outcomes only to the extent they do not compromise other uses of water or conflict with wider policy or legislation. For example, a government may decide that there is an overriding national interest to generate a given level of hydropower that limits provision of environmental water. However, the government may also have obligations under international agreements to protect environmental assets of significance. We argue that the vision statement is best obtained through the stakeholder engagement processes and identification of ecosystem services (outlined in Chapters 6, 7, and 8), set within the context of wider constraints and national priorities, laws, and international commitments.

10.3 ESTABLISHING OBJECTIVES FOR ENVIRONMENTAL WATER

A vision statement is not sufficiently detailed to support the many decisions required in an environmental water program. To support planning, a vision should be elaborated in multiple subordinate objectives. Objectives remain high-level statements but unbundle the vision for different environmental components, functions, or ecosystem services. Objectives may also be set for different sectors or institutions, such that when combined they deliver the wider vision. As with the vision, the process of development of objectives is best achieved through engagement between managers and stakeholders about the purposes for managing water (Carter et al., 2015; European Parliament, 2000; US Army Corps of Engineers, 2004). Objectives need to be underpinned by best scientific knowledge and data, hence the level of external scientific input (e.g., from environmental scientists, economists, social scientists) is likely to be greater than for vision setting. Objectives form the foundation for the development of quantitative and operational goals and targets (Rogers and Biggs, 1999).

Objectives may address regional, national, or international concerns, laws, or agreements particularly in large, transboundary systems (Canada and the United States of America, 1964; Council on Environmental Quality, 2013; Mekong River Basin Commission, 1995; Murray Darling Basin Authority, 2014a; Sudan and the United Arab Republic, 1959). They identify assets or species of importance. Where there is little external guidance on objectives, a process of examining preferred scenarios may be required to help define objectives.

There is a distinction that can be made in the approach taken to establishing objectives in a system that is: (1) highly constrained with extensive existing infrastructure, water use, and flow regulation where ecological objectives are being retroactively *fitted*; and one that is (2) less constrained where environmental policy provides strong levers and ecological objectives are set a priori. Much of the experience to date has been in systems that are highly constrained, however, newer planning and policy development around the world may see an increase in a priori setting of objectives in less constrained systems.

In highly constrained systems, a pragmatic approach to developing objectives begins with an understanding of the amount of water available at specific times for ecological purposes (Hall et al., 2011). In these systems, water availability is limited by the allocations of water to other uses and regulation of flows by infrastructure. In some cases, different objectives may be appropriate during different periods; for example, environmental water allocations may be constrained in the short-term due to high allocations to irrigated agriculture that may be reduced in the longer-term future as more water-use efficient technology is implemented. In the United Kingdom, dam operations are controlled by specific Acts of Parliament, thus while water has increasingly been allocated for environmental use to meet the European Water Framework Directive, full implementation must await law repeal (Acreman and Ferguson, 2010). In addition, objectives may vary spatially such that they combine to achieve a broader, for example, national vision such as numbers of breeding waterbirds or fish. The ability to implement management actions to support these objectives may be limited by the physical conditions in the system and credible objectives must consider the influence of factors other than flow regime. Water temperature is an influential factor for many aquatic organisms but water managers may have limited means for its management (Olden and Naiman, 2010; Vinson, 2001). Similarly, infrastructure constraints such as capacity of dam release valves may limit the ability to provide high-flow events in addition to constraints imposed by avoiding flooding risks to private property (Murray Darling Basin Authority, 2014b).

10.4 SETTING TARGETS

Evaluation of environmental water delivery programs should generally report performance in relation to measureable targets that indicate progress toward the objectives. We suggest SMART targets as they both measure the outcome of the project and define what is realistic in terms of available time and resources. SMART stands for targets that are *Specific* (concrete, detailed, well-defined), *Measurable* (quantity, comparable), *Achievable* (feasible, actionable), *Realistic* (considering resources), and *Time-bound* (a defined time line) (Hammond et al., 2011).

A useful process to identify targets is to use conceptual models to link management decisions (in this case, environmental water regime or flow components) to the objective. This depicts the known and proposed causal relationships that govern the response of the objective and helps to identify targets that need to be achieved and monitored to assess progress toward the objective (Horne et al., accepted). Conceptual models are also useful for communication with stakeholders in establishing a shared understanding of the translation from objectives through to implemented flow regime and monitoring outcomes.

Each objective is assessed by monitoring one or (usually) more targets, which are specific and measurable (the SM of SMART), drawn from the nodes or links of the conceptual model of the objective. The targets can occur at a range of temporal and spatial scales.

Targets should be informed by the available underpinning freshwater science, and in particular knowledge of the causal processes that might lead to the intended objectives. However, aquatic ecosystems are complex with multiple, interacting factors where individual components are controlled directly or indirectly by factors over a range of spatial and temporal scales (Allan and Castillo, 2007; Hynes, 1975; Poole, 2002; Strauss, 1991; Wiens, 2002). Ecological responses can have different temporal responses to water management actions:

1. Immediate responses such as physical conditions (Strydom and Whitfield, 2000) or behavior of organisms (King et al., 1998) coincident with water management actions;

2. Lagged responses such as increased reproduction or growth after an event due to changes in habitat or cascading effects at multiple trophic levels (Melis et al., 2012); or

3. Integrated response such as community shifts (Robinson and Uehlinger, 2008) or recruitment through multiple age classes in a population as a result of a sequence of streamflows or flow regime encompassing multiple flow components (Rood et al., 2003).

In general, short-term targets are more likely to relate to outcomes from a specific water provisioning action, such as salinity gradients in an estuary or the movement behavior of fishes and availability of prey (Bate and Adams, 2000; Cambray et al., 1997; Melis et al., 2012). These targets can be measured at specific locations and respond relatively quickly. These might be used for designing individual environmental water delivery actions (e.g., wetland filling or a flow fresh). In contrast, targets that relate to longer-term resource condition objectives represent higher-level ecological responses to water management (e.g., the spatial distribution of habitat patches, age structure of populations, and trophic or taxonomic structure of communities) and typically are affected by multiple factors integrated over timescales longer than those of individual management actions (Konrad, 2013; Robinson and Uehlinger, 2008; Rood et al., 2003; Snow et al., 2000; Townsend, 1989). Measured responses may thus not provide definitive evidence about the effects of specific environmental water actions or how specific environmental water prescriptions can be improved and thus how well targets are being met.

Although a small set of targets may be most sufficient, it is desirable to define targets that span multiple spatial and temporal scales. Such a hierarchical approach can address the immediate results of water management actions on single components of ecosystems, along with responses of populations, communities, or functions that are influenced by multiple factors integrated over time (Fig. 10.2). Ecosystem components (e.g., organism behavior, nutrient flux, habitat availability, population size) can be arrayed hierarchically based on the spatial scale of their controlling factors (Frissell et al., 1986; Snelder and Biggs, 2002). Adopting a hierarchical ecosystem framework to organize and identify targets can represent both the differences in the timescales of ecosystem (e.g., the time required to deposit a sand bar vs the time required to grow an established forest) and their dependency on multiple, interacting factors. Short temporal-scale responses can be used to rapidly demonstrate provision of conditions necessary to achieving the objective, whereas long-scale responses demonstrate integrative effects of management actions over time. They are closer in substance to the objective statement, and are more likely to resonate with stakeholders and the public.

Shifts in management strategy or combinations of management actions intended to address multiple objectives (environmental and social) become very problematic when evaluating progress toward long-term objectives: targets cannot be linked to specific actions and actions may have confounding effects (Korman et al., 2011). Each management target alone will have limited ability to inform the evaluation of progress in achieving long-term objectives; however, a suite of targets may be used to verify assumptions about how management actions will lead to longer-term objectives (Conservation Measures Partnership, 2014). For example, no single target is defined for assessing salmon recovery in the Columbia River Basin, United States and Canada. Instead, targets have been developed for each life stage of individual stocks at many different locations throughout the basin (National Oceanic and Atmospheric Administration, 2014).

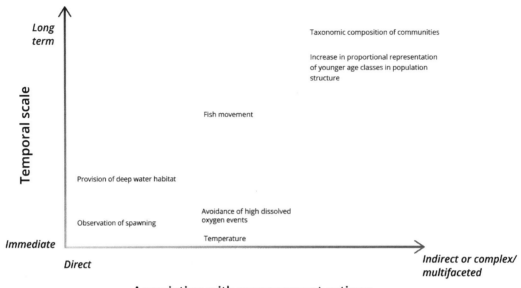

FIGURE 10.2

Hierarchical ecosystem framework to link objectives and targets—demonstration for a fish species.

10.5 GOALS

Goals have the properties of being Achievable, Realistic, and Time-bound (the ART in SMART). These are the statements of where a target needs to get to in order to indicate success of an implemented environmental water regime.

Ideally, goals should be measurable and quantitative: they should be a statement of the amount or rate of an ecosystem property that can be achieved by a certain time. However, setting such achievable goals relies on a solid knowledge of the quantitative relationship between the target and the management action. In the absence of this knowledge base, goals are often set in a qualitative or directional manner. They aim for an improvement in condition within a certain time frame, but stop short of specifying how much that improvement needs to be. The National Oceanic and Atmospheric Administration (2014), for example, requires that salmon populations increase over time for their recovery, but the rate of increase remains a point of contention with some stakeholders (Save Our Wild Salmon, 2014). Similarly, water managers in the Cedar River have the target of increasing salmon population, and the number of adult Chinook salmon returning to the Cedar River is indeed increasing, but water managers have not specified a goal in terms of the numbers of fish, which depends on ocean conditions and other factors beyond their control (Boxes 10.1 and 10.2).

BOX 10.1 BASIN-WIDE ENVIRONMENTAL WATERING STRATEGY, MURRAY–DARLING BASIN, AUSTRALIA

The Murray–Darling Basin Plan aims to restore a significant volume of water to the environment. To help inform the effective use of this water, a basin-wide environmental watering strategy was developed to define clear objectives and targets for the use of this water. The basin-wide strategy will then be implemented through a series of finer-scale plans for each region, with more clearly defined goals and linked monitoring programs.

The strategy identifies four important components of the basin's water-dependent ecosystems: river flows and connectivity, vegetation, waterbirds, and native fish (Fig. 10.3). These four elements were selected as (MDBA, 2014a):

1. They are good indicators of the health of a river system and are measurable;
2. They are important components of healthy functioning water-dependent ecosystems;
3. They are responsive to environmental water;
4. They are highly valued by people;
5. They have declined appreciably as a result of water resource development;
6. They require a basin-wide approach to be managed effectively.

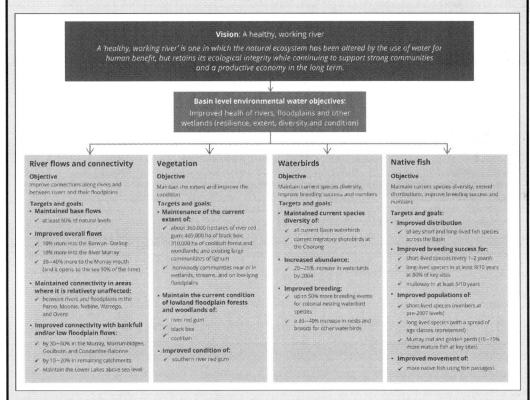

FIGURE 10.3

Summary of the vision, objectives, targets, and goals identified for the Murray–Darling Basin watering strategy.

Source: Adapted from Table 1, MDBA (2014a)

BOX 10.2 CEDAR RIVER, WASHINGTON, UNITED STATES

The Cedar River flows from its headwaters in the central Cascade Range, Washington, United States, to Lake Washington in the Puget Sound basin. The river has a drainage area of 470 km^2 and flows down 1 km over 74 km through steep, subalpine forest and a relatively confined lowland valley. Annual mean streamflow ranges from less than 10 m^3/s in dry years to more than 28 m^3/s in wet years. The Cedar River is the primary source of water for the City of Seattle, which protects its upper watershed to assure high-quality water. Seattle Public Utilities operates a dam constructed in 1914 at the outlet of a large natural lake in the upper watershed to regulate streamflow for a downstream diversion.

Fig. 10.4 shows the vision, objectives, and targets and goals for the river. The two targets around habitat condition are measurable but have not been translated into goals and quantified for a number of reasons: (1) their relation to streamflow is less direct depending on the condition of the river corridor that affects hydraulic conditions during high flows and the supply of wood and sediment to the river; (2) peak flows are not fully controlled by dam operation; (3) ecological goals must be balanced by flood risk reduction as part of managing peak flows; and (4) there may be trade-offs between redd scour and habitat maintenance that are easily determined. In the case of redd scour, the Instream Flow Commission has guidelines for limiting peak flows when possible, to less than a threshold based on field research (Cascades Environmental Services, 1991; Gendaszek et al., 2013). The longer-term objective of maintaining riverine habitats is not linked to any specific management action, but is the subject of ongoing monitoring and the development of quantitative targets (Gendaszek et al., 2013; Magirl et al., 2012).

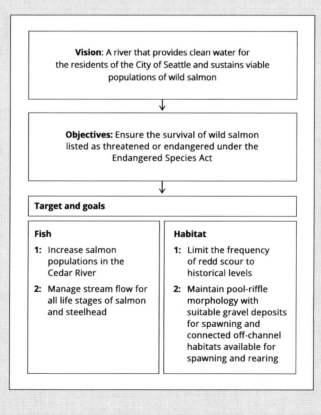

FIGURE 10.4

Vision, objectives, and targets and goals for the river.

10.6 MONITORING, EVALUATION, AND ADAPTIVE MANAGEMENT

The success of an environmental water program can be measured by the progress toward achieving the objectives, targets, and goals. Monitoring is designed to assess the environmental water program against the identified targets and goals (Chapter 25).

Visions, objectives, targets, and goals are not static, and will require updating over time. This may be due to, for example, changes in social preferences, alterations to overriding legislation, or improved knowledge of the scientific links between actions and outcomes. For example, in India increased demand for environmental water for the Ganges was driven by a sudden appreciation that the river was no longer in an acceptable condition to permit religious ceremonies. Updating should occur through an adaptive management framework (Chapter 25). Importantly, consideration of objectives or longer-time-scale targets should only be considered following major changes in the knowledge base or in societal values. Accommodating such changes may involve changes to practices, institutions involved, and/or in monitoring methods (and consequent redirection of funds away from some monitoring). This corresponds to the outer loop of adaptive management. In contrast, goals may be updated with new knowledge and understanding of how targets respond to management actions. This is a relatively short-term process that most closely corresponds to the *inner loop* of adaptive management (Chapter 25).

10.7 CONCLUSION

A clearly defined vision for a river basin will inform all other aspects of environmental water management. This process of achieving the vision is best translated into clear objectives that relate to the delivery of environmental water based on best available science and data. Involving water stakeholders in the definition of the vision and its objectives builds ownership and responsibility and forms part of a wider social inclusion process. However, some objectives may be governed by preset legislation or national good for which there is little room to incorporate local stakeholder views, in which case the process is limited to appreciation of the overriding laws and implications for environmental water delivery. Implementation of these objectives is best achieved through the identification and assessment of SMART targets and goals.

Targets for environmental water management can be arranged hierarchically based on the temporal and spatial scale of responses. The hierarchy ranges from targets that are defined narrowly in terms of a single component that responds rapidly to changes in response to flow management, to broader targets cast in terms of populations, communities, or systemic functions integrating the effects of many factors over longer timescales. The hierarchy does not presume greater value of either type of target, but does affect how the target informs management decisions.

REFERENCES

Acreman, M.C., 2001. Ethical aspects of water and ecosystems. Water Policy 3 (3), 257–265.
Acreman, M.C., Ferguson, A., 2010. Environmental flows and European Water Framework Directive. Freshw. Biol. 55, 32–48.
Allan, D.J., Castillo, M.M., 2007. Stream Ecology. Springer, Dordrecht, The Netherlands.

Bate, G.C., Adams, J.B., 2000. The effects of a single freshwater release into the Kromme Estuary. 5. Overview and interpretation for the future. Water SA 26, 329–332.

Cambray, J.A., King, J.M., Bruwer, C., 1997. Spawning behaviour and early development of the Clanwilliam yellowfish (*Barbus capensis*; Cyprinidae), linked to experimental dam releases in the Olifants River, South Africa. Regul. Rivers Res. Manage. 13, 579–602.

Canada and the United States of America, 1964. The Columbia River Treaty.

Carter, N.T., C.R. Seelke, and D.T. Shedd, 2015. U.S.-Mexico water sharing: background and recent developments. Congressional Research Service, Library of Congress. Available from: https://www.fas.org/sgp/crs/row/R43312.pdf (accessed 24.03.16).

Cascades Environmental Services, 1991. Cedar River Instream Flow and Salmonid Habitat Utilization Study. Seattle, Washington: Final Report Submitted to Seattle Water Department.

Conservation Measures Partnership, 2014. Open Standards for the Practice of Conservation.

Council On Environmental Quality, 2013. Principles and Requirements for Federal Investments in Water Resources.

European Parliament, 2000. A framework for community action in the field of water policy. Available from: http://eur-lex.europa.eu/legal-content/EN/TXT/HTML/?uri = CELEX:32000L0060&from = EN (accessed 24.03.16).

Frissell, C.A., Liss, W.J., Warren, C.E., Hurley, M.D., 1986. A hierarchical framework for stream habitat classification: Viewing streams in a watershed context. Environ. Manage. 10, 199–214.

Gendaszek, A.S., Magirl, C.S., Czuba, C.R., Konrad, C.P., 2013. The timing of scour and fill in a gravel-bedded river measured with buried accelerometers. J. Hydrol. 495, 186–196.

Hall, A.A., Rood, S.B., Higgins, P.S., 2011. Resizing a river: A downscaled, seasonal flow regime promotes riparian restoration. Restor. Ecol. 19, 351–359.

Hammond, D., Mant, J., Holloway, J., Elbourne, N., Janes, M., 2011. Practical river restoration appraisal guidance for monitoring options (PRAGMO). The River Restoration Centre (RRC), Bedfordshire, UK.

Harnish, R.A., Sharma, R., Mcmichael, G.A., Langshaw, R.B., Pearsons, T.N., 2014. Effect of hydroelectric dam operations on the freshwater productivity of a Columbia River fall Chinook salmon population. Can. J. Fish. Aquat. Sci. 71, 602–615.

Horne, A., Szemis, J., Webb, J.A., Kaur, S., Stewardson, M., Bond, N., et al (accepted). Informing environmental water management decisions: using conditional probability networks to address the information needs of planning and implementation cycles. Environ. Manage.

Hynes, H.B.N., 1975. The stream and its valley. Verhandlungen Internationale Vereinigung für Theoretische und Angewandte Limnologie 19, 1–15.

King, J., Cambray, J.A., Dean Impson, N., 1998. Linked effects of dam-released floods and water temperature on spawning of the Clanwilliam yellowfish *Barbus capensis*. Hydrobiologia 384, 245–265.

Konrad, C.P. 2013. Flow Experiment Attributes Data Set. Working group evaluating responses of freshwater ecosystems to experimental water management. National Center for Ecological Analysis and Synthesis.

Korman, J., Kaplinski, M., Melis, T.S., 2011. Effects of fluctuating flows and a controlled flood on incubation success and early survival rates and growth of age-0 rainbow trout in a large regulated river. Transact. of the Am. Fish. Soc. 140, 487–505.

Magirl, C. S., Gendaszek, A.S., Czuba, C.R., Konrad, C.P., Marineau, M.D., 2012. Geomorphic and hydrologic study of peak-flow management on the Cedar River, Washington. U.S. Geological Survey Open-File Report 2012-1240.

Mekong River Basin Commission, 1995. Agreement on the Cooperation for the Sustainable Development of the Mekong River Basin.

Melis, T.S., Korman, J., Kennedy, T.A., 2012. Abiotic & biotic responses of the Colorado River to controlled floods at Glen Canyon Dam, Arizona, USA. River Res. Appl. 28, 764–776.

Murray Darling Basin Authority, 2014a. Basin-wide environmental watering strategy. Canberra, Australia.

Murray Darling Basin Authority, 2014b. Constraints Management Strategy 2013–2014. Canberra, Australia.

National Oceanic and Atmospheric Administration, 2014. Endangered Species Act Section 7(a)(2) Supplemental Biological Opinion, Consultation on Remand for Operation of the Federal Columbia River Power System, National Marine Fisheries Service Northwest Region, NWR-2013-9562. Seattle, Washington.

Olden, J.D., Naiman, R.J., 2010. Incorporating thermal regimes into environmental flows assessments: modifying dam operations to restore freshwater ecosystem integrity. Freshw. Biol. 55, 86−107.

Olden, J.D., Konrad, C.P., Melis, T.S., Kennard, M.J., Freeman, M.C., Mims, M.C., et al., 2014. Are large-scale flow experiments informing the science and management of freshwater ecosystems? Front. Ecol. Environ. 12, 176−185.

Poff, N.L., Richter, B.D., Arthington, A.H., Bunn, S.E., Naiman, R.J., Kendy, E., et al., 2010. The ecological limits of hydrologic alteration (ELOHA): a new framework for developing regional environmental flow standards. Freshw. Biol. 55, 147−170.

Poole, G.C., 2002. Fluvial landscape ecology: addressing uniqueness within the river discontinuum. Freshw. Biol. 47, 641−660.

Richter, B.D., Warner, A.T., Meyer, J.L., Lutz, K., 2006. A collaborative and adaptive process for developing environmental flow recommendations. River Res. Appl. 22, 297−318.

Robinson, C.T., Uehlinger, U., 2008. Experimental floods cause ecosystem regime shift in a regulated river. Ecol. Appl. 18, 511−526.

Rogers, K., Biggs, H., 1999. Integrating indicators, endpoints and value systems in strategic management of the rivers of the Kruger National Park. Freshw. Biol. 41, 439−451.

Rood, S.B., Gourley, C.R., Ammon, E.M., Heki, L.G., Klotz, J.R., Morrison, M.L., et al., 2003. Flows for floodplain forests: A successful riparian restoration. BioScience 53, 647−656.

Save Our Wild Salmon, 2014. The Salmon Community's Analysis of the Columbia-Snake River Salmon Plan.

Snelder, T.H., Biggs, B.J.F., 2002. Multiscale river environment classification for water resources management. J. Am. Water Resour. Assoc. 38, 1225−1239.

Snow, G.C., Bate, G.C., Adams, J.B., 2000. The effects of a single freshwater release into the Kromme Estuary. 2: Microalgal response. Water SA 26, 301−310.

Strauss, S.Y., 1991. Indirect effects in community ecology: Their definition, study and importance. Trends Ecol. Evol. 6, 206−210.

Strydom, N., Whitfield, A., 2000. The effects of a single freshwater release into the Kromme Estuary. 4: Larval fish response. Water SA 26, 319−328.

Sudan and the United Arab Republic, 1959. Agreement (With Annexes) For The Full Utilization of the Nile Waters.

Townsend, C.R., 1989. The patch dynamics concept of stream community ecology. J. North Am. Benthol. Soc. 8, 36−50.

U.S. Army Corps of Engineers, 2004. Hanford reach fall chinook protection program. Available from: http://www.nwd-wc.usace.army.mil/tmt/documents/wmp/2011/draft/app5.pdf (accessed 12.09.15).

U.S. Department of the Interior, 2012. Environmental assessment: Development and implementation of a protocol for high-flow experimental releases from Glen Canyon Dam, Arizona, 2011 through 2020. Salt Lake City, Utah: Bureau of Reclamation.

Vinson, M., 2001. Long-term dynamics of an invertebrate assemblage downstream from a large dam. Ecol. Appl. 11, 711−730.

Wiens, J.A., 2002. Riverine landscapes: taking landscape ecology into the water. Freshw. Biol. 47, 501−515.

HOW MUCH WATER IS NEEDED: TOOLS FOR ENVIRONMENTAL FLOWS ASSESSMENT

IV

EVOLUTION OF ENVIRONMENTAL FLOWS ASSESSMENT SCIENCE, PRINCIPLES, AND METHODOLOGIES

11

N. LeRoy Poff[1,2], Rebecca E. Tharme[3], and Angela H. Arthington[4]

[1]*Colorado State University, Fort Collins, CO, United States* [2]*University of Canberra, Canberra, ACT, Australia*
[3]*Riverfutures, Derbyshire, United Kingdom* [4]*Griffith University, Nathan, QLD, Australia*

11.1 INTRODUCTION

The science underlying environmental water allocations (environmental flows assessment science or environmental water science) is based in the quantification of the linkages between hydrological processes and components and various ecological variables. This understanding supports the recommendations and establishment of a water regime needed to manage rivers (and other wetland systems) in a more ecologically and socially sustainable fashion (Chapter 1). Principles based in hydroecological (or ecohydrological) theory and concepts (Acreman, 2016; Bunn and Arthington, 2002; Dunbar and Acreman, 2001; Hannah et al., 2007), combined with empirical literature and stakeholder or indigenous knowledge guide and inform environmental water science and its practical application. Achieving a more quantitative understanding of ecological and social responses to both natural flow variability and its alteration by humans is one of the fundamentally challenging goals of environmental water science and assessment in order to achieve more precise and effective environmental water regimes. The application of environmental water has largely been in the context of managing existing water resources infrastructure (e.g., dams, diversion weirs, direct abstraction points), which are globally ubiquitous causes of hydrologic alteration (Richter and Thomas, 2007). However, environmental water science is also increasingly needed to contribute to achieving conservation objectives through protection of flow regimes in rivers and basins under pressure from proposed developments (Acreman et al., 2014a; Poff, 2014). Typically, the focus is on the ecological response to purely hydrologic change, and less so on the other important environmental factors that can be strongly modified by large infrastructure (Chapter 21), primarily thermal and sediment regimes (Chapter 12). Further, environmental water science has disproportionately emphasized the biophysical aspects of ecosystems (Tharme, 1996). There is growing attention, however, on the social dimensions of environmental water, from inclusion of ecosystem services and their

Water for the Environment. DOI: http://dx.doi.org/10.1016/B978-0-12-803907-6.00011-5

203

flow-linked dynamics (Chapter 8) through to the broader areas of culture, indigenous water, and social justice (Finn and Jackson, 2011; Wantzen et al., 2016; Chapter 9).

Environmental flows science and assessment have been driven primarily by concerns over the extent and rapid pace of deterioration in the biodiversity, ecological condition, and ecosystem function of rivers where natural flow regimes have been partially or completely regulated by humans (Chapter 4). The scientific understanding of flow—ecological relationships is continually evolving, as is the technical ability to characterize and analyze the biophysical features of rivers. We now have the ability to ask sophisticated, complex questions that guide the setting of meaningful flow targets to restore regulated rivers or to sustain the conservation values of rivers where new water developments are planned. However, achieving successful environmental water implementation requires application of appropriate tools and methods (including modeling) that can lead to successful outcomes. This process of linking predictive science to successful ecological and societal outcomes creates an important tension in environmental water, one that drives the further evolution of science and methods.

Fundamentally, the framing of a particular environmental water management question determines which methods and tools are required to adequately describe flow—ecology relationships (Chapter 14) that help address that question. A wide range of frames exist: from single species to entire ecosystem management, from short-term experimental flows to long-term flow regime modification, from rapid desktop estimation to guidance for planning of new water infrastructure, and from detailed field-based assessment of flows to achieve river restoration at individual sites, to assessments spanning systems across a large landscape or region. Over the more than 50-year history of environmental water science and assessment, the questions, approaches, and tools have advanced and diversified, in response to changing societal objectives, world views, and values (Chapters 2, 7, and 9), an increasing knowledge base (including indigenous knowledge), and significant advances in modeling capabilities (Chapters 13 and 14).

As the global human population continues to increase and place more demand on freshwater resources, and as climate change simultaneously imposes unprecedented challenges for sustainable resource management, new questions and frameworks for managing environmental water are emerging. These will be increasingly transdisciplinary and drive the development of novel approaches that will move forward as hydroecological and social understanding increase through research and monitored implementation of environmental water management actions.

11.1.1 FOUNDATIONS AND TYPES OF ENVIRONMENTAL FLOWS ASSESSMENT METHODOLOGIES

Environmental flows assessments began in earnest in the late 1940s in snowmelt streams and rivers of the western United States, where their main objective was to protect valuable cold-water fisheries (Poff and Matthews, 2013; Tharme, 2003). Here, and in Europe, early minimum flow recommendations were also being made to mitigate the problems of poor water quality at low flows and for pollution control. Rapid progress in the 1970s was primarily a result of new environmental and freshwater legislation coupled with demands for quantitative assessment of flows to protect aquatic species impacted by dam construction, then at a peak in the United States. The US *Clean Water Act 1972* set the objective of restoring and maintaining the chemical, physical, and biological integrity of that nation's waters. These foundations gave rise to more formalized hydrological, hydraulic rating, and habitat simulation methods, which collectively stimulated global awareness and developments of the science and implementation of environmental water. England, Australia, South Africa, and New Zealand began to engage strongly in the topic in the 1980s, followed by Brazil, Japan, and several countries of continental Europe (Arthington and Zalucki, 1998; Dyson et al., 2003; Tharme, 2003). Each application brought fresh perspectives on the challenges and solutions to streamflow management in different hydroclimatic, biophysical, and socio-political realms (Acreman and Dunbar, 2004).

From the 1990s onward, within a putative *modern* era (Poff and Matthews, 2013; Fig. 11.1), the evolution and expansion of the science and practice of environmental water were explosive. By 2002, a global review of environmental flows assessment literature and practice (Tharme, 2003) identified over 207 methods and several broader frameworks used to assess the water requirements of aquatic species, habitats, or ecosystem features, and support flow management practices to meet ecological and, increasingly, also social targets. These methods were recognized to differ widely in the extent to which they were capable of adequately representing the water needs of ecosystems (Tharme, 2003). Several of the earlier methods, particularly, were largely carryover methods that addressed water pollution and were, subsequently, seriously limited in their ability to satisfactorily reflect the diverse water quantity needs of ecosystems and deliver truly sustainable ecological and social flow outcomes.

Despite various strengths and limitations (Table 11.1), by the year 2000 these diverse environmental flows assessment approaches were in use in at least 44 countries within six broad regions: Australasia (Australia and New Zealand), the rest of Asia, Africa, North America, Central and South America (including Mexico and the Caribbean), Europe, and the Middle East. Places that had previously shown little or no activity in this area began to take environmental flows assessment principles on board and to tailor existing approaches for local application, including additional countries in Latin America, eastern Europe, southern and eastern Asia, and various African countries and river basins (Tharme, 2003). Recent developments in China, Kenya, Tanzania, Colombia, and Brazil, among others, are now extending the influence of environmental water as a major plank of water management, although implementation is not without challenges (Le Quesne et al., 2010; Chapter 27). The concept of environmental water has a firm place in many intergovernmental agreements (e.g., the Convention on Biological Diversity signed by 168 countries, https://www.cbd. int/information/parties.shtml). Environmental water concerns are now integrated into water policy and legislation in countries and regions as diverse as Australia, South Africa, Lesotho, New Zealand, Costa Rica, Tanzania, Pakistan, Mozambique, China, the Philippines, the United States, and the European Union (Acreman and Ferguson, 2010; Arthington, 2012; Hirji and Davis, 2009; Le Quesne et al., 2010; Chapter 27). Speed et al. (2011) provide an interesting analysis of the ways in which some environmental water concepts and frameworks have been linked with those of water management in various countries.

The four main categories of methods that were evident early on (Tharme, 1996, 2003) remain relevant today—hydrological, hydraulic rating, habitat simulation, and holistic methods (Table 11.1), but with expansion of holistic (ecosystem) approaches to include the kinds of regional frameworks discussed in Section 11.5. The more recent methods continue to improve with regular applications on the ground and through ongoing advances in the science. Although these types of approaches continue to focus largely on rivers, many are applicable, with some modification, to other water bodies such as standing waters (e.g., marshes, lakes) that also experience natural spatial and seasonal patterns in water-level fluctuations, wetting and drying, and connections with groundwater (Arthington, 2012). Methods tend to be applied hierarchically (Tharme, 1996), from hydrology-based approaches common and more appropriate in a precautionary, low-resolution framing of environmental water requirements at a water resources planning level, to increasingly comprehensive assessments using holistic methods. The broad types of methods described by Tharme (2003) were also typically used for single or multiple river reaches within a river system. Recent method developments to support more advanced river basin and broader landscape water management are the subject of Section 11.5.

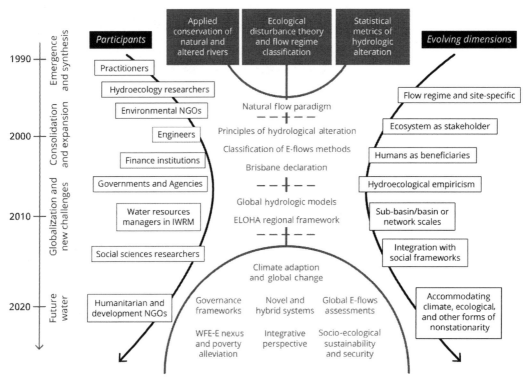

FIGURE 11.1

Historical timeline for "modern" environmental water, showing emerging directions in the principles and concepts that underpin the science and growth in the number and diversity of engaged institutions and practitioners. Timelines are shown that fall into relatively discrete periods of types of activities. Timelines for participants engaged in environmental water over time are shown to the left, for benchmark achievements in the center, and for evolving dimensions of environmental water on the right. ELOHA, ecological limits of hydrologic alteration; IWRM, Integrated Water Resources Management; NGO, nongovernmental organization; WFE-E, water, food, and energy-environment nexus.

Source: Adapted from Poff and Matthews (2013)

Table 11.1 Generalized Comparison of the Four Main Types of Methods and Frameworks Used Worldwide to Estimate Environmental Water Regimes for Rivers from Site to Regional Levels

Method Type	River Ecosystem Attributes/ Components Addressed	Knowledge and Expertise Required	Resource Intensity	Resolution of Output (Environmental Flow)	Appropriate Level(s) of Application
Hydrological	Whole ecosystem condition/ health, or nonspecific. Some include specific components (e.g., physical habitat, fish).	Primarily desktop, with low data needs. Use virgin/naturalized (or other reference state) historical flow records (daily, monthly, or annual). Single flow indices (often low-flow metrics), or more commonly multiple ecologically relevant flow metrics characterizing flow regime/whole hydrograph. Some use historical ecological data, hydraulic habitat data, or meta-analysis of results of multiple environmental water assessments to derive rules. Require expertise of a hydrologist. Few require ecological or geomorphological expertise, but such expertise is highly advantageous.	Low time and cost, and low or moderate technical capacity.	Mostly simple, flow targets for maintaining river health, based on estimates of the percentage of annual, seasonal, or monthly volume (often termed the minimum flow) that should be left in a river to maintain acceptable habitat or varying levels of river condition. Often expressed as % of monthly or annual flow (median or mean); or as limits to change in vital flow parameters, commonly low-flow indices. Low resolution, complexity, flexibility and confidence, or moderate and dynamic in a few more recent regime-focused methods.	Reconnaissance/planning level of water resource developments. Unsuitable for high-profile, negotiated cases, or where whole flow regime dynamics are critical. As a tool within habitat simulation or holistic methods. For highly data-deficient systems with limited ecological information. Regionalization potential for different river ecotypes.

Used widely in many developed and developing countries/basins. Simple single index, rule-of-thumb, and look-up table approaches (e.g., Montana method, Tennant, 1976; flow percentiles derived from Flow Duration Curve Analysis; Tharme, 2003, provides examples) becoming less common. Shift toward ecologically relevant flow metrics addressing multiple aspects of hydrological regime (e.g., Range of Variability approach, Richter et al., 1996; Environmental Flow Duration Curve, Smakhtin and Anputhas, 2006) and use of desktop models derived from meta-analyses of multiple environmental flows assessments (e.g., Desktop Reserve Model, Hughes and Hannart, 2003; Hughes et al., 2014).

(Continued)

Table 11.1 Generalized Comparison of the Four Main Types of Methods and Frameworks Used Worldwide to Estimate Environmental Water Regimes for Rivers from Site to Regional Levels *Continued*

Method Type	River Ecosystem Attributes/ Components Addressed	Knowledge and Expertise Required	Resource Intensity	Resolution of Output (Environmental Flow)	Appropriate Level(s) of Application
Hydraulic rating	Aquatic (instream) physical habitat for target species or assemblages.	Low to moderate data needs. Desktop analysis and limited field surveys. Historical flow records. Discharge linked to hydraulic variables, typically single river cross-section/transect. Single or multiple hydraulic variables. Require moderate expertise (hydrologist, field hydraulic habitat assessment, and modeling). Few require ecological or geomorphological expertise.	Mostly low, sometimes moderate time, cost, and technical capacity.	Hydraulic variables (e.g., wetted perimeter, depth) used as surrogate for habitat flow needs of target species or assemblages. Low, sometimes moderate, resolution, complexity, flexibility, and confidence.	Water resource developments where little negotiation is involved. As a tool within habitat simulation or holistic methods.

Used widely historically, mostly in developed countries (see Annear et al., 2004; Arthington, 2012; Tharme, 2003), but nowadays largely superseded or used as one of several integrated habitat modeling tools in habitat simulation or holistic methods (e.g., used within DRIFT, Arthington et al., 2003; King et al., 2003).

Method Type	River Ecosystem Attributes/ Components Addressed	Knowledge and Expertise Required	Resource Intensity	Resolution of Output (Environmental Flow)	Appropriate Level(s) of Application
Habitat simulation	Primarily instream physical habitat for target species, guilds, or assemblages. Some also consider channel form, sediment transport, water quality, riparian vegetation, wildlife, recreation, and esthetics.	Moderate to high data needs. Desktop, and field surveys. Historical flow records, typically average daily discharge. Few to many hydraulic variables are modeled at a range of discharges at multiple river cross-sections. Physical habitat availability, utilization, and preference data, or similar models, for target biota. A few use statistical summary methods based on results of multiple physical habitat studies.	High to sometimes moderate time, cost, and technical capacity.	Output in the form of weighted usable area (WUA) or similar habitat metrics for target biota (fish, invertebrates, plants). Often includes comparative analyses of time series of habitat availability. and duration and use. Moderate to high resolution, complexity, and confidence, moderate flexibility.	Water resource developments, often large scale, involving rivers of moderate to high strategic importance, often with complex, negotiated trade-offs among users. Commonly used as a method within holistic approaches and frameworks. Useful to examine a variety of alternative environmental water regime scenarios for several species/life stages/ assemblages.

	High level of expertise, with hydrologist, hydraulic habitat modeler. May use hydrodynamic modeling, GIS/remote sensing, ecological or geomorphological expertise			

Move away from single-species focus to increasing use for needs of species, guilds, and assemblages (IFIM, Bovee, 1982; see examples in Annear et al., 2004; Arthington, 2012; Tharme, 2003). Primarily applied in developed countries, using increasingly sophisticated and multidimensional (eco)hydraulic habitat modeling (e.g., Lamouroux and Jowett, 2005). Less commonly used in developing countries/basins, and then tending to be one of a suite of tools used to set environmental water within holistic approach (e.g., USAID, 2016).

Holistic (ecosystem) methods and frameworks	Entire ecosystem, all or several ecological components. Most consider instream and riparian components, some also consider groundwater, wetlands, floodplains, deltas, estuaries, lagoons, coastal waters. Few consider geomorphic processes (e.g., sediment dynamics, channel adjustments), or ecological functions/processes (e.g., nutrient dynamics, food web structure). Several explicitly address social and economic (e.g., livelihoods of rural subsistence users, human health) dependencies on species, ecosystem resources, and processes (i.e., ecosystem services, e.g., fisheries).	Typically, moderate to high knowledge and expertise, but several used in data-poor contexts. Desktop and often field studies (seasonal or more intensive). Many reliant on mix of data and expert judgment, using expert panels. Some use both scientific and traditional knowledge to develop or infer flow—ecology—social relationships. Use virgin/naturalized historical flow records, or rainfall records/other data for ungauged sites. Several use hydraulic habitat variables from multiple cross-sections. Typically use biological data on flow—ecology relationships for lifecycle stages of aquatic and riparian species, assemblages and components (e.g., fish migration and spawning cues, riparian water quality tolerances, exotic species requirements).	Moderate to high time, cost, and technical capacity.	Recommended hydrological regime linked to explicit quantitative or qualitative ecological, geomorphological, and sometimes, social and economic responses and consequences. Some address environmental water regimes for dry or wet years. Moderate to high complexity and confidence. Typically, high resolution and flexibility. Several with potential to generate outputs for multiple scenarios (past, future). Some explicitly address probabilities, interaction effects, risk, and/or uncertainty. A few incorporate climate change.	Water resource developments, typically large scale, involving rivers of high conservation and/or strategic importance, and/or with complex, negotiated trade-offs among stakeholders. Simpler approaches (e.g., expert panels) often used in basin contexts where flow—ecology knowledge is limited, and limited trade-offs exist among users, and/or time, resources, and capacity constraints exist. Used in planning stage of new developments to protect high conservation values. Also used in highly modified or novel ecosystems, with focus on flow regime to deliver specific restoration objectives, or to address socio-ecological values and services in novel ecosystems.

(Continued)

Table 11.1 Generalized Comparison of the Four Main Types of Methods and Frameworks Used Worldwide to Estimate Environmental Water Regimes for Rivers from Site to Regional Levels *Continued*

Method Type	River Ecosystem Attributes/Components Addressed	Knowledge and Expertise Required	Resource Intensity	Resolution of Output (Environmental Flow)	Appropriate Level(s) of Application
		Range of experts from different disciplines, including ecologists, hydrologists, and often a geomorphologist. Several include social scientists, other specialists (e.g., water chemistry, health), water managers.		Quantified environmental water release rules or standards for rivers of contrasting hydrological type or ecotype and points of management interest, at user-defined regional scale(s). Flow alteration-ecological/ social response relationships by river type. As for other holistic methods.	As for other holistic methods. Large systems/basins or aggregations of smaller ones, regions, entire states, or multiple projects. May be integrated with water management systems.
Regional and landscape-level holistic approaches.	As for other holistic methods, but for large-scale system(s).	Designed to use existing data sets and knowledge. In some cases, includes collection of new data, or modeling for system locations of interest for which hydrological and/or ecological data are absent.	As for other holistic methods.		

Increasingly common in developing and developed countries (e.g., BBM, King and Louw, 1998; Benchmarking, Brizga et al., 2002). Recent attention in developed regions focused on in-depth analysis of ecosystem components and, less commonly, functions/processes. Used regularly in developing countries, including for capacity development, and in complex basins with development pressures and, in many cases, communities with clear dependencies on aquatic systems (e.g., DRIFT, Arthington et al., 2003, 2007; Blake et al., 2011; King and Brown, 2010; King et al., 2000, 2014; Lokgariwar et al., 2014; McClain et al., 2014; Speed et al., 2011; Thompson et al., 2014; USAID, 2016). At regional scale, most applications are adaptations of a single framework, the Ecological Limits of Hydrologic Alteration (ELOHA; Poff et al., 2010; e.g., Arthington et al., 2012; James et al., 2016; McManamay et al., 2013; Rolls and Arthington, 2014; Solans and de Jalón, 2016) or similar approaches (e.g., Kendy et al., 2012). Expansion underway from applications in a few developed countries, to pilots in several developing countries, and increasing numbers of applications in large developed basins, with explicit links to water management tools and decision support systems (e.g., PROBFLO, Dickens et al., 2015).

Current practice is summarized below each method type, with select application examples from various world regions (for additional details of methods and case studies, see Acreman et al., 2014b; Arthington, 2012).

Source: Adapted from Tharme (2003)

The first *hydrological methods* were simple, low-resolution estimates of the percentage of annual, seasonal, or monthly flow volume (often termed the *minimum flow*) that should be left in a river to maintain minimal fish habitat and/or acceptable stream condition (e.g., single figure flow recommendations based on low-flow indices, derived from flow duration curves such as Q95; Tharme, 2003). Hydrological methods are often called fixed-percentage or look-up table methods, as they rely on formulae linked to historical flow records to estimate desirable discharges. Unusual for its time, the Montana method (Tennant, 1976) stands apart from such desktop approaches inasmuch as the look-up table of percentages of average annual flow, which correspond to different degrees of desired river condition, was derived from an empirical base of field-level flow habitat and ecological (fish) studies of many small US streams of specific biophysical character. Without appropriate validation for streams in new geographic regions or of different types, the use of the tabulated flow levels carries the risk of setting environmental water recommendations that are unsuitable (e.g., too high or too low) for local conditions; untested extrapolations of this kind remain a challenge common to many environmental flows assessment methods (Arthington, 2012). Hydrology-based methods have been variously elaborated over the years, and in the last decade or so have substantively advanced by taking a more regime-based approach that estimates a range of ecologically relevant streamflow characteristics such as magnitude, frequency, timing, and the duration of specific flood and low-flow events (e.g., Hughes and Hannart, 2003; Hughes et al., 2014; National Water Commission of Mexico, 2012; Richter et al., 2012).

Hydraulic rating methods emerged in parallel, with the intent to quantify how flowing water interacted with channel boundaries to create aquatic habitats of varying depth, velocity, substrate, and cover characteristics that varied over time with discharge pattern. The best known of these wetted perimeter methods define a minimum acceptable discharge that maintains wetted aquatic habitat for selected (*representative* and/or *critical*) channel cross-sections or stream reaches. These foundational hydraulic habitat methods paved the way for *habitat simulation* methods and associated tools (e.g., the *Physical Habitat Simulation [PHABSIM]* component of the *Instream Flow Incremental Methodology [IFIM]*; Bovee, 1982) and later, sophisticated innovations focused around two- and three-dimensional habitat modeling supported by spatially referenced habitat mapping techniques (Chapter 13). Habitat simulation methods, by virtue of their quantitative data and analytical demands, have been applied primarily to the description of habitat flow conditions for a few species (usually valued fish). Although efforts to expand these methods to the broader contexts of entire fish and macroinvertebrate communities have occurred, including through modeling of a range of habitat types occupied by different functional guilds (e.g., King and Tharme, 1994; Leonard and Orth, 1988; Parasiewicz, 2007), the many other facets of the flow regime and complex ecosystem requirements of rivers (Bunn and Arthington, 2002; Poff et al., 1997) require different kinds of methods (Section 11.3). Habitat simulation methods remain a vital and well-utilized tool, however, in desktop level (e.g., Hughes et al., 2014), habitat (with a focus on individual species threatened by loss of critical habitats), and holistic methods (Tharme, 2003). For examples of use in holistic frames, see Arthington et al. (2003), Blake et al. (2011), Illaszewicz et al. (2005), King et al. (2000), O'Keeffe et al. (2002), and USAID (2016).

11.2 HOLISTIC ENVIRONMENTAL FLOWS ASSESSMENT METHODS

11.2.1 EVOLUTION OF PRINCIPLES AND APPROACHES

The *holistic* or whole ecosystem perspective emerged conceptually from the growth in scientific understanding of flow–ecology relationships (see Section 6.3 and Chapter 14), as well as from a pressing need to reflect the flow conditions necessary to maintain the structure and function of entire ecosystems and the local communities and livelihoods they support (as was the case in South Africa; Tharme and King, 1998). By the late 1980s, the focus of river restoration and conservation or protection had begun to broaden well beyond individual species. Informed by advances in ecological theory, community and ecosystem perspectives explicitly incorporated the principle that hydrologic variability and disturbance are key to maintaining a dynamic aquatic environment that provides varying conditions under which many species can coexist over time (e.g., Resh et al., 1988). It similarly drew on new directions in geomorphological understanding of channel-forming processes over the same period (e.g., Hill et al., 1991; Newson and Newson, 2000; Petts and Calow, 1996; Rowntree and Wadeson, 1998).

This rapid transition to a new category of methods can be attributed to three separate strands of research and application that emerged in this period and eventually coalesced into the foundation of contemporary environmental water science (Fig. 11.1). One strand emerged simultaneously and collaboratively in Australia (Arthington, 1998; Arthington et al., 1992) and South Africa (King and Tharme, 1994; Tharme and King, 1998), where resource scientists were pursuing environmental water assessment and allocations based on multiple ecological targets, using modern hydroecological principles applied through expert judgment in specific, site-based river applications. These researchers and practitioners made fundamental contributions to the conceptual framing and elaboration of holistic environmental flows assessment frameworks to guide management of environmental water allocations (e.g., benchmarking and flow restoration methods, Arthington and Pusey, 2003; hierarchical application of methods and inclusion of social flow dependencies, King et al., 2003; Tharme, 1996) that are still widely used today.

A second source of activity was more academic, developing from investigation of how hydrologic disturbance in free flowing rivers could be analyzed statistically and classified across large hydroclimatic gradients. Following on Resh et al.'s (1988) articulation of the critical role of disturbance (extreme hydrologic events) in shaping the structure and function of lotic ecosystems, Poff and Ward (1989, 1990) developed a streamflow classification scheme based on flow metrics defined specifically to be ecologically relevant in covering the full range of hydrologic variation (from zero flow to peak flows). Similar regime-based ideas emerged in the United Kingdom (Gustard, 1979), Australia (Hughes and James, 1989), New Zealand (Biggs et al., 1990), and South Africa (Joubert and Hurly, 1994), as well as at the global scale (Haines et al., 1988). Flow metrics were extracted from long-term hydrographic data for unimpaired rivers to reveal geographically variable patterns of streamflow variability characterized by magnitude, frequency, duration, timing, and predictability of flow levels judged to be functionally important to riverine community and ecosystem processes. The focus on the theoretical role of natural disturbance regimes in regulating species performance and shaping whole communities and ecosystem states laid a strong foundation for future environmental water science and its application to rivers where hydrologic regimes, particularly their temporal variability and disturbance characteristics, were modified by dams or extensive water abstraction.

A third strand arose from the general perspective of river conservation and a desire to provide a method to enable managers to easily understand and more effectively manage the interrelationships between the types of flow alteration by dams and the ecological impairment downstream of them. The *Indicators of Hydrologic Alteration (IHA)* software program was developed by The Nature Conservancy and partners (Mathews and Richter, 2007; Richter et al., 1996) as a tool that calculated a diverse range of simple, but meaningful statistical metrics to characterize the alteration (in magnitude, frequency, duration, timing, rate-of-change, and variability) of the ecologically relevant components of an impaired flow regime. This readily available tool could be employed simply and universally in methods such as the *range of variability approach (RVA;* Richter et al., 1997), wherever there are discharge time series of pre- and/or postimpact hydrographs. The desktop IHA program has been highly successful, providing a common technical platform for assessment and comparison of flow alteration (as well as trend and seasonal comparative analyses) in rivers across the globe.

These three strands coalesced under the *natural flow regime* paradigm (Poff et al., 1997; Richter et al., 1997; Stanford et al., 1996), which later formed one of the foundational principles of the Brisbane Declaration (2007), an influential call for universal implementation of environmental water (see Arthington et al., 2010). *The natural flow regime (NFR)* paradigm provided a theory-based approach to understand the ecological roles of dynamic flow regimes, and a basis for evaluating the ecological consequences of flow alteration in any particular hydroclimatic setting (Bunn and Arthington, 2002; Lytle and Poff, 2004). It also supported frameworks for guiding environmental water efforts of two main types—environmental water to support the conservation of minimally impaired rivers and environmental water to guide restoration of highly flow-altered rivers (Arthington et al., 2010). The former circumstance requires maintaining components of an NFR to conserve biodiversity and ecosystem functions and services, whereas the latter is concerned with the restoration of key ecologically and socially relevant flow characteristics that have been lost through regulation or water abstraction.

A fundamental aspect of holistic approaches has been to use a long-term hydrologic time series of daily or monthly flows to derive a set of static flow metrics that quantify various aspects of the magnitude, frequency, timing, duration, and rate-of-change in discharge. The question as to which of many hundreds of flow metrics to use to characterize flow alteration in rivers has been approached in many ways, with the option to identify a final smaller, nonredundant suite through statistical filtering of metrics that are highly intercorrelated or convey redundant information (Olden and Poff, 2003). However, final selection of metrics is often largely arbitrary or highly context-dependent. Another approach to constrain metric selection has been to focus on key *functional* aspects of the flow regime that have clear ecological significance (e.g., Arthington and Pusey, 2003; Mathews and Richter, 2007; Yarnell et al., 2015). Regardless of the number of metrics used, the basic approach to quantifying hydrologic alteration has been to compare a current, modified hydrograph against a preimpact *baseline* hydrograph, which has generally been considered to reflect the *natural range of variation* (Auerbach et al., 2012; Richter et al., 1996) in ecologically relevant flow dimensions for a particular stream or river.

This general approach has allowed great ability to statistically compare river flow regimes in different hydroclimatic and geologic settings across large geographic extents. This facilitated classification of flow regime *types*, which in turn provided a template for how to stratify streams and rivers into hydrological classes that could be treated as similar management units (Arthington, et al., 2006; Poff et al., 2010). Flow typologies for unregulated (*natural*) rivers have proliferated over the years, and are available nationally for, among others, the United States (McManamay et al., 2014; Poff, 1996; Poff and Ward, 1989), Australia (Hughes and James, 1989; Kennard et al., 2010), New Zealand (Biggs et al., 1990), and individual large river basins such as the Huai River, China (Zhang et al., 2012), the Ebro River, Spain (Solans and Poff, 2013), and the Magdalena River Basin, Colombia (Walschburger et al., 2015). Some international comparisons of flow differences have also been pursued (e.g., Poff et al., 2006a,b). Extensive stream gauge networks have also allowed for the quantification of hydrologic alteration of regulated rivers (Carlisle et al., 2010; Mackay et al., 2014) and calls for a regulated river classification (Poff and Hart, 2002). However, streamflow classification systems for regulated rivers have not been developed as successfully, in part because of the somewhat idiosyncratic nature of flow alteration below dams (Mackay et al., 2014; McManamay et al., 2012; Poff et al., 2007).

11.2.2 A SURVEY OF HOLISTIC METHODS

The ease with which hydrologic data can be manipulated and placed into an *ecologically relevant* context led to new tools that could potentially guide environmental water assessment. For example, the RVA (Richter et al., 1996) was proposed as a simple rule to sustain a *normative* pattern of hydrologic variability (Stanford et al., 1996) in regulated rivers in the absence of strong empirical flow—ecology relationships. A similar hydrological approach of using *sustainability boundaries* around the natural (historical or other baseline) flow regime to establish presumptive guidelines for environmental water assessment has also been proposed, though rarely applied to date. There has, however, been a politically mobilized operational rule curve for Owen Falls Dam on the White Nile in Uganda, based on similar principles and in place since 1954 (M. McClain, pers. comm.). The approach uses risk-based envelopes reflecting decreasing levels of ecosystem protection with greater departure from the reference flow regime (Brizga et al., 2002; Richter et al., 2012). One disadvantage to this approach is that it is based on a statistical comparison of long hydrological time series and therefore not designed to guide short-term operations of dams to deliver environmental water. A similar precautionary, regime-based approach to setting environmental water regime standards is under development in the Magdalena Basin, Colombia (R. Tharme, pers. comm.).

The *building block methodology (BBM;* King and Louw, 1998; King et al., 2000) has been applied numerous times and at various levels of resolution in South Africa, as one of the standard methods for ecological reserve determination (King and Pienaar, 2011). It has also been adapted for local application in eastern African river basins (e.g., Rufiji Basin, Tanzania; USAID, 2016). In South Africa, environmental water regime prescriptions from available BBM applications have been grouped according to river hydrology, to extract more general hydroecological principles to guide assessment in rivers of contrasting hydrology such as the desktop reserve model (Hughes and Hannart, 2003). Its successor, the habitat flow-stressor response method, which is well aligned with the local holistic methods, is now widely applied for desktop-level reserve determination (Hughes et al., 2014). Some holistic methods, such as the BBM (e.g., Alfredsen et al., 2012) and Savannah method (Warner et al., 2014), have been developed for more proximate operational use for dynamic flow allocation.

The *Downstream Response to Imposed Flow Transformation (DRIFT)* methodology (King et al., 2003) evolved out of the BBM as a top-down, scenario-based alternative founded on a similar basis of flow—ecology relationships. King et al. (2003) developed DRIFT to meet the demand for a rigorous and transparent approach with the capacity to evaluate alternative scenarios of water management in data-rich and data-limited situations. The DRIFT methodology involves a robust collection of field and desktop procedures, database systems, and decision support tools for proactive planning of environmental water regimes for rivers (King et al., 2003). The DRIFT framework has been refined through many applications in major river basins in Africa and southeast Asia. It merged into a platform for Integrated Basin Flow Assessment, which can assess future development options in terms of potential changes in a wide range of river characteristics (e.g., channel configuration, riparian forests, estuarine water quality, and invasive species) and flow-related ecosystem services (e.g., food fish; Arthington et al., 2003), societal health, and well-being (Brown and Watson, 2007; King and Brown, 2010; King et al., 2014).

Another widely applied holistic framework is the Benchmarking Methodology (Brizga et al., 2002), used to set provisional limits on levels of hydrological alteration in many relatively undeveloped rivers in the coastal catchments of Queensland, Australia (see Arthington, 2012). Comparisons are made between near-natural *reference* reaches and a set of *benchmark* reaches that have experienced differing levels of impact resulting from existing water resource development (e.g., dams, weirs, pumped abstraction, or interbasin transfers of water). It can generate many scenarios of hydrologic alteration that are used to forecast likely ecological consequences and thus inform recommendations for flow regime implementation (Arthington, 2012). A limitation of this methodology is that it cannot be applied in places with few existing water infrastructure developments, because they are required to provide the negative benchmarks and ecological condition ratings that underpin risk assessment of future flow regime changes.

A generic framework for environmental water assessment that is holistic in scope, scenario-based, and that integrates scientific and social processes, has been developed and piloted in China (Gippel and Speed, 2010; Speed et al., 2011). It utilizes conceptual models of flow interrelationships with ecology, geomorphology, and water quality, to help establish environmental water release rules to maintain riverine and riparian ecological assets in a state of health that matches the agreed management objectives of stakeholders. Addressing river flows using a consistently applied environmental flows assessment methodology (e.g., some form of holistic ecosystem approach, Acreman and Ferguson, 2010) is also a goal of the *Water Framework Directive (WFD)* of the European Union (Council of the European Communities, 2000). The WFD sets the policy framework for restoration of aquatic ecosystems to *good ecological status* (referenced to aquatic biology) in most European rivers, lakes, and wetlands by 2015 (see also recent guidance under the European Commission, 2015).

A common challenge in the application of any environmental flows assessment method is how to characterize and account for uncertainty in relationships between flow alteration and ecological (or social) responses (Arthington et al., 2007; Stewart-Koster et al., 2010). Risk-based and probabilistic methods such as Bayesian methods and networks (Chapter 14) are a clear growth avenue for holistic methods to improve decision making in the face of such uncertainty. For example, the holistic method, Probflo, incorporates probabilistic elements of Bayesian network modeling and ecological risk assessment to evaluate probable socio-ecological consequences of altered flows and establish environmental water requirements for rivers, and it has been applied in Lesotho (Dickens et al., 2015). Efforts are also underway in several regions (e.g., in the Mara River Basin of Kenya and Tanzania) to further tailor Probflo for landscape-level application, borrowing from the architecture of *Ecological Limits of Hydrologic Alteration (ELOHA)* (Section 10.5). Bayesian network models that effectively integrate indigenous and scientific knowledge (Jackson et al., 2014) have supported environmental water decision making in the Daly River, Australia, showing, for example, that if current extraction entitlements were to be fully utilized, significant impacts on the populations of two iconic fish species would occur (Chan et al., 2012).

11.3 FLOW—ECOLOGY AND FLOW ALTERATION—ECOLOGICAL RESPONSE RELATIONSHIPS

At the core of environmental water science and assessment is the relationship between attributes of the flow regime and ecological responses to natural flow variability and to flow alterations, so-called flow—ecology relationships or flow alteration—ecological response relationships, respectively (Arthington et al., 2006; Poff et al., 2010). A tension that has persisted in environmental flows assessment science and application is the question of how much empirical testing of these hydroecological relationships is needed to justify curtailment of water development or flow restoration actions, both of which are often costly and socially challenging to implement. The first principles argument that *ecologically relevant* components of a flow regime should be retained or restored is generally supported by ecological theory (Bunn and Arthington, 2002; Monk et al., 2007; Poff and Ward, 1989; Poff et al., 1997). In the absence of quantitative understanding of how specific levels of magnitude, frequency, duration, and timing can be combined to achieve a desired level of ecological response, the principle of mimicking aspects of the NFR is often invoked from the conservation perspective of the so-called precautionary principle (*sensu* Myers, 1993).

However, a full restoration of natural flow pattern and timing is hardly feasible or necessarily desirable (Chapter 10) below all dams; therefore, the question arises: how much flow conservation (protection) or restoration is needed to effectively preserve or restore the ecosystem to some stated or desired level of condition? This question is long-standing in environmental water science (Richter et al., 1997), but human preferences are now acknowledged as a critical component of setting environmental water objectives (Acreman et al., 2014a; Brisbane Declaration, 2007; Poff et al., 2010). Setting specific restoration targets requires adequate empirical understanding of the relationship between flow alteration and ecological response (Arthington et al., 2006; Davies et al., 2014). Of importance here is whether there are critical thresholds above or below which key functions or elements of the ecosystem are impaired or lost (Arthington et al., 2006). These may include sufficient high flow to connect the river with its floodplain, or move sediment (Yarnell et al., 2015), or low flows that maintain within-channel connectivity between shallow (e.g., riffle) areas. Many flow—ecology relationships are better expressed as continuous curves and, therefore, pose greater challenges in identifying potential thresholds of change to guide management decisions. Irrespective of the nature of the flow—ecology relationship(s), the level of environmental water allocated should be decided using informed judgment or through a negotiated socio-political trade-off among all resource uses and users, including the river ecosystem (Acreman et al., 2014b; Arthington, 2012).

A persisting challenge for environmental water science is to develop robust and, if possible, transferrable flow—ecology relationships (Chapters 13 and 14). Environmental water knowledge generally derives from empirical flow—ecology relationships discovered through both extensive observational studies and manipulating and monitoring dam releases (see Poff et al., 2003), and a large body of environmental water knowledge has accumulated through these avenues

(Arthington, 2012; Olden et al., 2014; Chapter 14). However, environmental water knowledge can also be cogenerated by including local or indigenous perspectives on hydroecological relationships (Jackson et al., 2014). Given the many ways that flow variables and ecological response variables can be (and have been) defined and associated with one another at numerous spatial and temporal scales, generalization and transferability have proven a challenge. To date, generalized flow–ecology relationships that apply across hydroclimatic gradients and complex types of hydrologic alteration caused by dams have proven difficult to extract (Carlisle et al., 2011; Lloyd et al., 2003; Olden et al., 2014; Poff and Zimmerman, 2010; Webb et al., 2015). One incompletely explored avenue for developing a framework for generalization and transferability is the notion that *rules* could be developed for different hydrological types of streams (Arthington et al., 2006; Kennard et al., 2010; Poff and Ward, 1989; Reidy Liermann et al., 2012). Alternatively, different stream types may respond to dams and forms of flow regulation in particular ways (McManamay et al., 2012; Mackay et al., 2014; Poff and Hart, 2002; Poff et al., 2007; Rolls and Arthington, 2014). These avenues toward generalization and transferability merit more research.

A key working premise (assumption) of environmental water assessment has been that flow is a master variable and alteration of the flow regime leads directly to significant ecological responses. Environmental water has primarily and deliberately focused on flow management, because preventing or reversing flow alteration is a necessary condition to sustain or restore the ecological integrity of riverine species and ecosystems. Moreover, reregulation of the flow regime below a dam is relatively easily achieved compared to other types of environmental modification such as altered thermal and sediment regimes. However, from the fuller perspective of riverine ecology, we would expect to only partially explain flow–ecology relationships if we do not include key partially independent variables such as temperature (Olden and Naiman, 2010), sediment (Wohl et al., 2015), and species interactions (Shenton et al., 2012). Thus, the *noisy* flow–ecology relationships that emerge from literature reviews (e.g., Olden et al., 2014; Poff and Zimmerman, 2010) are to be expected.

Expanding the environmental water domain to include other key drivers of ecological process and pattern is not a new idea (e.g., Olden and Naiman, 2010; Poff et al., 2010; Chapter 22). Relationships between flow, thermal, and sediment alteration are likely to be site-specific (or perhaps stream class-specific), raising further challenges in how to adequately characterize environmental change in a generalized, quantitative, and transferrable framework that can guide effective ecological conservation or restoration. Comodeling of thermal and flow conditions is not well developed, but could be improved to help guide reoperation of flow depth releases to improve downstream temperature regimes for ecological recovery (e.g., see Maheu et al., 2016; Olden and Naiman, 2010).

Incorporation of sediment dynamics and channel structure has proven more challenging for environmental water science and implementation, and the need has been long recognized (Hill et al., 1991; Newcombe and Jensen, 1996; Poff et al., 2006 a,b; Rowntree and Wadeson, 1998; Tharme, 1996). Modeling the *natural sediment regime* is difficult (Wohl et al., 2015) because sediment dynamics depend on many catchment characteristics (soils, geology, topography, vegetation) and regional precipitation regime. It is important, however, because in regulated rivers where sediment dynamics are severely out of balance due to upstream storage of sediment by a dam, the *hydraulics* of habitat can change dramatically, such that flow restoration can lead to further habitat modification downstream rather than habitat restoration (see Wohl et al., 2015; Yarnell et al.,

2015). Thus, hydrology-based models that fail to account for flow interactions with the channel boundaries and transport of sediment will incompletely capture habitat dynamics and ecological response potential below many dams due to the sediment–water disequilibria.

There is a need to integrate hydraulic and hydrologic models into environmental water applications (Chapter 13); however, a particular challenge is the issue of scale. Hydraulic habitat models are data hungry and best applied at local (reach) scales. The regional nature of many environmental water applications (see below) places logistical and cost constraints on the extent to which hydraulic models and data can be employed in multisite models. Simple geomorphic classification of stream reaches across stream networks (based on slope, geology, etc.) can act as first-order hydraulic surrogates (Poff et al., 2010), but a better approach will be to develop hydraulic regionalization models (e.g., Wilding et al., 2014) that can be combined with hydrologic models to provide insight into habitat structure and dynamics across many sites simultaneously.

11.4 FROM LOCAL TO REGIONAL SCALES OF ENVIRONMENTAL WATER ASSESSMENT

In the past, environmental water methods have generally focused on management of flow regimes in the highly regulated sections of rivers downstream of major water storages. These reaches are often in need of significant flow management to attain some desired ecological condition; however, the consensus among riverine scientists is that ecological impairment from flow alteration is not restricted to river reaches immediately below dams (Acreman et al., 2014b; Bunn and Arthington, 2002; Dudgeon et al., 2006; Nilsson et al., 2005; Poff and Matthews, 2013). Consequently, a broader scope for environmental water management would extend to entire river systems in a regional context, where the network may include a mix of regulated and still undeveloped streams/rivers. Developing a regional (or multisite) environmental water approach is made difficult by the lack of detailed hydrologic and/or ecological information at poorly studied localities, and by the reality that most unstudied sites do not have records of known reference (prealteration) conditions (Flotemersch et al., 2015). Arthington et al. (2006) addressed this challenge by proposing to classify streams in a region into similar streamflow regime types based on unimpaired hydrology; the range of variation in the flow components defining the NFRs across the replicate, unimpaired streams for that class type could be considered as the bounds for *reference* hydrologic conditions. Localities with ecological data could be assigned to a particular flow class and then flow–ecology relationships generalized for unstudied streams in that same class. Through modeling, flow-altered streams could be assigned to the most appropriate unimpaired flow regime type, such that even unstudied, flow-altered streams could be statistically assigned a reference condition. For example, highly seasonal snowmelt-driven streams have different hydrologic conditions of magnitude, frequency, timing, duration, and rate-of-change than do stable groundwater-dominated streams (see Arthington et al., 2006; Poff and Ward, 1989) and the ecological effects of flow alteration by dams on these different types of rivers should manifest differently. This approach has an explicit empirical, hypothesis-driven foundation.

The ELOHA (Poff et al., 2010) extended the Arthington et al. (2006) approach by formalizing a scientific and social framework that can be used to guide the setting of flow prescriptions at the stream reach scale throughout entire stream networks (Fig. 11.2). ELOHA is a framework that represents a coalescence of precursor methods and approaches in environmental water science and assessment. It provides a multistep process of building a hydrologic foundation (modeled hydrographs for ungauged stream segments), streamflow classification, a further geomorphic substratification, assigning flow-altered streams to preimpact reference stream types, developing flow—ecology and flow alteration—ecological response relationships for stream types, and incorporating social preferences in the environmental water ecological targets. A guiding principle of ELOHA is that ecological responses to particular features of the altered flow regime can be interpreted most robustly and usefully when there is some mechanistic or process-based relationship between the ecological (or social) response and the particular flow regime component (Poff et al., 2010). Flow alteration—ecological response relationships can be compiled from existing data, or new data collected along a flow regulation gradient, and tested statistically to determine the form (e.g., threshold, linear) and degree of ecological change (positive or negative) associated with a particular type of flow regime alteration (Arthington et al., 2006).

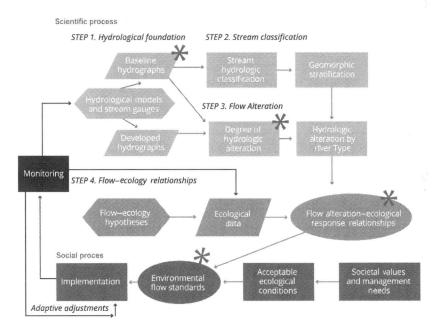

FIGURE 11.2

Framework for determining environmental water requirements for multiple sites throughout a region (river network[s]) as presented by the *Ecological Limits of Hydrologic Alteration (ELOHA)*. Stars indicate steps in the scientific process where the assumption of stationarity is implicit. Implementation, monitoring, and adaptive refinement of the recommended flows form part of the intimately linked social process, which can trigger additional scientific analysis or generation of new knowledge.

Source: Modified from Poff et al. (2010)

The ELOHA framework can be applied at various degrees of resolution in hydrologic data, hydroecological understanding, and ecosystem types. For example, Zhang et al. (2012) demonstrate how flow classification based on monthly flow records, although not ideal, can provide useful insights to support environmental water assessments and restoration in rivers where daily flow records are typically unavailable. Similarly, Sanderson et al. (2012) used monthly hydrologic data to assess potential risk to socially valued ecosystem components due to river damming and stream diversion throughout a small river basin. Studies following the ELOHA framework are developing new hydroecological knowledge for hydrological river classes in several regions (e.g., James et al., 2016; Mackay et al., 2014; McManamay et al., 2013; Rolls and Arthington, 2014).

The ELOHA framework can be used under widely differing governance and management systems (Pahl-Wostl et al., 2013) and more explicitly include indigenous knowledge, cultural services, and other social dimensions (Finn and Jackson, 2011; Jackson et al., 2014; Martin et al., 2015). In southern Mexico, for instance, ELOHA is being piloted for a collection of small coastal basins in Chiapas State, reflecting a diversity of river types, socio-economic contexts, and institutional arrangements for water management (Haney et al., 2015). Finn and Jackson (2011) propose a modified form of ELOHA with the potential capability to address the qualitative elements of indigenous people's ideas, beliefs, and socio-cultural relationships to water and rivers developed over many centuries of close interaction with aquatic ecosystems and dependence upon aquatic resources.

The framework has been adopted by US regulatory agencies (Kendy et al., 2012), and used as a foundation for regional river basin planning (e.g., see de Philip and Moburg, 2010, 2013) and multisite analysis of hydroecological impairment and prioritization of flow restoration (Martin et al., 2015; Sanderson et al., 2012). Applications of ELOHA in flow regime restoration in the Cheoah River (Upper Tennessee River Basin) demonstrated that, whereas fish richness did not increase as expected, recommended increases in flow magnitude reduced riparian encroachment within 4 years after flow restoration, as predicted from relationships between flow alteration and riparian responses (McManamay et al., 2013). In Colombia's Magdalena-Cauca Basin, a decision support system for basin water management has been evolving over the last few years, at both whole basin and lower administrative scales, that integrates ELOHA with elements of the Water Evaluation and Planning System software tool (http://www.weap21.org, see an example in Thompson et al., 2012) and IHA. It allows preliminary analyses to be made of future cumulative socio-ecological impacts by river type and of different scenarios of hydrological regime alteration across the river network (e.g., from urban water abstractions, planned and existing hydropower projects and irrigation). If implemented, it will enable environmental water to be set and adaptively monitored at management points of interest basin-wide (R.E. Tharme, pers. comm.).

The application of environmental water science at both the regional and local scales highlights the importance of appropriate tool selection in environmental water assessment and management (Fig. 11.3). Local, at-a-site application typically employs detailed hydrologic measurement (from streamflow gauges), high-detail hydraulic models, and potentially fine-scale ecological responses (e.g., species population dynamics or changes in abundance) supported by ecological modeling and with monitoring over time. Experimentation has occurred at this scale to test flow—ecology relationships and ecological responses to flow management (see review by Olden et al., 2014).

By contrast, regional methods involve modeling at unstudied locations and therefore generally rely on statistical tools and remote sensing to transfer information from studied reaches to locales

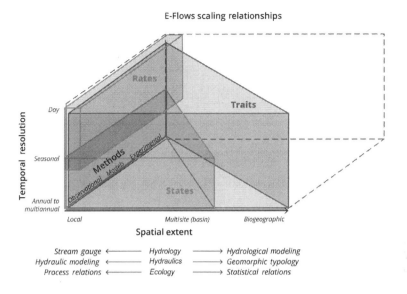

FIGURE 11.3

General scaling relationships in environmental water. Spatial grain of environmental water application (X-axis) ranges from local (single-site reach) to regional (multiple sites or whole stream networks) to biogeographic (where species replacements and shifting species pools naturally occur). Along this axis the tools that are available to characterize hydrologic setting, hydraulic conditions, and ecological relations vary. The Y-axis reflects methods of environmental and ecological data acquisition, from observation of existing conditions relationships, to modeling of relationships, to experimental manipulations of systems to observe responses. Observations can be collected at all spatial extents, and modeling can extend over large extents, but experiments are local in scope due to expense and logistics. The Z-axis shows the timescales over which different types of ecological responses occur. Ecological states (small triangular polygon) are species-based response variables that are limited to seasonal to annual turnover or response times within species' geographic ranges. Ecological rates (small rectangular polygon) are high-frequency response variables that can be measured over short timescales (days) (e.g., population growth or death rates, biogeochemical exchange rates) but that are applied at the local scale given data requirements. Ecological traits (large triangular polygon) are integrative measures of species function or performance (e.g., functional groups, life-history modes, dispersal ability) that can be generalized across species and hence biogeographic boundaries. Traits can be selected to reflect a range of temporal scales, from annual (e.g., inundation regime required) to daily (e.g., tolerance of transient extreme conditions). See text for further discussion.

that lack site-specific data. The broader extent is associated with lower information availability. Coarser hydrologic characterization (based on model calibration to a few gauges in the region) leads to the need for flow classification. Detailed hydraulic modeling is infeasible, so some kind of regionalization based on geomorphic surrogates is needed to generate surrogates of fine-scale hydraulics (e.g., Wilding et al., 2014; Chapter 13). Coarse-grained ecological metrics are also more appropriate, given the prohibitive cost of monitoring multiple sites.

11.5 NEW CHALLENGES FOR ENVIRONMENTAL WATER: MOVING FROM STATIC SYSTEMS TO NONSTATIONARY DYNAMICS

Environmental water science and assessment, like the broader supporting fields of water resources and conservation ecology, have long operated under the assumption of climatic stationarity, for instance precipitation and runoff processes occur within some statistically describable range of variation. The presumption of a stable climate created a scientific and management perspective that conveniently allowed environmental water to treat hydrologic variation and hydroecological dynamics in relatively static terms. Dynamic flow regimes from reference sites could be characterized by time-averaged statistical properties of ecological relevance such as average values (or standard deviations) of the frequency, duration, and timing of particular flow magnitudes. Similarly, ecological response variables could be treated in terms of alteration from some dynamically equilibrial, historical (reference) state (Poff, 2014).

In the last decade it has become clear that the global climate system is *nonstationary* (Milly et al., 2008) and that hydrologic *baselines* are dynamically changing, such that future hydrologic regimes will, in many locations, be significantly different from those of today (e.g., Laize et al., 2014; Reidy Liermann et al., 2012). Reliance on the assumption of historical flow conditions is no longer feasible for long-term planning in most systems due to rapid climate change and other sources of change, including population growth and increasing water demand (Acreman et al., 2014a; Kopf et al., 2015; Poff et al., 2016). In addition, natural ecological communities are rapidly changing with the spread of nonnative species and human-caused degradation (Humphries and Winemiller, 2009), making ecological restoration targets dynamic and variable (Moyle, 2014; Rahel and Olden, 2008).

Climatic and other sources of nonstationarity have thrust the science and management of environmental water into a new era. There is now a need to transition from a largely static to a more dynamic perspective, what has been referred to as a *process-based* understanding and framework for management of rivers (e.g., Beechie et al., 2010). This fundamentally means moving beyond strong reliance on simple correlations between flow variables and ecological responses to a more nuanced understanding of how ecological systems respond to dynamic hydrologic variation. New forms of ecological modeling that emphasize hydroecological process relationships (Chapter 14) can help guide targeted management in a more uncertain future.

Of course, even in highly modified and transient conditions, ecological principles of what might be called *natural flows thinking* will remain important (Acreman et al., 2014a; Auerbach et al., 2012; Kopf et al., 2015) and historical pattern—process relationships will continue to provide a key understanding of the historical and evolutionary context of species adaptations and ecosystem responses to environmental change (see Wiens et al., 2012 for a broad, comprehensive treatment of this issue). Understanding pattern—process relationships of the past helps us to manage the challenges of the present and predict possible futures under changing climates and novel ecological, social, and political environments. Existing frameworks such as ELOHA will continue to guide regional environmental water assessments; however, adjustments will need to be made due to current reliance on stationarity assumptions (see Fig. 11.2 notations), but the required evolution of environmental water in a nonstationary world will pose challenges because environmental water management is likely to become more locally contingent and to entail the loss of the relative comfort of fixed management targets that could be presumed under the long-held assumptions of climatic and ecological stationarity.

A key challenge associated with nonstationarity is to broaden the scope and scale and put environmental water into a larger landscape context of stream connectivity (Poff and Matthews, 2013). This challenge is driven by several factors, including the global biodiversity crisis and degradation of freshwater ecosystems (Dudgeon et al., 2006; Strayer and Dudgeon, 2010) and the competing societal demands for a shrinking freshwater resource. Environmental flows assessments will increasingly be applied beyond a multisite perspective, embodied in ELOHA, to a multibasin-scale perspective where water infrastructure's impact on regional diversity becomes part of a broader conservation or adaptation planning framework that accounts for habitat connectivity and species movements in hydrologically altered and fragmented stream networks (e.g., Jager et al., 2015; Poff, 2009; Winemiller et al., 2016). Conservation (or preservation) and restoration goals will need to be articulated in the context of competing human needs for freshwater, as exemplified in the contemporary basin-scale framing of water resources management as the *nexus* of food, water, and energy.

Environmental water is likely to evolve rapidly over the coming years, due to increasing human demand for water coupled with degrading aquatic ecosystems. Two fundamental challenges face the science of environmental water assessment in its quest to provide sound guidance and support to society on the management of human-impacted rivers to sustain or promote ecological integrity. These challenges focus on a necessary broadening of traditional approaches in environmental water science, with respect both to hydrologic characterization and ecological (and increasingly, also social) responses.

11.5.1 HYDROLOGY: REGIMES AND EVENTS

The hydrologic *regime* represents a template that influences how well species will persist and be capable of accommodating natural (and manmade) hydrologic change (Poff et al., 1997; Puckridge et al., 1998). Species have morphological, behavioral, or physiological traits that allow them to flourish under different types of hydrologic variation. Large differences in magnitude, frequency, duration, and timing of flow events among streams lead to the expectation of patterns of species abundance and persistence in hydrologically distinct streams (Poff et al., 1997). The relative stability of the global climate over the last approximately 10,000 years of the Holocene period has allowed hydrologists to consider the historical (instrumental) record to capture reasonably well the long-term, climatically driven flow regime, although pervasive land-use change has modified rainfall-runoff processes in long-settled regions (Acreman et al., 2014a).

Regime-based thinking has dominated environmental water over the last decades, but nonstationarity is presenting new challenges and alternative modes of framing how hydrologic alteration drives ecological responses. Individual hydrologic events often act as critical disturbances that may individually exert large and lasting effects on local populations or habitat structure. Thus, a focus on patterns of more immediate antecedent flows is arguably just as important as regime-based metrics in understanding and managing flow—ecology relationships. This perspective is increasingly relevant for management of nonstationary regimes where increasing variance in extremes is expected, exposing species to more frequent or intense extreme hydrologic events and perhaps simultaneously exacerbating extinction risk for small populations living in fragmented habitats

(Olden et al., 2007). Natural flow regimes are already shifting (e.g., snowmelt timing in western US streams; Stewart et al., 2005) and are projected to do so at an accelerated pace in the coming decades (e.g., Laize et al., 2014; Reidy Liermann et al., 2012). These trends suggest that environmental water management in the future will need to pay particular attention to novel sequences of hydrologic events that may shape ecosystems in new ways.

In some locations, environmental water applications are being specifically cast in the context of restoring specific elements of the flow regime that support the life-history needs of one or more target species, for example timed flow releases to aid cottonwood tree recruitment (Rood et al., 2003), or to give advantage to native fishes (Kiernan et al., 2012), or to eliminate recruitment bottlenecks for aquatic insects below hydropeaking dams (Kennedy et al., 2016). In situations where water is highly contested, environmental water releases are likely to be judged not only in terms of their measurable ecological benefits but also in terms of economic efficiency and other social considerations (Chapter 27; Poff and Schmidt, 2016). Thus, climate change that brings heightened uncertainty about future water availability will place new constraints on environmental water. However, new opportunities for environmental water management are likely to arise as well, such as the use of dams to provide cool water for downstream fishes (Thompson et al., 2012), or to provide flow refuges for species of interest (Shenton et al., 2012), or to manage against extreme low-flow conditions that may disadvantage native versus nonnative fishes (Ruhi et al., 2015). These kinds of strategic releases may be adjusted to achieve varying management in response to natural interannual variability of flow events, such as *dry* versus *wet* years, as occurs frequently in South African holistic reserve determinations (R.T. Tharme, pers. obs.) and has been proposed for some environmental water projects in the United States (Muth et al., 2000).

11.5.2 ECOLOGY: STATES, RATES, AND TRAITS

Nonstationarity applies to the ecological domain as well, insofar as riverine species distributions and local composition of communities are being reshuffled in response to human-assisted spread of nonnative species (Olden et al., 2004) and to species range changes in response to climate change (Rahel and Olden, 2008).

In the practice of environmental water, the broad question is what are the appropriate ecological *performance indicators* that can be used to gauge success of flow management? *Environmental water* science and practice have been mostly framed in terms of relatively static ecological endpoints such as ecosystem *states*, for example community composition or maintenance of key ecosystem features (riparian structure and condition) or target species populations (Arthington et al., 2006; Poff et al., 2010). This state-based approach has employed the implicit assumption that habitat is limiting and flow restoration improves ecological condition by providing more habitat for species that are often assumed to be *adapted* to the local flow regime (see Lytle and Poff, 2004). The dynamic processes that underlie the assembly of ecosystem states (e.g., connectivity to source populations and variable species' dispersal abilities, differential growth and mortality rates among species, species interactions of competition, predation, and facilitation) are typically ignored in environmental water science and assessment. However, increasing hydrological and ecological variance under nonstationarity will, inevitably, present novel conditions outside of historical, reference-based understanding. Consequently, predicting ecosystem states will require a more process-based (or *mechanistic*) understanding of key hydroecological linkages to guide flow management toward particular ecological endpoints (Beechie et al., 2010; Poff and Matthews, 2013).

Additional ecological endpoints beyond ecosystem states are needed to achieve a greater process basis in environmental water science and practice. One would be underlying biological and ecological *rates*, which can be conceived at the ecosystem level (i.e., rates of nutrient cycling or of production, e.g., Doyle et al., 2005) or at the individual or species population level such as environmentally sensitive and dynamic changes in growth, mortality, and colonization rates (e.g., Auerbach, 2013; Bond et al., 2015; Lancaster and Downes, 2010; Lytle and Merritt, 2004; Yen et al., 2013). These dynamic ecological representations correspondingly demand more attention to dynamic hydrologic processes such as the magnitudes and sequencing of extreme events that may create nonlinearities in ecological response beyond the simple *averaging* approach of traditional hydrologic statistical metrics (e.g., Ruhi et al., 2015; Shenton et al., 2012). Process-based environmental water management has been proposed (Anderson et al., 2006) but not widely employed. One significant challenge with a purely process-based approach is that it is often difficult to collect detailed information necessary to model population dynamics for more than a few species (e.g., Shenton et al., 2012). Thus, this approach may be most suitable for local applications or for particular species of concern where financial resources and scientific capacity are available.

One possible way to suffuse environmental water science with more process-based understanding at the community and ecosystem scales would be to take a more general, functional approach to species characterization, for instance *traits* of species rather than identities of species as ecological response units. Traits are organismal attributes that can be effectively related functionally (*mechanistically*) to environmental conditions (Poff, 1997; Verberk et al., 2013; Webb et al., 2010). Trait assignments have been assembled for various types of species including stream insects (Poff et al., 2006 a,b), fish (Mims et al., 2010, 2012; Olden et al., 2008), and riparian flora (Merritt et al., 2010). There are a growing number of traits-based or *functional* analyses of ecological response to hydrology, hydrologic alteration (or to climate change) that perhaps provide a middle ground of complexity between state-based and process-based analyses.

11.6 CONCLUSION: GUIDING ELEMENTS FOR BEST PRACTICE IN ENVIRONMENTAL FLOWS ASSESSMENT

We can view the development and evolution of environmental flows assessment as one of increasing technical sophistication and expanding focus, from single species to whole ecosystems to socio-ecological sustainability, as well as to more strongly empirically based flow—ecology relationships (Acreman et al., 2014b; Arthington, 2015; Poff and Matthews, 2013). Advances have helped promote the widespread recognition of the importance of environmental water for sustainable water management policies and water allocation plans, but schemes for active implementation lag behind on a global scale (Le Quesne et al., 2010; Chapter 27; Table 11.1). In this process of advancement and innovation, several critical elements of a full understanding of human alteration of aquatic systems have been underdeveloped. Among them are the inclusion of other key environmental drivers (sediment and temperature regimes) and hydraulic modeling, as well as a greater appreciation of the importance of flow sequences (not just long-term averages). From an ecological perspective, development of a more pluralistic and regionally calibrated approach to a range of ecological responses (states, rates, and traits) is now needed to improve the scientific foundations of environmental water. This is especially as pressures on limited freshwater resources increase with expanding human population growth and water demand and as systems becoming increasingly nonstationary hydrologically and ecologically with rapid climate change. The social sciences dimensions of environmental water, including an appreciation of the major social drivers, the conceptual foundation and empirical knowledge of important relationships between flow regime change and social responses, and methodology integration, also remain weakly developed.

Addressing these various, diverse elements is necessary to achieve more effective environmental water science and practice and to redress the persistent gap in implementation. To this end, we believe that environmental water can build on its past accomplishments, continue to innovate, and be successfully applied in a variety of settings. Here, we offer a set of general guiding principles or elements that are most likely to promote the successful progression and implementation of environmental water, for instance those that will facilitate achieving desirable outcomes for river ecosystems and for people and societies.

Guiding Element 1: *Clearly and quantitatively describe flow−ecology and flow−social relationships.* Flow−ecology relationships will continue to provide the foundation of environmental water applications (Chapter 14). Clear specification of flow−ecology relationships is key to shaping justifiable management expectations and ecological restoration objectives and hence to environmental water success (Davies et al., 2014; Poff and Schmidt, 2016). There are innumerable possible relationships that can form the foundation of an environmental water application. Selection of the most appropriate relationships for each environmental water application reflects a balance of ecological understanding, hydrologic modeling capacity, and management potential (Poff et al., 2010). Further, the selection and modeling of flow−ecology relationships must be understandable to managers, government agencies, and the public, as should the uncertainty contained in these relationships (Chapter 14).

Guiding Element 2: *Engage stakeholders to collaboratively set environmental water vision and objectives.* It is increasingly clear that environmental water restoration (or conservation) objectives are most likely to be achieved when they are collaboratively generated and agreed upon by multiple stakeholders, not just scientists (Chapter 7; Rogers, 2006). Science alone cannot objectively identify what the endpoints of environmental water restoration should be, because endpoints must reflect social values and appropriate socio-ecological goals of concern (Acreman et al., 2014b; Jackson et al., 2014; Roux and Foxcroft, 2011). This will require, in many instances, an explicit integration of social sciences and indigenous or stakeholder knowledge into the descriptions of relationships with flow (see Guiding Element 1).

Guiding Element 3: *Carefully identify what can be attained (and what cannot) from an implementation of environmental water regimes.* Flow manipulations alone are limited in terms of full ecosystem restoration, and other factors (e.g., highly altered thermal or sediment regimes, increasing urbanization or demographic shifts) may limit the extent to which flow management alone can achieve desirable outcomes. Objective assessment of the feasibility of achieving a stated environmental water objective is critical to building credibility of environmental water science as it is applied in ever more complex water resources settings (Acreman et al., 2014b), including climate change (Poff et al., 2016). Appreciating the basin management context and understanding the opportunities for integrated land and water resources management are also important when defining the scope and potential for environmental water to meet ecological targets (King and Brown, 2010).

Guiding Element 4: *Clearly identify at what spatial and temporal scale environmental water applications are appropriate and intended.* A range of possible environmental water frameworks exist that are appropriate across a range of spatial and temporal scales (Table 11.1; Figs. 11.2 and 11.3). Trade-offs in modeling precision and ecological prediction are associated with scale of application; therefore, it is critical to clearly identify scale and to identify which specific methods and tools can best be employed. Similarly, the social spheres of influence and factors for inclusion differ with such changes in scale.

Guiding Element 5: *Make efforts to engage with the proponents and engineers of new water infrastructure developments or proposed relicensing opportunities for existing infrastructure.* Early engagement with water resources developers helps create a bridge between impact assessments (project or strategic planning) and the environmental water process, helps target necessary baseline data collection on potential risks to regime change, and calls attention to the implications for downstream people and ecosystems of water development and management. Early engagement has the potential to more effectively influence project design (e.g., Richter and Thomas, 2007), operational regimes, and of course also decisions on the placement of infrastructure in river networks (e.g., King and Brown, 2010; Opperman et al., 2015; Poff et al., 2016), all of which play pivotal roles in securing environmental water.

Guiding Element 6: *Consider how environmental water goals and applications embed within and interact with other realms of influence that emerge with water governance and management at system scales.* Environmental water frameworks are increasingly of interest to larger spheres of management, social, and economic influence (see Fig. 11.1), such as systematic conservation planning (Nel et al., 2011), *Integrated Water Resources Management (IWRM*; Overton et al., 2014), and associated water allocation systems (Chapter 17). Socio-economic dimensions and governance contexts are increasingly recognized as critical to incorporate into environmental water assessments and effective management (Arthington et al., 2010; King et al., 2014; Pahl-Wostl et al., 2013). Environmental water applications should seek points of connection to these broader spheres of interest, in order to increase the social relevance and visibility of the environmental water activities.

Guiding Element 7: *Incorporate nonstationarity and process-based understanding into environmental water science and implementation to meet a new future.* A key challenge to environmental water science and an emerging and enduring context for environmental water applications is the issue of nonstationarity in hydrologic and ecological (and probably socio-political) conditions. Reference hydrologic and ecological conditions are increasingly impractical to establish, raising the need to shift many environmental water projects from a restoration focus to an adaptation focus (Palmer et al., 2009; Poff and Matthews, 2013; Poff et al., 2016; Thompson et al., 2014). Environmental water implementation can be a major element of more broad-based climate change adaptation in water resources (Matthews et al., 2009). Critical to meeting this challenge is the need for environmental water science to develop a better *process-based* understanding of hydroecological relationships, to allow for more dynamic management of regulated and highly modified river systems in order to meet, often by design, socially desired ecological targets that may be shifting through time as human population growth and climate change continue to accelerate the alteration of hydrologic systems into the future.

REFERENCES

Acreman, M., 2016. Environmental flows — basics for novices. WIREs Water 3, 622—628. Available from: http://dx.doi.org/10.1002/wat2.1160.

Acreman, M.C., Dunbar, M.J., 2004. Methods for defining environmental river flow requirements — a review. Hydrol. Earth Syst. Sci. 8, 121—133.

Acreman, M.C., Ferguson, A.J.D., 2010. Environmental flows and the European Water Framework Directive. Freshw. Biol. 55, 32—48.

Acreman, M., Arthington, A.H., Colloff, M.J., Couch, C., Crossman, N.D., Dyer, F., et al., 2014a. Environmental flows for natural, hybrid, and novel riverine ecosystems in a changing world. Front. Ecol. Environ. 12, 466—473.

Acreman, M.C., Overton, I.C., King, J., Wood, P.J., Cowx, I.G., Dunbar, M.J., et al., 2014b. The changing role of ecohydrological science in guiding environmental flows. Hydrol. Sci. J. 59, 433—450.

Alfredsen, K., Harby, A., Linnansaari, T., Ugedal, O., 2012. Development of an inflow-controlled environmental flow regime for a Norwegian river. River Res. Appl. 28, 731—739.

Anderson, K.E., Paul, A.J., McCauley, E., Jackson, L.J., Post, J.R., Nisbet, R.M., 2006. Instream flow needs in streams and rivers: the importance of understanding ecological dynamics. Front. Ecol. Environ. 4, 309—318.

Annear, T., Chisholm, I., Beecher, H., Locke, A., Aarrestad, P., Burkhard, N., et al., 2004. Instream Flows for Riverine Resource Stewardship. Instream Flow Council, Cheyenne, Wyoming.

Arthington, A.H., 1998. Comparative evaluation of environmental flow assessment techniques: review of holistic methodologies. LWRRDC Occasional Paper 26/98. LWRRDC, Canberra, Australia.

Arthington, A.H., 2012. Environmental Flows: Saving Rivers in the Third Millennium. University of California Press, Berkeley, California.

Arthington, A.H., 2015. Environmental flows: a scientific resource and policy tool for river conservation and restoration. Aquat. Conserv. Mar. Freshw. Ecosyst. 25, 155—161.

Arthington, A.H., Pusey, B.J., 2003. Flow restoration and protection in Australian rivers. River Res. Appl. 19, 377—395.

Arthington, A.H., Zalucki, J.M. (Eds.), 1998. Comparative evaluation of environmental flow assessment techniques: review of methods. LWRRDC Occasional Paper 27/98, LWRRDC, Canberra, Australia.

Arthington, A.H., Bunn, S.E., Pusey, B.J., Blühdorn, D.R., King, J.M., Day, J.A., et al., 1992. Development of an holistic approach for assessing environmental flow requirements of riverine ecosystems. In: Pigram, J.J., Hooper, B.P. (Eds.), Proceedings of an International Seminar and Workshop on Water Allocation for the Environment. Centre for Water Policy Research, Armidale, Australia, pp. 69—76.

Arthington, A.H., Bunn, S.E., Poff, N.L., Naiman, R.J., 2006. The challenge of providing environmental flow rules to sustain river ecosystems. Ecol. Appl. 16, 1311—1318.

Arthington, A.H., Baran, E., Brown, C.A., Dugan, P., Halls, A.S., King, J.M., et al., 2007. Water requirements of floodplain rivers and fisheries: existing decision support tools and pathways for development. Comprehensive Assessment of Water Management in Agriculture Research Report 17. International Water Management Institute, Colombo, Sri Lanka.

Arthington, A.H., Naiman, R.J., McClain, M.E., Nilsson, C., 2010. Preserving the biodiversity and ecological services of rivers: new challenges and research opportunities. Freshw. Biol. 55 (1), 1—16.

Arthington, A.H., Mackay, S.J., James, C.S., Rolls, R.J., Sternberg, D., Barnes, A., et al., 2012. Ecological limits of hydrologic alteration: a test of the ELOHA framework in south-east Queensland. National Water Commission Waterlines Report 75. Canberra, Australia.

Arthington, A.H., Rall, J.L., Kennard, M.J., Pusey, B.J., 2003. Environmental flow requirements of fish in Lesotho Rivers using the DRIFT methodology. River Res. Appl. 19 (5—6), 641—666.

Auerbach, D.A., 2013. Models of *Tamarix* and riparian vegetation response to hydrogeomorphic variation, dam management and climate change. PhD Dissertation. Colorado State University, Colorado.

Auerbach, D.A., Poff, N.L., McShane, R.R., Merritt, D.M., Pyne, M.I., Wilding, T.K., 2012. Streams past and future: fluvial responses to rapid environmental change in the context of historical variation. In: Wiens, J.A., Hayward, G.D., Safford, H.D., Giffen, C. (Eds.), Historical Environmental Variation in Conservation and Natural Resource Management. John Wiley & Sons, Ltd, Chichester, UK, pp. 232−245.

Beechie, T.J., Sear, D.A., Olden, J.D., Pess, G.R., Buffington, J.M., Moir, H., et al., 2010. Process-based principles for restoring river ecosystems. BioScience 60, 209−222.

Biggs, B.J.F., Duncan, M.J., Jowett, I.G., Quinn, J.M., Hickey, C.W., Daviescolley, R.J., et al., 1990. Ecological characterization, classification, and modeling of New Zealand rivers: an introduction and synthesis. N. Z. J. Mar. Freshw. Res. 24, 277−304.

Blake, D.J.H., Sunthornratana, U., Promphakping, B., Buaphuan, S., Sarkkula, J., Kummu, M., et al., 2011. E-flows in the Nam Songkhram River Basin. Final Report. M-POWER Mekong Program on Water, Environment and Resilience, IUCN, IWMI, CGIAR Challenge Program on Water and Food, and Aalto University, Finland.

Bond, N.R., Balcombe, S.R., Crook, D.A., Marshall, J.C., Menke, N., Lobegeiger, J.S., 2015. Fish population persistence in hydrologically variable landscapes. Ecol. Appl. 25, 901−913.

Bovee, K.D., 1982. A guide to stream habitat analysis using the Instream Flow Incremental Methodology. Instream Flow Information Paper 12. U.S.D.I. Fish and Wildlife Service, Office of Biological Services FWS/OBS−82/26.

Brisbane Declaration, 2007. Environmental flows are essential for freshwater ecosystem health and human well-being. 10th International River Symposium and International Environmental Flows Conference, 3−6 September 2007, Brisbane, Australia. Available from: http://www.eflownet.org.

Brizga, S.O., Arthington, A.H., Pusey, B.J., Kennard, M.J., Mackay, S.J., Werren, G.L., et al., 2002. Benchmarking, a 'top-down' methodology for assessing environmental flows in Australian rivers. Environmental Flows for River Systems. An International Working Conference on Assessment and Implementation, incorporating the 4th International Ecohydraulics Symposium. Southern Waters, Cape Town, South Africa.

Brown, C.A., Watson, P., 2007. Decision support systems for environmental flows: lessons from Southern Africa. Int. J. River Basin Manage. 5, 169−178.

Bunn, S.E., Arthington, A.H., 2002. Basic principles and ecological consequences of altered flow regimes for aquatic biodiversity. Environ. Manage. 30, 492−507.

Carlisle, D.M., Falcone, J., Wolock, D.M., Meador, M.R., Norris, R.H., 2010. Predicting the natural flow regime: models for assessing hydrological alteration in streams. River Res. Appl. 26, 118−136.

Carlisle, D.M., Wolock, D.M., Meador, M.R., 2011. Alteration of streamflow magnitudes and potential ecological consequences: a multiregional assessment. Front. Ecol. Environ. 9, 264−270.

Chan, T.U., Hart, B.T., Kennard, M.J., Pusey, B., Shenton, W., Douglas, M.M., et al., 2012. Bayesian network models for environmental flow decision making in the Daly River, Northern Territory, Australia. River Res. Appl. 28, 283−301.

Council of the European Communities, 2000. Directive 2000/60/EC of the European Parliament and of the Council of 23 October 2000 establishing a framework for community action in the field of water policy. OJEC L327 1−73.

Davies, P.M., Naiman, R.J., Warfe, D.M., Pettit, N.E., Arthington, A.H., Bunn, S.E., 2014. Flow-ecology relationships: closing the loop on effective environmental flows. Mar. Freshw. Res. 65, 133−141.

de Philip, M., Moberg, T., 2010. Ecosystem flow recommendations for the Susquehanna River Basin. The Nature Conservancy, Harrisburg, Pennsylvania. Available from: http://www.nature.org/media/pa/tnc-final-susquehanna-river-ecosystem-flows-study-report.pdf.

de Philip, M., Moberg, T., 2013. Ecosystem flow recommendations for the Upper Ohio River Basin in Western Pennsylvania. The Nature Conservancy, Harrisburg, Pennsylvania. Available from: http://www.nature.org/media/pa/ecosystem-flow-recommendations-upper-ohio-river-pa-2013.pdf.

Dickens, C., O'Brien, G.C., Stassen, J., Kleynhans, M., Rossouw, N., Rowntree, K., et al., 2015. Specialist consultants to undertake baseline studies (flow, water quality and geomorphology) and Instream Flow Requirement (IFR) Assessment for phase 2: instream flow requirements for the Senqu River: final report. Prepared by the Institute of Natural Resources NPC on behalf of the Lesotho Highlands Development Authority (LHDA). Maseru, Lesotho.

Doyle, M.W., Stanley, E.H., Strayer, D.L., Jacobson, R.B., Schmidt, J.C., 2005. Effective discharge analysis of ecological processes in streams. Water Resour. Res. 41.

Dudgeon, D., Arthington, A.H., Gessner, M.O., Kawabata, Z.I., Knowler, D.J., Leveque, C., et al., 2006. Freshwater biodiversity: importance, threats, status and conservation challenges. Biol. Rev. 81, 163–182.

Dunbar, M.J., Acreman, M.C., 2001. Applied hydro-ecological science for the twenty-first century. In: Acreman, M.C. (Ed.), Hydro-ecology: Linking Hydrology and Aquatic Ecology. International Association of Hydrological Sciences Publication No., 266.. IAHS Press, Centre for Ecology and Hydrology, Wallingford, UK, pp. 1–17.

Dyson, M., Bergkamp, G., Scanlon, J. (Eds.), 2003. Flow. The Essentials of Environmental Flows. IUCN, Gland, Switzerland.

European Commission, 2015. Ecological flows in the implementation of the WFD. CIS Guidance Document no. 31, Technical Report 2015–086, Brussels, Belgium.

Finn, M., Jackson, S., 2011. Protecting Indigenous values in water management: a challenge to conventional environmental flow assessments. Ecosystems 14, 1232–1248.

Flotemersch, J.E., Leibowitz, S.G., Hill, R.A., Stoddard, J.L., Thoms, M.C., Tharme, R.E., 2015. A watershed integrity definition and assessment approach to support strategic management of watersheds. River Res. Appl. Arena paper. Available from: http://dx.doi.org/10.1002/rra.2978.

Gippel, C.J., Speed, R., 2010. Environmental flow assessment framework and methods, including environmental asset identification and water re-allocation. ACEDP Australia-China Environment Development Partnership, River Health and Environmental Flow in China. International Water Centre, Brisbane.

Gustard, A., 1979. The characterisation of flow regimes for assessing the impact of water resource management on river ecology. In: Ward, J.W., Stanford, J.A. (Eds.), The Ecology of Regulated Streams. Plenum Press, New York and London, pp. 53–60.

Haines, A.T., Finlayson, B.L., McMahon, T.A., 1988. A global classification of river regimes. Appl. Geogr. 8, 255–272.

Haney, J., González, A., Tharme, R.E., 2015. Estudio de caudales ecológicos en las cuencas costeras de Chiapas, México. Technical Report. The Nature Conservancy, Mexico and North Central America.

Hannah, D.M., Sadler, J.P., Wood, P.J., 2007. Hydroecology and ecohydrology: a potential route forward? Hydrol. Process. 21, 3385–3390.

Hill, M.T., Platts, W.S., Beschta, R.L., 1991. Ecological and geomorphological concepts for instream and out-of-channel flow requirements. Rivers 2, 198–210.

Hirji, R., Davis, R., 2009. Environmental Flows in Water Resources Policies, Plans, and Projects. World Bank, Washington, DC.

Hughes, D.A., Hannart, P., 2003. A desktop model used to provide an initial estimate of the ecological instream flow requirements of rivers in South Africa. J. Hydrol. 270, 167–181.

Hughes, D.A., Desai, A.Y., Birkhead, A.L., Louw, D., 2014. A new approach to rapid, desktop-level, environmental flow assessments for rivers in South Africa. Hydrol. Sci. J. 59 (3), 1–15.

Hughes, J.M.R., James, B., 1989. A hydrological regionalization of streams in Victoria, Australia, with implications for stream ecology. Aust. J. Mar. Freshw. Res. 40, 303–326.

Humphries, P., Winemiller, K.O., 2009. Historical impacts on river fauna, shifting baselines, and challenges for restoration. BioScience 59, 673–684.

Illaszewicz, J., Tharme, R., Smakhtin, V., Dore, J. (Eds.), 2005. Environmental Flows: Rapid Environmental Flow Assessment for the Huong River Basin, Central Vietnam. IUCN Vietnam, Hanoi, Vietnam.

Jackson, S.E., Douglas, M.M., Kennard, M.J., Pusey, B.J., Huddleston, J., Harney, B., et al., 2014. "We like to listen to stories about fish": integrating indigenous ecological and scientific knowledge to inform environmental flow assessments. Ecol. Soc. 19 (1), 43.

Jager, H.I., Efroymson, R.A., Opperman, J.J., Kelly, M.R., 2015. Spatial design principles for sustainable hydropower development in river basins. Renew. Sustain. Energy Rev. 45, 808−816.

James, C., Mackay, S.J., Arthington, A.H., Capon, S., 2016. Does flow structure riparian vegetation in subtropical south-east Queensland? Ecol. Evol. Available from: http://dx.doi.org/10.1002/ece3.2249.

Joubert, A.R., Hurly, P.R., 1994. The use of daily flow data to classify South African rivers. Chapter 11: p. 286−359. In: King, J.M., Tharme, R.E. Assessment of the Instream Flow Incremental Methodology and Initial Development of Alternative Instream Flow Methodologies for South Africa. Water Research Commission Report No., 295/1/94. Water Research Commission, Pretoria, South Africa.

Kendy, E., Apse, C., Blann, K., 2012. A Practical Guide to Environmental Flows for Policy and Planning. The Nature Conservancy, Arlington, Virginia.

Kennard, M.J., Pusey, B.J., Olden, J.D., MacKay, S.J., Stein, J.L., Marsh, N., 2010. Classification of natural flow regimes in Australia to support environmental flow management. Freshw. Biol. 55, 171−193.

Kennedy, T.A., Muehlbauer, J.D., Yackulic, C.B., Lytle, D.A., Miller, S.W., Dibble, K.L., et al., 2016. Flow management for hydropower extirpates aquatic insects, undermining river food webs. BioScience 66, 561−575.

Kiernan, J.D., Moyle, P.B., Crain, P.K., 2012. Restoring native fish assemblages to a regulated California stream using the natural flow regime concept. Ecol. Appl. 22, 1472−1482.

King, J., Beuster, H., Brown, C., Joubert, A., 2014. Pro-active management: the role of environmental flows in transboundary cooperative planning for the Okavango River system. Hydrol. Sci. J. 59, 786−800.

King, J.M., Brown, C.A., 2010. Integrated basin flow assessments: concepts and method development in Africa and south-east Asia. Freshw. Biol. 55, 127−146.

King, J.M., Louw, M.D., 1998. Instream flow assessments for regulated rivers in South Africa using the Building Block Methodology. Aquat. Ecosyst. Health Manag. 1, 109−124.

King, J.M., Pienaar, H. (Eds.), 2011. Sustainable use of South Africa's inland waters: a situation assessment of resource directed measures 12 years after the 1998 National Water Act. Water Research Commission Report No. TT 491/11. Water Research Commission, Pretoria, South Africa.

King, J.M., Tharme, R.E., 1994. Assessment of the instream flow incremental methodology and initial development of alternative instream flow methodologies for South Africa. Water Research Commission Report No., 295/1/94. Water Research Commission, Pretoria, South Africa.

King, J.M., Tharme, R.E., de Villiers, M.S. (Eds.), 2000. Environmental flow assessments for rivers: manual for the Building Block Methodology. Water Research Commission Report TT 131/00, Pretoria, South Africa.

King, J.M., Brown, C.A., Sabet, H., 2003. A scenario-based holistic approach for environmental flow assessments. River Res. Appl. 19, 619−639.

Kopf, R.K., Finlayson, C.M., Humphries, P., Sims, N.C., Hladyz, S., 2015. Anthropocene baselines: assessing change and managing biodiversity in human-dominated aquatic ecosystems. BioScience 65, 798−811.

Laize, C.L.R., Acreman, M.C., Schneider, C., Dunbar, M.J., Houghton-Carr, H.A., Floerke, M., et al., 2014. Projected flow alteration and ecological risk for pan-European rivers. River Res. Appl. 30, 299−314.

Lamouroux, N., Jowett, I.G., 2005. Generalized instream habitat models. Can. J. Fish. Aquat. Sci. 62, 7−14.

Lancaster, J., Downes, B.J., 2010. Linking the hydraulic world of individual organisms to ecological processes: putting ecology into ecohydraulics. River Res. Appl. 26, 385−403.

Le Quesne T., Kendy, E., Weston, D., 2010. The Implementation Challenge: Taking Stock of Government Policies to Protect and Restore Environmental Flows. WWF-UK and The Nature Conservancy.

Leonard, P.M., Orth, D.J., 1988. Use of habitat guilds of fishes to determine instream flow requirements. N. Am. J. Fish. Manage. 8, 399−409.

Lloyd, N., Quinn, G., Thoms, M., Arthington, A., Gawne, B., Humphries, P., et al., 2003. Does flow modification cause geomorphological and ecological response in rivers? A literature review from an Australian perspective. Cooperative Research Centre for Freshwater Ecology Technical Report 1/2004. Canberra, Australia.

Lokgariwar, C., Chopra, V., Smakhtin, V., Bharati, L., O'Keeffe, J., 2014. Including cultural water requirements in environmental flow assessment: an example from the upper Ganga River, India. Water Int. 39, 81–96.

Lytle, D.A., Merritt, D.M., 2004. Hydrologic regimes and riparian forests: a structured population model for cottonwood. Ecology 85, 2493–2503.

Lytle, D.A., Poff, N.L., 2004. Adaptation to natural flow regimes. Trend. Ecol. Evol. 19, 94–100.

Mackay, S.J., Arthington, A.H., James, C.S., 2014. Classification and comparison of natural and altered flow regimes to support an Australian trial of the Ecological Limits of Hydrologic Alteration framework. Ecohydrology 7, 1485–1507.

Maheu, A., Poff, N.L., St-Hilaire, A., 2016. A classification of stream water temperature regimes in the conterminous United States. River Res. Appl. 32, 896–906.

Martin, D.M., Labadie, J.W., Poff, N.L., 2015. Incorporating social preferences into the ecological limits of hydrologic alteration (ELOHA): a case study in the Yampa-White River basin, Colorado. Freshw. Biol. 60, 1890–1900.

Mathews, R., Richter, B.D., 2007. Application of the indicators of hydrologic alteration software in environmental flow setting. J. Am. Water Resour. Assoc. 43 (5), 1–14.

Matthews, J., Aldous, A., Wickel, B., 2009. Managing water in a shifting climate. Am. Water Works Assoc. 101 (8), 29, 99.

McClain, M.E., Sabalusky, A.L., Anderson, E.P., Dessu, S.B., Melesse, A.M., Ndomba, P.M., et al., 2014. Comparing flow regime, channel; hydraulics and biological communities to infer flow-ecology relationships in the Mara River of Kenya and Tanzania. Hydrol. Sci. J. 59, 801–819.

McManamay, R.A., Orth, D.J., Dolloff, C.A., 2012. Revisiting the homogenization of dammed rivers in the southeastern US. J. Hydrol. 424, 217–237.

McManamay, R.A., Orth, D.J., Dolloff, C.A., Mathews, D.C., 2013. Application of the ELOHA framework to regulated rivers in the Upper Tennessee River Basin: a case study. Environ. Manage. 51, 1210–1235.

McManamay, R.A., Bevelhimer, M.S., Kao, S.C., 2014. Updating the US hydrologic classification: an approach to clustering and stratifying ecohydrologic data. Ecohydrology 7, 903–926.

Merritt, D.M., Scott, M.L., Poff, N.L., Auble, G.T., Lytle, D.A., 2010. Theory, methods and tools for determining environmental flows for riparian vegetation: riparian vegetation-flow response guilds. Freshw. Biol. 55, 206–225.

Milly, P.C.D., Betancourt, J., Falkenmark, M., Hirsch, R.M., Kundzewicz, Z.W., Lettenmaier, D.P., et al., 2008. Stationarity is dead: whither water management? Science 319, 573–574.

Mims, M.C., Olden, J.D., 2012. Life history theory predicts fish assemblage response to hydrologic regimes. Ecology 93 (1), 35–45.

Mims, M.C., Olden, J.D., Shattuck, Z.R., Poff, N.L., 2010. Life history trait diversity of native freshwater fishes in North America. Ecol. Freshw. Fish 19, 390–400.

Monk, W.A., Wood, P.J., Hannah, D.M., Wilson, D.A., 2007. Selection of river flow indices for the assessment of hydroecological change. River Res. Appl. 23, 113–122.

Moyle, P.B., 2014. Novel aquatic ecosystems: the new reality for streams in California and other Mediterranean climate regions. River Res. Appl. 30, 1335–1344.

Muth, R., Crist, L.W., LaGory, K.E., Hayse, J.W., Bestgen, K.R., Ryan, T.P., et al., 2000. Flow and temperature recommendations for endangered fishes in the Green River downstream of Flaming Gorge Dam. Final report FG-53 to the Upper Colorado River Endangered Fish Recovery Program. US Fish and Wildlife Service, Denver, Colorado. Larval Fish Laboratory Contribution 120.

Myers, N., 1993. Biodiversity and the precautionary principle. Ambio 22, 74–79.

National Water Commission of Mexico, 2012. Norma Mexicana 2012. NMX-AA-159-SCFI-2012. Que establece el procedimiento para la determinacioìn del caudal ecoloìgico en cuencas hidroloìgicas. Establishing the procedure for environmental flow determination in hydrological basins. Comisión Nacional del Agua, Conagua.

Nel, J.L., Turak, E., Linke, S., Brown, C., 2011. Integration of environmental flow assessment and freshwater conservation planning: a new era in catchment management. Mar. Freshw. Res. 62, 290–299.

Newcombe, C.P., Jensen, J.O.T., 1996. Channel suspended sediment and fisheries: a synthesis for quantitative assessment of risk and impact. N. Am. J. Fisheries Manage. 16, 693–727.

Newson, M.D., Newson, C.L., 2000. Geomorphology, ecology and river channel habitat: mesoscale approaches to basin-scale challenges. Prog. Phys. Geogr. 24 (2), 195–217.

Nilsson, C., Reidy, C.A., Dynesius, M., Revenga, C., 2005. Fragmentation and flow regulation of the world's large river systems. Science 308 (5720), 405–4088.

O'Keeffe, J., Hughes, D., Tharme, R., 2002. Linking ecological responses to altered flows, for use in environmental flow assessments: the Flow Stressor-Response method. Verh. Internat. Verein. Limnol. 28, 84–92.

Olden, J.D., Naiman, R.J., 2010. Incorporating thermal regimes into environmental flows assessments: modifying dam operations to restore freshwater ecosystem integrity. Freshw. Biol. 55, 86–107.

Olden, J.D., Poff, N.L., 2003. Redundancy and the choice of hydrologic indices for characterizing streamflow regimes. River Res. Appl. 19, 101–121.

Olden, J.D., Hogan, Z.S., Vander Zanden, M.J., 2007. Small fish, big fish, red fish, blue fish: size-biased extinction risk of the world's freshwater and marine fishes. Glob. Ecol. Biogeogr. 16, 694–701.

Olden, J.D., Kennard, M.J., Pusey, B.J., 2008. Species invasions and the changing biogeography of Australian freshwater fishes. Glob. Ecol. Biogeogr. 17, 25–37.

Olden, J.D., Konrad, C.P., Melis, T.S., Kennard, M.J., Freeman, M.C., Mims, M.C., et al., 2014. Are large-scale flow experiments informing the science and management of freshwater ecosystems? Front. Ecol. Environ. 12, 176–185.

Olden, J.D., Poff, N.L., Douglas, M.R., Douglas, M.E., Fausch, K.D., 2004. Ecological and evolutionary consequences of biotic homogenization. Trends Ecol. Evol. 19, 18–24.

Opperman, J., Grill, G., Hartmann, J., 2015. The power of rivers: finding balance between energy and conservation in hydropower development. Technical report. Available from: http://dx.doi.org/10.13140/RG.2.1.5054.5765. The Nature Conservancy, Washington, DC.

Overton, I.C., Smith, D.M., Dalton, J., Barchiest, S., Acreman, M.C., Stromberg, J.C., et al., 2014. Implementing environmental flows in integrated water resources management and the ecosystem approach. Hydrol. Sci. J. 59, 860–877.

Pahl-Wostl, C., Arthington, A., Bogardi, J., Bunn, S.E., Hoff, H., Lebel, L., et al., 2013. Environmental flows and water governance: managing sustainable water uses. Curr. Opin. Environ. Sustain. 5, 341–351.

Palmer, M.A., Lettenmaier, D.P., Poff, N.L., Postel, S., Richter, B., Warner, R., 2009. Climate change and river ecosystems: protection and adaptation options. Environ. Manage. 44, 1053–1068.

Parasiewicz, P., 2007. The MesoHABSIM model revisited. River Res. Appl. 23, 893–903.

Petts, G.E., Calow, P., 1996. River Flows and Channel Forms. Blackwell Science, Oxford, UK.

Poff, N.L., 1996. A hydrogeography of unregulated streams in the United States and an examination of scale-dependence in some hydrological descriptors. Freshw. Biol. 36, 71–91.

Poff, N.L., 1997. Landscape filters and species traits: towards mechanistic understanding and prediction in stream ecology. J. N. Am. Benthol. Soc. 16, 391–409.

Poff, N.L., 2009. Managing for variability to sustain freshwater ecosystems. J. Water Resour. Plan. Manage. 135, 1–4.

Poff, N.L., 2014. Rivers of the Anthropocene? Front. Ecol. Environ. 12, 427.

Poff, N.L., Hart, D.D., 2002. How dams vary and why it matters for the emerging science of dam removal. BioScience 52, 659−668.

Poff, N.L., Matthews, J.H., 2013. Environmental flows in the Anthropocene: past progress and future prospects. Curr. Opin. Environ. Sustain. 5, 667−675.

Poff, N.L., Schmidt, J.C., 2016. How dams can go with the flow: small changes to water flow regimes from dams can help to restore river ecosystems. Science 353 (6304), 7−8.

Poff, N.L., Ward, J.V., 1989. Implications of streamflow variability and predictability for lotic community structure: a regional analysis of streamflow patterns. Can. J. Fish. Aquat. Sci. 46, 1805−1818.

Poff, N.L., Ward, J.V., 1990. Physical habitat template of lotic systems: recovery in the context of historical pattern of spatiotemporal heterogeneity. Environ. Manage. 14, 629−645.

Poff, N.L., Zimmerman, J.K.H., 2010. Ecological responses to altered flow regimes: a literature review to inform the science and management of environmental flows. Freshw. Biol. 55, 194−205.

Poff, N.L., Allan, J.D., Bain, M.B., Karr, J.R., Prestegaard, K.L., Richter, B.D., et al., 1997. The natural flow regime. BioScience 47, 769−784.

Poff, N.L., Allan, J.D., Palmer, M.A., Hart, D.D., Richter, B.D., Arthington, A.H., et al., 2003. River flows and water wars: emerging science for environmental decision making. Front. Ecol. Environ. 1, 298−306.

Poff, N.L., Brown, C.M., Grantham, T.E., Matthews, J.H., Palmer, M.A., Spence, C.M., et al., 2016. Sustainable water management under future uncertainty with eco-engineering decision scaling. Nat. Clim. Change 6, 25−34.

Poff, N.L., Olden, J.D., Merritt, D.M., Pepin, D.M., 2007. Homogenization of regional river dynamics by dams and global biodiversity implications. Proc. Natl. Acad. Sci. USA 104, 5732−5737.

Poff, N.L., Olden, J.D., Pepin, D.M., Bledsoe, B.P., 2006a. Placing global streamflow variability in geographic and geomorphic contexts. River Res. Appl. 22, 149−166.

Poff, N.L., Olden, J.D., Vieira, N.K.M., Finn, D.S., Simmons, M.P., Kondratieff, B.C., 2006b. Functional trait niches of North American lotic insects: traits-based ecological applications in light of phylogenetic relationships. J. N. Am. Benthol. Soc. 25, 730−755.

Poff, N.L., Richter, B.D., Arthington, A.H., Bunn, S.E., Naiman, R.J., Kendy, E., et al., 2010. The ecological limits of hydrologic alteration (ELOHA): a new framework for developing regional environmental flow standards. Freshw. Biol. 55, 147−170.

Puckridge, J.T., Sheldon, F., Walker, K.F., Boulton, A.J., 1998. Flow variability and the ecology of large rivers. Mar. Freshw. Res. 49, 55−72.

Rahel, F.J., Olden, J.D., 2008. Assessing the effects of climate change on aquatic invasive species. Conserv. Biol. 22, 521−533.

Reidy Liermann, C.A., Olden, J.D., Beechie, T.J., Kennard, M.J., Skidmore, P.B., Konrad, C.P., et al., 2012. Hydrogeomorphic classification of Washington State rivers to support emerging environmental flow management strategies. River Res. Appl. 28, 1340−1358.

Resh, V.H., Brown, A.V., Covich, A.P., Gurtz, M.E., Li, H.W., Minshall, G.W., et al., 1988. The role of disturbance in stream ecology. J. N. Am. Benthol. Soc. 7, 433−455.

Richter, B.D., Thomas, G.A., 2007. Restoring environmental flows by modifying dam operations. Ecol. Soc. 12 (art. 12).

Richter, B.D., Baumgartner, J.V., Powell, J., Braun, D.P., 1996. A method for assessing hydrologic alteration within ecosystems. Conserv. Biol. 10, 1163−1174.

Richter, B.D., Baumgartner, J.V., Wigington, R., Braun, D.P., 1997. How much water does a river need? Freshw. Biol. 37, 231−249.

Richter, B.D., Davis, M.M., Apse, C., Konrad, C., 2012. A presumptive standard for environmental flow protection. River Res. Appl. 28, 1312−1321.

Rogers, K.H., 2006. The real river management challenge: integrating scientists, stakeholders and service agencies. River Res. Appl. 22, 269−280.

Rolls, R.J., Arthington, A.H., 2014. How do low magnitudes of hydrologic alteration impact riverine fish populations and assemblage characteristics? Ecol. Indic. 39, 179−188.

Rood, S.B., Gourley, C.R., Ammon, E.M., Heki, L.G., Klotz, J.R., Morrison, M.L., et al., 2003. Flows for floodplain forests: a successful riparian restoration. BioScience 53, 647−656.

Roux, D.J., Foxcroft, L.C., 2011. The development and application of strategic adaptive management within South African National Parks. Koedoe 53(2), Art. # 1049, 5. Available from: http://dx.doi.org/10.4102/koedoe.v53i2.1049.

Rowntree, K., Wadeson, R., 1998. A geomorphological framework for the assessment of instream flow requirements. Aquat. Ecosyst. Health Manage. 1, 125−141.

Ruhi, A., Holmes, E.E., Rinne, J.N., Sabo, J.L., 2015. Anomalous droughts, not invasion, decrease persistence of native fishes in a desert river. Glob. Change Biol. 21, 1482−1496.

Sanderson, J.S., Rowan, N., Wilding, T., Bledsoe, B.P., Miller, W.J., Poff, N.L., 2012. Getting to scale with environmental flow assessment: the watershed flow evaluation tool. River Res. Appl. 28, 1369−1377.

Shenton, W., Bond, N.R., Yen, J.D.L., Mac Nally, R., 2012. Putting the "ecology" into environmental flows: ecological dynamics and demographic modelling. Environ. Manage. 50, 1−10.

Smakhtin, V.U., Anputhas, M., 2006. An assessment of environmental flow requirements of Indian river basins. IWMI Research Report 107. International Water Management Institute, Colombo, Sri Lanka.

Solans, M.A., de Jalón, D.G., 2016. Basic tools for setting environmental flows at the regional scale: application of the ELOHA framework in a Mediterranean river basin. Ecohydrology 9, 1517−1538. Available from: http://dx.doi.org/10.1002/eco.1745.

Solans, M.A., Poff, N.L., 2013. Classification of natural flow regimes in the Ebro Basin (Spain) by using a wide range of hydrologic parameters. River Res. Appl. 29, 1147−1163. Available from: http://dx.doi.org/10.1002/rra.2598.

Speed, R., Binney, J., Pusey, B., Catford, J., 2011. Policy measures, mechanisms, and framework for addressing environmental flows. ACEDP Project Report. International Water Centre, Brisbane, QLD.

Stanford, J.A., Ward, J.V., Liss, W.J., Frissell, C.A., Williams, R.N., Lichatowich, J.A., et al., 1996. A general protocol for restoration of regulated rivers. Regul. Rivers Res. Manage. 12, 391−413.

Stewart, I.T., Cayan, D.R., Dettinger, M.D., 2005. Changes toward earlier streamflow timing across western North America. J. Climatol. 18, 1136−1155.

Stewart-Koster, B., Bunn, S.E., Mackay, S.J., Poff, N.L., Naiman, P.J., Lake, P.S., 2010. The use of Bayesian networks to guide investments in flow and catchment restoration for impaired river ecosystems. Freshw. Biol. 55, 243−260.

Strayer, D.L., Dudgeon, D., 2010. Freshwater biodiversity conservation: recent progress and future challenges. J. N. Am. Benthol. Soc. 29, 344−358.

Tennant, D.L., 1976. Instream flow regimens for fish, wildlife, recreation and related environmental resources. Fisheries 1 (4), 6−10.

Tharme, R.E., 1996. A review of international methodologies for the quantification of the instream flow requirements of rivers. Water law review report for policy development. Department of Water Affairs and Forestry, Pretoria, South Africa.

Tharme, R.E., 2003. A global perspective on environmental flow assessment: emerging trends in the development and application of environmental flow methodologies for rivers. River Res. Appl. 19, 397−441.

Tharme, R.E., King, J.M., 1998. Development of the Building Block Methodology for instream flow assessments, and supporting research on the effects of different magnitude flows on riverine ecosystems. Water Research Commission Report No. 576/1/98. Water Research Commission, Pretoria, South Africa.

Thompson, J.R., Laizé, C.L.R., Green, A.J., Acreman, M.C., Kingston, D.G., 2014. Climate change uncertainty in environmental flows for the Mekong River. Hydrol. Sci. J. 59 (3–4), 935–954.

Thompson, L.C., Escobar, M.I., Mosser, C.M., Purkey, D.R., Yates, D., Moyle, P.B., 2012. Water management adaptations to prevent loss of spring-run Chinook salmon in California under climate change. J. Water Resour. Plan. Manage. 138, 465–478.

USAID, 2016. Environmental flows in Rufiji River Basin assessed from the perspective of planned development in Kilombero and Lower Rufiji sub-basins. Technical assistance to support the development of irrigation and rural roads infrastructure project (IRRIP2). United States Agency for International Development, final report.

Verberk, W.C.E.P., van Noordwijk, C.G.E., Hildrew, A.G., 2013. Delivering on a promise: integrating species traits to transform descriptive community ecology into a predictive science. Freshw. Sci. 32, 531–547.

Walschburger, T., Angarita, H., Delgado, J., 2015. Hacia una gestión integral de las planicies inundables en la Cuenca Magdalena-Cauca. In: Rodríguez Becerra, M. (Ed.), ¿Para Dónde va el río Magdalena? Riesgos Sociales, Ambientales y Económicos del Proyecto de Navegabilidad. Friedrich Ebert Stiftung, Foro Nacional Ambiental, Bogotá, Colombia.

Wantzen, K.M., Ballouche, A.Z., Longuet, I., Bao, I., Bocoum, H., Cissé, L., et al., 2016. River culture: an eco-social approach to mitigate the biological and cultural diversity crisis in riverscapes. Ecohydrol. Hydrobiol. (in press). Available from: http://dx.doi.org/10.1016/j.ecohyd.2015.12.003.

Warner, A.T., Bach, L.B., Hickey, J.T., 2014. Restoring environmental flows through adaptive reservoir management: planning, science, and implementation through the Sustainable Rivers Project. Hydrol. Sci. J. 59, 770–785.

Webb, C.T., Hoeting, J.A., Ames, G.M., Pyne, M.I., Poff, N.L., 2010. A structured and dynamic framework to advance traits-based theory and prediction in ecology. Ecol. Lett. 13, 267–283.

Webb, J.A., De Little, S.C., Miller, K.A., Stewardson, M.J., Rutherfurd, I.D., Sharpe, A.K., et al., 2015. A general approach to predicting ecological responses to environmental flows: making best use of the literature, expert knowledge, and monitoring. River Res. Appl. 31, 505–514.

Wiens, J.A., Hayward, G.D., Safford, H.D., Giffen, C. (Eds.), 2012. Historical Environmental Variation in Conservation and Natural Resource Management. John Wiley & Sons, Ltd, Chichester, UK.

Wilding, T.K., Bledsoe, B., Poff, N.L., Sanderson, J., 2014. Predicting habitat response to flow using generalized habitat models for trout in Rocky Mountain streams. River Res. Appl. 30, 805–824.

Winemiller, K.O., McIntyre, P.B., Castello, L., Fluet-Chouinard, E., Giarrizzo, T., Nam, S., et al., 2016. Balancing hydropower and biodiversity in the Amazon, Congo, and Mekong. Science 351, 128–129.

Wohl, E., Bledsoe, B.P., Jacobson, R.B., Poff, N.L., Rathburn, S.L., Walters, D.M., et al., 2015. The natural sediment regime in rivers: broadening the foundation for ecosystem management. BioScience 65, 358–371.

Yarnell, S.M., Petts, G.E., Schmidt, J.C., Whipple, A.A., Beller, E.E., Dahm, C.N., et al., 2015. Functional flows in modified riverscapes: hydrographs, habitats and opportunities. BioScience 65, 963–972.

Yen, J.D.L., Bond, N.R., Shenton, W., Spring, D.A., Mac Nally, R., 2013. Identifying effective water-management strategies in variable climates using population dynamics models. J. Appl. Ecol. 50, 691–701.

Zhang, Y., Arthington, A.H., Bunn, S.E., Mackay, S., Xia, J., Kennard, M., 2012. Classification of flow regimes for environmental flow assessment in regulated rivers: the Huai River Basin, China. River Res. Appl. 28, 989–1005.

TOOLS FOR SEDIMENT MANAGEMENT IN RIVERS

12

David E. Rheinheimer[1] and Sarah M. Yarnell[2]
[1]*University of Massachusetts, Amherst, MA, United States* [2]*University of California, Davis, CA, United States*

12.1 INTRODUCTION

The goal of sediment management in the context of developing and implementing environmental water regimes is to support the creation and maintenance of geomorphic forms and processes that, along with the hydraulic characteristics of a river, provide the physical template needed to support river ecosystems (Chapter 5). Ecosystem management objectives might not be achieved if geomorphic forms and processes are omitted in environmental water assessments (Heitmuller and Raphelt, 2012; Yarnell et al., 2015).

Historically, however, environmental flows assessment methods have not explicitly incorporated sediment management, particularly in comparison with the wide range of hydrology-centric assessments and approaches (Meitzen et al., 2013). A major reason for this is the complexity of geomorphic processes and data acquisition, and conversely, the ease of obtaining flow data and its subsequent availability. Only two holistic environmental flows assessment methods, *Downstream Response to Impost Flow Transformation* (*DRIFT*; King et al., 2003) and to some degree *Ecological Limits of Hydrologic Alteration* (*ELOHA*; Poff et al., 2010), explicitly include maintenance of an ecologically supportive sediment regime as a requisite management objective. As a result of this historical deficiency, the need to explicitly account for sediment in environmental water regime design has been increasingly emphasized, whether for natural, highly altered, or novel river ecosystems (Acreman et al., 2014; Escobar-Arias and Pasternack, 2010; Meitzen et al., 2013; Wohl et al., 2015; Yarnell et al., 2015).

This chapter reviews a range of concepts and tools for providing the sediment regime needed to support the ecological functioning of a river. The focus is on in-river processes and sediment management, though with some mention of upland sediment management. First, the functional relationship between flow and sediment is reviewed with an overview of sediment transport theory. Second, several methods for assessing environmental water requirements for geomorphic processes are reviewed. Third, complementary tools are reviewed, including sediment budgets/models, land management, and options in and around dams. The chapter concludes with key points and research needs.

Managing sediment in rivers begins with recognizing the complexity of geomorphic processes in both space and time, and the basin context within which sediment management occurs. As a

Water for the Environment. DOI: http://dx.doi.org/10.1016/B978-0-12-803907-6.00012-7

237

result of the complex nature of sediment and its interactions with river morphology and ecology, each geomorphic system may have its own unique challenges. For this reason, sediment management issues, objectives, and tools vary widely, depending on location in catchment, relative degree of disturbance of the sediment system, the climate/geology/topography of the river, and timescales of interest. Thus, although sediment is important from headwaters to estuary in rivers of all sizes, specific approaches that may work well in one location may be poorly suited for another location, even within the same region. Regardless, there are several principles that are broadly relevant to most sediment regimes and several tools useful in many instances. Here we focus on the most common tools for reach-scale sediment management, and we refer the reader to Apitz and White (2003) and Owens (2008) for review of basin-scale sediment management in the context of the European Sediment Network, which is beyond the scope of this chapter. However, we also highlight some catchment-scale tools such as sediment budget assessments.

In addition, a wide range of human influences on geomorphic processes exist (Brookes, 1994), including indirect (land use and drainage) and direct (river regulation and channel management) changes, as listed in Table 12.1 and described in Chapter 5. Managing sediment therefore requires complementary actions well beyond environmental water management alone. Complementary actions include basin management, sediment management in and around reservoirs, and direct management of in-channel sediments and geomorphic features, some of which are reviewed here.

Table 12.1 Human Influences on Sediment Dynamics in Rivers	
Indirect Change	
Land-use change	Deforestation
	Afforestation
	Agricultural conversion
	Urbanization
	Mining
Land drainage	Agricultural drainage
	Surface water sewers
Direct Change	
Regulation	Impoundment of water
	Water diversions
Channel management	Gravel extraction
	Cattle grazing
	Straightening
	Flood control
	Bank erosion protection
	Dredging
Source: Adapted from Brookes (1994)	

12.2 **SEDIMENT MOBILIZATION THEORY**

As we move toward an increasingly mechanistic approach to managing rivers, such as espoused by Poff et al. (2010) for environmental water and Church and Ferguson (2015) for river morphology specifically, a mechanistic understanding of sediment mobilization is increasingly important compared to empirically based approaches. The complexity of sediment mobilization has resulted in many different characterization approaches. As such, the following is a decidedly cursory treatment, intended to provide a general overview of concepts useful for environmental water management. Many reviews of river sediment theory exist. Useful examples include Yang (2003), Fryirs and Brierley (2013), and Dey (2014); the latter is particularly thorough.

The sediment regime in a river is fundamentally driven by hydraulic forces of flowing water acting on particles in a river, primarily via lift and drag forces, which may be counteracted by other forces related to vegetation, channel form, and sediment supply. Sediment mobilization caused by these forces consists of *entrainment* (aka *erosion*), whereby particles are picked up or moved from a resting state; *transport*, whereby particles are moved some distance along the river bed or in the water column; and *deposition* (aka *sedimentation*), whereby there is no longer sufficient stream power to keep the particle in motion and the particle settles. For any given particle on the river bed, there is a specific threshold stream power above which the particle will become entrained. Once entrained, the stream power needed to continue to transport the particle decreases as turbulent forces in the water column contribute to motion. Overall, greater power (velocity) is needed to entrain sediment than is needed to continue to transport it. Both fine and coarse particles require more energy to entrain than intermediate sizes, the former due to physical and chemical cohesion mechanisms, and the latter due to the strength of gravitational forces relative to fluid forces. Once entrained, fine-grained sediments (fine sediments) may be easily transported as *suspended load*. Some very fine particles become *wash load*, never settling in typical flowing conditions. Coarse-grained sediments (coarse sediments), however, once entrained, will tend to drag along the bottom of the river as *bed-load*, and require much more energy to transport. Additional complexities affecting mobilization processes and sediment movement include the shape of the river bed, called *bed form*, and the presence of riparian vegetation and large woody debris. Hydraulic forces at the particle-scale are also influenced by larger-scale flow dynamics such as those resulting from channel meanders, drag through vegetation, and scour around large wood. Thus, in any particular location, channel formation and load transport is a function of inflow sediment characteristics, local river bed sediment size distribution, hydraulic characteristics of the river, presence of vegetation, and existing local bed form. These various interactions result in both nonlinearities and complexities that while sometimes may be in the form of *deterministic uncertainty* (Gomez and Phillips, 1999; Phillips, 2006), more commonly have made direct, theoretical assessments for environmental water requirements impossible in practice. As a result, semi-theoretical approaches or entirely empirical approaches are generally more useful.

12.2.1 **SEDIMENT ENTRAINMENT**

The concept of a *threshold stream power* as used above is a simplification of the complex process of sediment entrainment, and thus is difficult to quantify in practice. Two interrelated concepts are

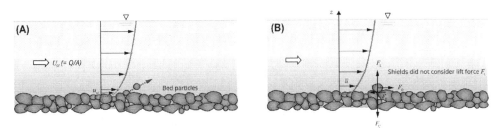

FIGURE 12.1

Conceptual illustration of (A) threshold velocity and (B) threshold bed shear stress for bed-load mobilization.

Source: Dey (2014)

particularly useful in understanding sediment transport, though they differ in their utility. *Threshold velocity* refers to local near-bed or depth-averaged velocity above which a particle of a given size is mobilized (Fig. 12.1A). However, although depth-averaged velocity may appear appealing as it is easy to calculate from discharge and channel geometry measurements, it is of limited practical use. Actual mobilization occurs due to local velocities, not average velocities, necessitating the introduction of various simplifying assumptions and resulting in imprecise results. Similarly, calculating near-bed velocity at the particle-scale requires simplifying assumptions that can be hard to estimate in practice.

Threshold bed shear stress refers to the tractive force per unit area on the bed floor above which a particle of a given size is mobilized (Fig. 12.1B), and, in contrast to threshold velocity, is the standard parameter used to predict the onset of sediment motion. Bed shear stress is based on an analysis of the hydraulic forces on bed particles in a flowing river, including a stabilizing resistive force (F_R), a driving force (drag force, F_D), gravitational force (F_G), and lift force (F_L; Fig. 12.1B). Although several studies have attempted to correlate sediment properties with a threshold (critical) shear stress (Dey, 2014), it is the semi-theoretical work of Shields (1936) that led to the development of the sediment threshold theory used today.

By analyzing the various forces on a particle, Shields (1936) developed a parameter Θ, now called the Shields parameter, that incorporates a range of factors affecting forces acting on a particle. (Note that Shields [1936] did not consider lift force, which can be important.) The Shields parameter is defined as:

$$\Theta_c = \frac{u_*^2}{\Delta gd} = \frac{\tau_0}{\Delta \rho gd}$$

where u_* is the shear velocity, Δ is the submerged relative density, ρ is the mass density of water, g is the acceleration due to gravity, d is the particle depth, and τ_0 is the Reynolds shear stress ($\tau_0 = \rho u_*^2$). The critical (threshold) Shields parameter Θ_c defines the onset of motion of a particle, and occurs at a critical shear stress, or when $u_* = u_{*c}$ or $\tau_0 = \tau_{0c}$. This formulation means that the critical Shields parameter Θ_c, and the onset of motion, is a function of fundamental properties of the water and particle.

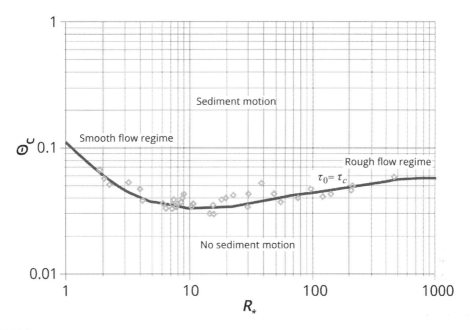

FIGURE 12.2

Shields diagram, where threshold (critical) Shields parameter Θ_c is represented as a function of shear Reynolds number R_*. τ_0 is the shear stress and τ_c is the critical shear stress.

Source: Shields (1936) adapted by Dey (2014)

Furthermore, the critical Shields parameter is a function of the shear Reynolds number R_*, which is a specific form of the more general particle Reynolds number R_p, an important and widely used parameter that indicates the hydraulic properties of a parcel of water around a particle (i.e., hydraulically smooth, hydraulically turbulent or transitional). R_p is defined as $R_p = uk_s/v$, where u is the water velocity, k_s is the bed particle size (diameter), and v is the viscosity; R_* is when velocity u equals shear velocity u_*. The relationship between R_* and Θ_c, which is nonlinear, can be plotted in what is known as the Shields diagram (Fig. 12.2). This diagram can help estimate the flow conditions needed for the onset of particle motion for different hydraulic characteristics: to mobilize any given particle size, flows should yield a shear Reynolds number such that $\Theta > \Theta_c$.

The Shields diagram, which is based on empirical data, also illustrates the inherent uncertainty in estimating Θ_c. This uncertainty is caused by two issues at the local scale. First, in a turbulent flow field such as a river, there is only a probability of entrainment of any given particle at the threshold bed shear stress. Second, the entrainment of a particle is influenced by the local particle size distribution and particle shape, which tend to be heterogeneous along a river bed. In practice, then, estimating the critical Shields parameter requires field measurements or empirical relationships, rendering its application to river management cumbersome or imprecise.

Despite these challenges, use of the Shields parameter has stood the test of time and remains an important tool for estimating flows needed to mobilize sediment in a river (Cao et al., 2006; Coleman and Nikora, 2008; Comiti and Mao, 2012). The challenge for environmental flows assessments is identifying the threshold Shields parameter for streams with complex sediment size distributions and varying bed structures. Much progress has been made, but this remains an active area of research (Bunte et al., 2013; Church et al., 2012; Mueller et al., 2005; Patel et al., 2013; Wilcock, 1993).

12.2.2 SEDIMENT TRANSPORT

Analysis of the sediment moving through a stream network typically focuses on bed-load and suspended load, which together form the *total load* of a stream. Although total load also includes wash load, wash load is often excluded from analyses as, by definition, it does not settle under typical flow conditions and can hence be treated as a conservative substance in the water column. Fig. 12.3 shows each of the three modes of particle transport. The processes and measurements involved in transport of bed-load versus suspended load are different enough that practitioners commonly assess the two fractions separately.

Bed-load transport

Bed-load consists of particles that, when mobilized, remain in near-continual contact with the bed. These particles move by rolling, sliding, or saltating (hopping) along the river bed (Fig. 12.3). The size range of bed-load material depends on the magnitude and turbidity of flow, but typically is dominated by coarser size fractions such as gravels and cobbles.

Numerous bed-load transport functions have been developed over the past century, ranging from simple models of unidirectional uniform-sized particle transport to more complex mixed

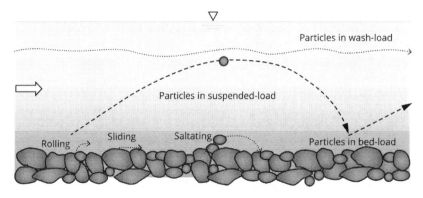

FIGURE 12.3

Different modes of sediment transport in a river.

Source: Dey (2014)

grain size models based on three-dimensional flow momentum equations (Dey, 2014). However, reflecting the challenge of sediment mobilization theory generally, they have been notoriously unreliable for predicting transport in either flumes or rivers due to the inherent complexities of flow and sediment interactions at multiple scales (Gomez and Church, 1989). Most transport models rely on calculations of excess shear stress, or the difference between the bed shear stress provided by a flow and the critical bed shear stress required to mobilize sediment, to determine the volume of sediment that can be moved through time. Complexities relating to estimates of grain size distribution, bed shear stress, degree of bed armoring, channel cross-sectional shape, and bed morphology, among other factors, are typically not accounted for, and thus contribute to error in sediment transport predictions.

Yet, even sediment transport estimates within an order of magnitude error can be useful for assessing some environmental water requirements such as determining which flows are large enough to mobilize and transport the majority of the bed sediment. Similarly, if management goals include a future change in channel form and sediment transport regime, then the *difference* between the sediment transport rate under current conditions versus future conditions is of more importance than the actual rate of transport. This is particularly important as changes in sediment transport rate along a stream reach are balanced by local erosion or deposition of bed and bank material. As a result, environmental water practitioners may seek to determine sediment transport rates based on management objectives rather than absolute accuracy.

Several computer programs are freely available to help estimate sediment transport. One such computer program is Bedload Assessment in Gravel-bedded Streams (http://www.stream.fs.fed.us/publications/bags.html), which provides the user a choice of different sediment transport formulas depending on the availability of stream data and information (Pitlick et al., 2009). The accompanying user's guide provides background information on sediment transport theory, common errors, and assumptions in sediment transport models, and aids the user in more accurately interpreting the results provided from the software (Wilcock et al., 2009).

Owing to the complexities inherent to geomorphic processes, most sediment transport calculations and models focus on the mobility of individual grain movements as discussed above. However, the shape and type of bed forms present within the channel can have profound effects on the mobility of sediment. Channel bars can create larger-scale form drag that compounds drag forces calculated solely from particle size, thus inducing potentially orders of magnitude error to calculations of sediment mobility and transport rates (Lawless and Robert, 2001). Similarly, erosion and deposition of bed forms during a single flood event can alter measured sediment transport rates during the flood by an order of magnitude, resulting in large discrepancies when compared with calculations of sediment transport based on particle size alone (Dinehart, 1992; Fryirs and Brierley, 2013). Currently, most readily available sediment transport models cannot account for bed morphology or channel bed forms, although the *US Hydrologic Engineering Center's River Analysis System (HEC-RAS)* (http://www.hec.usace.army.mil/software/hec-ras/) can simulate one-dimensional longitudinal sediment transport under mobile channel bed conditions. In the context of environmental flows assessments, some flow management objectives may be achieved with the more cursory estimates of sediment movement that particle-based calculations can provide. If more accurate calculations of sediment transport are needed for a particular geomorphic or flow objective, then consideration of bed form features and advanced modeling techniques may be required.

Suspended load transport

Suspended load consists of particles that move within the water column when mobilized but they interact and exchange with particles on the bed. Although in some extremely turbid debris floods very large particles may be picked up and carried within the water column, most suspended load material in streams is comprised of smaller particles such as sands and silts.

Unlike bed-load transport data that are more often estimated from various transport functions as discussed above, suspended sediment data are more commonly measured directly on site. Grey et al. (2000) (Section 5.3 in Diplas et al., 2008) provide descriptions of manual and automatic suspended sediment samplers, methods for their deployment, and a summary of equipment used in collections of subsamples. In general, samples are taken routinely during moderate- and high-flow events, and analyzed in a laboratory to determine the particle size distribution and volume transported over time. Laboratory methods for determining *suspended sediment concentrations* (*SSC*) or *total suspended solids* (*TSS*), both measured in mg/L, are similar but can result in differences due to the subsampling inherent to the methodology of TSS data collection (Grey et al., 2000). TSS data from subsamples tend to underestimate SSC data by 25%−35%, which can lead to large errors in calculations of annual sediment yields. In some instances, relationships between SSC/TSS as measured from samplers and turbidity as measured from optical sensors in *Nephelometric Turbidity Units* (*NTUs*) can be developed in an effort to provide a method for long-term monitoring of fine sediment discharge and water quality. However, use of turbidity alone as an analogue for suspended sediment concentration without development of a rating curve is fundamentally unreliable in streamflows (Grey et al., 2000).

Total load transport

The total load is the entirety of sediment moved downstream and is the combination of bed-load and suspended load. Most calculations of total load are made indirectly, by simply summing the bed-load and suspended load volumes over a common time frame. However, there may be instances where only total load is of importance or where there is no clear distinction between bed-load and suspended load. The latter is particularly true when suspended sediment loads are high. In this case, it may be more desirable to estimate total load directly. Numerous methods have been developed for this purpose, and are reviewed by Dey (2014).

12.2.3 SEDIMENT DEPOSITION

After particles are in motion, size-selective deposition on the channel bed begins once flow velocity falls below the *settling velocity* of a particle. The settling velocity of a particle is a function of grain size, grain density, water viscosity, and water density. Essentially, when the drag on a particle equals the submerged weight of the particle, the particle will begin to fall or settle to the bed. Alternatively, deposition occurs when the bed shear stress is less than the critical shear stress. As deposition is a grain size-selective process, as flow velocity decreases, coarser sediment is deposited first, followed by finer-grained particles as flows continue to decrease, resulting in sorting of bed material throughout the channel. Although settling velocity is closely related to particle size, it is greatly affected by sediment composition and associated drag forces. Spherical-shaped particles

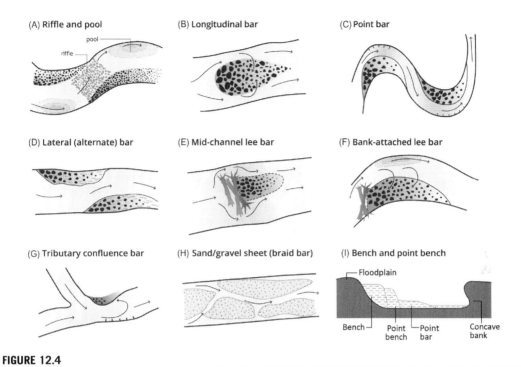

(A) Riffle and pool
pool
riffle

(B) Longitudinal bar

(C) Point bar

(D) Lateral (alternate) bar

(E) Mid-channel lee bar

(F) Bank-attached lee bar

(G) Tributary confluence bar

(H) Sand/gravel sheet (braid bar)

(I) Bench and point bench
Floodplain
Bench   Point   Point   Concave
          bench   bar   bank

FIGURE 12.4

(A—I) Some common geomorphic units in a river. Managing rivers for healthy ecosystems should include enabling hydraulic and geomorphic processes to create these features where appropriate.

Source: Fryirs and Brierley (2013)

have a higher settling velocity than angular particles, and higher concentrations of sediment in motion will support higher settling velocities as particle-to-particle interactions become more frequent (Fryirs and Brierley, 2013).

Sediment deposition can occur not only following a large sediment-moving flow event as velocities decrease, but also in any location within a channel where flow velocities locally decrease, resulting in numerous ecologically significant geomorphic units within a river (Fig. 12.4). Obstructions in the channel provide decreased velocity zones on the downstream or lee side promoting sediment deposition and the formation of *lee bars*. Meander bends create differential velocity zones with lower velocities on the inside of the bend, promoting deposition and the formation of *point bars*. Decreased velocities in eddies along the lateral sides of the channel create *eddy bars* (coarse-grained) or *benches* (fine-grained; sensu Vietz et al., 2006). Expansion (widening) and contraction (narrowing) zones in the stream create increased velocity and scour of sediment in constricted areas followed by decreased velocities as the channel widens and deposition of *mid-channel bars*. In each of these cases, local features and morphology create zones of differential velocities that locally promote deposition of fine or coarse material, creating the various channel habitat features that are so ecologically important.

12.3 ENVIRONMENTAL FLOWS ASSESSMENT METHODS FOR GEOMORPHIC MAINTENANCE

Determining environmental water requirements for geomorphic maintenance entails two steps: (1) defining specific management objectives and (2) estimating the flows needed to support those objectives. Historically, management objectives were fairly specific, and fell into three broad categories: *flushing*, whereby the primary objective is to remove (flush) finer sediment from coarser substrate; *channel maintenance*, whereby the objective is to maintain overall channel structure and geometry (e.g., scour sediment from pools, clear banks of encroaching vegetation); and *channel forming*, whereby the objective is to reshape the channel morphology. As river complexity and heterogeneity have been increasingly recognized as important for river ecosystem health (Naiman et al., 2008; Ward et al., 2001), management objectives have broadened to include maintaining the natural sediment regime (sensu Wohl et al., 2015). However, recognizing the importance of maintaining a natural flow or sediment regime does not necessarily entail simply mimicking these natural regimes. In highly regulated rivers in particular, flows should be tied to specific functional objectives (*functional flows*), where channel heterogeneity is one of many objectives (Yarnell et al., 2015).

Below we describe two general approaches to developing geomorphically effective flows, organized as threshold-based approaches and field/laboratory methods. Additionally, we describe how these methods relate to holistic environmental flows assessment methods. In practice, there is no clear distinction between these approaches, as field/laboratory methods may be used to estimate flow thresholds. Some of these methods depend specifically on quantifying the threshold shear stresses discussed above, whereas others rely on determining flow frequency and exceedances. Each of these methods may be applicable to one or more specific geomorphic process to be maintained. This organization differs somewhat from others. Kondolf and Wilcock (1996) used the terms self-adjusted channel methods, sediment entrainment methods, and direct calibration methods, whereas Gippel (2001) organized methods by statistical (natural hydrology) methods, desktop calculation methods, and field methods. These differences are mostly a matter of organizational convenience and semantics; specific methods and approaches are the same.

12.3.1 THRESHOLD-BASED APPROACHES

The earliest methods developed to provide geomorphic functions within environmental water regimes were designed to flush fine sediments from coarser river substrate, primarily to support salmon spawning in the United States (Gippel, 2001; Reiser et al., 1985). Using empirical data collected from multiple rivers within a broad region, these methods determine a flow threshold above which sediments are mobilized. Generally, they are based on a fixed flow statistic such as a percentage of natural runoff either averaged over a period of record, averaged for specific times of the year, or averaged for a specific flow return interval (Gippel, 2001). One such flow threshold is the *effective discharge*, which is the discharge that mobilizes more sediment per flow volume than any other discharge. Wolman and Miller (1960), who pioneered this concept, found the effective discharge occurred approximately every 1−2 years for many alluvial streams in the western United States. Subsequent studies have found that the frequency of effective discharge in any given stream depends on the variability of streamflow and basin characteristics, with rivers with greater variability—and smaller basins—mobilizing more sediment with less frequent, higher-flow events (Sholtes et al., 2014).

The general approach of using streamflow frequency to provide specific geomorphic functions has been used in numerous instances for both flushing fine sediments and channel maintenance, particularly in the United States in the 1970s and 1980s. Examples of flushing flows include: discharge near the mean daily flow averaged over the high-flow period for flushing for cold and warm water rivers in the northern United States (Northern Great Plains Resource Program, 1974); the 17% flow exceedance for 48 h for flushing flows in Colorado, United States (Hoppe, 1975); the 5% flow exceedance for Oregon coastal streams (Beschta and Jackson, 1979); 200% of the mean daily unregulated flow to flush fine sediments to maintain spawning gravels for salmonids (Tennant, 1976); and 3- or 7-day average maximum flows for flushing sediments in Oregon, United States, corresponding to about 400–1000% of the mean daily flow (Reiser et al., 1985).

Similar flow frequency approaches have been used for channel-forming flows that require significant bed mobilization. Parker (1979) assumed 80% of the bankfull flow depth for sediment mobilization, whereas Montana Department of Fish (1981) used a 24-h bankfull discharge with a recurrence interval of 1.5 years as the effective discharge needed for channel maintenance. Bankfull discharge is often assumed to be the effective discharge in low-gradient alluvial streams and thus is commonly applied as a channel-forming flow (Hill et al., 1991). Although bankfull flows are often used to maintain channel form, there is widespread recognition that infrequent larger floods and flow variability are also necessary for diversity in channel habitat (Arthington et al., 2006; Naiman et al., 2008; Tockner et al., 2000). In urban streams, where urban disturbance regimes tend to increase the rate and extent of channel erosion, Hawley and Vietz (2016) found that controls on stormwater discharge are needed to maintain channel stability, with limits of $> \sim 0.1-1$ times 2-year peak discharge for gravel/boulder streams and < 0.01 times 2-year peak discharge for sand-dominated streams.

Flow frequency approaches have the advantage of recognizing some degree of regional geomorphic homogeneity. However, specific values cited for one region are not generally applicable to other regions. Even within a region, the assumption of geomorphic homogeneity may not be appropriate. For this reason, statistical methods, whether for flushing or channel-forming, may be useful for preliminary regional planning, though insufficient for specific rivers/locations.

Semi-theoretical approaches can be tailored to individual streams as they estimate flows needed for sediment mobilization based on threshold velocities or shear stress such as with the critical Shield's parameter discussed above. For example, Newbury and Gaboury (1994) equated bed shear stress (flow depth in meters $\times$ slope $\times$ 10^3, kg/m) to incipient particle diameter in centimeters, which, compared with the bed particle size distribution and channel dimensions, allowed for estimation of the discharge needed to mobilize the bed material.

Although the application of semi-theoretical approaches has historically been limited by the complexity encountered in real-world settings (Bleed, 1987; Buffington and Montgomery, 1997), this type of mechanistic approach is often relied upon in highly regulated rivers, particularly when time or funding for more accurate field-based methods is limited. Despite the limitations of threshold-based approaches, estimating sediment mobilization flows needed for specific geomorphic functions is increasingly recognized as both necessary and possible, particularly for restoration of highly modified rivers, which tend to require more engineered approaches (i.e., with less reliance on the natural flow regime) to achieve management objectives (Yarnell et al., 2015).

12.3.2 **FIELD/LABORATORY METHODS**

Field and laboratory experiments offer the most direct method of estimating flows for sediment mobilization. Field experiments entail in situ characterization of sediment mobilization using tracers (e.g., Vázquez-Tarrío and Menéndez-Duarte, 2015), acoustic Doppler (Kostaschuk et al., 2005; Reichel and Nachtnebel, 1994), or other sediment monitoring techniques. The obvious appeal of field methods is the development of location-specific knowledge. However, the trade-off is that such methods are typically complex, requiring extensive reliance on field personnel, requisite equipment, and logistical support. In contrast, laboratory experiments such as with scaled river models or flumes, are much simpler, providing a very controlled setting. Laboratory experiments are more commonly used to better understand fundamental hydraulics and sediment transport theory, but may be tailored to better understand location-specific geomorphic processes (e.g., Warburton and Davies, 1994). In some instances, laboratory methods may be combined with field methods such as in one study of geomorphic maintenance flows for Colorado River squawfish where recommendations were provided for flushing and bankfull flows to maintain the required fish habitat (O'Brien, 1984).

Some field experiments may involve active regulated flow trials such as the test floods in the Colorado River below Glen Canyon Dam (Olden et al., 2014; Stevens et al., 2001). Results from a series of experimental flow releases designed to mobilize sediment and create channel bars showed that high-magnitude short-duration floods coinciding with natural floods from tributaries were most effective in creating desired channel bar habitat (Melis et al., 2012; Schmidt and Grams, 2011). Although these types of large-scale flow experiments are highly valuable for determining actual responses of sediment to various flows, appropriate study design, modeling, and monitoring must be incorporated to ensure effective outcomes for managers (Konrad et al., 2011).

12.3.3 **SEDIMENT MANAGEMENT IN HOLISTIC ENVIRONMENTAL FLOW METHODS**

Sediment management may occur as part of a comprehensive, holistic environmental water assessment method/framework, though the degree to which geomorphic processes and their ecological significance are considered depends on the method. The two most prominent holistic methods are DRIFT (King et al., 2003) and the ELOHA (Poff et al., 2010) described further in Chapter 11. DRIFT includes four *modules*: biophysical, sociological, scenario development, and economic. In DRIFT, geomorphology and sediment management are explicit components of the biophysical module, recognizing the critical role of geomorphology in river ecosystems. However, DRIFT is only a framework, and does not prescribe any specific method to determine components for environmental water requirements. To integrate appropriate sediment management into a DRIFT application, methods specific to geomorphic processes, such as described above, and complementary tools, such as described below, are needed. ELOHA focuses more exclusively on flows, through the development of data-driven flow alteration—ecological response relationships within a region. ELOHA considers geomorphology primarily in the context of hydrologic classification (as *geomorphic subclassification*), which is used to help define the importance of a particular flow regime type in a particular river reach. However, ELOHA recognizes that the development of mechanistic or process-based flow alteration—ecological response relationships often requires an understanding of the various ecosystem processes involved, including geomorphic processes.

Table 12.2 Example Functional Flows, Their Geomorphic Importance, and Suggested Magnitude, Timing, and Duration to Support Geomorphic Processes for Highly Modified Rivers in the Western United States (Yarnell et al., 2015). All Functional Flows Should Incorporate Interannual Variability

Flow Type	Geomorphic Function	Magnitude	Timing	Duration
Wet season initiation flows	Flushing fine sediments accumulated during dry season	High enough to establish connectivity with riparian zone and flush organic matter and fine sediments from the substrate	Onset of wet season precipitation	Sufficient to cue species migrations or initiate nutrient exchange
Peak magnitude flows	Mobilization of larger sediments for channel maintenance and forming	High enough to connect to floodplain, mobilize bed material, maintain pools, and maintain in-channel bar forms	Any time within the natural high-flow season	Long enough to support geomorphic objectives (floodplain deposition, channel bar formation, etc.), but short enough to prevent excessive sediment depletion
Spring recession flows	Redistribution of sediments mobilized by peak flows	Starting at spring snowmelt peak, decreasing at a rate typical of local unimpaired recession	Natural spring snowmelt period	As needed to enable desired rate of recession
Dry season low flows	Support riparian vegetation for geomorphic heterogeneity	Low enough to provide limiting conditions, but still maintain dry season perenniality or ephemerality	Natural low-flow period	Consistent with premodified conditions

In addition to these holistic environmental flows assessment methods, the functional flows paradigm described by Yarnell et al. (2015), though not a method per se, provides an integrative approach to identifying environmental water requirements for geomorphic and ecological functions in highly degraded rivers. Yarnell et al. (2015) highlight four functional flow types to support geomorphic and other processes in rivers in the western United States, each of which inherently achieves objectives such as flushing or channel maintenance. These flows, their geomorphic importance, and suggestions for their magnitude, timing, and duration are listed in Table 12.2; we emphasize that these flows are more broadly supportive of habitat diversity and maintenance, not strictly for sediment mobilization.

12.4 COMPLEMENTARY OPTIONS FOR SEDIMENT MANAGEMENT

The flow regime alone may be insufficient to achieve geomorphic objectives, as a river's geomorphic characteristics also depend directly on a range of other factors. This section provides an overview of complementary options for sediment management, including catchment-scale sediment conceptual models and budgets to help prioritize local options, off-river options to support appropriate rates of sediment supply, through- and below-dam options to help maintain appropriate rates of sediment mobilization continuity within the channel, and stream restoration to encourage sediment retention, distribution, and heterogeneity within the channel.

12.4.1 CATCHMENT-SCALE CONCEPTUAL SEDIMENT MODELS AND BUDGETS

Catchment-scale quantitative understanding of sources, sinks, stores, and movement pathways of sediment is needed for local sediment management strategies to be ecologically meaningful in river systems, particularly if nonpoint sources are of concern. Conceptual models and sediment budgets can help identify the most important sources of sediment, and consequently, targets for management intervention (Owens, 2005; Walling and Collins, 2008). The main purpose of a conceptual sediment model—or, more broadly, a conceptual river basin model—is to identify key subsystems of a river basin and map the relationship between these subsystems in terms of sediment fluxes (Owens, 2005). Once the conceptual model is created, a sediment budget may be performed, whereby fluxes of sediments from sources to sinks are quantified along each source-to-sink pathway. A variation of the sediment budget is to include sediment reservoirs and associated residence times (i.e., behind dams). Sediment budgets are typically presented in both tabular and graphical form, with an example of the latter shown in Fig. 12.5 for the Kaleya catchment in Zambia. Sediment budgets can also help show how management interventions affect catchment sediment fluxes over time (Nyssen et al., 2009).

Several tools have been used to develop catchment sediment models (Walling and Collins, 2008). One method entails frequent monitoring of in-river sediment. This can be accomplished using manual or automatic direct sampling of sediment in the water column, or by using turbidity sensors (Walling et al., 2001). One drawback to monitoring is the need for high-frequency sampling during high-sediment load events, which requires significant human resources for both monitoring and subsequent data processing. Another method uses fallout radionuclides such as cesium-137, which accumulated in the stratosphere and soil as a result of testing thermonuclear weapons in the 1950s and 1960s, and may be useful for estimating medium-term (about 50-year) rates of erosion, deposition, and loss of fine sediments in upland source and floodplain areas (Walling, 2004). Finally, fingerprinting, whereby the geochemical signature of sediment is linked to its source (Collins et al., 1997), is also particularly useful in a management context, as the sediment sources can then be targeted for management intervention. In general, an integrated approach, whereby more than one method is used, is preferable to using any single method. Walling et al. (2001) used an integrated approach to develop the suspended sediment budget for the upper Kaleya catchment in Zambia (Fig. 12.5).

Once developed, a sediment budget may be used to strategically prioritize sediment management actions. In the upper Kaleya example (Fig. 12.5), management actions to reduce the supply of suspended sediments should focus on communal cultivation and bush grazing; actions targeting commercial cultivation would be less impactful.

12.4.2 BASIN LAND MANAGEMENT

Terrestrial development and activity may either increase or decrease the quantity and quality of sediment supplied to a river. Historically, the increase in fine sediment caused by land management practices in rural areas has been the primary concern worldwide (Owens et al., 2005). However, the

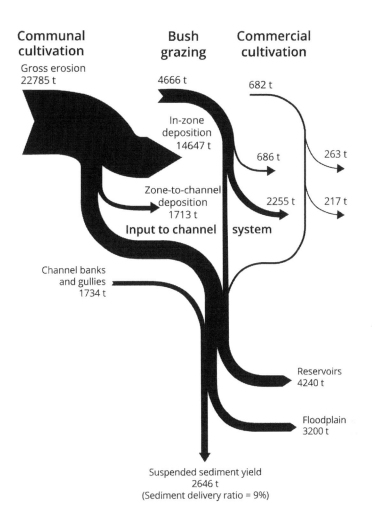

FIGURE 12.5

Suspended sediment budget of the upper Kaleya catchment, Zambia (where 1 ton is equal to 1000 kg).

Source: Walling et al. (2001)

effects of urbanization on geomorphic processes can also be significant, particularly the short-term increase in erosion rates during construction activities, later reduction in coarser sediment supply due to increased imperviousness, and increased in-stream sediment disturbance by increased runoff magnitude and flashiness (often as mediated by stormwater drainage systems) both during and after initial urbanization activities (Bledsoe, 2002; Russell et al., 2016; Vietz and Hawley, 2016; Wolman, 1967). Wolman (1967) eloquently expressed sediment yield over time as a function of typically observed land-use change (Fig. 12.6).

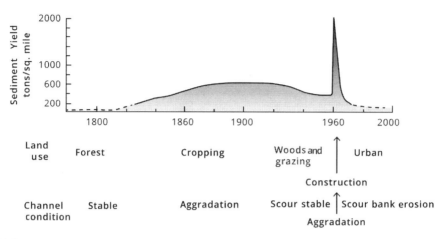

FIGURE 12.6

The cycle of land-use change, sediment yield, and channel behavior in Piedmont, Maryland, United States (where 1 ton/square mile is equal to 0.392 tonnes/square kilometer).

Source: Wolman (1967)

In rural areas, dominant anthropogenic sources of fine sediment in rivers include soil erosion resulting from crop cultivation, overgrazing, and deforestation, and land disturbance resulting from forestry activities, construction, and mining (Owens et al., 2005). Many management options exist in each of these areas to reduce excessive sediment supply. In crop cultivation, agronomic options include covering soil surface, increasing surface roughness, increasing surface depression storage, increasing infiltration, soil management, fertilizers/manures, and subsoiling/drainage, whereas mechanical options include contouring/ridging, terraces, and waterways (Mekonnen et al., 2015; Montgomery, 2007; Morgan, 2005). Reforestation has been widely demonstrated as a means of reducing excessive sediment yield (e.g., Fang et al., 2016; Marden, 2012; Venkatesh et al., 2014). In forestry, soil erosion can be reduced through a variety of best management practices—particularly well-designed, constructed, used, and maintained roads. However, their efficacy can vary significantly depending on a range of factors related to design, implementation, and socio-political factors (Cristan et al., 2016). Many options similarly exist to reduce sediment from construction and mining activities. In each of these areas, one dominant principle is to use vegetative cover as a means of minimizing erosion, whether through cover crops in agriculture or reforestation.

In urban areas, management options similarly exist to help protect and restore stream morphology in urbanizing catchments, as reviewed by Vietz et al. (2016). Options that specifically target the sediment regime include: compensating for sediment reduction in urban areas with protecting coarse sediment sources in headwaters; removing sediment barriers such as weirs; designing and using stormwater control measures such as gross pollutant traps that allow noncontaminated sediments—which often include coarser sediments—to pass through; enabling sediment entrainment from some streams; daylight piped streams; and sediment augmentation similar to below dams (described below). The implementation of any of these measures likely needs to be accompanied with other infrastructure and noninfrastructure options to address other challenges in managing urban river morphology: excess stormwater runoff, limited riparian space legacy impacts to streams, and social and institutional impediments (Vietz and Hawley, 2016). As with rural areas, vegetative cover (e.g., vegetation strips) is also widely used in urban areas to reduce excessive soil erosion.

12.4.3 SEDIMENT MANAGEMENT THROUGH DAMS

The disruption to sediment regimes caused by dams can be partly mitigated with sediment pass-through systems, whereby sediment is routed through or around a dam. Although typically used to help manage sediment accumulation in reservoirs, these systems are increasingly recognized as beneficial for maintaining sediment regimes in a river and supporting river ecosystems. Four such methods include sediment bypass, clear water diversion, sluicing, and turbidity current venting (Kondolf et al., 2014; Morris, 2015).

Sediment bypass systems (Fig. 12.7A) divert sediment-laden water around a reservoir via a weir and channel or tunnel (Auel and Boes, 2011; Boes, 2015). The bypass gradient must be high enough to support movement, so sediment bypassing works best at sharp river bends, or when the river is relatively steep such as at higher elevations. Sediment bypass systems have been demonstrated to help improve downstream channel morphology (Fukuda et al., 2012) and sediment respiration, fine particulate organic matter, periphyton biomass, and macroinvertebrate density and richness (Martín et al., 2015). The main drawback of sediment bypass systems is their historically high costs, though these may be small compared to benefits (Sumi, 2015), and maintenance costs are decreasing (Nakajima et al., 2015; Vischer et al., 1997).

Clear water diversion (Fig. 12.7B) is a variation of the sediment bypass concept and entails diverting clear water to an off-stream reservoir (Morris, 2010). This minimizes river fragmentation and maintains natural sediment loads downstream, though the flow regime is still altered. This type of scheme is useful for relatively small-capacity off-stream reservoirs.

Sediment sluicing (Fig. 12.7C), also called drawdown routing, consists of passing sediment-laden water directly through the reservoir during high-flow events (Kondolf et al., 2014). Sluicing requires the reservoir level to be lowered before the flood event to enable a higher flow gradient, and therefore a higher flow velocity, for adequate sediment transport. Still, reservoir velocities typically remain too low for transporting course sediment, so sluicing is more effective in large mid- to low-elevation reservoirs where coarse sediments are minimal (e.g., Wang et al., 2005). Sluicing may result in poor downstream river quality (e.g., lower dissolved oxygen) if not carefully designed.

Turbidity currents occur when a sediment-laden river enters a reservoir with a distinctly higher density than the water in the reservoir. Turbidity current venting entails releasing a turbidity current through a low-level outlet (Fig. 12.7D) or diverting it to a mid-level outlet. In contrast with sluicing, there is no need to lower the reservoir level prior to a high-turbidity event. However, turbidity current venting requires a unique set of hydraulic conditions (Morris and Fan, 1998).

Although in each of these cases infrastructure still results in alteration of the sediment regime, if managed properly they nonetheless help enhance downstream conditions overall compared to no pass-through system (Facchini et al., 2015; Martín et al., 2015; Oertli and Auel, 2015), resulting in a co-benefit to maintaining reservoir capacity. However, there are relatively few case studies in the literature focused on their use in managing river ecosystems; significant work remains to better design and operate sediment pass-through systems for environmental water management.

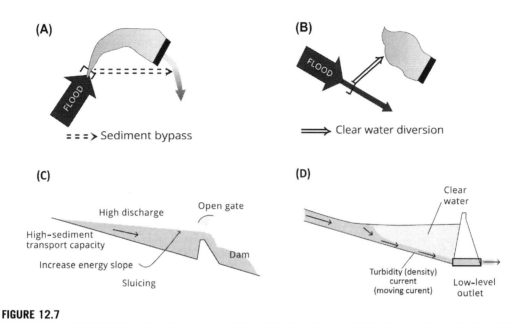

FIGURE 12.7

Conceptual diagram of sediment pass-through systems: (A) sediment bypass, (B) clear water diversion, (C) sediment sluicing, and (D) turbidity current venting.

Source: Kondolf et al. (2014)

12.4.4 SEDIMENT AUGMENTATION BELOW DAMS

In the context of river health, *sediment augmentation* or *replenishment* (or *gravel augmentation* for gravel and coarse sediments only) entails supplying sediment to rivers below dams by artificial methods to help mitigate the geomorphic effects of dams. Historically, sediment augmentation efforts focused on adding gravel to recreate habitat lost to incision, beginning with the creation of riffles for salmon spawning below Lewiston Dam on the Trinity River, California, in 1976 (Kondolf and Matthews, 1993). However, the last couple of decades have seen the increasing use of coarse sediment augmentation to support restoration of geomorphic processes such as channel migration and the formation of complex bar features (Gaeuman, 2014). Gravel augmentation has been widely used in Northern California (including in the Trinity River; www.trrp.net/restore/gravels), has been introduced to other parts of the United States, and is increasingly used in Japan (Kantoush et al., 2010; Kondolf and Minear, 2004; Ock et al., 2013; Okano et al., 2004; Smith, 2011). Gravel augmentation is also implemented in Europe, such as below the Barrage Iffezheim on the Rhine River (Kuhl, 1992), though in that case for nonecosystem purposes.

A wide range of factors must be considered when designing a sediment augmentation scheme, including characteristics related to sediment, flow regime, river morphology, vegetation, augmentation implementation (placement, frequency, etc.), monitoring, and economics (Bunte, 2004; Kantoush and Sumi, 2010). Bunte (2004) provides a particularly useful review of gravel augmentation design.

Implementation method is a particularly key design element, though all design and operational aspects are important for effective results. Four general implementation methods have been used, varying by placement in the channel and delivery method (Bunte, 2004; Ock et al., 2013) (Fig. 12.8.). An *in-channel bed stockpile* provides gravel directly to the low-flow channel (Fig. 12.8A), creating immediately usable habitat and typical of earlier gravel augmentation schemes. A *high-flow stockpile* entails placing sediment at the side of the bank, above the low-flow level but below the bankfull flow level (Fig. 12.8B). The purpose of this method is to let the river redistribute the sediment downstream during higher flow events to support geomorphic processes generally, and therefore includes a range of coarse sediment sizes. This method is common in Japan (Ock et al., 2013). A *point bar stockpile* augments or creates a point bar on a bend (Fig. 12.8C). Like the in-channel stockpile method, this involves direct placement where the gravel is needed, but like the high-flow stockpile method can include a range of sediment sizes for more complex geomorphology. As a point bar may erode quickly during a high-flow event, typically more sediment is added than may be needed in the short term. Finally, in the *high-flow direct injection* method, heavy equipment is used to deliver sediment continuously during a high-flow event (Fig. 12.8D). This has

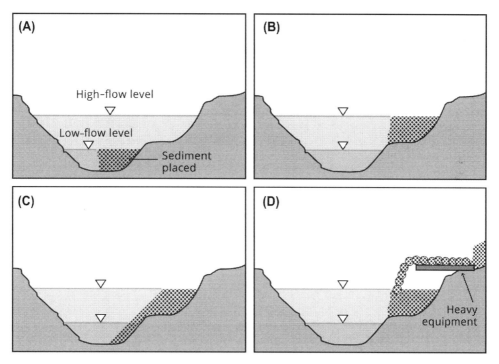

FIGURE 12.8

Sediment augmentation methods. (A) In-channel bed stockpile, (B) high-flow stockpile, (C) point bar stockpile, and (D) high-flow direct injection.

Source: Ock et al. (2013)

the major advantage of immediately replacing sediment eroded by the event. Each of these methods has several specific advantages and disadvantages, reviewed by Ock et al. (2013).

12.4.5 STREAM RESTORATION

In some cases, stream restoration may be needed to support geomorphic processes. Stream restoration generally can refer to a wide variety of different activities, depending on the stream management goal (Shields et al., 2003). For this reason, activities should be selected to be consistent with management objectives: either reversing existing poor stream management practices such as those listed in Table 12.1, or facilitating the restoration of geomorphic processes caused by previous or existing poor management practices upstream. As with other sediment management activities, restoration is increasingly focused on supporting catchment-scale ecosystem restoration, rather than narrowly focused on local sites (Mant et al., 2016).

A wide range of restoration options have been developed and implemented to achieve these objectives. Some examples include: dam removal to restore longitudinal connectivity (of water, sediment, nutrients, etc.), levee breaching to restore lateral connectivity, vegetative methods to control stream bank erosion, riparian road improvements, and physical in-stream restoration (Roni et al., 2002; Shields et al., 2003). In-stream restoration that focuses on creating immediately usable habitat such as adding boulders or large woody debris may be useful pending the full restoration of process-supporting activities. Numerous options also exist for restoring geomorphic characteristics and functions diminished by upstream activities such as where channels have been incised due to changes in flow and sediment regimes upstream. Shields Jr et al. (2003) reviews some of these options from a hydraulic and sediment design perspective. Useful recent practice-oriented stream restoration handbooks include Roni and Beechie (2012) and Simon et al. (2013), among others.

12.5 CONCLUSION

Sediment management, including supply and continual redistribution of sediments for channel maintenance and channel-forming, is critical for effective environmental water management. Although this is generally widely acknowledged, sediment management, particularly for geomorphic processes, historically has not been prioritized in practice. This chapter reviewed the options for improving sediment management in an environmental water context, which are summarized here.

Sediment management is complex, such that there is generally no single unifying approach that is suitable for all situations. Regardless, understanding hydraulic and geomorphic processes, including entrainment, transport, and deposition of sediments, is critical to effective sediment management. Entrainment of sediments can be quantified using the concept of threshold (critical) shear stress, which depends on streamflow, the degree of turbulence in a river, sediment characteristics, and the morphologic characteristics of the river, each of which dynamically interacts with each other. Sediment mobilization processes result in the specific geomorphic features of a river and, subsequently, the quality of habitat for aquatic species.

A wide range of methods have been developed and used to identify the flows needed to achieve sediment management objectives. Most broadly, the objective of these methods is to specify the flow ranges needed to entrain, transport, and deposit sediments, and are based on flow statistics and/or field/laboratory experimentation. The method selected or otherwise developed should be based on broader environmental flows assessment considerations. These may occur within a holistic environmental flows assessment method.

Several modeling tools exist to help understand and predict sediment flux and mobilization in rivers. These include conceptual river basin models and catchment sediment budgets, as well as physically based sediment transport models. The latter are particularly important, and an important area of research.

Environmental water regimes may need to be complemented with other management options for effective sediment management. These may include upland, terrestrially based options (e.g., forest or agriculture management) or in-river options (e.g., reservoir sediment bypass systems, gravel augmentation, local stream restoration). Despite their potential importance, some options such as stream restoration are necessarily underdiscussed here.

Sediment management tools complementary to environmental water regimes often offer co-benefits with other management domains, whether for ecosystem maintenance/improvement or other objectives. For example, well-designed sediment pass-through systems in dams both improve downstream river conditions and reduce sediment accumulation in the dam.

This chapter only summarizes the dominant sediment management tools related to environmental water assessments and management. Many more specific tools exist for detailed questions related to fluvial geomorphic processes at multiple spatial and temporal scales (e.g., Kondolf and Piégay, 2003), which may be important when designing and ultimately implementing environmental water regimes. For example, many tools exist to estimate sediment budgets at the reach scale, which may be necessary for monitoring the need for and efficacy of environmental water releases for sediment management. Importantly, this chapter also omits economic, social, and political tools in sediment management. These tools, including policies, governance structures, and management frameworks, are just as important as the specific technical tools discussed here.

Lastly, we note three broad research needs. First, the coordination of environmental water regimes for sediment management with structural options is a relatively new endeavor and an area of ongoing research. Specifically, a better understanding of how environmental water releases can be integrated with sediment bypass systems in/around hydraulically disruptive infrastructure (dams, weirs, stormwater systems, etc.) is needed. Second, theoretical approaches, while promising, remain elusive as a reliable means of estimating environmental water requirements for various geomorphic processes; continued research is needed to develop theoretical or semi-theoretical approaches that can be scaled to complex, real-world situations. The continued development of computer models of sediment transport processes that integrate these advances will help improve sediment management. Finally, there is a general continual need for additional research and experience in environmental flows assessments for sediment management. The practitioner should feel encouraged to innovate as needed and able, and to report innovations and outcomes to the broader river management community. Similarly, researchers are encouraged to continually engage with practitioners both to help ensure that the latest science is available in real management settings and to ensure that research efforts are aimed at better understanding practical challenges.

ACKNOWLEDGMENTS

The authors thank Christopher Gippel for initiating this chapter, and Angus Webb and Geoff Vietz for reviewing and improving this chapter.

REFERENCES

Acreman, M., Arthington, A.H., Colloff, M.J., Couch, C., Crossman, N.D., Dyer, F., et al., 2014. Environmental flows for natural, hybrid, and novel riverine ecosystems in a changing world. Front. Ecol. Environ. 12 (8), 466–473.

Apitz, S., White, S., 2003. A conceptual framework for river-basin-scale sediment management. J. Soils Sediments 3 (3), 132–138.

Arthington, A.H., Bunn, S.E., Poff, N.L., Naiman, R.J., 2006. The challenge of providing environmental flow rules to sustain river ecosystems. Ecol. Appl. 16 (4), 1311–1318.

Auel, C., Boes, R., 2011. Sediment bypass tunnel design–review and outlook. In: Schleiss, A.J., Boes, R.M. (Eds.), Proc. 79th Annual Meeting of ICOLD: Dams and Reservoirs under Changing Challenges. CRC Press/Balkema, Lucerne, Switzerland, pp. 403–412.

Beschta, R.L., Jackson, W.L., 1979. The intrusion of fine sediments into a stable gravel bed. J. Fish. Board Can. 36 (2), 204–210.

Bledsoe, B.P., 2002. Stream erosion potential and stormwater management strategies. J. Water Resour. Plan. Manage. 128 (6), 451–455.

Bleed, A.S., 1987. Limitations of concepts used to determine instream flow requirements for habitat maintenance. J. Am. Water Resour. Assoc. 23, 1173–1178.

Boes, R.M. (Ed.), 2015. Proceedings of the First International Workshop on Sediment Bypass Tunnels, "VAW-Mitteilungen 232". Laboratory of Hydraulics, Hydrology and Glaciology (VAW), ETH Zurich.

Brookes, A., 1994. River channel change. In: Calow, P., Petts, G.E. (Eds.), The Rivers Handbook: Hydrological and Ecological Principles. Blackwell Science, Oxford, UK.

Buffington, J.M., Montgomery, D.R., 1997. A systematic analysis of eight decades of incipient motion studies, with special reference to gravel-bedded rivers. Water Resour. Res. 33 (8), 1993–2029.

Bunte, K., 2004. Gravel mitigation and augmentation below hydroelectric dams: a geomorphological perspective. Engineering Research Center, Colorado State University.

Bunte, K., Abt, S.R., Swingle, K.W., Cenderelli, D.A., Schneider, J.M., 2013. Critical Shields values in coarse-bedded steep streams. Water Resour. Res. 49 (11), 7427–7447.

Cao, Z., Pender, G., Meng, J., 2006. Explicit formulation of the shields diagram for incipient motion of sediment. J. Hydraul. Eng. 132 (10), 1097–1099.

Coleman, S.E., Nikora, V.I., 2008. A unifying framework for particle entrainment. Water Resour. Res. 44 (4), 10.

Collins, A., Walling, D., Leeks, G., 1997. Source type ascription for fluvial suspended sediment based on a quantitative composite fingerprinting technique. Catena 29 (1), 1–27.

Comiti, F., Mao, L., 2012. Recent advances in the dynamics of steep channels. Gravel-bed rivers: processes, tools. Environments 351–377.

Cristan, R., Aust, W.M., Bolding, M.C., Barrett, S.M., Munsell, J.F., Schilling, E., 2016. Effectiveness of forestry best management practices in the United States: literature review. Forest Ecol. Manage. 360, 133–151.

Church, M., Ferguson, R.I., 2015. Morphodynamics: rivers beyond steady state. Water Resour. Res. 51 (4), 1883–1897.

Church, M., Biron, P., Roy, A., 2012. Gravel Bed Rivers: Processes, Tools, Environments. John Wiley & Sons.

Dey, S., 2014. Fluvial Hydrodynamics: Hydrodynamic and Sediment Transport Phenomena. GeoPlanet: Earth and Planetary Sciences. Springer, Berlin, Germany.

Dinehart, R.L., 1992. Gravel-bed deposition and erosion by bedform migration observed ultrasonically during storm flow, North Fork Toutle River, Washington. J. Hydrol. 136 (1), 51−71.

Diplas, P., Kuhnle, R., Gray, J., Glysson, D., Edwards, T., 2008. Sediment transport measurements. Sedimentation Engineering: Theories, Measurements, Modeling, and Practice. ASCE Manuals and Reports on Engineering Practice 110, 165−252.

Escobar-Arias, M., Pasternack, G.B., 2010. A hydrogeomorphic dynamics approach to assess in-stream ecological functionality using the functional flows model, part 1—model characteristics. River Res. Appl. 26 (9), 1103−1128.

Facchini, M., Siviglia, A., Boes, R.M., 2015. Downstream morphological impact of a sediment bypass tunnel − preliminary results and forthcoming actions. In: Boes, R. (Ed.), First International Workshop on Sediment Bypass Tunnels. ETH Zurich, pp. 137−146.

Fang, N., Chen, F., Zhang, H., Wang, Y., Shi, Z., 2016. Effects of cultivation and reforestation on suspended sediment concentrations: a case study in a mountainous catchment in China. Hydrol. Earth Syst. Sci. 20 (1), 13−25.

Fryirs, K.A., Brierley, G.J., 2013. Geomorphic Analysis of River Systems: An Approach to Reading The Landscape. John Wiley & Sons, Chichester, West Sussex, UK.

Fukuda, T., Yamashita, K., Osada, K., Fukuoka, S., 2012. Study on Flushing Mechanism of Dam Reservoir Sedimentation and Recovery of Riffle-Pool in Downstream Reach by a Flushing Bypass Tunnel, International Symposium on Dams for a Changing World, Kyoto, Japan, pp. 6.

Gaeuman, D., 2014. High-flow gravel injection for constructing designed in-channel features. River Res. Appl. 30 (6), 685−706.

Gippel, C.J., 2001. Geomorphic issues associated with environmental flow assessment in alluvial non-tidal rivers. Austr. J. Water Resour. 5 (1), 3−19.

Gomez, B., Church, M., 1989. An assessment of bed load sediment transport formulae for gravel bed rivers. Water Resour. Res. 25 (6), 1161−1186.

Gomez, B., Phillips, J., 1999. Deterministic uncertainty in bed load transport. J. Hydraul. Eng. 125 (3), 305−308.

Grey, J., Glysson, G., Turcios, L., Schwarz, G., 2000. Comparability of total suspended solids and suspended sediment concentration data. US Geological Survey Water Resources Investigations Report 00-4191. US Geological Survey, Reston, VA.

Hawley, R., Vietz, G., 2016. Addressing the urban stream disturbance regime. Freshw. Sci. 35 (1), 278−292.

Heitmuller, F.T., Raphelt, N., 2012. The role of sediment-transport evaluations for development of modeled instream flows: Policy and approach in Texas. J. Environ. Manage. 102, 37−49.

Hill, M.T., Platts, W.S., Beschta, R.L., 1991. Ecological and geomorphological concepts for instream and out-of-channel flow requirements. Rivers 2 (3), 198−210.

Hoppe, R., 1975. Minimum streamflows for fish, Soilshydrology workshop. US Forest Service. Montana State University.

Kantoush, S., Sumi, T., Kubota, A., 2010. Geomorphic response of rivers below dams by sediment replenishment technique. In: Proceedings of the River Flow 2010 Conference, Braunschweig, Germany, pp. 1155−1163.

Kantoush, S.A., Sumi, T., 2010. River morphology and sediment management strategies for sustainable reservoir in Japan and European Alps. Ann. Disas. Prev. Res. Inst., Kyoto Univ. 53, 821−839.

King, J., Brown, C., Sabet, H., 2003. A scenario-based holistic approach to environmental flow assessments for rivers. River Res. Appl. 19 (5–6), 619–639.

Kondolf, G.M., Gao, Y., Annandale, G.W., Morris, G.L., Jiang, E., Zhang, J., et al., 2014. Sustainable sediment management in reservoirs and regulated rivers: experiences from five continents. Earth's Future 2, 256–280.

Kondolf, G.M., Matthews, W.V.G., 1993. Management of Coars Sediment in Regulated Rivers of California. 80. University of California Water Resources Center, Davis, California.

Kondolf, G.M., Minear, J.T., 2004. Coarse Sediment Augmentation on the Trinity River Below Lewiston Dam: Geomorphic Perspectives and Review of Past Projects, Final Report. Trinity River Restoration Program, Weaverville, Calif.

Kondolf, G.M., Piégay, H. (Eds.), 2003. Tools in Fluvial Geomorphology. John Wiley & Sons, Ltd.

Kondolf, G.M., Wilcock, P.R., 1996. The flushing flow problem: defining and evaluating objectives. Water Resour. Res. 32 (8), 2589–2599.

Konrad, C.P., Olden, J.D., Lytle, D.A., Melis, T.S., Schmidt, J.C., Bray, E.N., et al., 2011. Large-scale flow experiments for managing river systems. BioScience 61 (12), 948–959.

Kostaschuk, R., Best, J., Villard, P., Peakall, J., Franklin, M., 2005. Measuring flow velocity and sediment transport with an acoustic Doppler current profiler. Geomorphology 68 (1–2), 25–37.

Kuhl, D., 1992. 14 Years of artificial grain feeding in the Rhine downstream the barrage Iffezheim. In: Proceedings of the 5th international symposium on river sedimentation: sediment management. University of Karlsruhe, Karlsruhe, Germany, pp. 1121–1129.

Lawless, M., Robert, A., 2001. Scales of boundary resistance in coarse-grained channels: turbulent velocity profiles and implications. Geomorphology 39 (3), 221–238.

Mant, J., Large, A., Newson, M., 2016. River restoration: from site-specific rehabilitation design towards ecosystem-based approaches. River Science. John Wiley & Sons, Ltd, pp. 313–334.

Marden, M., 2012. Effectiveness of reforestation in erosion mitigation and implications for future sediment yields, East Coast catchments, New Zealand: a review. N. Z. Geogr. 68 (1), 24–35.

Martín, E.J., Doering, M., Robinson, C.T., 2015. Ecological effects of sediment bypass tunnels. In: Boes, R. (Ed.), First International Workshop on Sediment Bypass Tunnels. ETH Zurich, pp. 147–156.

Meitzen, K.M., Doyle, M.W., Thoms, M.C., Burns, C.E., 2013. Geomorphology within the interdisciplinary science of environmental flows. Geomorphology 200, 143–154.

Mekonnen, M., Keesstra, S.D., Stroosnijder, L., Baartman, J.E.M., Maroulis, J., 2015. Soil conservation through sediment trapping: a review. Land Degrad. Dev. 26 (6), 544–556.

Melis, T., Korman, J., Kennedy, T.A., 2012. Abiotic & biotic responses of the Colorado River to controlled floods at Glen Canyon Dam, Arizona, USA. River Res. Appl. 28 (6), 764–776.

Montana Department of Fish, W.a.P., 1981. Instream flow evaluation for selected waterways in western Montana.

Montgomery, D.R., 2007. Soil erosion and agricultural sustainability. Proc. Natl. Acad. Sci. 104 (33), 13268–13272.

Morgan, R.P.C., 2005. Soil Erosion and Conservation. Blackwell Publishing, Oxford, United Kingdom.

Morris, G.L., 2010. Offstream reservoirs for sustainable water supply in Puerto Rico. Am. Water Resource Assn., Summer Specialty Conf.

Morris, G.L., 2015. Management Alternatives to combat reservoir sedimentation. First International Workshop on Sediment Bypass Tunnels. ETH Zurich.

Morris, G.L., Fan, J., 1998. Reservoir Sedimentation Handbook: Design and Management of Dams, Reservoirs, and Watershed For Sustainable Use. McGraw-Hill, New York.

Mueller, E.R., Pitlick, J., Nelson, J.M., 2005. Variation in the reference Shields stress for bed load transport in gravel-bed streams and rivers. Water Resour. Res. 41 (4).

Naiman, R.J., Latterell, J.J., Pettit, N.E., Olden, J.D., 2008. Flow variability and the biophysical vitality of river systems. C. R. Geosci. 340 (9), 629−643.

Nakajima, H., Otsubo, Y., Omoto, Y., 2015. Abrasion and corrective measures of a sediment bypass system at Asahi Dam. In: Boes, R. (Ed.), First International Workshop on Sediment Bypass Tunnels. ETH Zurich, pp. 21−32.

Newbury, R.W., Gaboury, M.N., 1994. Stream Analysis and Fish Habitat Design: A Field Manual. Newbury Hydraulics Ltd., Gibsons, British Columbia.

Northern Great Plains Resource Program, 1974. Instream needs sub-group report, Work Group C, Water.

Nyssen, J., Clymans, W., Poesen, J., Vandecasteele, I., De Baets, S., Haregeweyn, N., et al., 2009. How soil conservation affects the catchment sediment budget−a comprehensive study in the north Ethiopian highlands. Earth Surf. Process. Landf. 34 (9), 1216−1233.

O'Brien, J., 1984. Hydraulic and sediment transport investigation, Yampa River, Dinosaur National Monument. WRFSL Rept 83, 8.

Ock, G., Sumi, T., Takemon, Y., 2013. Sediment replenishment to downstream reaches below dams: implementation perspectives. Hydrol. Res. Lett. 7 (3), 54−59.

Oertli, C., Auel, C., 2015. Solis sediment bypass tunnel: first operation experiences. In: Boes, R. (Ed.), First International Workshop on Sediment Bypass Tunnels. ETH Zurich, pp. 223−233.

Okano, M., Kikui, M., Ishida, H., Sumi, T., 2004. Reservoir sedimentation management by coarse sediment replenishment below dams. In: Proceedings of the Ninth International Symposium on River Sedimentation, Yichang, China.

Olden, J.D., Konrad, C.P., Melis, T.S., Kennard, M.J., Freeman, M.C., Mims, M.C., et al., 2014. Are large-scale flow experiments informing the science and management of freshwater ecosystems? Front. Ecol. Environ. 12 (3), 176−185.

Owens, P., 2005. Conceptual models and budgets for sediment management at the River Basin Scale (12 pp). J. Soils Sediments 5, 201−212.

Owens, P., Batalla, R., Collins, A., Gomez, B., Hicks, D., Horowitz, A., et al., 2005. Fine-grained sediment in river systems: environmental significance and management issues. River Res. Appl. 21 (7), 693−717.

Owens, P.N. (Ed.), 2008. Sustainable Management of Sediment Resources: Sediment Management at the River Basin Scale, 4. Elsevier.

Parker, G., 1979. Hydraulic Geometry of Active Gravel Rivers. J. Hydraul. Eng. ASCE 105 (9), 1185−1201.

Patel, S.B., Patel, P.L., Porey, P.D., 2013. Threshold for initiation of motion of unimodal and bimodal sediments. Int. J. Sediment Res. 28 (1), 24−33.

Phillips, J.D., 2006. Deterministic chaos and historical geomorphology: a review and look forward. Geomorphology 76 (1−2), 109−121.

Pitlick, J., Cui, Y., Wilcock, P., 2009. Manual for computing bed load transport using BAGS (Bedload Assessment for Gravel-bed Streams) Software.

Poff, N.L., Richter, B.D., Arthington, A.H., Bunn, S.E., Naiman, R.J., Kendy, E., et al., 2010. The ecological limits of hydrologic alteration (ELOHA): a new framework for developing regional environmental flow standards. Freshw. Biol. 55 (1), 147−170.

Reichel, G., Nachtnebel, H.P., 1994. Suspended sediment monitoring in a fluvial environment: advantages and limitations applying an acoustic doppler current profiler. Water Res. 28 (4), 751−761.

Reiser, D., Ramey, M., Lambert, T., 1985. Review of Flushing Flow Requirements in Regulated Streams. Pacific Gas and Electric Co., Department of Engineering Research, San Ramon, California.

Roni, P., Beechie, T., 2012. Stream and Watershed Restoration: A Guide to Restoring Riverine Processes and Habitats. John Wiley & Sons.

Roni, P., Beechie, T.J., Bilby, R.E., Leonetti, F.E., Pollock, M.M., Pess, G.R., 2002. A review of stream restoration techniques and a hierarchical strategy for prioritizing restoration in Pacific Northwest watersheds. North Am. J. Fish. Manage. 22 (1), 1−20.

Russell, K., Vietz, G., Fletcher, T.D., 2016. Not just a flow problem: how does urbanization impact on the sediment regime of streams? In: 8th Australian Stream Management Conference, Leura, New South Wales, Australia, pp. 661−667.

Schmidt, J.C., Grams, P.E., 2011. The high flows—Physical science results. Effects of three high-flow experiments on the colorado river ecosystem downstream from Glen Canyon Dam, Arizona. US Department of the Interior, US Geological Survey. Circular 1366, 53−91.

Shields, A., 1936. Application of similarity principles and turbulence research to bed-load movement. Soil Conservation Service.

Shields, F.D., Cooper, C., Knight, S.S., Moore, M., 2003. Stream corridor restoration research: a long and winding road. Ecol. Eng. 20 (5), 441−454.

Shields Jr, F.D., Copeland, R.R., Klingeman, P.C., Doyle, M.W., Simon, A., 2003. Design for stream restoration. J. Hydraul. Eng. 129 (8), 575−584.

Sholtes, J., Werbylo, K., Bledsoe, B., 2014. Physical context for theoretical approaches to sediment transport magnitude-frequency analysis in alluvial channels. Water Resour. Res. 50 (10), 7900−7914.

Simon, A., Bennett, S.J., Castro, J.M., 2013. Stream Restoration in Dynamic Fluvial Systems: Scientific Approaches, Analyses, and Tools, 194. John Wiley & Sons.

Smith, C.B., 2011. Adaptive management on the central Platte River—Science, engineering, and decision analysis to assist in the recovery of four species. J. Environ. Manage. 92 (5), 1414−1419.

Stevens, L.E., Ayers, T.J., Bennett, J.B., Christensen, K., Kearsley, M.J.C., Meretsky, V.J., et al., 2001. Planned flooding and colorado river riparian trade-offs downstream from Glen Canyon Dam, Arizona. Ecol. Appl. 11 (3), 701−710.

Sumi, T., 2015. Comprehensive reservoir sedimentation countermeasures in Japan. First International Workshop on Sediment Bypass Tunnels. ETH Zurich.

Tennant, D.L., 1976. Instream Flow Regimens for fish, wildlife, recreation and related environmental resources. Fisheries 1, 6−10.

Tockner, K., Malard, F., Ward, J., 2000. An extension of the flood pulse concept. Hydrol. Process. 14 (16−17), 2861−2883.

Vázquez-Tarrío, D., Menéndez-Duarte, R., 2015. Assessment of bedload equations using data obtained with tracers in two coarse-bed mountain streams (Narcea River basin, NW Spain). Geomorphology 238, 78−93.

Venkatesh, B., Lakshman, N., Purandara, B., 2014. Hydrological impacts of afforestation—a review of research in India. J. Forestry Res. 25 (1), 37−42.

Vietz, G., Hawley, R.J., 2016. Urban streams and disturbance regimes: a hydrogeomorphic approach. In: 8th Australian Stream Management Conference, Leura, New South Wales, Australia.

Vietz, G., Stewardson, M., Finlayson, B., 2006. Flows that Form: The Hydromorphology of Concave-Bank Bench Formation in the Ovens River, Australia, 306. Iahs Publication, p. 267.

Vietz, G.J., Rutherfurd, I.D., Fletcher, T.D., Walsh, C.J., 2016. Thinking outside the channel: challenges and opportunities for protection and restoration of stream morphology in urbanizing catchments. Landsc. Urban Plan. 145, 34−44.

Vischer, D., Hager, W., Casanova, C., Joos, B., Lier, P., Martini, O., 1997. Bypass tunnels to prevent reservoir sedimentation. Trans. Int. Congr. Large Dams, 605−624.

Walling, D., Collins, A., Sichingabula, H., Leeks, G., 2001. Integrated assessment of catchment suspended sediment budgets: a Zambian example. Land Degrad. Dev. 12 (5), 387−415.

Walling, D.E., 2004. Using environmental radionuclides to trace sediment mobilization and delivery in river basins as an aid to catchment management. In: Proceedings of the Ninth International Symposium on River Sedimentation, pp. 18−21.

Walling, D.E., Collins, A.L., 2008. The catchment sediment budget as a management tool. Environ. Sci. Policy 11, 136−143.

Wang, G., Wu, B., Wang, Z.Y., 2005. Sedimentation problems and management strategies of Sanmenxia Reservoir, Yellow River, China. Water Resour. Res. 41, 9.

Warburton, J., Davies, T., 1994. Variability of bedload transport and channel morphology in a braided river hydraulic model. Earth Surf. Proc. Land. 19 (5), 403–421.

Ward, J., Tockner, K., Uehlinger, U., Malard, F., 2001. Understanding natural patterns and processes in river corridors as the basis for effective river restoration. Regul. Rivers Res. Manage. 17 (4–5), 311–323.

Wilcock, P., Pitlick, J., Cui, Y., 2009. Sediment transport primer: estimating bed-material transport in gravel-bed rivers.

Wilcock, P.R., 1993. Critical shear stress of natural sediments. J. Hydraul. Eng. 119 (4), 491–505.

Wohl, E., Bledsoe, B.P., Jacobson, R.B., Poff, N.L., Rathburn, S.L., Walters, D.M., et al., 2015. The natural sediment regime in rivers: broadening the foundation for ecosystem management. BioScience 65, 358–371.

Wolman, M.G., 1967. A cycle of sedimentation and erosion in urban river channels. Geogr. Annal. A Phys. Geogr. 385–395.

Wolman, M.G., Miller, J.P., 1960. Magnitude and frequency of forces in geomorphic processes. J. Geol. 54–74.

Yang, X., 2003. Manual on Sediment Management and Measurement. World Meteorological Organization, Geneva, Switzerland.

Yarnell, S.M., Petts, G.E., Schmidt, J.C., Whipple, A.A., Beller, E.E., Dahm, C.N., et al., 2015. Functional flows in modified riverscapes: hydrographs, habitats and opportunities. BioScience biv102.

PHYSICAL HABITAT MODELING AND ECOHYDROLOGICAL TOOLS 13

Nicolas Lamouroux[1], Christoph Hauer[2], Michael J. Stewardson[3], and N. LeRoy Poff[4,5]

[1]Irstea UR Maly, Villeurbanne, France [2]BOKU, Vienna, Austria [3]The University of Melbourne, Parkville, VIC, Australia [4]Colorado State University, Fort Collins, CO, United States [5]University of Canberra, Canberra, ACT, Australia

13.1 INTRODUCTION: PRINCIPLES OF ECOHYDROLOGICAL AND HYDRAULIC-HABITAT TOOLS

Environmental flows assessment approaches in rivers increasingly consider environmental water regimes as complex compromises between water uses and ecosystem functions (see Chapters 1 and 11). These compromises generally result from negotiations that involve engineering constraints, economic aspects, and societal objectives within a governance context (Pahl-Wostl et al., 2013). However, biophysical issues remain central in the process because the knowledge of biological responses to flow changes helps define the trade-offs being negotiated. Ideally, negotiations and management decisions should be based on reliable quantitative predictions of the ecological effects of flow management (Lamouroux et al., 2015). Anthropogenic impacts on flow regime and resulting ecological effects are well documented (see Chapters 3 and 4; Poff and Zimmerman, 2010); however, challenges remain in defining and quantifying the relationships between flow management and ecological outcomes (see Chapter 15).

Ecohydrological methods and hydraulic-habitat models are the tools most commonly used to quantify the expected ecological responses to flow changes (Linnansaari et al., 2013). Ecohydrological tools describe the alterations of many components of the flow regime (i.e., statistics describing discharge variations) that have been empirically shown to relate to ecological characteristics of aquatic species and communities (Poff et al., 2010). Ecohydrological components include: the characteristics of hourly discharge variations (MetCalfe and Schmidt, 2014); the magnitude, timing, frequency, duration, and rate of change of daily flows (Poff and Ward, 1989; Poff et al., 1997; Richter et al., 1996, 1997; e.g., Box 13.1); and the timing, frequency, and duration of low-flow spells or high-flow pulses (Gordon et al., 2004). Acceptable ecohydrological alterations can be negotiated either: (1) in consideration of the natural range of variability of ecohydrological variables (Richter et al., 1997) and/or (2) in consideration of empirical relationships identified between flow and ecology alterations (Arthington et al., 2006; Poff et al., 2010).

Hydraulic-habitat tools are based on the idea that hydraulic descriptions (e.g., water depths, current velocities, and bed shear stresses) are fundamental characteristics of the physical habitat of aquatic taxa. Hydraulic variables provide more direct descriptors of habitat conditions relevant to biota than discharge statistics. Accordingly, many empirical relationships have been established

Water for the Environment. DOI: http://dx.doi.org/10.1016/B978-0-12-803907-6.00013-9

BOX 13.1 PRINCIPLES OF ECOHYDROLOGICAL TOOLS

Ecohydrological tools (Fig. 13.1) describe the alterations of many components of the flow regime. In Fig. 13.1A (Richter et al., 1996), 33 variables are used to describe the degree of alteration, compared to a natural flow regime, of daily flow variations. *Flow—ecology* relationships can be used to quantify the expected ecological effects of flow regime alteration. In Fig. 13.1B (after Webb et al., 2015), the empirical relationship between the percentage of alluvial vegetation in the floodplain and the annual duration of floods can be used to predict the effect of flood duration alteration.

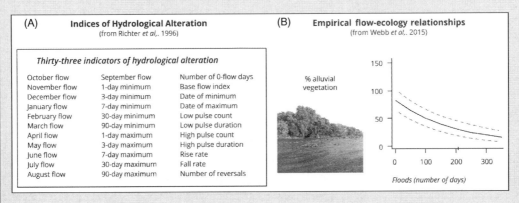

FIGURE 13.1

Ecohydrological tools.

Source: Adapted from Lamouroux et al. (in press)

between individual taxa densities and the hydraulic characteristics of their microhabitats (e.g., Borchardt, 1993; Gibbins et al., 2010; Lamouroux et al., 2013; Mérigoux et al., 2009). As the same discharge or discharge percentile in different rivers may be associated with different velocities or depths depending on local channel morphology (e.g., slope, substrate, and cross-section shape), similar discharge patterns in morphologically different rivers can translate into different expected ecological responses than those anticipated from discharge alone (Poff et al., 2006).

Most hydraulic-habitat models focus on low to mean annual flow levels and are based on *preference* models relating aquatic taxa densities to their microhabitat hydraulics (Box 13.2). The principle of these models is to link hydraulic models that represent microhabitat hydraulics in stream reaches (e.g., local depths, velocities, and bed shear stresses) with biological models reflecting the preferences of taxa for these microhabitat variables. Many types of models have been developed based on this principle (see Dunbar et al., 2012), although they differ, for example, in the habitat variables included and in the spatial scale at which microhabitats are defined.

Ecohydrological tools and hydraulic-habitat models have both been used at the scale of stream reaches (e.g., for defining the minimum flow regime in a specific river reach bypassed for hydroelectricity production) and at the scale of whole catchments (e.g., for defining water abstraction rules). Whatever the scale considered, there is a growing consensus of views recommending the combination of ecohydrological and hydraulic-habitat tools when comparing the ecological effects of flow management scenarios (Dunbar et al., 2012; King et al., 2003; Poff et al., 2010).

BOX 13.2 PRINCIPLES OF HYDRAULIC-HABITAT MODELS

Most hydraulic-habitat models (Fig. 13.2) are based on the observation of *hydraulic-habitat preferences* of species (or specific life stages, or groups of species or *guilds*). These preferences are defined by quantifying variations in species density or occurrence within a stream reach, as a function of the hydraulic characteristics of the microhabitats where the organisms are sampled (e.g., by electrofishing or snorkeling). In the preference model of the figure (drawn using the EVHA software; Ginot, 1998), the relative density of a fish life stage is expressed as a joint function of the current velocity and water depth of microhabitats (a value of 1 corresponds to the maximum observed density).

Hydraulic-habitat models link a hydraulic model of the stream reach of interest with such preference models. The hydraulic model is used to map microhabitat hydraulics at different discharge rates in the reach, and the preference models are used to translate this information into maps of habitat values for the species. In the habitat map of the figure, the color of the reach microhabitat units indicates habitat values (from low to high relative density, or preference) for a given stream discharge. As discharge changes over time, the amount of low-quality or high-quality habitat changes at the stream reach scale. The reach-averaged habitat value (typically a score ranging between 0 and 1) is used to quantify the expected effect of discharge changes. It is frequently multiplied by the wetted area of the reach in order to provide the reach scale *usable area*, which accounts for changes in both habitat quality (habitat value) and quantity (wetted area).

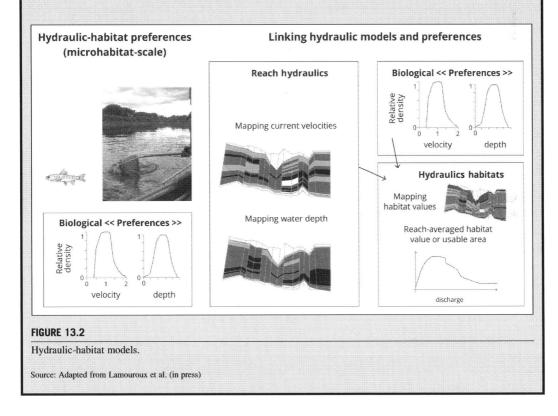

FIGURE 13.2

Hydraulic-habitat models.

Source: Adapted from Lamouroux et al. (in press)

Environmental flows assessment studies should include some sort of ecohydrological analysis, generally involving a description of natural, actual, and future flow regimes (see Chapter 11). When possible and relevant (in particular for low-flow values), hydraulic-habitat models can help by translating the main hydrological alterations into habitat alterations, which can be expressed in terms of empirically derived hydraulic preferences of organisms. Inevitably, involving hydraulic-habitat models within environmental flows assessments has a cost: these models require more input data than discharge estimates, including some information on river morphologies for translating discharge values into hydraulic values. However, some habitat approaches use easily obtained hydraulic metrics (e.g., the average depth and width of reaches at various discharge rates), which facilitates their implementation (Lamouroux and Capra, 2002; Snelder et al., 2011).

In this chapter, we select four case studies to illustrate the potential of ecohydrological tools and hydraulic-habitat models to guide environmental flows assessments at the scale of stream reaches or whole catchments. We first provide an overview of the field validations of these tools that justify their use in supporting management decisions. The first case study shows the potential of ecohydrological classifications for regional-scale environmental flows assessments. Two other case studies show that spatially explicit hydraulic-habitat models are particularly adapted to some situations, such as the quantification of the ecological impacts of hydropeaking (i.e., rapid changes in discharge caused by hydropower plants) or morphological restoration. The last case study shows that simplified hydraulic-habitat models can also inform environmental water planning at the catchment-scale. We use these examples to discuss how ecohydrological and hydraulic approaches could be further developed and combined.

13.2 PREDICTIVE POWER OF ECOHYDROLOGICAL AND HYDRAULIC-HABITAT TOOLS

The literature provides much evidence of ecological effects of flow regime alteration, but literature reviews have uncovered very few transferable quantitative models relating ecological response to metrics of hydrological alteration. Such reviews have indicated that the alteration of many aspects of the flow regime influence aquatic communities (Murchie et al., 2008; Poff and Zimmerman, 2010; Poff et al, 1997; Webb et al., 2013), providing a large body of evidence that fully justifies the importance of environmental water regimes. Flow alteration has consistent effects, across multiple studies, on both fish (abundance, assemblage structure, and diversity) and vegetation (diversity and mortality; Webb et al., 2013). However, few general quantitative relationships can be drawn from these reviews (see Chapter 15). An example of a quantitative result is that changes in the magnitude of extreme flows of a few dozen percent frequently have ecological impacts on abundance and diversity metrics (Carlisle et al., 2011; Poff and Zimmerman, 2010).

The limited success in establishing general, quantitative models of the ecological effects of flow alterations can be attributed to several factors: (1) the diversity and natural variability of flow regimes and their alterations; (2) the varying hydraulic effects of flow alterations, depending upon the morphology of stream reaches; (3) the complexity of biological responses that are influenced by many biotic and abiotic factors; (4) limitations of ecological monitoring; and (5) inconsistent design and reporting of environmental water monitoring studies. Ideally, experimental flow regimes or flow restoration case studies should contribute to a better identification of the ecological effects of flow changes. Unfortunately, long-term flow experiments that clearly disentangle the effects of natural and anthropogenic variations are still uncommon, poorly coordinated, and often not comparable (Gillespie et al., 2015; Olden et al., 2014). Many evaluations of flow experiments are based on monitoring durations that are too short; others use space-for-time substitutions, for instance comparisons between sites with varying levels of hydrological alteration rather than monitoring changes through time as flow regimes are restored. Therefore, such evaluations must be interpreted with care.

Tests of the ecological effects of hydraulic-habitat alterations on aquatic communities also suffer from limited generality (e.g., Sabaton et al., 2008). However, a few studies involving habitat models have generated convincing predictions of the ecological effects of low-flow changes, providing support for efforts to include hydraulic information in environmental water assessments (Poff et al., 2010). For example, Jowett and Biggs (2006) reported that predictions of fish population response to low-flow settings based on hydraulic-habitat models were qualitatively consistent with observed effects on fish populations in three of four case studies in New Zealand. In the Rhône River in France, where minimum flows were intentionally increased downstream of four dams, habitat models generated accurate predictions of changes in many fish and invertebrate taxa densities (as well as changes in community traits) in restored reaches (Lamouroux et al., 2015). These tests of habitat model predictions were particularly convincing because predictions were developed before flow restoration, at another spatial scale and in other reaches than those used for the tests.

In summary, there is no doubt concerning the significant effects of hydrologic and hydraulic changes on biological communities in streams, but establishing quantitative models of these effects remains elusive and unlikely to be resolved quickly. Accordingly, an important degree of expertise is still needed when using ecohydrological tools and habitat models (see Chapter 11). In particular, expertise is needed to assess the likelihood and severity of influence from multifaceted hydrologic alterations on aquatic communities in a given environmental context. Similarly, expertise is needed to appreciate the relevance of previously published hydraulic preference models at a particular site. Nevertheless, the potential of combining hydrological and hydraulic approaches is obvious: hydrologic approaches consider all aspects of flow regime and their potential effects on all aspects of the ecosystem; hydraulic approaches are more focused on some flow characteristics, and in particular low flows that are frequent and persistent in most managed systems. They can provide useful predictions of biological changes corresponding to alternative flow scenarios. These approaches are further complementary in that ecohydrological approaches are well suited to characterize temporal variations in the aquatic environment, whereas hydraulic-habitat models allow analysis of spatial habitat heterogeneity within reaches.

13.3 EXAMPLES OF ECOHYDROLOGICAL TOOLS

13.3.1 OVERVIEW

A number of ecohydrological software tools are frequently used to derive statistics reflecting the intensity, duration, frequency, timing, and rate of change of daily flow rates (*Indicators of Hydrologic Alteration [IHA]* software that calculates 33 ecohydrological indices, Richter et al., 1996; Hydrological Index Tool software that calculates 171 hydrological indices, as described in Olden and Poff, 2003; the River Analysis Package in Marsh et al., 2006; see a review in Rinaldi et al., 2013). Although the hydrological indices obtained from these tools may be partly redundant, they are particularly useful for describing natural flow regimes and their alteration. Fewer tools enable descriptions of subdaily flow patterns, which are necessary for describing the impacts of flow alterations such as hydropeaking (see SAAS software; MetCalfe and Schmidt, 2014). For other situations such as the definition of environmental water regimes in secondary channels of large rivers, other statistics may be particularly relevant for assessing ecological responses. For example, Castella et al. (2015) identified the frequency of physical hydrologic connection of secondary channels of the Rhône River as the major predictor of invertebrate community structure in these channels, and they used estimates of lateral connectivity to make quantitative predictions of the effects of floodplain restoration on invertebrate community structure.

Applications of ecohydrological methods generally require discharge series simulated for different environmental water scenarios, often including an *unimpacted* case as a baseline reference. In addition, catchment-scale approaches require spatial extrapolation of discharge series or flow metrics, using either statistical methods or catchment hydrological modeling (Singh and Frevert, 2006). Many methods exist for these purposes, making hydrological approaches particularly suited to catchment-scale approaches. Nevertheless, the uncertainty associated with hydrological extrapolations should be recognized and taken into account in environmental flows assessments (see Chapter 15). Extrapolating hydrological characteristics is particularly uncertain in systems such as headwater streams, intermittent rivers, or karstic networks. Such situations are more frequent than generally thought, for example, Snelder et al. (2013) found that more than 30% of the length of the French stream network is made of intermittent reaches despite a temperate climate. Therefore, the accuracy of spatial and temporal extrapolation of flow regimes should be assessed in all environmental water assessments before hypotheses of ecological responses to flow alteration are developed.

13.3.2 CASE STUDY 1: ECOHYDROLOGICAL ASSESSMENT OF A LARGE RIVER CATCHMENT

The *Murray—Darling Basin (MDB)* is economically and environmentally important in Australia (Davies et al., 2010) with an area of 1,040,000 km^2. In 2000, with increasing investment in river and catchment restoration, the then MDB Commission began work on the *Sustainable Rivers Audit (SRA)* to report on status and trends in river condition across the basin (Davies et al., 2010). The SRA report is focused on detecting and reporting the signs of change in ecological condition. The second SRA report (for the period 2008—2010) includes a comprehensive ecohydrological assessment of flow alteration across the basin (Davies et al., 2012).

This case study highlights how ecohydrological methods can be applied at the basin scale as an initial screening tool to identify flow-stressed rivers that should be prioritized for more detailed environmental assessments. This ecohydrological assessment considered the temporal and spatial pattern of streamflow (or *flow regime*) at sites throughout the MDB covering a total of 191,000 km of river length. The assessment focuses on sensitive measures of hydrological alteration from an ecological viewpoint, to compare hydrological condition between rivers. Hydrological alterations are assessed by comparing streamflows subject to current water resource arrangements (*Current scenario*) with streamflows that would have occurred in the absence of European settlement in Australia (*reference scenario*). Reference scenario flows are modeled using water resource models (where these are available) to remove the effects of flow regulation and diversion. In addition, hydrological models were used to remove the effects of farm dams and tree clearing in the reference scenario. An overview of hydrological modeling methods is provided in Davies et al. (2012) with more details available in several background technical reports (e.g., SKM, 2011). An important feature of this assessment is the separate treatment of mainstem rivers ($\sim$10% of the total river length) and headwater streams ($\sim$90% of the total river length) and the different data sources used for each of them, reflecting data availability. The existing water resource models (commonly developed for water planning or accounting) could be used for the mainstem rivers. However, as is commonly the case, coverage of hydrological data and models is patchy for the smaller headwater streams and regression equations were used to extrapolate effects of farm dams and land clearance across these headwaters. This required a significant project over 2$-$3 years with much work required to compile the necessary spatial data. Many of the mid-size rivers ($\sim$100$-$1000 km^2 catchment area) could not be included in the analysis because extrapolation techniques were not applicable with the presence of substantial private diversions but limited data on their location and magnitude.

The ecohydrological assessment was based on a set of metrics that characterized hydrological alteration from the reference condition described as the *Flow Stress Ranking (FSR)* procedure (SKM, 2005). The FSR uses metrics calculated from analysis of daily or monthly streamflow series (depending on temporal resolution of the available modeled data) representing both the flow regime being assessed and reference conditions. The selected metrics characterized alteration in flow seasonality, mean and variation of annual flows, flow distribution, high- and low-flow spells, flow intermittency, and overbank flows.

The FSR metrics were calculated from model runs using the climate sequence for the period 1895$-$2009 for the mainstem rivers and approximately the last 40 years for the headwater streams. The resulting FSR metrics are based on a comparison of flow regime under current and reference conditions. It was necessary to agree on the *current conditions* to be used in this analysis given that catchment development (including water policy settings, infrastructure, and use patterns) will continue to change during the assessment project. These current conditions are assessed using the full climatic sequence. The actual *historic* flow regime status over the period of record may differ from this current scenario, particularly where major changes in water management arrangements have occurred in recent decades.

The FSR metrics were combined to calculate a single hydrological index representing flow alteration including all aspects of the regime represented by individual metrics. Component metrics were combined sequentially. In turn, the final hydrologic index is based on a combination of two indicators related to in-channel and overbank flows. The in-channel hydrology indicator was calculated from subindices related to: flow volume, high flows, and low flows, and these indices were each calculated from relevant FSR metrics. *Expert rules* were used at each calculation step based on ecological considerations rather than using a weighted average. Indicators and indices are calculated from metrics by *integration* using these expert rules, using a computational process based on *fuzzy logic* (e.g., Negnevitsky, 2002). This method allows for the integration of assessment results in ways that reflect ecological insight, but that cannot be achieved by simple mathematic methods such as averaging. This integration process is also transparent and open to review in the light of new knowledge (more details are provided in Davies et al., 2012). To illustrate these expert rules, Fig. 13.3A shows how two FSR metrics related to alteration in the amplitude and timing of seasonal flow variations are combined to calculate a flow seasonality subindex. A metric value of 1 indicates no change from reference condition. Note that the amplitude can increase (metric $>$ 1) or decrease (metric $<$ 1) but changes in timing are unidirectional. Contour lines in this diagram indicate the seasonality subindex score (out of a maximum of 100) derived from these two variables. The shape of these contour lines has been informed by expert knowledge. For example, the subindex score is more sensitive to reductions in seasonal amplitude (top left quadrant) than increases (top right quadrant) because of the importance of seasonality in triggering a range of ecological processes. Similarly, reductions in amplitude combined with reductions in amplitude (bottom left corner) are scored lower than reductions in period alone (middle-bottom region of this figure).

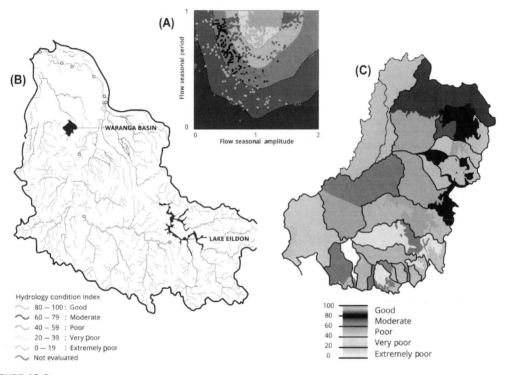

FIGURE 13.3

Ecohydrological assessment of hydrological alteration showing: (A) expert rules for combining flow amplitude and timing metrics to calculate the flow seasonality subindex; (B) results for the Goulburn River subcatchment showing an aggregate measure of hydrological alteration (good to extremely poor indicated by color) for river reaches where data are available; and (C) results for the Murray—Darling Basin showing aggregated results for lowland, midland, and upland zones in each of the 23 subcatchments.

Source: Davies et al. (2012)

Results for all stream reaches are aggregated, separately for the mainstem and headwater river reaches using a length-weighted average index, where length refers to the length of the river reach (Fig. 13.3B). This aggregation was used to calculate an aggregated measure of hydrological alteration for lowland, midland, and upland zones for each subcatchment of the basin (Fig. 13.3C). Final hydrological index scores varied from 0 to 100 and were assigned descriptive ratings of good (80—100), moderate (60—80), poor (40—60), very poor (20—40), and extremely poor (0—20). Across the basin, the hydrological condition of 56% of the mainstem river length is rated as poor or worse. Modifications to all aspects of the flow regime are widespread across the mainstem river network. The greatest human impacts are on flow seasonality and flow variability. However, alterations to high- and low-flow events as well as the total volume of flow are also widespread and severe in many cases. In headwater streams, the SRA only considers impacts of farm dams and altered woody cover. Based on this assessment, 99% of the basin's headwater streams are rated in good condition. There are some restricted areas (less than 5% of the total headwater stream length) where there are moderate alterations to flow seasonality and variability relative to reference conditions. The SRA results can be reported at different levels of detail depending on the audience for this information. The overall length-averaged index score provides a useful summary of the state of the MDB, but the component indices at the reach scale provide a rich data source for examining the spatial pattern and characteristics of flow alterations.

13.4 EXAMPLES OF HYDRAULIC-HABITAT APPROACHES

13.4.1 OVERVIEW

Many tools are available for implementing hydraulic-habitat approaches (Dunbar et al., 2012). Although they follow the same basic principle (Box 13.2), they differ by the nature of their hydraulic and biological components. We can identify three major axes of development of habitat models, related to their hydraulic components (Table 13.1). First, the development of numerical habitat models (i.e., models involving a numerical, deterministic hydraulic component) has benefitted from technological advances (field measurements and computing times). Numerical models provide spatially explicit descriptions of within-reach hydraulic variability. Such descriptions are particularly useful when assessing the effects of morphological restoration on habitats, when conducting behavioral studies, or when studying the effects of hydropeaking on habitats. However, their application is costly and cannot be reasonably applied over whole catchments. In particular, numerical models require a three-dimensional (3D) topography of the reach, as well as extensive current velocity and water depth measurements for model calibration. Second, *mesohabitat* models have been developed, based on the idea that the frequency of habitats at the mesoscale (e.g., a riffle or a pool) can be modeled from surveys at several discharge rates and related to biological responses (Parasiewicz, 2007). Although the identification of mesohabitats is partly subjective (Jowett, 1993), these approaches enable applications at larger spatial scales than numerical models (Vezza et al., 2014). Third, statistical habitat models have been proposed as simple alternatives to deterministic ones (Lamouroux and Capra, 2002; Wilding et al., 2014). These simplified approaches are based on the observation that the frequency distribution of microhabitat hydraulic variables in reaches depends on the average hydraulic geometries of reaches (i.e., depth- and width-discharge relationships; Girard et al., 2014). They provide estimates of statistical distributions of within-reach hydraulics but do not provide spatially explicit results. These models apply to reaches with morphologies close to natural.

Hydraulic-habitat models involve a large variety of biological *preference* models that describe the hydraulic suitability of microhabitats for aquatic organisms (Dunbar et al., 2012; Linnansaari et al., 2013). Most biological models reflect changes in occurrence or abundance of aquatic taxa, at low to medium flows, as a function of the hydraulic characteristics of their micro- or mesohabitats (e.g., Box 13.2). Fewer models focus on characteristics of population dynamics, for example mortality estimates due to drift of aquatic invertebrates during high flows (Borchardt, 1993) or to

Table 13.1 Overview of the Advantages and Limits of Ecohydrological and Hydraulic-Habitat Tools

Tools	Input Data and Catchment-Scale Applications	Output Data			
		Consideration of All Aspects of Flow Regime	Consideration of Hydraulics	Representation of Spatial 2D Patterns (e.g., for Restoration, Hydropeaking)	Generality/ Availability
Ecohydrological	+ +	+ +	− −	− −	+ +
	Needs discharge data only	Yes	No	No	Yes
Hydraulic-Habitat					
Numerical	− −	+	+ +	+ +	+
	Needs detailed topo- hydro-measures	Depends on biological models and calibration	Yes	Yes	Unadapted to complex 3D/ torrential flows
Mesohabitats	+	−	+	+	−
	Needs mapping at several discharge rates	Depends on the biological models and measurement discharges	Via mesohabitats	Simplified representations	Includes some subjectivity
Statistical	+	−	+ +	− −	−
	Needs estimates of reach hydraulic geometries	Depends on biological models, applies below bankfull	Yes	No	Developed in regional contexts and for natural morphologies

stranding of fish during rapid flow decreases (Hauer et al., 2014). The principles of the biological components are largely shared by all types of hydraulic-habitat models, indicating that models are generally classified according to their hydraulic modeling approach. Biological preference models have been criticized for lacking realism, for being static (i.e., not taking into account temporal habitat variations), and for relying on correlative approaches (Lancaster and Downes, 2010). In addition, habitat preferences vary with many environmental attributes such as temperature and water quality, as well as biological factors including the type of activity or the presence of predators. Despite the complexity of animal behavior, the analyses of responses of fish and invertebrates to hydraulics has been observed across many taxa and in a wide variety of streams (Lamouroux et al., 1999, 2010, 2013) and has led to two general results that support the use of hydraulic-habitat models. First, the density of most aquatic taxa varies strongly and significantly with hydraulics. Second, average preference models of aquatic taxa developed in a region transfer well to independent streams (i.e., not used to develop the models) in approximately 60% of the cases on average across taxa. Transferable preference models are suited for the application of hydraulic-habitat models over whole catchments, although the uncertainties of such exercises should be taken into account in communicating results. Catchment-scale applications may be most useful as a desktop screening exercise, to identify impacted sites where more detailed field-based assessment is warranted.

Numerical models can describe the effects of low flows, flow pulses, and flood pulses on habitat characteristics (Junk et al., 1989; Tockner et al., 2000). Flow pulses may cause movements of individuals for improving their energetic trade-offs (e.g., feeding versus swimming effort). The time and duration of flood pulses influence the access to spawning, feeding, or refuge habitats. When changes in flow are rapid, unsteady models can take into account the complex effects of rapid flow changes on hydraulics at a given discharge. Two-dimensional (2D) depth-averaged numerical models represent a good compromise between effort and accuracy for studying unsteady hydraulic patterns (Hauer et al., 2013). Compared with simpler one-dimensional (1D) models (e.g., the example in Box 13.2) they improve the spatial quantification of habitat distribution and habitat quality. They are easier to implement than more sophisticated 3D models (e.g., Sinha et al., 1998). Moreover, specific 2D models enable consideration of the impact of ice formation on the habitat distribution (Morse and Hicks, 2005) or the effects of engineering works (e.g., bridges and culverts; Ahmed and Rajaratnam, 1998). With digital terrain models of sufficient quality, unsteady 2D models can be used to assess both environmental hazards (e.g., floods) and habitat issues (Acreman and Dunbar, 2004; Hauer and Habersack, 2009; Krapesch et al., 2011). Overall, Hauer et al. (2009) suggested that 2D depth-averaged numerical models can provide an accurate description of the physical environment (point water depth, depth-averaged flow velocity) if bed particle size d_{90} (90% finer) is below 20 cm.

In rivers with coarser substrates such as step-pool and cascade geomorphic units, the complex instream hydraulics (e.g., turbulence) due to an overtopping of boulders in and along the river bed cannot be modeled using a numerical model in an ecologically relevant way. Statistical approaches are an attractive alternative for such complex flows (Girard et al., 2014). They are also attractive for reach scale applications of habitat models that do not require spatial descriptions of habitats and for a number of catchment-scale applications. In the next sections, we describe three case studies that illustrate the potential of these different hydraulic-habitat tools, two focused on situations where numerical models are developed, and one illustrating the potential of simplified, statistical habitat models for catchment-scale habitat analyses.

13.4.2 CASE STUDY 2: UNSTEADY 2D-NUMERICAL HABITAT MODELING FOR ASSESSING THE EFFECT OF HYDROPEAKING IN AUSTRIAN ALPINE RIVERS

This case study illustrates the potential of unsteady 2D depth-averaged numerical modeling to assess the impacts of hydropeaking on aquatic ecology. Hydropeaking refers to rapid changes in discharge downstream of hydropower dams, for which steady-state modeling approaches cannot be recommended as they neglect the important dynamic component in hydrology. Hydropeaking leads to pronounced dewatering areas between peak and base flow along gravel bars and channel banks. This generates a high risk of stranding of benthic invertebrates and fish. In this case study, the unsteady 2D modeling approach has been applied in different stream reaches to assess possible differences in the risk of stranding for young of the year brown trout (0 +). Interestingly, the biological component of the habitat model considered here is dynamic: it describes the potential mortality of the early life stage of brown trout. Variation in stranding risk was expected among reaches due to their different morphological features (e.g., presence of gravel bars and groin fields). More generally, the unsteady 2D modeling approach could also be used to study peak flow attenuation and the effects of additional pressures such as channel embankment for flood protection. Therefore, this approach is particularly useful in alpine rivers where multiple pressures often co-occur.

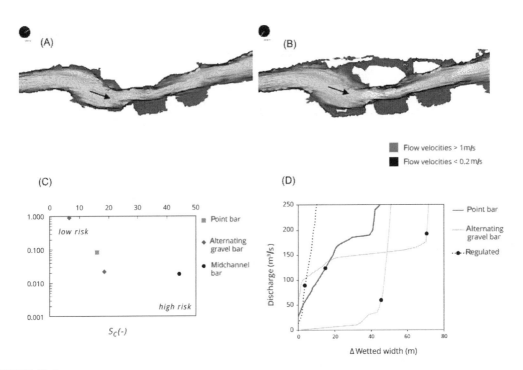

FIGURE 13.4

Unsteady 2D depth-averaged modeling results for the environmental assessment of hydropeaking impacts: flow velocity distribution during (A) base flow ($Q = 33$ m³/s) and (B) peak flow ($Q = 99$ m³/s) at the Inn River (a river with heterogeneous morphology, gravel bars, and groins); (C) conceptual stranding risk model, for 0 + brown trout, relating the stranding risk to the habitat suitability of the reach at peak flow (weighted usable area), the dewatered surface area between peak and base flow (ΔA) and the grain size variability (Sc, an index that increases with particle size heterogeneity) of the dewatered area; the stranding risk varies across four reaches dominated by different morphologies; (D) rate of change in wetted width as a function of discharge for the four different reaches; black dots indicate mean flow.

Source: (C and D) Adapted from Hauer et al. (2014)

Fig. 13.4A,B shows, in one reach, how the area dewatered during the flow reduction phase can be quantified by the 2D model for specific hydropeaking scenarios. Flow velocity maps for base and peak flows help to identify areas less sensitive to artificial flow fluctuations due to (1) minor extent of the dewatering zone and (2) low-flow velocities for both base and peak flow. Fig. 13.4C indicates how the stranding risk was conceptually linked, in the different reaches, to (1) the habitat value of 0+ brown trout in the entire reach; (2) the dewatered areas between high and peak flow; and (3) the grain size heterogeneity of the dewatered area. It indicates that rivers with different morphologies and grain size exhibit different stranding risks for aquatic organisms, with heterogeneous grain size increasing the risk of stranding (Hauer et al., 2014). Finally, Fig. 13.4D illustrates that the rate of change in wetted area varies among rivers and as a function of base flow values. Such relationships guide the negotiation of ramping ratios (ratios between base and peak discharge). For example, in one of the four rivers considered, it was estimated that similar dewatered areas would be obtained with ramping ratios of 1:5 (when estimated relative to winter base flow conditions) and 1:2 when estimated relative to summer base flow conditions).

13.4.3 CASE STUDY 3: 2D-NUMERICAL HABITAT MODELING FOR ASSESSING THE EFFECT OF MORPHOLOGICAL RESTORATION

This case study involving 2D depth-averaged numerical habitat models deals with the evaluation of an ecologically orientated flood protection project at the Sulm River, Austria, carried out between 1998 and 2000. Here, several 2D models corresponding to different morphologies observed during a 3-year monitoring program (2001−2003) helped to assess the effects of river widening (for flood protection) and the artificial construction of a new meander (for ecological purposes) inspired by historical reference conditions. These 2D hydrodynamic numerical models were combined with habitat suitability data of fish species.

The results showed a strong influence of river bed changes (Fig. 13.5) on the habitat quality and quantity for the young stages of nase (0+, 1+, 2+ age classes of *Chondrostoma nasus*, a key fish species of the Sulm River). Although the restoration was globally judged as successful, the morphological conditions modified by the combined influence of restoration and floods had decreased suitable areas for 0+ nase (Fig. 13.5). The primary factor was river bed aggradation, especially along the inner bend of the artificial meander. The higher flow velocities and shallower depths after restoration, combined with the steeper bank angle, reduced the area suitable for 0+ *C. nasus* (calculated at low-flow: 2.66 m^3/s) from 325 m^2 before restoration to 110 m^2 after (entire reach length: 500 m). The reduction of suitable area at mean flow (6.3 m^3/s) was less pronounced, with values ranging from 111 m^2 in 2001 to 58 m^2 in 2003. These figures show the influence of morphological changes on habitat suitability at similar discharge rates.

The results of the 2D depth-averaged habitat modeling were consistent with the results of electrofishing surveys, which showed a decline in 0+ nase after restoration (Fig. 13.5). In this case study, habitat changes resulted from the combined effects of floods and restoration, and a full assessment of the restoration project will be only possible when a dynamic equilibrium in river morphology has been achieved.

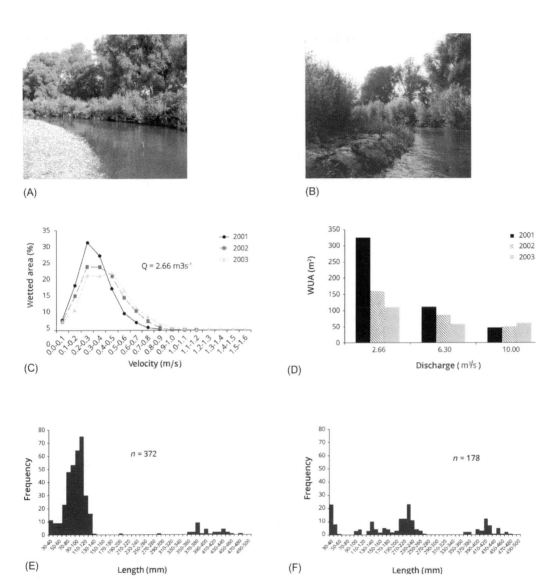

FIGURE 13.5

Evaluation of river restoration measures (creation of an artificial meander) in a river reach in Austria, based on 2D depth-averaged numerical modeling; (A) and (B) show the reach before (2001) and after (2003) a combined effect of restoration and floods; (C) changes in depth-averaged flow velocity frequency distributions; and (D) changes in habitat suitability (usable area WUA) of 0 + nase (*Chondrostoma nasus*); (E) and (F) changes in the population structure (length distribution) of nase.

Source: Adapted from Hauer et al. (2008)

13.4.4 CASE STUDY 4: CATCHMENT-SCALE DISTRIBUTED HABITAT MODELING USING STATISTICAL HABITAT MODELS

Simplified statistical hydraulic-habitat models may be used to translate hydrological alterations into habitat alterations at the scale of a large catchment (from Miguel et al., 2016). The simulation described here addressed a request of the Seine River Catchment Water Agency (France) to evaluate the ecological impact of groundwater abstraction in the catchment. The study used an existing distributed hydrological model of the Seine Catchment that accounted for surface–groundwater interactions and allowed an assessment of daily discharge alterations under different management scenarios. The hydrological model provided estimates of the alteration of low-flow values (Q95, the daily discharge exceeded 95% of the time) due to groundwater abstraction (for all river reaches in the catchment, including first-order reaches). The low-flow percentile Q95 was chosen because it reflected well the influence of water abstraction on annual low flows.

A series of models was used to translate Q95 hydrologic alterations into habitat alterations for fish at low flows, for all reaches in the catchment. First, hydraulic geometry relationships (depth- and width-discharge relationships for reaches), calibrated in a large range of French streams (Lamouroux and Capra, 2002), allowed translation of Q95 alterations into hydraulic alterations in reaches. Second, published fish species distribution models were used to identify fish species (or groups of species: guilds) that potentially occurred in each reach. These distribution models accounted for the effects of several environmental variables (e.g., catchment area, distance to source, air temperature, and reach slope) on the probability of occurrence of species. They were used to run habitat simulations only for fish species/guilds likely occurring in a given reach. Third, statistical hydraulic models were used to translate hydraulic alterations into habitat alterations for the fish species that were potentially present (Lamouroux and Capra, 2002). For each reach, the highest habitat alteration (across fish species and guilds considered) was mapped (Fig. 13.6).

Groundwater abstraction rate in the Seine catchment represents on average approximately 15% of the Q95 at the catchment mouth but is heterogeneous across the catchment. Results of simulations showed that groundwater abstraction causes a moderate median alteration of low flows (median reduction of Q95: 4.1%) and a weak alteration of usable habitats for fish (median reduction of usable area at Q95: 1.6%). However, the spatial distribution of habitat alterations helped to identify regions where impacts were much stronger (habitat reduction around 15% at Q95; Fig. 13.6). High habitat alteration reflected a combined effect of strong abstraction, specific hydraulic geometries and/or the presence of sensitive fish species. An uncertainty analysis (Miguel et al., 2016) further indicated that results should be interpreted at the scale of subbasins or for groups of reaches rather than individual reaches, due to very high uncertainties at the reach scale. This uncertainty analysis was made by propagating numerically the errors associated with the hydrological models (estimations of Q95 values and their alteration) and the hydraulic geometry models (estimations of reach hydraulic characteristics).

This case study and others (e.g., Snelder et al., 2011) show that hydraulic-habitat models can be applied in catchments using available data and models to estimate their input data (i.e., hydraulic geometries of reaches). This opens the possibility to translate a number of hydrological alterations into hydraulic alterations, with high ecological relevance. However, these simulations involve several models applied at the catchment-scale that rely on a number of simplifications and hypotheses.

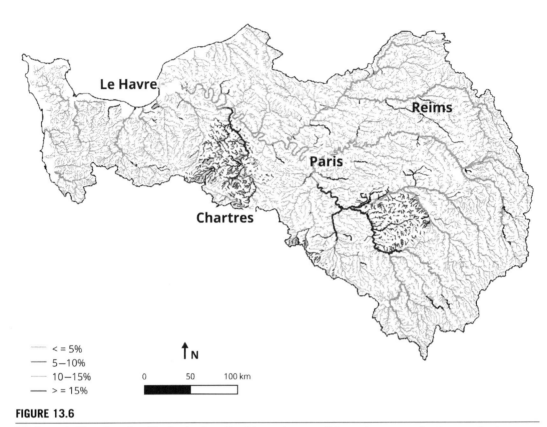

FIGURE 13.6

Maximum habitat alteration (% of reduction of usable area), across fish species groups, due to groundwater abstraction. Estimations are distributed over the whole Seine Catchment, France, using a combination of hydrological models, hydraulic geometry models, fish species distribution models, and statistical habitat models.

Source: Adapted from Miguel et al. (2016)

Therefore, addressing uncertainties that arise from all models and extrapolations in such habitat simulations is important.

13.5 CONCLUSION: COMBINING HYDROLOGICAL AND HYDRAULIC-HABITAT TOOLS WITHIN MODULAR LIBRARIES

Although they have often been considered opposing approaches, our case studies suggest that eco-hydrological and hydraulic-habitat tools should ideally be combined in environmental water studies. As summarized in Table 13.1, ecohydrological tools describe hydrological alterations for all aspects of the flow regime. Among hydraulic-habitat models, numerical models enable the description of complex spatial and temporal flow patterns that are particularly relevant in restoration studies, hydropeaking, or fish passage studies (Table 13.1). Simplified or statistical hydraulic approaches are interesting alternatives for some reach scale studies where spatially explicit results are not required; they also enable catchment-scale habitat applications (Table 13.1).

Improved combinations of these tools could be achieved, for example by involving hydraulic-habitat descriptions (e.g., as in Case Study 4 concerning low flows) in catchment-scale assessments of physical alteration (e.g., Case Study 1). This could improve the ecological relevance of large-scale assessments given the influence of morphology on stream habitat suitability (e.g., Case Study 3). In the same vein, repeated detailed habitat studies (e.g., hydropeaking; Case Study 2) can suggest important ecologically relevant hydraulic characteristics of stream reaches that would be possible to consider in large-scale assessments. Among these are the dewatering areas associated with rapid flow variations (which could be estimated using generalized width-discharge relationships, e.g., Booker, 2010) or the grain size heterogeneity of stream banks (which could be estimated by remote sensing; Mandlburger et al., 2015). In practice, however, hydraulic approaches (and particularly numerical approaches) require more complex input data than ecohydrological tools. Therefore, the degree of involvement of the different tools in environmental water assessments will depend on many criteria such as the objective and scale of the study, the availability of data, the degree of flow alteration, the degree of ecological sensitivity or the available funds (see, e.g., Table 7.2 in the European guidance on ecological flows; European Commission, 2015).

Our case studies also illustrate that a single technical tool cannot address all environmental flows assessment issues. Depending on environmental flows assessment studies objectives, scales, and stakes, it will be appropriate to combine different ecohydrological tools, and different types of hydraulic and biological models (see Chapter 11). Therefore, it would be desirable to move toward integrated management platforms that facilitate the combination of these different tools in flexible ways. Examples of model combinations addressing environmental flows assessment issues at the catchment-scale exist locally, among which are the *Estimkart* toolbox used in Case Study 4 (Miguel et al., 2016), the continuing development of a River Environmental Classification in New Zealand (Snelder et al., 2011), the Water Flow Evaluation Tool of Sanderson et al. (2012), or the River Analysis Package (Stewardson and Marsh, 2004). As catchment-scale environmental flows assessments often require using catchment-scale hydrologic and if possible morphodynamic models (see Case Study 3), a better integration of environmental flows assessment approaches and physical catchment-scale models (e.g., SedNet, Wilkinson et al., 2006; NetMap, Benda et al., 2007) is also desirable.

Future technical tools in the environmental flows assessment domain should both be more integrated (i.e., combining hydrology, unsteady hydraulics at various flow levels, morphodynamics, and water quality issues) and address uncertainty issues (see Chapter 15). In addition, available techniques in the field of Digital Elevation Models acquisition, hydraulic modeling, and spatial data representation are evolving rapidly. For example, airborne LiDAR (Light Detection And Ranging) bathymetry has improved in recent years, allowing high-resolution maps of fluvial topography (>20 points/m^2 and height accuracy <10 cm) to be created for both the aquatic and the riparian area (Mandlburger et al., 2015). Therefore, future tools should be designed to be as modular as possible, to allow for rapid evolution of their individual components.

We emphasize that the degree of involvement of technical tools in environmental flows assessments depends on the degree of confidence in flow—ecology or hydraulic—ecology relationships. The knowledge of quantitative ecological responses to hydrological and hydraulic patterns remains a priority in the environmental flows assessment arena (see Chapter 15). In particular, a better understanding of the role of habitat variations on aquatic population dynamics is needed (Shenton et al., 2012). Advanced technical tools can provide quantitative descriptions of the physical habitat and thus improve our understanding of biological responses to flow alterations at reach to catchment scales (see Chapter 15).

REFERENCES

Acreman, M.C., Dunbar, M.J., 2004. Defining environmental river flow requirements: A review. Hydrol. Earth Syst. Sci. Discuss. 8, 861–876.

Ahmed, F., Rajaratnam, N., 1998. Flow around Bridge Piers. J. Hydraul. Eng. 124, 288–300.

Arthington, A.H., Bunn, S.E., Poff, N.L., Naiman, R.J., 2006. The challenge of providing environmental flow rules to sustain river ecosystems. Ecol. Appl. 16, 1311–1318.

Benda, L., Miller, D.J., Andras, K., Bigelow, P., Reeves, G., Michael, D., 2007. NetMap: a new tool in support of watershed science and resource management. Forest Sci. 52, 206–219.

Booker, D.J., 2010. Predicting width in any river at any discharge. Earth Surf. Process. Landf. 35, 828–841.

Borchardt, D., 1993. Effects of flow and refugia on the drift loss of benthic macroinvertebrates: implications for lowland stream restoration. Freshw. Biol. 29, 221–227.

Carlisle, D.M., Wolock, D.M., Meador, M.R., 2011. Alteration of streamflow magnitudes and potential ecological consequences: a multiregional assessment. Front. Ecol. Environ. 9, 264–270.

Castella, E., Beguin, O., Besacier-Monbertrand, A.-L., Hug Peter, D., Lamouroux, N., Mayor Siméant, H., et al., 2015. Changes in benthic invertebrates and their prediction after the restoration of lateral connectivity in a large river floodplain. Freshw. Biol. 60, 1131–1146.

Davies, P.E., Harris, J.H., Hillman, T.J., Walker, J.F., 2010. The Sustainable Rivers Audit: assessing river ecosystem health in the Murray-Darling Basin, Australia. Mar. Freshw. Res. 61, 764–777.

Davies, P.E., Stewardson, M.J., Hillman, T.J., Roberts, J., Thoms, M., 2012. Sustainable Rivers Audit Report 2: The ecological health of rivers in the Murray-Darling Basin at the end of the Millennium Drought (2008–2010). Murray-Darling Basin Authority, Canberra, Australia. Available from: http://www.mdba.gov.au/what-we-do/mon-eval-reporting/sustainable-rivers-audit.

Dunbar, M.J., Alfredsen, K., Harby, A., 2012. Hydraulic-habitat modelling for setting environmental river flow needs for salmonids. Fish. Ecol. Manage. 19, 500–517.

European Commission, 2015. Ecological flows in the implementation of the Water Framework Directive. CIS Guidance Document No. 31. Technical Report, 2015–086.

Gibbins, C., Batalla, R.J., Vericat, D., 2010. Invertebrate drift and benthic exhaustion during disturbance: response of mayflies (Ephemeroptera) to increasing shear stress and river bed instability. River Res. Appl. 26, 499–511.

Gillespie, B.R., Desmet, S., Kay, P., Tillotson, M.R., Brown, L.E., 2015. A critical analysis of regulated river ecosystem responses to managed environmental flows from reservoirs. Freshw. Biol. 60, 410–425.

Ginot, V., 1998. Logiciel EVHA. Evaluation de l'habitat physique des poissons en rivière (version 2.0). Cemagref Lyon et Ministère de l'aménagement du Territoire et de l'Environnement, Direction de l'Eau, Paris, France.

Girard, V., Lamouroux, N., Mons, R., 2014. Modeling point velocity and depth statistical distributions in steep tropical and alpine stream reaches. Water Resour. Res. 50, 427–439.

Gordon, N.D., McMahon, T.A., Finlayson, B.L., Gippel, C.J., Nathan, R.J., 2004. Stream Hydrology: an Introduction for Ecologists. John Wiley & Sons, Chichester, UK.

Hauer, C., Habersack, H., 2009. Morphodynamics of a 1000-year flood in the Kamp River, Austria, and impacts on floodplain morphology. Earth Surf. Process. Landf. 34, 654–682.

Hauer, C., Mandlburger, G., Habersack, H., 2009. Hydraulically related hydro-morphological units: description based on a new conceptual mesohabitat evaluation model (MEM) using LiDAR as geometric input. River Res. Appl. 25, 29−47.

Hauer, C., Schober, B., Habersack, H., 2013. Impact analysis of river morphology and roughness variability on hydropeaking based on numerical modelling. Hydrol. Process. 27, 2209−2224.

Hauer, C., Unfer, G., Holzapfel, P., Haimann, M., Habersack, H., 2014. Impact of channel bar form and grain size variability on estimated stranding risk of juvenile brown trout during hydropeaking. Earth Surf. Process. Landf. 39, 1622−1641.

Hauer, C., Unfer, G., Schmutz, S., Habersack, H., 2008. Morphodynamic effects on the habitat of juvenile cyprinids (*Chondrostoma nasus*) in a restored Austrian lowland river. Environ. Manage. 42, 279−296.

Jowett, I.G., 1993. A method for objectively identifying pool, run, and riffle habitats from physical measurements. N. Z. J. Mar. Freshw. Res. 27, 241−248.

Jowett, I.G., Biggs, B.J.F., 2006. Flow regime requirements and the biological effectiveness of habitat-based minimum flow assessments for six rivers. Int. J. River Basin Manage. 4, 179−189.

Junk, W.J., Bayley, P.B., Sparks, R.E., 1989. The flood pulse concept in river floodplain systems. Can. Spec. Publ. Fish. Aquat. Sci. 106, 110−127.

King, J., Brown, C., Sabet, H., 2003. A scenario-based holistic approach to environmental flow assessments for rivers. River Res. Appl. 19, 619−639.

Krapesch, G., Hauer, C., Habersack, H., 2011. Scale orientated analysis of river width changes due to extreme flood hazards. Nat. Hazards 11, 2137−2147.

Lamouroux, N., Augeard, B., Baran, P., Capra, H., Le Coarer, Y., Girard, V., et al. (in press) Ecological flows: the role of hydraulic habitat models within an integrated framework. Hydroécol. Appl. [in French].

Lamouroux, N., Capra, H., 2002. Simple predictions of instream habitat model outputs for target fish populations. Freshw. Biol. 47, 1543−1556.

Lamouroux, N., Capra, H., Pouilly, M., Souchon, Y., 1999. Fish habitat preferences at the local scale in large streams of southern France. Freshw. Biol. 42, 673−687.

Lamouroux, N., Gore, J.A., Lepori, F., Statzner, B., 2015. The ecological restoration of large rivers needs science-based, predictive tools meeting public expectations: an overview of the Rhône project. Freshw. Biol. 60, 1069−1084.

Lamouroux, N., Mérigoux, S., Capra, H., Dolédec, S., Jowett, I.G., Statzner, B., 2010. The generality of abundance-environment relationships in microhabitats: a comment on Lancaster and Downes (2009). River Res. Appl. 26, 915−920.

Lamouroux, N., Mérigoux, S., Dolédec, S., Snelder, T.H., 2013. Transferability of hydraulic preference models for aquatic macroinvertebrates. River Res. Appl. 29, 933−937.

Lancaster, J., Downes, B.J., 2010. Linking the hydraulic world of individual organisms to ecological processes: putting ecology into ecohydraulics. River Res. Appl. 26, 385−403.

Linnansaari, T., Monk, W.A., Baird, D.J., Curry, R.A., 2013. Review of approaches and methods to assess Environmental Flows across Canada and internationally. Research Document 2012/039. Canadian Science Advisory Secretariat. Fisheries and Oceans, Canada, p. 75.

Marsh, N.A., Stewardson, M.J., Kennard, M.J., 2006. River Analysis Package, Cooperative Research Centre for Catchment Hydrology. Monash University, Melbourne, Australia.

Mandlburger, G., Hauer, C., Wieser, M., Pfeifer, N., 2015. Topo-bathymetric LiDAR for monitoring river morphodynamics and instream habitats − a case study at the Pielach River. Remote Sens. 7, 6160−6195.

Mérigoux, S., Lamouroux, N., Olivier, J.-M., Dolédec, S., 2009. Invertebrate hydraulic preferences and predicted impacts of changes in discharge in a large river. Freshw. Biol. 54, 1343−1356.

Metcalfe, R.A., Schmidt, B., 2014. Streamflow Analysis and Assessment Software (SAAS). Available from: http://people.trentu.ca/rmetcalfe/SAAS.html.

Miguel, C., Lamouroux, N., Pella, H., Labarthe, B., Flipo, N., Akopian, M., et al., 2016. Hydraulic habitat alteration at the catchmentscale: impacts of groundwater abstraction in the Seine-Normandie catchment. La Houille Blanche 3, 65−74 [in French].

Morse, B., Hicks, F., 2005. Advances in river ice hydrology 1999−2003. Hydrol. Process. 19, 247−263.

Murchie, K.J., Hair, K.P.E., Pullen, A.C.E., Redpath, A.T.D., Stephens, A.H.R., Cooke, S.J., 2008. Fish response to modified flow regimes in regulated rivers: research methods, effects and opportunities. River Res. Appl. 24, 197−217.

Negnevitsky, M., 2002. Artificial Intelligence, a Guide to Intelligent Systems. Pearson Education, Frenchs Forest, New South Wales, Australia.

Olden, J.D., Konrad, C.P., Melis, T.S., Kennard, M.J., Freeman, M.C., Mims, M.C., et al., 2014. Are large-scale flow experiments informing the science and management of freshwater ecosystems? Front. Ecol. Environ. 12, 176−185.

Olden, J.D., Poff, N.L., 2003. Redundancy and the choice of hydrologic indices for characterizing streamflow regimes. River Res. Appl. 19, 101−121.

Pahl-Wostl, C., Arthington, A.H., Bogardi, J., Bunn, S.E., Hoff, H., Lebel, L., et al., 2013. Environmental flows and water governance: managing sustainable water uses. Curr. Opin. Environ. Sustain. 5, 341−351.

Parasiewicz, P., 2007. The MesoHABSIM model revisited. River Res. Appl. 23, 893−903.

Poff, N.L., Ward, J.V., 1989. Implications of streamflow variability and predictability for lotic community structure: a regional analysis of streamflow patterns. Can. J. Fish. Aquat. Sci. 46, 1805−1818.

Poff, N.L., Zimmerman, J.K.H., 2010. Ecological responses to altered flow regimes: a literature review to inform the science and management of environmental flows. Freshw. Biol. 55, 194−205.

Poff, N.L., Allan, J.D., Bain, M.B., Karr, J.R., Prestegaard, K.L., Richter, B.D., et al., 1997. The natural flow regime. A paradigm for river conservation and restoration. BioScience 47, 769−784.

Poff, N.L., Olden, J.D., Pepin, D.M., Bledsoe, B.P., 2006. Placing global streamflow variability in geographic and geomorphic contexts. River Res. Appl. 22, 149−166.

Poff, N.L., Richter, B.D., Arthington, A.H., Bunn, S.E., Naiman, R.J., Kendy, E., et al., 2010. The ecological limits of hydrologic alteration (ELOHA): a new framework for developing regional environmental flow standards. Freshw. Biol. 55, 147−170.

Richter, B.D., Baumgartner, J.V., Powell, J., Braun, D.P., 1996. A method for assessing hydrologic alteration within ecosystems. Conserv. Biol. 10, 1163−1174.

Richter, B.D., Baumgartner, J.V., Wigington, R., Braun, D.P., 1997. How much water does a river need? Freshw. Biol. 37, 231−249.

Rinaldi, M., Belletti, B., Van de Bund, W., Bertoldi, W., Gurnell, A., Buijse, T., et al., 2013. Review on eco-hydromorphological methods. In: Friberg, N., O'Hare, M., Poulsen, A., (Eds.), Deliverables of the EU FP7 REFORM Project. Available from: http://www.reformrivers.eu

Sabaton, C., Souchon, Y., Capra, H., Gouraud, V., Lascaux, J.-M., Tissot, L., 2008. Long-term brown trout populations responses to flow manipulation. River Res. Appl. 24, 476−505.

Sanderson, J.S., Rowan, N., Wilding, T., Bledsoe, B.P., Miller, W.J., Poff, N.L., 2012. Getting to scale with environmental flow assessment: the Watershed Flow Evaluation Tool. River Res. Appl. 28, 1369−1377.

Shenton, W., Bond, N.R., Yen, J.D.L., Mac Nally, R., 2012. Putting the "Ecology" into environmental flows: ecological dynamics and demographic modelling. Environ. Manage. 50, 1−10.

Sinha, S., Sotiropoulos, F., Odgaard, A., 1998. Three-dimensional numerical model for flow through natural rivers. J. Hydraul. Eng. 124, 13−24.

Singh, V.P., Frevert, D.K., 2006. Watershed Models. CRC Press, Boca Raton, Florida.

SKM, 2005. Development and application of a flow stressed ranking procedure. Final Report to Department of Sustainability and Environment, Victoria. Sinclair Knight Merz, Armadale, Australia.

SKM, 2011. Sustainable rivers audit hydrology theme background report: integration of stress scores for farm dams, land use change, and regulation. Report to Murray Darling Basin Authority. Sinclair Knight Merz, Armadale, Australia.

Snelder, T.H., Booker, D., Lamouroux, N., 2011. A method to assess and define environmental flow rules for large jurisdictional regions. J. Am. Water Resour. Assoc. 47, 828–840.

Snelder, T.H., Datry, T., Lamouroux, N., Larned, S.T., Sauquet, E., Pella, H., et al., 2013. Regionalization of patterns of flow intermittence from gauging station records. Hydrol. Earth Syst. Sci. 17, 2685–2699.

Stewardson, M., Marsh, N., 2004. River Analysis Program (RAP) brochure. Available from: http://www.toolk-it.net.au/tools/RAP.

Tockner, K., Malard, F., Ward, J.V., 2000. An extension of the flood pulse concept. Hydrol. Process. 14, 2861–2883.

Vezza, P., Parasiewicz, P., Spairani, M., Comoglio, C., 2014. Habitat modeling in high-gradient streams: the mesoscale approach and application. Ecol. Appl. 24, 844–861.

Webb, J.A., Miller, K.A., King, E.L., De Little, S.C., Stewardson, M.J., Zimmerman, J.K.H., et al., 2013. Squeezing the most out of existing literature: a systematic re-analysis of published evidence on ecological responses to altered flows. Freshw. Biol. 58, 2439–2451.

Webb, J.A., De Little, S.C., Miller, K.A., Stewardson, M.J., Rutherfurd, I.D., Sharpe, A.K., et al., 2015. A general approach to predicting ecological responses to environmental flows: making best use of the literature, expert knowledge and monitoring data. River Res. Appl. 31, 505–514.

Wilding, T.K., Bledsoe, B., Poff, N.L., Sanderson, J., 2014. Predicting habitat response to flow using generalized habitat models for trout in Rocky Mountain streams. River Res. Appl. 7, 805–824.

Wilkinson, S.N., Prosser, I.P., Hughes, A.O., 2006. Predicting the distribution of bed material accumulation using river network sediment budgets. Water Resour. Res. 42, 17. Available from: http://dx.doi.org/10.1029/2006WR004958.

MODELS OF ECOLOGICAL RESPONSES TO FLOW REGIME CHANGE TO INFORM ENVIRONMENTAL FLOWS ASSESSMENTS

14

J. Angus Webb[1], Angela H. Arthington[2], and Julian D. Olden[3,2]

[1]*The University of Melbourne, Parkville, VIC, Australia*
[2]*Griffith University, Nathan, QLD, Australia* [3]*University of Washington, Seattle, WA, United States*

14.1 INTRODUCTION

Environmental water regimes are designed and delivered for the expressed benefit of protecting or restoring freshwater ecosystems. Although aquatic environments include physical, chemical, and ecological dimensions, it is the ecological characteristics that are most often cited as the targets for restoration and conservation through implementation of environmental water regimes. The foundational papers in environmental water research stress the linkages between flow regime attributes and ecological response (Bunn and Arthington, 2002; Lytle and Poff, 2004; Poff et al., 1997), and more recently, attention has focused on responses to flow regime change.

It is striking that the majority of environmental flows assessments attempt to forecast ecological responses to flow regime change using physical habitat models alone (Tharme, 2003). Early methods for environmental flows assessment were focused predominantly on hydrological endpoints (e.g., Tennant, 1976), with later methods using physical habitat models to relate hydrology to habitat for particular species of interest, particularly fish (Thomas and Bovee, 1993). Even the more comprehensive *holistic* (sensu Tharme, 2003) methods that make direct predictions of ecological responses to flow regime attributes rely heavily on modeled associations between river discharge and physical habitat. These are then combined with knowledge from the literature and expert opinion about how flora and fauna will likely respond to improvements in flow and habitat.

This approach to quantifying ecological response is perhaps not surprising for frameworks such as the Building Block Methodology (King and Louw, 1998) and *Downstream Response to Flow Transformation (DRIFT*; King et al., 2003) that were designed to cater for situations of limited ecological knowledge and modeling capabilities (see King and Brown, 2010 for case studies). Indeed, the main reason given for this reliance on physical models coupled with expert opinion is that we lack sufficient empirical knowledge to model relationships between particular flow events or alterations and ecological outcomes directly (Davies et al., 2014; Stewardson and Webb, 2010). Attempts to synthesize multiple studies to identify generic relationships have failed, at least in part because of inconsistent approaches, endpoints, analytical techniques, and environmental/climatic contexts, and therefore ecological responses (Poff and Zimmerman, 2010); a problem that has also emerged in global reviews of large-scale experimental water releases below dams (Gillespie et al., 2015; Konrad et al., 2011; Olden et al., 2014).

Against this background of inconsistent outcomes from empirical studies and consequent uncertainty concerning ecological responses to flow variability or flow regime change, the perceived confidence associated with hydrological and habitat models has proved appealing and somewhat comforting to scientists and water managers. Nevertheless, considerable uncertainty still exists regarding the ecological benefits predicted from models of physical processes (see Chapter 15), causing many practitioners to be mindful of the *Field of Dreams* hypothesis (Palmer et al., 1997); that provision of habitat is sufficient to elicit a desired ecological response. However, numerous studies inform us that although provision of suitable habitat in the right place and at the right time of year is essential, it is not by itself sufficient to achieve desirable ecological outcomes (Bunn and Arthington, 2002). Rather, other environmental attributes are required to allow organisms to take advantage of the physical habitat provided. These can include recruitment processes, and attributes such as longitudinal or lateral habitat connectivity to provide for fish and invertebrate movements and dispersal (Colloff and Baldwin, 2010), water quality characteristics such as appropriate temperatures for spawning and growth (Olden and Naiman, 2010), primary production and energy flows along systems to provide food resources (Bunn et al., 2003), and community-level attributes such as the presence of multiple species and robust food webs (Naiman et al., 2012).

Recent decades have witnessed considerable publication of research relating ecological responses to natural flow variability and different facets of altered flow regimes (Arthington, 2012; Naiman et al., 2008; Stewardson and Webb, 2010). In many cases, it can no longer be argued that we have insufficient empirical data to derive useful ecological models. How can we better employ this huge and underutilized resource to improve the modeling of ecological responses to flow variability and flow regime alterations in environmental flows assessments?

A broad range of modeling tools can be used to improve our understanding of how ecosystems and their components respond to flow variability and flow regime alteration, and then applied to support environmental water planning, assessment, and management. Ecological modeling tools can be used to explore a range of questions, from broad patterns across flow regime types, local relationships between particular flow events and ecological attributes (e.g., species richness, abundance, and recruitment), to long-term dynamic population or ecosystem outcomes in response to particular flow sequences. These different questions need different types of methods and different data sets that may or may not be available for individual environmental flows assessments.

In this chapter, we focus on the middle range of these questions and the statistical modeling approaches that can be used to explore and quantify relationships between flow events and ecological attributes of interest to environmental water scientists and managers. This focus is consistent with understandings from foundational papers in environmental water research, which largely address the relationships between flow regime attributes and ecological response (Bunn and Arthington, 2002; Lytle and Poff, 2004; Poff et al., 1997), and ecological responses to flow regime change (Poff and Zimmerman, 2010; Webb et al., 2013). Our aim is to bring different kinds of flow—ecology models to the attention of practitioners who may be unfamiliar with the options available, and to step through the processes involved in model conceptualization and development, illustrated by case studies. We believe that this will promote and facilitate the use of flow—ecology models in future environmental flows assessments, and thereby support improvements in the rigor, transparency, and repeatability of recommendations.

14.2 CONCEPTUAL BOUNDS

We restrict our discussion to statistical model types and case studies that have been used, or we believe have the potential to be used, to inform environmental flows assessments and implementation. Therefore, we assume that broad ecological goals of the environmental water regime have been established by stakeholders and that the environmental flows assessment panel has been convened to develop flow recommendations. The primary delineation of what constitutes a useful model will be based on a combination of whether the model has predictive capacity, and on its degree of complexity and utility. We do not expect that every reader will agree with where we draw the line. By *model* we mean any statistical relationship between environmental drivers (and here we concentrate on flow drivers) and one or more ecological variables. Our definition of *ecological variable* encompasses any response at any level of ecological organization, from genetics to ecosystems. This treatment excludes all models that predict habitat only; and also excludes water quality models that assess the effects of physical causes (e.g., instream nutrient levels enriched by fertilizer runoff). For a comprehensive summary of ecological endpoints that may be used in developing flow–ecology relationships to inform environmental flows assessments, see Table 2 of Poff et al. (2010). In summary, our goal is to explore the major statistical model types, and illustrate each of these with an example that is emblematic of the broad challenges and opportunities that exist. However, first we introduce the concept of model parsimony as being the primary consideration when developing or using flow–ecology models.

14.3 STRIVING FOR PARSIMONY IN ENVIRONMENTAL FLOWS ASSESSMENTS

Everything should be made as simple as possible, but no simpler.

Albert Einstein

Parsimony is the principal criterion by which one should assess whether a model is useful for environmental flows assessment. In the simplest sense, a parsimonious model includes the principal environmental drivers (e.g., descriptors of the flow regime and other flow-influenced factors), while omitting other drivers that explain little variation in the ecological endpoints of interest. The important practical argument for parsimony is that models need to be complex enough to capture flow–ecology relationships, yet simple enough to be used as part of a multistakeholder environmental flows assessment. Practitioners need to be able to engage with and understand the model so that they can assess and trust the predicted ecological responses to different environmental water delivery scenarios. Models that are unnecessarily complex, or that take an excessively long time to parameterize and implement, are unlikely to be broadly useful in environmental flows assessments, especially when management decisions must be made within short time frames. Moreover, excess complexity may limit model transferability in both space and time.

There are also more theoretical reasons for considering model parsimony as the principal determinant of whether or not a model may be useful for environmental flows assessment. Complex models will almost certainly have many parameters, all of which must be estimated during model development. This can be done by fitting data to the model, employing values from the literature or expert opinion, or a combination of both. If the parameter values are estimated by fitting the model to data, we run the risk of having insufficient information to reliably estimate the parameter values. Models can become over-fit; resulting in *torturing* of the data to the point where model generality is unachievable. Such models may predict the data to which they have been fitted, but have little potential to predict new data (Olden and Jackson, 2002). If values are taken from the literature for physically based parameters, this typically fails to take into account the specific context of the studies that would have affected these values.

Although simple models may demonstrate greater predictive capacity because of their focus on the primary driving variables (MacNally, 2000), they must still be able to account for sufficient variation in the data, or will be of limited use as a predictive tool. Information criteria such as *Aikake's Information Criterion (AIC)*, the *Deviance Information Criterion (DIC)*, and the *Bayesian Information Criterion (BIC)* provide an assessment of the trade-off between model fit to the data (complexity) and number of parameters (simplicity). They can be used to identify parsimonious models from a set of candidate models (Burnham and Anderson, 2002).

Model parsimony is linked to utility of the models for decision making. Any model used in an environmental flows assessment must be able to inform decision making. Parsimonious models are more likely to be useful for decision making because they have fewer parameters that focus on the primary drivers, and hence have fewer management *levers* that need to be considered by decision makers when developing flow recommendations. To inform decisions, a model must be able to predict ecological responses under different flow scenarios. All of the model types described below can make such predictions, although this is more easily done for some methods than others.

14.4 WHAT CHOICES EXIST FOR ECOLOGICAL MODELING TO ASSIST ENVIRONMENTAL FLOWS ASSESSMENTS?

All too often, practitioners undertaking environmental flows assessments will not have access to preexisting models (see Section 14.6). In these situations, a multitude of statistical modeling approaches exists, but choosing the most appropriate method is challenging.

The choice of what method to use will be influenced by the choice of response variables and predictor variables. A response variable will be a measurable expression (a *measurement endpoint*) of an ecological value that is chosen as an objective through the environmental flows assessment process (*assessment endpoint*). For example, an assessment endpoint may be "a sustainable population of golden perch," with measurement endpoints for this including evidence of breeding response, and subsequent growth and recruitment (King et al., 2009). In more general terms, a response variable should have the following characteristics:

1. A response is conceptually linked to the *treatment*, for instance provision of a particular environmental water event;
2. It is economic to measure, using well-established techniques;
3. The response variable is not affected by high levels of measurement uncertainty;
4. A response is expected within an expected time frame;
5. The ecological response is considered valuable by stakeholders.

The choice of predictor variables will be affected by conceptual understanding of the influence of flow on the measurement endpoint chosen, and upon the availability of data. Consistent with our overarching call for parsimony in ecological models, predictor variables should be those believed to have the greatest effect upon the response variable, with the statistical model using this ecological knowledge to make predictions of response under different flow scenarios. For the golden perch example above, spring *fresh* (high-flow event) volume and rate of change in flow rate, along with water temperature, are understood to be the most important variables for predicting the occurrence of the breeding response (Webb et al., 2016). This example illustrates the two types of predictor variable that should be considered.

1. Variables with a known relationship to the management levers available. Here, water managers can manipulate the peak discharge volume of the spring fresh, and the rate of change in flow (both subject to infrastructure and operational constraints).
2. Variables not directly related to the management levers, but which are so important to the response variable that they can modify, or even override, the effect of the management lever. Sufficiently high water temperatures are a necessary but not sufficient condition for golden perch spawning. Ideal fresh volumes and rates of change delivered during a period of cold water temperatures will not elicit a spawning response. If water temperature is critical to fish spawning, managers may be able to release water from an appropriate depth in a storage (Olden and Naiman, 2010).

When possible, predictor variables should be expressed in terms that will translate across different rivers. This maximizes managers' abilities to transfer learnings from one river to a novel context (Chapter 25). The peak discharge volume and rate of change described above would not fulfill this criterion, as different rivers will often have different geomorphic settings and channel characteristics (Poff et al., 2010). However, hydraulic models could be used to translate discharge and rate of change into hydraulic variables such as depth, average velocity, and rate of change in velocity (e.g., Shafroth et al., 2010). Such variables are more likely to be directly causally related to the ecological response, as it is typically the hydraulic conditions rather than the hydrology that are experienced by the biota (Chapter 13). Variables with a direct causal linkage are more likely to be transferable across different settings.

The nature of the hypothesized relationship between response and predictor variables, the number of predictor variables, and the way they are expected to interact with one another, will all affect the choice of the most appropriate modeling approach. In the remainder of this section, we outline major approaches we believe can be applied to modeling ecological responses to flow variation as part of environmental flows assessments, including the characteristics that may suit them to particular applications. We do not exhaustively cover all modeling approaches, but seek to cover the range of complexities of statistical approaches that we believe will be useful to environmental flows assessments. The description starts with the simplest approaches, with simple functional forms and single (or few) predictor variables linking response to predictor variables, through to more complex modeling approaches. In Section 14.5, we propose that some of the more complex modeling approaches, while feasible for environmental flows assessments, may require the inclusion of an ecological modeling specialist on the assessment panel.

14.4.1 SIMPLE LINEAR MODELS

We may expect to see some ecological responses being overwhelmingly determined by a single aspect of the flow regime, with a continuous response. In these circumstances, a simple linear regression could be used to make predictions of ecological outcomes from a managed flow event, together with uncertainties. Driver et al. (2005) present data that describe such a relationship between flood duration and the number of ibis (*Threskiornis molucca*) nests in the Booligal Swamp in New South Wales, Australia (Fig. 14.1). Although this is probably the simplest of applicable statistical models, the R^2 of 0.76 demonstrates that the majority of variation is explained by fitting just two parameters (intercept, slope).

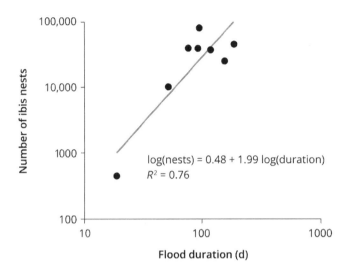

FIGURE 14.1

Simple linear relationship between bird breeding (number of nests) and flood duration (days).

Source: Data from Driver et al. (2005)

14.4.2 GENERALIZED LINEAR AND NONLINEAR MODELS

Various refinements upon the basic linear modeling approach have allowed the incorporation of greater complexity within this general modeling framework (McCullagh and Nelder, 1989). This allows analyses to incorporate such features as nonlinear relationships, non-Gaussian residuals, noncontinuous and nonnormal response data (e.g., logistic regression), and random effects to avoid pseudoreplication. These features are handled through a variety of approaches that we do not attempt to separate here; practitioners need only be aware that these options are available for modeling ecological responses.

Arthington et al. (2012) used generalized least squares regression, which allows for correlated errors and unequal variances (Bolker, 2008; Pinheiro, 2011), to assess relationships between descriptors of flow regime and indicators of biotic responses of riparian vegetation, aquatic plant and fish assemblages, and measures of species abundance. Two significant relationships discovered for riparian vegetation are provided in Fig. 14.2: an inverse quadratic relationship between species richness of riparian vegetation and the coefficient of variation in annual flows, with lowest values of species richness found at intermediate values of variation (A); and an inverse linear relationship between the log-transformed density of native vegetation and the coefficient of variation of daily flows during the dry season (B). Although the data in Fig. 14.2B, in particular, do not explain a large proportion of the variation in density of native species, they could still be used to inform decision making around acceptable levels of dry season flow variability.

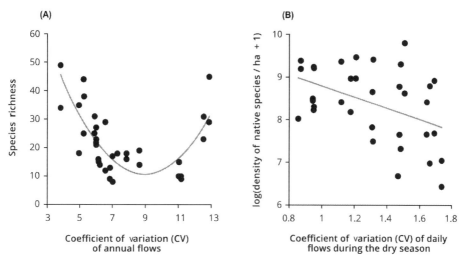

FIGURE 14.2

Nonlinear (A) and linear (B) models of relationships between discharge and biotic effects in a test of the *Ecological Limits of Hydrologic Alteration (ELOHA)* framework in subtropical Australia.

Source: Redrawn from Arthington et al. (2012)

14.4.3 **HIERARCHICAL MODELS**

Hierarchical models are being used increasingly in environmental research because of their great flexibility and consequent ability to be fitted to the complex, incomplete, and *messy* data that characterize many environmental investigations (Clark, 2005). In particular, the hierarchical aspect allows data from multiple *sampling units* (e.g., rivers) to be combined within the same statistical model to improve inferential strength. Hierarchical models can be implemented within a classical *frequentist* framework using approaches such as data cloning (Lele et al., 2007), but this is complex. Bayesian approaches to hierarchical models are more straightforward, and we restrict the remainder of this section to the consideration of Bayesian methods.

Bayesian methods provide a mathematical framework for using data to update belief. A *prior probability distribution* for each parameter in a model is the expression of belief in the parameter value prior to collecting new data. The data are expressed in the *likelihood function* and combined with the prior to produce a *posterior probability distribution*, which reflects the updated belief (McCarthy, 2007).

The prior probability provides an opportunity to incorporate existing knowledge into the analysis to improve outcomes (McCarthy and Masters, 2005), but is also an ongoing source of controversy as it can be seen as making Bayesian analyses too subjective (McCarthy, 2007). However, prior probabilities only have a large effect on the posterior when there are few data available. Moreover, in the absence of prior knowledge (or in an attempt to avoid perceptions of subjectivity), *minimally informative* priors are often used. These priors have little impact upon the posterior probability distributions gained at the end of the analysis.

In a hierarchical Bayesian model, estimates of model parameters from different sampling units are linked via their priors (Gelman et al., 2004). The practical effect of the hierarchical approach is that predictive uncertainties within individual sampling units are reduced—*borrowing strength*, and parameter values are drawn some way toward a global average—*shrinkage*. Like other forms of prior information, the hierarchical model structure has the greatest effect on posterior estimates when data are few and uncertainty within sampling units is high (Webb et al., 2010b).

Webb et al. (2010b) proposed the use of hierarchical Bayesian models for predicting ecological effects of environmental water regimes. Using data for Australian smelt (*Retropinna semonii*) in the Thomson River, Victoria, Australia, they demonstrated that uncertainties in regression slopes that related smelt abundance to summer flows were substantially reduced when a hierarchical model was used, as opposed to when the five river reaches were analyzed separately (Fig. 14.3).

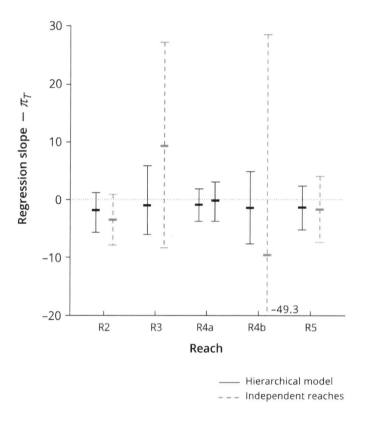

FIGURE 14.3

Improvements in predictive uncertainty gained from using a hierarchical Bayesian modeling approach. The graph shows the median value of a regression slope linking abundance of Australian smelt (*Retropinna semoni*) to summer flows, with values shown for each of five river reaches. The central ticks are median values and the error bars encompass the 95% credible interval for the estimate. Results from the hierarchical model are shown using solid lines, with those from the independent reaches analysis using dashed lines, as per the inset key.

Source: Webb et al. (2010b)

14.4.4 **FUNCTIONAL LINEAR MODELS**

There are recognized challenges in using hydrologic metrics for establishing flow—ecology relationships. The range of hydrological metrics is vast (Olden and Poff, 2003) and they can have biases in their calculation (Kennard et al., 2007, 2010). Recent statistical advances have sought to provide a more holistic characterization of the flow regime in an attempt to develop flow—ecology relationships without the need to select and quantify hydrologic metrics (e.g., Sabo and Post, 2008). Functional linear models aim to decompose the mathematical functions that underlie the observed time series for use in subsequent statistical analyses. The resulting *functional* data typically consist of continuous, or effectively continuous, observations of a process over space or time that are derived from mathematical functions (Ramsay and Silverman, 2002) such as temperature or river discharge (e.g., Ainsworth et al., 2011).

The functional data series may be used as predictor variables in a functional linear model (Muller and Stadtmuller, 2005; Ramsay and Silverman, 2002). In such models, the actual time series, in this case the hydrograph, becomes the predictor variable in a linear model as opposed to a set of hydrologic metrics that aim to describe it. Stewart-Koster et al. (2014) demonstrated the utility of this approach for quantifying the relationships between fish species densities and the timing, magnitude, and duration of flows at both local and regional spatial scales in the United States. By using a mathematical decomposition of the actual discharge, a functional approach removed the uncertainty that can be associated with calculating hydrologic metrics from flow records of different lengths (e.g., Kennard et al., 2007, 2010).

The functional predictor variable and, therefore, the regression coefficient do not readily quantify the influence of variability of flows, although Stewart-Koster et al. (2014) combined functional predictors and metrics to remedy this challenge. In summary, functional methodologies may provide a valuable tool to overcome some, but not necessarily all, of the challenges associated with the use of hydrologic metrics in ecological models.

14.4.5 **MACHINE LEARNING APPROACHES**

Machine learning (ML) is a rapidly growing area of ecoinformatics that is concerned with identifying structure in complex, often nonlinear data and generating accurate predictive models. ML algorithms can be organized according to a diverse taxonomy that reflects the desired outcome of the modeling process. A number of ML techniques have been promoted in ecology as powerful alternatives to traditional modeling approaches. These include supervised learning approaches that attempt to model the relationship between a set of inputs and known outputs such as *artificial neural networks (ANNs)*, cellular automata, classification and regression trees, fuzzy logic, genetic algorithms and programming, maximum entropy, support vector machines, and wavelet analysis (Olden et al., 2008). The growing use of these methods in recent years is the direct result of their ability to model complex, nonlinear relationships in ecological data without having to satisfy the restrictive assumptions required by conventional, parametric statistical approaches.

In one example, Kennard et al. (2007) used multiresponse artificial neural networks to model fish community composition of two coastal Australian rivers as a function of both environmental and hydrological attributes (Fig. 14.4). In addition to the flexibility of neural networks to model multiple response variables, neural networks have the advantage of modeling nonlinear associations with a variety of data types, require no specific assumptions concerning the distributional characteristics of the independent variables, and can accommodate interactions among predictor variables without any a priori specification (Olden et al., 2006). In addition to achieving high predictive performance, the artificial neural networks produced by Kennard et al. (2007) also demonstrated that landscape- and local-scale habitat variables and characteristics of the long-term flow regime were generally more important predictors of fish assemblage structure than variables describing the short-term history of hydrological events. Moreover, the modeling approach revealed the importance of interactions among environmental and hydrologic factors operating at multiple spatial and temporal scales.

ML methods are powerful tools for prediction and explanation, and they will enhance our ability to model ecological systems. They are not, however, a solution to all ecological modeling problems. Over the past decade there has been considerable research on the development of methods for understanding the explanatory contributions of the independent variables in ANNs (Giam and Olden, 2015; Olden and Jackson, 2002). This was, in part, prompted by the fact that neural networks were considered a *black box* approach to modeling ecological data because of the perceived difficulty in understanding their inner workings. This is no longer the case. No one ML approach will be best suited to addressing all problems nor will ML approaches always be preferable to traditional statistical approaches. Although ML methods are generally more flexible with respect to modeling complex relationships and messy data sets, the models they produce are often more difficult to interpret, and the modeling process itself is often far from transparent. Although ML technologies strengthen our ability to model ecological phenomena, advances in understanding the fundamental processes underlying those phenomena are clearly critical as well.

14.4.6 BAYESIAN NETWORKS

Bayesian networks (BNs) are graphical probabilistic models that link drivers to outcomes via a series of *conditional probability tables (CPTs*; Pearl, 2000). They have several advantages that make them appealing (Horne et al., 2017), and which have seen an explosion in their use in natural resource management problems since the early 2000s.

1. They are quite easy to construct, with software platforms such as Netica having a drag-and-drop method for building network structures.
2. The model has its basis in an explicit graphical conceptual model that shows how the different driving variables interact to affect the final outcome. This facilitates communication and understanding of the model without restricting it to be overly simple in terms of structure.
3. The stochastic nature of relationships is fully specified via the CPTs, with this stochasticity propagating through the model to provide an explicit representation of uncertainty in the final prediction.
4. BNs can be used to rapidly test the likely ecological outcomes from different scenarios of driving variables.

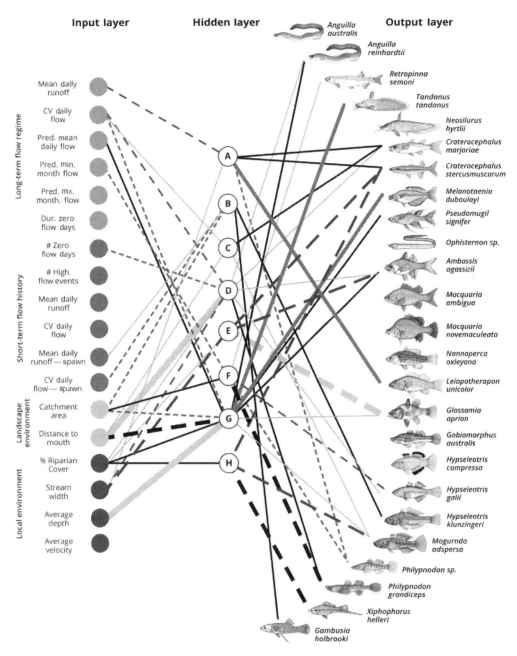

FIGURE 14.4

Structure and link weights of the multiresponse neural network relating different fish species to long-term flow characteristics, short-term changes in hydrological environment, and landscape-scale attributes.

Source: Kennard et al. (2007). © Canadian Science Publishing or its licensors

However, perhaps the major reason for the rapid uptake of BNs in natural resource management is their capacity to employ different data types in the derivation of CPTs. Within a model, some relationships might be specified by data, others by model outputs, and others by expert opinion. Data of different types can also be combined in the same CPT through the application of Bayes' rule to update *prior* CPTs. This capacity has seen many networks created that are largely expert driven.

For example, Shenton et al. (2011, 2014) used BNs to build models of ecological response to changing flow regimes for the Australian grayling (*Prototroctes maraena*) and blackfish (*Gadopsis marmoratus*) in two southeastern Australian rivers. The models were built using an expert consultation and elicitation process, and then used to predict the probable outcomes of different environmental water regimes.

14.5 COMPARING MODELING APPROACHES

14.5.1 CLASSIFYING MODELING APPROACHES: DATA REQUIREMENTS VERSUS NEED FOR KNOWLEDGE OF ECOLOGICAL PROCESSES

The range of methods outlined above can be loosely classified according to the need for ecological expertise for specifying the model structure, and the amount and type of data that are needed or can be employed (Fig. 14.5). Again, we do not expect that all readers will agree with this classification, but it serves as a useful tool for raising awareness of the needs of different types of models. For a more detailed classification of the strengths and weaknesses of several ML methods versus linear models, see Olden et al. (2008).

Of course, no modeling approaches are found in the bottom left corner of Fig. 14.5—models that would require neither knowledge of ecological process nor extensive data sets.

At the top left of the graph lie the ML approaches. These approaches are appealing because they allow for a largely data-driven outcome, with less need for a priori knowledge of model structure. Moreover, a model developed using one of these approaches is guaranteed to be no worse in its fit than any other approach (but more probably superior), if fitted to the same data set (Elith et al., 2006). However, some ML approaches rely on having large sets of empirical data (e.g., Boosted Regression Trees; Elith et al., 2006). The ability of these approaches to model nonlinear and nonsmooth functions for the relationships between drivers and endpoints is one of their great strengths, but the lack of any assumption (e.g., linearity) regarding the shape of the relationship between two variables makes extrapolation impossible.

Bayesian methods and, in particular, hierarchical models as described above (on the right-hand side of Fig. 14.5), offer tremendous flexibility in model structure (Clark, 2005). This allows models to be built that emulate physical and biological processes, for example instream primary production and ecosystem respiration as part of the *BASE (BAyesian Single-station Estimation)* model (Grace et al., 2015), rather than simply testing for relationships among empirical data sets. However, with this flexibility comes the requirement for extensive knowledge, or at least hypotheses, about the causal linkages that describe the processes that drive the ecological response of interest. Bayesian methods have the ability to assimilate different kinds of data such as those derived from experts or

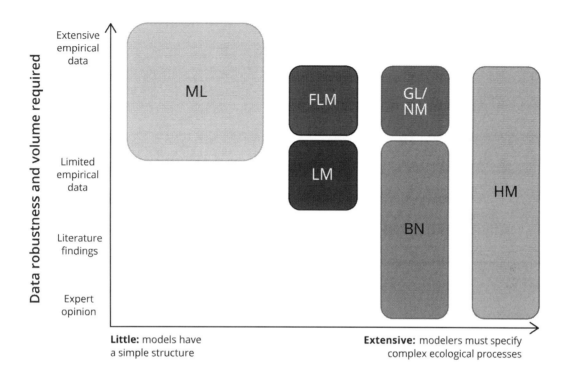

Ecological expertise required for model specification

FIGURE 14.5

How do the modeling approaches outlined differ in their need for ecological expertise in model specification and the amount and type of data required? BN, Bayesian networks; FLM, functional linear models; GL/NM, generalized linear and nonlinear models; HM, hierarchical models; LM, linear models; ML, machine learning.

literature as prior probability distributions to combine with empirical data. Bayesian networks (Pearl, 2000) have a similar ability to directly combine data of different kinds and are especially useful for modeling of relationships based on expert opinions (Shenton et al., 2014). The use of prior information reduces uncertainty, potentially meaning that informative conclusions can be reached with fewer data than alternative approaches.

In the mid-range of the *x*-axis of Fig. 14.5 lie the generalized linear and nonlinear modeling approaches. The reduced requirement for ecological understanding is caused by the fact that these modeling approaches are more restrictive of the types of relationships that can be specified. In general, more empirical data are required to successfully parameterize such relationships, especially for generalized modeling approaches that may seek to improve inferential strength by including data from multiple sampling units (e.g., sites), and must include random effects to account for between-sampling unit differences and to avoid pseudoreplication in the estimates of main effects.

14.5.2 IMPLEMENTING MODELING APPROACHES: THE NEED FOR TECHNICAL EXPERTISE

The modeling approaches could also be classified by the amount of technical expertise required to implement them, and by the requirement for specialist software. These characteristics will affect the degree to which one or more types of modeling are feasible choices for environmental flows assessments.

Approaches such as linear modeling and even generalized linear and nonlinear modeling should be familiar to most ecological specialists employed on environmental flows assessment teams, and also to managers responsible for these assessments. These types of approaches are taught in many undergraduate and postgraduate courses, and can be implemented through many different kinds of statistical software (e.g., Minitab, R, SPSS).

Bayesian network modeling is also an accessible approach. The basics of modeling may be taught within a short (1–3 days) course, and do not require any specialized mathematical knowledge. Software is required to implement BNs, with the most popular package being Netica by Norsys. This is an inexpensive software by commercial standards, but could still be a budgetary consideration for an environmental flows assessment.

ML approaches pose a greater degree of difficulty. They can be implemented using free software libraries through the R statistical language that are increasingly widely available and make implementation straightforward. However, the rarity of these approaches means that few people are versed in the methods and their implementation in software (Olden et al., 2008). Thus, specification of an ML model requires specialist knowledge that is not found in most environmental flows assessment teams or management agencies.

Similarly, the highly mathematical statistical approaches such as Bayesian hierarchical models and functional linear models are, as yet, the domain of a small number of specialists. Like the ML-based approaches, specialist software is available on free platforms. However, reasonably high-level coding skills are required to implement the models.

Does this mean that the more technical modeling approaches described here are not relevant to environmental flows assessment as currently carried out? For reasons outlined below, we do not believe so. Nearly all environmental flows assessments already make use of modeling specialists for hydrologic and hydraulic modeling. It would therefore only be a slight extension of current practice to engage ecological modeling specialists responsible for implementing more advanced modeling approaches to aid the assessment. Such individuals would be additional to the ecological specialists already engaged on environmental flows assessment teams; their primary job would be to translate the ecological specialist's knowledge into relationships and models that could inform the assessment. Such an approach would add a little more complexity (and therefore expense) to the current practice of environmental flows assessment (see Fig. 14.6 in Box 14.1), but it is a feasible pathway to taking advantage of the latest research developments in ecological response modeling to inform the assessments.

However, the wisdom of investing effort into developing statistical models to guide the environmental flows assessment, and indeed the choice of model type, will depend upon the amount of data available relevant to the system being assessed. Data specific to the study system will often be rare or completely absent, and few environmental flows assessments have the time or resources to collect new data to develop models. However, data from comparable systems may be available. As outlined in Section 14.1, there is often an assumption that no data exist to inform environmental flows assessments. Against this, *Ecological Limits of Hydrological Alternation (ELOHA)* assessments

BOX 14.1 EMBEDDING STATISTICAL MODEL SELECTION AND MODIFICATION INTO AN ENVIRONMENTAL FLOWS ASSESSMENT: EXAMPLE USING THE FLOWS METHOD

The FLOWS method (DEPI, 2013) is the most common approach used for determining environmental flows requirements in the state of Victoria, Australia. It is a Building Block method (sensu King and Louw, 1998), and hence

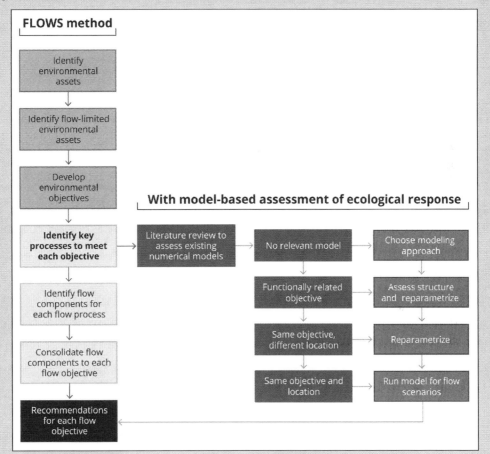

FIGURE 14.6

Incorporating ecological response models into the FLOWS method. The first column of the figure has been reproduced from the FLOWS method manual. The top three boxes present the process of identifying important ecological endpoints; the next three boxes involve the process of determining their responses to flow; and the final box is the output of environmental flows assessment—water regime recommendations to meet the desired ecological objectives. Here, we suggest an alternative path for determining the ecological responses, adding formal assessment of models in the literature (columns 2 and 3), and subsequent modeling (the final column), to improve the rigor of the environmental water regime recommendations produced.

Source: DEPI (2013). © State of Victoria, Department of Environment Land Water and Planning 2015, www.delwp.vic.gov.au

(Continued)

BOX 14.1 (CONTINUED)

shares a great deal of features with many other approaches to environmental flows assessment being used around the world (see Tharme, 2003 and Chapter 11). FLOWS is a panel-based process that employs a stakeholder-driven approach to identifying ecological assets (e.g., species) of priority for conservation/restoration, uses hydrologic and hydraulic models to assess the provision of physical habitat under different discharge scenarios, and then uses largely expert-based assessments of what the ecological response is likely to be to different habitats and flow regimes. The method has been implemented successfully for dozens of environmental flows assessments in Victoria (e.g., Cottingham et al., 2014; EarthTech, 2003), and is well-accepted by stakeholders. It has also been applied in an international setting (Bond et al., 2012).

Of most relevance to this chapter is the process used to predict ecological responses to different flow regimes, which ultimately informs the environmental water regime recommendations. This is illustrated in the left column of Fig. 14.6 (modified from DEPI, 2013). The method is not prescriptive about the process used to "identify key flow processes to meet each objective," and this is where we believe that ecological response models could be used to improve the rigor of environmental water regime recommendations.

The right-hand side of Fig. 14.6 illustrates how we believe the process outlined above (Section 14.6.1) could be used within the environmental flows assessment process. The *identify key processes* box of the current FLOWS specification involves the development of a preliminary *conceptual ecological model (CEM)*. The literature is then reviewed (columns 2 and 3) to assess the extent to which statistical models already exist that capture the processes embedded in the CEMs. Depending on the degree to which models and data already exist, related to the ecological objectives and location under consideration, different levels of effort need to be invested in the modeling process outlined in Section 14.6 (final column). The final parameterized model can then be used to set the flow recommendations for each ecological objective. This may range from a data-driven model based upon relevant empirical data, to (if models and data are lacking for the endpoint being assessed) an expert-parameterized model (e.g., Bayesian network).

For example, an environmental flows assessment seeking to manage species richness of riparian vegetation could employ the relationship developed by Arthington et al. (2012) and displayed in Fig. 14.2A to set operational recommendations for interannual flow variability. If the assessment were being carried out in southeast Queensland, the relationship could be used in its current form. If being undertaken elsewhere, additional data may be needed to reparameterize the inverse quadratic relationship.

We have used FLOWS as an example here in order to use the workflow diagram, but the process described above should be compatible with any Building Block method, and indeed is already implicitly specified within some methods (e.g., ELOHA; Poff et al., 2010). This compatibility is important because it means that adopting the process would not disrupt existing successful methods for setting environmental water regimes. It would increase the amount of work required to identify the responses of priority environmental endpoints to flow variation, and possibly also require the involvement of ecological modeling specialists in the environmental flows assessment panel (see Section 14.5.2), but we feel that the improvement in rigor and transparency/repeatability of the environmental flows assessment would justify this extra investment in many instances.

(Poff et al., 2010) require data to develop their flow–response relationships, and the increasing number of ELOHA case studies (e.g., Arthington et al., 2012; Kendy et al., 2012; McManamay et al., 2013) being published puts lie to this assumption. In the case that no relevant data are truly available, expert-driven models (e.g., BNs) may still be employed. These considerations feed into the process outlined below for arriving at a parsimonious model to inform environmental flows assessment.

14.6 ARRIVING AT A PARSIMONIOUS MODEL
14.6.1 IDENTIFYING EXISTING MODELS OR GAPS

It is not possible to be entirely prescriptive about the types of models that will be useful for environmental flows assessments. Model complexity and components will necessarily be context-dependent. Instead, we can propose a process through which a parsimonious model can be: identified from existing research, modified from such research so that it matches the intended purpose, or built from scratch.

Conceptual ecological models (CEMs) specify our a priori knowledge or beliefs of what we consider to be the most important hydrologic factors driving ecological responses. Here, we consider CEMs as node-link graphical models linking system variables via functional pathways (e.g., Fig. 14.7). The linkages in such a CEM can be thought of as a series of testable hypotheses, the relationships among which can then be quantified using data or some other source of information. These conceptual models underpin statistical models. Other types of CEMs may be more difficult to translate into statistical hypotheses (e.g., written descriptions, or spatial *mud maps* of systems), but these models still portray functional linkages between system components. The most important consideration for parsimony is restricting the model to include only the most important variables.

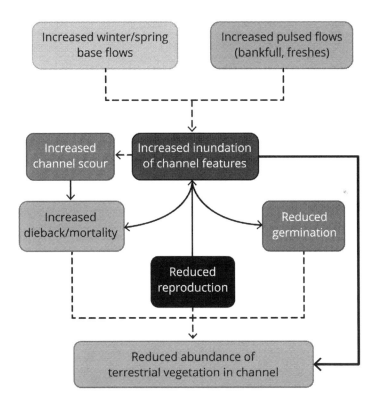

FIGURE 14.7

Conceptual ecological model linking model drivers (flow components, top row) to proximate stressors (second row) to ecological processes (third row) and an ultimate ecological outcome (bottom cell). The linkages between nodes can be tested and quantified by the type of statistical models outlined in this chapter.

Source: Reprinted from Miller et al. (2013)

However, our experience in workshops building CEMs is that there is a tendency to include many variables, making any resulting statistical model too complex to use and difficult or impossible to parameterize.

It is our contention that far too often, CEMs are built from scratch with little reference to existing models or the scientific literature. In an environmental flows assessment, the CEMs often derive from the knowledge of expert panel members. This constant reinvention of the (water) wheel misses the opportunity to leverage existing knowledge and research, and can lead to a proliferation of CEMs that all purport to address the same (or at least similar) issues.

We believe that it is necessary for environmental flows assessments to develop preliminary conceptual models based upon the considerations and context for the system being assessed. However, once developed, these models should be assessed against the existing literature to determine whether important driving variables have been missed, or proposed driving variables may be less important than initially believed. A CEM does not necessarily have to conform to those used previously, but there should be sound system- or application-specific reasons for differences.

At this point, the CEM can be assessed against existing statistical models to determine the extent to which existing models may be useful to inform the environmental flows assessment. There are four potential outcomes at this stage:

1. A suitable statistical model exists for the species or processes in question and in the system being studied. This can be used without modification. Such an outcome is likely to be extremely rare, but is obviously very welcome.
2. A statistical model exists for the species or processes in question, but not for the system being studied. The model structure may be able to be retained without modification, but reparameterization of the model will be required to account for the different geomorphic and climatic conditions (e.g., the relationship between discharge and extent of riverine habitat, or differences in the seasonal patterns of flow) that characterize the system being assessed.
3. A statistical model exists for a functionally related species. The model structure may need to be modified, but the core elements of it can be retained, such as the principal driving responses. Reparameterization will be necessary.
4. No statistical model exists that can be suitably modified. This is probably the norm. To employ an ecological response model in the environmental flows assessment, practitioners need to select an appropriate modeling approach to translate their CEM into a statistical model or series of models with predictive power.

Box 14.1 illustrates how the above process could fit within an environmental flows assessment compared to existing practice that typically relies on expert-based interpretations of ecological responses to the predictions of hydraulic-habitat models. For Outcomes 2–4, it will also be necessary to assess the amount of relevant data available to parameterize any model modified or developed. In an extremely data-poor environment, expert-driven models may be required.

14.6.2 **IMPROVING MODEL PARSIMONY OVER TIME**

It should be recognized that models are not static entities, and that they also evolve over time as new information and knowledge become available. There may also be an evolution of model parsimony over time, with improved models being used in future environmental flows assessments. Ward (2008) describes a *simplicity cycle*, whereby in first tackling a problem, a simple solution is proposed that does not perform as well as desired. Over time, the system is refined and builds in complexity to the point where it performs well, but is complex and potentially difficult to use. However, as understanding improves, opportunities to remove complexity without compromising performance emerge (Fig. 14.8). The eventual result is an elegant solution of *requisite simplicity* (Rogers, 2007) that could not have emerged without taking the full journey from naïve simplicity, and via informed complexity.

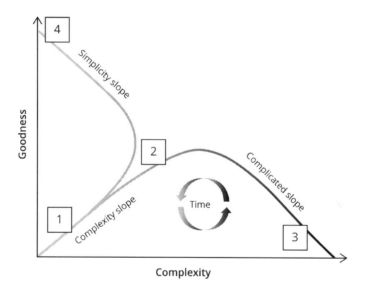

FIGURE 14.8

The Simplicity Cycle. Solutions to problems may be classified by their degree of complexity versus their utility for solving a problem (Goodness). Over time, solutions to problems—here we are discussing models of ecological responses to flow variation—proceed from the Simplistic (1) to the region of Complexity (2). At that point, solutions may proceed to the unnecessarily Complicated (3), or the Simple (4), where elegant solutions reduce Complexity while maximizing Goodness.

Source: Modified from Ward (2005)

We believe that these concepts apply equally to models of environmental responses to flow variation. For example, many analyses attempt to relate ecological responses to simple flow metrics such as changes in flow magnitude; 71 of 165 studies analyzed by Poff and Zimmerman (2010) and Webb et al. (2013) examined ecological effects of changes in flow magnitude (naïve simplicity). Over time, researchers have developed diverse flow metrics, often aimed at slightly different characterizations of the flow regime, to attempt to better explain ecological responses (informed complexity; e.g., Marsh et al., 2012). However, other analyses have shown that the vast majority of variation in flow regimes can be captured with just a few of these metrics (Olden and Poff, 2003), allowing an elegantly simple approach to statistical analysis (requisite simplicity).

14.7 UPTAKE

As described in the introduction to this chapter, there are as yet few examples where statistical flow—response models have been directly used in environmental flows assessments (see Kendy et al., 2012). Below, we explore potential reasons for this, and possible ways forward.

There are many statistical models relating ecological responses to changes in flow regimes published in the peer-reviewed literature (e.g., King et al., 2016; Shafroth et al., 2010; Webb et al., 2010b). Why are these models not being used in environmental flows assessments? The answer is that while some of them are being used (e.g., McManamay et al., 2013; Overton et al., 2014; Shafroth et al., 2010), many are research studies on models and methods that *could* be used to guide management, but have yet to be taken up.

Above, we discuss the overwhelming need for flow—ecological response models to be parsimonious for them to be useful for environmental flows assessments. Simple models that describe sufficient variation in ecological endpoints will generally be more useful for management than complex models. However, for publication, the opposite is true. All journals, and especially the higher impact journals, stress the need for contributions to be novel and innovative. In describing ecological responses to changing flow regimes, this translates to models that use new statistical approaches known to few practitioners, combine data in new ways not previously seen, or develop an intricate understanding of the complex causal pathways underpinning ecological responses. There is thus a mismatch between the modeling needs of environmental flows assessments, and the drive to explore new approaches and get novel ecological models published.

One environmental flows assessment framework that is designed to produce and make use of statistical (and other) flow—ecological response models is the ELOHA approach (see Chapter 11). Within this flexible framework, flow—ecological response models can be compiled from existing data, or new data collected along a flow regulation gradient, and tested statistically to determine the form (e.g., threshold, linear) and degree of ecological change (positive or negative) associated with a particular type of flow alteration (Arthington et al., 2006). The expectation is that robust flow—ecological response models developed for different hydrologic classes of river can be transferred to less well-studied rivers of the same class, and used to guide environmental water release decisions for members of each distinctive river type (Poff et al., 2010).

In line with the wide range of methods for deriving flow—response models outlined in this chapter, methods being used in ELOHA assessments are also diversifying. Kendy et al. (2012) provide a useful summary of US case studies and methods for mining existing data sets to produce ecological response models. Several of these studies applied quantile regression to isolate the potential response of fish, invertebrates, and riparian vegetation to changes in selected flow parameters (using Blossom statistical software; Cade and Richards, 2005). For example, Wilding and

Poff (2009) applied quantile regression to develop relationships between percentage change in riparian vegetation (relative to reference condition) and percentage change in peak flow magnitude (via sediment supply, disturbance, seedling establishment, etc.), and a relationship for biomass of flannelmouth sucker to peak flow magnitude.

There is no reason why the approaches employed in these ELOHA studies cannot also be applied within other environmental flows assessment frameworks (Box 14.1). The difference is perhaps the emphasis that ELOHA places upon deriving statistical flow–ecological response models for rivers of particular hydrological and geomorphic character, compared to other frameworks. The successes above demonstrate our contention at the start of this chapter that we often do now have sufficient empirical data to derive useful ecological models. Where we do not have sufficient ecological data, the ELOHA approach can be used to guide the collection of useful *reference* data and dam-impacted data on a wide range of ecological response variables (see Table 2 in Poff et al., 2010). Alternatively, we can employ expert-based models, as is customary in applying the DRIFT framework (King et al., 2003).

Consistent with our observation above that the type of models that will be useful for environmental flows assessments do not often find their way into the peer-reviewed literature, many of the models being used in ELOHA-based assessments are generated specifically for their context, and are either not published beyond the gray literature (e.g., Arthington et al., 2012; Wilding and Poff, 2009), or publication is delayed. As applications of ELOHA increase, useful publications are increasing. McManamay et al. (2013) published an ELOHA study in the upper Tennessee River basin, which included a summary of the flow–ecological response models developed, followed by outcomes from the implementation and monitoring of flow restoration recommendations. Similarly, Arthington et al. (2014) and Rolls and Arthington (2014) published a collection of models for fish responses to flow patterns and flow regulation in subtropical Queensland.

This type of compendium publication may offer the best way forward for dissemination of parsimonious flow–ecological response models in the peer-reviewed literature. In these cases, the models were not the focus of the papers, but how they had been formulated, populated, and could be applied to guide flow management. Further publications of this type, especially of collections of models that may be readily extrapolated beyond the case study for which they were derived, offer the best chance for rapid proliferation of models useful for environmental flows assessment.

This proliferation will be an important step toward increasing uptake of these approaches because it will mean that environmental flows assessment teams do not need to develop as many models from scratch. Currently, models are mostly confined to the gray literature, where they may be poorly documented (i.e., the underpinning conceptual model and/or rationale for different choices is not easily discoverable). Better documentation and dissemination will greatly reduce the workload associated with using flow–ecological response models in environmental flows assessments (see also Chapter 25).

Beyond this, uptake is a question of change of attitude and practice. Existing methods have served us well, and thus have a great deal of inertia associated with them. A change in practice to more fully embrace flow–ecological response modeling in environmental flows assessment will only come about through sustained demonstration of the benefits of doing so through case studies such as those outlined above, and through the involvement of managers and regulators able to see the benefits of applying modeling approaches within the scope of their activities.

14.8 BUILDING KNOWLEDGE AND IMPROVING ECOLOGICAL RESPONSE MODELS

14.8.1 ADAPTIVE LEARNING, THE IMPORTANCE OF MAINTAINING CLOSE LINKS TO RESEARCH

The ultimate goal of the development of statistical ecological models in environmental flows assessments is a capacity to make quantitative predictions of ecological response to a variety of potential flow scenarios. These predictions must include estimates of their uncertainty, and should also be able to be made for rivers that have not had extensive histories of research or monitoring (Poff et al., 2010). This is a major challenge, but is already within our reach for some endpoints (Box 14.2).

Over time, we expect to see an improvement in ecological modeling capacity. This will be driven by the acquisition of more and better data on responses to environmental water regimes, both through targeted research and through monitoring programs in well-described contexts; through improved understanding of ecological processes, driven by research efforts; and through improved modeling approaches such as some of the more recently developed approaches described above (Section 14.4).

These improvements in capacity and consequent predictive capability will come about through a cycle of adaptive learning embedded within the adaptive management cycle for environmental water (Chapter 25). Within this cycle, statistical ecological models will be used to make predictions of ecological outcomes to various environmental water release scenarios (management decision), one or more of which is then implemented. Monitoring and evaluation of the ecological outcomes of the environmental water release decisions will improve understanding and lead to either minor (e.g., parameter update) or major (e.g., model structure changed) updates of the model. These continuously updated models will be available for practitioners to use in more refined environmental flows assessments (Section 14.6.2).

For this optimistic future to be realized, it will require strong engagement of the research community with managers responsible for conducting environmental flows assessments. The research community must be actively engaged to ensure that the outcomes of monitoring and evaluation can be used to improve models of ecological response to flow variation (see Poff et al., 2003; Shafroth et al., 2010). A failure to engage will perpetuate the existing processes for environmental flows assessment that often isolate the implementation phase from the research process, and in doing so fail to take advantage of adaptive learning and heightened understanding of ecosystem responses over time. The existing approaches have served us well over many years, but we feel can be improved by the greater use of ecological modeling within many environmental flows assessments.

14.8.2 WHAT MIGHT THE FUTURE HOLD?

Some of the more recent developments in ecological modeling that we have discussed (e.g., functional linear models) demonstrate that the field is evolving rapidly. New approaches are likely to be proposed on a regular basis. Although it is impossible to predict with certainty what the field of ecological response modeling for environmental water regimes will look like in 20 or even 10 years, we believe that two features will become more prominent in the near future.

BOX 14.2 QUANTITATIVE PREDICTIVE CAPACITY FOR ENVIRONMENTAL FLOWS ASSESSMENTS: MODELING EFFECTS OF INUNDATION ON ENCROACHMENT OF TERRESTRIAL VEGETATION INTO REGULATED RIVER CHANNELS

Webb et al. (2015) describe a multistage process used to model the effects of inundation on the encroachment of terrestrial vegetation into regulated river channels. The process was established to make best use of all sources of information that can be brought to bear on the issue of detecting and predicting the effects of flow variation. The process employed the following steps:

1. A systematic literature review to develop an evidence-based conceptual model of the main driving processes for terrestrial vegetation response to inundation (Miller et al., 2013).
2. A formal expert elicitation process (see also Chapter 15) to provide a prior quantification of the links within the conceptual model (de Little et al., 2012).
3. Collection of data through a large-scale coordinated monitoring program (Webb et al., 2010a, 2014). See Chapter 25 for general principles relating to monitoring.
4. Translation of the conceptual model structure into a statistical model structure for analyzing data (de Little et al., 2013).
5. Development of a hierarchical Bayesian model of this structure that combined the expert-based priors with the monitoring data (de Little et al., 2013).

The resulting model had the capacity to make predictions (Fig. 14.9) of average terrestrial vegetation cover for different inundation scenarios (days inundated, frequency of inundation, and season of inundation), and at three different scales (site, river, and region) likely to be of greatest relevance to different stakeholders (individual landowner, local catchment management authority, and state river manager, respectively). The model also had the capacity to make predictions (but with greater uncertainties) for any site or river with the necessary hydrological (and other driving) data, but for which vegetation data had not been collected.

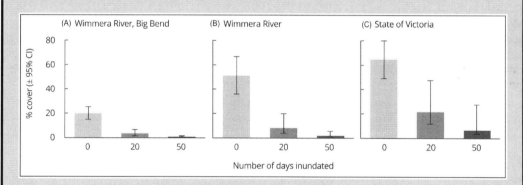

FIGURE 14.9

Prediction of terrestrial vegetation cover under three different inundation scenarios (0, 20, 50 days), with inundation being a single continuous inundation event, and with all inundation occurring during the water year "summer" (November–April). Bars show median predicted cover, with the error bars encompassing 95% of the credible interval of the estimate. Scale of the y-axis is the same for each panel (0–80%). Results are presented at three scales: (A) one site in the Wimmera River, western Victoria, Australia; (B) the Wimmera River as a whole; and (C) the state of Victoria as a whole.

The results illustrate that, for some environmental endpoints, we have the capacity to make robust quantitative predictions that can inform environmental water assessments. The process used in this case was quite involved and intensive; however, we were able to show that each of the processes contributed to improved precision of posterior estimates of terrestrial vegetation cover (i.e., lower uncertainty), underscoring the value of all the components.

First, hydraulic modeling has increased in sophistication over recent years. Although many environmental flows assessments are still informed by the outputs from one-dimensional hydraulic models (e.g., *Hydrologic Engineering Center's River Analysis System [HEC-RAS];* http://www.hec.usace.army.mil/software/hec-ras/), the use of two- and three-dimensional hydraulic models to estimate hydraulic-habitat is increasing, with an accompanying greater level of sophistication (Shafroth et al., 2010; Webb et al., 2016). We believe that ecological models will begin to take more advantage of these developments in hydraulic modeling, moving away from the current dominance of using static hydrologic metrics to describe flow regimes. After 20 years of interaction of the two separate disciplines of ecology and hydraulics, the interdisciplinary field of ecohydraulics is increasing in profile, and is beginning to develop general principles to guide future research (Nestler et al., 2016a; Nestler et al., 2016b). Although we hold to our primary contention of this chapter, that models must be parsimonious, we see the closer ties between hydrology, hydraulics, and ecology as being a major area of development.

Second, we are moving into the era of *big data* for many scientific disciplines (Frankel and Reid, 2008), and there is no reason to expect that environmental water research will not be among these. Hydrologic data, converted via models to hydraulic data, may soon be collected via remote sensing, providing continuous descriptions of flow regimes and habitat along rivers, rather than just at selected gauging points. Similarly, data on ecological variables may become far more diverse, abundant, and informative, through approaches such as remote sensing, environmental genomics, and the measurement of *environmental DNA* (Chariton et al., 2016; Hampton et al., 2013). These sorts of advances could completely reverse the present situation regarding the availability of useful ecological data. From having small data sets, but of high accuracy, at a few points along the river; we could soon be faced with large amounts of spatially contiguous data, although potentially with much greater uncertainty around any data point. This deluge of data may drive ecological response modeling toward more data-driven approaches such as the ML-based approaches (Chariton et al., 2016). Such approaches rely far less heavily on individual scientists and managers having an understanding of ecological processes and the causal links between flow regime attributes and species responses. Rather, the data can be analyzed to speak for themselves.

In this data-rich future, does specialist understanding, experience, and ecological knowledge become an anachronism? We certainly do not believe so. In general, unless there is a clear process being modeled, no models can be used to reliably predict beyond the experience of the data set used to create them. However, for an environmental water delivery program seeking to restore flows to a heavily regulated river, we may need to do just that, for instance predict ecological responses to flow regimes that have not been seen in the system for many years and for which we have no empirical data. Moreover, as we move further into the Anthropocene (sensu Steffen et al., 2007), environments will continue to change under numerous and potentially novel driving forces. We will see more and more *no analogue* or novel ecosystems driven by environmental changes such as climate change (Acreman et al., 2014). This raises the prospect, not only of needing to retain and indeed expand ecological expertise, but also moving beyond the statistical models presented within this chapter to more explicitly embrace process-based representations such as dynamic population models of ecological responses to changing flow regimes (e.g., Jager and Rose, 2003; Shenton et al., 2012). Thus, we can expect more (and more complex) ecological *surprises* and *novel* ecosystems (Acreman et al., 2014; Doak et al., 2008). The increasing amounts of empirical information promised by the era of big data can help us to elucidate the causal pathways that led to previously expected ecological outcomes, but we still require strong basic knowledge of ecological processes to interpret this information (Olden et al., 2006, 2008) and predict alternative futures.

14.9 CONCLUSION

Environmental flows assessments have long made use of quantitative hydrologic and hydraulic models to predict the physical habitat created by flow releases. In contrast, ecological responses to flow and habitat have largely been predicted, based upon expert opinion. However, huge amounts of published research on ecological responses to changing flow regimes, coupled with an ever-expanding suite of modeling tools, means that we are now well able to use statistical and other types of models to predict ecological trends and outcomes, and to guide environmental flows assessments. The increased use of ecological models for making predictions of ecological responses to changes in flow regime holds the promise of improving the rigor, transparency, and repeatability of the recommendations from environmental flows assessments.

In this chapter, we have outlined a process by which practitioners may locate and adapt existing statistical models for use in flow assessments. We have also outlined several of the major groups of modeling tools available in the likely event that no model exists, and classified these tools by their needs for data and ecological expertise (Box 14.3).

BOX 14.3 IN A NUTSHELL

1. Environmental flows assessments have not generally used statistical ecological models of response to changing flow regimes. Rather, ecological responses have been predicted using expert opinion to elicit anticipated ecological responses to changes in flow and habitat. Statistical models can provide greater robustness of predictions and confidence in environmental water regime recommendations.
2. Large (and increasing) amounts of empirical research are being published concerning ecological responses to flow variability and changing flow regimes. Greater use can be made of this resource.
3. Environmental flows assessments may be able to employ existing models of ecological response, either without change, or with reparameterization and/or structural changes.
4. However, often there will be no existing models, and environmental flows assessment teams will need to develop models for each study system. There is a wide variety of approaches available for modeling ecological responses to changes in flow regime.
5. For model reparameterization or development, data relevant to the case study system are required. If no data are available, an expert-derived model may be able to be used.
6. Modeling methods vary in the amount and type of data needed to run them, the level of ecological expertise required to compose and populate the models, and the level of technical expertise required to construct them. There should be a more prominent role for ecological modeling specialists in future environmental flows assessments.
7. We are as yet short of a general ability to make quantitative predictions concerning ecological responses to changing flow regimes, but the tools and data exist for the much greater uptake of statistical ecological modeling into environmental flows assessments.

Although we are some way from achieving the goal of having the ability to make robust quantitative predictions of all ecological responses to changing flow regimes, the models and principles outlined above demonstrate that ecological response modeling is sufficiently advanced to be incorporated directly into many environmental flows assessments. Just as important will be processes by which those models are incorporated into decision making, both for the environmental flows assessment process (Chapter 11) and subsequent adaptive management of environmental water in light of changing knowledge (Chapter 25). Simple or complex modeling techniques may be useful to make predictions of ecological responses to restored flow regimes, but practitioners must be able to communicate those predictions so that they can inform decisions by those who ultimately decide how to deliver environmental water. Equally important is that implementation is followed by monitoring to document the success or otherwise of flow regime restoration based on quantitative flow–ecology response models (e.g., McManamay et al., 2013; Shafroth et al., 2010).

REFERENCES

Acreman, M., Arthington, A.H., Colloff, M.J., Couch, C., Crossman, N.D., Dyer, F., et al., 2014. Environmental flows for natural, hybrid, and novel riverine ecosystems in a changing world. Front. Ecol. Environ. 12, 466–473.

Ainsworth, L.M., Routledge, R., Cao, J., 2011. Functional data analysis in ecosystem research: the decline of Oweekeno Lake Sockeye Salmon and Wannock River flow. J. Agric. Biol. Environ. Stat. 16, 282–300.

Arthington, A.H., 2012. Environmental Flows: Saving Rivers in the Third Millenium. University of California Press, Berkely and Los Angeles, CA.

Arthington, A.H., Bunn, S.E., Poff, N.L., Naiman, R.J., 2006. The challenge of providing environmental flow rules to sustain river ecosystems. Ecol. Appl. 16, 1311–1318.

Arthington, A.H., Mackay, S.J., James, C.S., Rolls, R.J., Sternberg, D., Barnes, A., et al., 2012. Ecological-Limits-of-Hydrologic-Alteration: A Test of the ELOHA Framework in South-East Queensland. National Water Commission, Canberra.

Arthington, A.H., Rolls, R.J., Sternberg, D., Mackay, S.J., James, C.S., 2014. Fish assemblages in subtropical rivers: low-flow hydrology dominates hydro-ecological relationships. Hydrol. Sci. J. 59, 594–604.

Bolker, B.M., 2008. Ecological Models and Data in R. Princeton University Press, Princeton.

Bond, N., Gippel, C., Catford, J., Lishi, L., Hao, L., Bin, L., et al., 2012. River Health and Environmental Flow in China Project: Preliminary Environmental Flows Assessment in the Li River. International Water Centre, Brisbane.

Bunn, S.E., Arthington, A.H., 2002. Basic principles and ecological consequences of altered flow regimes for aquatic biodiversity. Environ. Manage. 30, 492–507.

Bunn, S.E., Davies, P.M., Winning, M., 2003. Sources of organic carbon supporting the food web of an arid zone floodplain river. Freshw. Biol. 48, 619–635.

Burnham, K.P., Anderson, D.R., 2002. Model Selection and Multimodel Inference. Second edition. Springer, New York.

Cade, B.S., Richards, J.D., 2005. User manual for Blossom statistical software. U.S. Geological Survey.

Chariton, A., Sun, M., Gibson, J., Webb, J.A., Leung, K.M.Y., Hickey, C.W., et al., 2016. Emergent technologies and analytical approaches for understanding the effects of multiple stressors in aquatic environments. Mar. Freshw. Res. 67, 414–428.

Clark, J.S., 2005. Why environmental scientists are becoming Bayesians. Ecol. Lett. 8, 2–14.

Colloff, M.J., Baldwin, D.S., 2010. Resilience of floodplain ecosystems in a semi-arid environment. The Rangeland J. 32, 305−314.

Cottingham, P., Brown, P., Lyon, J., Pettigrove, V., Roberts, J., Vietz, G., et al., 2014. Mid Goulburn River FLOWS Study—Final Report: Flow Recommendations. Peter Cottingham and Associates, Melbourne.

Davies, P.M., Naiman, R.J., Warfe, D.M., Pettit, N.E., Arthington, A.H., Bunn, S.E., 2014. Flow-ecology relationships: closing the loop on effective environmental flows. Mar. Freshw. Res. 65, 133−141.

DEPI, 2013. FLOWS—A Method for Determining Environmental Water Requirements in Victoria. Second edition. Department of Environment and Primary Industries, Melbourne.

de Little, S.C., Webb, J.A., Patulny, L., Miller, K.A., Stewardson, M.J., 2012. Novel methodology for detecting ecological responses to environmental flow regimes: using causal criteria analysis and expert elicitation to examine the effects of different flow regimes on terrestrial vegetation. Proceedings of the 9th International Symposium on Ecohydraulics. International Association for Hydro-Environmental Engineering and Research (IAHR), Vienna, Austria, paper 13886−2.

de Little, S.C., Webb, J.A., Miller, K.A., Rutherfurd, I.D., Stewardson, M.J., 2013. Using Bayesian hierarchical models to measure and predict the effectiveness of environmental flows at the site, river and regional scales. Proceedings of MODSIM2013, 20th International Congress on Modelling and Simulation. Modelling and Simulation Society of Australia and New Zealand, Adelaide, pp. 359−365.

Doak, D.F., Estes, J.A., Halpern, B.S., Jacob, U., Lindberg, D.R., Lovvorn, J., et al., 2008. Understanding and predicting ecological dynamics: Are major surprises inevitable? Ecology 89, 952−961.

Driver, P., Chowdhury, S., Wettin, P., Jones, H., 2005. Models to predict the effects of environmental flow releases on wetland inundation and the success of colonial bird breeding in the Lachlan River, NSW. Proceedings of the 4th Australian Stream Management Conference: Linking Rivers to Landscapes. Tasmanian Department of Primary Industries, Water and Environment, Launceston, Tasmania, pp. 192−198.

EarthTech., 2003. Thomson River environmental flow requirements & options to manage flow stress. Report to West Gippsland Catchment Management Authority, Dept. of Sustainability and Environment, Melbourne Water Corporation and Southern Rural Water, Earth Tech Engineering Pty Ltd, Melbourne, Australia.

Elith, J., Graham, C.H., Anderson, R.P., Dudik, M., Ferrier, S., Guisan, A., et al., 2006. Novel methods improve prediction of species' distributions from occurrence data. Ecography 29, 129−151.

Frankel, F., Reid, R., 2008. Big data: distilling meaning from data. Nature 455, 30.

Giam, X., Olden, J.D., 2015. A new R-2-based metric to shed greater insight on variable importance in artificial neural networks. Ecol. Model. 313, 307−313.

Gillespie, B.R., Desmet, S., Kay, P., Tillotson, M.R., Brown, L.E., 2015. A critical analysis of regulated river ecosystem responses to managed environmental flows from reservoirs. Freshw. Biol. 60, 410−425.

Gelman, A., Rubin, J.B., Stern, H.S., Rubin, D.B., 2004. Bayesian Data Analysis. Chapman & Hall/CRC, Boca Raton, FL.

Grace, M.R., Gilling, D.P., Hladyz, S., Caron, V., Thompson, R.M., Mac Nally, R., 2015. Fast processing of diel oxygen curves: estimating stream betabolism with BASE (BAyesian Single-station Estimation). Limnol. Oceanogr. Methods 13, 103−114.

Hampton, S., Strasser, C., Tewksbury, J., Gram, W., Budden, A., Batcheller, A., et al., 2013. Big data and the future of ecology. Front. Ecol. Environ. 11, 156−162.

Horne, A., Kaur, S., Szemis, J., Costa, A., Webb, J.A., Nathan, R., 2017. Using optimization to develop a "designer" environmental flow regime. Env. Model. Soft. 88, 188−199.

Jager, H.I., Rose, K.A., 2003. Designing optimal flow patterns for fall Chinook salmon in a central valley, California, river. N. Am. J. Fish. Manage. 23, 1−21.

Kendy, E., Apse, C., Blann, K., 2012. A Practical Guide to Environmental Flows for Policy and Planning. The Nature Conservancy, Arlington, VA.

Kennard, M.J., Olden, J.D., Arthington, A.H., Pusey, B.J., Poff, N.L., 2007. Multiscale effects of flow regime and habitat and their interaction on fish assemblage structure in eastern Australia. Can. J. Fish. Aquat. Sci. 64, 1346–1359.

Kennard, M.J., Mackay, S.J., Pusey, B.J., Olden, J.D., Marsh, N., 2010. Quantifying uncertainty in estimation of hydrologic metrics for ecohydrological studies. River Res. Appl. 26, 137–156.

King, A.J., Tonkin, Z., Mahoney, J., 2009. Environmntal flow enhances native fish spawning and recruitment in the Murray River, Australia. River Res. Appl. 25, 1205–1218.

King, A.J., Gwinn, D.C., Tonkin, Z., Mahoney, J., Raymond, S., Beesley, L., 2016. Using abiotic drivers of fish spawning to inform environmental flow management. J. Appl. Ecol. 53, 34–43.

King, J., Brown, C., 2010. Integrated basin flow assessments: concepts and method development in Africa and South-east Asia. Freshw. Biol. 55, 127–146.

King, J., Louw, D., 1998. Instream flow assessments for regulated rivers in South Africa using the Building Block Methodology. Aquat. Ecosyst. Health Manag. 1, 109–124.

King, J., Brown, C., Sabet, H., 2003. A scenario-based holistic approach to environmental flow assessments for rivers. River Res. Appl. 19, 619–639.

Konrad, C.P., Olden, J.D., Lytle, D.A., Melis, T.S., Schmidt, J.C., Bray, E.N., et al., 2011. Large-scale flow experiments for managing river systems. BioScience 61, 948–959.

Lele, S.R., Dennis, B., Lutscher, F., 2007. Data cloning: easy maximum likelihood estimation for complex ecological models using Bayesian Markov chain Monte Carlo methods. Ecol. Lett. 10, 551–563.

Lytle, D.A., Poff, N.L., 2004. Adaptation to natural flow regimes. Trends Ecol. Evol. 19, 94–100.

MacNally, R., 2000. Regression and model-building in conservation biology, biogeography and ecology: the distinction between – and reconciliation of – 'predictive' and 'explanatory' models. Biodivers. Conserv. 9, 655–671.

Marsh, N., Sheldon, F., Rolls, R., 2012. Synthesis of Case Studies Quantifying Ecological Responses to Low Flows. National Water Commission, Canberra, Australia.

McCarthy, M.A., 2007. Bayesian Methods for Ecology. Cambridge University Press, Cambridge, UK/New York.

McCarthy, M.A., Masters, P., 2005. Profiting from prior information in Bayesian analyses of ecological data. J. Appl. Ecol. 42, 1012–1019.

McCullagh, P., Nelder, J.A., 1989. Generlized Linear Models. second edition Chapman and Hall/CRC Press, London.

McManamay, R.A., Orth, D.J., Dolloff, C.A., Mathews, D.C., 2013. Application of the ELOHA framework to regulated rivers in the upper Tennessee River basin: a case study. Environ. Manage. 51, 1210–1235.

Miller, K.A., Webb, J.A., de Little, S.C., Stewardson, M.J., 2013. Environmental flows can reduce the encroachment of terrestrial vegetation into river channels: a systematic literature review. Environ. Manage. 52, 1201–1212.

Muller, H.G., Stadtmuller, U., 2005. Generalized functional linear models. Ann. Stat. 33, 774–805.

Naiman, R.J., Latterell, J.J., Pettit, N.E., Olden, J.D., 2008. Flow variability and the biophysical vitality of river systems. Compte Rendus Geosci. 340, 629–643.

Naiman, R.J., Alldredge, J.R., Beauchamp, D.A., Bisson, P.A., Congleton, J., Henny, C.J., et al., 2012. Developing a broader scientific foundation for river restoration: Columbia River food webs. Proc. Natl. Acad. Sci. USA. 109, 21201–21207.

Nestler, J.M., Stewardson, M., Gilvear, D., Webb, J.A., Smith, D.L., 2016a. Does ecohydraulics have guiding principles. Proceedings of the 11th International Symposium on Ecohydraulics. The University of Melbourne, Melbourne, Australia, paper 26780.

Nestler, J.M., Stewardson, M.J., Gilvear, D., Webb, J.A. and Smith, D.L., 2016b. Ecohydraulics exemplifies the emerging "paradigm of the interdisciplines". J. Ecohydraul. 1, 5–15.

Olden, J.D., Jackson, D.A., 2002. A comparison of statistical approaches for modelling fish species distributions. Freshw. Biol. 47, 1976−1995.

Olden, J.D., Naiman, R.J., 2010. Incorporating thermal regimes into environmental flows assessments: modifying dam operations to restore freshwater ecosystem integrity. Freshw. Biol. 55, 86−107.

Olden, J.D., Poff, N.L., 2003. Redundancy and the choice of hydrologic indices for characterizing streamflow regimes. River Res. Appl. 19, 101−121.

Olden, J.D., Poff, N.L., Bledsoe, B.P., 2006. Incorporating ecological knowledge into ecoinformatics: An example of modeling hierarchically structured aquatic communities with neural networks. Ecol. Inform. 1, 33−42.

Olden, J.D., Lawler, J.J., Poff, N.L., 2008. Machine learning methods without tears: a primer for ecologists. Q. Rev. Biol. 83, 171−193.

Olden, J.D., Konrad, C.P., Melis, T.S., Kennard, M.J., Freeman, M.C., Mims, M.C., et al., 2014. Are large-scale flow experiments informing the science and management of freshwater ecosystems? Front. Ecol. Environ. 12, 176−185.

Overton, I.C., Pollino, C.A., Roberts, J., Reid, J.R.W., Bond, N.R., McGinness, H.M., et al. 2014. Development of the Murray Darling basin plan SDL adjustment ecological elements method. CSIRO Land and Water Flagship, Adelaide.

Palmer, M.A., Ambrose, R.F., Poff, N.L., 1997. Ecological theory and community restoration ecology. Restor. Ecol. 5, 291−300.

Pearl, J., 2000. Causality: Models, Reasoning, and Inference. Cambridge University Press, Cambridge, UK.

Pinheiro, J., 2011. The nlme package, Version 3.1−98.

Poff, N.L., Zimmerman, J.K.H., 2010. Ecological responses to altered flow regimes: a literature review to inform the science and management of regulated rivers. Freshw. Biol. 55, 194−205.

Poff, N.L., Allan, J.D., Bain, M.B., Karr, J.R., Prestegaard, K.L., Richter, B.D., et al., 1997. The natural flow regime. BioScience 47, 769−784.

Poff, N.L., Allan, J.D., Palmer, M.A., Hart, D.D., Richter, B.D., Arthington, A.H., et al., 2003. River flows and water wars: emerging science for environmental decision making. Front. Ecol. Environ. 1, 298−306.

Poff, N.L., Richter, B.D., Arthington, A.H., Bunn, S.E., Naiman, R.J., Kendy, E., et al., 2010. The ecological limits of hydrologic alteration (ELOHA): a new framework for developing regional environmental flow standards. Freshw. Biol. 55, 147−170.

Ramsay, J.O., Silverman, B.W., 2002. Applied Functional Data Analysis: Methods and Case Studies. Springer, New York.

Rogers, K.H., 2007. Complexity and simplicity: complementary requisites for policy implementation. Oral presentation at the 10th International Riversymposium and Environmental Flows Conference. The Nature Conservancy, Brisbane, Australia.

Rolls, R.J., Arthington, A.H., 2014. How do low magnitudes of hydrologic alteration impact riverine fish populations and assemblage characteristics? Ecol. Indic. 39, 179−188.

Sabo, J.L., Post, D.M., 2008. Quantifying periodic, stochastic, and catastrophic environmental variation. Ecol. Monogr. 78, 19−40.

Shafroth, P.B., Wilcox, A.C., Lytle, D.A., Hickey, J.T., Andersen, D.C., Beauchamp, V.B., et al., 2010. Ecosystem effects of environmental flows: modelling and experimental floods in a dryland river. Freshw. Biol. 55, 68−85.

Shenton, W., Hart, B.T., Chan, T., 2011. Bayesian network models for environmental flow decision-making: 1. Latrobe River Austrlaia. River Res. Appl. 27, 283−296.

Shenton, W., Bond, N.R., Yen, J.D.L., Mac Nally, R., 2012. Putting the "Ecology" into environmental flows: Ecological dynamics and demographic modelling. Environ. Manage. 50, 1−10.

Shenton, W., Hart, B.T., Chan, T.U., 2014. A Bayesian network approach to support environmental flow restoration decisions in the Yarra River, Australia. Stoch. Environ. Risk Assess. 28, 57−65.

Steffen, W., Crutzen, P.J., McNeill, J.R., 2007. The Anthropocene: are humans now overwhelming the great forces of nature. Ambio 36, 614–621.

Stewardson, M.J., Webb, J.A., 2010. Modelling ecological responses to flow alteration: making the most of existing data and knowledge. In: Saintilan, N., Overton, I. (Eds.), Ecosystem Response Modelling in the Murray-Darling Basin. CSIRO Publishing, Melbourne, Australia, pp. 37–49.

Stewart-Koster, B., Olden, J.D., Gido, K.B., 2014. Quantifying flow-ecology relationships with functional linear models. Hydrol. Sci. J. 59, 629–644.

Tennant, D.L., 1976. Instream flow regimens for fish, wildlife, recreation, and related environmental resources. Proceedings of the Symposium and Specialty Conference on Instream Flow Needs, May 3-6. American Fisheries Society, Boise, ID, pp. 359–373.

Tharme, R.E., 2003. A global perspective on environmental flow assessment: emerging trends in the development and application of environmental flow methodologies for rivers. River Res. Appl. 19, 397–441.

Thomas, J.A., Bovee, K.D., 1993. Application and testing of a procedure to evaluate transferability of habitat suitability criteria. Regul. Rivers Res. Manage. 8, 285–294.

Ward D., 2005. The simplicity cycle: simplicity and complexity in design. *Defence AT&L, Defense Acquisitions University,* November-December, 18–21.

Ward, D., 2008. The Simplicity Cycle: An Exploration of the Relationship Between Complexity, Goodness, and Time. Rouge Press.

Webb, A., Casanelia, S., Earl, G., Grace, M., King, E., Koster, W., et al., 2016. Commonwealth Environmental Water Office Long Term Intervention Monitoring Project: Goulburn River Selected Area Evaluation Report 2014-15. University of Melbourne Commercial, Melbourne.

Webb, J.A., Stewardson, M.J., Chee, Y.E., Schreiber, E.S.G., Sharpe, A.K., Jensz, M.C., 2010a. Negotiating the turbulent boundary: the challenges of building a science-management collaboration for landscape-scale monitoring of environmental flows. Mar. Freshw. Res. 61, 798–807.

Webb, J.A., Stewardson, M.J., Koster, W.M., 2010b. Detecting ecological responses to flow variation using Bayesian hierarchical models. Freshw. Biol. 55, 108–126.

Webb, J.A., Miller, K.A., King, E.L., de Little, S.C., Stewardson, M.J., Zimmerman, J.K.H., et al., 2013. Squeezing the most out of existing literature: a systematic re-analysis of published evidence on ecological responses to altered flows. Freshw. Biol. 58, 2439–2451.

Webb, J.A., Miller, K.A., de Little, S.C., Stewardson, M.J., 2014. Overcoming the challenges of monitoring and evaluating environmental flows through science-management partnerships. Int. J. River Basin Manage. 12, 111–121.

Webb, J.A., de Little, S.C., Miller, K.A., Stewardson, M.J., Rutherfurd, I.D., Sharpe, A.K., et al., 2015. A general approach to predicting ecological responses to environmental flows: making best use of the literature, expert knowledge, and monitoring data. River Res. Appl. 31, 505–514.

Wilding, T.K., Poff, N.L., 2009. Flow-Ecology Relationships for the Watershed Flow Evaluation Tool. Colorado Water Conservation Board, Denver, CO.

UNCERTAINTY AND ENVIRONMENTAL WATER

15

Lisa Lowe[1], Joanna Szemis[2], and J. Angus Webb[2]

[1]*Department of Environment, Land, Water, and Planning, Melbourne, VIC, Australia*
[2]*The University of Melbourne, Parkville, VIC, Australia*

15.1 WHY CONSIDER UNCERTAINTY?

There are known knowns. These are things we know that we know. There are known unknowns. That is to say, there are things that we know we don't know. But there are also unknown unknowns. There are things we don't know we don't know.

Donald Rumsfeld (2002)

Uncertainty is an unavoidable characteristic of environmental water management. Decades of research continues to show that the ecological health of our rivers is reliant on sufficient water being available. This is a known known. The greatest challenge arises from the aspects of environmental water management where "we know we don't know." Although much progress has been made in recent years to better understand environmental water requirements, this is not yet something that practitioners understand with certainty (Poff and Zimmerman, 2010).

Uncertainties are widely acknowledged, but they are rarely quantified. Some academic authors have investigated individual sources of uncertainty in environmental flows assessments (e.g., Caldwell et al., 2015; Fu and Guillaume, 2014; Stewardson and Rutherfurd, 2006; Van der Lee et al., 2006; Warmink et al., 2010), but there are no comprehensive assessments of uncertainty. A formal assessment of uncertainty is rarely incorporated into environmental flows assessments and is not a formal step in the methods outlined in Chapter 11. However, better understanding of these uncertainties would provide many benefits.

In situations where there is competition for water, it can be difficult to convince a community to accept tangible economic losses when they cannot be certain that the desired environmental benefits will be achieved (Clark, 2002; Ladson and Argent, 2002). A shared understanding of the uncertainties will allow a more open dialogue between the various stakeholders, whereas ignoring uncertainty can diminish public trust (Ascough et al., 2008).

Understanding and considering uncertainty can also improve the operational decisions made by environmental water managers. For example, these managers may make judgments about how much water to release from a reservoir to achieve a desired environmental benefit. If they release too little water the environmental benefit may not be realized and the water will have been wasted; however, it will also be wasted if more water is released than is needed (Stewardson and Rutherfurd, 2006).

Water for the Environment. DOI: http://dx.doi.org/10.1016/B978-0-12-803907-6.00015-2

A better understanding of the uncertainty in the environmental water requirements will allow the manager to assess the probability that a flow of a given magnitude will provide the environmental benefit, and compare this with the benefit of using the water for another purpose.

The most obvious way to reduce uncertainty is to increase monitoring and research efforts. Again, a better understanding of uncertainty is beneficial. Funding for monitoring and research is limited. Identifying the major sources of uncertainty and considering how these influence management decisions can guide future research and monitoring investments to most effectively improve our knowledge. However, even with increased research and monitoring efforts, uncertainty will remain.

The purpose of this chapter is to provide practitioners who are not involved in undertaking uncertainty assessments with an appreciation for the concepts involved and show how they relate to water for the environment. It provides a high-level overview of how to assess uncertainty in the context of environmental water. Determining and addressing all uncertainties is well beyond what can be covered within a single book chapter. As such, the focus of this chapter is on the uncertainty in environmental flows assessments. It begins by considering the different types of uncertainty and how these are applicable in the context of environmental flows assessments (Section 15.2). There are many sources of uncertainty and, again, it is well beyond the scope of this chapter to provide a comprehensive treatment. Rather, Section 15.3 presents a framework to show how different sources of uncertainty can be identified. An introduction to methods available to quantify the magnitude of uncertainty is given in Section 15.4. Once uncertainty is better understood, steps can be taken to address this uncertainty and these are discussed in Section 15.5. More detailed descriptions of the methods available to undertake uncertainty assessments are provided by Burgman (2005), Quinn and Keough (2002), and Cullen and Frey (1999).

15.2 THE NATURE OF UNCERTAINTY

Uncertainty is an all-encompassing term and may take on a different meaning for different people. There is thus even uncertainty about the meaning of *uncertainty*! For some, the term *uncertainty* may relate to a lack of knowledge of the volume of water needed to achieve an environmental benefit. Others may associate the term *uncertainty* with future events such as not knowing if there will be sufficient water available in the coming months to meet environmental requirements. Others again may be uncertain about how the community will react to a large environmental water release. A very general definition of uncertainty is needed to cover all these meanings. Uncertainty can be defined as "any departure from the unachievable ideal of complete determinism" (Walker et al., 2003).

The uncertainties in environmental flows assessments fall into three main classes: epistemic uncertainty, ambiguity, and natural variability. The classes are also known by alternate names (Fig. 15.1).

Epistemic uncertainty is related to our lack of knowledge. This type of uncertainty can be reduced with improved knowledge. For example, we may be uncertain about what species of fish inhabit a particular river, what environmental processes are important for affecting a certain outcome for fish, or the value of a coefficient to use in a predictive model of fish abundance. We can reduce this type of uncertainty by undertaking research or monitoring.

Ambiguity is due to multiple interpretations of a situation (Brungnah et al., 2008). It is not a lack of information, but differences in how the information is interpreted or understood. For example, the language used in an environmental flows assessment may be vague or ambiguous and may be interpreted differently. Different flow recommendations are sometimes made for dry and wet years, and unless these are well-defined, there is room for differing interpretations.

Natural variability is the inherent randomness in systems that cannot be predicted (Ascough et al., 2008; Walker et al., 2003). Although natural variability can be better understood and more reliably estimated, it cannot be reduced with improved knowledge (Beven, 2016). Despite the possibility that at some point in the future our knowledge of natural systems may have improved to such an extent that we understand and can predict processes that now appear to be random, it does seem useful to distinguish between the uncertainty that we can conceivably reduce with available resources and those that we cannot (Warmink et al., 2010).

All three classes of uncertainty play an important role in environmental water management; epistemic uncertainty is the focus of this chapter, although natural variability is also considered. Ambiguity is not discussed further and readers are referred to Burgman (2005) and Beven (2016) for a more detailed discussion, including approaches to address this class of uncertainty.

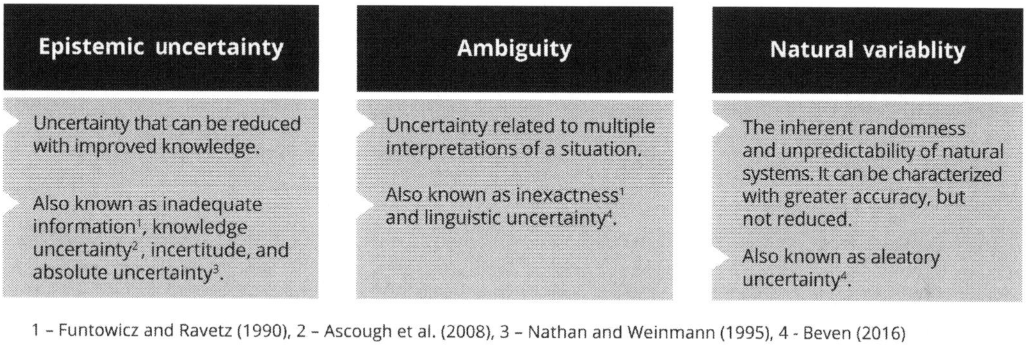

Epistemic uncertainty	Ambiguity	Natural variablity
Uncertainty that can be reduced with improved knowledge. Also known as inadequate information[1], knowledge uncertainty[2], incertitude, and absolute uncertainty[3].	Uncertainty related to multiple interpretations of a situation. Also known as inexactness[1] and linguistic uncertainty[4].	The inherent randomness and unpredictability of natural systems. It can be characterized with greater accuracy, but not reduced. Also known as aleatory uncertainty[4].

1 – Funtowicz and Ravetz (1990), 2 – Ascough et al. (2008), 3 – Nathan and Weinmann (1995), 4 - Beven (2016)

FIGURE 15.1

Classes of uncertainty.

15.3 SOURCES OF UNCERTAINTY IN ENVIRONMENTAL FLOWS ASSESSMENTS

There are many different sources of epistemic uncertainty and natural variability (referred to as *uncertainty* for the remainder of this chapter) related to environmental flows assessments. A systematic process can be used to locate the different sources of uncertainty within these assessments. Classes of uncertainty (as defined in the previous section) can be a useful tool in this process as they organize ideas about uncertainty (Ascough et al., 2008; Huijbregts et al., 2001; Maier et al., 2008; Walker et al., 2003). Unfortunately, a generic classification of uncertainty does not exist (Walker et al., 2003) for environmental flows assessments. Rather we can draw upon available classes that are relevant (namely, Ascough et al., 2008; Burgman, 2005; Walker et al., 2003) to help us better understand the uncertainties introduced during each phase of an environmental flows assessment. We can thus categorize five aspects of environmental flows assessments where potential sources of uncertainty exist. The first begins at the very start of environmental water assessments where the boundaries of the problem being assessed are specified, along with relevant and important processes and variables (context). The data and information (inputs) required are then determined, after which the representation of processes (structure) and the quantification of these relationships (parameters) are chosen. Finally, the combined uncertainties from the previous features ultimately impact the results of the environmental flows assessment (output).

To clearly explain the different sources of uncertainties associated with these groups in the following subsections, we use a hypothetical ecological response function (Fig. 15.2). An ecological response function defines the relationship between an ecological outcome and different flow conditions. The ecological response may be a direct function of some descriptor of the flow regime (a hydrological statistic), or an indirect function through, for example, the hydraulic habitat conditions created by flows (see Chapter 14).

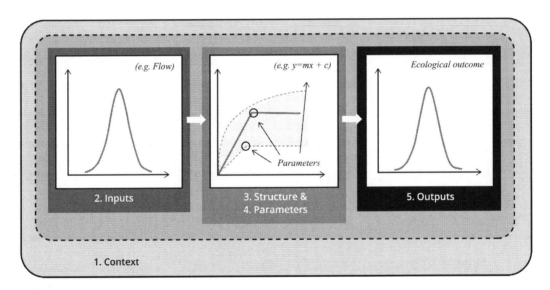

FIGURE 15.2

Sources of uncertainty in an ecological response function.

15.3.1 CONTEXT

In an ideal word, when selecting an environmental flows assessment method, we would have a complete understanding of the nature of the problem, the governing processes, and a comprehensive set of information, data, and models. However, this is never the case as ecological responses are complex. It is in the initial stages that the *context* of the problem is defined, specifying the boundary of the problem as well as the relevant and important processes and variables. Even though we may be seeking understanding of the effects of a certain driver of ecological responses, these responses are mediated through complex causal webs that we do not fully understand (Norton et al., 2008). An environmental response to a watering action may be affected by interactions with other causal factors of which we are not aware. As a result, an incorrect representation of the ecological response may be selected or the current environmental water assessment practices may not be suited, thus introducing uncertainties to the *context* of the environmental flows assessment (as presented in Fig. 15.2).

For instance, an environmental water practitioner may use a single ecological response curve, as shown in the box in Fig. 15.2. The context in this example is associated with the choice of x (input) and y (output) variables on each axis. The relevance of the selected variables for the given purpose should be discussed and specified; such as whether the use of discharge, velocity, or wetted perimeter to predict an ecological response is appropriate. Other factors relevant to the chosen processes should also be considered such as selecting a single process and using a curve as in Fig. 15.2, or multiple curves that represents complex interactions between the inputs (e.g., flow events) and ecological outcome (e.g., Young et al., 2003). This is discussed in more detailed in Section 15.3.3.

The context uncertainty described above arises because of a lack of knowledge (Ascough et al., 2008). It can also be related to uncertainty in environmental management priorities introduced by targeting species and locations according to external factors such as funding decisions that focus on priority rivers (Roberts, 2002). In addition to this, the environmental flows assessment choice may largely be informed by previous practice, with individual methods entrenched in various jurisdictions. Defining the context is the most critical stage of any environmental flows assessment method as it identifies the main issues being addressed and inevitably the modeling and resources required to answer it (Walker et al., 2003).

Generally, the more complexity considered in the environmental flows assessment methodology, the greater the certainty we have about the context because the risk of excluding processes that potentially impact the ecological response prediction is reduced. Hydrological methods, the simplest of environmental flows assessment methods, only use hydrological data (Tharme, 2003). As such, they do not directly consider the ecological processes associated with these data, and in fact only infer ecological effects via the *field of dreams* (Palmer et al., 1997) hypothesis that the provision of water will be enough to elicit a biological response. Consequently, this method has greater context uncertainty versus the more complex holistic methods, which also include hydraulic, geomorphic, and biological processes (Arthington et al., 2007). Although this reduces context uncertainty, using complex environmental models can introduce other sources of uncertainties associated with the other components, namely structure and parameters. This is discussed in more detail later in the section.

15.3.2 **INPUTS**

Before the environmental flows assessment is undertaken, data or inputs are obtained. Inputs are essentially the data and information that feed into the environmental flows assessment. These are the input variables to the ecological response model in Fig. 15.2 and could include spatial data (i.e., vegetation maps and wetland polygons), bathymetry, or monitoring data (i.e., hydrological, meteorological, and biological). Both natural variability and epistemic uncertainties impact the input variables. Uncertainty associated with these inputs occurs in the early stages when the initial raw data are measured, collected, and later processed for specific use within the environmental flows assessment. Uncertainty due to natural variability has most relevance during the measurement of the raw data. For example, natural variability will lead to measurement error (Kennard et al., 2010) and affects the uncertainty introduced by the sampling intensity. Stewardson and Rutherfurd (2008) suggested that by simply obtaining extra field measurements, the uncertainty in hydraulic model estimates can be significantly reduced (see Box 15.1). Epistemic uncertainty is more significant during the processing of the raw data to generate inputs. Uncertainty may arise as a result of infrequent instrument recalibration; or a change in channel dimensions affecting the rating curve (see Box 15.2); or the sampling approach used when collecting ecological data (see Box 15.3); or

---

BOX 15.1 HYDRAULIC MODELS

Hydraulic models are commonly used in environmental flows assessments to describe and predict the hydraulics of water flow in river channels and adjacent wetlands and floodplains using mathematical equations. Generally, hydraulic variable predictions are used to estimate habitat for different parts of the biota (e.g., minimum depth for fish connectivity, wetted perimeter for large woody debris as fish habitat). If hydraulic models are employed in isolation, they rely on the assumption that the provision of the hydraulic habitat will sufficiently achieve the required ecological endpoint. It should be noted that these assumptions can be tested by using monitoring data.

Many environmental water practitioners assume that the hydraulics/hydrology of river systems is well known and that uncertainty in predictions is low (Stewardson and Rutherfurd, 2008). However, this is not the case and there are many different sources of uncertainty associated with the inputs used to develop models, the model structure, and the methods used to define or calibrate the parameters. This can introduce major uncertainties that have a significant impact on the final model predictions. Table 15.1 presents the sources of uncertainty for a one-dimensional hydraulic model of the mid-Goulburn River in northern Victoria, Australia. Stewardson and Rutherfurd (2008) assessed these uncertainties using Monte Carlo simulations and determined that channel hydraulics estimations (e.g., Manning's n) contributed most of the error. The study concluded that to reduce the overall uncertainty it would be more beneficial to obtain extra field measurements rather than use a complex bed-load threshold function.

Table 15.1 Sources of Uncertainty in a Hydraulic Model. The Sources of Uncertainty Have Been Categorized in Terms of Inputs (Both Sampling and Measurements), the Model Structure, and the Three Key Physical Aspects of the Model (Including the River Channel, Shear Stress, and Streamflow)

Model Component	Source of Uncertainty	River Channel	Shear	Streamflow
Inputs	Sampling	Cross-section sampling	Number of particles in sample size	Length of data
	Measurement	Equipment and/or technique used for measurement	Grain size measurement	Error in rating curve
Model (i.e., model structure and parameter)	Model error	Manning's n	Bed shear stress equation (Shield's entrainment function)	Flow routing and estimating flow from ungauged catchments

Source: Adapted from Stewardson and Rutherfurd (2008)

BOX 15.2 UNCERTAINTY IN STREAMFLOW GAUGING

Streamflow measurements are commonly based on water level measurements that are converted to a discharge rate using a rating curve (see Fig. 15.3). The rating curve is constructed based on a sample of streamflows measured using the (time-consuming) velocity–area method and their corresponding water level (concurrent streamflow and water level data sample is termed *gauging*). The rating curve is then fit to the gaugings using statistical techniques, but may also take into account other factors such as the influence of the shape of the river cross-section.

The major sources of uncertainty in streamflow records are associated with the measurement of water level and the rating curve. The degree of measurement uncertainty for the water level depends on the type of recorder used. The number of gaugings undertaken, and their accuracy, dictates the uncertainty of the rating curve.

In Australia, there is a specified method for quantifying the uncertainty associated with streamflow measurements (Standards Australia, 1990). This method was used to assess 14 streamflow gauges within the Werribee River catchment located in Victoria, Australia. The uncertainty in the annual streamflows during 2005/2006 ranged from ±4% to ±41% of the reported flow (Lowe et al., 2009a). As the extent of the uncertainty depends on several site-specific factors, it is difficult to make generalizations about the magnitude of uncertainty associated with streamflow gaugings. However, this single result illustrates that discharge records available for stream gauges are subject to reasonable levels of uncertainty.

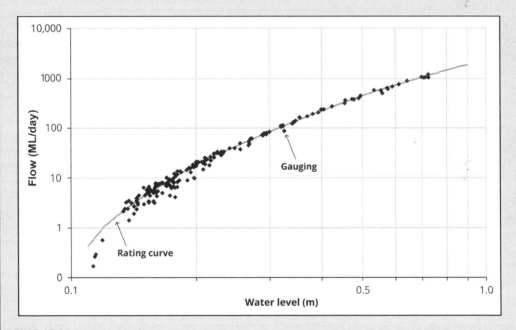

FIGURE 15.3

Gaugings and rating curve. The scatter in the gaugings around a given water level indicates the extent of the uncertainty in the rating curve.

Source: Lowe (2009)

BOX 15.3 ECOLOGICAL DATA

Ecological data are often quite imprecise. Many kinds of data are affected by high sampling uncertainty; the individual data points that we measure depart by random amounts from the true value that they are attempting to measure. For example, if we were measuring abundance of a particular species in a river reach, then there is a true value of abundance that is unknown to us. We conduct electrofishing and netting surveys to estimate this value, but our estimate will almost always be wrong to some extent. Sampling uncertainty is often not calculated, but affects all kinds of data collection. For example, the SIGNAL index (Chessman, 1995) is used to summarize macroinvertebrate data in Australian systems. The index ranges from 1 to 10, with higher values indicative of better ecological conditions. Metzeling et al. (2004) assessed uncertainty in SIGNAL scores derived from independent pairs of macroinvertebrate samples. A synthesis of these data found an average uncertainty of approximately 0.5 points for *edge* habitat samples (Webb and King, 2009). This 5% error rate is large when considering that a difference in SIGNAL score of just 2 points (from 6 to 4) can imply the difference between very good ecological condition and severe environmental impacts (Chessman, 1995).

Some ecological sampling approaches also have variable performance under different environmental conditions. For example, electrofishing is widely used to assess fish assemblages. This technique involves using electrical fields to stun fish, which are then collected and measured before being released. Under high-flow conditions, electrofishing can underestimate fish abundances, with stunned fish being washed downstream before they can be retrieved, and with fish able to flee from the electrofisher altogether. Conversely, under very low-flow conditions, electrofishing has a high likelihood of efficiently sampling the fish present. Such differences in sampling efficiency may be able to be accounted for in statistical models (e.g., Webb et al., 2010), but are a major impediment to assessing ecological responses.

---

during the pre- and postprocessing stage where uncertainties can occur from data aggregation (e.g., from daily to monthly streamflow), infilling missing data (Kennard et al., 2010), or simply when delineating polygons in geographic information system software (Fortin and Edwards, 2001).

15.3.3 STRUCTURE

All models, whether those used in hydrological methods (e.g., Gippel, 2001; Richter et al., 1996), habitat simulations methods (e.g., Maughan and Barrett, 1992), or holistic methods (e.g., King et al., 2008), are approximations of real-world processes and will not provide an exact replication of what occurs in reality (Burgman, 2005). Uncertainties arise for a number of reasons, including: (1) the use of equations as simplified representations of complex processes; (2) the use of variables such as depth, wetted perimeter, and velocity as measures of flow stressor; and (3) the omission of nonflow variables that drive ecological response (Ascough et al., 2008).

In terms of the ecological response function in Fig. 15.2, the way in which the relationship between the input variable (*x*-axis) and the ecological outcomes (*y*-axis) is defined introduces uncertainties in the model structure. In this case, a simple linear relationship ($y = mx + c$) is defined until the break of slope. Above this there is no further improvement in ecological response as the *x*-variable increases. Bounds are also shown to indicate the uncertainty associated with this given ecological response function. Considerations of whether: (1) this is an appropriate representation of the ecological outcomes; (2) the relationship is linear or nonlinear; or more importantly (3) the appropriate structure is adequate in estimating beyond the range of observed events used in model formulation, should be made. Also, are there dependencies or important interactions between variables that need to be taken and how they should be represented? For example, consider the maintenance of a strong self-sustaining fish population as the ecological endpoint. We know that the particular species requires a fresh event to trigger spawning, but it also needs low flows year round for habitat provision (Shenton et al., 2014). We can represent this as either a single response function relating daily flow to fish population, or as multiple curves relating summer low-flow to adult condition, and fresh magnitude and timing to recruitment (Shenton et al., 2014). The challenge then arises as to the relative importance and interaction between these flow components and the ultimate outcome for the fish community. An appropriate conceptual model to represent these complex interactions and how best to combine these curves needs to be selected (Chapter 14). This is no easy task and input from experts is vital.

It is also worth noting that there may be multiple and competing model structures available, usually developed for different purposes by different agencies using a different knowledge base (Caldwell et al., 2015). The manner as to how ecological response is predicted either via approximations, simple equations, or complex models inevitably results in different degrees of uncertainty.

15.3.4 PARAMETERS

Another component of environmental flows assessments is the parameters. These are different from inputs; instead parameters are constant values that are selected based on the context and problem being investigated (Walker et al., 2003). In terms of the ecological response function example in Fig. 15.2, the choice of parameter values includes the breakpoints on the curve as well as slope (m) and intercept (c) in the linear equation. The aim is to select parameter values that are accurate, that is, the model estimates reasonably closely the average of the true observations.

There are a number of different methods for selecting appropriate parameter values such as direct or indirect measurements (e.g., Manning's roughness coefficient), expert elicitation, or through a comprehensive calibration process (Ascough et al., 2008). Inevitably, no matter the selection process, uncertainties will be present and their sources will be different. For example, parameters that are selected subjectively or by expert judgment occur when there are insufficient measurements/data to make reliable estimates (Burgman, 2005). In this case, experts make recommendations based on observation and experience (Arthington et al., 2006), which can lead to misclassification of parameters as a result of bias and overconfidence (see Box 15.4). Alternatively,

---

BOX 15.4 EXPERT OPINION

Expert opinion can be a useful source of information for environmental assessments when data are lacking (Burgman, 2005; Martin et al., 2005). Expert opinion is frequently used in environmental flows assessments, partly because of the perception that there are insufficient data to develop empirical models of ecological responses to flow alteration (Stewardson and Webb, 2010). That perception is being challenged by the huge amounts of work being published in the literature (see Chapter 11), but for the present, expert opinion (most likely involving the engagement of expert panels) is likely to remain a key feature of environmental flows assessments.

When expert opinions are used in environmental flows assessments, there is often no attempt to specify the level of uncertainty associated with those estimates; that is, they are provided as a single best estimate. One exception is the revised FLOWS method used in Victoria, Australia. That method now requires a "clear statement of confidence or uncertainty in any relationship" (Department of Environment and Primary Industries, 2013) to be included as supporting

(Continued)

BOX 15.4 (CONTINUED)

information for the conceptual model used to predict ecological responses and hence derive the flow recommendations. The assessment also needs to include a comment on the information required to reduce the level of uncertainties for key flow—ecology relationships.

Assessment of the uncertainty of expert opinions is important because these estimates are likely to be affected by bias and overconfidence (Lin and Bier, 2008), and group think can affect the estimates (Lakoff, 1987). Bias emerges because the opinion of the most acknowledged (or most senior) expert takes on undue weight in the group-level estimate. Similarly, acknowledged experts are more likely to unconsciously overstate their confidence in an estimate (Speirs-Bridge et al., 2010).

There is little guidance in environmental flows assessment methods as to how to overcome these problems. However, researchers in cognitive psychology have devised methods of *structured expert elicitation* that reduce the sources of uncertainty in the elicited opinions (see Section 15.4.5).

the uncertainties related to parameters selected using calibration strategies (e.g., optimization) are closely associated with data uncertainties, where the length, quality, and type of data have an impact on the parameter selection (Ascough et al., 2008). Ascough et al. (2008) adds that the type of calibration strategy (e.g., trial and error vs optimization) can have an impact on the parameter uncertainty as the most optimal parameters may not be selected.

15.3.5 OUTPUTS

The uncertainties associated with context, input, structure, and parameters combine and propagate through the environmental flows assessment to impact the final outputs or the ecological outcomes (see output box in Fig. 15.2). Very few environmental flows assessments attempt to quantify the uncertainty in outputs. There are numerous methods available to combine these different sources of uncertainty, but this in itself is a difficult process. Section 15.4.6 discusses these methods and their applicability to environmental flows assessments.

15.4 QUANTIFYING UNCERTAINTY

Quantifying uncertainty is inherently difficult. In essence, it is attempting to quantify the unknown. Even so, it is a subject that has been studied extensively and there is a vast body of literature addressing this very question. Just as there are many and varied sources of uncertainty, there are numerous methods that can be used to quantify uncertainty. The most appropriate method to adopt depends on factors such as the source of uncertainty and the level of information available to quantify it (Fig. 15.4).

The level of information available can fall along a continuum between having complete knowledge (although this is not possible in practice) and total ignorance, where we have no information and are unaware that a lack of knowledge even exists (Refsgaard et al., 2005; Walker et al., 2003; Warmink et al., 2010). Between these extremes, environmental water requirements may be based on many observations, few observations, or qualitative information.

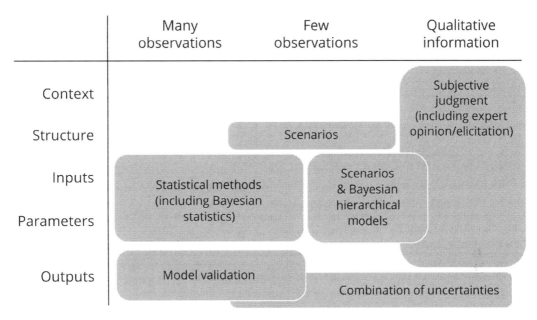

FIGURE 15.4

Relationship between the level of available information (top row), source of uncertainty (left column), and the approach that may be used to quantify uncertainty (boxes).

Few environmental flows assessments are in the enviable position of having *many observations* upon which to base environmental water recommendations. Inputs or parameters that can be readily sampled and measured are more likely to have many observations such as bed particle sizes (Stewardson and Rutherfurd, 2006). Relationships between hydrological variables and ecological response variables are another example where many observations may be available if good monitoring data exist. For example, Pettit et al. (2001) observed relationships between aspects of flooding and variables related to riparian vegetation. Statistical methods are available to quantify uncertainty in these cases (Section 15.4.2).

In other cases, only a *few observations* may be available, generally because it is either impractical or too costly to collect more data within the required time frame. In these cases, it may be reasonable to compare the range of observations to make an assessment of uncertainty (see Section 15.4.3). If the few available observations can be combined with prior knowledge, Bayesian modeling may be an appropriate approach to quantifying uncertainty (Section 15.4.4). For example, it may take years to collect data related to fish spawning in a specific river; however, this can be combined with the knowledge of spawning in other river systems in order to characterize uncertainty.

On occasions, *qualitative information* may be the only type of information available. For example, an expert may be required to estimate the environmental water requirements of a species that has not been comprehensively studied. In such circumstances the uncertainty is quantified using subjective judgment (Section 15.4.5).

Determination of an environmental water requirement involves several steps, and uncertainties will be introduced at each step. The total uncertainty will be a combination of the uncertainty introduced at each step of an environmental flows assessment. In some cases, the total uncertainty may be quantified by comparing model outputs with observations. In the absence of observations, the total uncertainty can be quantified by combining all of the uncertainties using one of the approaches presented in Section 15.4.6.

The following section provides a brief overview of the methods that are either most commonly used or have potential to be more widely adopted. Before these methods are discussed, an introduction to probability distributions is given in Section 15.4.1 as this concept flows through the various quantification methods that are available.

15.4.1 PROBABILITY DISTRIBUTIONS

Probability distributions assign probabilities to all possible values of a variable. They provide a comprehensive picture of uncertainty and enable more complex statistical methods to be used to combine uncertainties (see Section 15.4.6). In the context of environmental water requirements, it is useful to consider probability distributions for both discrete and continuous variables.

Discrete variables have a finite number of values. The number of flushing flows required per year is a discrete variable because it must be a whole number that is greater than (or equal to) zero. The probability distribution representing the uncertainty in the number of flushing flows (Fig. 15.5A) shows that the most likely number of flushing flows required is three with a probability of 60%. The probability that the number of flushing flows required is equal to either one, two, three, or four is 100%.

Continuous variables can take any value within a given interval. A minimum flow requirement is an example of a continuous variable as it can take any value greater than zero. The probability of a given value cannot be calculated for a continuous variable because there are an infinite number of possible values. Instead, the uncertainty can be described as the probability that a given interval contains the true value. This probability can be calculated from the *probability density function* (*PDF*) and is equal to the area that falls under the PDF and within the interval. In Fig. 15.5B, the best estimate of the minimum flow requirement is 6 ML/day for both probability distributions shown. However, the uncertainty surrounding this best estimate varies between the two probability distributions. The 95% confidence interval for the distribution represented with a solid line is ± 2 ML/day. This interval is much larger for the probability distribution represented with a dashed line and there is more uncertainty surrounding the estimate.

A number of probability distributions are useful for describing uncertainties. The distribution of discrete events can be shown using a probability histogram (Fig. 15.5A), but may also be approximated using discrete probability distributions such as the binomial distribution. A normal distribution is shown in Fig. 15.5B. Simpler distributions can also be useful, particularly if they are derived using subjective judgment (see Section 15.4.5). A triangular distribution is used to show the uncertainty associated with the minimum flow requirement in Fig. 15.5C. Again, the best estimate of the minimum environmental flow is 6 ML/day. Unlike the normal distribution, the triangular distribution shows the lowest and highest possible values of the minimum flow requirement. Unsurprisingly, the lowest possible minimum flow is zero. In this example, there is no possibility

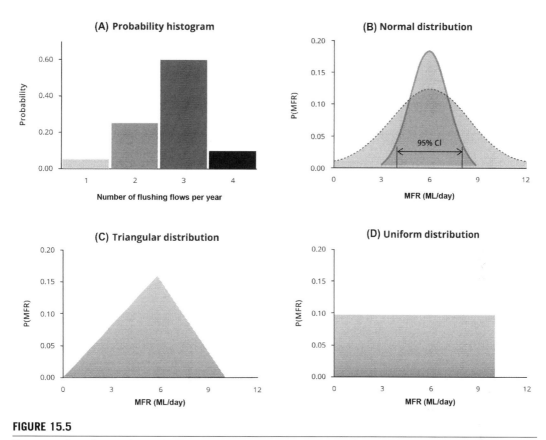

FIGURE 15.5

Representation of uncertainty using probability distributions: (A) probability histogram, (B) normal distribution, (C) triangular distribution, and (D) uniform distribution.

Source: Lowe (2009)

that the true value of the minimum environmental water release exceeds 10 ML/day. In Fig. 15.5D, the only information known about the minimum flow requirement is that the true value falls in the range from 0 ML/day to 10 ML/day. There is no information to show which values within this range are more likely. This is represented with a uniform probability distribution.

15.4.2 STATISTICAL METHODS

Statistical methods are available to estimate probabilities based on observed frequencies that are obtained from repeated sampling. Sampling is used when it is difficult to measure a variable precisely. For example, it would be time-consuming and costly to measure the percentage of an entire river bank that is covered by annual herbs. Instead, a sample is taken and statistical methods are used to make inferences about the value of annual herb cover for the entire river bank, or full population in statistical terms (Box 15.5).

BOX 15.5 CONFIDENCE INTERVALS

An input to an environmental flows assessment may be the percentage of the river bank that is covered in annual herbs. The bank is surveyed using 30 quadrats and the results are used to estimate the average for the entire river bank. The average percentage cover is 5% and the standard error (the standard deviation of the mean) is 0.49%. This means that there is a 95% probability that the true mean percentage cover is between 4.8% and 5.2%. This is equivalent to ±20% of the sample mean or best estimate. Increasing sampling effort will reduce the size of the confidence interval, but only in proportion to the *square root of the sample size*. Thus there is a diminishing return on increasing effort as sample size increases. In this example, doubling the number of quadrats to 60 only reduces the standard error to ±19.6% of the estimated mean. Conversely, reducing sampling effort does not increase uncertainty greatly. If the number of quadrats was halved, the standard error would increase from ±20% to ±21%. Of course, if there was less variability between the quadrats, the uncertainty would be reduced.

The sample mean ($\bar{x}$) is the best estimate of the population mean (μ) of the variable. In the example above, the sample of annual herb cover can be used to estimate the average cover across the entire river bank. The uncertainty associated with the average cover can be represented using a normal distribution where the precision will depend on both the number of samples (n) and the variation between the samples (i.e., the standard deviation [s]). The confidence interval for the mean can be calculated using the following equation:

$$\bar{x} \pm t^*_{(n-1)} \frac{s}{\sqrt{n}}$$

The value of $t^*_{(n-1)}$ follows the $t_{(n-1)}$ distribution and its value will depend on both the degrees of freedom (which is $n-1$) and the size of the confidence interval (i.e., 95%). The cumulative probabilities associated with different t values can be found in tables contained in most statistical textbooks or obtained from relevant software.

Confidence intervals can also be used to show the uncertainty in the relationship between two variables (Freund et al., 2006). The confidence intervals reflect the uncertainty in model parameters. Consider an example where the percentage cover of exotic species can be estimated from the duration of a flooding event (Fig. 15.6). For a flood that lasts for 30 days the mean cover of exotic species is expected to be 25% and the 95% confidence interval equivalent to ±2.9% (or ±12% of the best estimate).

The statistical methods introduced in this section are covered in more detail in many statistics textbooks such as Quinn and Keough (2002) and Cullen and Frey (1999).

15.4.3 SCENARIO APPROACH

The scenario approach has been widely used to characterize the uncertainty associated with model structure (Refsgaard et al., 2006), but can also be used to characterize the uncertainty associated with inputs and parameters, particularly where they are obtained using expert opinion (Fig. 15.4). In situations where only a few observations are available, the most appropriate approach to quantify uncertainty may be to consider the range of the observations. For example, where there is more than one available model, each model will make different predictions and the variability will reflect the extent of the model's structural uncertainty.

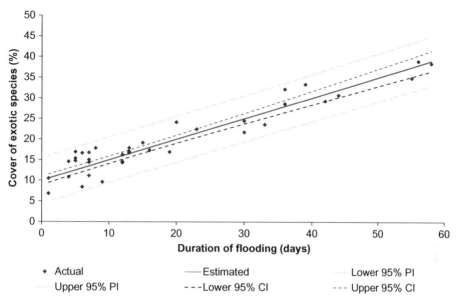

FIGURE 15.6

Linear regression plot (the observations are shown using black diamonds, the mean of the slope of the relationship is shown with a solid black line, and the 95% confidence interval of that mean is shown using dashed black lines).

Consider a hypothetical example where an environmental water manager wants to predict the average daily flow and the magnitude of a high-flow event (exceeded only 25% of days) and a low-flow event (exceeded on 75% of days) for a given catchment. The environmental water manager could have access to more than one hydrological model that are able to predict these flow statistics, but each model may provide the environmental water manager with a different prediction. Fig. 15.7 illustrates one approach for comparing the results obtained from different models. It shows that for this hypothetical example the models predict a large range for the magnitude of the high-flow event, meaning that there is substantial uncertainty in this result, and that the models disagree in their assessment of how the catchment responds to high-rainfall events. All of the models give a very similar value for the magnitude of a low-flow event, indicating that we can have more confidence in this estimated value than for the high-flow event. However, it is possible that all models are giving the same, but wrong, results.

A similar approach was used by Caldwell et al. (2015) who compared the ecologically relevant flow statistics generated by six hydrological models at five sites in the southeastern United States. Their analysis did not identify one superior model, rather that models performed better for different flow statistics, and all models performed poorly (>30% difference from observed) for at least one flow statistic.

The scenario approach can also be applied to quantify the uncertainty in inputs or parameters that are quantified using qualitative information such as expert opinion (Fig. 15.4). For example, a few fish biologists may provide an estimate of the depth of water required to maintain an adequate

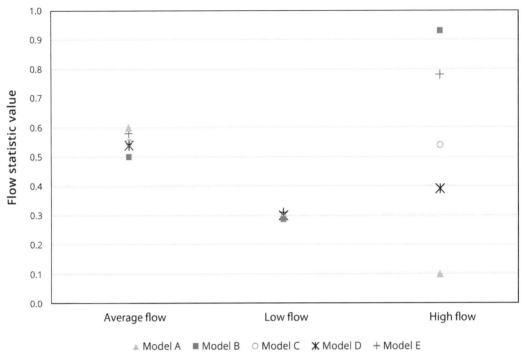

FIGURE 15.7

Illustration of the scenario approach. The uncertainty in the magnitude of three flow events (low, average, and high) predicted by hydrological models is assessed by comparing the predictions made by five different models.

habitat for a particular species of fish. The range in estimates provided will give some indication of the uncertainty in the estimates.

15.4.4 BAYESIAN STATISTICS

Bayesian statistical methods warrant a special mention because of their potential applicability to environmental flows assessments, both for assessing uncertainty in assessments and also reducing uncertainty through the analysis of monitoring and assessment data. Several features make Bayesian methods appropriate for use in assessing uncertainty in environmental flows assessments.

Model flexibility Bayesian methods are very flexible (Clark, 2005). This provides the ability to write models that more closely emulate the processes leading to observed outcomes. This flexibility will reduce unexplained variation in models, providing more precise estimates of parameters, and more precise predictions.

Use of prior data Bayes theorem provides a formal method for updating our knowledge as new knowledge becomes available. The *prior probability distribution* of, for example, a regression parameter represents existing knowledge before we collect new data. The prior is then updated to a *posterior probability distribution* with new data. The use of prior information is a source of controversy, as it is sometimes seen as making Bayesian methods too subjective (McCarthy, 2007). However, human thought processes generally work in a manner similar to Bayesian reasoning—we have a belief concerning something, and can update that belief when new information becomes available. When empirical data are scarce, prior data improve the precision of posterior estimates; when data are plentiful, the prior will have little influence (Webb et al., 2010). *Minimally informative* priors can also be used when we have no particular beliefs concerning a model outcome. One class of Bayesian model, *hierarchical* Bayesian models, effectively uses data from other sampling units (e.g., sites, reaches, or rivers) as prior information to inform posterior probability distributions for other sampling units. All types of prior information reduce the uncertainty of posterior estimates, improving the precision of model predictions.

Webb et al. (in prep) described the use of Bayesian modeling for quantifying the relationship between stream bank inundation and encroachment of terrestrial vegetation into the river channel. They demonstrated that through the use of hierarchical models and expert-based prior information, credible intervals (the Bayesian equivalent of a confidence interval) model predictions were reduced by a factor of approximately three.

The application of Bayesian statistical methods to ecological data is covered in more detail in textbooks such as McCarthy (2007) and Kéry (2010). Their specific application to environmental water monitoring and evaluation was first proposed by Webb et al. (2010).

15.4.5 SUBJECTIVE JUDGMENT

It is often the case that decisions around environmental water need to be made when there are little, or even no, data available. The statistical methods above are clearly not applicable in these cases. Rather than ignore the uncertainty, subjective judgment can be used to quantify uncertainty.

Just as expert opinion is often used when estimating environmental water requirements, expert opinion may be also used to quantify the uncertainty around that estimate. However, this is seldom done in environmental flows assessments. A recent exception is the revised FLOWS environmental flows assessment method used in Victoria, Australia (Department of Environment and Primary Industries, 2013). The FLOWS method now requires the sources of uncertainty and implications for the recommendations to be documented, with a recommendation to use a qualitative approach relying predominantly on subjective judgment.

Even if experts cannot estimate the entire probability distribution, they may be able to provide an estimate of the best estimate and an upper and lower bound. This information is enough to generate a uniform or triangular probability distribution (as shown in Fig. 15.5). However, as previously discussed expert opinion is also a source of uncertainty (see Box 15.4). Expert opinions are prone to bias and overconfidence (Lin and Bier, 2008).

Considerable research has resulted in methods for structured expert elicitation that can help minimize such uncertainty. These methods explicitly require experts to state their uncertainty by nominating upper and lower bounds for an estimate, as well as the *best* estimate. They reduce bias and overconfidence by seeking multiple independent opinions, trying to ensure that experts do not *anchor* one estimate to a previous estimate for an alternative scenario, and by allowing group-level discussion of estimates only after all members of a panel have provided initial estimates (e.g., Speirs-Bridge et al., 2010). These methods rely, to some degree, on the *wisdom of the crowd* (Surowiecki, 2004)—the (faintly astonishing) fact that a group of nonexperts are collectively more likely to come up with a good estimate of an endpoint (e.g., a parameter in a response model) than a single out-and-out expert in the field.

Although there are as yet no environmental flows assessment methods that formally include structured expert elicitation in their methods, the use of this approach would be consistent with the ELOHA framework (Poff et al., 2010). ELOHA provides a great deal of flexibility in how flow—response relationships are derived, and even though it does not mandate the consideration of uncertainty as part of this process, it allows for the possibility. de Little et al. (2012) describe a method that was used to develop probability distributions of ecological responses to flow variation. That approach was based heavily on the method of Speirs-Bridge et al. (2010), and would be feasible for use in environmental flows assessments.

15.4.6 COMBINING UNCERTAINTIES

Environmental flows assessments generally involve several steps (refer to Chapter 11) and uncertainties will be introduced at each step. The preceding sections have introduced some methods available to quantify uncertainty associated with individual inputs, parameters, and model structure. However, the uncertainty in the environmental water requirement is the combination of the uncertainties generated in each step.

The combined uncertainty can be assessed by comparing predictions with observations. Residuals are the difference between the model predictions and the observed values. The magnitude of these residuals reflects the uncertainty in individual predictions and a nonzero average indicates that there is a bias in the predictions.

Observations are not always available to compare with predictions. A number of methods are suitable for combining uncertainties in environmental flows and these include First-order approximation methods, Monte Carlo Simulations, Fuzzy Bounds, and Bayesian Networks. A brief introduction to each of these methods is presented below.

First-order approximation methods provide an analytical approach for combining uncertainties. Equations are used to add, multiply, and divide the mean and standard deviations (that represent uncertainty) associated with each of the inputs (Cullen and Frey, 1999). These methods are applicable when all the inputs are independent, have the same distribution shape, and no one input is dominant.

Monte Carlo simulation is a widely used numerical method. Using this approach the environmental water requirement for a target endpoint (e.g., a species of fish) is calculated N times. In each of the N simulations each input is randomly selected from its probability distribution and used to calculate the environmental water requirement. The environmental water requirement calculated in each simulation is recorded and the variation in these results is used to characterize the uncertainty (see Fig. 15.8). Generally, thousands of simulations are required to adequately represent the uncertainty. A more thorough explanation of this technique is provided by Cullen and Frey (1999).

Another method to combine uncertainties in environmental flows is fuzzy logic. Unlike binary logic where values can equal only 0 or 1, a fuzzy logic value can range between 0 and 1, and is represented by its fuzzy set and associated membership function, which can have linear or a trapezoidal relationship. For instance, consider an environmental variable X and A—a fuzzy set of acceptable values. If x is a potential value of X, then $A(x)$ represents the membership degree

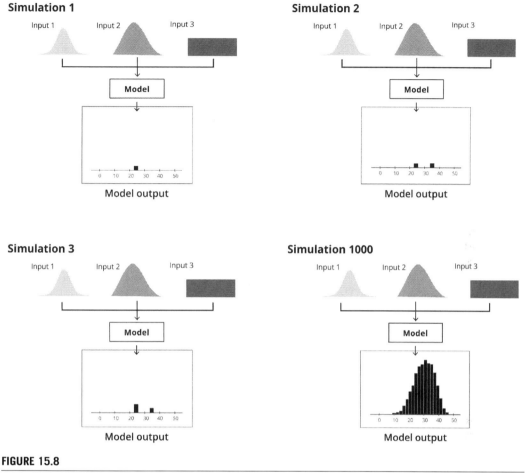

FIGURE 15.8

Illustration of the Monte Carlo method.

Source: Lowe (2009)

of the fuzzy set, *A* (Adriaenssens et al., 2004). The advantage of using a membership function is that it can take into account imprecise data or incomplete knowledge (Zadeh, 1965). This is especially useful when accounting for linguistic uncertainty, that is when knowledge or ecological data are expressed in written form (Adriaenssens et al., 2004; Ascough et al., 2008). For example, Jorde et al. (2001) determined fuzzy logic suitable for habitat modeling to allow expert knowledge from a fish specialist to be developed into data sets by forming fuzzy-based rules (i.e., using if–then statements). Fukuda (2009), however, developed fuzzy sets by considering different hydraulic and biological variables (i.e., water depth, velocity, cover, and vegetation coverage), and combining these to predict the likely habitat in a river reach.

Bayesian networks models are extensively used in environmental management; in particular, they have been used to inform environmental flows assessments (Chan et al., 2012; Shenton et al., 2011, 2014; Webb et al., 2013). They are graphical statistical models that consist of three elements: (1) nodes that represent the state variables (e.g., discharge and temperature); (2) links that define the cause-and-effect relationship (e.g., impact of temperature on an important ecological process); and (3) conditional probability tables that quantify the dependencies between connecting nodes (Chan et al., 2012). Bayesian networks models have the advantage of being able to combine different data sources (e.g., expert opinion and empirical data), explicitly represent uncertainty via conditional probabilities, and provide a fast response (Chan et al., 2012; Uusitalo, 2007; Webb et al., 2013). They can also be readily updated as new knowledge becomes available. They are potentially useful for environmental water assessments and informing management strategies, because they meet the challenges not only in predicting ecological response with limited data and understanding, but also quantifying this uncertainty.

15.4.7 CHALLENGES FOR QUANTIFYING UNCERTAINTY

There are numerous methods available to quantify uncertainty and they vary in their complexity. Some methods require specialist skills and possibly considerable time to undertake. In some situations the use of complex methods will be warranted, whereas in others a simpler approach will suffice. Practitioners need to understand how the results of the analysis will be used and make a trade-off between the complexity and comprehensiveness of an analysis and the effort required.

For many rivers a scarcity of data poses a challenge for the environmental flows assessment, including the uncertainty assessment. The most appropriate method to adopt to quantify uncertainty will depend on the level of information (or data) available (Fig. 15.4). However, uncertainties will not be quantifiable if they are not well understood (Ascough et al., 2008). For example, there may be a situation where experts recognize that they cannot identify all causal factors, and it is not possible to quantify the impact this has on the estimate of an environmental water requirement.

15.5 ADDRESSING UNCERTAINTY

The most obvious way to address uncertainty is to seek to reduce it (Section 15.5.1). Indeed, many authors identify uncertainty analysis as a tool to identify where more research or better data are needed (e.g., Brungnah et al., 2008; Fu and Guillaume, 2014; Stewardson and Rutherfurd, 2006; Van der Lee et al., 2006). Even with increased research efforts, uncertainty will remain and so must be addressed. Addressing uncertainty involves first reporting the uncertainty (Section 15.5.2) and then incorporating it into decision making (Section 15.5.3).

15.5.1 IMPROVED KNOWLEDGE

Epistemic uncertainty (a focus of this chapter) is related to our lack of knowledge and can be reduced with improved knowledge (Section 15.2). Any new knowledge will reduce uncertainty, but the benefit of understanding the sources and magnitudes of relevant uncertainties is that further investigations can be focused on the areas where the greatest gains can be made (see Box 15.6).

BOX 15.6 UNCERTAINTY IN LOSSES ATTRIBUTED TO ENVIRONMENTAL WATER DELIVERIES IN THE CAMPASPE RIVER, VICTORIA, AUSTRALIA

With increasing recognition of environmental water requirements, creative ways of manipulating the operation of irrigation systems operation are being investigated to achieve environmental benefits at the same time as delivering consumptive water. In theory, this provides environmental outcomes without a specific environmental water right. However, in reality, changing the operation of a system may increase the losses in the system due to evaporation, channel seepage, and groundwater interactions.

An uncertainty analysis was undertaken to quantify the uncertainty associated with the additional losses incurred by rerouting water to achieve environmental benefits along the Campaspe River (Lowe et al., 2009b). The analysis considered uncertainty due to measurement of streamflow, and the metering or estimation of water extractions. The total expected monthly losses are shown using a boxplot in Fig. 15.9A. The largest losses occurred in April 2007 when the best estimate of the losses was 339 ML, but the 95% confidence interval ranged between 179 ML and 499 ML. The uncertainty was in the order of ± 47%. The relative contribution of each source of uncertainty was assessed by removing each source in turn and comparing the result to the combined uncertainty (shown using a spider plot in Fig. 15.9B). The results showed that the uncertainty could be drastically reduced by improving the accuracy of streamflow measurements at the upstream and downstream locations, but little improvement would be gained by increasing metering of water extractions or inflows from drains.

To avoid third-party impacts, the losses incurred due to the changed system operation would need to be accounted for against the environmental right. The uncertainty analysis allows the risk of third-party impacts to be explicitly considered. If the onus is on the environmental manager to demonstrate that there will be no third-party impacts, a conservatively high estimate of the losses should be adopted (i.e., 499 ML). However, if the risk is shared equally between the environment and consumptive users, the best estimate of losses (339 ML) can be deducted from the environmental account. Without an assessment of uncertainty the best estimate would be adopted without the risks of third-party impacts being apparent.

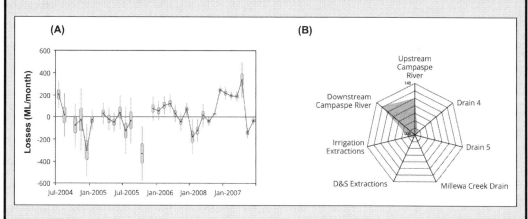

FIGURE 15.9

Illustration of the uncertainty in losses along the Campaspe River, Victoria, Australia: (A) the expected range of monthly river losses represented using the interquartile range (shaded box) and the 95% confidence interval (vertical lines); (B) relative influence of inputs to the uncertainty in losses. Each input is represented by an axis and the shaded area crosses the axis at the value of the coefficient of variation used to describe the uncertainty.

Source: Lowe et al. (2009b)

Unfortunately, it is difficult to make universal statements about where improved knowledge is needed the most. Sources of uncertainty with the greatest influence in an environmental flows assessment will vary. The source of the greatest uncertainty will depend on factors such as the extent of monitoring, the models developed, and how well the ecology of the system is understood. An uncertainty assessment at Lake Ijsselmeer in the Netherlands showed that there was not one single variable that contributed most to the uncertainty in the habitat suitability index for pondweed (Van der Lee et al., 2006). Rather, different sources of uncertainty had the greatest influence at different locations.

There are two potential approaches to tackling the problem of uncertainty in data and responses. First, monitoring and research can focus on endpoints that are deemed to be important, but for which we have limited understanding. Such a focus is consistent with an adaptive management approach, which highlights that decisions must be made under uncertainty and seeks to reduce those uncertainties (Chapter 25). Second, monitoring can focus on endpoints that are well understood, can be measured with high precision, and are expected to respond predictably to flow variation. Such an approach is consistent with tenets of restoration monitoring, where the focus is on demonstrating an outcome for stakeholders.

Adaptive management is one way to reduce the uncertainty in environmental flows assessments as it provides an opportunity to learn from management decisions (see Chapter 25). It is most suited to rivers where the release of environmental water can be controlled (and therefore varied) and where there is enough interest and resources to justify ongoing monitoring.

Although uncertainties can be reduced with further research and monitoring they are unlikely to ever be eliminated. Rather, uncertainty is increasingly being seen as an inherent challenge that needs to be built into the management approach (Brungnah et al., 2008; Clark, 2002).

15.5.2 REPORTING UNCERTAINTIES

If environmental water managers are to consider uncertainty in environmental flows assessments, they need to be provided with meaningful information. Many different ways to present information about uncertainty exist; however, not all of these can be easily understood by nonexperts (Wardekker et al., 2008); communicating what an estimate of uncertainty means can be difficult (Beven, 2016). To be effective, reporting needs to be simple and tailored to the decisions that are likely to be made so that the implications of the uncertainty are clear (Lowell, 2007; Wardekker et al., 2008). For example, a manager may use the outcomes from an environmental flows assessment to decide how much water to release from a reservoir to meet environmental needs. In this case, the uncertainty could be reported as the probability that the environmental needs will be met for different release volumes.

Of course, an environmental flows assessment may be used to make decisions that cannot be foreseen at the time it is undertaken, and therefore more detailed reporting of uncertainty may also be warranted. It may be more appropriate to include such information in supplementary information such as an appendix to the main report. Decision makers are likely to be interested in what information was used in the assessment, the magnitude of the uncertainty, the major sources of uncertainty, and steps that can be taken to reduce it (Wardekker et al., 2008).

15.5.3 INCORPORATING UNCERTAINTY IN DECISION MAKING

Decisions made with active consideration of the uncertainty of outcomes are more robust than those that assume deterministic outcomes. It is possible that an *uncertain* decision may not achieve the best possible outcome, but it is much more likely to achieve an acceptable outcome through consideration of how uncertainty impacts possible outcomes (Ben-Haim, 2005; Warmink et al., 2010).

Consider an environmental water manager who is planning to release a flushing flow to turn over a gravel bed downstream of a reservoir. If the uncertainty in the magnitude of the flushing flow required is known, the manager can assess the probability that a flow of a given magnitude will turn over the bed (Stewardson and Rutherfurd, 2006). This in turn allows an open dialogue between various stakeholders and the release decision can reflect the level of risk the community is prepared to accept versus the amount of water to be used (Heaney et al., 2012). Often managers must make trade-off decisions between competing demands for water. The manager who is planning to release a flushing flow may also be planning to make a subsequent release to trigger fish spawning. The likelihood that a release of water will provide the intended environmental benefit could factor into the decision about which flow event is prioritized by the manager.

Clearly, the decisions become more complicated when multiple competing demands for water and their associated uncertainties are considered. Software-driven decision support systems play an important role in allowing these complexities to be incorporated into decision making (Ascough et al., 2008; Clark, 2002). One such tool is the *seasonal environmental water decision support (SEWDS)* tool (Horne et al., 2015), which aims to optimize environmental water, drawing on seasonal streamflow forecasts and ecological response predictions. Decision support systems such as the SEWDS tool ensure that decisions made are consistent and transparent, improve our understanding of decision drivers, as well as being able to incorporate uncertainties associated in the decision process and formalize an adaptive management approach.

Rather than consider the probability that a particular management decision (i.e., a flow release) will be a success, uncertainty assessments can be used to identify what decisions can be made with confidence, given the uncertainty that exists (Fu and Guillaume, 2014). For example, a manager may need to know what environmental water releases should be prioritized in the upcoming season. However, they may have available to them several ecological response models. If all models identify the same priorities, they can make their decision with a higher degree of confidence. If, however, the models do not agree, the manager may need to seek better information before a reliable decision can be made.

15.5.4 CHALLENGES FOR ADDRESSING UNCERTAINTY

Uncertainty does not need to be a reason for inaction, rather consideration of uncertainty can improve decision making. However, exactly how, and to what extent, uncertainty influences decision making will depend on a range of factors, including consideration of social values. Consider again an environmental water manager who must decide how much water to release for an environmental purpose. If the release is to benefit an endangered species that is highly valued by the community, they are more likely to release a volume of water that provides high confidence that the environmental response will be observed.

The focus of this chapter has been on the analysis and treatment of epistemic uncertainty in environmental flows assessments. However, in making decisions, managers and policy makers may also take into account sources of uncertainty in addition to those associated with the environmental flows assessments. A large source of uncertainty is future climate conditions, both short- and long term. In deciding whether to make an environmental release, a manager may need to consider the likelihood that the flow event will be met by a natural event anyway, which means the water could have been used for another purpose such as being saved to help alleviate drying conditions in future years (Heaney et al., 2012). The relative contribution of this type of uncertainty may be large compared with our understanding of environmental water requirements.

15.6 CONCLUSION

Uncertainty remains a challenge for sustainable water management. Consideration of uncertainty leads to more robust decision making; however, few environmental flows assessments identify and quantify uncertainties. We have presented a simple typology that allows practitioners to use a structured approach to identify the different sources of uncertainty associated with environmental flows assessments and describe some of the main methods to quantify these uncertainties. Any uncer-

---

BOX 15.7 IN A NUTSHELL

1. Uncertainty is "any departure from the unachievable ideal of complete determinism" (Walker et al., 2003).
2. Uncertainty is introduced into environmental flows assessments because we do not completely understand which factors influence the environmental watering requirements, how these different factors interact, and because there are uncertainties associated with the measurement and estimation of these factors.
3. There are numerous methods available to quantify uncertainty in environmental water requirements.
4. The most obvious way to reduce uncertainty is to increase monitoring and research efforts. An uncertainty assessment can identify the major sources of uncertainty and help identify the most effective way to reduce the uncertainties.
5. Uncertainty is likely to remain, and consideration of this uncertainty can improve decision making. Decision makers need to be provided with meaningful information and, in more complex situations with decision support tools, to help them understand the implications of the uncertainty on the decisions they make.

---

tainty analysis should be tailored toward the decisions that rely on the outcomes of the analysis. This will influence the level of detail included in the analysis, the selection of an appropriate quantification method, and how the uncertainty is reported. An uncertainty analysis can also help identify the greatest source of uncertainty and thereby focus future monitoring and research efforts (Box 15.7).

REFERENCES

Adriaenssens, V., De Baets, B., Goethals, P.L.M., De Pauw, N., 2004. Fuzzy rule-based models for decision support in ecosystem management. Sci. Total Environ. 319, 1—12.

Arthington, A.H., Bunn, S.E., Poff, N.L., Naiman, R.J., 2006. The challenge of providing environmental flow rules to sustain river ecosystems. Ecol. Appl. 16, 1311−1318.

Arthington, A.H., Baran, E., Brown, C.A., Dugan, P., Halls, A.S., King, J., et al., 2007. Water requirements of floodplain rivers and fisheries: Existing decision support tools and pathways for development. International Water Management Institute, Colombo, Sri Lanka.

Ascough, J.C., Maier, H.R., Ravalico, J.K., Strudley, M.W., 2008. Future research challenges for incorporation of uncertainty in environmental and ecological decision-making. Ecol. Model. 219, 383−399.

Ben-Haim, Y., 2005. Info-gap Decision Theory For Engineering Design. Or: Why 'Good' is Preferable to 'Best'. In: Nikolaidis, E., Ghiocel, D.M., Singhal, S. (Eds.), Engineering Design Reliability Handbook. CRC Press.

Beven, K., 2016. Facets of uncertainty: epistemic uncertainty, non-stationarity, likelihood, hypothesis testing, and communication. Hydrol. Sci. J. 61, 1652−1665.

Brungnah, M., Dewulf, A., Cpahl-Wostl, C., Taillieu, T., 2008. Toward a relational concept of uncertainty: about knowing too little, knowing too differently, and accepting not to know. Ecol. Soc. 13, 30.

Burgman, M.A., 2005. Risks and decisions for conservation and environmental management. Cambridge University Press, Cambridge.

Caldwell, P.V., Kennen, J.G., Sun, G., Kiang, J.E., Butcher, J.B., Eddy, M.C., et al., 2015. A comparison of hydrologic models for ecological flows and water availability. Ecohydrology.

Chan, T.U., Hart, B.T., Kennard, M.J., Pusey, B.J., Shenton, W., Douglas, M.M., et al., 2012. Bayesian network models for environmental flow decision making in the Daly River, Northern Territory, Australia. River Res. Appl. 28, 283−301.

Chessman, B.C., 1995. Rapid assessment of rivers using macroinvertebrates: a procedure based on habitat-specific sampling, family-level identification and a biotic index. Aust. J. Ecol. 20, 122−129.

Clark, J.S., 2005. Why environmental scientists are becoming Bayesians. Ecol. Lett. 8, 2−14.

Clark, M.J., 2002. Dealing with uncertainty: adaptive approaches to sustainable river management. Aquat. Conserv. Mar. Freshw. Ecosyst. 12, 347−363.

Cullen, A., Frey, C.H., 1999. Probabalistic Techniques in Exposure Assessment: A Handbook for Dealing with Variability and Uncertainty in Models and Inputs. Plenum Press, New York.

De Little, S.C., Webb, J.A., Patulny, L., Miller, K.A., Stewardson, M.J., 2012. Novel methodology for detecting ecological responses to environmental flow regimes: using causal criteria analysis and expert elicitation to examine the effects of different flow regimes on terrestrial vegetation. In: Mader, H., Kraml, J. (Eds.), 9th International Symposium on Ecohydraulics 2012 Proceedings, Sep 17-21. International Association for Hydro-Environmental Engineering and Research (IAHR), 13896_2, Vienna, Austria.

Department Of Environment And Primary Industries, 2013. Flows − a method for determining environmental water requirements in Victoria. Edition 2 ed Victorian Government Department of Environment, Melbourne.

Fortin, M.-J., Edwards, G., 2001. Delineation and Analysis of Vegetation Boundaries. In: Hunsaker, C., Goodchild, M., Friedl, M., Case, T. (Eds.), Spatial Uncertainty in Ecology. Springer, New York.

Freund, R.J., Wilson, W.J., Sa, P., 2006. Regression Analysis: Statistical Modelling of a Response Variable. Elsevier, USA.

Fu, B., Guillaume, J.H.A., 2014. Assessing certainty and uncertainty in riparian habitat suitability models by identifying parameters with extreme outputs. Environ. Model. Softw. 60, 277−289.

Fukuda, S., 2009. Consideration of fuzziness: Is it necessary in modelling fish habitat preference of Japanese medaka (*Oryzias latipes*)? Ecol. Model. 220, 2877−2884.

Funtowicz, S.O., Ravetz, J.R., 1990. Uncertainty and quality in science for policy. Springer Science & Business Media.

Gippel, C.J., 2001. Australia's environmental flow initiative: filling some knowledge gaps and exposing others. Water Sci. Tech. 43, 73−88.

Heaney, A., Beare, S. & Brennan, D.C., 2012. Managing environmental flow objectives under uncertainty: The case of the lower Goulburn River floodplain, Victoria. 2012 Conference (56th), February 7–10, 2012, Freemantle, Australia, 2012. Australian Agricultural and Resource Economics Society.

Horne, A., Costa, A., Boland, N., Kaur, S., Szemis, J.M. & Stewardson, M. 2015. Developing a seasonal environmental watering tool. 36th Hydrology and water resource symposium. Hobart, Tasmania.

Huijbregts, M.A., Norris, G., Bretz, R., Ciroth, A., Maurice, B., Von Bahr, B., et al., 2001. Framework for modelling data uncertainty in life cycle inventories. Int. J. Life Cycle Assess. 6, 127–132.

Jorde, K., Schneider, M., Peter, A., Zoellner, F., 2001. Fuzzy based models for the evaluation of fish habitat quality and instream flow assessment. Proceedings of the 2001 International Symposium on Environmental Hydraulics 27–28.

Kennard, M.J., Mackay, S.J., Pusey, B.J., Olden, J.D., Marsh, N., 2010. Quantifying uncertainty in estimation of hydrologic metrics for ecohydrological studies. River Res. Appl. 26, 137–156.

Kéry, M., 2010. Introduction to WinBUGS for Ecologists. Elsevier, Chennai, India.

King, J.M., Tharme, R.E., De Villiers, M.S., 2008. Environmental Flow Assessments for Rivers: Manual for the Building Block Methodology (Updated Edition). Water Research Comission, Republic of South Africa.

Ladson, A., Argent, R., 2002. Adaptive management of environmental flows: lessons for the Murray-Darling Basin from three large North American Rivers. Aust. J. Water Resour. 5, 89–101.

Lakoff, G., 1987. Women, Fire, and Dangerous Things. University of Chicago Press, Chicago.

Lin, S.W., Bier, V.M., 2008. A study of expert overconfidence. Reliab. Eng. Syst. Safe. 93, 711–721.

Lowe, L., 2009. Addressing Uncertainties Associated with Water Accounting. PhD thesis. Department of Civil and Environmental Engineering. University of Melbourne. March 2009.

Lowe, L., Etchells, T., Malano, H., Nathan, R. & Potter, B. 2009a. Addressing uncertainties in water accounting. 18th World IMACS/MODSIM Congress. Cairns, Australia.

Lowe, L., Horne, A. & Stewardson, M. 2009b. Using irrigation deliveries to achieve environmental benefits: accounting for river losses. International conference on implementing environmental flow allocations. Port Elizabeth, South Africa.

Lowell, K.E. 2007. At what level will decision-makers be able to use uncertainty information? Modelling and Simulation Society of Australia and New Zealand. New Zealand.

Maier, H., Ascough II, J., Wattenbach, M., Renschler, C., Labiosa, W., Ravalico, J., 2008. Chapter five uncertainty in environmental decision making: issues, challenges and future directions. Developments in Integrated Environmental Assessment 3, 69–85.

Martin, T.G., Kuhnert, P.M., Mengersen, K., Possingham, H.P., 2005. The power of expert opinion in ecological models using Bayesian methods: Impact of grazing on birds. Ecol. Appl. 15, 266–280.

Maughan, O.E., Barrett, P.J., 1992. An evaluation of the Instream Flow Incremental Methodology (IFIM). J. Ariz.-Nev. Acad. Sci. 24/25, 75–77.

Mccarthy, M.A., 2007. Bayesian Methods for Ecology. Cambridge University Press, Cambridge, UK; New York.

Metzeling, L., Wells, F., Newall, P., Tiller, D., Reid, J., 2004. Biological Objectives for Rivers and Streams – Ecosystem Protection. Victorian Environment Protection Authority, Melbourne.

Nathan, R., Weinmann, P., 1995. The estimation of extreme floods–the need and scope for revision of our national guidelines. Aust. J. Water Resour. 1 (1), 40–50.

Norton, S.B., Cormier, S.M., Suter, G.W., Schofield, I.K., Yuan, L., Shaw-Allen, P., et al., 2008. CADDIS: the causal analysis/diagnosis decision information system. In: Marcomini, A., Suter, I.G.W., Critto, A. (Eds.), Decision Support Systems for Risk-Based Management of Contaminated Sites. Springer, New York.

Palmer, M.A., Hakenkamp, C.C., Nelson-Baker, K., 1997. Ecological heterogeneity in streams: why variance matters. J. N. Am. Benthol. Soc. 16, 189–202.

Pettit, N.E., Froend, R.H., Davie, P.M., 2001. Identifying the natural flow regime and the relationship with riparian vegetation for two contrasting Western Australian Rivers. Regul. Rivers Res. Manage. 17, 201–215.

Poff, N.L., Zimmerman, J.K.H., 2010. Ecological responses to altered flow regimes: a literature review to inform the science and management of regulated rivers. Freshw. Biol. 55, 194−205.

Poff, N.L., Richter, B.D., Arthington, A.H., Bunn, S.E., Naiman, R.J., Kendy, E., et al., 2010. The ecological limits of hydrologic alteration (ELOHA): a new framework for developing regional environmental flow standards. Freshw. Biol. 55, 147−170.

Quinn, G.P., Keough, M.J., 2002. Experimental Design and Analysis for Biologists. Cambridge University Press, Cambridge.

Refsgaard, J.C., Nilsson, B., Brown, J., Klauer, B., Moore, R., Bech, T., et al., 2005. Harmonised techniques and represenative river basin data for assessment and use of uncertainty information in integrated water management (HarmoniRiB). Environ. Sci. Pol. 8, 267−277.

Refsgaard, J.C., Van Der Sluijs, J.P., Brown, J., Van Der Keur, P., 2006. A framework for dealing with uncertainty due to model structure error. Adv. Water Resour. 29, 1586−1597.

Richter, B.D., Baumgartner, J.V., Powell, J., Braun, D.P., 1996. A Method for Assessing Hydrologic Alteration within Ecosystems. Conserv. Biol. 10, 1163−1174.

Roberts, J., 2002. Species-level knowledge of riverine and riparian plants: a constraint for determing flow requirments in the future. Aust. J. Water Resour. 5 (1), 21−31.

Shenton, W., Hart, B., Chan, T., 2011. Bayesian network models for environmental flow decision-making: 1. Latrobe River Australia. Rivers Res. Appl. 27, 283−296.

Shenton, W., Hart, B.T., Chan, T.U., 2014. A Bayesian network approach to support environmental flow restoration decisions in the Yarra River, Australia. Stoch. Environ. Res. Risk Assess. 28, 57−65.

Speirs-Bridge, A., Fidler, F., Mcbride, M., Flander, L., Cumming, G., Burgman, M., 2010. Reducing overconfidence in the interval judgments of experts. Risk Anal. 30, 512−523.

Standards Australia 1990. Measurement of Water Flow in Open Channels.

Stewardson, M., Rutherfurd, I., 2006. Quantifying uncertainty in environmental flow assessments. Aust. J. Water Resour. 10, 151−160.

Stewardson, M., Rutherfurd, I., 2008. Conceptual and Mathematical Modelling in River Restoration: Do We Have Unreasonable Confidence? In: Darby, S., Sear, D. (Eds.), River Restoration Managing the Uncertainty in Restoring Physical Habitat. John Wiley & Sons Ltd, West Sussex, England.

Stewardson, M.J., Webb, J.A., 2010. Modelling ecological responses to flow alteration: making the most of existing data and knowledge. In: Saintilan, N., Overton, I. (Eds.), Ecosystem Response Modelling in the Murray-Darling Basin. CSIRO Publishing, Melbourne, Australia.

Surowiecki, J., 2004. The Wisdom of Crowds. Doubleday; Anchor, USA.

Tharme, R.E., 2003. A Global Perspective on Environmental Flow Assessment: Emerging Trends in the Development and Application of Environmental Flow Methodologies for Rivers. River Res. Appl. 19, 397−441.

Uusitalo, L., 2007. Advantages and challenges of Bayesian networks in environmental modelling. Ecol. Model. 203, 312−318.

Van Der Lee, G.E.M., Van Der Molen, D.T., Van Den Boogaard, H.F.P., Van Der Klis, H., 2006. Uncertainty analysis of a spatial habitat suitability model and implications for ecological management of water bodies. Landscape Ecol. 21, 1019−1032.

Walker, W.E., Harremoes, P., Rotmans, J., Van Der Sluijs, J.P., Van Asselt, M.B.A., Janseen, P., et al., 2003. Defining uncertainty: A conceptual basis for uncertainty management in model-based decision support. Integrated Assessment 4, 15−17.

Wardekker, J., Van Der Sluijs, J.P., Janssen, P.H.M., Kloprogge, P., Petersen, A.C., 2008. Uncertainty communication in environmental assessments: views from the Dutch science-policy interface. Environ. Sci. Pol. 11, 627−641.

Warmink, J.J., Janssen, J.A.E.B., Booij, M.J., Krol, M.S., 2010. Identification and classification of uncertainties in the application of environmental models. Environ. Model. Softw. 25, 1518–1527.

Webb, J.A., King, E.L., 2009. A Bayesian hierarchical trend analysis finds strong evidence for large-scale temporal declines in stream ecological condition around Melbourne, Australia. Ecography 32, 215–225.

Webb, J.A., De Little, S.C., Miller, K.A. & Stewardson, M.J. in prep. Quantifying the benefits of environmental flows: combining large-scale monitoring data within hierarchical Bayesian models. Freshw. Biol.

Webb, J.A., Stewardson, M.J., Koster, W.M., 2010. Detecting ecological responses to flow variation using Bayesian hierarchical models. Freshw. Biol. 55, 108–126.

Webb, J., De Little, S., Miller, K., Stewardson, M., Rutherfurd, I., Sharpe, A., et al., 2013. Modelling ecological responses to environmental flows: making best use of the literature, expert knowledge, and monitoring data. The 3rd Biennial ISRS Symposium: Achieving Healthy and Viable Rivers, 221–234.

Young, W.J., Scott, A.C.T., Cuddy, S.M., Rennie, B.A., 2003. Murray Flow Assessment Tool – a technical description. CSIRO Land and Water, Canberra.

Zadeh, L.A., 1965. Fuzzy sets. Information and Control 8, 338–353.

ENVIRONMENTAL WATER WITHIN WATER RESOURCE PLANNING

WATER BUDGETS TO INFORM SUSTAINABLE WATER MANAGEMENT

Brian Richter[1] and Stuart Orr[2]

[1]*Sustainable Waters, Crozet, VA, United States* [2]*WWF International, Gland, Switzerland*

16.1 INTRODUCTION

Two-thirds of the freshwater sources on earth—rivers, lakes, and aquifers—have been only lightly exploited to provide water supplies for cities, farms, and industries (Brauman et al., 2016). That is very good news for those concerned with leaving enough water flowing through the planet's freshwater and estuarine ecosystems to sustain their health.

The bad news is that the other one-third has been very heavily tapped (Fig. 16.1). In these water basins, three-fourths or more of the natural replenishment of water from precipitation, runoff, and recharge is being depleted for human use, either on an ongoing basis or at least during drier months or years. That means that present-day water flow regimes in these basins bear little resemblance to the historic, naturally varying flows that previously existed.[1]

As a consequence of this heavy water use in many regions, freshwater ecosystems have been heavily degraded and thousands of aquatic species have become extinct. The populations of species living in freshwater have declined by an estimated 76% since 1970 (WWF, 2014).

Formulating strategies to restore or protect environmental water regimes depends on a clear understanding of not only depletion levels (present or future), but also the particular uses for the water being removed. For rivers that are already being heavily depleted, we can assume that recovering targeted species populations or some degree of overall ecological health will require restoring some of the original volume and timing of flow. This means that the consumptive use of water in one or more water-use sectors will need to be reduced, and some portion or all of the unused water will need to be reallocated to environmental use. This does not necessarily mean that productive uses need to be curtailed or eliminated, however; it may be possible to gain substantial reductions in consumptive use by reducing nonbeneficial losses such as evaporation from soils and leaking canals in irrigated farming (Richter et al., in press).

---

[1]We acknowledge that river flow regimes can be altered by dam operations and other human activities as well (see Chapter 3), but this chapter focuses primarily on volumetric depletion and ways to mitigate it. Additionally, without diminishing the importance of proactive protection of environmental water regimes in two-thirds of global basins where water supplies are lightly used, this chapter focuses on one-third of global basins where renewable water supplies are being fully used and competition among water-use sectors is high.

Water for the Environment. DOI: http://dx.doi.org/10.1016/B978-0-12-803907-6.00016-4

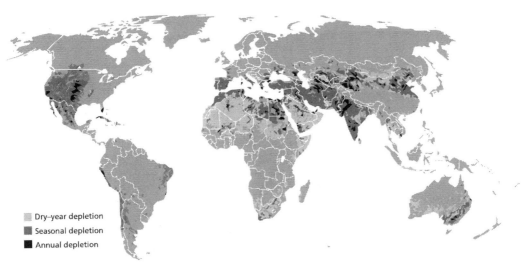

FIGURE 16.1

This map highlights water sources that are being heavily depleted by water use, placing communities and freshwater ecosystems at risk of water shortages. Annual depletion means that the renewable water supply is depleted by more than 75% on an average annual basis. Seasonal depletion means that heavy depletion occurs primarily in certain months, and dry-year depletion means that water shortages occur during drier years or droughts.

Source: Adapted from Brauman et al. (2016)

Accurate water accounting data—including estimation of how much water is being consumed by each water-use sector—can greatly aid in identifying the sectors of water use primarily responsible for river depletions, facilitating better-honed ideas for reducing water consumption in those sectors. This information can also help in identifying which parties might be cooperative, or alternatively resistant to, your efforts.

In this chapter we discuss the nature of the water accounting, or *water budgeting*, necessary to develop an effective flow restoration or protection strategy. We then discuss the needs, concerns, and motivations that drive decision making by both water users and governments, as context for development of flow restoration strategies.

16.2 CONSTRUCTING A WATER BUDGET

A *water budget*, also commonly referred to as a *water balance*, is simply an accounting for the water inputs and extractions from a particular water source such as a river. In this sense, a water budget is much like a financial budget, or a personal bank account. These budgets help you to understand how much is going into the account, how much is coming out, and how much remains. With both water and financial budgets, it can also be helpful to further understand where the outputs or extractions are going, that is, how the water is being spent.

A water budget can be constructed at many different levels of spatial and temporal resolution, so it will be important to consider how the water budget will be used in developing a flow restoration or protection strategy. For instance, do you need to know which sectors are using water during each month, or do you only need a general characterization of how much water gets used in total over the course of a year?

With respect to spatial resolution, the water budget accounting should ideally focus on a particular point or reach along a river such as the location where the environmental water regime is to be improved. The accounting for inputs would include all basin runoff entering the river system above the location of interest, any groundwater discharge entering the river system, and any water imported into the system such as through interbasin water transfers or desalination. The outputs would include all extractions and losses of water from the river system upstream of the location of interest, including transfers out of the basin and any natural losses; note that if some of the water extracted for use is returned to the river upstream of your point of interest, those return flows will need to be accounted for as well. Fig. 16.2 for the Colorado River Basin in Texas, United States—

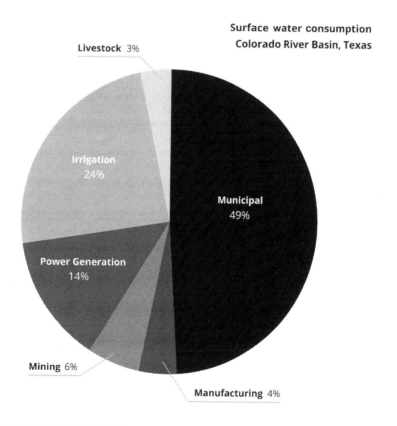

FIGURE 16.2

This pie chart diagram provides insights into the overall consumptive use of water within a basin, indicating that municipal, power generation, and irrigation uses are the biggest water consumers. Consumptive water use in cities (municipal use) is high in this region because of extensive outdoor landscape watering.

Source: Texas Water Development Board, Texas Commission on Environmental Quality

only representing the water-use side of the water budget—is an example of the most common way that water-use information is reported; such tallies provide insight into which sectors are consuming the greatest volumes of water.

In many countries, some types of water budget estimates have already been prepared for most or all river basins, usually by governmental entities such as ministries or departments of water resources. For example, the UK Environment Agency has prepared detailed water budgets for all catchments in England and Wales, and set limits on water extractions that are intended to protect environmental water regimes (Environment Agency, 2013).

However, readily available water budget information may not yet have been prepared or summarized at the spatial or temporal scale most useful for determining potential impacts of water depletions at your location of interest. For instance, water budgets specific to a particular river basin or catchment may not be available (e.g., summaries may be only available for the entire state or country), or only available for the entirety of a river basin, as shown in Fig. 16.2. That may or may not provide information useful to the location where the environmental water regime is of interest. In the case of the Colorado River, river conservationists in Texas are quite interested in restoring freshwater inflows into Matagorda Bay, located at the mouth of the Colorado River Basin; therefore, this summary of the overall water use in the river basin helps them to understand that municipal, power generation, and irrigation are the largest sectors of water use.

Most water budget information (i.e., Fig. 16.2) has been prepared on an average annual basis, meaning that water use is estimated over the course of multiple years and then reported as annual averages. This annual resolution can in many instances be quite helpful to the development of flow restoration strategies because it offers a high-level overview of the water inputs or outputs, or both. However, water budgets prepared for each year (as opposed to averaged annual results), or summarized on average monthly intervals, may reveal interannual, seasonal, or shorter-term flow issues that are not apparent in annually averaged budgets.

For example, Fig. 16.3 illustrates the annual estimates of water consumption used to create the average annual pie chart in Fig. 16.2. This graph provides some important additional information. For instance, it appears that overall water use has remained relatively stable over time. However, among the individual sectors of water use, we see that municipal use is growing and irrigation use is declining. Both interannual and within-sector variations and trends can be quite useful in formulating environmental water regime restoration strategies.

Although Figs. 16.2 and 16.3 provide useful information about water use, anyone interested in restoring environmental water regimes will want to know more. For instance, it will be important to know what portion of the natural or historical river flow is being consumptively used (removed) from the river. This information can then inform assessments of whether or not that volume of river depletion is damaging freshwater and estuarine ecosystems or increasing the likelihood of water shortages for some water users (Richter, 2016).

Given that most environmental water assessments will be developed at a monthly or shorter timescale, water budget information summarized at a similar timescale will be most valuable in determining how best to meet the environmental water requirements. Unfortunately, such information at monthly or shorter time frames is rarely available from measurements of the volumes of water availability and use, and must therefore be simulated using hydrologic models, as explained in Chapter 3.

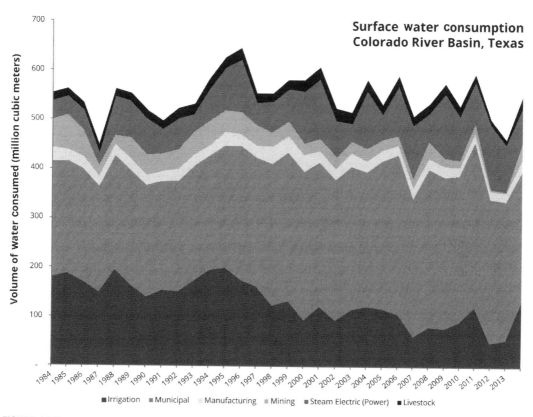

FIGURE 16.3

This graph of the changing water consumption by sector in the Colorado River Basin of Texas, United States (where 1 million cubic meters = 1 GL) provides important additional information about year-to-year variations and trends within individual sectors of water use.

Source: Texas Water Development Board, Texas Commission on Environmental Quality

Fig. 16.4 illustrates the results from a hydrologic simulation model developed for the Colorado River in Texas.[2] This particular model simulates the water inputs into the river for each month, over multiple decades, and then accounts for the water extracted from the river and any water returned to the river after use. The graph suggests that in many months over the model-simulated time series, water consumption depleted a large portion of the originally available river flow. This depletion took place in all months of the year (Fig. 16.4); water flows in the typically wetter winter months (November–February) are captured and stored in large reservoirs for later use during the drier summer (May–September) months, but summer flows can be heavily depleted as well, when

---

[2]Note that the period simulated by this model ends in 1998, and more recent simulation results are not available. This is another challenge of developing water budgets: the requisite data for compiling a complete budget may not be available for the time period(s) of interest.

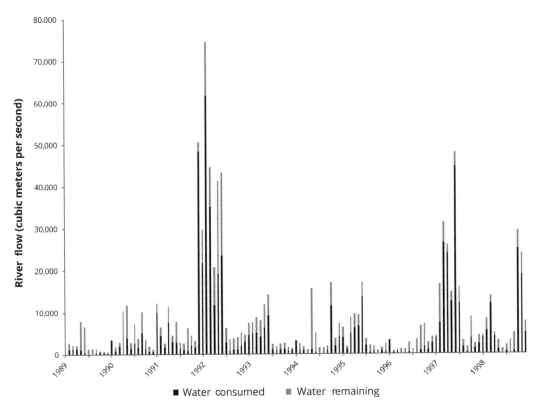

FIGURE 16.4

This graph was generated using the output of the water availability model created by the state water agencies in Texas, United States. This model simulates month-by-month water inflows into, and extractions from, the Colorado River system over a 60-year simulation period ending in 1998 (where 1 cubic meter per second = 86.3 ML/d). Only a portion of the available output is shown here.

irrigation is taking place on farms and residential lawns are being watered (more than half of all urban use in this area of Texas goes to lawn watering).

At this point, we know how much water is being removed from the Colorado River, and for what purposes. However, we still do not know whether that volume of water removal is too much, given environmental water requirements. The answer to that question should be based on a comparison of the water remaining in the river with an environmental flows assessment, as discussed in other chapters in this book. The difference between the water regime remaining in the river and environmental water requirements will tell us how much water will need to be restored, and at what time.

In Texas, teams of river scientists have been developing environmental flows assessments for rivers across the state. Multiple environmental water target locations have been evaluated in the Colorado River Basin, including an assessment of the monthly freshwater inflows necessary to sustain the ecological health of the lower river and the Matagorda Bay estuary located at the river's

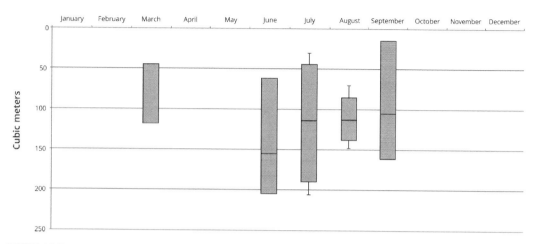

FIGURE 16.5

This graph summarizes the volume of flow restoration needed to meet base-flow environmental water requirements on a monthly basis (where 1 cubic meter = 0.001 ML). It is based on a comparison of the environmental flows assessment prepared for this location with actual measured river flows during the recent 20 years. Each vertical bar in this box-and-whisker plot illustrates the minimum, 25th percentile, median, 75th percentile, and maximum volumes of flow restoration that would have been necessary in each month of the year to fully meet environmental water requirements. Note that in the computer simulation of Matagorda Bay inflows used to evaluate these environmental water shortfalls, only 7 of 20 years assessed had any environmental water gaps at all; therefore, the box-and-whisker diagram here is based on the 7 years with shortfalls. Additionally, shortfalls did not exist in some months of the year.

Source: Computer simulation from Texas Commission on Environmental Quality; assessment of environmental water gaps by The Nature Conservancy

mouth. Scientists have compared their environmental flows assessments with the actual measured river flows entering Matagorda Bay. A summary of the results is presented in Fig. 16.5.

16.3 REBALANCING A WATER BUDGET FOR LONG-TERM SUSTAINABILITY

In this section we discuss some of the political, governance, economic, and social considerations relevant in developing strategies for restoring river flows. We begin by discussing two challenges pertinent to flow protection and restoration worldwide: (1) setting limits on consumptive use and (2) reallocating water to environmental use. These concepts are explored in further detail in Chapters 17 and 18.

16.3.1 SETTING LIMITS ON CONSUMPTIVE USE

As discussed previously, it is highly likely that freshwater species populations and ecosystems have been heavily impacted in at least one-third of all river basins globally, given that water flows have been depleted by 75% or more in these basins on an ongoing or episodic basis (Brauman et al., 2016; Richter, 2016; Richter et al., 2011). A primary concern in all of these depleted basins is the question of whether river depletions could continue to worsen in the future. If the relevant water management agencies or ministries continue to allow water consumption to increase in the future, the challenges of restoring river flows to meet environmental water requirements will intensify.

Appropriate water-use restrictions can be imposed in a number of ways. For example, a river basin or groundwater aquifer can be *legally closed* to any additional uses in the future such as by refusing to issue any new surface water-use rights or permits to construct new wells. By setting a maximum limit or *cap* on allowable extractions of water, a governing entity can begin to address the problems of overallocation and design or facilitate more optimal water-use scenarios (e.g., see Environment Agency, 2013; Richter, 2014a; Richter et al., 2016). The establishment of a cap clarifies and institutionalizes the goal to provide environmental water protection as an integral part of sustainable water management (Postel and Richter, 2003; Richter, 2014a; Richter et al., 2011); such a cap should specify the maximum volume of water consumption that would be compatible with maintaining environmental water regimes at the targeted level. The preparation of a water budget can be of great help in determining the level at which a consumptive use cap should be set, and regular recalculations of a water budget over time can help to determine whether the cap is being attained.

As with water budget information, a cap that is based upon and varies on a monthly basis (i.e., see *cap and flex* approach in Richter 2014a), or that sets limits on the degree to which natural daily flows can be altered (i.e., see *percent of flow approach* or *presumptive standard* in Richter et al., 2011), will be most effective in ensuring that environmental water requirements are met.

However, simply preventing *increases* in consumptive use may be only half the solution if a gap in environmental water requirements already exists, meaning that total consumptive water use is already too great and will need to be *reduced* at certain times to achieve the desired environmental water regime. For example, Fig. 16.5 suggests that existing consumptive uses need to be reduced in the months of March and June through September in order to meet environmental water requirements in the Colorado River of Texas. In this situation, the setting of a targeted cap at a level lower than existing consumption levels can be immensely beneficial in building political and financial support for environmental water regime restoration, particularly if the cap is agreed to through an open, inclusive stakeholder process (Chapter 7).

We openly acknowledge that any suggestion that existing levels of consumptive use need to be reduced can be expected to be extremely controversial, and politically challenging to implement, particularly if pursued through regulatory reductions. Not surprisingly, regulatory reductions in consumptive water use have been achieved in only a limited number of countries and river basins to date. Some of the keys to success in such endeavors include: (1) providing strong financial incentives for voluntary (rather than mandatory) reductions in water use; (2) providing sufficient time for water users to make adjustments; and (3) actively engaging the water users themselves in designing the most favorable ways to accomplish the desired reductions (Chapters 7 and 18).

Once the level of the cap is determined and adopted, strategies can be formulated for bringing existing water uses into line with these limitations on consumptive use. Chapter 18 provides an in-depth discussion of the strategies available for reducing consumptive use, but it is worth noting that both public (governmental) and private (especially nongovernmental) initiatives have success-fully reduced consumptive uses for the benefit of environmental water regime restoration. For example, in the Murray–Darling Basin of Australia, the Commonwealth government is pursuing a program of acquiring existing water entitlements from irrigation farmers to attain a goal of 20% reduction in overall consumptive use (Hart, 2015; Richter, 2014a). Those acquired water entitle-ments are held by the Commonwealth Environmental Water Holder for use in maintaining the envi-ronmental water regime. Similarly, in the western United States, many nongovernmental conservation organizations have used private funds to acquire water rights from consumptive users and then designated those rights to environmental water purposes.

16.3.2 REALLOCATING WATER TO ENVIRONMENTAL USE

Each of these public and private initiatives has had to overcome the second challenge mentioned above, that of explicitly reallocating water to environmental use. Simply reducing the consumptive use of water by individuals or water-use sectors does not guarantee that the water savings will be made available for the intended purpose. An all-too-common problem with water-saving initiatives around the world is that saved water is quickly taken up for other uses, instead of being allowed to contribute to environmental water regime restoration. For instance, most investments in improving water-use efficiency in irrigated agriculture simply result in more irrigation or other consumptive use taking place, because the improved efficiencies have made more water available to farmers. Similarly, urban water conservation programs rarely result in leaving more water in a river; instead, the reduction in water use by city residents or industries simply enables the population to grow or other water uses to expand. There is a real challenge in preventing new or expanded uses from con-suming water that is being saved and intended for environmental water restoration.

A highly desirable solution from an environmental perspective is to formally (legally) reallocate the saved water to the environment. This can be accomplished by reserving all saved or protected water flows from nonenvironmental uses such as by reducing or retiring some portion of a water user's rights to use water, or by formally changing the purpose of a water right to environmental water purposes.

As discussed above, the setting of a cap at a level commensurate with environmental water pro-tection can be immensely helpful in institutionalizing and socializing such flow restoration efforts, because it sets a formal target against which flow restoration efforts can be measured. For example, in the Murray–Darling Basin, the Australian government has set *sustainable diversion limits* (caps) within each of 32 subbasins, and calculated the reductions in existing consumptive uses necessary to attain those limits. Those targeted reductions represent goals for environmental water regime restora-tion, to which the government is holding itself responsible.

16.3.3 THE ECONOMIC AND POLITICAL PROSPECTS FOR ENVIRONMENTAL WATER REGIME RESTORATION

Although it is very important to formulate environmental water restoration plans on the basis of sound scientific and technical analysis, the ultimate success of those plans will likely hinge much more heavily on economics and politics rather than on environmental science. In reality, environmental values and even ecosystem service arguments seldom carry much weight in real-world water allocation and management decisions. Around the world, clean and healthy river basins and environmental water regimes are undervalued and regularly ignored in planning for economic growth and social well-being. The importance of the services that freshwater ecosystems render, and the risk incurred when these systems are disturbed, are typically given little weight in economic and political trade-offs involving water.

To be successful, advocates for protecting environmental water regimes need to consider the social and economic values and trends that are most strongly influencing water allocation and use in the river basin of interest, and find ways to align and integrate environmental water restoration and protection initiatives into development plans that will likely be dominated by other values and objectives.

In most river basins, the economics and politics associated with agricultural water use will be of paramount importance. Although opportunities to harvest water savings in municipal and industrial water use should never be ignored in an environmental water restoration effort, the dominance of irrigation as the largest water consumer in most water-stressed basins (84% of consumptive use globally; Brauman et al., 2016) suggests that the greatest opportunities for reallocation of water to the environment—both in terms of volume and cost—will usually be gained from irrigated agriculture (Richter et al., 2013). Also of interest is the fact that at the global scale the economic importance of agriculture has been diminishing; there has been a widespread reduction in the relative contribution of agriculture in most national economies (as measured by change in agriculture's contribution to gross domestic product; Quandl, 2016). Most countries are now experiencing a gradual, but in some cases a rapid shift toward industrial production (including mining, utilities, construction, and manufacturing) and service-oriented economies, with associated changes in water use (Debaere and Kurzendoerfer, 2016). Although we undeniably need to produce enough food to feed the world, the diminishing contribution of irrigated agriculture in many national economies suggests that opportunities may exist to *make the case* for improvements in environmental water as the sectoral uses of water are changing.

In many regions of the industrialized world, and more specifically in water-stressed basins within industrialized regions, the ongoing expansion of urban and industrial water uses will come primarily from reallocation of water from irrigation. For example, in the Colorado River Basin of Texas, as evidenced by Fig. 16.3, economic production and associated water use have been markedly shifting away from agriculture toward service industries and power generation. Along with this reallocation of water use—which in this case is accomplished through voluntary and compensated trading of water-use rights—have come notable increases in the economic productivity of water use. The economic output per unit of water consumed has more than doubled in the Colorado River Basin since 1990 (Debaere and Hamilton, 2016), due to the comparatively high economic value of water in urban and industrial uses as compared with irrigated agriculture.

We can expect similar reallocations of water use among sectors in many countries in coming decades as the economic structure of industrialized nations evolves (Debaere and Kurzendoerfer, 2016). Given that approximately half of all 60 + nations experiencing water scarcity manage their water use through some type of water rights-based allocation system (Richter, 2016), reallocation of water use in these countries will typically involve a transfer of water-use rights, either through governmental reductions in permitted extractions or through market-based transactions such as water rights purchases, as exemplified by Australia (e.g., Murray—Darling Basin) and the western United States. Successful restoration of environmental water regimes in these settings will depend upon the ability of environmental water advocates to intervene in policy reforms or market trading. This can result from convincing governments to reallocate more water to the environment or from purchases of water rights for the environment in water market settings. For example, the Israeli government has decided to use ocean desalination more heavily in order to allow more water to flow in the Jordan River (Richter, 2014b), and the Australian government has allocated substantial public funding for the *buyback* of water rights from farmers in the Murray—Darling Basin (Hart, 2015; Richter, 2014a).

The prospects and strategies for environmental water restoration in developing regions look quite different. The regulatory systems governing water use are much stronger in industrialized regions, populations are more stable, and income levels are higher and rising. In contrast, the vast majority of projected global population growth will occur in developing countries that typically lack adequate water management capacity and effective institutional frameworks (Orr and Cartwright, 2010). As these countries develop and standards of living improve, we can expect that their diets will become more water-intensive, and overall commodity consumption will increase; some projections forecast a doubling in irrigation volume will be required to feed future population growth (IWMI, 2007).

Unfortunately, many of these countries are already approaching the limits of water availability, as evidenced by high levels of water depletion depicted in Fig. 16.1. As a common pool resource, the sustainable and equitable management of water in these developing countries will require continual reconciliation of trade-offs between private interests and values associated with collective well-being (Hepworth and Orr, 2013).

The ability of countries to adapt to changing and increasing water needs, and the success of environmental water advocates in protecting or restoring flows, will be largely dependent upon the development of adequate organizational capacity at the national and river basin levels. Equally important will be support for the creation and implementation of laws, rights, and recourse mechanisms to reconcile trade-offs and meet rising demands. Ensuring these as a basis for sustainable water management remains a daunting task in many developing countries.

The dominant priority in many countries with high population growth is for the delivery of water supply and sanitation services to poor and marginalized segments of the population, which has been given renewed emphasis under the recently formulated global *Sustainable Development Goals* (UNDP, 2016). Development priorities such as building water infrastructure to provide water for economically productive purposes—mainly energy and agricultural expansion—are high on many countries' wish lists. Therefore, any arguments for environmental water protection or restoration will need to resonate with business continuity and government priorities, and perhaps with higher degrees of community participation.

The good news is that strong water policy exists, at least on paper, in many developing countries, and there is growing commitment to allocate water with a clear recognition of the many important social and economic services that rivers provide. An important characteristic of the livelihoods and well-being of populations in developing regions is a strong dependence upon natural ecosystems as a foundation of food security, that is, through fisheries, floodplain farming, and so on. Importantly, there is also a growing appreciation for the need for private sector and civil society stakeholders to become engaged in water-related decision making. This includes management partnerships, particularly where government capacity for delivering and managing water is lacking. These trends offer expanding opportunities for environmental advocates to engage in water dialogues.

Recent discussions of water management priorities in the Kafue River Basin, a tributary to the Zambezi River in Zambia, help to illustrate these challenges and the need for water budgets to guide economic development and environmental protection decisions. The Kafue Flats is an immense floodplain area (6500 km^2) situated between two dams on the Kafue River. Although it is widely understood that a large population depends directly upon fishing, hunting, and floodplain farming in the Flats, decisions over dam operations and water allocations for out-of-stream diversions are being made without adequate understanding of the environmental water regime necessary to sustain their ecological productivity. Stakeholder consultation, led by the *World Wildlife Fund (WWF)*, an international, nongovernmental conservation organization, was undertaken to clarify risks and opportunities related to water use in the area (WWF, 2016). Although not framed as an environmental water assessment per se, an understanding of existing and potential water use helped to characterize water demands for energy generation, sugar and beef production, export crops, Lusaka's water supply, and ecosystem-dependent livelihoods and food security (WWF, 2016).

Alteration of the natural flow regime was identified as an overarching risk to the local economy dependent upon ecosystem benefits and services. The river flow through the Kafue Flats is now largely controlled by the operations of the Itezhi-Tezhi Dam, which in turn are dictated by the need to release water upon demand for hydropower generation downstream of the Flats at the Kafue Gorge Dam. Stakeholders in the Kafue Flats have recognized the need for better understanding of the hydrology of the Flats, particularly the influence of the dams and diversions on ecosystem function, so that the dam operating rules can be optimized. Although extensive research exists on the hydrology of the Flats, awareness of the needs of different users dependent upon the Kafue Flats ecosystem was not clear (WWF, 2016).

The Zambian capital city of Lusaka is already dependent on an offtake pipe from the Kafue Flats for 44% of its water supply, with domestic demand projected to grow considerably in the coming decades. The food and energy needs of this growing urban population are tied to the Kafue Flats as well. Electricity is generated through the Kafue Gorge Dam, whereas food staples such as maize, beef, fish, milk, and sugar consumed in Lusaka are mostly sourced from the Kafue Flats region. Agriculture accounts for 73% of total water withdrawals in Zambia, the majority of which takes place within the Kafue Flats, where large tracts of sugar cane are irrigated. In addition to sugar, this is the home of the largest concentration of cattle in Zambia, the largest area of maize planted (mostly by smallholders), and barley production (WWF, 2016).

Even though a tacit understanding of the qualitative connections between flow alteration and livelihoods in the Kafue Flats is held by many of the water users and the government through the stakeholder engagement process, there is considerably less understanding—particularly quantitative—of the potential consequences of ongoing water allocations to support economic expansion. Water budgets have been calculated for the Kafue Basin in the past (Wamulume et al., 2011), but the steep increase in water demands and the continued growth of the Zambian economy mean that changes are rapid, and decisions impacting the complex hydrology of the region need to be revisited and updated regularly to reflect the changing socio-economic landscape.

Given the rates of change in much of the developing world, the need for water budgets to serve as the basis of water governance could not be more urgent. Water budgets can serve as the basis for stakeholder dialogue, providing a means for formulating proactive agreements around desired levels of environmental water protection and appropriate allocations to consumptive use. Although the ability of developing nations to establish effective governance systems remains highly uncertain in many instances, nongovernmental and other private entities can help focus attention and resources on establishing water budgets that can lay the foundation of a sustainable water future in these countries.

16.4 CONCLUSION

This chapter has illustrated the value of preparing a water budget as an essential foundation for developing environmental water regime restoration strategies. By gaining a better understanding of how and when water is used in the basin of interest, and the impact of this water use on the natural or historical flow regime, environmental water advocates will be better able to identify and prioritize water-saving strategies that could free up water for reallocation to environmental water.

However, such reallocation of saved water back to the environment is by no means a guaranteed outcome, and environmental water advocates will need to find ways to legally reallocate and protect any saved water to remain instream, and not to be taken up for other consumptive uses. The political opportunities and challenges incumbent in this endeavor should be well understood such as by gaining understanding of the economics and politics of the water use in the basin of interest, as well as development priorities that may align, or obstruct, the restoration of water for the environment.

REFERENCES

Brauman, K., Richter, B.D., Postel, S., Malsy, M., Florke, M., 2016. Water depletion: an improved metric for incorporating seasonal and dry-year water scarcity into water risk assessments. Elem. Sci. Anth. 4, 83. Available from: http://dx.doi.org/10.12952/journal.elementa.000083.

Debaere, P., Hamilton, B., 2016. Unpublished data, University of Virginia.

Debaere, P., Kurzendoerfer, A., 2016. Decomposing U.S. water withdrawal since 1950. J. Assoc. Environ. Resour. Econ. in press.

Environment Agency, 2013. Managing Water Abstraction. Environment Agency, Bristol, England.

Hart, B.T., 2015. The Australian Murray−Darling Basin Plan: challenges in its implementation. Int. J. Water Resour. Dev. Available from: http://dx.doi.org/10.1080/07900627.2015.1083847 (accessed 24.03.16).

Hepworth, N., Orr, S., 2013. Corporate water stewardship: new paradigms in private sector water engagement. In: Lankford, B.A., Bakker, K., Zeitoun, M., Conway, D. (Eds.), Water Security: Principles, Perspectives and Practices. Earthscan, London.

IWMI, 2007. Water for Food, Water for Life: a Comprehensive Assessment of Water Management in Agriculture. International Water Management Institute. Earthscan, London and International Water Management Institute, Colombo, Sri Lanka.

Orr, S., Cartwright, A., 2010. Water scarcity risks: experience of the private sector. In: Martinez-Cortina, L., Garrido, A., Lopez-Gunn, E. (Eds.), Re-thinking Water and Food Security. CRC Press, London.

Postel, S., Richter, B., 2003. Rivers for Life: Managing Water for People and Nature. Island Press, Washington, DC.

Quandl, 2016. Agricultural share of GDP by country. Available from: https://www.quandl.com/collections/economics/agriculture-share-of-gdp-by-country (accessed 15.08.16).

Richter, B., 2014a. Chasing Water: a Guide for Moving from Scarcity to Sustainability. Island Press, Washington, DC.

Richter, B., 2014b. Can desalination save a holy river? Water currents. National Geographic. Available from: http://voices.nationalgeographic.com/2014/11/02/can-desalination-help-save-a-holy-river/ (accessed 02.11.14).

Richter, B., 2016. Water Share: Using Water Markets and Impact Investing to Drive Sustainability. The Nature Conservancy, Arlington, Virginia.

Richter, B.D., Davis, M., Apse, C., Konrad, C., 2011. A presumptive standard for environmental flow protection. River Res. Appl. 28, 1312–1321.

Richter, B.D., Abell, D., Bacha, E., Brauman, K., Calos, S., Cohn, A., et al., 2013. Tapped out: how can cities secure their water future? Water Policy 15 (2013), 335–363.

Richter, B.D., Powell, E.M., Lystash, T., Faggert, M., 2016. Protection and restoration of freshwater ecosystems. Chapter 7. In: Miller, K.A., Hamlet, A.F., Kenney, D.S., Redmond, K.T. (Eds.), Water Policy and Planning in a Variable and Changing Climate. CRC Press, Boca Raton, FL.

Richter, B.D., Brown, J.D., DiBenedetto, R., Gorsky, A., Keenan, E., Madray, C., et al. Opportunities for saving and reallocating agricultural water to alleviate water scarcity. Water Policy, in press.

UNDP, 2016. UNDP support to the implementation of the 2030 agenda for sustainable development. United Nations Development Programme, New York.

Wamulume, J., Landert, J., Zurbrügg, R., Nyambe, I., Wehrli, B., Senn, D.B., 2011. Exploring the hydrology and biogeochemistry of the dam-impacted Kafue River and Kafue Flats (Zambia). Phys. Chem. Earth 36, 775–788.

WWF, 2014. Living planet report: species and places, people and places. World Wildlife Fund International, Gland, Switzerland.

WWF, 2016. Kafue Flats shared risks assessment report. World Wildlife Fund Zambia, Lusaka.

MECHANISMS TO ALLOCATE ENVIRONMENTAL WATER

17

Avril C. Horne[1], Erin L. O'Donnell[1], and Rebecca E. Tharme[2]

[1]The University of Melbourne, Parkville, VIC, Australia [2]Riverfutures, Derbyshire, United Kingdom

17.1 INTRODUCTION

Many countries have now embraced a policy of environmental water provision, but the implementation of these policies on the ground in highly diverse and complex socio-political contexts presents significant challenges (Le Quesne et al., 2010). Taking an environmental water regime from policy to actual delivery of water to rivers and other wetlands, once the environmental water requirements have been determined (see Chapter 11 for a discussion of the science and methodologies involved) and an appropriate water regime agreed, requires a mechanism to allocate the water to the environment. Effective management and enforcement of this mechanism is also dependent on the broader institutional settings (see Chapter 19).

In this chapter, we identify and discuss the choices available to policy makers in terms of selecting one or more *allocation mechanisms* for environmental water. We use the term *allocation mechanism* to encompass the legal or policy tools available to provide water for environmental purposes. Allocation mechanisms can be given effect through legislation, water resource management plans, water abstraction licenses, and water infrastructure operation licenses (Speed et al., 2013). In this chapter, the array of allocation mechanisms that have been implemented around the world are classified into a finite number of representative and distinct choices, based on the legal characteristics, primary function, and operation of the mechanism. From a legal perspective, the allocation mechanisms can be considered as two distinct categories:

1. Mechanisms that impose conditions on other water users (such as a cap); and
2. Mechanisms that establish a legal right to water for the environment itself (such as an environmental water right).

Within these two broad legal categories, we have further classified these allocation mechanisms based on their function and operation (described in detail below). There are three distinct mechanisms that impose conditions on other water users: (1) a cap on total water abstraction; (2) license conditions on water abstractors; and (3) conditions on storage operators. There are two further mechanisms that create a distinct legal right to water for the environment: (1) an ecological or environmental water reserve and (2) environmental water rights.

As the case studies below demonstrate, these allocation mechanisms can be applied in combination within a single system (Speed et al., 2013), or used in isolation. Distinguishing among each of the allocation mechanisms is important, as they afford the environment different levels of

Water for the Environment. DOI: http://dx.doi.org/10.1016/B978-0-12-803907-6.00017-6

protection relative to human water uses, they have varying levels of flexibility, and they have different operational challenges.

In this chapter, *environmental water* encompasses all water available to the environment through the array of possible allocation mechanisms. Environmental water may or may not equate to the environmental water requirements—the volume and pattern of water (i.e., the magnitude, timing, and quality) estimated to meet the flow needs of the ecosystem, and the human livelihoods and well-being that directly depend on it (The Brisbane Declaration, 2007).

In this chapter, we explore contemporary water management approaches to provide environmental water, while acknowledging that customary (or indigenous) management systems for water resources (Chapter 9) can be part of, or exist alongside, these approaches. We begin by describing the different allocation mechanisms that are currently used worldwide. Choosing a mechanism to allocate environmental water requires consideration of the current water management policy and practice, as well as the biophysical state of a river system. We then outline a process to help select an appropriate allocation mechanism in context. The discussion of this process draws on the work of earlier chapters, which provide insights into the vision and objectives for environmentally sustainable water resource management (Chapters 7−10), the balance between consumptive water use and the needs of the environment (Chapter 16), and the technical and social engagement tools that are available and used to identify the required environmental water regimes ahead of their allocation (Chapters 11−15 and Chapter 7, respectively). The chapter then concludes with a series of case studies that highlight some of the adopted allocation mechanisms currently used across the globe.

17.2 TYPES OF ALLOCATION MECHANISMS: LEGAL BASIS, FUNCTION, AND OPERATION

The mechanisms for allocating water to the environment can be classified into five broad types, based on their legal characteristics, primary function, and operation of the mechanism. As discussed above, the most fundamental difference among these allocation mechanisms relates to the way they use the law to protect environmental water. Some allocation mechanisms provide environmental water by imposing conditions on the way that other users access water such as imposing a cap on the water that can be used for consumptive purposes. Alternatively, allocation mechanisms can create specific legal rights to water for the environment itself. This distinction is important as the different legal rights created by the allocation mechanisms can give rise to different legal remedies, supporting policies, and regulatory frameworks, and the need for different organizational capacities (see Chapter 19).

Within the two broad groups defined earlier, the differences between these allocation mechanisms relate to their function and operation. Identifying specific mechanisms in use in a particular situation is complicated by the fact that they are often implemented in combination, and terminology is extremely variable, so our discussion focuses on the function of each mechanism. The diversity of the applications of these mechanisms is demonstrated by the case studies later in the chapter. Each allocation mechanism is discussed briefly below, with more detail provided in Table 17.1.

Table 17.1 A Comparison of Mechanisms for Environmental Water Allocation

	Cap on Consumptive Water Use	Conditions on Other Water Users		Environmental Rights	
		License Conditions for Water Abstractors	Conditions on Storage Operators or Water Resource Managers	Ecological or Environmental Reserve	Environmental Water Rights
Description	Limit on the total volume of licenses issued and/or the extraction/abstraction of water against these licenses.	Conditions listed on the license of individual water users that restrict the volume and/or timing of extractions.	Conditions on a storage operator prescribing releases from storage for downstream ecological needs.	Legally establishes environmental water as a prior right to consumptive water use.	Individual water access rights held by an environmental water manager.
Implementation options	A limit on gross or net extractions and/or volumes held in storage. Limit can be implemented as a percentage of total available resource or as a volumetric limit. More complex caps can be applied but will be more difficult to enforce.	Minimum instream flow required before extraction. Variable instream flow regime protected before extraction. Seasonal license (e.g., wet season pumping only).	Passing a fixed daily release from storage. Passing a percentage of storage inflow. Passing a prescribed environmental water regime.	A fixed annual volume based on minimum needs. An adaptive volume updated annually based on given years' water needs.	In a system where water is stored for use in a dam, environmental water can be released to create any component of flow regime. In a system where water rights are linked to a share of river flow environmental water rights can protect existing flows.
Establishment	Legislative changes or water resource plan development to implement a cap.	Reissuing of licenses with changed conditions. Changes to a water allocation plan.	Infrastructure upgrades required to make releases possible. New licenses or storage operator contracts with changed conditions. Changes to a water allocation plan.	Legislative changes or water resource plan development to create the reserve. Secondary implementation mechanisms are likely to be required if creating the reserve after water has been allocated to other users.	Establishment of an organization to hold the water rights. A process to manage the water rights. Acquisition or creation of the water rights.
Operation	Compliance and enforcement—often many individuals.	Compliance and enforcement—many individuals. Compliance difficulty increases with seasonal or flow-related limits.	Compliance and enforcement of storage license or contract conditions.	Planning and implementation of reserve on an annual basis. Compliance of individual license holders to ensure that reserve is protected.	Ongoing planning, use, and performance reporting of environmental water rights. Enforcement of water rights to prevent others taking environmental water.
Ability for adaptive management	Difficult to retrospectively adjust limit once consumptive rights have already been allocated. Easier to apply as a precautionary limit prior to allocation.	Can be difficult to adjust once license conditions are established unless process for review incorporated into water resource plans.	Can be difficult to adjust once license conditions are established unless process for review incorporated into water resource plans.	Flexible on a planning scale through revised determinations of environmental water needs, depending on the specific measures used for implementation.	Highly flexible on a continuous basis through change in portfolio of water rights held by the environment.
Protection of environment in times of scarcity	A volumetric cap will provide little environmental water during times of scarcity, however, if set as a percentage of available water allocated it will continue to provide some environmental outcome.	A license condition that limits pumping at times of low-flow will protect the remaining flows in the river.	A flow release condition on a dam will depend on the volume stored in the dam and the priority given to the environmental release.	A reserve may still protect the environment as one of the recognized priorities, but this depends on where it sits in the priority list.	Environmental water rights are often reduced in line with other users, rights. In a priority-based system (i.e., prior appropriation), the priority of the right determines the protection, highlighting the need for high-priority rights.
Security of the right (i.e., legal procedure for removal)	Likely to require legislative change to remove the cap (although altering the specific level of the cap may be less difficult).	Can be changed at the planning level (unlikely to require a bylaw, regulation, or legal change).	Can be changed at the planning level (unlikely to require a bylaw, regulation, or legal change).	Depending on where reserve is specified, removal would require constitutional or legislative change. However, the specific implementation measures used may be more vulnerable individually.	Would require legislative change to remove the capacity of the environment to hold water rights.

17.2.1 ALLOCATION MECHANISMS THAT IMPOSE CONDITIONS ON OTHER WATER USERS

Cap on water abstraction

A cap sets a limit on the total volume of water that can be extracted from a system by consumptive users, and can be expressed as an absolute volume or as a share of the available volume of water in a given period. This is seen by many as a key step in the sustainable allocation of water (Le Quesne et al., 2010; Mekonnen and Hoekstra, 2016; Richter, 2014). A cap typically operates at the bulk level of water allocation, and will need to be supported by accurate accounting of water use by individual users. A cap is more than the sum of limits to individual water use (usually stipulated as conditions on water licenses, see below), as it prevents the issuing of new water licenses once the cap has been reached.

When used to limit water consumption at the current level of water use (or higher), it is in many ways the easiest mechanism to implement and can rapidly limit further water extraction (Speed et al., 2013). When combined with the capacity to transfer water rights between users, it may also stimulate, in a more sustainable manner, the further development and efficient use of available water. However, once water has been fully allocated to consumptive water users (or indeed over-allocated relative to actual system water availability), it is difficult and costly to reduce these rights and return water to the environment at a later stage (Hirji and Davis, 2009a). For example, the US state of Michigan amended its water law to protect environmental water from future water withdrawals, but had to exempt existing users due to political pressure (Le Quesne et al., 2010). Although the recent reforms in the *Murray−Darling Basin (MDB)*, Australia, have seen the existing cap lowered further with the implementation of new *sustainable diversion limits* (MDBA, 2012), this has required significant political will and substantial investment in water recovery (Connell and Grafton, 2011; Hart, 2015). Where possible, it is far preferable to use caution when allocating consumptive water rights in the first instance, although history would show that this is not often the case (Hirji and Davis, 2009a).

Importantly, caps are usually applied and audited annually, with compliance based on long-term average consumption. This allows for variable water consumption between years, to reflect varying water availability. As an example, the original MDB cap on diversions explicitly considered the *climatic and hydrological conditions* of the year the cap was set (MDB Ministerial Council, 2000), and the new sustainable diversion limit (SDL) relies on the concept of long-term average yield (MDBA, 2012).

Although a cap is an important mechanism for limiting overuse of water resources, a cap has limited ability to address the full suite of hydrological alteration of the flow regime caused by consumptive extractions and/or other forms of flow regime alteration. A simple cap on diversions most effectively limits extraction of the higher portions of a hydrograph, such that low-flow conditions may still be inadequate from a social or ecological standpoint, and does not provide any requirement around the flow regime itself. Although caps can be constructed in more complex ways to protect low-flow requirements or other features of the hydrograph, they become more challenging to implement. Further, where a cap is a volumetric limit (rather than set as a proportion of flow), the environmental water sitting outside the cap will be the first impacted should total inflows to the system reduce due to climate change (or in periods of extended drought; Speed et al., 2013). For these important reasons, caps are quite often implemented in combination with other mechanisms to allocate environmental water.

License conditions for water abstractors

Water abstraction licenses enable water users to pump or divert water directly from the water source such as a river or aquifer. Typically, these licenses will impose conditions on those water users. These conditions may be specified on the water license itself, or a generic condition may require compliance with the relevant water management plan, which would impose a range of specific conditions. These conditions can include daily, seasonal, or annual abstraction limits, as well as specifying the level of water availability (as an instream flow or aquifer depth) that must be met prior to abstraction, or some combination. Conditions on water licenses are often used to provide environmental water in systems that do not have major onstream storages (in Australia and elsewhere, these are often called *unregulated* systems).

Two Australian examples help to illustrate the importance of these extraction limits for protecting environmental water. In the Olinda Catchment in Victoria, water extraction licenses are issued either as (1) all-year licenses or (2) dam-filling licenses. The dam-filling licenses can only be used during the high-flow period (July–October), thus reducing extractions during low-flow periods. All-year licenses face further seasonal limits, so that diversions must cease when streamflows fall below a set level. In this instance, the level was predetermined through an environmental flows assessment process, but in other cases it may simply reflect historical water-sharing arrangements. This instream flow level varies depending on the location in the catchment, and is intended to ensure that the river cannot be completely dewatered by water abstractors. In the Condamine Balonne Catchment in New South Wales, water licenses include conditions to limit water extraction during the winter bird-breeding months. In particular, during a bird-breeding event, water extraction must be reduced by 10% for a 10-day period (Speed et al., 2011).

Where water licenses are issued in perpetuity, it is difficult and often controversial to implement new limitations on existing users, particularly without some form of compensation to users (although seldom to the environment). Such alterations are usually most effective when licenses or water allocation plans come up for renewal on a regular basis, creating an expectation that license conditions can change. Changes to licenses can also be achieved through the use of highly consultative approaches, but these may be lengthy and complex (State of Victoria, 2004).

Conditions on water licenses that protect environmental water can also be overridden by governments, most notably in times of water scarcity. For example, during the Millennium drought in Australia, the newly established water resource management plans of New South Wales were suspended to enable drought-affected irrigators to continue to access water that would otherwise have been allocated to the environment (NWC, 2009).

Setting conditions on the licenses held by private users can be used to protect minimum instream flows (or aquifer depths), as well as seasonal flow variations and the high-flow portions of the hydrograph, depending on the specific conditions applied. However, where such conditions have not been applied when water licenses were issued, it requires significant political will and community support to change existing licenses. Moreover, as private water licenses are a relatively low level of legal instrument (see Box 17.1), their conditions can be difficult to protect from arbitrary change. A further implementation challenge with this allocation mechanism is the requirement to check compliance and enforce license conditions across what is potentially a large number of individual license holders in any given river basin.

BOX 17.1 LEGAL INSTRUMENT HIERARCHY

Each of the allocation mechanisms described in Table 17.1 uses a specific legal or regulatory tool to either impose conditions on other users, or create specific legal rights for the environment. The power of the legal tool will vary depending on the local context, based on the power of the rule of law and the capacity to enforce legal rights. A short primer on the nature of these legal instruments is provided to help guide the choice. However, each jurisdiction is different, so any instrument will need to be adapted based on the requirements of the relevant location, including case law. In transboundary situations, this will be further complicated by the need to find an acceptable international legal instrument, and to embed this instrument within the legal hierarchies of each state (Lenaerts and Desomer, 2005).

In most jurisdictions, the legal instruments exist in a hierarchy of legal power (see Fig. 17.1). Typically, a national constitution creates the legal power for the legislature to pass laws (Endicott, 2011). This legislation, in turn, supports the creation of regulations and the decisions of the executive and their delegates. These regulations often enable the creation of additional rules and policies to guide the implementation of the law (Killingbeck and Charles, 2011). As a result, for example, "a statute enacted by the directly elected parliament generally takes precedence over any regulation" (see discussion in Lenaerts and Desomer (2005), p. 745).

Each of the legal mechanisms outlined in Table 17.1 can be established at one of the several levels within this legal hierarchy. For example, the Ecological Reserve has been enacted within statute (Godden, 2005), whereas conditions on licenses in Victoria, Australia, are usually set by a water corporation as a sublegal tool. In choosing a legal instrument, there is a core trade-off that needs to be navigated: the choice between legal power and ease of implementation (Fig. 17.1).

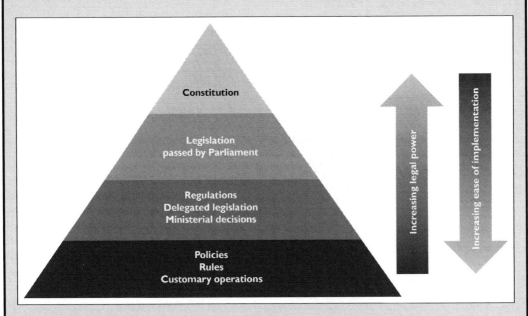

FIGURE 17.1

Legal instrument hierarchy in a constitutional Parliamentary democracy, such as Australia.

(*Continued*)

BOX 17.1 (CONTINUED)

Legal power increases as one moves up the legal instrument hierarchy. Choosing a legal instrument that is embedded in legislation has much greater legal power than a policy or rule. In the event of a challenge to a particular allocation mechanism, this legal power matters. This is most relevant if the environmental water allocation is likely to be challenged in court, but can also be helpful in setting new norms around the importance of environmental water in the local context (Fisher, 2010).

However, ease of implementation increases as one moves down the legal instrument hierarchy. It is usually much easier to alter a regulation than it is to change legislation. Changing legislation requires the support of the government of the day, and usually the support of opposition parties, so that such changes are not immediately undone following a change in government. If the environmental water allocation is still at the pilot stage, it may be easier to use a subordinate legal instrument (i.e., a regulation or bylaw) to implement it in the first instance.

Finally, the balance between power and ease of implementation is further exacerbated by the need to respond to changing circumstances. Once the environmental water allocation has been made, it will be progressively more difficult to change the mechanism to allocate environmental water as you move up the legal hierarchy. This can be helpful in order to protect the environmental water allocation against arbitrary changes, but will create barriers to changing in response to changing environmental conditions. Although some environmental water allocation mechanisms include the capacity to review the volume or share of water that is set aside for the environment, in practice this is often hard to do (unless there is adequate compensation). For example, the Murray–Darling Basin Plan in Australia is a legislative instrument, adopted by the minister, and endorsed by a majority of Parliament under the *Water Act 2007*. Although Parliament was not required to vote on the Basin Plan, Parliament could vote to disallow it. The Basin Plan has specified a cap on water use (the sustainable diversion limit), and although the capacity for amending the cap is embedded in the legislation (sections 23A and 23B, *Water Act 2007*), any changes to this cap would require amendments to the Basin Plan, requiring approval by the minister and acceptance by Parliament.

Environmental water allocations, like other aspects of environment and natural resource management, can benefit from a combined approach. A high-level legal instrument such as legislation or the constitution can be used to set a commitment to providing environmental water, with the details of the implementation left for regulations and policies (e.g., Wade, 2010). Targeting the right legal instrument for the specific goals of the environmental water allocation mechanism will be essential to success.

Conditions on storage operation or water resource management

When there is an onstream storage, particularly large infrastructure, conditions may be imposed on the storage operator to provide environmental water. These conditions may include compliance with a cap (see above), as well as requiring water releases at particular times or volumes to meet specific environmental water requirements (Richter, 2009; Warner et al., 2014). In contrast to conditions for water abstractors, conditions on storage operation (including operational rules for reservoirs and other water infrastructure) apply to the water resource manager or storage operator, as opposed to individual water abstractors.

This approach has been most commonly used to ensure release of a minimum flow downstream of a dam. It can, however, also be used in a regime-focused way (nowadays considered best practice, as compared with a constant or only seasonally varying minimum flow; see Chapter 11), to create high-flow events such as freshes and floods, require reasonable rates of rise and fall in any water releases for consumptive or hydropower use, or to impose limits on flow releases during naturally low-flow seasons (thereby avoiding potentially damaging seasonal reversals in the hydrograph; Speed et al., 2011). These conditions on operation can also affect the quality of water released from storage, for example, by requiring multilevel offtakes to ensure that environmental water releases comply with required temperatures or dissolved oxygen levels downstream.

A key challenge when implementing this approach is to ensure that the storage infrastructure is capable of delivering flow releases to meet the specified release conditions. Existing water infrastructure may often be constrained by its physical character and outlet works. For example, a dam's outlet may not have sufficient capacity to release a *controlled flood* of the magnitude needed to restore ecological processes downstream. This allocation mechanism can best be achieved through a design of appropriate infrastructure in the project planning stage (e.g., multiple release outlets, or an array of different-sized power-generating turbines). However, although it is likely to be more complex and costly, it is also possible to retrofit the structures needed for environmental water releases (Opperman et al., 2011; Richter and Thomas, 2007). Dam relicensing programs provide an opportunity to undertake these modifications (Chapter 21).

As with conditions on individual licenses, conditions on storage operators are unlikely to be specified in statute, and are usually established as a regulation or policy. This means that they can more easily be altered, but where these changes will affect the reliability of supply to other users, altering these conditions is likely to be controversial (Ward and Ward, 2004).

17.2.2 ALLOCATION MECHANISMS THAT CREATE LEGAL WATER RIGHTS FOR THE ENVIRONMENT

An ecological or environmental water reserve

A reserve represents an allocation that is a *legally established prior right to water* for the environment, typically ahead of economic uses of water. A reserve is a legal right specified in legislation, which gives it significant legal power, in contrast to the conditions on existing license holders or storage operators, which do not provide a formal environmental right. The Ecological Reserve of South Africa is the best-known example (see the case study below), and demonstrates the importance of both legal power and effective implementation (which may depend on the use of a range of other allocation mechanisms in combination with the reserve).

This mechanism can work in a variety of ways, but generally depends on a well-developed understanding of the environmental water regime for a river (rather than a simple annual volume of water for the environment, or minimum flow), explicitly linked to a desired future condition for the system, defined through stakeholder consultation. It may require a significant adjustment to consumptive water use, especially in instances where existing water use is high. Catchment management strategies or water resource plans, or a similar policy vehicle, may be required to make the reserve operational in a particular basin. This may necessitate a compulsory relicensing process to adjust historical water rights so that the desired social and ecological objectives can be achieved (Le Quesne et al., 2010).

Although there are a number of examples of the calculation of Ecological Reserves, in southern Africa, their implementation continues to present real challenges (Le Quesne et al., 2010). The case study discussion of the South African Ecological Reserve below gives insights into some of these challenges and the difficulties inherent in the pragmatic application of the reserve within large-scale water reform.

Finally, the term *reserve* is used in a wide variety of contexts. In South Africa, the reserve relates to a statutory priority right to water, including both a Basic Human Needs Reserve and an Ecological

Reserve (see below). Similarly, in Mexico, the National Water Reserve Program (*Programa Nacional de Reservas de Agua [PNRA]*) has as its primary objective the legal establishment of water reserves for the conservation or restoration of vital ecosystem services, such that the reserve water volume is excluded from the total amount that can be allocated in concessions (Comisión Nacional del Agua, 2011). The first phase of the PNRA aims to ensure the issue, by 2018, of reserve decrees for 189 priority river basins where conservation potential is high but water allocation to users is presently low; a second phase will focus on basins already facing intensive pressure on their water resources. In Victoria, reserve is a statutory term used to bring together other mechanisms for allocating environmental water under one legal umbrella (Foerster, 2007; Godden, 2005).

Environmental Water Rights

The environment may be issued with water rights of the same structure and legal properties as those of consumptive water users. An organization of some kind is required to hold and manage these environmental water rights on behalf of the environment (further discussed in Chapter 19). These water rights can be used to protect low flows, high flows, or both, depending on the nature of the river system and the volume of rights held. Where these water rights are held in storage, or include the right to be extracted from the river, they can be actively managed, for instance to create flushing flow events or modified overbank flows. In Victoria, these water rights are held by the *Victorian Environmental Water Holder (VEWH)*, which uses the water available each year to generate specific flow regimes in wetlands across the state (VEWH, 2015). The VEWH prioritizes how best to use the environmental water rights, providing different aspects of the flow regime at different locations, depending on the environmental needs for a particular season and year (further discussed in Chapter 23).

Environmental water rights can be allocated in bulk as part of the initial water allocation in an underutilized water system. Significantly, they are one of the most effective mechanisms that presently exist to address over allocation, as environmental water rights can be created by increasing water use efficiency (and transferring savings to the environment), or by purchasing water rights from existing water users (see Chapter 18).

The precise role of environmental water rights in providing environmental water will depend on the water rights framework. Where water rights are largely homogeneous, environmental water rights can be used most flexibly. However, in situations such as the western United States, which uses a water rights framework based on prior appropriation, the specific nature of the water rights will be much more important. Prior appropriation creates a wide array of heterogeneous water rights, and means that the seniority of the water right held effectively defines the aspect of the flow regime that the license can protect (see the Columbia Basin case study below). When a junior water right is held for the environment, it effectively establishes a partial limit on new users to the system, as it has priority over all future issued licenses. It does not, however, have priority over existing water users. To provide baseflow(s), the water right held by the environment would need to be the most (or one of the most) senior water rights in that section of stream, so that it has priority over all existing users. It is perhaps unlikely that the prior appropriation framework would be adopted for use in other locations, but understanding the different roles of senior and junior environmental water rights remains crucial in systems where this framework is in place.

Environmental water rights can be used creatively to allocate water for the environment. For example, in the Colorado River delta in Mexico, the existing legal framework did not explicitly allow for the transfer of water away from irrigation use. However, a parcel of water was transferred from irrigated cropland to natural floodplain wetlands, for the specific purpose of *irrigating* riparian vegetation and wetlands (Le Quesne et al., 2010). This eliminated the need to change the official use associated with the water right. There is now a direct right to allocation for the environment (O. Hinojosa, Pronatura Noreste, pers. comm.) as well as a recently delivered environmental water release (flood) to the delta (Daesslé et al., 2016; Tarlock, 2014).

17.3 SELECTION OF AN APPROPRIATE MECHANISM FOR ENVIRONMENTAL WATER ALLOCATION

We propose a process to help explicitly set the terms for the selection of the mechanism for environmental water allocation in a context-sensitive way (refer to Fig. 17.2). The steps are expressed as three critical questions for environmental water policy makers and practitioners who will guide the selection of the most appropriate mechanism(s) to allocate environmental water:

1. What are the initial conditions and fundamental constraints of the river system?
2. What philosophy underpins the value of the environment?
3. How responsive does the approach need to be to biophysical variability and change?

Once these questions are answered, the appropriate allocation mechanism can be selected to implement the environmental water allocation.

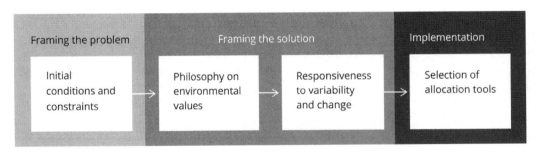

FIGURE 17.2

Four-step process to set the terms of allocation of environmental water.

The discussion of allocation mechanisms is generally considered in the context of a river basin. The same principles are applicable at a larger landscape scale, but implementation must then recognize the need to engage across borders (as is the case for a number of the case studies discussed later in this chapter). Importantly, environmental water allocation should be considered in the context of the whole basin (including linked groundwater). This is discussed further in Chapter 16, which describes the challenges of accounting for catchment processes that may alter water availability.

Framing the problem (understanding the initial conditions and system constraints) and setting the philosophy on environmental values are significant questions discussed in more detail in

Chapters 7, 8, and 9. The following is a brief analysis of these issues as they pertain to the selection of environmental allocation mechanisms.

17.3.1 FRAMING THE PROBLEM: INITIAL CONDITIONS AND FUNDAMENTAL CONSTRAINTS

The first step is to define the problem to be solved by allocating (or recovering) environmental water. This includes understanding:

1. The types of water bodies that constitute the resource (e.g., Are the rivers permanent or ephemeral? Are there floodplain wetlands? Is there an important delta or estuary?).
2. The current levels of allocation for consumptive use (is the system near natural with no or minimal detectable socio-ecological impacts or is it already heavily utilized with extractions causing ongoing environmental deterioration?).
3. The specific socio-ecological needs that the environmental water allocation will address (capturing the aspirations of stakeholders and identified vision for the river system).
4. The legal, institutional, and technical capacity to allocate water to the environment and enforce such allocations.

The level of development in a catchment significantly influences the range of appropriate options for implementing environmental water across the continuum from flow *protection* to flow *restoration*. It is arguably simpler and more cost-effective longer term to protect existing flows in situ rather than having to restore absent flows. It is also easier to implement mechanisms that place requirements on new users than to require changes to the entitlements or activities of existing users (Le Quesne et al., 2010). The objectives and, hence, the approach used, to establish environmental water to *protect* a river system with little or no existing water resource development may differ considerably from those where flows need to be restored in a system that is heavily modified for human use and already has high competing demands for water (Acreman et al., 2014).

Where the aim is to protect a system against future additional use and associated detrimental effects on its integrity and ecosystem services to people, it is also possible to use some of the mechanisms discussed in this chapter as a precautionary step to protect existing environmental water (e.g., a cap on consumptive use or an environmental reserve). Although best practice may involve significant investment in determining the environmental water requirements, these mechanisms can be implemented as a first step toward delivering a more complete environmental water regime when capacity is developed to invest in this additional scientific work.

In heavily used systems, the emphasis shifts to recovering water for the environment and redressing the imbalance between consumptive and environmental water uses and improving ecosystem health to a societally agreed state, by restoring environmental water. The allocation of environmental water under these circumstances is arguably more challenging. This chapter focuses on the mechanisms or tools to allocate environmental water, which can be applied in all water systems. Chapter 16 discusses the broad formation of strategies for restoring flows, and Chapter 18 follows with a discussion of the market-based approaches to acquire water, which is one such recovery strategy.

The types of hydrological impact and the elements of the flow regime required to meet social and ecological objectives will also influence the selection of environmental allocation mechanisms. As discussed in Chapter 3, impacts to the hydrological regime can include changes in the characteristics of flow magnitude, timing, duration, frequency, and rate of change, as well as other aspects of intra- and interannual variability. There must be a match between the elements of the flow regime that are most affected and most relevant for ecosystem condition and services, and the capacity of the allocation mechanism to provide these elements of the flow regime. Addressing some of these ecohydrological impacts may require a change in the allocation of water between the environment and consumptive water users, whereas others (i.e., seasonal inversions in flow) may be improved through other management mechanisms that do not alter the relative distribution of water between consumptive and environmental uses (as discussed later in Chapter 21; e.g., see Hillman, 2008).

Importantly, when allocating environmental water, the complete water cycle needs to be considered, taking into account interrelationships between surface water and groundwater (Hirji and Davis, 2009b), as well as the entire ecosystem and its functional network (e.g., from headwater streams to coastal estuary, or a river and its associated complex of floodplain lakes). This is discussed in more detail in Chapters 20 and 22.

Finally, there are a wide range of existing legislative and institutional arrangements already in place as part of the existing system of water resources management for all water users. There are enabling conditions required for implementation and these will vary depending on the allocation mechanism. Speed et al. (2011) identified these enabling conditions broadly as:

1. High-level government support (creation of a legal mandate and policy framework);
2. Institutional capacity (human resources, regulatory and management systems, system modeling, monitoring, enforcement, and science);
3. Basin-wide planning.

The existing arrangements do not necessarily limit the options available in terms of allocation mechanisms for environmental water, but they will make some allocation mechanisms easier, quicker, and more cost-effective to implement (Garrick et al., 2009). Le Quesne et al. (2010) identify three broad pathways to the adoption and implementation of environmental water policies: comprehensive water policy reform, incremental water policy reform, and interstate treaties. They emphasize that it is important to commence the process of allocating environmental water early and in a precautionary manner, but also to recognize that quickly and easily implemented options may need to be followed by longer-term, substantive reforms. Allocating water to the environment requires recognition of path dependency, to ensure that immediate steps do not preclude or limit future options in providing water to the environment. For example, creating property rights to water in a bid to increase the efficiency of water use may limit the power to alter conditions on water licenses without compensation in the future.

17.3.2 PHILOSOPHY ON ENVIRONMENTAL VALUES

Central to determining how to provide environmental water is the societally driven process of agreeing on a vision and objectives for the river system (Rogers and Biggs, 1999; Roux and Foxcroft, 2011): What is the value system that is going to define the solution? What are the roles of cultural values and belief systems? Is the aim to maintain or return the system to as natural a state as possible, or to develop a flow regime that supports multiple values and benefits including economic growth and community outcomes, or some state in between? Earlier chapters provide detail on the setting of objectives for environmental water (Chapter 7) and the sharing of water with consumptive users (Chapter 16).

The underlying value system informs the priority the environment has relative to other water users, influences stakeholder perceptions and the negotiation process involved. This is a crucial element in selecting an appropriate allocation mechanism as it informs the required priority, security, and flexibility of the allocation.

In brief, there are four distinct environmental philosophies that underpin current water allocation systems:

1. *Ecocentric*: An ecocentric philosophy values nature for its own sake, judging that it requires protection because of its intrinsic value (O'Riordan, 1991). Given the difficulty and, often, the lack of desirability of restoring rivers to natural states, this philosophy is perhaps most relevant when considering rivers that are still in near natural condition and/or where conservation and biodiversity values are top priorities. After a survey of Australian rivers in *near pristine condition*, the Queensland government introduced legislation (*Wild Rivers Act 2005 (QLD)*) that restrict mining or intensive aquaculture in declared high preservation areas (Neale, 2012a). The aim was to ensure that the integrity of these natural values of rivers remained preserved. Conservation groups campaigned for this Act, but it was opposed by local indigenous landowners and industry (Neale, 2012b), namely those groups that had social and economic stakes in potential developments. Notably, this legislation was repealed in 2014.

2. *Anthropocentric*: An anthropocentric view is that the environment should be protected because of the ecosystem services it provides to society. Under this philosophy, nature should be protected because human comfort and survival are dependent on it (Gagnon Thompson and Barton, 1994).

3. *Triple bottom line*: Originating in the business world in the late 1990s, the term *triple bottom line* refers to three areas of potential impact when assessing outcomes: economic, social, and environmental (Gray and Milne, 2014). When applied to natural resource management, a triple bottom line philosophy accepts a relationship and balance between economic, social, and environmental outcomes. The term *working river* has been used in the MDB to describe "aquatic ecosystems harnessed as a natural resource" (Hillman, 2008). This acknowledges the need to understand the ecological processes that determine the health of the resource, the processes by which the resource can support production (economic outcomes) into the future, and the social preferences and implications of the resource management approach (Hillman, 2008).

4. *Minimum protection*: This suggests a *minimal* allocation of environmental water that acts as a limit on the economic uses of water. Chile's commitment to minimum instream flow in 2005 is an example of the way that minimum protection is often applied (O'Donnell and Macpherson, 2013). The notion of environmental water can also be conflated with *conveyance water*, which is the water required for running the river system for economic purposes (e.g., see State of Victoria, 2009).

17.3.3 RESPONSIVENESS TO VARIABILITY AND CHANGE

Environmental water allocations should reflect the inherent variability in water availability (both within and between years), as well as having the capacity to change in response to long-term shifts in water availability (i.e., climate change), community preferences, and technology.

In many river systems around the world, there is significant natural variability in streamflow both within years and between years. Environmental water recommendations and flow studies may identify a suite of flow components that are ecologically significant and reflect within-year variability. Different mechanisms to allocate environmental water protect different elements of the flow regime (refer to Table 17.1 for more detail), and multiple mechanisms may be required to protect the full suite of flows required. For example, conditions on private diverters may protect low or baseflow(s), whereas conditions on storage operators that define dam operating rules may help sustain the necessary high flows. The detail of how a particular allocation mechanism is used will also determine how effective particular mechanisms will be at protecting the complete recommended flow regime (e.g., a condition on a storage operator may require release of a constant flow, or variable event-based flow releases).

Similarly, many environmental flows assessment methods account for interannual variability by generating different recommended flow regimes for different year types. Matching the allocation mechanism to provide this variability is not always straightforward. Successfully providing variability typically depends on designing the allocation mechanism as a proportion of impacted catchment inflows. For example, a storage operator may be required to pass a percentage of dam inflows rather than a fixed-flow regime, so that more environmental water releases are delivered in wetter years, and less in dry years. Similarly, a cap may be implemented as a percentage of the available resource rather than a volumetric limit, which ensures that a river will not be completely dewatered during dry years.

Along with the flexibility to respond to climate variability, the allocation approach should also acknowledge the uncertain climate future. An environmental water allocation approach should be devised for our current best state of knowledge, but also needs to be robust and adaptive under plausible future scenarios (Walker et al., 2001). Climate change will exacerbate variability with changed rainfall patterns and altered magnitudes and patterns of streamflow (Jiménez Cisneros et al., 2014; Ukkola et al., 2015). Importantly, climate change is likely to increase the frequency and duration of droughts in many already dry regions (IPCC, 2014).

Where environmental allocation mechanisms are designed based on historical streamflow patterns and current social values and objectives, these will reflect an agreement on the appropriate balance between consumptive and instream water use. However, in many systems where existing environmental allocation mechanisms are in place (i.e., caps or conditions on license holders), if there is a step shift in the overall water available in the system, the reduction in water availability will be felt more by the environment than by consumptive water users. The simplest of examples is to consider a system with a cap on consumptive use. For simplicity, let us assume that the average water availability under the current climate is 150 GL/year, and the cap on extractions is set at 100 GL/year. If climate change causes a *reduction* in the mean annual streamflow of 25 GL/year, consumptive users may still consume up to 100 GL/year, leaving only 25 GL/year for the environment. By setting the cap as a volumetric limit (rather than a proportion), the loss will be felt solely by the environment and not by consumptive water users. To demonstrate this using real data, consider the catchments of the MDB. Fig. 17.3 shows projected reductions in water availability and surface water diversions under a median 2030 climate scenario for river catchments in the MDB (CSIRO, 2008). Slightly different allocation mechanisms for environmental water, and often a combination of multiple allocation mechanisms, are employed for each of these catchments. The majority of river catchments show that surface water diversions do not decrease at the same rate as the climate impacted reduction in water availability. The Ovens Basin, for example, shows a 12% reduction in water availability, with no reduction in surface water diversions. This means that the 12% reduction is borne by the environment rather than shared across consumptive water users. The Gwydir Basin is the closest to having an equal reduction in water availability and consumptive use (around 10%), and it is the only catchment where, under current allocation mechanisms, the environment and consumptive water users would equally feel the impacts of climate change.

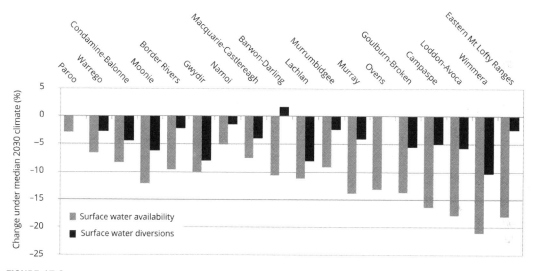

FIGURE 17.3

Impact of changed climate on water availability and surface water diversions for river catchments in the Murray−Darling Basin, Australia, for the median 2030 climate scenario.

Source: CSIRO (2008)

There is a strong link between the responsiveness of the mechanism to allocate environmental water and the philosophy that underpins the allocation of environmental water. Under current climate conditions, there may be some agreed share of water between the environment and consumptive water users, based on the initial philosophy underpinning environmental water requirements. However, under a climate change scenario there will be reduced water availability. The way that the reduction in water availability is shared between the environment and consumptive users should *continue* to reflect this philosophy of how water is *currently* allocated between the environment and the consumptive users. For example, we consider a system where, under current climate conditions, the total water availability in the system is 100 GL/year, with 60 GL/year allocated to consumptive users and 40 GL to the environment (Fig. 17.4A). If there was a long-term drop in water availability, an allocation mechanism that meets the environment's needs first (under an ecocentric philosophy) would require that the consumptive users' allocation would need to decline, to ensure that the environment continues to receive the same volume (but a higher proportion of available water; Fig. 17.4B). Alternatively, a minimum protection approach might well result in the environment bearing the risk of long-term water reductions and consumptive water users maintaining access to their current volume for as long as possible (Fig. 17.4C). Other approaches (an anthropocentric or triple bottom line approach) would see the risks of climate change shared between the environment and consumptive water users (Fig. 17.4D). Such a division of climate risk should be a conscious consideration in the selection of mechanisms to allocate environmental water and be made explicit.

(A) Allocated share under current climate

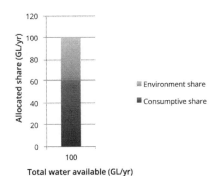

(B) Allocated share under climate change, consumptive user takes climate risk

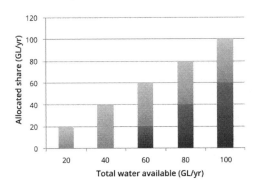

(C) Allocated share under climate change, environment takes climate risk

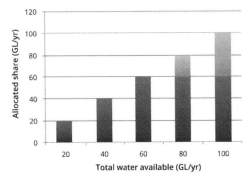

(D) Allocated share under climate change, climate risk shared

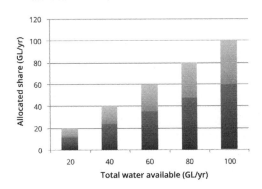

FIGURE 17.4

Options for sharing of climate risks between consumptive water users and the environment. (A) Allocated share under current climate. (B) Allocated share under climate change, consumptive user takes climate risk. (C) Allocated share under climate change, environment takes climate risk. (D) Allocated share under climate change, climate risk shared.

Source: Adapted from CIRES (2014)

Policy also needs to adapt in response to social, economic, and technological drivers of change. The reality is that as science improves, our understanding of environmental water needs and water resource system operation will change. Environmental flows assessments and policy documents often refer to the concept of adaptive management and the ability to use knowledge gained from trial flow releases, monitoring, and evaluation to then inform future watering decisions and needs (Olden et al., 2014). Environmental water allocation mechanisms should be implemented in a way that allows for adaptive management and periodic review as new information becomes available. Similarly, the values of a society may change over time, and the ability to refine allocation mechanisms and reflect current objectives is an important consideration. However, the mechanisms to allocate environmental water should still recognize the environment as an important use of available water, and any alteration should be the result of considered debate. Environmental water should not be perceived as a political tool, and should be buffered against arbitrary alterations (see discussion in O'Donnell [2012]).

Once these three questions have been addressed, the next step is selecting the appropriate allocation mechanism to meet these needs. This is a decision that is heavily influenced by the specific requirements of the local jurisdiction, but each jurisdiction can learn from the efforts of others. This chapter presents a review of mechanisms to allocate environmental water as a series of case studies, highlighting the specific factors contributing to success (or not), so that these lessons can inform future environmental water allocation choices. Importantly, there is no one right answer: the correct allocation mechanism will be the one that works in context, to meet the needs of the people and ecosystems concerned.

17.4 CURRENT MECHANISMS FOR ENVIRONMENTAL WATER ALLOCATION: CASE STUDIES ILLUSTRATING FACTORS FOR SUCCESS IN IMPLEMENTATION

In this section, using Table 17.1 as a guide, we examine case studies of the environmental water allocation mechanisms and how they were used in context to protect existing flows, and to restore environmental water regimes in dewatered rivers. The case studies focus on the mechanisms to allocate environmental water and only briefly introduce the catchments and the environmental attributes of the system for context. Each case study can help guide the selection of future mechanisms to allocate environmental water in a particular basin context, by identifying some of the strengths and limitations of the mechanisms employed, and factors relevant to the relative success of their implementation.

17.4.1 THE MURRAY–DARLING BASIN, AUSTRALIA

The history of environmental water allocation in Australia dates back to 1967, with the creation of the first environmental allocation (NWC, 2012), but significant and ongoing changes in the allocation mechanisms have occurred over the last 20 years. The MDB now has a combination of allocation mechanisms, including conditions on storage operators, conditions on individual license holders, a cap on consumptive use, and environmental water rights.

Placing conditions on the storage operators within the Basin was one of the initial water planning instruments adopted to protect environmental water in regulated river systems (rivers with storages that regulate flows). In many cases, the environmental water releases also act as conveyance flows providing the water to keep the river operational for delivery of consumptive water. These releases primarily provide baseflow(s) in the river, with some provisions for event-based releases to protect water quality (Goulbourn Broken Catchment Management Authority, 2013).

Conditions on individual license holders are used in unregulated systems, implemented through water resource plans or as part of the license itself. These conditions restrict pumping at times of low-flow, limit pumping to particular seasons, and set maximum daily pumping limits. This approach was initiated early in the water policy reform process and still remains in many unregulated systems across the Basin. Unregulated systems are discussed further in Box 17.2.

BOX 17.2 UNREGULATED RIVERS IN THE MURRAY–DARLING BASIN, AUSTRALIA, AND THE CHALLENGE OF ENVIRONMENTAL WATER RIGHTS

There are large portions of the Murray–Darling Basin, Australia, where river flow remains largely unimpacted by onstream water storages. There are specific challenges in effectively using environmental water entitlements as an allocation mechanism in these catchments, particularly in the northern Basin where distances are extreme and the floodplains extensive. There are two particular water purchases that the *Commonwealth Environmental Water Holder (CEWH)* considered that demonstrate these challenges.

In September 2008, Toorale Station in northwest *New South Wales (NSW)* was purchased to transfer the water rights to the CEWH. The property included a 14 GL entitlement to extract water from the Darling and Warrego Rivers, along with rights to harvest water from the floodplain (Senator the Hon. Penny Wong and Carmel Tebbutt, 2008). Under the usual water-sharing arrangements, there was no mechanism in place that allowed water held in the unregulated Barwon–Darling river valleys to be transferred for use as a licensed entitlement in the Murray or Lower Darling River valleys, where the CEWH wished to make use of the water (DWE, 2009). The river is unregulated with downstream license holders able to extract water as it passes their properties, according to their license conditions. The CEWH wished to forgo extracting the Toorale Station water and instead protect and *shepherd* this water downstream into the regulated Murray system. To do so, the states had to agree on how much water would be lost along the way, as the Darling is extensive in length and there are significant system losses. During a large rainfall event in February and March 2009, a trial was run where the NSW government calculated the estimated losses associated with the various stages of the transfer. Of the 11.4 GL that would have been available at Toorale Station for extraction, 8.7 GL was deemed to reach the Murray regulated system. Finally, the states also had to agree to forgo their rights to this water, as this water is usually distributed through the state water-sharing arrangements once it reaches the Murray River regulated system. A special agreement was made to maintain the CEWH as the owner of this water.

The second example is the large water holding of Cubbie Station in Queensland, just north of the NSW border. There was high publicity of the Cubbie Station offer to sell half of its entitlement (70 GL) to the CEWH, which was promoted as the solution to problems caused by the "water-sucking Cubbie station" (ABC News, 2009). The MDBA calculations indicated that as little as 20% of this water would reach the Lower Lakes in South Australia, and that under the existing water-sharing arrangements, NSW irrigators would be permitted to extract this water from the river due to their individual license conditions in the unregulated river. These factors would severely limit the effectiveness of purchasing this water right for the environment. This highlights the difference in allocation mechanisms and how they operate within existing policy and legislation, particularly where there are governance boundaries within the Basin. Here, the environmental water would be in the form of an environmental water right and thus not protected in an unregulated river. If the environmental water right was instead in the form of a condition on water abstractors in the form of a passing flow prior to pumping (under NSW policy), it would be protected instream.

These examples demonstrate the challenge of identifying and using environmental water rights as an allocation mechanism in unregulated river systems, but that increased monitoring and a willingness to negotiate can help.

The Basin water resources were initially capped at 1993/1994 levels of development. Importantly, at this time, groundwater and catchment interception activities were not considered within the cap, which meant that there was substantial capacity to increase water extraction from the Basin, even after the cap was implemented. Although the driver for the implementation of the cap was in part environmental concern, it was also part of a reform agenda to address water security in a highly variable system (MDBC, 1998). This included reforms to property right structures that changed them from a right to a volume to a right to a share of the resource (to allow for a variable total resource availability each year) and to enable water rights to be traded separately from land, thus allowing for an increase in water trade. A further complication of the cap implementation was that the development level was based on licensed volume rather than actual historical use. With the increase in water trade, a number of previously dormant licenses were sold onto the market, thus increasing actual water use, despite license volume remaining capped (Hussey and Dovers, 2007).

The Millennium drought (van Dijk et al., 2013) and the ongoing environmental degradation in the MDB came to a head in 2007 (Hart, 2015). The ultimate outcome was an increased role for the federal government in managing water resources in the Basin. The individual states referred their legal powers to the Commonwealth, enabling the federal government to pass the *Water Act 2007*, a broad piece of water legislation that required a binding management plan for the MDB. The new Basin Plan, which came into effect in 2012, lowered the cap on consumptive water use, now referred to as the SDL. The *Water Act 2007* required the SDL to reflect a sustainable level of water resource development, and was set as a share of the available water each year. This was made feasible by previous work under the National Water Initiative, which converted water licenses in the Basin to a share of the resource, rather than an absolute volumetric right (COAG, 2004). The SDL also includes interception activities (i.e., farm dams and forestation) and groundwater. Implementation of the SDL, however, required a significant reduction in consumptive water use (MDBA, 2010), which was politically challenging and unpopular. The communication of, and community involvement in developing, the Basin Plan and SDL, received criticism and opposition from the powerful irrigation lobby. The government eventually integrated the SDL policy with a commitment to environmental water buyback to ensure that compensation was paid to existing license holders (Commonwealth of Australia, 2010). There was also a greater emphasis and commitment to stakeholder engagement that led to a successful completion of the Basin Plan. The challenges experienced in the MDB demonstrate the difficulties in retrospectively limiting extraction—rather than setting a precautionary cap up front—and the need to engage the community in the transition process and allow time within policy development for this process to occur. However, importantly, it also demonstrates that in a heavily used system with intensive economic uses for water it is still possible to achieve a significant redistribution of water to the environment.

The MDB is unique in the extent to which environmental water rights exist across the Basin and the use of this allocation mechanism to provide water to floodplain assets and wetlands. In 2003, the federal government and southern Basin states agreed to the Living Murray Program. This program included a AU$650 million contribution to recover water for the environment and build infrastructure to deliver this water to floodplains in a highly targeted way (MDBA, 2015). The environmental water is held as bulk environmental water rights managed by the states and the former Murray–Darling Basin Commission (the decision-making forum for basin jurisdictions). In 2006, the process for recovering this water also included a tender process to buy water rights from individual irrigators (Grafton and Hussey, 2006).

A second program for creating environmental water rights across the Basin began in 2007. The *Water Act 2007* created the *Commonwealth Environmental Water Holder (CEWH)* to hold and manage environmental water entitlements to achieve environmental outcomes. This allows the CEWH to hold water rights with the same legal basis as consumptive water rights (for more detail on the process to acquire this water, see Chapter 18 and for more on the institutional arrangements of managing this water, see Chapter 19). Although this clearly provides a mechanism to allocate environmental water, government documents initially indicated confused messages about the objectives of the acquisition of the water rights through the Water for the Future program, which committed AU$3.2 billion to buy environmental water entitlements (Horne et al., 2011). The initial objectives were "multiple, poorly defined and at times conflicting" (Productivity Commission, 2010), and included both easing the transition for consumptive users to a future with less water availability, as well as providing short-term environmental outcomes (Productivity Commission, 2010).

As this chapter has argued, the selection of mechanisms to allocate environmental water depends on the underlying objectives. In the MDB, the design of the environmental water rights and the interaction with the SDL (a cap) are fundamental in terms of how water is provided to the environment. If water rights held by the environment are defined by the SDL, they can provide a structural adjustment mechanism (a form of compensation to assist water users to adjust to the new cap) and some flexibility of water delivery, but they are not able to provide additional volumes of water to the environment over and above that established through the SDL. In contrast, if environmental water rights were in addition to the SDL (as they are nonconsumptive, and therefore arguably not defined by a limit on consumptive water use), they would provide additional environmental benefit on top of the cap and some ability to effectively readjust the balance between consumptive and environmental water on an ongoing basis (Horne et al., 2011). The Murray–Darling Basin Authority has determined that the environmental water rights in the MDB will be defined by the SDL (not in addition to it). In terms of environmental outcome, although this approach does allow flexibility of delivery, it also requires a more complex accounting system to track compliance with the SDL if environmental water is moved between catchments (or between the environment and consumptive users) using the water market (Horne et al., 2011).

The Murray–Darling case study demonstrates how a range of allocation mechanisms are combined to provide environmental water (Table 17.2). Although these allocation mechanisms are now used in parallel, the conception of these policies lacked an integrated vision of the role and interaction of each mechanism.

Importantly, it has been a 20-year process with incremental steps to reach the current level of environmental water and this process has happened in parallel with other major water reforms including clearer definitions of water rights, improved monitoring, and improved water resource planning. It has also been an expensive process with the government investing large amounts in water recovery. The process of allocating environmental water demonstrates the trade-offs required in a system of cooperative federalism between comprehensiveness and pragmatism (MacDonald and Young, 2001). Each step in the process included limitations, but also significant progress.

17.4.2 COLUMBIA RIVER BASIN, UNITED STATES

The Columbia Basin is located in the northwest of North America, extending from Alberta, across the Canadian border and into the states of Washington, Montana, Oregon, Idaho, with some fringes

Table 17.2 Mechanisms to Allocate Environmental Water and Success Factors for the Murray−Darling River Basin, Australia

Mechanisms for Environmental Water Allocation	Success Factors
Conditions on storage operators	Legislation: Enabling the setting of conditions through water resource management plans.
	Physical availability: Opportunities for conjunctive management as environmental water needs often met through releases to consumptive water users or for conveyance water, but the trade-off is that it is difficult to define (and therefore protect) releases that are solely for environmental outcomes.
	River type: used primarily on regulated rivers.
Environmental water rights	Legislation: Provided for through a federal water act (*Water Act 2007*) and the establishment of a formal water holder. Also provided through various state legislation, which enabled water rights to be created and acquired for environmental purposes.
	Water markets: Allow the environment to acquire water from consumptive water users.
	Costs: Require significant up-front costs in acquiring water and ongoing management costs.
License conditions for water abstractors	Legislation: Enables conditions on individual licenses to be set by the minister or the relevant water authority (i.e., a winter-fill only license) or modified by a water resource plan.
	River type: Used primarily in unregulated rivers.
Cap on water extraction	Legislation: Enabled through state and federal water acts and intergovernmental agreements.
	Enforcement: Jurisdictions announce water availability for individual license holders and keep allocations below cap levels on a long-term average.
	Monitoring: Extensive monitoring of extraction by consumptive users and water accounts balanced on annual usage.

of the basin reaching Wyoming, Utah, and Nevada (see Maps section). Over the past 200 years, the flow regime of the Columbia River and its tributaries has been greatly altered by the combined effects of flood control, hydropower generation, water extraction for irrigation, and use of the rivers for navigation (Ward and Ward, 2004). This case study focuses on the Columbia Basin within the United States.

In response to water scarcity, and as part of a program to mitigate the effects of hydropower production on the aquatic ecosystems of the Columbia, Bonneville Power Administration and the *National Fish and Wildlife Foundation (NFWF)* came together in 2002 to build a program to restore the water regime in dewatered sections of the Columbia River Basin (Columbia Basin Water Transactions Program, 2011). To understand how environmental allocation mechanisms have been implemented in the Columbia Basin, it is important to recognize that water rights have evolved under a system called prior appropriation (Wiel, 1911). A short primer on prior appropriation is provided in Box 17.3.

BOX 17.3 PRIMER ON PRIOR APPROPRIATION IN THE WESTERN UNITED STATES

Under prior appropriation, individual rights to water are based on the initial extraction and ongoing use of water, so the first person to extract and use the water will continue to have the first call on that water (MacDonnell, 2009a). As a legal doctrine, prior appropriation has four significant elements (Tarlock, 2000). First, the date of the initial extraction defines the seniority of the right, known as "first in time, first in right." Second, the use to which the water is put must be considered a *beneficial use*, which requires that the water is actually used on the land, and excludes any water that reenters the stream (as drainage or through groundwater). Third, prior appropriation rights can be transferred, but any transfer must comply with a *no injury* rule, which requires the assessment of any proposed transfer by a state administrator or water court to confirm that the change in location or use of the water will not impact other water users (Committee on Western Water Management, 1992). Fourth, the water must have been in continuous use, where continuous is defined by the state in question. Any water left unused for the defined period (typically 5–10 years) is considered to be relinquished by the water user, who then has no right to claim that water in future (Donohew, 2009). In the western United States, this requirement to calculate the water that has actually been used (and not returned to the river or left unused for the necessary period) creates the difference between the *paper* water right (the water claimed by the water user) and the *wet* water, which is the volume that can be demonstrated as having been used and not returned (Sax et al., 1991). Most significantly, under this system of water allocation, it is feasible for all the water available in a stream to be allocated to a beneficial use, which means that the stream can literally run dry.

The Columbia Basin is a good example of the use of a range of environmental water allocation tools; and also for highlighting those that were not available. Two tools were not used: conditions on existing water users' licenses and the creation of a reserve. The water allocation framework in the Columbia Basin does not support intervention by the states to limit existing use by imposing conditions on water licenses beyond the requirements of the prior appropriation doctrine. Although the states of the Columbia Basin retain the ownership of the water itself (Gillilan and Brown, 1997), the role of the state in regulating water use by individuals is largely restricted to administering the existing rights (Tarlock, 2000). Unlike jurisdictions where water rights are based in statute, such as Australia, the western US states have very little capacity to alter the conditions of the existing users' water rights in a unilateral manner. Similarly, there has been no capacity to set an environmental water reserve as part of the existing legal framework. Although legal change is always a possibility, there has been no widespread support for creating an environmental water reserve, and past attempts to unilaterally act to prioritize environmental water releases have been extremely controversial and ultimately ineffective (see, e.g., Doremus and Tarlock, 2008).

However, the capacity to set conditions on the operation of onstream dams has been a powerful tool when imposed at the federal level (for more on the dam relicensing program, see Bowman [2002]). The *Endangered Species Act 2000* (16 U.S.C. §§ 1531–1544; ESA) required that all federal projects be operated in a manner to avoid causing harm to species listed as endangered or threatened. When salmon and steelhead species were listed under the ESA, this created a federal power for altering the operation of the hydropower dams, and other storages that were operated by federal agencies (see, e.g., Doremus and Tarlock, 2003). The situation in the Columbia Basin was further enhanced by the *Pacific Northwest Electric Power Planning and Conservation Act 1980* (16 U.S.C. §839(6); Power Act), which established an interstate agency, the *Northwest Power and Conservation Council (NPCC)*:

...protect, mitigate and enhance the fish and wildlife, including related spawning grounds and habitat, of the Columbia River and its tributaries, particularly anadromous fish which are of significant importance to the social and economic well-being of the Pacific Northwest and the Nation and which are dependent on suitable environmental conditions substantially obtainable from the management and operation of the Federal Columbia River Power System and other power generating facilities on the Columbia River and its tributaries.

The combination of the ESA and the Power Act has been effective in providing tools to reduce the impact of the hydropower in the Columbia Basin (Leonard et al., 2015). Initially, the NPCC focused on the mainstem of the Columbia, investing in fish passage around dams and the provision of adequate water to maintain instream fish habitat (see the NPCC programs 1982–1994, at Northwest Council [2015]). In the early 2000s, the program was expanded to focus on the tributaries as well (discussed below).

The creation of a cap has been a challenge in the Columbia River Basin. Often, the only cap on water use was the physical availability of the water itself (Aylward, 2008). The states of Washington and Oregon have closed some streams to new appropriations, effectively capping water extraction at the existing levels (Cronin and Fowler, 2012). However, this moratorium on new licenses has been overruled by the legislature in some instances, enabling new licenses to be issued (Mistry, 2015).

What has been more successful throughout the Columbia River Basin is the appropriation of streamflows for the environment, effectively the creation of environmental water rights that have the same legal recognition as other users' water rights. These environmental water rights have been appropriated to protect existing flows, as well as to restore flows in dewatered systems.

To protect existing flows from future impact, states recognized the environment as a beneficial use, which required changes to the water laws in the states, and the definition of the use for which instream flows could be appropriated (see MacDonnell, 2009b). Following this change, states began to appropriate water for the environment in high-priority river systems where some water remained unallocated (e.g., Byorth, 2009). This appropriation of water for the environment effectively placed a cap on future development, and several states combined this appropriation program with the creation of a water bank or a water market, so that future development of water resources could continue (Cronin and Fowler, 2012). Although this program has seen some success, it faced two significant problems. First, there had to be unallocated water remaining in the river, and many rivers were already dry during summer. Second, the environmental water right thus created was a junior right: so it only protected the flow against new appropriations, and did not stop senior water users from extracting their water (Malloch, 2005).

The restoration of environmental water regimes to historically dewatered systems attempted to address these specific problems by obtaining existing senior water rights and transferring them to the environment. One of the lessons learned from the water wars of the Klamath Basin is that top-down regulatory demands that take water from one group and transfer it to another are extremely controversial, even when that recipient is the environment. If one group (i.e., irrigators) is bearing the cost of the transfer whereas others share in the benefit, this is unlikely to be politically acceptable. However, using a market to achieve the same outcome can significantly reduce this tension (Ferguson et al., 2006; Thompson, 2000). As this chapter describes, market-based transfers rely on willing buyers and sellers, so that each transaction is voluntary.

The restoration of environmental water regimes in the Columbia Basin required further legal change: each state had to enable the transfer of existing water rights to an environmental organization, and the conversion of those rights into instream flows, without losing their reliability (set by the priority date). Some states focused on enabling long-term, renewable leasing arrangements, whereas others supported permanent transfers as well (Garrick et al., 2009). In the Columbia, the transactions to acquire water for the environment began with highly targeted, small acquisitions to restore flows to dry stream beds, particularly during the summer low-flow periods. In many cases, these transactions were one-on-one arrangements between the irrigator(s) and the environmental purchaser, with a negotiated outcome depending on the capacity of the existing user to make alternative arrangements that were within the capacity of the environmental purchaser to fund. These transactions included investment in water use efficiency as well as altering the farming practices to alleviate the need for water at particular times. In most cases, the early transactions sought to maintain the farmer on the land, as well as the water in the stream (Columbia Basin Water Transactions Program, 2011). Although this is still a priority for the water acquisition programs in the Columbia, as the program has matured and state investments in establishing water banks have increased, there is more capacity to undertake the environmental transactions through a functioning water market rather than just the one-on-one approach (Garrick et al., 2009).

The Columbia Basin Water Transactions Program is the ongoing, overarching program for environmental water restoration in the Columbia. It is funded by the Bonneville Power Administration, as part of its obligations under the Power Act and the ESA, and managed by the NFWF (Columbia Basin Water Transaction Program, 2015). NFWF has identified the *qualified legal entities* in each state or subbasin of the Columbia, which then work with the state agencies to identify the priority reaches for streamflow restoration, engage with local water users, and identify potential water transactions, complete the due diligence to ensure that the water is in fact available, secure the funding, and complete the legal arrangements for the transaction (Malloch, 2005; see Chapter 18). Once the transfer is complete, the streamflows are then monitored to ensure that the environmental water right has been provided. In 2011, the Columbia Basin Water Transactions Program had enhanced the streamflow in 974 km of tributaries to the Columbia River, and restored approximately 7155 GL of flows (over the lifetime of the transactions).

In summary, the Columbia River Basin is an example of how a small initial recovery process for environmental water can grow. It highlights the need for a strong legal framework that recognizes the environment as a user of water, and the necessary funding to complete the environmental water recovery (see Table 17.3). The Columbia River Basin also demonstrates the challenges of attempting to scale up from a small beginning, and how to adjust the model to reduce the costs of each environmental water right purchase (Garrick and Aylward, 2012).

17.4.3 SOUTH AFRICA

South Africa is a semi-arid, water-scarce country, with a history of intensive water use and with an increasing number of river basins in major deficit (Pollard and Du Toit, 2014). Major political reform (with the 1994 democratic elections and new 1996 Constitution) coincided with an urgent need for statutory reform of the national water management system in response to exploitable water resources nearing their limits (Muller, 2014). Fundamental revision of water legislation, guided by a White Paper on national water policy (DWAF, 1997), led to the promulgation of a new *National*

Table 17.3 Environmental Water Allocation Tools and Success Factors for the Columbia River Basin, Australia

Environmental Water Allocation Tool	Success Factors in the Columbia River Basin
Conditions on storage operators (Bowman, 2002; Leonard et al., 2015)	Legislation: The ESA and the Power Act both set powerful legal requirements to alter the operation of hydropower and other federal water storages to protect endangered species.
	Funding: The actions under the Power Act were funded by the Bonneville Power Administration from the sale of hydropower electricity on the federal US market.
Environmental water rights: Protection of existing flows (MacDonnell, 2009b)	Legislation: Each state legally recognized the environment as a beneficial use and identified who can hold the environmental water right.
	Physical availability: Water was still available in some streams to benefit from this protection.
	Water markets: Setting a de facto cap on water extraction is much more palatable to existing and future water users when they can purchase water from other users so that development can still proceed, and states supported the creation of water transfers through water markets and water banks.
Environmental water rights: Acquisition of new flows (Garrick et al., 2009)	Legislation: Enabling an existing water right to be transferred to an instream use without losing seniority and identifying who can hold the environmental water right.
	Water purchasers: A range of organizations with the capacity to engage potential sellers, facilitate the transaction process, and pay for the water right were established in the Columbia River Basin, including private and government agencies.
	Funding: The water purchase programs in the Columbia River Basin received funding from the Bonneville Power Administration, the National Fish and Wildlife Foundation, various state-funding sources, and a range of philanthropic donations.

Water Act 1998 (*NWA*; Republic of South Africa, 1998). The NWA aims to ensure that all water resources are used, managed, and controlled in such a way that they benefit all users, striking a balance between pressing economic development and sustainability and intergenerational rights (Muller, 2014).

The overarching tool for allocating environmental water in South Africa is the Ecological Reserve, which is defined in the NWA. The Reserve comprises the amount of water needed to provide for basic human needs (the Basic Human Needs Reserve) together with the water regime required to sustain the ecosystem (the Ecological Reserve). The legislation defines the Ecological Reserve as: "the quantity and quality of water required to protect ecosystems in order to secure ecologically sustainable development and use of the relevant resource" (*National Water Act 1998* §1 (10)(xviii)(b)). The Reserve is the only water right specified as inviolable in the law, with water for basic human needs having the highest allocation priority, followed by ecosystems (DWA, 2011). Accordingly, no water can be allocated to other uses prior to setting the Reserve for a water resource (Pollard and Du Toit, 2014). Implementation of this high-level legal allocation of water to the environment remains a real challenge (Le Quesne et al., 2010) and is contingent on effective operationalization through the additional mechanisms of conditions on storage operators and license conditions for water abstraction (McLoughlin et al., 2011).

Reserve implementation requires two processes—determination of the Reserve requirements and operationalization, including "operational planning, monitoring, enforcement, reflection and learning" (Pollard et al., 2011); each of these will be discussed in turn.

The determination of the Ecological Reserve involves a number of steps. This includes, among other things, the identification of manageable water resource units (be that surface water, wetlands, groundwater, or an estuary), through a consultative process, and classification using a national system where the chosen *Management Class (MC)* defines the nature and extent of permissible and sustainable resource use (Pienaar et al., 2011). For example, a system that is identified as currently in the class "Unacceptably degraded resource," although it must be restored to a greater level of health, may only have the objective of being returned to the class "Heavily used/impacted," where the system is still significantly changed from its natural state. Thus, the definition of the MC helps establish the catchment vision and stakeholder objectives (McLoughlin et al., 2011). The appropriate *resource quality objectives (RQOs)* are then established (Palmer et al., 2004), including the quantity, pattern, timing, water level, and assurance of flow determined using environmental flows assessment methods selected according to the scale and level of resolution required, resource constraints, and the basin context (Chapter 11). The RQOs establish clear goals relating to the quality of the relevant water resources, and in addition to instream flow conditions, also describe water quality, instream and riparian habitat, and biota (DWA, 2011). After the Reserve and associated RQOs are determined for each of the water resource units, they are gazetted and become legally binding. Importantly, the NWA allows for preliminary determinations of the Reserve to be applied, and later be replaced by more detailed, higher confidence studies. This allows for progressive implementation of the Reserve (Pienaar et al., 2011). It acknowledges that the Reserve will require revision as new knowledge becomes available, or modification to ensure that it is indeed meeting its objectives.

The practicalities of operationalizing the Reserve from this point onward present perhaps the greatest challenge to its implementation, as illustrated in an analysis of Reserve implementation in the three *Water Management Areas (WMAs)*, the Luvuvhu/Letaba, Olifants, and Inkomati, in the eastern (lowveld) region of South Africa. The three WMAs, for which only the Inkomati has an established *Catchment Management Agency (CMA*; see below), share many similar biophysical and socio-economic characteristics and are in water deficit (Pollard and Toit, 2011; Pollard et al., 2011). Most of the rivers traverse densely populated rural areas (former bantustans) with high levels of poverty and unemployment, flow through a major protected area, the Kruger National Park, and all form part of international river systems.

In 2011, none of the flow regimes of the eight rivers assessed met the Reserve requirements, with the incidence of noncompliance between 18% and 85% (Pollard et al., 2011). With the sole exception of the Sabie River, this situation has deteriorated over the last decade since the new NWA, despite changes in policy and/or specific management interventions (Fig. 17.5; note, the lighter shaded bars at the left of each pair are based on monthly flow data before the NWA, and the darker bars are based on daily flow data after the Act; see Pollard and Du Toit, 2014; Pollard et al., 2011). Implementation is expected to improve in the Crocodile River, the most stressed of all the rivers within the Inkomati WMA, due to the development of real-time management tools (Pollard and Toit, 2011).

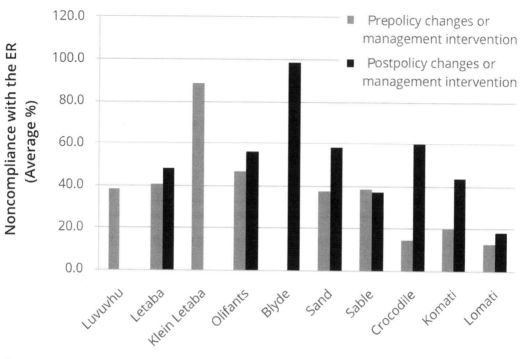

FIGURE 17.5

A comparison of the incidence of noncompliance with the Ecological Reserve for South Africa's eastern rivers before and after policy changes or management intervention.

Source: Pollard and Du Toit (2011)

Presently, perhaps the greatest obstacle to implementation is the lack of a clear, pragmatic, and well-established approach to give the needed support to water managers to integrate the Ecological Reserve into water resources planning and operation (Pollard and Du Toit, 2011; Table 17.4). The outcomes from the lengthy Reserve determination process are often complex and not readily translatable into an operational plan. To illustrate, in the Olifants WMA the Reserve was established in 2004, significant water resources were classified in 2012, and the determination of RQOs concluded in 2014 (DWS, 2014). A total of 494 RQOs of varying complexity were generated for river, wetland, dam, and groundwater resources (DWS, 2014). The Reserve determination process is often linked to natural flow, which presents challenges in forecasting the likely Reserve requirements for a given time period (Pollard et al., 2011). Hydrological modeling could be used to compare projected daily river flows at locations key to Ecological Reserve requirements, but quality rainfall data are limited in many catchments. In the absence of such an approach for making the Ecological Reserve operational through a real-time water resources management model, water managers have indicated that they simply interpret it as a minimum flow, rather than as a spatially and temporally variable flow regime (magnitude, duration, and frequency of low and high flows linked

Table 17.4 Mechanisms to Allocate Environmental Water and Success Factors for South Africa

Environmental Water Allocation Tool	Success Factors
Environmental reserve	Legislation: The National Water Act established the Reserve, comprising the Basic Human Needs Reserve and the Ecological Reserve as the highest priority water allocations.
	Methods for environmental flows assessment have been highly effective at identifying the environmental water needs of South Africa's aquatic ecosystems.
	Remaining challenges: Implementation and enforcement of the Reserve at the operational level.
Conditions on water abstractors	Used to operationalize the Reserve, but requires improved methods, framework, and monitoring.
	Potential for self-organization and self-regulation.
Conditions on storage operators	Used to operationalize the Reserve, but requires improved methods, framework, and monitoring.

to prevailing climatic conditions) (Pollard et al., 2011). Pollard et al. (2011) observed, however, that *Department of Water Affairs (DWA)* are in the process of developing a framework to operationalize the Reserve. To this end, McLoughlin et al. (2011) describe Reserve modeling tools and procedures under development for the eastern rivers (e.g., a real-time, rapid response system for adaptive management of dam flow releases to achieve the Ecological Reserve in the Groot Letaba River).

A further challenge exists around the practicalities of implementing complex license conditions for water abstractors or storage operators. The NWA decoupled water rights from land ownership (Gowlland-Gualtieri, 2007) and required that water abstraction licenses be cognizant of the Ecological Reserve (Muller, 2014). Further, the licensing process has been hindered by the delays in establishing CMAs, the agencies effectively operationalizing the Reserve (Pienaar et al., 2011). Although the DWA maintains oversight of water resource management, the NWA provides for the administration of certain water management functions to be delegated to CMAs (Muller, 2014). The NWA provides for the establishment of 19 WMAs, to be managed by CMAs, but these have recently been collapsed into nine. Only three of these WMAs have been established nationally (Pollard and Toit, 2011) and only two have been devolved to management by a CMA (DWA, 2013). Delays in establishing the CMAs have caused resultant delays in the generation of functional catchment management strategies, and in license implementation and monitoring, perpetuating a climate of uncertainty (Muller, 2014; Pollard and Du Toit, 2014). Consequently, most abstractions still do not take place on the basis of licenses under the Water Act.

In practice, the conditions on abstractive users—particularly where these are small-scale users—must not be overly cumbersome for a water user to implement, understand, and adhere to. Similarly, conditions on storage operators are complicated by infrastructure and operational practices that are often geared toward more consistent releases to meet downstream consumptive demands, rather than the variable flow patterns required to meet the Environmental Reserve (Pienaar et al., 2011). This is more readily achieved when determining release conditions for a new storage (e.g., the Berg River Dam; Pienaar et al., 2011).

Monitoring and enforcement of license conditions are also limited (Pienaar et al., 2011). Unlawful water use is an issue in some areas, as are the lack of incentives for compliance and of political will to enforce (Pollard et al., 2011). This highlights the importance of stakeholder engagement and support for the allocation process. There is an ongoing perception problem for the Reserve in South Africa, with van Wyk et al. (2006) observing considerable stakeholder frustration directed at the legal requirement to allocate water to ecosystems in a country with such high water demands and where rights of use of available water have already been allocated. The dominantly held belief is that "the needs of people compete with the needs of river ecosystems and that the Reserve strengthens this dichotomy" (van Wyk et al., 2006). This situation has confounded attempts to reallocate water for development among marginalized groups (van Wyk et al., 2006), including in the lowveld rivers. Stakeholder fatigue is also a recognized factor. Positively, the Groot Letaba River (Olifants WMA) provides an example of where a first step toward formal conditions on abstractive licenses is emerging in a system developed at the local level involving stakeholders in managing their water use patterns below Tzaneen Dam (McLoughlin et al., 2011). This demonstrates the potential and importance of self-organization and self-regulation (Pollard and Toit, 2011). It is hoped that these kinds of efforts will be accompanied with improved regulation and enforcement at higher levels.

In practice, with full or over allocation already an issue in many catchments, safeguarding the environmental water allocation is necessarily a complex and protracted process of negotiated trade-offs. Different sectors within the same WMA continue to see very different priorities for managing the shared water resource, although the situation is improving (Pollard and Du Toit, 2014). Muller (2014) argues that the challenge of bringing together diverse interest groups, ranging from major corporations through to poor, largely illiterate rural communities, in a process that is effective, efficient, fair, and equitable has been the critical difficulty behind the delay in establishing the CMAs. Where draft catchment management strategies have been developed, the process has been further detrimentally impacted by a lack of delegation of powers and associated budget (Pejan, 2013, cited in Pollard and Du Toit, 2014). This governance challenge is a common problem for the implementation of environmental water allocation mechanisms, but is likely to be exacerbated in developing countries and emerging economies.

Pollard and Du Toit (2011) observe that "securing the Reserve in reality relies on the collective contribution and synergies of a number of strategies, plans and practices... that make up IWRM." There is a real need for the government to lead the way through a cohesive strategic plan that includes frameworks, tools, and management systems to operationalize the Reserve (Pollard and Du Toit, 2011). Greater knowledge exchange on how different mechanisms for environmental water allocation can be used in combination with the Reserve to facilitate implementation is likely to become as vital for adaptive learning in the future, particularly as similar governance frameworks and allocation mechanisms for environmental water management are in evolution in other countries in the region (e.g., Kenya [Mumma, 2007], Lesotho [Le Quesne et al., 2010], and Tanzania [LVBC and WWF-ESARPO, 2010]). Building technical capacity to stem the observed decline in South Africa since the 1990s (Ashton et al., 2012) will be fundamental to long-term success in the region (Table 17.4).

17.4.4 YELLOW RIVER, PEOPLE'S REPUBLIC OF CHINA

Growing concerns over water scarcity, altered flow regimes, and poor water quality have spurred new policies and approaches to improve the environmental condition of China's rivers. As with many of the policy processes in China, water policy is established centrally (by the *Ministry of Water Resources [MWR]*); however, increasing governance and administrative powers of provincial government and regional institutions called *Water Conservancy Commissions (WCC)* provide implementation (Moore, 2014; Silveira, 2014).

The overarching water legislation is the Water Law (1988, amended 2002), which builds on the 1982 Constitution defining water as a property of the State (Moore, 2014). It also allows for the establishment of a water rights system that assigns individual rights; however, in many cases, these individual rights remain ambiguous (Moore, 2014). Shen and Speed (2009) highlight three key components of the water allocation system: (1) the development of river basin and regional allocation undertaken by the WCC to allocate water to administrative regions; (2) administrative regions (provincial governments) allocate water to abstractive users (effectively bulk users) through a permit system; and (3) allocation of the abstractive water permit to individual users. The Law on Environmental Protection (1979 and amended 1989) and Law on the Prevention and Control of Water Pollution (1984 and amended 1996 and 2008) also provide a basis for environmental water (Jian, 2014). Importantly, however, there is "no tradition of governing according to the rule of law. ... enactment of national laws or the passing of central decisions is believed to be the beginning of a bargaining process between central government and local authorities concerning their implementation" (Silveira, 2014). Although the laws themselves have been praised, limitations in enforcement have been highlighted as a key area for improvement (Silveira, 2014).

The Yellow River spans seven provinces and two autonomous regions in a heavily populated and arid region of Northern China. It has been managed as a whole basin since the inception of the *Yellow River Conservancy Commission (YRCC)* in 1950 (Moore, 2014). Historically, the river has provided significant agricultural water to users in the lower reaches, but there are increasing demands for urban and industrial (including coal and gas) water use higher in the catchment (Moore, 2014). There are a number of large storages in the upper Basin. The heavy use of water resources in the Basin, coincident with a period of extended drought in both the Yellow River Basin and the entire North China Plain, led to the river failing to reach the sea for an average of 120 days per year between 1995 and 1998 (Moore, 2014). In response, significant efforts have been made to address environmental water requirements in the river and its coastal delta. Minimum flow requirements to meet critical ecosystem needs have been identified at key flow gauging stations. In China, as in many countries, the definition of environmental water includes not only the maintenance of biodiversity, but also of ecosystem services. For this reason, in the Yellow River, sediment flushing is a major component of the recommended flow regime (Giordano et al., 2004). Meeting the environmental water requirement is challenging as the target represents one-third of the average annual flows and nearly half of the flow volumes observed during the 1990s drought (Zhu et al., 2003). There are two allocation mechanisms used in the Yellow River that provide environmental water: a cap on consumptive use and conditions on storage and water resource operators.

There have been a number of government processes to implement a cap in the Yellow River. Indeed, capping water use is now a central government policy introduced as the "Three Red Lines" (defined through the water policy document "Accelerating the Water Conservancy Reform and Development," often referred to as Central No. 1 Policy Document 2011) that limits water usage, and sets efficiency and quality targets (Moore, 2014). This high-level centralized document limits national water consumption to less than 700,000 GL/year. A core challenge in cap implementation has been around governance and the often entwined sense of economic activities with local governments, leading to vested interests and a lack of trust and coordination between regions (Moore, 2014). As water scarcity issues in the Yellow River have increased, there have been increasing disputes between provincial governments and the YRCC and MWR have been challenged to manage these (Moore, 2014). The challenges of the multilayered governance of water in China and the need for clear delineations of responsibility are well documented (e.g., Fang, 2014; Silveira, 2014). In the Yellow River, a number of attempts have been made to define water demand levels and cap extraction, with the water resources allocation plan determining the water available for consumption, as well as prescribing instream flow requirements at locations in the system (Svensson, 2014). However, the 2011 YRCC reports that despite policies being in place, there remain "some localities which do not put into practice the water quantity allocation and dispatch plan, and exceed the allocation limits in using water resulting from interprovincial flows not according with control limits" (Huangshuihui [YRCC] quoted in Moore, 2014). This overextraction is achieved by extracting large volumes from tributaries rather than the mainstem of the river where sophisticated monitoring systems are in place (Moore, 2013). These enforcement challenges effectively make the cap limited to only the reaches where monitoring is in place, and not in the extensive tributaries. A further challenge in implementing a cap on extraction is that groundwater is currently managed as a separate resource (Barnett et al., 2006).

The South−North Water Transfer Project has been introduced to ease water scarcity in the north through a large-scale reallocation of water between river basins (Fang, 2014). This effectively increases the volume of water available under the cap, and also the instream flow, by sourcing water from outside of the river system and will allow more water to remain instream. If successful, restoration in the Yellow River stands to be the world's largest-scale reallocation of water to the environment to date (Le Quesne et al., 2010). Such interbasin transfers require consideration of the environmental impacts and allocation mechanisms across all donor and recipient river systems impacted (negatively or positively) by the transfer project.

Another important mechanism for environmental water allocation is the modified operation rules for Xiaolangdi Dam. In 2008, the YRCC amended its policy to require storage releases targeting floods and sediment transport to preserve ecological functions. The storage operating conditions are targeted at maintaining continuous flow in the lower Yellow River and managing sediment transport (The Natural Heritage Institute, 2013).

There has been measurable progress in providing instream water in the Yellow River and more broadly in supporting sustainable development (Table 17.5). This has been mostly possible through the centralized governance arrangements in China. This is a significant achievement alongside the parallel process of economic development (Zhang and Wen, 2008). However, layered governance arrangements and limited enforcement have hindered the implementation of these policies (Silveira, 2014).

Table 17.5 Mechanism to Allocate Environmental Water and Challenges and Success Factors for the Yellow River Basin, China

Mechanism to Allocate Environmental Water	Challenges and Success Factors
Cap on consumptive use (Moore, 2014)	Legislation: Directive from the No. 1 policy documents (a centralized legislation) but implemented through YRCC water plans. Allocations centrally announced for different areas of the river system (annual, monthly, or daily announcements).
	Enforcement: Requires enforcement across the basin, to date there has been limited ability to enforce plan so overextraction continues. Limited monitoring in tributaries.
	Institutional arrangements: Centrally controlled.
Conditions on storage or water resource operators (Speed et al., 2011)	Legislation: Through the annual regulation plan developed by YRCC.
	Effectiveness: Targeted releases for sediment transport and other objectives introduced.

17.5 CONCLUSION

The mechanisms to provide environmental water are varied and may be employed in different combinations. However, by classifying these allocation mechanisms based on their legal characteristics, function, and operation, we can make the process of selecting an appropriate mechanism in a particular context much more explicit. Even in systems where allocation mechanisms have already been implemented, this classification helps to frame the analysis of the efficacy of existing mechanisms and target any potential improvements.

There are five specific allocation mechanisms, in two groups:

1. Mechanisms that impose conditions on the rights of other water users:
   a. A cap on total consumptive water use;
   b. Conditions on water abstraction licenses;
   c. Conditions on water storage operators.
2. Mechanisms that create specific legal rights for the environment itself:
   a. Ecological or Environmental Water Reserve;
   b. Environmental water rights.

These allocation mechanisms have been examined in situ in a range of case studies from Australia, the United States, South Africa, and China. By focusing on the specific allocation mechanisms in this way, it is possible to understand and analyze the way that environmental water has been allocated in very different water law frameworks.

The implementation of each of these approaches will require government to play a significant role in developing and supporting the environmental water allocation, as well as the institutions and structures for monitoring, compliance, and enforcement. As Chapter 18 highlights, nongovernmental organizations can also play an important role in developing and managing environmental water allocation mechanisms. Transitioning from the current situation to new environmental water arrangements will require different levels of political will. Significant changes to environmental water provision have tended to occur when the political climate supports such changes (e.g., constitutional reform coupled with water scarcity in South Africa; and the Millennium drought in the MDB).

The implementation and practice of environmental water has more chance of being successful if the objectives and allocation mechanisms recognize the need to integrate social and economic values and outcomes (Poff and Matthews, 2013). Although each of the underpinning philosophies identified in Section 17.3 places value on the environment in different ways, implementing environmental water allocation mechanisms will require an acknowledgment of the social and economic impacts of a changed water regime. This will need to incorporate, where necessary, some form of mitigation, including compensation for any impacts that cannot be avoided or fully mitigated.

Importantly, the selection of an appropriate allocation mechanism should explicitly include an assessment of flexibility and ability to adapt to a changing and variable climate (see also Chapter 11). A discussion of the balance between environmental and consumptive water users should be part of the overall water planning process, including under historical patterns of water availability, as well as under extreme drought and future climatic change.

REFERENCES

ABC News. 2009. Water-sucking Cubbie Station for sale.

Acreman, M., Arthington, A.H., Colloff, M.J., Couch, C., Crossman, N.D., Dyer, F., et al., 2014. Environmental flows for natural, hybrid, and novel riverine ecosystems in a changing world. Front. Ecol. Environ. 12, 466−473.

Ashton, P.J., Roux, D.J., Breen, C.M., Day, J.A., Mitchell, S.A., Seaman, M.T., et al., 2012. The freshwater science landscape in South Africa, 1900−2010. Overview of research topics, key individuals, institutional change and operating culture. Water Research Commission, Pretoria, South Africa.

Aylward, B. 2008. Water markets: a mechanism for mainstreaming ecosystem services into water management?: Briefing Paper, Water and Nature Initiative, IUCN.

Barnett, J., Webber, M., Wang, M., Finlayson, B., Dickinson, D., 2006. Ten key questions about the management of water in the yellow river basin. Environ. Manage. 38, 179−188.

Bowman, M., 2002. Legal perspectives on dam removal. BioScience 52, 739−747.

The Brisbane Declaration, 2007. Brisbane declaration and call to action, in 10th international river symposium. 3−6 September 2007: Brisbane, Australia.

Byorth, P.A., 2009. Conflict to compact: Federal reserved water rights, instream flows and native fish conservation on national forests. Montana. Pub. Land Resour. L. Rev. 30, 35−55.

CIRES. 2014. CIRES webcast: Presentation by Tony McLeod [Online]. Available from: https://cirescolorado. adobeconnect.com/_a1166535166/p2v7b7okxek/?launcher=false&fcsContent=true&pbMode=normal (accessed 27.10.15).

COAG, 2004. Intergovernmental agreement on a national water initiative, Council of Australian Governments.

Columbia Basin Water Transactions Program, 2011. Annual Report Financial Year 2011. Portland, Oregon: National Fish and Wildlife Foundtion.

Columbia Basin Water Transaction Program, 2015. The Program [Online]. Available from: http://www.cbwtp. org/jsp/cbwtp/program.jsp (accessed 19.10.15).

Comisión Nacional Del Agua, 2011. Identificación de reservas potenciales de agua para el medio ambiente en México. México: CONAGUA.

Committee On Western Water Management, 1992. Water Transfers in the West: Efficiency, Equity and the Environment. National Academy Press, Washington, D.C.

Commonwealth Of Australia, 2010. Securing Our Water Future. Australian Government Department of the Environment, Water, Heritage and the Arts, Canberra, Australia.

Connell, D., Grafton, R.Q., 2011. Water reform in the Murray Darling Basin. Water Resour. Res. 47 (12), 1–62.

Cronin, A.E., Fowler, L.B., 2012. The Water Report #102: Northwest Water Banking. Seattle: Envirotech Publications.

CSIRO, 2008. Water availability in the Murray-Darling Basin. A report to the Australian Government from the CSIRO Murray-Darling Basin Sustainable Yields Project.

Daesslé, L.W., Van Geldern, R., Orozco-Durána, A., Barth, J.A.C., 2016. The 2014 water release into the arid Colorado River delta and associated water losses by evaporation. Sci. Total Environ. 542, 586–590.

Donohew, Z., 2009. Property rights and western United States water markets. Aust. J. Agric. Resour. Econ. 53, 85–103.

Doremus, H., Tarlock, A.D., 2003. Fish, farms and the clash of cultures in the Klamath Basin. Ecol. Law Quart. 30, 279–350.

Doremus, H., Tarlock, A.D., 2008. Water War in the Klamath Basin: Macho Law, Combat Biology and Dirty Politics. Island Press.

DWA, 2011. Directorate Water Resource Planning Systems: Water Quality Planning. Resource Directed Management of Water Quality. Planning Level Review of Water Quality in South Africa. Department of Water Affairs, Republic of South Africa, Pretoria, South Africa.

DWA, 2013. Strategic overview of the water sector in South Africa. Department of Water Affairs, Republic of South Africa, Pretoria, South Africa.

DWAF, 1997. White Paper on a National Water Policy for South Africa. Department of Water Affairs and Forestry, Pretoria, South Africa.

DWE, 2009. Proposal to enable environmental water entitlements acquired in the Darling River at Toorale Station, to be diverted downstream of the Menindee Lakes.

DWS, 2014. Determination of Resource Quality Objectives in the Olifants Water Management Area (WMA4): Resource quality objectives and numerical limits report. Department of Water and Sanitation, Prepared by the Institute of Natural Resources (INR) NPC, Pietermaritzburg, South Africa.

Endicott, T., 2011. Administrative Law. Oxford University Press, Oxford, UK.

Fang, L., 2014. China's federal river management - the example of the Han River. In: Garrick, D. E., Anderson, G. R. M., Connell, D. & Pittock, J. (Eds.), Federal rivers. Managing water in multi-layered political systems.

Ferguson, J., Chilcott-Hall, B., Randall, B., 2006. Private water leasing: working within the prior appropriation system to restore streamflows. Pub. Land Resour. L. Rev. 27, 1–13.

Fisher, D.E., 2010. Australian Environmental Law: Norms, Principles and Rules. Thomson Reuters, Sydney.

Foerster, A., 2007. Victoria's new Environmental Water Reserve: what's in a name? Australas. J. Nat. Resour. Law Policy 11, 145.

Gagnon Thompson, S.C., Barton, M.A., 1994. Ecocentric and anthropocentric attitudes toward the environment. J. Environ. Psychol. 14, 149–157.

Garrick, D., Aylward, B., 2012. Transaction costs and institutional performance in market-based environmental water allocation. Land Econ. 88, 536–560.

Garrick, D., Siebentritt, M.A., Aylward, B., Bauer, C.J., Purkey, A., 2009. Water markets and freshwater ecosystem services: Policy reform and implementation in the Columbia and Murray-Darling Basins. Ecol. Econ. 69, 366–379.

Gillilan, D.M., Brown, T.C., 1997. Instream Flow Protection: Seeking a Balance in Western Water Use. Island Press, Covelo, California.

Giordano, M., Zhu, Z., Cai, X., Hong, S., Zhang, X. & Xue, Y., 2004. Water management in the Yellow River Basin: Background, current critical issues and future research needs: Research Report 3. Colombo, Sri Lanka.

Godden, L., 2005. Water law reform in Australia and South Africa: sustainability, efficiency and social justice. J. Environ. Law 17, 181.

Goulbourn Broken Catchment Management Authority, 2013. Seasonal Watering Proposal.

Gowlland-Gualtieri, A., 2007. South Africa's Water Law and Policy Framework — Implications for the Right to Water. International Environmental Law Research Centre.

Grafton, R.Q., Hussey, K., 2006. Buying back the Living Murray: At What Price? Economics and Environmetn Network Working Paper EEN0606. Australian National University.

Gray, R., Milne, M.J., 2014. Explainer: what is the triple bottom line? The Conversation February 6, 2014.

Hart, B., 2015. The Australian Murray-Darling Basin Plan: challenges in its implementation (part 1). Int. J. Water Resour. Dev.

Hillman, 2008. Ecological Requirements: Creating a working River in the Murray-Darling Basin. In: Crase, L. (Ed.), Water Policy in Australia — The Impact of change and uncertainty. Resources for the future, Washington, USA.

Hirji, R., Davis, R., 2009a. Envirnmental flows in water resources policies plans, and projects: case studies. Washington DC: World Bank Environemnt Department Papers, Natural Resource Managmenet Series, Paper No 117.

Hirji, R., Davis, R., 2009b. Environmental flows in water resources policies, plans and projects: findings and recommendations. The World Bank, Washington DC.

Horne, A., Freebairn, J., O'Donnell, E., 2011. Establishment of environmental water in the Murray-Darling Basin — an analysis of two key policy initiatives. Aust. J. Water Resour. 15, 7.

Hussey, K., Dovers, S., 2007. Managing Water for Australia: The Social and Insitutional Challenges. CSIRO Publishing.

IPCC, 2014. Climate Change 2014 — Synthesis Report. Contribution of Working Groups I, II and III to the Fifth Assessment Report of the Intergovernmental Panel on Climate Change [Core Writing Team, R.K. Pachauri and L.A. Meyer (Eds.)]. Geneva, Switzerland.

Jian, K., 2014. Watershed management in Tai Lake Basin in China. In: Garrick, D. E., Anderson, G. R. M., Connell, D. & Pittock, J. (Eds.) Federal rivers. Managing water in multi-layered political systems.

Jiménez Cisneros, B.E., Oki, T., Arnell, N.W., Benito, G., Cogley, J.G., Döll, P., et al., 2014. Freshwater resources. In: Field, C.B., Barros, V.R., Dokken, D.J., Mach, K.J., Mastrandrea, M.D., Bilir, T.E., Chatterjee, M., Ebi, K.L., Estrada, Y.O., Genova, R.C., Girma, B., Kissel, E.S., Levy, A.N., Maccracken, S., Mastrandrea, P.R., White, L.L. (Eds.), Climate Change 2014: Impacts, Adaptation, and Vulnerability. Part A: Global and Sectoral Aspects. Contribution of Working Group II to the Fifth Assessment Report of the Intergovernmental Panel on Climate Change. Cambridge University Press, Cambridge, United Kingdom and New York, NY, USA.

Killingbeck, S., Charles, M.B., 2011. The place of legislation and regulation and the role of policy: lessons from the CPRS. South. Cross Univ. Law Rev. 14, 93—118.

Lenaerts, K., Desomer, M., 2005. Towards a hierarchy of legal acts in the European Union? Simplification of legal instruments and procedures. Eur. Law J. 11, 744—765.

Leonard, N.J., Fritsch, M.A., Ruff, J.D., Fazio, J.F., Harrison, J., Grover, T., 2015. The challenge of managing the Columbia River Basin for energy and fish. Fish. Manag. Ecol. 22, 88—98.

LE Quesne, Kendy, T.E., Weston, D., 2010a. The Implementation Challenge: Taking stock of government policies to protect and restore environmental flows. World Wildlife Fund UK, Godalming, Surrey.

LVBC and WWF-ESARPO, 2010. Assessing Reserve flows for the Mara River. Nairobi and Kisumu, Kenya: Lake Victoria Basin Commission (LVBC) and WWF Eastern & Southern Africa Regional Programme Office (WWF-ESARPO).

Macdonald, D.H. & Young, M., 2001. A case study of the Murray-Darling Basin. Prepared for the International Water Management Institute.

Macdonnell, L., 2009a. Environmental flows in the rocky mountain west: a progress report. Wyo. Law Rev. 9, 335–396.

Macdonnell, L., 2009b. Return to the River: Environmental Flow Policy in the United States and Canada. J. Am. Water Resour. Assoc. 45, 1087–1099.

Malloch, S., 2005. Liquid Assets: Protecting and Restoring the West's Rivers and Wetlands through Environmental Water Transactions. Trout Unlimited, Arlington, VA.

McLoughlin, C., Mackenzie, J., Rountree, M., Grant, R., 2011. Implementation of strategic adaptive management for freshwater protection under the South African national water policy. Water Research Commission, South Africa.

MDBA (Murray-Darling Basin Authority), 2010. Guide to the proposed Basin Plan: overview. Murray-Darling Basin Authority, Canberra.

MDBA (Murray-Darling Basin Authority) Murray-Darling Basin Plan, 2012. Commonwealth of Australia, Canberra, Australia.

MDBA, 2015. Ten years of The Living Murray Program – restoring the health of the River Murray [Online]. Available from: http://www.mdba.gov.au/what-we-do/working-with-others/ten-years-of-tlm-program (accessed 10.11.15).

MDBC (Murray-Darling Basin Commission), 1998. Murray-Darling Basin Cap on Diversions Water Year 1997/98: Striking the Balance. Murray-Darling Basin Commission, Canberra.

Mekonnen, M.M., Hoekstra, A.Y., 2016. Four billion people facing severe water scarcity. Science Advances 2.

Mistry, K., 2015. Columbia River flows to be protected [Online]. Available from: http://www.celp.org/tag/celp/ (accessed 16.10.15).

Moore, S., 2013. Issue Brief: Water Resource Issues, Policy and Politics in China. Brookings Institute. Available from: https://www.brookings.edu/research/issue-brief-water-resource-issues-policy-and-politics-in-china/.

Moore, S., 2014. The politics of thirst: Managing water resources under scarcity in the Yello River Basin, People's Republic of China. Discussion Paper #2013-08. Harvard Kennedy School, Belfer Center for Science and International Affairs.

Muller, M., 2014. Allocating powers and functions in a federal design: the experience of South Africa. Part IV: South Africa. In: Garrick, D., Anderson, G.R.M., Connell, D., Pittock, J. (Eds.), Federal rivers. Managing water in multi-layered political systems. Edward Elgar Publishing Limited, Cheltenham, UK and Northampton, MA, USA, and IWA Publishing, London, UK.

Mumma, A., 2007. Kenya's new water law: an analysis of the implications of Kenya's Water Act, 2002, for the rural poor. In: Van Koppen, B., Giordano, M., Butterworth, J. (Eds.), Community-based water law and water resource management reform in developing countries. Comprehensive Assessment of Water Management in Agriculture Series 5. CAB International, Wallingford, UK.

Murray Darling Basin Ministerial Council, 2000. Review of the Operation of the Cap - Overview Report of the Murray-Darling Basin Commisssion. Canberra, Australia.

Neale, T., 2012a. Contest and Contest: The legacy of the Wild Rivers Act 2005 (QLD). Indigenous Law Bull. 8.

Neale, T., 2012b. The Wild Rivers Act Controversy. The Conversation.

Northwest Council, 2015. Reports [Online]. Available from: https://www.nwcouncil.org/reports/ (accessed 16.10.15).

NWC (National Water Commission), 2009. Australian Water Reform 2009: Second biennial assessment of progress in implementation of the National Water Initiative. National Water Commission, Canberra.

NWC, 2012. Australina environmental water management: framework criteria. National Water Commission, Canberra.

O'Donnell, E., 2012. Institutional reform in environmental water management: the new Victorian Environmental Water Holder. J. Water Law 22, 73–84.

O'Donnell, E., Macpherson, E., 2013. Challenges and opportunities for environmental water management in Chile: an Australian perspective. J. Water Law 23, 24–36.

O'Riordan, T., 1991. The new environmentalism and sustainable development. Sci. Total Environ. 108, 5.

Olden, J.D., Konrad, C.P., Melis, T.S., Kennard, M.J., Freeman, M.C., Mims, M.C., et al., 2014. Are large-scale flow experiments informing the science and management of freshwater ecosystems? Front. Ecol. Environ. 12, 176–185.

Opperman, J.J., Royte, J., Banke, J., Day, L.R., Apse, C., 2011. The Penobscot River, Maine, USA: a basin-scale approach to balancing power generation and ecosystem restoration. Ecol. Soc. 16, 7.

Palmer, C., Muller, W., Gordon, A., Scherman, P., Davies-Coleman, H., Pakhomova, L., et al., 2004. The development of a toxicity database using freshwater macroinvertebrates, and its application to the protection of South African water resources. South Afr. J. Sci. 100, 643–650.

Pienaar, H., Belcher, A., Grobler, D.F., 2011. Protecting Aquatic Ecosystem Health for Sustainable Use. In: Schreiner, B., Hassan, R. (Eds.), Transforming Water Management in South Africa: Designing and Implementing a New Policy Framework. Springer.

Poff, N.L., Matthews, J.H., 2013. Environmental flows in the Anthropocene: past progress and future prospects. Curr. Opin. Environ. Sustain. 5, 667–675.

Pollard, S., Du Toit, D., 2011. Towards the Sustainability of Freshwater Systems in South Africa: An Exploration of Factors That Enable and Constrain Meeting the Ecological Reserve Within the Context of Integrated Water Resources Management in the Catchments of the Lowveld. Report to WRC.

Pollard, S., Du Toit, D., 2014. Meeting the challenges of equity and sustainability in complex and uncertain worlds: the emergence of integrated water resources management in the eastern rivers of South Africa. In: Garrick, D., Anderson, G.R.M., Connell, D., Pittock, J. (Eds.), Federal rivers. Managing water in multi-layered political systems. Edward Elgar Publishing Limited, Cheltenham, UK and Northampton, MA, USA, and IWA Publishing, London, UK.

Pollard, S., Mallory, S., Riddell, E., Sawunyama, T., 2011. Towards improving the assessment and implementation of the Reserve: real-time assessment and implementation of the Ecological Reserve. Water Research Commission, South Africa.

Productivity Commission, 2010. Market mechanisms for recovering water in the Murray-Darling basin (Final Report, March). Productivity Commission, Canberra.

Republic of South Africa, 1998. National Water Act. In: Government Gazette, V., No. 19182. (Ed.) Act No. 36 of 1998. Cape Town.

Richter, B., 2009. Re-thinking environmental flows: from allocations and reserves to sustainability boundaries. River Res. Appl. 25, 1–12.

Richter, B.D., 2014. Chasing water: A guide for moving from scarcity to sustainability. Island Press, Washington.

Richter, B.D., Thomas, G.A., 2007. Restoring environmental flows by modifying dam operations. Ecol. Soc. 12, 12.

Rogers, K., Biggs, H., 1999. Integrating indicators, endpoints and value systems in strategic management of the rivers of the Kruger National Park. Freshw. Biol. 41, 439–451.

Roux, D., Foxcroft, L., 2011. The development and application of strategic adaptive management within South African National Parks. Koedoe 53.

Sax, J.L., Abrams, R.H., Thompson, B.H., 1991. Legal Control of Water Resources. West Publishing Company, St Paul, Minnesota.

Senator the Hon. Penny Wong and Carmel Tebbutt, 2008. Joint Media release, Commonwealth and NSW purchase Toorale.

Shen, D., Speed, R., 2009. Water Resources Allocation in the People's Republic of China. Int. J. Water Resour. Dev. 25, 209–225.

Silveira, A., 2014. China's political system, economic reform and the governance of water quality in the Pearl River Basin. In: Garrick, D. E., Anderson, G. R. M., Connell, D., Pittock, J. (Eds.) Federal rivers. Managing water in multi-layered political systems.

Speed, R., Binney, J., Pusey, B., Catford, J., 2011. Policy measures, mechanisms, and framework for addressing environmental flows. International WaterCentre, Brisbane.

Speed, R., Li, Y., Le Quesne, T., Pegram, G., Zhiwei, Z., 2013. Basin water allocation planning — principles, procedures and approaches for basin allocation planning. UNESCO, Paris.

State Of Victoria, 2004. Victorian Government White Paper: Securing Our Water Future Together. Department of Sustainability and Environment.

State Of Victoria, 2009. Northern Region Sustainable Water Strategy. Department of Sustainability and Environment, Melbourne.

Svensson, J., 2014. Development of Water Markets in the Yellow River Basin: A Case-Study of the Ningxia Hui Autonomous Region. Lund University.

Tarlock, A.D., 2000. Prior appropriation: rule, principle or rhetoric? N. D. Law Rev. 76, 881—910.

Tarlock, A.D., 2014. Mexico and the United States assume a legal duty to provide Colorado River Delta restoration flows: An important International Environmental and Water Law Precedent. Rev. Eur. Commun. Int. Environ. Law 23, 76—87.

The Natural Heritage Institute, 2013. Concept Paper: Reoptimization of Xialangdi Dam on the Yellow River.

Thompson, J.B.H., 2000. Markets for Nature. William Mary Environ. Law Policy Rev. 25, 261.

Ukkola, A.M., Prentice, I.C., Keenan, T.F., Van Dijk, A.I., Viney, J.M., Myneni, N.R., et al., 2015. Reduced streamflow in water-stressed climates consistent with CO_2 effects on vegetation. Nat. Clim. Change. 6, 75—78.

Van Dijk, A.I.J.M., Beck, H.E., Crosbie, R.S., De Jeu, R.A.M., Liu, Y.Y., Podger, G.M., et al., 2013. The Millennium Drought in southeast Australia (2001—2009): Natural and human causes and implications for water resources, ecosystems, economy, and society. Water Resour. Res. 49, 1040—1057.

Van Wyk, E., Breen, C.M., Roux, D.J., Rogers, K.H., Sherwill, T., Van Wilgen, B.W., 2006. The Ecological Reserve: towards a common understanding for river management in South Africa. Water SA 32, 403—409.

VEWH, 2015. Reflections: Environmental Watering in Victoria 2014—15. Victorian Environmental Water Holder, Melbourne.

Wade, J.H.B., 2010. How national park service operations relate to law and policy. J. Interpret. Res. 15, 33—39.

Walker, W., Adnan Rahman, S., Cave, J., 2001. Adaptive policies, policy analysis, and policy-making. Eur. J. Oper. Res. 128, 282—289.

Ward, N.D., Ward, D.L., 2004. Resident fish in the Columbia River Basin: restoration, enhancement and mitigation for losses associated with hydroelectric development and operations. Fisheries 29, 10—18.

Warner, A.T., Lb, B., Jt, H., 2014. Restoring environmental flows through adaptive reservoir management: planning, science, and implementation through the Sustainable Rivers Project. Hydrol. Sci. J. 59, 770—785.

Wiel, S., 1911. Water Rights in the Western United States. Bancroft Whitney Company, San Francisco.

Zhang, K.M., Wen, Z.G., 2008. Review and challenges of policies of environmental protection and sustainable development in China. J. Environ. Manage. 88, 1249—1261.

Zhu, Z., Giordano, M., Cai, X., Molden, D., Shangchi, H., Huiyan, Z., et al., 2003. Yellow River Comprehensive Assessment: Basin Features and Issues. Collaborative Research between International Water Management Insitutite (IWMI) and Yellow River Conservancy Commission (YRCC).

REBALANCING THE SYSTEM: ACQUIRING WATER AND TRADE

18

Claire Settre[1] and Sarah A. Wheeler[1,2]

[1]*The University of Adelaide, Adelaide, SA, Australia* [2]*The University of South Australia, Adelaide, SA, Australia*

18.1 INTRODUCTION

Water shortages were named as the world's top global risk by the World Economic Forum's Global Risk Report in 2015. Ten years ago, it was not even on the list (WEF, 2015). As populations increase, the climate changes and scarcity factors intensify in arid and semi-arid regions, water-dependent ecosystems will endure this risk. This is especially true in fully appropriated and over-allocated river basins where little or no water is left to maintain riverine ecosystems. In these cases, the health and resilience of water-dependent ecosystems is a question of providing increased water to the environment. Restoring natural assets through increased provision of environmental water has gained recognition and varying degrees of consensus internationally (Lane-Miller et al., 2013). What is less agreed upon is how this can be achieved. Various mechanisms to reallocate water to the environment are discussed in Chapter 17. In some cases, improving environmental quality can be achieved by simply improving management of existing environmental water supplies. To a certain extent, new supplies can be created through desalination to replace water extracted from rivers, though at significant cost. At an urban scale, rain and stormwater harvesting and effluent reuse can also reduce urban pressure on riverine systems. On top of these practices, restoring water-stressed basins requires high-level policy that facilitates larger-scale water reallocation. There are a number of ways in which this can be achieved and the environmental water policy toolkit includes:

1. *Regulations and sanctions (command and control)*: Norms and standards for water quality (e.g., drinking water quality, ambient water quality for recreational water bodies, and industrial discharges); performance-based standards; charges to water rights; reducing water allocations; restrictions or bans on activities that have an impact on water resources (e.g., polluting activities in catchment areas and ban on phosphorus detergents); abstraction and discharge permits; water rights; land-use regulation and zoning (e.g., buffer zone requirements for pesticide application).

2. *Voluntary and education instruments*: Metering of water usage; eco-labeling and certification (e.g., for agriculture and water-saving household appliances); voluntary agreements between businesses and government for water efficiency; introduction of carryover for water holders to store water over multiperiods; water donation schemes; promotion of awareness raising and training in ecological farming practices or improved irrigation technologies; stakeholder

Water for the Environment. DOI: http://dx.doi.org/10.1016/B978-0-12-803907-6.00018-8

399

initiatives and cooperative arrangements seeking to improve water systems, for example between farmers and water utilities; planning tools (e.g., integrated river basin management plans).

3. *Economic instruments*: Charges (e.g., abstraction and pollution); user tariffs (e.g., for water services); payment for watershed services (e.g., for protection of catchment upstream); water pricing; reform of environmentally harmful subsidies (e.g., production-linked agricultural support and energy subsidies for pumping water); water markets; buyback of water entitlements; buyback of water allocations; subsidies (e.g., public investment in infrastructure and social pricing of water); tradable water allocations (temporary water) and entitlements (permanent water), options, and quotas; insurance schemes (Grafton and Wheeler, 2015).

In some cases, the relative lack of success of voluntary and regulation approaches in water management has meant that there has been an increased emphasis given to economic instruments over time (Griffin, 2016). Economic reallocation instruments, and in particular how water markets can be used to purchase water for the environment, is the focus of this chapter. Section 18.2 first discusses fundamentals about water markets.

18.2 WATER MARKET FUNDAMENTALS

A water market, either formal or informal, is the voluntary interaction of willing buyers and sellers with the purpose of exchanging a property right for water that is legally permitted to be traded. The exchanges undertaken in a water market are called water trades, transfers, or transactions and can be temporary (lease) or permanent (sale) in nature. The water right, or aspect thereof, that is sold on the market is called a water product and is typically characterized by the type of right (groundwater, surface water) and duration of the product it is exchanged for (permanent, short-term, split-season, option contact). Surface water rights are the most common water market products due to relative ease of measurement and mobility of the resource, although groundwater trade is also possible within hydrologically connected groundwater jurisdictions. Under ideal conditions, water markets provide a mechanism to distribute water among competing users to achieve a more socially and economically optimal solution by moving water to the highest value uses. Advocates for water marketing cite the dual benefits of increased water use efficiency and environmental conservation by signaling water users with the opportunity cost of water through the market (Chong and Sunding, 2006; Johnson et al., 2001; Rosegrant and Binswanger, 1994).

In general, water markets are a response to scarcity and where water is plentiful there will be no need for a market (a notable exception are markets for water quality, see Doyle et al., 2014). When a basin is under scarcity pressure, water markets can allow water users (e.g., farmers) to mitigate their supply risk by purchasing water when it is most needed (Zuo et al., 2014) or selling water if the price of water sale exceeds the use value derived from applying the water in irrigation. Water markets also benefit urban users by allowing cities to purchase reliable supply for critical human needs during drought or to support urban expansion. When a water market is used for reallocation from consumptive users to the environment, ecological conditions can also benefit when water is bought and left in the river (Section 18.3).

The characteristics of water markets are as diverse as the regions in which they operate. Market participants, products, and trades within the market are subject to the legal, institutional, and hydrological rules that govern the allocation and use of water of the region. Water markets exist in varying degrees of formality around the world. In an informal market the right to use an agreed-upon volume of water for a short time such as year or season is the product traded (Bjornlund, 2004). Informal water markets typically occur within irrigation districts between farmers and proceed with little administrative input. Formal water markets are less common and there are few examples of mature water markets. Formal water markets allow for the legal transfer of the water right on a perpetual basis (as opposed to the right to use for a period of time) and are significantly more institutionally complex and require a higher degree of administrative support (Bjornlund, 2004). Water markets of varying degrees of formality have been observed in India, Pakistan, Mexico, Chile, Spain, Australia, China, and the United States (Baillat, 2010; Easter et al., 1999; Moore, 2014; Palomo-Hierro et al., 2015; Rosegrant and Binswanger, 1994; Venkatachalam, 2015; Wheeler et al., 2014a). In Australia, for example, water marketing has progressed to a point of formality such that water can be bought and sold through water brokers or online trading platforms.

Markets for water, formal or informal, can effectively operate at a variety of geographic scales within agricultural districts, regions, basins, across basins, and across states. For example, in Australia, which hosts one of the most mature water markets in the world (see Box 18.1), the water market operates at a watershed scale within the Murray—Darling Basin (MDB) and covers five states and territories of southeastern Australia. By contrast, in China, quasi-market style mechanisms facilitate the transfer of water across the country in large south-to-north interbasin transfers (Shao et al., 2003).

As well as varying geographic scales, water markets can also operate at a range of institutional scales involving a variety of willing buyers and sellers. As agriculture typically uses the majority of water (up to and over 70% in many countries), farmers tend to be the sellers of water rights (Wheeler et al., 2013). Water sales can occur within an irrigation district from one farmer to another, or can occur from agriculture to urban, agriculture to industry, or agriculture to the environment, and back again. For example, in the western United States (see Box 18.2), the primary source of water in the market is farmers, who sell predominately to other farmers, cities, and the environment (Hanak and Stryjewski, 2012).

Establishing and regulating a water market has many challenges including strategies to minimize transaction costs and third-party impacts. At a more basic level, difficulties in water marketing arise due to the very nature of water itself. Assigning a monetary value to water as a commodity in a market framework is complicated not only by physical heterogeneity in quality, quantity, and location, but also by politically charged economics of water pricing and the symbolic nature of water as an essential life-giving human right (Hanemann, 2005). Water is a nonstandard, complex, and uncooperative commodity (Bakker, 2005; Chong and Sunding, 2006; Hanemann, 2005) that has the unique duality of being both finite and renewable. Further, the economic and environmental value of water is not only derived from its quantity, but also its quality, reliability, location, and timing. This adds a burden to the trading of water for optimal social and environmental benefit. Although laws permitting the formal transfer of water rights exist around the world, progress, and adoption of water markets remains highly contentious and underwhelming. For example, legislation passed in 1999 (Law 45/1999) incorporated formal water markets into the Spanish regulatory framework, but high transaction costs owing to administrative, legal, and cultural difficulties has led

BOX 18.1 WATER MARKETS IN THE MURRAY–DARLING BASIN, AUSTRALIA

There is a long history of informal water markets in Australia. Following scarcity and water quality pressures temporary water trading was first permitted in south-eastern MDB states in the 1960s and 1970s. In 1992, a Murray–Darling Basin Agreement and market-based instruments were developed. In 1997, a cap was established to limit water extraction and was critical in helping to promote markets by creating a cap and trade context. The most recent stage of water market development in Australia occurred with the introduction of a Water Act 2007 (Cwth), which sought to achieve large legislative, regulatory, and stakeholder reform through coordinated federal–state action. Importantly, these reforms included defining and securing water property rights; reforming markets to improve effectiveness; and removing barriers to trade water out of irrigation districts and between regions (Loch et al., 2013; Wheeler et al., 2014a). The northern and southern regions of the MDB are not hydrologically linked and therefore trade between them is not possible. In contrast, the southern Murray–Darling Basin (sMDB) is made up of a number of hydrologically connected water systems that transcend state boundaries. The sMDB accounts for most of the water used and traded, as well as most of the irrigated agricultural activity in the Basin. Water allocation trading is the most common form of trading (see Fig. 18.1), although entitlement trading has grown considerably over time, especially since the Millennium Drought (1997–2010) and the entry of the Commonwealth Environmental Water Holder into the market for the purpose of buying water entitlements for environmental use (see Section 18.3.3). By the end of 2010, at least 70% of all irrigators had made at least one water market trade, making it a common management tool used by irrigators in the Basin (Wheeler et al., 2014a).

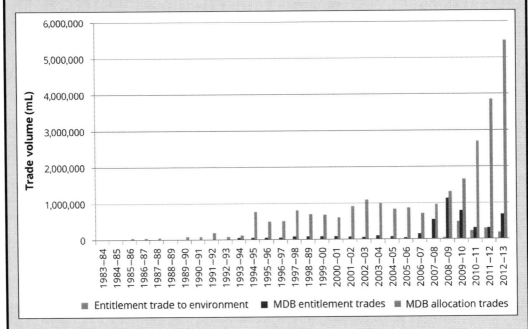

FIGURE 18.1

Water market trade volumes in the southern Murray–Darling Basin, Australia, from 1983 to 2013.

BOX 18.2 WATER MARKETS IN THE WESTERN UNITED STATES

Water markets in the United States are mostly limited to the semi-arid western states, where scarcity pressure necessitates the movement of water between agriculture, cities, industry, and the environment in response to changing supply-and-demand patterns. Generally, all types of water rights are legally transferable (Mooney et al., 2003), but not necessarily tradable separately from land. Water trade is concentrated among agricultural communities, where formal and informal reallocations within irrigation districts can be an important scarcity management tool (Debaere et al., 2014). The most common water rights in the United States are the appropriative rights. Water markets function within the wider context of western US water law, which rest on the pillars of prior appropriation and beneficial use. Prior appropriation assigns water to users on the basis of the seniority of their right, which is determined by the date the water was appropriated. This guarantees that senior beneficial users are last to lose supply during periods of low flow, so long as they maintain a beneficial use without prolonged interruptions (Garrick et al., 2011). Traditionally, beneficial use has referred to water used for productive means such as diversions for agriculture, mining, industry, domestic, and urban use (Clayton, 2009), although in many states legislation now recognizes and protects environmental water. Unlike the federal reforms in Australia, institutional development of water markets in the United States has progressed unevenly owing to differing state-based rules, local institutions, and limitations on water trading. In water management practice, many different formal and informal institutions interact across the landscape, governed by a myriad of contracts and case law.

---

to a very narrow water market (Palomo-Hierro et al., 2015). Proponents and critics of water markets often cite the same logic: water is vital and scarce (Ronald et al., 2013).

Proper accounting of water use and understanding of hydrological realities, as discussed in Chapter 3, is integral to the purposes of acquiring water for the environment. Young (2014) suggests six institutional principles when designing water markets:

1. Separate water access arrangements into their various component parts;
2. Assign any policy instruments for only specific purposes, and do not use multi-instruments;
3. Design instruments with hydrological integrity;
4. Keep transaction costs as low as possible;
5. Assign risk to one interest group; and
6. Ensure robustness of a system through proper accounting of water use.

In recognition of the need to provide water to the environment to maintain and restore fresh water ecosystems, water markets have been identified, and in some cases used, as a means to reallocate water to the environment.

18.3 ACQUIRING WATER FOR THE ENVIRONMENT

When legislation allows for it, water markets can be used as a platform to acquire water for the environment through the voluntary and compensated reallocation of consumptive water to the environment. When water rights are fully appropriated or where a cap on extractions is established, acquiring environmental water through voluntary and compensated market transactions is often a more politically palatable approach than administratively curtailing existing consumptive water rights.

18.3.1 CONDITIONS FOR THE ENVIRONMENT ENTERING THE WATER MARKET

As defined by Garrick et al. (2009), there are three conditions necessary for the establishment of environmental water markets. Without these conditions satisfied, markets for environmental water are infeasible (Aylward, 2008). The conditions are:

1. Establishment of rights to and limits on freshwater extraction and alteration;
2. Recognition of the environment as a legitimate water use; and
3. Authority to transfer existing water rights to an environmental purpose.

In addition to these conditions, acquiring water rights for the environment requires basic market principles to be satisfied such as well-defined, tradable, and legally defendable property rights. Without such conditions, allocating water to the environment will be a challenging task. Further discussion on natural resource property rights can be found in Schlager and Ostrom (1992) and Heltberg (2002).

Globally, the recognition of the environment as a legitimate water user remains on a relatively small scale. Growing environmental awareness and changes to water laws internationally signal a slow transition to legitimizing the environment as a beneficial user. Further discussion of this was provided in Chapter 2. The following select examples illustrate this point: the introduction of an ecological reserve flow in South Africa's 1998 water reform (Schreiner, 2013); the 2005 reforms to the Chilean Water Code allowing the state to reject new water right appropriations on the basis of ecological protection (Bitran et al., 2011); the New Start for Fresh Water reforms in New Zealand (Ministry for the Environment, 2009); the establishment of the Commonwealth Environmental Water Holder (CEWH), a national environmental water steward, in Australia (Wheeler et al., 2013); the provision of a pulse flow for ecological benefit in the Mexican reaches of the Colorado Basin (King et al., 2014); and the growing state-based and public recognition of beneficial instream use in the western United States (Loehman and Charney, 2011).

18.3.2 WATER PRODUCTS FOR THE ENVIRONMENT

When operating in an existing water market, there may be numerous ways for water to be acquired for the environment. The nature of environmental water acquisitions is dictated by the physical, institutional, and hydrological rules governing the system.

The legally defined nature of water (e.g., annual, short-term, seasonal, etc.) purchased for the environment can be described as a water product. The nature of the water product acquired for the environment influences the management of environmental water over spatial and temporal scales (Wheeler et al., 2013) and hence the environmental water use strategy. The management of environmental water use to optimize environmental objectives is discussed in Chapters 23 and 24. To highlight the varying types of products and innovations to acquire water for the environment we examine the approaches adopted in the western United States and Australia—two key examples of environmental water markets (see Section 18.3.4). Table 18.1 summarizes the main approaches used in these markets.

Column 3 of Table 18.1 lists entitlements and allocation trading. Water rights in Australia are defined as the right to access a share or entitlement of water from a consumptive pool. This water entitlement can be transferred (permanent trade). Each entitlement yields a seasonal volumetric allocation that can be traded, known as water allocation trade (temporary trade). The seasonal

Table 18.1 Water Rights Used for the Environment in Western United States and Australia

Contract Duration	Western United States	Australia
Permanent	Water right purchases Conserved water Groundwater–surface water source switch Change in point of diversion	Entitlement purchases Improving on- and off-farm efficiency of infrastructure
Temporary	Leases Split-season leases Groundwater–surface water source switch	Allocation trading Long-term lease agreements Counter-cyclical trading
Options	Dry-year agreements Minimum flow agreements Rotational pools	Options contracts

Source: Adapted from Wheeler et al. (2013, p. 430)

allocation varies annually depending on the reliability of the entitlement and climactic conditions—as scarcity increases volumetric allocations decrease. Consumptive water rights and environmental water rights are subject to the same allocation rules. Entitlement and allocation purchases are the same as water rights purchases and water leasing in the western United States, respectively. When water is leased to the environment, the water right holder legally retains the ownership of the water right, but allows it to be used instream for a specified duration.

In Australia, entitlement purchases (permanent trade) are the most common form of acquiring water for the environment. A novel alternative to this approach is yearly trade of allocations between consumptive users and the environment, called counter-cyclical trading. Consider, for example, a floodplain wetland that requires inundation every three years to maintain optimal ecological health. In this example, counter-cyclical trading is beneficial by allowing the environmental steward to lease allocations to farmers, two in every three years, and deliver the allocations to the floodplain in the third year, which can be supplemented by additional allocations purchased with funds raised by the previous year's leasing. In essence, when water is not needed for environmental watering, the allocation can be placed back onto the market to be purchased by a consumptive user. In this way, the environment generates an income from the sale of water when it is not needed in order to buy water at more crucial times (Connor et al., 2013; Kirby et al., 2006).

Water products in the United States are more varied than in Australia. A number of temporary options involve contractual agreements to change the characteristic of the water right such as split-season leasing or dry-year agreements. Split-season leases are agreements for consumptive water rights holders to use water for part of the season and lease water for the other part of the season. An ideal arrangement of this kind meets consumptive water demands during all or part of the irrigation season and meets environmental water demands during critical ecological times such as during fish migration or spawning. Conditional contracts, which stipulate the transfer of water to the environment on a temporary or permanent basis when some condition is met, are also viable products for the environment. In the United States, examples of these include minimum flow agreements and dry-year contracts, which are triggered by a hydroclimactic threshold stipulated in the contract. These types of agreements specify that a consumptive user will be compensated to forgo the use of water in years in which additional water is required for the environment.

BOX 18.3 ECOLOGICAL FLOW PROVISIONS OF MINUTE 319: UNITED STATES AND MEXICO BINATIONAL AGREEMENT

Minute 319 is an unprecedented agreement between the United States and Mexico to share shortages and surpluses, allow Mexico to store water in US dams, and provide environmental water to the parched Colorado River Delta. With the exception of unusually wet years, very little water has reached the Colorado River Delta since 1960, because up to 90% of water is diverted from major upstream dams in Arizona, Nevada, and California, United States, and the remaining 10% diverted in the Mexicali Valley (Mexico) before reaching the Delta (Flessa, 2001). This has had significant implications for the ecology of the Delta, which Minute 319 aims partly to address. The environmental regimes are made up of two components. First, a one-off pulse flow of 126 GL jointly provided by US and Mexican governments, including the intentionally created Mexican allocation generated by conservation activities in the Mexicali Valley, released from Lake Mead. The pulse flow was released in March 2014 and timed to mimic natural floods at the time of native cottonwood seed release. The pulse flow entered the Gulf of California in May 2014 (Flessa et al., 2014). Second, a base flow of 64 GL acquired by a coalition of nongovernmental organizations (see Section 18.3.4) was delivered to priority restoration areas along the Mexican reaches of the river. The environmental water deliveries were made with the aim of enhancing 9.3 km^2 of Delta habitat (Sonoran Institute, 2013). A range of additional restoration efforts were also carried out under Minute 319, including hydroseeding of native vegetation and removal of invasive species.

Negotiation of water storage rights is also becoming a prevalent aspect for both short- and long-term environmental water agreements in the western United States and beyond. For example, the pulse flow delivered to the Mexican reaches of the Colorado River implemented to aid in the Colorado Delta restoration (see Box 18.3) was derived from storage held in upstream Lake Mead in Nevada. The stored water was generated, in part, by the intentionally created Mexican allocation, which allowed Mexico to voluntarily conserve and store water in US storage for future consumptive and environmental use (King et al., 2014).

In addition to direct water purchases, incentivizing conservation through an increase in irrigation efficiency has been employed in both Australia and the United States. This approach benefits the environment because the water saved by the improvement in irrigation efficiency can be sold, donated, or purchased for the environment. In these cases, farmers benefit from often subsidized investments in on-farm infrastructure and added income from the sale of water. In the case of Australia (and also the United States), irrigators have expressed strong preferences for infrastructure improvement options over purchases of water entitlements as a means to recover water for the environment (Bjornlund et al., 2011, Loch et al., 2014). Alternative measures such as switching crops to reduce irrigation demand is also used to decrease consumptive water use and increase water available to be sold to the environment, although this is less common in Australia than irrigation efficiency upgrades.

Groundwater and surface water source switch, such that the previously used surface water is made available for the environment, is also a way of increasing local environmental water availability. The upper Deschutes Basin Groundwater Mitigation Program (GMP) is an instructive case where this approach has been implemented in a hydrologically connected groundwater–surface water. The GMP specified that the impact of new groundwater use must be offset by a mitigation project (or purchase of mitigation credits from established mitigation banks), which results in a quantity of water protected instream for environmental use (Water Resources Department, 2008). In essence, for every new groundwater permit issued water from a surface water right would be dedicated to instream flow to mitigate the impact of the new use for the life of the groundwater use

(Lieberherr, 2011). Surface water—groundwater switches are not formally practiced in Australia, although there is evidence that some irrigators who have sold their water rights substitute their water use with groundwater entitlements (Wheeler and Cheesman, 2013; Wheeler et al., 2016). In addition to source switches, changing the point of diversion from a tributary to a main channel can be implemented, typically by negotiation with private irrigators who manage their own infrastructure. This approach is often quite ecologically effective because the removal of water from a small tributary can have substantially larger environmental impacts than the same amount of volume diverted from a main river channel, therefore increasing the local environmental conditions in the tributary.

Typically, it is surface water rights that are purchased for the environment. However, there may be cases where it is ecologically beneficial and legally possible to transfer groundwater to the environment by paying a consumptive user to cease pumping operations. This approach is far less common than surface water transfers, although some arrangements of this kind have been pioneered by the Arizona Land and Water Trust in programs designed to protect groundwater-fed springs' riparian habitat of desert rivers (Citron and Garrick, 2010). Payments for ceasing pumping operations may gain greater traction as institutions governing groundwater markets mature and environmental water purchases become more common. Wheeler et al. (2016) provides greater discussion on groundwater market issues.

In addition to water products listed in Table 18.1, it is also possible for water to be donated to the environment from a consumptive user on a permanent or temporary basis. Incentivizing donations to the environment can be achieved by making water tax deductible, akin to a donation to a charity, as was initiated in Australia in 2010 (Wheeler et al., 2014b).

With many options available to be purchased, what water product should be bought for the environment? The answer not only hinges on the environmental demand for water but also on the political, economic, and social implications of environmental water acquisitions. In particular, the water acquisition strategy must reflect irrigator willingness to participate in reallocation programs.

There are a number of advantages of permanent acquisition strategies. Namely, permanent acquisition means that water is reallocated to the environment in the truest sense, as it can be used by the environment every year. This provides security and allows environmental water managers to adopt long-term watering plans. This is particularly useful when working to reestablish or restore an environmental asset that requires ongoing maintenance such as main channel recovery requiring an increase in base flow(s). By permanently acquiring water, the administrative burden of the transfer occurs only once at the time of sale. This reduces the ongoing transaction costs that would occur if annual leasing or contract negotiation was to be used. Conversely, in the case of the western United States, permanent acquisitions can attract significant transaction costs in the form of lengthy approval processes and extensive legal scrutiny of permanent transfer applications (Pilz, 2006).

However, permanently reallocated consumptive water use rights typically have even intra-annual supply profiles that are not necessarily consistent with environmental water demands (Connor et al., 2013). Further, the permanent water purchases, even when compensation is provided through the market, require considerable adjustment by farmers and regional communities. When water is sold permanently, land could be taken out of production if dry-land practices or input substitution are not adopted. This is known as buying and drying and can be damaging to rural farming communities if wide adjustment help is not provided. This may provide an incentive for temporary water to be purchased for the environment to aid in gradually transitioning farmers during reallocation programs and potentially reduce socio-economic impacts of water reallocation (Wheeler et al., 2013). Temporary purchases, however, require administrative input on a yearly basis and do not provide long-term security of water for the environment.

A portfolio approach to water acquisition is likely to be optimal to provide long-term security of supply and short-term ability to acquire or sell water as needed. For the case of Australia, Wheeler et al. (2013) suggest that a portfolio approach increases irrigator willingness to engage with reallocation programs and increases the volume of water available for the environment. Discussing water products for the environment quickly leads to the question of who can buy this water, which is discussed in the following section.

18.3.3 WHO CAN BUY WATER FOR THE ENVIRONMENT?

Globally, there are many options and an array of institutional collaborations that have been formed to address the question of who can buy water for the environment. These collaborations range from federal- and state-led solutions to private enterprise, as well as a number of collaborative partnerships. The institutional arrangements underpinning the management of environmental water are discussed in Chapter 19.

There are many trade-offs to federal-, state-, or privately led reallocation strategies. Federally led water purchases have the advantage of funding availability and supporting institutions to promote the economically efficient acquisition of water. Ongoing federal water reforms in Australia established robust institutions to support the function and regulation of the environmental water market. For example, the Murray—Darling Basin Authority (MDBA) was established to identify volumetric targets for reallocation, the CEWH was established to manage the purchased water, the Commonwealth Environmental Water Office was established to support decision making of the CEWH, and the National Water Commission (now dissolved) monitored the progress of reform. Without federal leadership in reallocation, the establishment of these vital institutions would likely be unrealized. Further, federal involvement in reallocation meant a considerable budget for purchasing water, around AU$3.1 billion, a sum that would likely be unmatched by state-led programs or private enterprise such as nongovernmental organizations (NGOs).

An additional advantage of federally led reallocation is that because water is purchased with public money (e.g., through tax revenue), the government is beholden to the public to use it in a way that maximizes social and environmental benefit for the public benefit. Theoretically, this results in the dual benefit of transparent government expenditure and optimal reallocation to the environment. Consequently, the CEWH's current policy is to make all its decision making as transparent as possible (e.g., the CEWH plans to upload all models and data on publicly available websites).

Similar to Australia, state and federal agencies remain the primary agencies acquiring instream water rights in the United States, and federal funding is the principal source of funding for NGO water acquisition (Scarborough, 2010). In recent years, however, there has been notable growth in philanthropic and fundraising organizations fueling environmental water market activity in the Colorado Basin. A notable example is the Raise the River campaign implemented to raise funds for instream water purchases and revegetation in the Mexican reaches of the Colorado River (Raise the River, 2016). The question of financial sustainability of philanthropically funded NGO water purchases is an ongoing one vital to the success of market-based programs. Significant innovations to develop sustainable business models for environmental water purchases are evident in a recent example in the MDB. In 2015, The Nature Conservancy launched an environmental water investment model, known as the Murray—Darling Basin Balanced Water Fund, and intends to purchase permanent water rights in the southern MDB. Funding for additional environmental acquisitions is generated by leasing the majority of water allocations (annual water assigned to the permanent entitlement) to consumptive users to derive a financial return and deliver remaining allocations to the environment to improve ecological condition (Kilter Rural, 2015).

There are significant advantages derived from NGO involvement in environmental water markets such as flexibility to develop relationships with farmers and build community trust. In situations where there is reluctance to sell water to the state or federal governments, there is an opportunity for private enterprise to enter the market and acquire water in place of the government. Flexibility in acquisition strategies and ability to work out case-by-case solutions such as split-season leasing or point of diversion changes with individual irrigators are also added advantages. Further, in cases when political will to reallocate water to the environment is lacking or environmental objectives are not recognized, NGOs can play a unique role by filling this gap.

The role of private individuals and NGOs participating in the market is governed by local institutions and laws. In the United States, for example, state-based legislation varies considerably regarding private entity acquiring rights for the environment. In the state of Colorado, a private entity is not allowed to acquire and manage water for environmental water rights, but rather the acquired right may be purchased and donated to the governing state body (Scarborough, 2012). Comparatively, in California, water rights may be purchased and managed for the environment by a private individual (Mooney et al., 2003; Scarborough, 2012). In Australia, private environmental water rights purchase occurs on a relatively small scale.

Although there are a number of water markets around the world and a growing amount of governments and private enterprise working on behalf of the environment, there are still only a few examples of water for the environment being acquired through the market, discussed in the following section.

18.3.4 EXAMPLES OF ENVIRONMENTAL WATER MARKETS

Water markets are not typically developed for the purpose of reallocating water to the environment, but are rather developed as a means for consumptive users to deal with scarcity. Regardless, a number of existing water markets have been used and expanded as vehicles to provide the transfer of water from consumptive uses to the environment (Debaere et al., 2014). Three key examples of this are found in Australia, the western United States, and Mexico. In these cases, the amount of water held by the environmental water holder is small in proportion to the hydrological volume of the basin. These examples are by no means exhaustive, and state- or regionally based reallocation programs certainly exist both within these countries and elsewhere, though documentation is scarce. A detailed discussion of the governance arrangements underpinning environmental water management in these case studies is provided in Chapter 24.

The MDB is the site of one of the most mature water markets in the world (Wheeler, 2014). As a recap, two main water markets exist in Australia: (1) a water entitlement market where the permanent rights to water are bought and sold. A water entitlement provides exclusive long-term access to a share of the water resources from a consumptive pool. An entitlement can be a low, general, or high-security entitlement, which dictates the likelihood of receiving a nonzero allocation each year; or (2) a temporary water allocation market where the right to use water awarded to an entitlement is traded on a yearly basis. Water allocations are announced at the beginning of each season as a percentage of the total water entitlement and are dependent upon changing water availability (and can sometimes be zero). Box 18.1 provided more detail on the evolution of water markets in Australia.

Using the water market to reallocate water to the environment in the MDB is a result of decades of ongoing water policy development, reform, and implementation. A key outcome is that the environment has become a statutorily recognized water user in its own right and has been awarded the same degree of security as water entitlements held for consumptive use (COAG, 2004). Most recently in 2007, the *Water Act* (Cth) was passed and created the MDBA, which was charged with the task of developing a plan to return the MDB to sustainable levels of water use. In 2012 the Murray–Darling Basin Plan was passed into law. The Plan is a blueprint for implementing sustainable diversion limits (SDLs), which are an upper limit on the amount of water that can be extracted for consumptive use and come into full effect in 2019. A reallocation target of 2750 GL was set to be reallocated to the environment by 2019, with an extra 450 GL recovered through infrastructure investment expenditure. At the time of writing, a two-pronged approach is currently underway to overcome the gap between current extraction levels and the SDLs, and includes investment in on-farm irrigation efficiency upgrades and purchasing water rights for the environment from consumptive users through the water market. Table 18.2 shows the amount of environmental water purchased by the Australian government as of February 29, 2016. Currently, around two-thirds of the target volume has been acquired.

Similar to Australia, the emergence of environmental water transactions in the western United States has been the result of ongoing legislative reform and increasing recognition of the importance of water for the environment (Box 18.2). Initially through avenues of common law such as the Public Trust Doctrine, the environment has been incorporated into the water rights system as a legitimate water user (Garrick et al., 2009). Based on the Public Trust Doctrine, private water rights holders do not have the right to infringe on the quality of public resources held in trust

Table 18.2 Water Rights and Average Annual Yield of Environmental Water Rights Held by the Australian Commonwealth Environmental Water Holder, February 29, 2016

State	Security	Water Right Volume (Entitlement) (ML)	Average Annual Yield (Allocation) (ML)
Queensland	Medium	15,585	5,284
New South Wales	High	30,357	29,060
	General	890,855	556,353
Victoria	High	605,499	575,006
	Low	55,851	23,369
South Australia	High	147,837	133,053
TOTAL	High	783,693	737,120
	General/ Medium/Low	962,291	585,006

Source: Department of Environment (2016)

by the state (Mooney et al., 2003). Administrative rulings with their roots in the Public Trust Doctrine have been used to appropriate water for the environment as a beneficial use. Federal level legislation, such as the Endangered Species Act 1973, and to a lesser extent the Clean Water Act 1972, have also driven progress in administrative- and market-based reallocation to provide environmental water to sustain critically endangered aquatic species.

The Columbia River Basin located in the northwest and shared with Canada is a key example where market-based water transactions have been used to improve habitat for the maintenance of an endangered species—in this case, salmon. The Columbia River Basin Water Transactions Program (CBWTP) specifies a set of qualified local entities (QLEs) that are able to buy water on behalf of the environment. As part of the CBWTP, extensive permanent acquisitions, irrigation efficiency projects, and temporary water transactions (e.g., annual or short-term leases) and split-season leases are used to return water to the river. The QLEs in the CBWTP are a mixture of state, river basin conservancies, and NGOs, and provide a successful example of nested governance arrangements in a market-based water management context. State-based legislation is also paramount for protecting the water transferred to the environment. The US states of the Columbia River Basin were among the first to take the legal leap to establish environmental water provisions (Neuman et al., 2006). For example, laws allowing administrative designations of minimum or instream flows were passed in Washington (Minimum Water Flows and Levels Act 1967) and Idaho (Minimum Stream-Flow Act 1978) followed by the Instream Water Rights Act in Oregon (1987), which provided an explicit basis for a market-based approach by allowing senior rights to be acquired, leased, or donated for instream use. Progress toward reallocation to the environment has been slower in states experiencing greater scarcity pressure such as the American reaches of the Lower Colorado Basin, where environmental rights can be both legally ambiguous and insecure.

In addition to the CBWTP, water transactions occur in many other instances. Overall, environmental water transactions made up 40% of total water traded and 7% of total value traded from 2003 to 2012 in the western United States (WestWater Research, 2014). The majority of this water was purchased by federal agencies, including the Bureau of Reclamation and the US Fish and Wildlife Service. Environmental water acquisitions, expenditure, volume, frequency of trade, and type of right vary significantly across states and between basins (Scarborough and Lund, 2007). Table 18.3 summarizes volumes traded to the environment in the most active subbasins in the Columbia River Basin from 2003 to 2010.

Also in the United States and shared with Mexico, the Colorado River Basin is the site of an internationally unique binational environmental watering program. The Colorado River Basin partly covers seven US states and two states of Mexico and is governed by the 1944 Water Treaty. The environmental watering program targets the Colorado River Delta located in northern Mexico and is part of a wider 5-year pilot agreement: Minute 319. At the core of Minute 319 is a binational partnership incorporating US and Mexican federal governments, binational NGOs and US water providers that aim to manage the river as a shared resource by sharing shortage and surplus (King et al., 2014). A key provision of Minute 319 is the delivery of a pulse and base flow to the Colorado River Delta (see Box 18.3). The pulse flow is to be provided by both US and Mexican governments, whereas the base flow is to be provided by a coalition of NGOs, known as the Colorado River Delta Water Trust. To acquire the water for the base flow, the Trust makes use of the active water market in the Mexicali Valley through which it buys water to be delivered to the environment. Progress toward achieving the target volume has been substantial and has been facilitated by the Trust, both purchasing and leasing water rights. Funding for the purchase of water rights is gained through philanthropic donations and raised funds. In the remaining years of the

Table 18.3 Water Recovered, Budget, and Expenditure for 10 Active Water Trading Subbasins in the Columbia River Basin, US, from 2003 to 2010

Subbasin, State	Average Annual Water Recovered (GL/year) 2003–2010	Total Water Recovered (GL/year) 2003–2010	Total Program Budget 2003–2010 (USD 2007)
Bitterroot, Montana	189.32	1515.42	$975,553
Blackfoot, Montana	87.51	701.01	$942,076
Deschutes, Oregon	294.69	2359.31	$7,121,465
Grande Ronde, Oregon	22.33	179.49	$829,389
John Day, Oregon	50.90	408.99	$1,337,186
Salmon, Idaho	151.81	1215.37	$1,066,464
Umatilla, Oregon	21.43	172.35	$820,454
Upper Columbia, Washington	57.15	459.00	$1,844,845
Walla Walla, Washington	69.65	557.23	$2,012,633
Yakima, Washington	118.77	952.83	$3,424,480
TOTAL	1063.56	8521.01	$20,374,545

Source: Adapted from Garrick (2015, pp. 143–145)

Minute 319 pilot period (ending 2017), the Trust will continue to purchase and lease water from farmers in the Mexicali Valley.

Although the MDB and Mexicali Valley environmental water markets are underpinned by existing and active irrigation water trade, the absence of active water trading does not preclude the possibilities of water transfers to the environment. For example, water market activity in the Columbia Basin is arguably driven by environmental demand and government mandates to protect species listed under the Endangered Species Act 1973. The Walker Basin, Nevada, is another key example of where an environmental water acquisition program to restore lake levels occurs in a setting where market activity is comparatively limited (Doherty and Smith, 2012).

18.4 CHALLENGES TO ENVIRONMENTAL WATER ACQUISITIONS

In many countries, there is a general recognition of the benefit of purchasing or leasing (e.g., conservation easements) land for conservation purposes. Property rights were established long ago and land is bought and sold like any other commodity. Land is also a stock resource that is more or less static. Conversely, buying water for the environment, even when all conditions are met (see Section 18.3.1), introduces significant challenges. Tension between water as a common or as a commodity provides a barrier to the sale of water separate from land. Water is also a flow resource and is dynamic across time and space—making ownership, price, and third-party impacts difficult to define and quantify. As a result, buying water for the environment is a challenging task. In

Table 18.4 Environmental Water Acquisition Criteria

Indicators	Water Acquirers	Water Sellers	Management
Efficiency	Least-cost path to buy environmental water (price)	Most beneficial path to increased welfare (price)	Irrigator: Sell when monetary benefits outweigh production profits Environment Holder: Sell water when monetary benefits outweigh environmental demand
Effectiveness	Secure adequate volume (quantity)	Retain right to water for future use if desired (alternative transactions)	Irrigator: Sell water in accordance with farm plan
	Willingness to participate in the market	Willingness to participate in the market	Environment Holder: Optimize portfolio of rights and delivery of water
	Transaction costs of acquisition	Transaction costs of sale	Irrigator, Environment Holder: Ability to operate in the market Transaction costs of management
Equity	Minimize third-party impacts	Minimize third-party impacts	Irrigator: No harm principle
	Avoid stranded assets		Environment Holder: Minimize market distortion in both acquisition and management phases
	Facilitate restructure		Contribute to irrigator infrastructure when shared for environmental deliveries

Source: Adapted from Wheeler et al. (2013, p. 431)

particular, quantifying the outcomes and impacts of environmental water purchases poses quite a scientific and monitoring challenge. Wheeler et al. (2013) provide the following performance criteria for assessing environmental water acquisitions, namely: (1) efficiency; (2) effectiveness; and (3) equity. Table 18.4 provides further detail.

18.4.1 GOVERNMENT INTERVENTION, INSTITUTIONAL, AND ECONOMIC CHALLENGES

Whenever governments of private entities enter the water market to acquire water for the environment, they are faced with concerns from irrigators that they may inadvertently influence water price and market behavior. For example, it is often claimed that the government buying water entitlements in Australia has driven up water entitlement and allocation prices, although this is yet to be fully investigated. In terms of the socio-economic impacts, a rise in water prices in general is a benefit for all irrigators who own water. It is also a benefit for those who choose to sell their water, but it involves an increased farming operating cost for those who buy water allocations or those first entering irrigation. Similar criticisms about price impacts have also been made in the Columbia Basin and the Mexican reaches of the Colorado Basin, although we are not aware of any quantitative evidence.

As described earlier, the successful acquisition of water for the environment rests on a number of conditions (see Section 18.3.1). A cornerstone of this is secure and well-defined property rights for water. Poorly defined property rights are a key limitation to instream flow acquisitions (Scarborough, 2012). In the MDB, water rights are volumetric, and the yearly amount received is reduced proportionally based on the seasonal availability of water. As such, burden and risk are shared equally between rights holders with the same level of entitlement security, including the environment. In the case of the United States, legal ambiguity regarding the volume able to be traded introduces significant barriers to environmental water trade. The volume able to be transferred may be a function of the consistent beneficial consumptive use of the right, less the return flows relied upon by downstream users, or the volume of conserved water made available (Mooney et al., 2003). This ambiguity is in part a function of the prior appropriation doctrine (see Box 18.2) and the lack of adjudication of rights within this system.

Other institutional challenges include the need to minimize transaction costs. Transaction costs in environmental policy refer to the resources that are required to define, establish, maintain, and transfer property rights (McCann, 2013; McCann et al., 2005). Transaction costs of transferring property rights of complex resources are inherently difficult to define and manage as they are affected by unique fundamental physical, technical, cultural, and institutional factors inherent in any system (McCann, 2013). The nature of transactions costs and the mechanisms put in place to minimize them will in part determine the long-run prospects of environmental water markets (Garrick et al., 2009). In the case of the western United States, the administrative requirements involved in the formal partitioning of a transfer for a water right cannot be overstated. The transaction costs associated with acquiring water for the environment is a combination of payment to the seller of the right, the completion of a water rights valuation, administrative efforts at a district and/or state level, and technical expertise for environmental impact investigations for long-term or perpetual transfers (Mooney et al., 2003).

As highlighted in this chapter, private or collaborative programs play a substantial role in water right acquisitions in the United States. Contrasting with the Australian experience where public environmental water holders are funded by state and federal coffers, private conservation groups in the United States can rely on a myriad of funding sources, including donations, private investments, subsidies, and government funding. This begs the question of the financial ability of NGOs, particularly not-for-profit organizations, to be able to consistently acquire water for the environment, especially when scarcity pressure increases the cost of water supply and urban users are able to pay a premium for rural to urban transfers.

18.4.2 COMMUNITY AND PARTICIPATORY CHALLENGES

As with any reforms and policy innovations, the success of environmental water acquisition is a function of public participation. Methods and case studies of stakeholder engagement in environmental water programs are discussed in Chapter 13. The importance of the role of community support and political will has been exemplified in the ongoing Australian water reform process. A challenge in environmental water acquisition internationally is that such an approach is contingent upon the existence of willing sellers. Even under ideal legal circumstances, acquiring environmental water will be unsuccessful without the participation of consumptive rights holders (Lane-Miller et al., 2013). There has been significant progress toward creative ways to engage rights holders in the sale of water rights. Private negotiations with irrigators are one such approach, although this has a number of ramifications for transaction costs. In addition, brokering deals with individual irrigators naturally results in smaller-volume transactions, compared with transfer arrangements entered into by government buyers and institutional sellers.

Alternative water market products that do not require the permanent sale of the water rights also provide a way to acquire environmental water while compensating irrigators through a market transaction. Examples of these include split-season leasing, tax-deductible water donations, and options contracts. Creative solutions such as these prove advantageous by allowing rights holders to make yearly decisions, removing the fear of stranded assets associated with permanent sale, and reducing administrative hurdles (Hardner and Gullison, 2007; Lane-Miller et al., 2013). Similar patterns have been observed in Australia's water market, where allocation (short-term) trade is more widely adopted by irrigators because of reduced risk and complexity. Although short-term leasing and alternative water products have benefits as outlined earlier, the lack of perpetual (entitlement) acquisitions can have negative impacts on long-term environmental protection and decision making that can be afforded by owning environmental water rights in perpetuity.

18.5 CONCLUSION

Environmental water governance has covered three main paradigms: state management, collective management, and more recently, water markets (Meinzen-Dick, 2007). Water markets are institutional innovations that can help allocate scarce water between consumptive water users by facilitating the trade of water rights. When legal conditions allow for it (see Section 18.3.1), water markets can also be a useful tool mechanism to purchase water for the environment. In river basins, where water rights are fully appropriated, water markets are more politically palatable than administrative reallocation as consumptive users are compensated for forgoing the use of their water right.

Although water markets exist in varying degrees of formality around the world, there are few examples where they have been used to reallocate water to the environment (see Section 18.3.4). In Australia, the federally led buyback of consumptive water rights provides a leading example of how markets can be used to achieve environmental water objectives. In the western United States (see Box 18.2), although there has been a lack of cohesive federal and state leadership on reallocation of water to the environment, there have been a number of projects that have delivered modest but encouraging hydrological and ecological benefits. Similar benefits can be observed in the Mexican reaches of the Colorado River, where although the volumes acquired through the market are comparably modest, the institutional innovation and ecological results are undeniable (see Box 18.3).

Depending on the institutional and hydrological underpinning of the water market, various types of water products can be purchased for the environment (see Section 18.3.2). In the Australian case study, these water products have typically been permanent acquisitions, though there is considerable opportunity to expand the federally led buyback system to include annual water purchases and options contracts. Adopting these innovations is likely to play a role in increasing environmental water availability and also increase irrigator willingness to participate in reallocation programs.

The role of who can buy water for the environment (see Section 18.3.3) has an evolving answer that is different across all contexts owing to the social and legal norms that support the water market. In the case of the western United States and Mexico, collaborative state, federal, and NGO partnerships have played a considerable role in facilitating water transfers to the environment.

BOX 18.4 IN A NUTSHELL

1. There are a number of conditions that must be met to facilitate successful environmental water trade.
2. When these conditions are met, purchasing water for the environment can be a viable way to increase instream flows.
3. Water markets have been successfully used in this way in the western United States, Mexico, and Australia.
4. There are a number of institutional, economic, and participatory challenges associated with markets for environmental water.
5. As scarcity pressures increase, market-based policies can play a role in future environmental water management.

This approach is gaining traction in Australia, where NGO participation in the water market is likely to increase.

When examining the three case studies presented in this chapter, it is clear that using water markets to reallocate water to the environment is a result of significant institutional reform and commitment from federal and state governments, and the private sector. These processes have highlighted considerable challenges to environmental water acquisitions (Section 18.4). Notably, market reallocation strategies are most effective when operating in the context of local institutions and with the support of irrigator communities—which remains a challenge in the case of both Australia and the western United States.

In cases where there is an active environmental water market, the amount of water held by the environmental stewards (government or private) is small in comparison to the hydrological volume of the basin. Acquiring water through trade is not a complete alternative to traditional water management paradigms but rather a supplementary tool. As scarcity pressure in arid to semi-arid regions of the world increases, a focus on demand-based policies such as purchasing water for the environment is likely to become a viable policy option for environmental water managers (Box 18.4).

REFERENCES

Aylward, B., 2008. Water markets: a mechanism for mainstreaming ecosystem services into water management? IUCN Briefing paper, Water and Nature Initiative. International Union for Conservation of Nature, Gland, Switzerland.

Baillat, A., 2010. International Trade in Water Rights: the Next Step. IWA Publishing, London.

Bakker, K., 2005. Neoliberalizing nature? Market environmentalism in water supply in England and Wales. Ann. Assoc. Am. Geogr. 95 (3), 542–565.

Bitran, E., Rivera P., Villena, M., 2011. Water management problems in the Copiapo Basin, Chile; markets, severe scarcity and the regulator. Global Forum on the Environment: Making Water Reform Happen. Paris.

Bjornlund, H., 2004. Formal and informal water markets: drivers of sustainable rural communities? Water Resour. Res. 40 (9).

Bjornlund, H., Wheeler, S., Cheesman, J., 2011. Irrigators, water trading, the environment, and debt: perspectives and realities of buying water entitlements for the environment. In: Grafton, Q., Connell, D. (Eds.), Basin Futures: Water Reform in the Murray-Darling Basin. ANU Press, Canberra, pp. 291–302.

Chong, H., Sunding, D., 2006. Water markets and trading. Ann. Rev. Environ. Resour. 31, 239–264.

Citron, A., Garrick, D., 2010. Benefiting Landowners and Desert Rivers: a Water Rights Handbook for Conservation Agreements in Arizona. Arizona Land and Water Trust, Tuscon, AZ.

Clayton, J., 2009. Market-driven solutions to economic, environmental, and social issues related to water management in the western USA. Water 1 (1), 19–31.

COAG, 2004. The intergovernmental agreement on a national water initiative. Council of Australian Governments, Canberra, Australia.

Connor, J., Franklin, B., Loch, A., Kirby, M., Wheeler, S., 2013. Trading water to improve environmental flow outcomes. Water Resour. Res. 49 (7), 4265–4276.

Debaere, P., Richter, B., Davis, K., Duvall, M., Gephart, J., O'Bannon, C., et al., 2014. Water markets as a response to scarcity. Water Policy 16, 625–649.

Department of Environment, 2016. Environmental water holdings. Available from: http://www.environment.gov.au/water/cewo/about/water-holdings (accessed 15.04.16).

Doherty, T., Smith, R., 2012. Water Transfers in the West: Projects, Trends and Leading Practices in Voluntary Water Trading. Western States Water Council, Murray, UT.

Doyle, M., Patterson, L., Chen, Y., Schnier, K., Yates, A., 2014. Optimizing the scale of markets for water quality trading. Water Resour. Res. 50 (9), 7231–7244.

Easter, W., Rosegrant, M., Dinar, A., 1999. Formal and informal markets for water: institutions, performance and constraints. World Bank Res. Obs. 14 (1), 99–116.

Flessa, K., 2001. The effects of freshwater diversions on the marine invertebrates of the Colorado River Delta and estuary. United States–Mexico Colorado River Delta Symposium, Mexicali, Mexico, September 11–12, 2001.

Flessa, K., Kendy, E., Schlatter, K., 2014. Minute 319 Colorado River Delta Environmental Flows Monitoring: Initial Progress Report. Available from: https://www.ibwc.gov/EMD/Min319Monitoring.pdf (accessed 26.04.17).

Garrick, D., 2015. Water Allocation in Rivers under Pressure: Water Trading, Transaction Costs and Transboundary Governance in the Western US and Australia. Elgar, Cheltenham, UK, Northampton, MA.

Garrick, D., Lane-Miller, C., McCoy, A., 2011. Institutional innovations to govern environmental water in the western United States: lessons for Australia's Murray Darling Basin. Econ. Pap. 30 (2), 167–184.

Garrick, D., Siebentritt, M., Aylward, B., Bauer, C., Purkey, A., 2009. Water markets and freshwater ecosystem services: policy reform and implementation in the Columbia and Murray-Darling Basins. Ecol. Econ. 69 (2), 366–379.

Grafton, R., Wheeler, S., 2015. Water economics. In: Halvorsen, R., Layton, D. (Eds.), Handbook on the Economics of Natural Resources. Edward Elgar Publishing.

Griffin, R., 2016. Water Resource Economics: The Analysis of Scarcity, Policies and Projects, second edition. The MIT Press, Cambridge, MA, USA, p. 496.

Hanak, E., Stryjewski, E., 2012. California's water market by the numbers: update 2012. Public Policy Institute of California, San Francisco, CA.

Hanemann, W.M., 2005. The economic conception of water. In: Rogers, P., Llamas, R., Martinez-Cortina, L. (Eds.), Water Crisis: Myth or Reality. Taylor & Francis, London, New York.

Hardner, J., Gullison, R., 2007. Independent external evaluation of the Columbia Basin water transactions (2003–2006). Gardner and Gullison Consulting.

Heltberg, R., 2002. Property rights and natural resource management in developing countries. J. Econ. Surv. 16 (2), 189–214.

International Boundary & Water Commission: United States and Mexico, 2012. Minute 319: Interim international cooperative measures in the Colorado River Basin through 2017 and extension of Minute 318 cooperative measures to address the continued effects of the April 2010 earthquake in the Mexicali Valley, Baja California. Available from: http://www.ibwc.gov/Files/Minutes/Minute_319.pdf (accessed 20.07.15).

Johnson, N., Revenga, C., Echeverria, J., 2001. Managing water for people and nature. Science 292 (5519), 1071–1072.

Kilter, Rural, 2015. Information memorandum: the Murray-Darling Basin balanced water fund. Kilter Rural, Bendigo, Australia.

King, J., Culp, P., Parra, C.D.L., 2014. Getting to the right side of the river. Denver Univ. Law Rev. 18 (36), 1−77.

Kirby, M., Qureshi, M., Mainuddin, M., Dyack, B., 2006. Catchment behaviour and counter-cyclical water trade: an integrated model. Nat. Resour. Modell. 19 (4), 483−510.

Lane-Miller, C., Wheeler, S., Bjornlund, H., Connor, J., 2013. Acquiring water for the environment: lessons from natural resource management. J. Environ. Pol. Plan. 15 (4), 513−532.

Lieberherr, E., 2011. Acceptability of the Deschutes groundwater mitigation program. Ecol. Law Curr. Available from: http://elq.typepad.com/currents/2011/06/currents38-04-lieberherr-2011-0607.html#_edn12 (accessed 28.07.16).

Loch, A., Wheeler, S., Boxall, P., Hatton-Macdonald, B., Adamowicz, V., Bjornlund, H., 2014. Irrigator preferences for water recovery budget expenditure in the Murray-Darling Basin, Australia. Land Use Policy 36, 396−404.

Loch, A., Wheeler, S., Bjornlund, H., Beecham, S., Edwards, J., Zuo, A., et al., 2013. The Role of Water Markets in Climate Change Adaptation. NCCARF, Gold Coast 126.

Loehman, E., Charney, S., 2011. Further down the road to sustainable environmental flows: funding, management activities and governance for six western US states. Water Int. 36 (7), 873−893.

McCann, L., 2013. Transaction costs and environmental policy design. Ecol. Econ. 88, 253−262.

McCann, L., Colby, B., Easter, W., Kasterine, A., Kuperan, K., 2005. Transaction cost measurement for evaluating environmental policies. Ecol. Econ. 52 (4), 527−542.

Meinzen-Dick, R., 2007. Beyond panaceas in water institutions. Proc. Natl Acad. Sci. USA 104 (39), 15200−15205.

Ministry for the Environment, 2009. Implementing the new start for freshwater: proposed officials work program. Available from: http://www.mfe.govt.nz/more/cabinet-papers-and-related-material-search/cabinet-papers/freshwater/implementing-new-start (accessed 04.04.16).

Mooney, D., Burch, M., Holland, E., 2003. California Water Acquisition Handbook. The Trust for Public Land, CA.

Moore, S., 2014. Water Markets in China: Challenges, Opportunities, and Constrains in the Development of Market-based Mechanisms for Water Resource Allocation in the People's Republic of China: Discussion Paper #2014-09. Harvard Kennedy School of Government, p. 20.

Neuman, J., Squier, A., Achterman, G., 2006. Sometimes a great notion: Oregon's instream flow experiments. Environ. Law 36 (4), 1125−1155.

Palomo-Hierro, S., Gomez-Limon, J., Riesgo, L., 2015. Water markets in Spain: performances and challenges. Water 7 (2), 652−678.

Pilz, R., 2006. At the confluence: Oregon's instream water rights law in theory and practice. Environ. Law 36, 1383−1420.

Raise the River, 2016. Historic change for the delta. Available from: http://raisetheriver.org/our-work/ (accessed 28.07.16).

Ronald, C., Griffin, D., Peck, E., Maestu, J., 2013. Introduction: myths, principles and issues in water trading. In: Maestu, J. (Ed.), Water Trading and Global Water Scarcity: International Experiences. Routledge, Abingdon, pp. 1−14.

Rosegrant, M., Binswanger, H., 1994. Markets in tradable water rights: potential for efficiency gains in developing country water allocation. World Dev. 22 (11), 1613−1625.

Scarborough, B., 2010. Environmental Water Markets: Restoring Streams through Trade. PERC Policy Series No. 46. Property and Environment Research Center, Bozeman, MT.

Scarborough, B., 2012. Buying water for the environment. In: Gardner, D., Simmons, R. (Eds.), Aquanomics: Water Markets and the Environment. The Independent Institute, Oakland, CA, pp. 75−105.

Scarborough, B., Lund, H., 2007. Saving Our Streams. Property and Environment Research Centre, Bozeman, MT.

Schlager, E., Ostrom, E., 1992. Property-rights regimes and natural resources: a conceptual analysis. Land Econ. 68 (3), 249–262.

Schreiner, B., 2013. Why has the South African National Water Act been so difficult to implement? Water Altern. 6 (2), 239–245.

Shao, X., Wang, H., Wang, Z., 2003. Inter-basin transfer projects and their implications: a China case study. Int. J. River Basin Manage. 1 (1), 5–14.

Sonoran Institute, 2013. Colorado River Delta restoration project. Available from: http://www.sonoraninstitute.org/component/docman/doc_details/1552-minute-319-factsheet-09152013.html?Itemid=3 (accessed 20.07.15).

Venkatachalam, L., 2015. Informal water markets and willingness to pay for water: a case study of the urban poor in Chennai City, India. Int. J. Water Resour. D 31 (1), 134–145.

Water Resources Department (State of Oregon), 2008. Deschutes groundwater mitigation program: five year program evaluation report. Water Resources Department, Salem, OR.

WestWater Research, 2014. Environmental Water Markets. WestWater Research, Boise, ID.

Wheeler, S., Cheesman, J., 2013. Key findings of a survey of sellers to the restoring the balance program. Econ. Papers 32 (2), 340–352.

Wheeler, S., Garrick, D., Loch, A., Bjornlund, H., 2013. Evaluating water market products to acquire water for the environment. Land Use Policy 30 (1), 427–436.

Wheeler, S., 2014. Insights, lessons and benefits from improved regional water security and integration in Australia. Water Resour. Econ. 8, 57–78. Available from: http://dx.doi.org/10.1016/j.wre.2014.05.006.

Wheeler, S., Loch, A., Zuo, A., Bjornlund, H., 2014a. Reviewing the adoption and impact of water markets in the Murray-Darling Basin, Australia. J. Hydrol. 518, 28–41.

Wheeler, S., Zuo, A., Bjornlund, H., 2014b. Australian irrigators' recognition of the need for more environmental water flows and intentions to donate water allocations. J. Environ. Plan. Manage. 57, 104–122.

Wheeler, S., Schoengold, K., Bjornlund, H., 2016. Lessons to be learned from groundwater trading in Australia and the United States. In: Jakeman, A.J., Barreteau, O., Hunt, R.J., Rinaudo, J.-D., Ross, A. (Eds.), Integrated Groundwater Management: Concepts, Approaches and Challenges. Springer.

WEF, 2015. Global Risks 2015, tenth ed. World Economic Forum, Geneva, Switzerland.

Young, M., 2014. Designing water abstraction regimes for an ever-changing and ever-varying future. Agric. Water Manage. 145, 32–38.

Zuo, A., Nauges, C., Wheeler, S., 2014. Farmers' exposure to risk and their temporary water trading. Eur. Rev. Agric. Econ. 42 (1), 1–24. Available from: http://erae.oxfordjournals.org/cgi/doi/10.1093/erae/jbu003.

ENVIRONMENTAL WATER ORGANIZATIONS AND INSTITUTIONAL SETTINGS

19

Erin L. O'Donnell[1] and Dustin E. Garrick[2]

[1]The University of Melbourne, Parkville, VIC, Australia [2]University of Oxford, Oxford, United Kingdom

19.1 INTRODUCTION

Establishing effective environmental water institutions is a necessary element in the implementation of environmental water policies (Le Quesne et al., 2010). Earlier chapters have explored the various drivers for environmental water policies, and the tools that have been developed to help set and implement environmental water regimes. This chapter focuses on the organizations tasked with the responsibility of implementing environmental water policies. These organizations can be (and often are) involved in helping to create the political will for an environmental water policy. However, this chapter focuses on the activities of that organization in helping to *implement* the environmental water policy.

Chapter 17 identified five mechanisms for allocating environmental water:

1. License conditions on consumptive users;
2. Conditions on storage operators or water resource managers;
3. An ecological reserve;
4. A cap on consumptive use;
5. Environmental water rights.

Chapter 17 linked these mechanisms to a consideration of whether the existing levels of environmental water available in the system are considered sufficient, or not. This chapter distills these environmental water policies into two distinct policy strategies: (1) protection and maintenance and (2) recovery and management. Where there is no attempt to enhance the existing environmental water regime, the protection and maintenance strategy is followed, setting limits on water use for consumptive purposes, usually by placing a cap on water extractions, establishing an ecological reserve or setting conditions on license holders or storage operators. This protection and maintenance policy includes:

1. Protecting the existing environmental water regimes into the future by establishing the tools and instruments that set water aside for the environment.
2. Maintaining the tools and instruments over time with adjustments as necessary to reflect changing environmental needs.

Water for the Environment. DOI: http://dx.doi.org/10.1016/B978-0-12-803907-6.00019-X

3. Ensuring compliance with those instruments so that the environmental water is actually received in the aquatic ecosystem.

Alternatively, it may be necessary to restore environmental water to previously dewatered or over-allocated systems by increasing its volume. As discussed in Chapters 17 and 18, environmental water can be increased in a variety of ways, including changing conditions on license holders and storage operators, lowering a cap, increasing the ecological reserve and investing in water savings or acquiring water rights from existing users. This recovery and management strategy includes:

1. Recovering additional water to restore environmental water regimes, by establishing and using the tools and instruments that increase water for the environment.
2. Making the necessary decisions to use, trade, or reacquire temporary access to the environmental water.
3. Adjusting the volume of environmental water as needed to reflect changing environmental needs.
4. Ensuring compliance with those instruments, so that the environmental water is actually received in the aquatic ecosystem.

It is important to note that both strategies incorporate the need to ensure that the environmental water releases are actually provided, and the need to adaptively manage their volume and application over time. Although many environmental water regimes around the world have been implemented using a set of rules that do not explicitly require evaluation and adjustment, this adaptive management in response to changing circumstances, both social and environmental, should always be part of environmental water policy (Foerster, 2011).

Environmental water organizations (EWOs) can have multiple objectives, but at least one of these will include implementing environmental water policies in a particular location. Many government EWOs are part of a broader government department or Ministry with overarching responsibility for water resource management. Many private EWOs also have broader environmental advocacy objectives, and helping to implement environmental water regimes may be only one of the activities they undertake. However, some of these EWOs are extremely narrow in remit, and have been created solely for the purpose of implementing environmental water regimes (for more on these EWOs, see O'Donnell, 2014).

It is unlikely that a single organization will be solely responsible for the entire set of actions required for environmental water implementation. The principle of subsidiarity focuses action at the lowest level it can be practically undertaken, but the variety of skills needed and the interaction of environmental water with other natural resource policies means that EWOs will almost always work in partnership with other organizations (O'Donnell, 2012). These partnerships can operate in parallel, across different policy areas such as water resource management, catchment, and land management, and across jurisdictional boundaries. Typically, EWOs are embedded in a scheme of nested governance arrangements, interacting with other organizations along boundaries of scale, legal powers, and responsibilities (Garrick et al., 2011, 2012). The partnerships that EWOs engage in are explored in more detail as part of the case studies for the Colorado River in the United States and Australia's environmental water managers. Importantly, the EWOs assessed in this chapter are focused on the *volume* and *timing* of water provided to the environment. Although water quality is an important element of the environmental water regime, this chapter addresses the organizational requirements for improving water quantity. One of the reasons for this limitation is that organizations with water quality responsibility often have an extremely broad pollution control remit (i.e., United States Federal Environment Protection Agency).

EWOs are *identifiable* organizations, agencies, or persons with at least some *responsibility* for *implementing* environmental water policies (see Box 19.1 for a discussion on the roles of organiza-

---

BOX 19.1 INSTITUTIONS AND ORGANIZATIONS

Nobel Laureate Douglass North defined institutions as the "rules of the game"—a set of humanly devised constraints and incentives that structure interactions between people, and between people and the environment, over time. Institutions involve rules codified in policies and regulations, as well as more informal norms and practices. In this context, organizations are distinct from institutions. Organizations can be thought of as the *teams* striving to meet their objectives given the rules in place. Sometimes organizations will strive to change the rules of the game (through legal reform, administrative measures, or courts) to tilt the playing field in their favor (North, 1990). Organizations may be public or private entities, or a partnership of the two, and can operate at a range of scales.

This sports analogy is useful for making sense of the complex institutional frameworks and organizational arrangements needed to meet environmental water needs. The distinction between institutions and organizations is important because the institutional frameworks for water planning and allocation will shape the types of organizations best placed to plan for, acquire, deliver, and/or monitor environmental water. There is a broad range of environmental watering functions involving distinct competencies and economies of scale (see Tables 19.1 and 19.2). Different organizations may be better suited for different tasks within a given institutional context.

---

tions and institutions). They have a clear objective to achieve environmental water provision, using at least one of the mechanisms identified in Chapter 17. Although the institutional settings differ and they may share this overall responsibility with other organizations, their identity as EWOs enhances transparency and legitimacy for environmental water implementation.

Selection of an EWO will be constrained by the history and politics of water resource management in each jurisdiction (Garrick et al., 2009), and will also depend on the environmental allocation mechanism(s) adopted (see Chapter 17). This chapter presents an approach to selecting (or modifying) EWOs in response to the requirements of the environmental water allocation mechanism(s) once a commitment to an environmental water policy has been made (although not necessarily implemented).

19.2 WHAT TYPE OF ENVIRONMENTAL WATER ORGANIZATION IS REQUIRED?

EWOs can be established before, during, or after the implementation of an environmental water allocation mechanism (see Chapter 17). Where the organization pre-dates the environmental water allocation, it can play a significant role in driving the development of the environmental water policies, including lobbying for necessary legal changes (Ferguson et al., 2006; Neuman, 2004).

The EWO can also be a modification to an existing organization. It is not necessary to create an entirely new organization to implement and manage environmental water, but the point this chapter emphasizes is that the organization needs to be capable of performing the necessary environmental water actions for which it will be responsible. Roles and responsibilities need to be clear and unambiguous to ensure that the relevant organization can be held accountable for environmental water implementation (Horne and O'Donnell, 2014).

EWOs are fundamentally designed to assist in implementing environmental water policy. They will need a significant technical skill set as well as the more general administration skills to perform this role. For example, in Victoria, Australia, the commissioners appointed to the *Victorian Environmental Water Holder (VEWH)* must have skills in at least one of environmental management, sustainable water management, economics, or public administration (*Water Act 1989* (Vic), section 33DF(2)). However, EWOs are also often operating on a tight budget. They can, and should, leverage the skills of other technical specialists by building links with specialist research organizations and other EWOs. For example, the Columbia Basin Water Transactions Program in the western United States actively promotes knowledge exchange between EWOs, researchers, and other government agencies (for more on the Columbia Basin, see Box 19.2 and Chapter 17). One of the ongoing challenges for successful environmental water management is to ensure sufficient independence within the water resource management framework. In some instances, the organization legally responsible for holding the environmental water rights is also the same organization responsible for ensuring compliance by all water users. When compliance with water extraction rights is largely complaint driven, this combined role can create a perceived conflict of interest (Garrick and O'Donnell, 2016). Potential conflicts can also arise where the organization or individual responsible for decisions on how and when to use environmental water is also part of the organization responsible for making water resource management policy (O'Donnell, 2012).

Many factors will affect the choice of an EWO. Context-specific factors such as political will, the societal importance placed on the rule of law, and the nature of the legal rights to water are all important (Dovers, 2005; Dovers and Connor, 2006). The choice of EWO cannot be separated from the politics and political economy of water allocation and management. Environmental water decisions are inherently political; they involve decisions about the social value of water, and who wins and who loses. Often environmental water decisions involve foreclosing future development or the reallocation of existing water rights. These decisions involve the access to and distribution of resources, and hence can involve conflicts. Therefore, politics matter and figure as a key element of the context shaping the range of institutional reforms and organization options suitable in a given place. Without overlooking the importance of these factors, they are beyond the scope of this chapter to discuss. Any new organizations will need to adapt to operating within the context of the state or region in which the environmental water is allocated.

19.2.1 ENVIRONMENTAL WATER ORGANIZATIONS: FUNCTIONS AND ACTIVITIES

Although there are many different ways to construct EWOs, there must be a match between the organizations, their powers and capacities, scale of operation, and the type of environmental water for which they are responsible (see discussion in Kunz et al., 2013). The activities of the EWO will be set by the policy and the environmental water allocation mechanism(s).

First, is the required policy one of protection or recovery? If environmental water is deemed adequate, the policy is that of protecting the current values of the aquatic system against future development, or other form of system change such as climate change (see Chapter 17 for more detail). The critical activities of the policy and maintenance project cycle are outlined in Table 19.1 (based on a table presented in Garrick and O'Donnell, 2016).

Table 19.1 Protection and Maintenance Project Cycle		
Protection	Policy	Long-term goals
	Planning	Identify values, assets, objectives, risks, and prioritization of actions
	Legal tool	Establish allocation mechanism to give effect to policy and planning
Maintenance	Monitoring	Environmental outcomes
	Compliance	Enforcement of allocation mechanism
	Evaluation and learning	Review and feedback into planning

The EWO needs to be capable of undertaking these critical tasks, either itself or via partnership with another organization. For example, EWOs often rely on other organizations to enforce compliance with the water allocation mechanism (which is most likely to be a cap or a condition on other water users or storage operators).

As Chapter 17 explains, the protection and maintenance policy means a focus on the more rules-based procedural environmental water allocation mechanisms such as a cap on extractions or conditions on license holders or storage operators. As discussed in the case study of Idaho, in prior appropriation systems, setting a cap may be achieved through allocating the most junior water rights to the environment (see Chapter 17 for a discussion of prior appropriation water rights).

If the current level of environmental water is inadequate, then additional environmental water is required, necessitating a recovery-oriented policy path. The critical activities of the recovery and management cycle are outlined in Table 19.2 (based on a table originally presented in Garrick and O'Donnell, 2016).

As discussed in Chapters 17 and 18, there are many ways to increase the environmental water available, depending on the allocation mechanism and process of water recovery. It may be possible to alter the conditions on a storage operator to improve environmental water management (see the case studies from Ghana and the Sustainable Rivers Program in the United States), or to use administrative mechanisms to alter conditions on consumptive water users or lower the overall cap on water extractions (see Chapter 17 and the discussion of streamflow management plans). However, in a fully allocated system, it is most likely that recovering environmental water will require some form of investment in redistributing water. One option is improving water efficiency by investing in water savings and returning the saved water to the environment (see the case study of the Deschutes River Conservancy, Oregon, US, below). This water can be legally allocated to the environment as an ecological reserve or environmental water rights. A second option is acquiring the water from existing users through a form of trade or compulsory acquisition. In this case, it

Table 19.2 Recovery and Management Project Cycle

Recovery	Planning and prioritization	Identify needs and prioritize projects
	Financing	Budgets (acquisition and ongoing administration)
	Find recovery opportunities (water sellers, water savings)	Outreach
	Regulation review	Consider investment impacts
	Conflict resolution	Engagement and consultation
	Conversion to environmental water allocation (may include holding water rights)	Legal conversion process (may include a transfer from user to environment)
	Monitoring	Contract compliance
	Evaluation and learning	Review and feedback into planning
Management	Policy	Long-term goals
	Planning	Identify values, assets, objectives, risks, and prioritization of actions
	Decision	Commitment to specific watering plan
	Implementation	Use of water in accordance with plan
	Monitoring and reporting	Monitor water delivered and report on objectives
	Evaluation	Monitor environmental response and adapt policy and planning

is most likely that the water recovered will be allocated to the environment as environmental water rights (see the case studies of the Oregon Freshwater Trust and the Commonwealth Environmental Water Holder (CEWH), Australia).

It is clear that the recovery and management policy requires a significantly larger set of skills than that of protection and maintenance. Most importantly, when this policy is implemented using a flexible allocation mechanism, there will be an ongoing need for active decision making about how and where to use the environmental water to achieve the greatest possible environmental benefits.

19.2.2 ACTIVE OR PASSIVE MANAGEMENT: WHAT DOES THIS MEAN?

EWOs occupy a spectrum of organizational types, depending on their specific activities in the environmental water management space:

1. Those at the more *passive* end of the spectrum, which are responsible for maintaining the legal and policy framework in which environmental water is acquired and managed (and in some instances, may be the legal *owner* of environmental water).
2. Those at the more *active* end of the spectrum, which are responsible for recovering environmental water rights and deciding how and where to use (or trade) water to achieve environmental outcomes.

Of course, not all institutions fit neatly into either category, and many jurisdictions have both active and passive EWOs.

In this context, active and passive refer to the decisions that are necessary in order to use environmental water. It is not a statement about the activity levels of any given organization, as it is well-recognized that protecting and maintaining the legal and policy frameworks that provide environmental water is a complex, time- and cost-intensive, ongoing role (Horne and O'Donnell, 2014). However, the critical difference relates to the mechanism by which the environmental water is provided, and the flexibility of this mechanism. More flexible mechanisms place greater emphasis on the need for a decision on how best to obtain or use the environmental water.

As discussed in Chapter 17, mechanisms that set a cap on consumptive water use or set conditions on consumptive users or storage operators do not require an active decision on if, how, and when the environmental water is used in any given year. Where there is a separate organization tasked with the responsibility of managing water resources, the passive EWO typically relies on that other body to enforce the conditions on water use that protect environmental water regimes. Although the EWO may seek to alter the institutional settings and policies over time to better protect environmental water, the procedures in place will deliver environmental water without requiring any additional ongoing input from the passive EWO.

However, active management by EWOs is necessary when the environmental water allocation mechanism is flexible and requires a decision on how and when to use water. This flexibility can be part of a condition on storage operators that requires a decision in response to changing environmental circumstances. For example, the Barmah–Millewa environmental water in Australia was initially created as a condition on bulk water managers in New South Wales and Victoria to make the water available to the environment no less than every 5 years, on the basis of decisions by environmental water managers on whether conditions were right for release (Dexter and Macleod, 2010).

However, this decision making is most necessary when the allocation mechanism is environmental water rights, and these rights are:

1. Of limited duration (i.e., a lease that requires renewal to maintain environmental water regimes);
2. Rights to water in storage, which are only released on demand;
3. Capable of being extracted for use in a wetland; and/or
4. Capable of being traded to another user.

In these circumstances, an active EWO will be required with the capacity to make decisions on where, when, and how this water will be used or traded each water year.

Where environmental water rights protect an instream flow, the critical decisions relate to the choice to reacquire that water in the future, and ensuring that the environmental water is protected from other users in the system. This is likely to require a combination of flow monitoring and, where necessary, complaint-based enforcement. The active EWO will need to choose when and what to acquire (or reacquire), and how and when to enforce the instream rights.

When environmental water is held in storage, active management is required to decide when and how to release the water from storage, and whether to deliver it as an instream flow or to extract it for use in a particular location (i.e., watering a wetland). This decision is made in a broader context of environmental water management (see the case study of the Victorian Environmental Water Holder).

Finally, when environmental water rights can also be traded, this creates the opportunity to use this water in yet another way—to generate funds. These funds can be invested in complementary works to enhance the impact of environmental water, or they can be used to invest in alternative environmental water rights (in a better location or of a higher reliability).

Active management of environmental water is an important concept that has emerged in response to the need for efficient and effective use of environmental water to achieve the maximum environmental benefits. EWOs that undertake active management typically possess some form of legal personhood: they can enter contracts (for the acquisition or lease of water rights) and they have legal standing to sue for the enforcement of their water rights in court (when necessary) (Stone, 1972).

19.2.3 CHOOSING YOUR ENVIRONMENTAL WATER ORGANIZATION: A GUIDE

For a policy maker or environmental water practitioner, establishing an EWO (or modifying an existing organization) can be approached using a step-by-step process (see Fig. 19.1). This guide builds on the work of Chapter 17, which focused on two critical choices: whether there is sufficient environmental water available in the system, and what mechanism is best suited to provide the necessary environmental water. This chapter links these two decision points with the question of how the environmental water can be maintained and/or managed into the future to guide the choice of EWO.

Chapter 17 (and the earlier chapters in this volume) provides an approach to guide practitioners through the process of determining whether current levels of environmental water are sufficient. If there is no desire to increase environmental water, then the appropriate policy is one of protection and maintenance. As Chapter 17 explains, this means a focus on the more rules-based procedural environmental water allocation mechanisms such as a cap on extractions or conditions on license holders or storage operators. As discussed in the case study of Idaho in prior appropriation systems, setting a cap may be achieved through the mechanism of allocating the most junior water rights to the environment (see Chapter 17 for a discussion of prior appropriation water rights). In this scenario, a passive environmental water manager will be desirable.

If the current environmental water regime is not sufficient, then a policy of recovery and management of environmental water will be required. As Chapters 17 and 18 discuss, there are many ways to increase the environmental water available, depending on the allocation mechanism and the process of water recovery. It may be possible to alter the conditions on a storage operator to improve environmental water (see the case studies from Ghana and the Sustainable Rivers Program) or to use administrative mechanisms to alter conditions on consumptive water users or lower the overall cap on water extractions (see Chapter 17 and the discussion of streamflow management plans). However, in a fully allocated system, it is most likely that recovering environmental water will require some form of investment in redistributing water. One option is improving water efficiency by investing in water savings and returning the saved water to the environment (see the case study of the Deschutes River Conservancy). This water can be legally allocated to the environment as an ecological reserve or environmental water rights. A second option is acquiring the water from existing users through a form of trade or compulsory acquisition. In this case, it is most likely that the water recovered will be allocated to the environment as environmental water rights (see the case studies of the Oregon Freshwater Trust and the CEWH). In this scenario, an active environmental water manager will be needed.

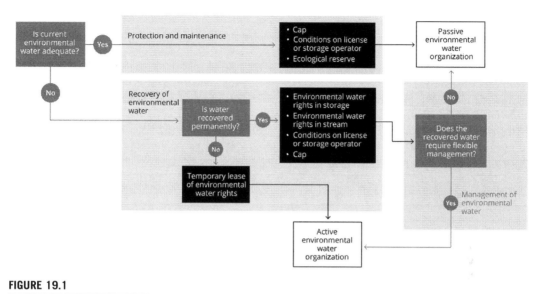

FIGURE 19.1

Choosing an environmental water organization.

Fig. 19.1 shows how to work through these choices when creating (or modifying) an EWO.

Importantly, an EWO may begin life as an active manager and transition into a passive organization. For example, where the water recovery is permanent and treated as a one-off adjustment in the water allocation between environment and consumptive users, it may be feasible to convert environmental water rights to an alternative mechanism (i.e., a cap) that does not require ongoing decision making. As Chapter 17 demonstrates, this may reduce flexibility in use, but may also cut the costs of managing the environmental water (effectively transitioning back into a maintenance policy after the water is recovered). In this case, the *active* EWO may be only needed during the water recovery and responsibility may shift to a *passive* organization once the recovery program is completed.

19.2.4 HYDROLOGY, SCALE, AND PARTNERSHIPS

EWOs rarely work alone, and are often embedded within multiple relationships with other organizations. These relationships help the EWO tailor its activities in response to the context-specific challenges posed by the hydrology and politics of the system in which the EWO will operate, and the jurisdictional boundaries and spatial scales of its operations.

The hydrology of the system is a fundamental constraint on selecting an EWO. As discussed above, a regulated river, with a large onstream dam, requires different environmental water implementation activities than an unregulated river, where water users have a right to a share of the available flow instream (e.g., the mechanisms discussed in Chapter 17). Ephemeral rivers and wetlands create different challenges to permanent rivers (see Chapter 11). A strongly interactive groundwater—surface water system will need a different response to one in which the different water sources are effectively isolated (Nelson, 2013).

EWOs can target their activities on a catchment or jurisdictional boundary scale. They are often established at the catchment scale (e.g., the Deschutes River Conservancy), but can also be designed to operate across an entire state (e.g., the VEWH). Catchment scale organizations can help the EWO coordinate their activities with other river basin management organizations, and this is often an appropriate scale for engaging with local stakeholders. The principle of subsidiarity requires environmental management decisions to be devolved to the lowest level where such decisions can be practically made, and the catchment scale is a natural fit.

However, the capacity to make decisions effectively and efficiently is also crucial (see Brinkerhoff and Morgan, 2010). This capacity is more likely to be found in larger organizations, so it operates as a constraint on the subsidiarity principle. The more complex the decisions, and the higher the transaction costs associated with novel activities, the more likely it is that the organization may need to be operating at a larger scale (Smith, 2008). Building and retaining capacity over time requires ongoing investment (Marshall, 2002; Watson, 2006), and there may be economies of scale to be achieved by operating the EWO on a regional or state-wide scale.

Sometimes, of course, river basins themselves demand a large-scale approach. The Murray—Darling Basin, Australia, and the Colorado River are two rivers that have had extensive investment in environmental water recovery (Garrick, 2015), and operate on a multistate (and in the case of the Colorado, multinational) scale. In each of these examples, however, EWOs operate at both the local scale (e.g., the Colorado Water Trust operates within the state of Colorado) and across the entire basin (in the case of the Murray—Darling, the CEWH holds and manages environmental water across the entire basin, and is supported by regional catchment managers). Ultimately, the decision about the scale of operations of an EWO needs to be made in context. The EWO's capacities and activities need to enable it to achieve its environmental goals, so it must be able to act at the scale required to achieve environmental improvements; while also making the most of local knowledge and local decision making.

Understanding who does what is critical for setting appropriate accountability measures for the EWOs. This is especially the case when they operate in partnership with other agencies (Garrick et al., 2011; Horne and O'Donnell, 2014; O'Donnell, 2013). Partnerships can be further complicated by the interaction between public and private agencies. Around the world, environmental water has been successfully implemented and managed by both public and private agencies to achieve environmental benefits that accrue to the population at large (see Table 19.3).

It is typically government agencies that play the important roles of: (1) creating the legal framework that recognizes the environment as a user of water and (2) setting large-scale policy directions on the use of environmental water. However, private agencies can be critical in creating the need for change, by changing public perceptions and creating the required political will (Neuman et al., 2006). In some jurisdictions, the law limits the ownership of environmental water to a single body, which is usually a government agency. In these cases, the government agency often works hand in hand with a private agency to acquire and manage environmental water (see Box 19.2). Private agencies can also be extremely effective in acquiring water for the environment, especially in circumstances when it may be difficult for government organizations to enter the water market (Ferguson et al., 2006; Malloch, 2005; Scarborough, 2010). Often there is no one correct answer, and public and private organizations can operate successfully together (Garrick et al., 2011; Garrick and O'Donnell, 2016).

Table 19.3 Examples of Environmental Water Organizations in the Public and Private Sectors

Sector	Organizational Form	Examples
Public (government)	Government department Water regulatory agency Statutory entity (e.g., Water bank)	Dirección General de Aguas (Chile) Idaho Water Resource Board (Idaho, US) Commonwealth Environmental Water Holder (Australia)
Private	Water Trust Watershed management not-for-profit Conservation not-for-profit with a recreational use origin (e.g., fishing, hunting) Conservation not-for-profit	Washington Water Trust (Washington, US) Deschutes River Conservancy (Oregon, US) Trout Unlimited (US), Ducks Unlimited (US and Canada) The Nature Conservancy (international)

BOX 19.2 PUBLIC AND PRIVATE ENVIRONMENTAL WATER ORGANIZATIONS

The Columbia Basin Water Transactions Program was established in 2002 to implement incentive-based water acquisition projects for salmon recovery in the four primary US states sharing the Columbia Basin: Idaho, Montana, Oregon, and Washington. The Bonneville Power Administration (a federal agency in the United States) solicited proposals to identify entities to assist in meeting its obligations under the Endangered Species Act (Action 151 under the 2000 Biological Opinion on the Operation of the Federal Columbia River Power System) and the Northwest Power Planning Council's Fish and Wildlife Program (Implementation Provision A.8). The National Fish and Wildlife Foundation (a not-for-profit organization) was selected to run the program, and in turn issued a call for *qualified local entities (QLEs)* working within tributaries of the four states (for more on the Columbia Basin, see Chapter 16).

Today, the program features 11 QLEs, including the state regulatory agencies in all four states and a mix of nongovernmental water trusts, conservation organizations, and watershed groups. The inclusion of regulatory agencies and conservation brokers has facilitated the development of complementary public and private roles across the recovery and management project cycles (Table 19.2). Federal and state agencies have led planning and financing tasks, whereas not-for-profit organizations have leveraged comparative advantages in community engagement to identify sellers and negotiate acquisition projects. However, it would be a mistake to consider a strict separation of functions between public and private entities. There is productive and necessary overlap and coordination for several functions. Such partnerships have proven critical in light of the cultural and economic importance of irrigated agriculture in the region, which triggered political resistance initially and prompted collaborative efforts to identify and respond to fears and concerns associated with environmental water projects. For example, state regulatory agencies have devised instream flow transfer and leasing rules with input from not-for-profit organizations and irrigation districts; in turn, not-for-profit organizations and irrigation districts have supplemented formal monitoring efforts of state regulatory and fish and wildlife agencies.

The program operates across different state allocation frameworks by providing financing and capacity-building services to support the QLEs, including regular meetings to share good practice and devise implementation strategies and new transactional tools. Since 2003, it has restored over 800 cubic feet per second (22.6 cubic m/s) to enhance fish habitat in tributaries throughout the Basin. Efforts in the Deschutes (Oregon), Salmon (Idaho), Upper Columbia (Washington), and Bitterroot (Montana) have illustrated the potential for this regional partnership to be adapted for local constraints and opportunities in all four states.

The following sections provide case study examples of the roles of passive and active EWOs in implementing both the protection and maintenance, and recovery and management policies, using a variety of environmental water allocation mechanisms across different hydrologies and scales.

19.3 PROTECTION AND MAINTENANCE

When there is deemed to be sufficient environmental water, the emphasis is on protecting the existing ecological functions, and maintaining that protection over time as water use or water availability changes. This policy and project cycle is the province of the passive EWO. These organizations, in addition to protecting and maintaining environmental water, often also play important roles in maintaining the broader policy framework in which the environmental water policies are situated.

It is worth noting that the protection and maintenance project cycle does not imply that the current levels of environmental water are ecologically adequate. Protection and maintenance can be pursued under two distinct water use scenarios. First, when the water ecosystems are at or near pristine condition, and very little water extraction has occurred, the protection and maintenance policy will help to protect this system for the future. As described in Chapter 17, a good example of this was the *Wild Rivers Act 2005* in Queensland, Australia.

Second, in scenarios where water extraction for consumptive use or hydropower has already occurred and the aquatic ecosystem has already been affected, the protection and maintenance project cycle is a method to protect the existing ecological function against further alteration to water use patterns. Under this scenario, the remaining environmental water regimes can be protected. This protection can operate as an end in itself, or as a precursor to a recovery and management policy to increase environmental water release.

The following case studies have been selected to give insight into the range of passive EWOs undertaking protection and maintenance policies in a diverse array of water systems.

19.3.1 PROTECTION IN ADVANCE: PROPOSED CONDITIONS ON OPERATORS OF THE PROPOSED PATUCA III DAM IN HONDURAS

As discussed in Chapter 17, it is always easier to protect environmental water regimes before water has been allocated to another use. In rivers where a large onstream dam is proposed, the best time to act to protect environmental water regimes is before the construction of the dam. These dams, whether used for extraction (i.e., irrigation or drinking water supplies) or hydropower, significantly affect the hydrology of river systems, with far-reaching effects on the ecological function and social values associated with the river (Millennium Ecosystem Assessment, 2005; World Commission on Dams, 2000). Since 1960, the volume of water stored in dams has quadrupled to between 6000 and 7000 km^3 stored in 45,000 large dams, with another estimated 800,000 small dams (Millennium Ecosystem Assessment, 2005). There is still an increasing interest in building dams for hydropower in many developing nations (e.g., Esselman, 2010).

In Honduras, a large new hydropower dam has been proposed for the Patuca River. Two organizations were involved in establishing the required environmental water management: the quasi-governmental agency Empresa Nacional de Energía Eléctrica (ENEE) and The Nature Conservancy

(TNC). ENEE was the national electricity organization, making it responsible for the planning and design of the dam (Patuca III) as an electricity generation scheme. TNC, an international environmental nongovernment organization, brought its experience and funding to the table to help prepare an environmental flows assessment for the Patuca River. Both organizations signed a memorandum of understanding, which agreed that the environmental flows assessment and the recommended environmental water regimes would form part of the Patuca III environmental impact assessment (Esselman, 2010). Importantly, neither of these organizations would become owners of environmental water, nor would TNC necessarily have an ongoing role in managing the system. However, TNC was able to focus public attention on the likely impacts of the dam, and make it both more important, and easier, for ENEE to embrace the requirement for environmental water management provisions as part of the conditions for the operations of the new dam. The ultimate outcome in this circumstance remains unclear, as the Patuca III dam has yet to be constructed and currently lacks financing (see Central America Data [April 8, 2015]).

19.3.2 PROTECTION AND MAINTENANCE ACROSS MULTIPLE STATES: THE MURRAY–DARLING BASIN CAP

Even after large onstream storages have significantly altered the hydrology of a system, it is still possible to protect the remaining ecological function. As described in Chapter 17, the Murray–Darling Basin includes parts of the states Queensland, New South Wales, Victoria, and South Australia, as well as the Australian Capital Territory. The Murray–Darling Basin had been the subject of interstate management arrangements since the River Murray Waters Agreement came into force in 1915 (Connell, 2007). In 1988, in a spirit of cooperative federalism, the Australian and basin state and territory governments established the Murray–Darling Basin Commission (MDBC) and a Ministerial Council, on which all the governments were represented. This situation was formalized in the Murray–Darling Basin Agreement in 1992, and passage of relevant state legislation by 1995.

The MDBC was a corporate body established by statute, and composed of commissioners appointed by the member states and the Australian government. The MDBC was empowered to advise and assist the Ministerial Council in the "equitable, efficient and sustainable use of water" (s17, *Murray–Darling Basin Agreement 1992*). The MDBC was funded to undertake investigations, measurements, and monitoring as required to provide assistance and advice.

In 1995, the MDBC released a report that showed that any continued growth in water use would undermine the reliability of water supply to existing users, as well as causing severe declines in ecosystem health and water quality (Murray-Darling Basin Commission, 1995). This report formed the basis for the Ministerial Council's moratorium on any new water licenses in the Murray–Darling Basin, and water resource development was capped at 1994 levels of water extraction (Murray-Darling Basin Commission, 1998). Importantly, the Ministerial Council operated by unanimous vote, so that the decision to cap water extraction reflected the consensus of the Murray–Darling Basin state governments. The Ministerial Council and the MDBC did not seek to recover any additional environmental water, but they did succeed in capping water extractions, thus protecting the environment of the Murray–Darling Basin from future extractions. Although this cap was not a perfect or complete solution, it was an extremely effective first step in protecting the remaining environmental water.

The MDBC is an example of a passive EWO that effectively provided the information and impetus to enable the responsible governments to agree to change the laws on water allocation. However, the capacity of the MDBC was severely limited by its constitution. It was unable to take direct responsibility for either setting the cap or maintaining it, as the states retained responsibility for making laws with respect to water resource management. Changing the water laws to reflect the cap on water extractions and maintenance of the cap became the province of the state governments. However, the MDBC and the Ministerial Council did provide a mechanism for the states to report on water extraction from the Murray—Darling Basin and their compliance with the Murray—Darling Basin cap.

19.3.3 PROTECTION IN FULLY OR OVERALLOCATED SYSTEMS: SETTING A CAP AND PROTECTING MINIMUM FLOWS

The environment is embedded within water resource management frameworks, and the protection and maintenance policy often targets existing water law and policy to recognize and maintain environmental water regimes. As a result, many of the passive EWOs are government departments or government agencies.

When protecting existing environmental water, the techniques vary depending on the nature of the existing water laws. The following examples show two different methods, both employed by public organizations (government departments and government agencies) to protect existing environmental water from future development.

The first example involves the setting of a cap by a state agency in Idaho. Western states in the United States operate under a *prior appropriation* water law framework, which means that the first person to divert water from the river for a beneficial purpose has the first call on that water into the future (for a brief primer on the prior appropriation system, refer to Chapter 17; see also Zellmer, 2008). As a result, a cap can effectively be established in these states by appropriating new instream flow rights to protect the environment. In the future, additional water can be only extracted once these instream rights have been met. In highly allocated water systems, it is extremely unlikely that any water will remain after meeting the environment's instream flows, thus preventing the further appropriation of water.

The Idaho Water Resource Board (IWRB) was established in 1965 in response to concerns within Idaho that "a more politically powerful state, federal government or other entity would gobble up Idaho water" (Idaho Water Resource Board, 2015). In addition to broad water resource development and planning powers, in 1978 the IWRB was also given the capacity to appropriate water rights for instream flows. The IWRB consists of eight board members appointed by the governor, and is located within the Idaho Water Resources Department, which provides the staff and other support as needed to assist the Board to carry out its duties. This relationship between the IWRB and the Department is a fairly typical governance arrangement for a government EWO.

In Idaho, instream flows protection under state law seeks to protect the minimum flow (or minimum lake level) necessary to "protect fish and wildlife habitat, aquatic life, navigation, transportation, recreation, water quality or aesthetic beauty" (*Idaho Code*, Chapter 15, Title 42). The IWRB is the only agency permitted to hold water rights for instream flow purposes, and it holds these rights in trust for all the citizens of Idaho (Idaho Water Resource Board, 2013). The Idaho instream flows program requires the appropriation of water rights, which is an activity more typically associated with an active environmental water manager. However, like many state instream flows programs in the western United States (Zellmer, 2008), the Idaho instream flow appropriations operate much more like a cap, rather than the recovery of additional environmental water. It is worth noting that the program operates differently within the Lehmi and Salmon River basins, where the program includes the capacity to transfer water rights from existing users. In these basins, the policy is one of recovery, which requires active environmental water management and an environmental organization with the capacity to purchase water rights.

In Idaho, an instream flow appropriation may be only made when there is sufficient unappropriated available water to meet the instream flow right (based on historical flow data), and where this appropriation is no more than the minimum necessary to preserve stream values (*Idaho Code* 42-1502, 42-1503). These requirements have severely limited the appropriation of instream flow rights in Idaho, and less than 2% of the stream miles are protected by an instream flow (Idaho Water Resource Board, 2013). Furthermore, the nature of the prior appropriation system means that these recently appropriated water rights operate to protect the existing flows against future water resource development. As a result, the appropriation of these rights, while using the mechanism of environmental water rights, operates much more like a cap on water extraction. In cases where there remains significant volumes of water above and beyond the level required by the appropriated instream flows, they may act as a form of minimum flow, preventing future water appropriations from dewatering the stream.

The second example comes from Chile, where the state water rights agency (Dirección General de Aguas [DGA]) is responsible for maintaining minimum stream flows. The DGA is responsible for water regulation and management, but its legal capacity is extremely limited by the Chilean constitution (Guiloff, 2012). Generally, the DGA must grant a water right when requested, whenever the water is physically available for use. If there is insufficient water to meet all requests, then the DGA must hold a public auction for the water rights requested, selling them to the highest bidder (Bauer, 1997). Once the water rights are issued, they are protected as private property under the Chilean constitution and the DGA cannot cancel or restrict these rights without purchasing them.

Although limited in legal power, the DGA does maintain and manage hydrologic data, as well as keeping records of water rights granted, and can conduct studies to inform the legislative branch of government (Bauer, 1997). Both these functions are important in supporting the extension to its legal powers under the 2005 amendment to the Chilean *Water Code*. Under article 129, the DGA must now establish a minimum streamflow every time it grants a new water right. This is defined as "the minimum flow that rivers must have in order to maintain the existing ecosystems and preserve ecological quality" (see Government of Chile, 2007, cited in Guiloff, 2012). Although this addition to the legal powers of the DGA has enabled it, for the first time, to protect existing environmental water regimes, the actual effect of the amendment has been small. Most of the rivers in Chile had effectively reached full allocation prior to the 2005 amendment, so very few new water rights have been granted with the minimum streamflow requirement (O'Donnell and Macpherson, 2014).

Both these examples demonstrate the challenge of using a passive EWO to protect environmental water regimes after significant volumes of water rights have already been granted. These examples also highlight the tension between a broad water resources management remit and the specific need to allocate environmental water to protect existing ecological functions. Multiple objectives can result in regulatory capture by existing water users, and can create conflicts of interest when enforcing the environmental water regime. For example, in the western United States, the state agency responsible for appropriating the instream flow rights is often also the agency responsible for enforcing compliance; when one arm of the agency complains to the other that instream flows are being extracted unlawfully, this can create a perceived conflict of interest when the agency acts to constrain the unlawful extraction by an irrigator. As a result, state agencies can be unwilling to enforce complaints about unlawful taking of instream flows (Garrick and O'Donnell, 2016).

19.4 RECOVERY AND MANAGEMENT

When there is not enough environmental water to support water-dependent ecosystems and their associated uses at the level deemed acceptable by the local community, it will be necessary to recover additional water for the environment. In some circumstances, this additional water can be obtained through investment in more efficient systems of water storage and delivery, which means that more water can be made available to the environment without reducing the water rights of existing water users (Australian Government, 2010; Deschutes River Conservancy, 2015; State of Victoria, 2009). In most cases, however, the recovery of additional water for the environment will necessarily require a transfer from existing consumptive water users to the environment.

The recovery and management policy can be undertaken by passive or active EWOs, depending on how the water is being recovered, and the nature of the mechanism used for allocating the water to the environment, in the short and long term (see Fig. 19.1). Passive EWOs can be highly effective at changing existing conditions on water storage operators to enable more water to be passed downstream at the appropriate time. Passive organizations can also be the recipient of water rights acquired by active EWOs, where this acquisition is permanent and where it does not require any additional decision making on the part of the water right holder to implement the environmental water regimes (Fig. 19.1).

Active EWOs are necessary when the water rights are acquired as a result of trade between the environmental organization and existing water users. There is a continued requirement for an active organization capable of making decisions and entering into contracts when the trades are of a temporary nature (and require additional contracts to renew the acquisitions over time; see Chapter 18 for a discussion of temporary trades and leases) and/or where the water rights are held in storage, requiring the organization to decide each year how and where to use the water to achieve the best overall result. This capacity for decision making is the crucial distinction between active and passive environmental organizations (e.g., O'Donnell, 2013). Decisions demand accountability and independence from external influences, and the need to demonstrate transparency (see Section 19.5). Active EWOs need to be able to demonstrate that their decisions on how and where to acquire and use environmental water have delivered environmental benefits, which can also include the need to show that those environmental water rights have been effectively enforced.

The following case studies have been selected to give insight into the range of active and passive EWOs undertaking the recovery and management of environmental water in different hydrologies and across different scales.

19.4.1 RECOVERY BY PASSIVE ORGANIZATIONS: CHANGING CONDITIONS FOR STORAGE OPERATIONS IN GHANA AND THE UNITED STATES

The effects of large onstream dams include not only those at the site of the dam construction and the new reservoir, but also the changes to flow regimes, sediment transport, and fish passage that affect the river ecosystem for many kilometers downstream of the dam (Bunn and Arthington, 2002; Krchnak et al., 2009). Although it is easier to protect river flows from the impact of dam operations in advance of the dam construction, the many large dams already in operation around the world means that this is not always possible (Millennium Ecosystem Assessment, 2005). Large dams are operated for a variety of reasons, including irrigation, hydropower generation, and flood control (Richter and Thomas, 2007). Importantly, these objectives can change over time, in response to changing environmental conditions or societal needs, and this creates an opportunity to alter the operations to improve environmental water regimes below the dam (Krchnak et al., 2009).

Passive EWOs can play a significant role in driving the process for changing the dam operations to release more environmental water at the times when it is most needed. This process is considered a recovery of environmental water, as it will ultimately lead to increased environmental water release, although the precise change in volume of environmental water can be small. Changing the operation of an onstream dam to increase environmental water is as often as much about changing the timing of water releases as it is about increasing the volume (Krchnak et al., 2009).

Two examples are examined here, to demonstrate the way that dam reoperation can unfold in both developed and developing nations. Both examples demonstrate that this is a long-term process, and heavily dependent on consultation with affected communities. For more on changing the operation of large infrastructure to improve environmental management, see Chapter 21.

The first example comes from the Volta River, in Ghana. Two large hydropower dams, the Akosombo and the Kpong, are located on the lower Volta River. These dams provide up to 95% of Ghana's electricity, as well as some flood protection. Upstream, the dam reservoirs provide a significant fishery, as well as supporting navigation (Natural Heritage Institute, 2014). However, since the dams were constructed (the Akosombo in 1965 and the Kpong in 1982), their combined operations have effectively eliminated the annual floods in the lower Volta River. The loss of these flow events has cut off the river from its floodplains, wetlands, and estuaries. The altered river flows have increased invasive species (devastating a shell fishery) and reduced sediment transport, causing beach erosion, loss of mangrove habitat, and lower productivity of the pelagic fishery. The annual flood pulse was also important for communities downstream who historically relied on it to support floodplain agriculture (Natural Heritage Institute, 2014).

The proposal to alter the dam operations to improve environmental water management originated with an international environmental organization, the Natural Heritage Institute (NHI). The NHI is a not-for-profit, nongovernmental, public interest law firm and conservation advocacy organization. The NHI aims to "restore and protect the natural functions that support water dependent ecosystems and the services they provide to sustain and enrich human life" (Natural Heritage

Institute, 2008). The NHI operates as a knowledge broker, and brings together the necessary technical and legal skills with the local decision makers and affected communities to improve the operation of major river basin infrastructure around the world. In Ghana, the NHI worked with the project partners in 2007 to develop and design a project to investigate the technical and economic feasibility of reoperating the Akosombo and Kpong dams to improve the downstream environmental water (Natural Heritage Institute, 2014). The program received funding from the African Water Facility in 2011 (African Water Facility, 2010), and work is underway to develop downstream environmental water targets that accurately reflect the various social and ecological water needs (Natural Heritage Institute, 2014).

Although the work on the dam reoperation in the Volta River is ongoing, it demonstrates the power of a passive EWO such as the NHI to work with local governments and local communities to drive change. The NHI brings global experience to local problems through its Global Dam Re-Operation Initiative. The NHI works with local partners in many locations around the world to develop and implement reoperation plans for major onstream dams (Natural Heritage Institute, 2015). Although the NHI has a substantial role in these projects, it does not aim to acquire rights to water for the environment, but to only change conditions for the storage operators. As a result, it remains on the passive end of the spectrum for EWOs.

The second example is the Sustainable Rivers Program. Unlike the situation in Ghana, where the development of a new set of operations for the dams remains in the early stages, the Sustainable Rivers Program has succeeded in implementing improved environmental water management below several dams in the United States (Richter and Thomas, 2007). The Sustainable Rivers Project emerged through an early collaboration between TNC and the US Army Corps of Engineers (the Corps) to restore environmental water to the Green River in Kentucky, US, below the Green River Dam (Hickey and Warner, 2005). TNC is a global nongovernment, not-for-profit corporation, and has been working to "conserve the lands and waters on which all life depends" since 1951 (The Nature Conservancy, 2015). The Corps is a federal agency with broad responsibility for constructing and operating dams on rivers throughout the United States.

The original project on Green River was identified by TNC as an opportunity to demonstrate that dam operations could be modified to improve environmental outcomes without compromising other objectives. Following the development of alternative management plans for Green River, TNC and the Corps signed a memorandum of understanding in 2000 to work together on the Sustainable Rivers Program (US Army Corps of Engineers and The Nature Conservancy, 2011b). The program officially began in 2002, when new environmental water regimes were implemented on Green River in 2002 (where dam operations were formally changed in 2006). In 2003, a Corps staff member was assigned to TNC, to continue deepening the partnership underpinning the Sustainable Rivers Program. The program has been used on eight rivers across the United States, and although it remains in the early stages in some rivers, it has delivered improved environmental water regimes to the Green, Savannah, and Bill Williams rivers (US Army Corps of Engineers and The Nature Conservancy, 2011a).

TNC, like NHI, is a passive EWO with the capacity to drive change in the way that dams are operated to improve environmental water management. It does not seek to create rights to water for the environment, but to work with the Corps to develop win—win strategies to change conditions of storage operation to improve environmental water regimes without undermining the other dam operation objectives.

19.4.2 RECOVERY AND MANAGEMENT ACROSS NATIONAL AND STATE BORDERS: ENVIRONMENTAL WATER RIGHTS IN THE COLORADO RIVER

The recovery and management of environmental water for downstream frequently involves coordination across state and national borders, particularly to deliver water to large deltas in closed or closing rivers (Grafton et al., 2013). The Colorado River is a prominent example of a large river where environmental water has been recovered and managed across state and international borders. The Colorado River Basin is shared by seven US and two Mexican states. It is governed by a complex set of laws, court cases, and operational criteria known as the "Law of the River." In the late 1990s, long-range demand and supply intersected for the first time, highlighting the historic over-allocation of the waters of the basin. The river's vast delta (the second largest in North America) is a fraction of its historic extent due to upstream development and diversions. Since 2000, the Basin has experienced an unprecedented sequence of dry years, which has paradoxically provided an opening and impetus for water allocation reforms.

These reforms include novel international commitments to acquire and manage environmental water to restore both base flows and periodic flood pulses (Gerlak, 2015). In 2012, the United States and Mexico negotiated *Minute 319*, an update to the 1944 International Treaty that included several provisions, including the sharing of shortages between the United States and Mexico. It also included the provision (Section III.6) to deliver water for the environment through a "pilot program ... to create 158,088 acre-feet (195 GL) of water for base flow and pulse flow for the Colorado River Limitrophe and its delta by means of the participation of the United States, Mexico, and nongovernmental organizations" (Flessa et al., 2014).

This unique public–private partnership was founded on a collaborative network that coalesced in mid-2005, with leading regional and international nongovernmental organizations in partnership with federal and state actors (Gerlak, 2015). Just under 10 years later, in 2014, over an 8-week period between March and May, an environmental water pulse was successfully delivered to the Colorado Delta, with water reaching the sea on 15 May 2014. The pulse flow event is being further enhanced by an ongoing commitment to a base flow in priority restoration areas. Minute 319 commits to base-flow delivery until 2017 (Flessa et al., 2014). The achievement of this environmental water release during a period of sustained drought in the southwestern United States speaks to the power of this international public–private partnership.

19.4.3 RECOVERY AND MANAGEMENT BY PRIVATE ORGANIZATIONS: ENVIRONMENTAL WATER RIGHTS IN THE WESTERN UNITED STATES

The western United States has grappled with water over allocation since the 1950s, when awareness of the extent of the problem was prompted by declines in freshwater fisheries due to surface water extractions. By the late 1980s, market-based policy reforms established the necessary enabling conditions needed for water trading to support environmental recovery in several river basins and states (see Chapter 18 for more detail on the use of water markets to enhance environmental water regimes, and refer to the case study of the Columbia Basin in Chapter 17 for more detail on the western United States, specifically).

Oregon provided the first regional model of legal reform to achieve these conditions in a system of prior appropriation (see Chapter 17) when it passed the *Instream Water Rights Act* in 1987. However, the first water rights lease to restore environmental water did not occur until 1994, 7 years later (Neuman, 2004). Private nongovernmental organizations were critical of the development and implementation of market-oriented environmental water transactions.

In Oregon, not-for-profit agencies catalyzed implementation under the *Instream Water Rights Act*'s dormant authority. The Oregon Water Trust, formed in 1993, was the first of its kind internationally. It is a nonprofit organization governed by a multistakeholder board of directors to acquire water rights for streamflow restoration. The Trust pioneered Oregon's first environmental water lease, permanent transfer, and a range of other water products compatible with the state's evolving water management framework (Neuman, 2004).

The Trust's work lent impetus and institutional capacity to the *Instream Water Rights Act* by: (1) contributing to rule making and policy implementation; (2) conducting outreach with existing water users and communities vulnerable to unintended consequences of water trading; (3) acquiring water rights for environmental water through leases, purchases, and other water products; and (4) contributing private monitoring and enforcement to uphold the public interest in environmental water rights. The Trust's influence has spread across the western United States, paving the way for water trusts in Washington, Colorado, Montana, Arizona, Northern California, New Mexico, and Texas.

The spread of the water trust model has been accompanied by an evolution toward both localism and regional integration. State-wide trusts have defined priority watersheds and, increasingly, have reformed their organizational structure to establish a local presence. The Washington Water Trust, for example, opened its first field office in 2008 to situate its work within the local watershed context of the Upper Columbia and Yakima River subbasins.

Basin organizations are another institutional innovation. Such organizations range from voluntary and decentralized to statutory and centrally established entities. The Deschutes River Conservancy, for example, grew out of a working group addressing water allocation, fisheries, and tribal water settlements. The result was a quasi-governmental basin organization run by a board of directors representing a diverse range of local stakeholders. The Conservancy conducted its first environmental water lease in 1998 and secured federal funding to support institutional capacity and pilot projects. In 2006, the Conservancy convened a water summit that culminated a multiyear process of basin planning to clarify water needs for instream flows, cities, and agriculture. The Conservancy has worked with local irrigation cooperatives to invest in water savings, and the saved water has been transferred as an instream flow right to the Conservancy (Deschutes River Conservancy, 2015). A mix of permanent and ad hoc arrangements has included a water bank, a groundwater mitigation bank and voluntary water-leasing programs. Importantly, feedback processes were included to reform the statutory and administrative framework on the basis of lessons gained through stakeholder forums. The Conservancy has institutionalized environmental water acquisitions and has begun to convert temporary water transactions into long-range protection.

19.4.4 MANAGEMENT OF RECOVERED ENVIRONMENTAL WATER RIGHTS IN AUSTRALIA

Government organizations are typically passive EWOs, with responsibility for the broader water resource management framework. However, they can also be active EWOs, with responsibility for acquiring and managing additional environmental water that is provided in the form of water rights for the environment (see Chapter 17). In Australia, active EWOs have been established at the federal and state levels, with the specific goal of efficient and effective management of recently recovered environmental water. These EWOs are responsible for *holding* (owning) the environmental water rights, and are termed *environmental water holders* (O'Donnell, 2013). Management in the Australian context is focused on the decision of where and how to use the environmental water, and occurs within a network of partnerships between the environmental water holders, catchment management authorities, and water authorities (see Fig. 19.2).

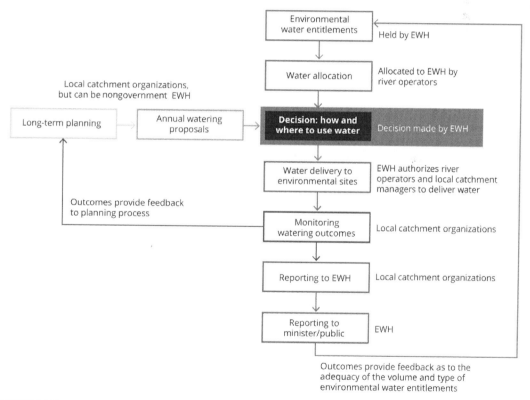

FIGURE 19.2

Using environmental water in Australia: Decision making by environmental water holders. EWH, environmental water holder.

Source: O'Donnell (2013)

At the federal level, the CEWH was created in 2007, under the new federal *Water Act 2007*. The CEWH is part of the new federal water policy for the Murray–Darling Basin, which included the acquisition of large volumes of environmental water to meet the new Sustainable Diversion Limit (Horne and O'Donnell, 2014). It is a statutory position to which an employee of the federal government can be appointed (*Water Act 2007*, ss104, 115). The federal government's environmental water rights, which have been obtained through purchase or investment in water efficiency savings, are managed by the CEWH (Australian Government, 2010). The CEWH is supported by staff in the Commonwealth Environmental Water Office, which is part of the Department of Environment (Commonwealth of Australia, 2015a).

In managing the environmental water rights, the CEWH must determine when and where to use water to achieve environmental outcomes, or whether to trade some of its water holdings. Importantly, the federal minister can direct the CEWH on where and how to use its water, but cannot issue directions on how and whether any water is to be traded (*Water Act 2007*, s107). Decisions on where and how to use the environmental water are made by the CEWH in the context of inputs from the state governments and the Murray–Darling Basin Authority (Commonwealth Environmental Water Office, 2013). The CEWH explicitly aims to maximize the environmental outcomes achieved by using its water efficiently and effectively each year, and between years (as water can be held over in storage for use in the future, or traded on the water market).

Decisions to trade water are highly constrained under the legislation (*Water Act 2007*, s106), unless the CEWH can demonstrate that the water was surplus to its current requirements and that the funds generated by the trade will improve the capacity of the CEWH to meet its obligations under the environmental watering plan. As of November 23, 2015, the CEWH holds 1649 GL of water rights (as long-term average annual yield; see Commonwealth of Australia, 2015b). In 2015, the CEWH completed the first sale of environmental water by selling almost 23 GL of temporary water allocations deemed in excess to its annual requirements (Commonwealth of Australia, 2015c). Interestingly, at the time of sale there was no firm commitment as to how the funds generated would be used to improve environmental outcomes.

At the state level, the VEWH operates under a similar model to the CEWH, with a focus on managing the state's environmental water rights (*Water Act 1989*, s33DC). In Victoria, these water rights (termed *holdings*) have been recovered over the past several decades, with significant investment in water recovery occurring since 2004 (State of Victoria, 2004). Unlike the CEWH, the VEWH was not established as part of an ongoing water recovery program, but was created solely to manage the recently recovered environmental water efficiently (O'Donnell, 2012). The VEWH has a significantly smaller set of water holdings than the CEWH, but they include a range of entitlements to both water in storage as well as instream flows (Victorian Environmental Water Holder, 2015b). The VEWH makes decisions each year on where and how to use water available under its water holdings, on the basis of inputs from the Victorian catchment management authorities (*Water Act 1989*, s33DX). The VEWH has also sold water in the past to generate funds that have then been used to purchase additional water holdings in other locations (Victorian Environmental Water Holder, 2015a).

The VEWH also departs from the CEWH in terms of its organizational structure. The VEWH is a statutory government-owned corporation, composed of three commissioners and a small staff

(*Water Act 1989*, ss33DB, 33DF). Importantly, it is not part of the relevant Victorian Government department, although the staff members are public service employees (*Water Act 1989*, s33DM). Significantly, the VEWH is also protected from ministerial direction on how and where to use water in any given year, including whether or not to trade water (*Water Act 1989*, s33DS). Although the CEWH was established to provide a single decision maker for environmental water across the Murray–Darling Basin (and, to a lesser extent, Australia), it was the need for some independence from the politics of the government of the day that was an important driver for the creation of the VEWH (O'Donnell, 2012).

The VEWH and the CEWH are only two of the several government-based active EWOs in Australia (O'Donnell, 2013), but they are the organizations with the clearest objectives when it comes to managing recently acquired environmental water rights. As statutory entities with legal personality that goes beyond that of the relevant government department, they are also more identifiable and thus more easily held accountable by the public for how they manage their environmental water.

19.5 GETTING THE INSTITUTIONAL SETTINGS RIGHT

EWOs operate within a much broader institutional context that defines their roles, responsibilities, and ultimately, their capacity for success in providing environmental water to the ecosystems and communities who demand it. Getting the institutional settings right is fundamental to creating an effective EWO. This chapter has explicitly examined the links between the policy intention (to protect existing environmental water regimes or to acquire additional environmental water), the mechanisms used to allocate the environmental water and the active or passive nature of the EWO. At the most basic level, there must be a match between the objectives and responsibilities of the EWO, and its legal powers and organizational capacity. This final section explores some of the other elements of the institutional framework in which the EWOs operate, from their partnerships to the capacity to hold these organizations to account for their actions.

EWOs typically work in partnership with other water resource managers, both up and down geographic scales and across the public–private sector divide. Enabling these partnerships to be effective without compromising the clarity of roles and responsibilities requires a great deal of planning, as well as the capacity to adapt to unexpected outcomes. When establishing the EWO, it needs to be clear what their responsibilities will be, and how they will interact with the other players. The case studies used in this chapter demonstrate the multitude of ways in which EWOs can work with others. TNC and the NHI are examples of international nongovernment organizations that have built context-specific relationships with local communities and government water resource decision makers. The Murray–Darling Basin has been managed collectively between the state and the Australian governments since 1915, and partnership arrangements are often explicit (although there is still room for improvement, see Horne and O'Donnell, 2014).

EWOs are tasked with the responsibility of implementing environmental water regimes, through the protection or recovery of environmental water. As other chapters in this book have made clear, environmental water belongs to all the people, and any EWO needs to be accountable for the actions it undertakes to deliver environmental water. This need for transparency and accountability is affected by the nature of the EWO as a public or private body. Government organizations will operate within the local public sector accountability frameworks of the relevant jurisdiction, and will benefit from greater specificity. When an EWO is merely part of a broad government agency or department, it can be very difficult for individuals to identify the precise environmental water responsibilities of that organization. Even when an organization operates from within a government department, ensuring that the organization has its own name and clearly brands its own activities, will enhance the capacity of interested citizens to engage and hold the organization to account. Two examples from this chapter include the IWRB (which sits within the Idaho Department of Water Resources) and the CEWH (a statutory position within the Department for Environment): they may be part of the overall department, but their decisions, activities, and responsibilities are clearly identified.

Private organizations often have their own transparency and accountability requirements as part of their tax-exempt status, and as part of their relationships to their own sources of funding. As a general rule, private organizations are extremely interested in publicizing their achievements as part of building ongoing support for their activities. However, these existing reporting arrangements may not be sufficient when a private organization plays a central role in the implementation of environmental water, which is fundamentally a public asset. By building strong links between the private organization and the responsible government agency or department, the actions of the private organization can be more transparent as part of the jurisdiction's overall environmental water program. Some jurisdictions have achieved this by requiring that only the government agency can legally hold environmental water rights, so that any acquired water is eventually transferred to the government (MacDonnell, 2009; Malloch, 2005). This can work in the opposite direction too: in Australia, the CEWH is entering into legal agreements with private organizations (e.g., the Water for Nature Trust of the Nature Foundation SA) who take on the responsibility of managing some of the environmental water for an agreed period. These arrangements are legally binding and reported on by the government body, as well as by the private organization.

Finally, there is the question of independence. This is a controversial issue when it comes to environmental water, and the need to be accountable to the public at large. In democracies, this accountability is often achieved through the electoral process, and attempts to weaken the link between spending public money and being accountable to the relevant government ministers are generally met with concern. However, this is where the institutional settings become paramount. Although the decision on whether and how much to invest in an environmental water program is political, the day-to-day decisions and activities should be shielded from the political spotlight to some extent. Drawing this line is not easy. For example, whenever an instream flow is appropriated for the environment in Idaho, in addition to jumping all the legal hurdles discussed above, the final appropriation must be approved by the state legislature (Zellmer, 2008), which means that each and every decision to appropriate water for the environment is open to political debate. This may well be another reason for the slow implementation of instream flows in Idaho (Idaho Water Resource Board, 2013). In Australia, although the CEWH is shielded from ministerial intervention in some of its decisions, the minister can intervene to direct the CEWH on where and how to use its water. Although no such directions have been given as yet, political parties have campaigned on the promise to do so in future (O'Donnell and Macpherson, 2014). In Victoria, it was the desire for independence from political intervention in when and how environmental water should be used in any given year that gave the VEWH its protection from ministerial directions (O'Donnell, 2012). In that case, the link back to the parliamentary democracy was maintained through the capacity of the minister to set long-term rules about the management of environmental water; just not the specifics of where and how water is used each year.

Private EWOs are not immune from the challenge of independence. In their case, it can be a delicate balance to maintain relationships with the relevant government decision makers, their funders, and the support of the community in which they operate. In the western United States, private EWOs have been broadly successful in acquiring additional environmental water (Scarborough, 2010), but their focus on win−win outcomes (where an irrigator can be paid to transition to a more efficient use of water and thus keep water instream while at the same time maintaining the irrigation business) means that this process is slow, and they have struggled to scale up their activities effectively (Garrick and Aylward, 2012).

EWOs of all kinds need to be able to demonstrate both accountability and independence. There is no one-size-fits-all solution to this challenge, and it will require the capacity to adjust relationships and expectations over time. However, being cognizant of the need for both accountability and independence at the outset will help to minimize future challenges. For more on how to build successful environmental water institutions and organizations, refer to the criteria in Chapter 26.

19.6 CONCLUSION

This chapter has focused on the role of EWOs in the implementation of environmental water management. EWOs are identifiable organizations that can be held accountable for specific tasks in implementing environmental water policies.

Choosing an EWO is critically dependent on:

1. The overall policy: Is it protection and maintenance or recovery and management?
2. The mechanism for allocating environmental water: Does it require an active EWO to make decisions about water use?
3. The scale of the problem: What is the appropriate scale for the EWO to achieve its goals? Can it be local or does it need a transboundary approach?
4. The partners it will work with: Who does what, and how is each organization held accountable?
5. The specific job requirements: Does the EWO have the required legal powers and organizational capacity to achieve its objectives?
6. The need to balance accountability and independence: To whom and how is the EWO accountable, and is it sufficiently independent from external interference to achieve its objectives?

Table 19.4 gives an overview of a range of EWOs from around the world. This is not an exhaustive list by any means, but it does demonstrate the wide range of organizational models available.

Table 19.4 Examples of Environmental Water Organizations

Environmental Water Organizations / Name and location	Policy and Project Cycle		Organization		Hydrology		Governance		Scale			
	Protection and maintenance	Acquisition and management	Active	Passive	Regulated	Unregulated	Public	Private	Local/ Community	Provincial/ State	National	International
United States												
Freshwater Trust		x	x		x	x		x		x		
Deschutes River Conservancy		x	x			x		x	x			
Washington Department of Ecology	x	x		x	x	x	x			x		
Washington Water Trust		x	x		x	x		x		x		
Colorado Water Conservation Board	x	x		x	x	x	x			x		
Colorado Water Trust		x	x		x	x		x		x		
Idaho Water Resource Board	x			x	x	x	x			x		
US Forest Service (Montana Compact)	x			x	x	x	x			x		
Montana Department of Fish, Wildlife, and Parks	x			x		x	x			x		
Trout Unlimited Montana		x	x			x		x		x		
The Nature Conservancy Sustainable Rivers	x	x		x								
Florida Water Management Districts	x	x		x	x	x	x		x			
Mexico												
Colorado River Delta Water Trust		x	x		x		x	x				x
Pronatura Noroeste		x	x		x		x	x	x			
CONAGUA	x			x	x	x	x				x	

Australia								
Commonwealth Environmental Water Holder	x	x		x	x			x
Murray–Darling Basin Authority	x	x	x	x	x			x
Victorian Environmental Water Holder	x	x		x	x		x	
Nature Foundation SA (Water for Nature Trust)	x	x		x		x	x	
Ghana								
Natural Heritage Institute	x	x		x		x	x	
Brazil								
National Water Agency	x	x		x	x		x	x
Canada								
British Columbia Consultative Committee	x	x		x	x		x	
Water Conservation Trust of Canada	x	x		x		x	x	
Alberta Minister for Environment	x		x	x	x		x	
Ducks Unlimited Canada (Yukon Territories)	x	x		x		x	x	
Chile								
Dirección General de Aguas	x	x	x	x	x			x

REFERENCES

African Water Facility, 2010. Reoptimisation and reoperation study of Akosombo and Kpong Dams: Project Appraisal Report. African Water Facility, Tunis Belvedere, Tunisia.

Australian Government, 2010. Water For the Future. Canberra: Department of Sustainability, Environment, Water, Population and Communities.

Bauer, C.J., 1997. Bringing Water Markets Down to Earth: The Political Economy of Water Rights in Chile, 1976–95. World Dev. 25, 639–656.

Brinkerhoff, D.W., Morgan, P.J., 2010. Capacity and capacity development: coping with complexity. Public Adm. Dev. 30, 2–10.

Bunn, S.E., Arthington, A.H., 2002. Basic principles and ecological consequences of altered flow regimes for aquatic biodiversity. Environ. Manage. 30, 492–507.

Central America Data. 8 April 2015. No Money for Hydroelectric Station Patuca III [Online]. Available from: http://en.centralamericadata.com/en/article/home/No_Money_for_Hydroelectric_Station_Patuca_III (accessed 9.10.15).

Commonwealth Environmental Water Office, 2013. Framework for Determining Commonwealth Environmental Water Use. Australia: Commonwealth of Australia, Canberra.

Commonwealth of Australia, 2015a. Commonwealth Environmental Water Office [Online]. Available from: https://www.environment.gov.au/water/cewo: Commonwealth of Australia (accessed 23.11.15).

Commonwealth of Australia, 2015b. Environmental Water Holdings [Online]. Available from: https://www.environment.gov.au/water/cewo/about/water-holdings: Commonwealth of Australia (accessed 23.11.15).

Commonwealth of Australia, 2015c. Media Release: Sale of Commonwealth environmental water from the Goulburn catchment benefits Victorian irrigators and the environment. Available from: https://www.environment.gov.au/water/cewo/media-release/sale-commonwealth-environmental-water-goulburn-catchment-benefits-victorian-irrigators.

Connell, D., 2007. Water Politics in the Murray-Darling Basin. The Federation Press, Sydney.

Deschutes River Conservancy, 2015. Water Conservation Program: Permanent Streamflow Protection [Online]. Available from: http://www.deschutesriver.org/what-we-do/streamflow-restoration-programs/water-conservation/ (accessed 30.10.15).

Dexter, B.D. & Macleod, D.J., 2010. The management of the Murray-Darling Basin: Barmah-Millewa Forest Hydrologic Indicator Site. A case study for effective and efficient environmental watering and the role of the community. Canberra, ACT: Parliament of Australia Senate Submission to Senate Standing Committee on Rural Affairs and Transport.

Dovers, S., Connor, R., 2006. Institutional and Policy Change for Sustainability. In: Richardson, B.J., Wood, S. (Eds.), Environmental Law for Sustainability. Hart Publishing, Portland, Oregon.

Dovers, S., 2005. Environment and Sustainability Policy: Creation, Implementation, Evaluation. The Federation Press, Sydney.

Esselman, P.C., Opperman, J.J., 2010. Overcoming information limitations for the prescription of an environmental flow regime for a Central American river. Ecol. Soc. 15 [online]. Available from: http://www.ecologyandsociety.org/vol15/iss1/art6/.

Ferguson, J., Chilcott-Hall, B., Randall, B., 2006. Private water leasing: working within the prior appropriation system to restore streamflows. Pub. Land & Resources L. Rev. 27, 1–13.

Flessa, K., Kendy, E. & Schlatter, K., 2014. Minute 319 Colorado River Delta Environmental Flows Monitoring: Initial Progress Report. Available from: http://www.ibwc.gov/EMD/Min319Monitoring.pdf (accessed 11.12.15).

Foerster, A., 2011. Developing purposeful and adaptive institutions for effective environmental water governance. Water Resour. Manage. 25, 4005–4018.

Garrick, D., Aylward, B., 2012. Transaction costs and institutional performance in market-based environmental water allocation. Land Econ. 88, 536–560.

Garrick, D., O'Donnell, E., 2016. Exploring private roles in environmental watering in Australia and the US. In: Bennett, J. (Ed.), Protecting the Environment, Privately. World Scientific Publishing.

Garrick, D., 2015. Water Allocation in Rivers Under Pressure: Water trading, transaction costs, and transboundary governance in the Western USA and Australia. Edward Elgar, Cheltenham, UK.

Garrick, D., Bark, R., Connor, J., Banerjee, O., 2012. Environmental water governance in federal rivers: opportunities and limits for subsidiarity in Australia's Murray-Darling River. Water Policy 14, 915–936.

Garrick, D., Lane-Miller, C., McCoy, A.L., 2011. Institutional innovations to govern environmental water in the Western United States: Lessons for Australia's Murray–Darling Basin. Econ. Pap. 30, 167–184.

Garrick, D., Siebentritt, M.A., Aylward, B., Bauer, C.J., Purkey, A., 2009. Water markets and freshwater ecosystem services: Policy reform and implementation in the Columbia and Murray-Darling Basins. Ecol. Econ. 69, 366–379.

Gerlak, A.K., 2015. Resistance and reform: Transboundary water governance in the Colorado River Delta. Rev. Policy Res. 32, 100–123.

Government of Chile, 2007. General Water Directorate handbook of rules and procedures: Conservation and Hydric Resources Protection Department of the General Water Directorate.

Grafton, R.Q., Pittock, J., Davis, R., Williams, J., Fu, G., Warburton, M., et al., 2013. Global insights into water resources, climate change and governance. Nat. Clim. Change 3, 315–321.

Guiloff, M., 2012. A pragmatic approach to multiple water use coordination in Chile. Water International 37, 121.

Hickey, J., Warner, A., 2005. River brings together Corps, The Nature Conservancy. The Corps Environment.

Horne, A., O'Donnell, E., 2014. Decision making roles and responsibility for environmental water in the Murray-Darling Basin. Aust. J. Water Resour. 18, 118–132.

Idaho Water Resource Board, 2013. Idaho Minimum Stream Flow Program. Idaho Department of Water Resources, Boise, Idaho.

Idaho Water Resource Board, 2015. About the Idaho Water Resource Board [Online]. Idaho State Government. Available from: https://www.idwr.idaho.gov/waterboard/About.htm (accessed 20.11.15).

Krchnak, K., Richter, B., Thomas, G., 2009. Integrating environmental flows into hydropower dam planning, design and operations. Water Working Notes. The World Bank, Washington, DC.

Kunz, N.C., Moran, C.J., Kastelle, T., 2013. Implementing an integrated approach to water management by matching problem complexity with management responses: a case study of a mine water site committee. J. Clean. Prod. 52, 362–373.

Le Quesne, T., Kendy, E., Weston, D., 2010. The Implementation Challenge: Taking stock of government policies to protect and restore environmental flows. World Wildlife Fund UK, Godalming, Surrey.

MacDonnell, L., 2009. Return to the river: Environmental flow policy in the United States and Canada. J. Am. Water Resour. Assoc. 45, 1087–1099.

Malloch, S., 2005. Liquid Assets: Protecting and Restoring the West's Rivers and Wetlands through Environmental Water Transactions. Trout Unlimited, Arlington, VA.

Marshall, G.R., 2002. Institutionalising cost sharing for catchment management: Lessons from land and water management planning in Australia. Water. Water, Sci. Tech. 45, 101–111.

Millennium Ecosystem Assessment, 2005. Ecosystems and human well-being: wetlands and water synthesis. World Resources Institute, Washington (DC).

Murray-Darling Basin Commission, 1995. An audit of water use in the Murray-Darling Basin: June 1995. Murray-Darling Basin Commission, Canberra.

Murray-Darling Basin Commission, 1998. Murray-Darling Basin Cap on Diversions Water Year 1997/98: Striking the Balance. Murray-Darling Basin Commission, Canberra.

Natural Heritage Institute, 2014. The Akosombo and Kpong Dams Reoptimization and Reoperation Study: Project Summary and Key Elements. Natural Heritage Institute, San Francisco CA.

Natural Heritage Institute, 2008. Natural Heritage Institute: About Us [Online]. Available from: http://www.n-h-i. org/about-nhi/about-us.html (accessed 23.11.15).

Natural Heritage Institute, 2015. NHI Global Dam Re-operation Initiative: Restoring Aquatic Ecosystems and Human Livelihoods [Online]. Available from: http://www.global-dam-re-operation.org/: Natural Heritage Institute (accessed 23.11.15).

Nelson, R., 2013. Groundwater, rivers and ecosystems: comparative insights into law and policy for making the links. Aust. Environ. Rev. 28, 558–566.

Neuman, J., 2004. The good, the bad and the ugly: the first ten years of the Oregon Water Trust. Neb. L. Rev. 83, 432–484.

Neuman, J., Squier, A., Achterman, G., 2006. Sometimes a great notion: Oregon's instream flow experiments. Envtl. L. 36, 1125–1155.

North, D.C., 1990. Institutions, institutional change and economic performance. Cambridge University Press.

O'Donnell, E., Macpherson, E., 2014. Challenges and Opportunities for Environmental Water Management in Chile: an Australian Perspective. J. Water Law 23, 24–36.

O'Donnell, E., 2012. Institutional reform in environmental water management: the new Victorian Environmental Water Holder. J. Water Law 22, 73–84.

O'Donnell, E., 2013. Australia's environmental water holders: who is managing our environmental water? Austr. Environ. Rev. 28, 508–513.

O'Donnell, E., 2014. Common legal and policy factors in the emergence of environmental water managers. In: Brebbia, C.A. (Ed.), Water and Society II. WIT Press, Southampton, UK.

Richter, B.D., Thomas, G.A., 2007. Restoring environmental flows by modifying dam operations. Ecology and Society 12, [online]. Available from: http://www.ecologyandsociety.org/vol12/iss1/art12/.

Scarborough, B., 2010. Environmental water markets: restoring streams through trade. In: Meniers, R. (Ed.), PERC Policy Series. Montana: PERC, Bozeman.

Smith, J.L., 2008. A critical appreciation of the "bottom-up" approach to sustainable water management: embracing complexity rather than desirability. Local Environ. 13, 353–366.

State of Victoria, 2004. Victorian Government White Paper: Securing Our Water Future Together. Department of Sustainability and Environment, Melbourne.

State of Victoria, 2009. Northern Region Sustainable Water Strategy. Department of Sustainability and Environment, Melbourne.

Stone, C.D., 1972. Should trees have standing? Towards legal rights for natural objects. S. Cal. L. Rev. 45, 450.

The Nature Conservancy, 2015. About Us: Vision and Mission [Online]. Available from: http://www.nature. org/about-us/vision-mission/about-vision-mission-main.xml (accessed 23.11.15).

US Army Corps of Engineers & The Nature Conservancy, 2011a. Sustainable Rivers Project: Improving the Health and Life of Rivers, Enhancing Economies, Benefiting Rivers, Communities and the Nation. US Army Corps of Engineers and The Nature Conservancy, Alexandria, Virginia.

US Army Corps of Engineers & The Nature Conservancy, 2011b. Sustainable Rivers Project: Understanding the Past, Vision for the Future. US Army Corps of Engineers and The Nature Conservancy, Alexandria, Virginia.

Victorian Environmental Water Holder, 2015a. VEWH water allocation trading strategy 2015−16. Victorian Environmental Water Holder, Melbourne, Victoria.

Victorian Environmental Water Holder, 2015b. Water Holdings [Online]. Available from: http://www.vewh.vic.gov.au/managing-the-water-holdings: Victorian Environmental Water Holder (accessed 23.11.15).

Watson, D., 2006. Monitoring and Evaluation of Capacity and Capacity Development. European Centre for Development Policy Management, Maastricht, The Netherlands.

World Commission on Dams, 2000. Dams and development: a new framework for decision-making. Earthscan, London, UK.

Zellmer, S., 2008. Chapter 12: Legal tools for instream flow protection. In: Locke, A., Stalnalter, C., Zellmer, S., Williams, K., Beecher, H., Richards, T., Robertson, C., Wald, A., Andrew, P., Annear, T. (Eds.), Integrated Approaches to Riverine Resource Stewardship: Case Studies, Science, Law, People, and Policy. WY: The Instream Flow Council, Cheyenne.

MANAGEMENT OPTIONS TO ADDRESS DIFFUSE CAUSES OF HYDROLOGIC ALTERATION

20

Avril C. Horne, Carlo R. Morris, Keirnan J.A. Fowler, Justin F. Costelloe, and Tim D. Fletcher

The University of Melbourne, Parkville, VIC, Australia

20.1 INTRODUCTION

The impacts of large dams and diversions on flow regimes have been widely documented (ICOLD, 2007; Lehner et al., 2011; Wada and Bierkens, 2014) and have received much attention in the environmental water literature. However, there are significant changes to flow regimes as a result of diffuse catchment changes. Addressing diffuse catchment changes is challenging as the sources of the problems may be scattered over wide areas. Environmental managers may not be able to isolate and address the numerous individual sources of the problems or they may not even know where the sources are located. Even if the multiple sources of the problems could be identified, addressing them all may prove financially prohibitive. Moreover, monitoring the problems and ensuring compliance with measures to solve them would be resource-intensive. A further challenge is that there is often a temporal and spatial gap between the cause of the problems and the resulting impact on the streamflow series.

This chapter discusses the challenges associated with managing streamflow alteration caused by diffuse catchment changes, and the potential methods to address these problems using three examples: farm dams, groundwater diversions, and land-use change (urbanization and reforestation) (Neal et al., 2001; O'Connor, 2001; Pokherl et al., 2015; Walsh et al., 2012).

20.2 THE NATURE OF DIFFUSE CATCHMENT CHANGES AND THEIR MANAGEMENT CHALLENGES

The economist Alfred E. Kahn (1966) termed the phrase "the tyranny of small decisions" to describe the situation where a number of individuals making small seemingly independent decisions, ultimately cumulate into big post hoc decisions, the outcomes of which are often not optimal or desired (Odum, 1982).

Diffuse catchment changes fit neatly into this category. Often there is significant hydrological alteration, not caused by the decisions of an individual but from a multitude of small independent choices across the landscape. It is often the case with these diffuse small decisions that larger encompassing management decisions or solutions are considered only once the cumulative impacts have resulted in significant ecological impacts (Odum, 1982).

Water for the Environment. DOI: http://dx.doi.org/10.1016/B978-0-12-803907-6.00020-6

Although there are many catchment processes that may lead to changed river hydrology, we can summarize diffuse causes of hydrologic alteration as broadly having the following aspects in common:

1. There are numerous impact points distributed across the landscape.
2. Impacts are cumulative (the effects of a single decision are small but the combined effect of many can be substantial).
3. Impacts are challenging to quantify:
   a. Data can be difficult and costly to obtain;
   b. Impacts are unique to each location (if the same disturbance occurs at two locations in the catchment, this will not necessarily result in the same hydrological alteration from each disturbance);
   c. There are often spatial or temporal delays in impacts.
4. There are often multiple actors.
5. They may be managed by different policy frameworks or institutions to those involved in river management.

These elements lead to challenges in understanding the nature of the impacts (quantifying the timing and magnitude of impacts) and also introduce specific management challenges.

We have identified three broad management approaches for diffuse causes of hydrological alteration: regulation, technical solutions, and collaborative management.

20.2.1 REGULATION AND LICENSING

A central aspect of water resource management is the concept of setting sustainable resource limits and then allocating water (licensing) within these limits. A key approach for managing diffuse causes of hydrological alteration is to bring these water uses within the resource limit and license these activities or their impacts. It is far easier to make the resource limits comprehensive from the outset rather than retrospectively regulate water use (Finlayson et al., 2008).

A key challenge in regulating diffuse activities is monitoring and compliance across large spatial scales and potentially a large number of entities. One approach that may be useful is the concept of responsive regulation, an approach developed by Ayres and Braithwaite (1992). Compliance with regulations is most likely to occur when a regulator uses an approach that Ayres and Braithwaite refer to as a regulatory pyramid. At the base of the pyramid are dialogue-based approaches for securing compliance such as persuasion and education. Ayres and Braithwaite suggest that these should be the most commonly used techniques. The higher levels of the pyramid consist of more demanding and punitive interventions such as formal warnings, penalties, and license revocation (Fig. 20.1). Braithwaite (1985, 2002) argues that regulators should always start at the base of the pyramid before reluctantly escalating to more severe approaches in cases when dialogue fails, and only employing more punitive approaches when the more modest forms of punishment have failed. This technique is used by Melbourne Water (2013), an Australian water agency, to enforce farm dam laws. As with many environmental policies, public education to change perceptions may contribute toward a successful policy (Gunningham and Sinclair, 2005; Ring and Schröter-Schlaack, 2011).

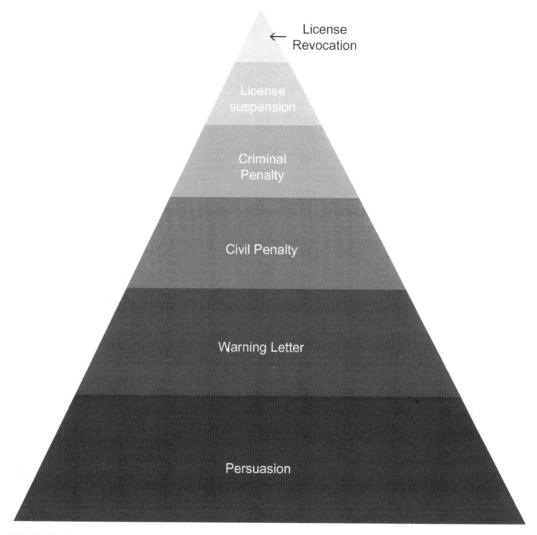

FIGURE 20.1

An example of a regulatory pyramid. The size of each layer suggests the appropriate proportion of activity at each level. The content of the pyramid will vary depending on the context of the particular regulation.

Source: Ayres and Braithwaite (1992)

20.2.2 **TECHNICAL SOLUTIONS**

There are often technical solutions that can be implemented to reduce the impact of catchment activities on river hydrology. These technical solutions can be targeted to specific areas of a catchment, or to manage impact on particular elements of the flow regime. A challenge in this approach is defining an appropriate funding model whereby the burden of costs is appropriately allocated.

20.2.3 COLLABORATIVE MANAGEMENT

Owing to the nature of the many individual landholders or institutions involved, addressing diffuse causes of hydrologic alteration requires cooperation and consent from a large number of stakeholders. Collaborative management is a strategy that goes beyond consent and allows decisions to be made at a local level. The targets or requirements can be established by government, but landowners can be given autonomy to decide how best to meet these requirements. More broadly, Ostrom (1990, 1993, 2011) among others, has advocated the feasibility of self-governing systems, appealing to numerous examples of long-running collaborative ventures, including irrigation systems in the Phillipines, Nepal, and Spain (Ostrom, 1993). Principles of community-led management may be well-suited to diffuse causes of hydrologic alteration as they are difficult to monitor and control centrally, which sometimes leads to unintended outcomes of centralized policy (Crase et al., 2012). In contrast, community-led management allows on-the-ground information to be incorporated into management practice in real time. Thus, community-led management strategies can be flexible and may be more effective. Collaborative management may also engage landowners in broader issues of environmental stewardship (Edeson, 2015), resulting in benefits beyond simply meeting environmental water management targets (see also Chapter 7).

The following sections discuss these concepts in relation to three diffuse catchment changes using three examples: farm dams, groundwater diversions, and urbanization and reforestation. The management approaches discussed are nonexhaustive, and must be chosen in the context of the specific resource, allocation framework, and institutional setting.

20.3 MANAGING FARM DAMS

Farm dams, also known as farm ponds, are small earth structures designed to store water, usually on private land (<100 ML—Nathan and Lowe, 2012). Some dams are located on defined waterways, whereas other dams are not located close to defined waterways and fill by capturing catchment runoff or by water pumped from streams, bores, or water tanks. As described in Chapter 3, farm dams cause reductions in river flows because, in addition to capturing rainfall, they intercept local surface runoff that may otherwise have flowed into streams and rivers (Schreider et al., 2002). This diverted water may then be used on-farm or *lost* to evaporation or seepage. The cumulative impact of many such water bodies across a landscape may be significant both on hydrology (Beavis and Howden, 1996; Fowler et al., 2016) and ecology (Chin et al., 2008; Mantel et al., 2010a, 2010b; Maxted et al., 2005). Note that this chapter does not discuss those dams constructed on floodplains to capture water during large floods for later consumptive use. Farm dams are common in numerous countries including Australia (MDBC, 2008), the United States (Ignatius and Stallins, 2011), New Zealand (Thompson, 2012), South Africa (Hughes and Mantel, 2010), Brazil (Souza Da Silva et al., 2011), and Pakistan (Ashraf et al., 2007).

20.3.1 CHARACTERIZING FARM DAM HYDROLOGICAL IMPACTS

The hydrologic impact of large reservoirs can be estimated by a comparison of upstream and downstream gauged flows, but this is difficult to extend to small private storages because they are so numerous and they often lie on ephemeral, ill-defined watercourses that are difficult to gage. How then can managers determine the severity of hydrological alteration by farm dams? Although a detailed answer is beyond the scope of this chapter, we provide a summary of the main steps of a commonly used approach here and refer the interested reader to the references herein.

The first step toward characterization of farm dam impacts is to identify and count the farm dams in the catchment of interest. If licenses are required to build farm dams, the records of the licensing agency provide a starting point. If no such records exist, farm dams can be identified using aerial photography, topographic maps, or satelite imagery. Larger farm dams may be visible in freely available photography such as LANDSAT. For smaller water bodies, higher-resolution aerial photography products may need to be acquired.

Having identified the farm dams in the catchment, the next step is to estimate their volumetric capacity. Some larger dams may have been surveyed as part of a formal construction process and the volume can be calculated from the survey documentation. Alternatively, high-quality terrain data such as *light detection and ranging (LiDAR)* data, may allow the volume of dams to be estimated on the basis of their internal geometry (Walter and Merritts, 2008), provided the dams were close to empty at the time of data acquisition.

However, the cost of acquiring and processing the necessary data can be expensive, especially for large areas (Sinclair Knight Merz, 2012). As an alternative, the volumetric capacity of a farm dam may be estimated using the dam's surface area. Several studies have derived relationships between surface area and volume. For example, Fowler et al. (2016) list five such equations for different regions of Australia and Hughes and Mantel (2010) derive three equations for different regions in South Africa. Habets et al. (2014) used a different approach when estimating the effects of farm dams on streamflows in western France. They assumed that all dams had an average depth of 3 m on the basis of a survey of dams in the region.

McMurray (2004) investigated several proposed equations. He found that farm dam geometry can vary greatly and consequently any given surface area can result in a wide range of possible volumes. He concluded that although an equation is not a good instrument for estimating the volume of an individual farm dam, an equation does provide a reasonable estimate when used to measure the combined volume of a number of dams (i.e., within an entire catchment).

The final step is to compare the estimated volume of the dams with the measured streamflow regime to determine their likely impact. For example, if the estimated volume of dams in a catchment was 1% of annual flows, then a sensible conclusion would be that the farm dams have negligible hydrological impact. However, if the conclusion is not so straightforward, further insight can be gained using a water balance model. Numerous examples of farm dam water balance models exist in the literature (e.g., Arnold and Stockle, 1991; Hughes and Mantel, 2010; Liebe et al., 2009; Nathan et al., 2005) and some are publically available windows-type programs (Cresswell et al., 2002; Fowler et al., 2012b; Jordan et al., 2011). Generally, water balance models take into account the various water fluxes into and out of individal farm dams, including rainfall, evaporation, inflows due to surface runoff, consumptive use, seepage, and downstream spills or releases. The models use these data to perform a water balance on each dam in the catchment. The results are combined to estimate the overall impact of dams in a catchment. Some models can incorporate spatial data to determine topological flow paths, which is important if there is any doubt which farm dams are upstream of an ecological asset (Fowler et al., 2012b).

Given that the proportion of streamflow harvested by farm dams is usually highest during dry periods (Fowler et al., 2016; Lett et al., 2009), models that characterize the seasonal and year-to-year variability of climate and streamflow, such as those listed above, may be preferable to simple calculations that characterize long-term averages. Like any hydrologic model, farm dam water balance models require numerous assumptions and simplifications, and decision makers should be aware of the resulting modeling uncertainty when basing decisions on model outputs (Fowler et al., 2016; Hughes and Mantel, 2010; Lowe and Nathan, 2008).

20.3.2 OPTIONS AVAILABLE TO ADDRESS HYDROLOGICAL IMPACTS OF FARM DAMS

Regulation and licensing

Licensing can be used to manage farm dams and limit their effects on water resources. For example, in 2000, the Victorian government in Australia proposed changes to the farm dam licensing system so that all dams used for irrigation or commercial purposes would require a license (DNRE, 2000). Some dam owners strongly opposed the proposed changes (Rochford, 2004). However, the Victorian government took steps to make the changes more acceptable. A committee was established to consult with dam owners, environmental groups, and farming lobby groups. The committee made several revisions to the initial proposal after consideration of written submissions, views expressed at specially convened public meetings, and after consultation with lobby groups (VFDRC, 2001). Furthermore, the government made concessions to the opposition political parties. One of the main changes to the original proposal was the inclusion of grandfathering, which gave people who owned existing irrigation dams the option of registering their dams at no charge (DNRE, 2001). The Victorian Parliament passed the legislation for a farm dam licensing system in 2002.

When it is politically unpalatable to impose actions on existing dams, it might still be possible to regulate the approval and design of new dams. Western Australia, faced with a high level of farm dam development in horticultural and viticultural regions in the southwest of the state, commissioned a decision support tool based on water balance modeling of farm dams (Fowler et al., 2012a, 2012b) to assist with the approval of new licenses. When a license application is received, the proposed new farm dam is simulated using the decision support tool and is only approved if the tool indicates that: (1) the impact on downstream environmental assets is judged to be acceptable and (2) the reliability of supply of other dam owners located downstream is not unduly compromised.

Another method of limiting the hydrological impact of dams is for agencies to encourage the construction of winter-fill dams. These dams are often located offstream from a watercourse and collect water via pumping, only during wet periods of the year when the rivers are at high flow. The water is stored in dams for use during dry months of the year. The benefit of winter-fill dams is that they provide a secure source of water and reduce demand for water during dry periods when rivers are under most stress.

Another method to reduce the effects of dams on streamflows is to encourage their removal. Although a compulsory decommissioning program would likely be seen as heavy-handed, a voluntary program, whereby landowners are requested (and possibly incentivized) to decommission their farm dams could be a politically palatable alternative. Landowners may be willing to decommission farm dams if: (1) land ownership has changed and the new owners use the land differently, resulting in *legacy* farm dam infrastructure that is passed on from the previous owners and is not providing any value or considered a safety hazard by the new owners; (2) dams were built during previous dry periods and are no longer needed; (3) dams are old and managing the risks of the aging infrastructure outweighs the benefits; and/or (4) landowners wish to respond to the environmental imperative to return water to streams and rivers, or the economic imperative of a financial incentive.

An advantage of this approach is that compliance is generally assured: once a dam is decommissioned (e.g., by bulldozing part of the wall) it is quasi-permanent and expensive to reverse. A disadvantage is that landowners may require significant financial compensation as an incentive to act. Furthermore, it is difficult to control which dams are decommissioned, so this approach may not be useful to target a specific ecological asset, compared with low-flow bypasses (discussed in the following subsection). The authors are aware of cases where voluntary decommissioning has been considered as a management option, but to our knowledge it is yet to be trialed in practice.

Technical solutions: low-flow bypasses

As mentioned in Chapter 3, farm dams have the biggest impact on streamflows during the dry months of the year (McMurray, 2006; Nathan and Lowe, 2012). That is, during times of low flow, farm dams generally harvest less water in absolute terms, but this water accounts for a greater proportion of the catchment discharge, which would have occurred had the farm dams never been built (Lett et al., 2009). In times of drought, these diversions may deprive refuge habitats of vital water. In such cases, management actions focusing specifically on times of low flow may be the best management response.

Low-flow bypasses are devices that ensure that a minimum flow is passed downstream of a farm dam. When the flow rate of water entering the dam is less than the design threshold of the bypass, 100% of the incoming flow is passed downstream. When the inflow rate exceeds the threshold flow rate, water passed downstream is capped to the threshold value, and the remainder flows into the dam (Lee, 2003).

A number of different designs can fulfill this purpose. Fig. 20.2 shows a schematic for a design using an upstream weir and bypass pipeline (State of Victoria, 2008). Another design is a contour channel that diverts all flow into a distribution pit. During high flows, water in the pit is directed into the dam and during low flows water is passed around the dam. Another bypass design is a floating pipe that can release water from within the dam (for long dams located on streams where bypass pipes are uneconomical).

An advantage of this approach is that bypasses can be installed just for those farm dams that matter most to environmental water delivery. For example, Fowler and Morden (2009), working in a catchment with a small number of large farm dams, identified which farm dams to target using topography and dam location data to infer flow pathways. Cetin et al. (2013) similarly assessed the relative benefit of focusing low-flow bypasses on subsets of farm dams based on location and licensing status.

Disadvantages include the high cost of retrofitting existing dams with low-flow bypasses, perceptions of unfairness if only certain dams are targeted whereas others are left alone, perceptions that the reliability of supply of water might be compromised, and the vulnerability of bypasses to manual tampering, for example by blocking up bypass pipes. Depending on the number of farm dams with bypasses installed, significant resources may be required to conduct compliance checks with sufficient regularity. Thus, the cooperation and goodwill of landowners may be essential to positive outcomes.

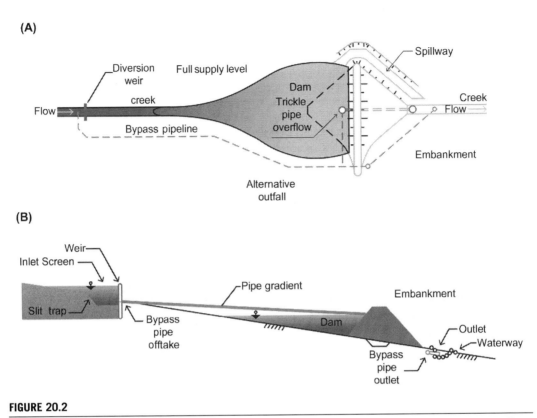

FIGURE 20.2

Example design of a low-flow bypass for a farm dam, involving an upstream diversion weir and bypass pipeline. Such designs are useful when the inflows to the dam are confined to a well-defined channel. (A) Plan view and (B) longitudinal cross-section.

Source: Adapted from State of Victoria (2008)

The inclusion of low-flow bypasses could be encouraged or required for all new dams in a region. For example, a policy was introduced in Victoria that all new farm dams used for commercial purposes must be installed with low-flow bypasses (State of Victoria, 2003). The cost of incorporating a bypass at the design stage is usually significantly less than a retrofit. The Victorian government produced a state-wide map of the passing flow requirements (Fig. 20.3) and associated technical guidelines on how to incorporate the requirements into dam design (State of Victoria, 2008).

Collaborative management

Collaborative management between government and water users may also assist in minimizing the effect of dams on streamflows. For example, Edeson (2015) describes how a government department worked collaboratively with farm dam owners across a river catchment in Tasmania, Australia. The government set environmental water requirements. If flows dropped below a critical level (40 ML/day), it would place a ban on diverting water. In an experiment in collaborative

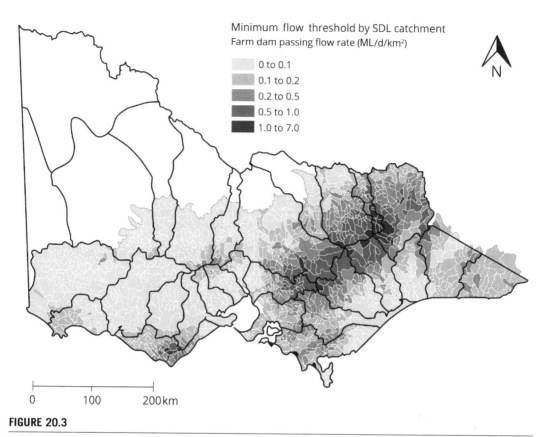

FIGURE 20.3

Map of the State of Victoria, Australia, showing the minimum flow thresholds for low-flow bypasses for new commercial farm dams. The rate for a given dam can be calculated (in ML/day) as the product of the value from the map and the upstream contributing area of the dam (in km^2). Darker shades correspond to wetter areas with higher flow requirements.

Source: State of Victoria (2008)

management, farmers were given access to real-time data about rivers flows. When the farmers noticed that flows were about to drop below the critical level, they cooperated to release water from their dams and avoid the ban. The result was that farmers kept access to irrigation water and the effect of low flows was mitigated.

20.4 MANAGING HYDROLOGICAL ALTERATION DUE TO GROUNDWATER CHANGES

Groundwater–surface water interactions can lead to two distinct effects on streamflow. Most commonly, groundwater extractions lead to reduced streamflow by decreasing the groundwater discharge to streams or by inducing increased recharge from streams. Somewhat less common is that increases in groundwater levels, due to excessive irrigation or increased recharge from land-use change, can lead to increased *base flows* and water quality issues in the stream and riparian zone. Both effects pose significant technical challenges in quantifying the degree of hydrologic alteration and subsequent management.

20.4.1 CHARACTERIZING GROUNDWATER—SURFACE WATER IMPACTS

There are a number of key technical challenges in understanding groundwater and surface water interactions. These include spatiotemporal scale issues for management, understanding the fluxes between groundwater and surface water systems, and the monitoring of groundwater extractions.

The temporal scale of the management decisions may be different to the timescale at which streamflow impacts are observed as there are lag times and uncertainty around fluxes between groundwater and surface water (Evans, 2007). Unlike surface water systems, where low-flow requirements can typically be managed on a daily basis by reservoir releases and low-flow threshold conditions on river diversions, the effects of groundwater pumping restrictions on streamflow- and groundwater-dependent ecosystems may take months to decades (or even centuries) to become apparent (Bredehoeft and Young, 1970). In addition to the temporal scale issues, the spatial extents of groundwater and surface water systems are often not aligned, which can lead to institutional management complexities. For example, large groundwater systems such as the High Plains Aquifer of mid-west United States (Kustu et al., 2010) and the Great Artesian Basin of Australia (Habermehl, 1980) cover a number of river basins and state jurisdictions. Many other large groundwater systems with significant groundwater withdrawals, such as the Guarani Aquifer in South America and aquifers underlying the Tigris-Euphrates Basin, occur over a number of different countries (Famiglietti, 2014; Voss et al., 2013).

The effects of groundwater extraction on surface water systems are typically diffuse and difficult to determine. The estimation of the groundwater contribution to streamflow remains a scientific challenge despite decades of research (Gonzales et al., 2009). Most commonly, base-flow filter methods are applied to streamflow records to estimate the base-flow contribution to streamflow. These methods are simple and inexpensive to apply, but separate streamflow into quick and slow flow components, and groundwater discharge may be only one component of the slow flow (Cartwright et al., 2014). Tracer separation methods are powerful but expensive, and can only be applied with confidence at the spatial and temporal scales of the tracer data collection (McGlynn and McDonnell, 2003). These scales are typically quite limited and rarely are time series of tracer data available with the record length of groundwater and streamflow monitoring. Groundwater models provide another powerful technique but are expensive to produce, require adequate groundwater and streamflow data for model calibration and validation, and are reach-specific (Brunner et al., 2010). The advantage of groundwater models is that they are the only method that can use the entire length of the observation record and also be used for forward modeling (i.e., scenario testing). As such, groundwater models allow managers to examine the effects of different groundwater extraction scenarios, management options, and climate change (Mulligan et al., 2014).

There are significant monitoring and knowledge acquisition costs to manage groundwater extractions. The monitoring is required to both evaluate the state of the system (i.e., water levels in the aquifer) and to measure actual volumes and patterns of water extraction. An adequate groundwater monitoring bore network is required to evaluate the groundwater dynamics resulting from pumping, and subsequent effects on surface water systems. This is a significant cost imposition on the management of the resource, given the high cost of installing monitoring bores. Enforcement of sustainable groundwater allocation requires monitoring of groundwater pumping bores to ensure compliance. Initial regulation of groundwater resources in Australia typically involved allocation of maximum volumes for extraction, but the determination of actual volumes utilized was only measured over long time frames (e.g., annually), limiting the accuracy and effectiveness of this information for management (Skurray, 2015). High temporal frequency monitoring of extraction by meters is required for management to be able to separate the effects of climate variation (i.e., natural drivers) to those of pumping (i.e., anthropogenic drivers) on the aquifer water levels (Shapoori et al., 2015). Metering of multiple bores in an aquifer is expensive, subject to tampering, and requires that the locations of all pumping bores are known to the water management authority (Ross and Martinez-Santos, 2010). In Australia, the requirement for metering of bores used for substantial water extraction (e.g., for irrigation, town water supply, and salinity alleviation) is becoming more common and this is the case in many countries with a high reliance on groundwater. For example, Jordan has a high proportion of private bores metered for extraction ($>90\%$), but only 61% were reported to be operational in 2002 (reported in Venot and Molle, 2008). However, in many countries, monitoring of groundwater extraction is rare and enforcement is difficult or not applied. For example, monitoring of groundwater occurs in many parts of India, but regulation of groundwater extraction for irrigation is not enforced in most states (Kumar, 2005; Rodell et al., 2009). California only brought in legislation requiring the metering of groundwater extraction at the basin scale in 2014 and this continues to face resistance from landholders (Wee, 2015), despite strong advocacy from scientific groups (Christian-Smith and Abhold, 2015).

20.4.2 OPTIONS AVAILABLE TO ADDRESS HYDROLOGICAL IMPACTS FROM GROUNDWATER USE

Regulation and licensing

The ideal paradigm is to consider groundwater—surface water as integrated systems that need to be managed conjunctively, recognizing that groundwater extractions could have an impact on streamflow but typically of a poorly defined magnitude and lag time. Water policy reforms in the last two decades have sought to implement this paradigm, such as the European Water Framework Directive (Griffith, 2002), Council of Australian Governments groundwater reforms (Nevill, 2009), and the Central Ground Water Authority in India (Rodell et al., 2009). However, this approach is far from universal. For instance, California, as the state with the largest groundwater use in the United States, lacked a state-wide framework for groundwater management (Nelson, 2012) until only recently when regulation (*Sustainable Groundwater Management Act*) has commenced in response to the current Californian drought (McNutt, 2014). Conjunctive management and licensing of groundwater and surface water allocations by a regional manager (e.g., regional water authority or state government department) provides the desired model for managing streamflow effects from groundwater extraction. Methods for such conjunctive management have been explored for decades (e.g., Young and Bredehoeft, 1972), and although implementing this policy approach is becoming more accepted, groundwater resources are still commonly managed in isolation (Famiglietti, 2014) and the effects of nonconjunctive management can be felt for decades later (Bredehoeft and Young, 1970).

Two examples illustrate the need to consider policies in surface water and groundwater resources simultaneously. In the Murray–Darling Basin, groundwater use increased by over 50% since 1994, following the introduction of a *cap* on surface water extraction due to concerns about water overallocation from the river systems (Nevill, 2009). The increase in groundwater use also coincided with a long-term drought, resulting in further decreases in groundwater levels, and likely contributed to reductions in river low flows in this period (Leblanc et al., 2012; Nevill, 2009). In the Upper Guadiana Basin of central Spain, the area under intensive irrigation using groundwater increased fivefold between the 1960s and 2000s, resulting in the groundwater table dropping by >20 m and severe degradation of the large groundwater-dependent wetland, Tablas de Daimiel (Martinez-Santos et al., 2008). A groundwater management plan was introduced in 1992, but has not successfully halted the decline in the groundwater table, at least in part due to the large number of unregulated, illegal wells in the area and irrigation methods that were in place before management was introduced (Martinez-Santos et al., 2008). There have been numerous modeling studies that have identified the value of conjunctive management of groundwater and surface water (e.g., Singh, 2014; Zhang, 2015), but the Australian, Spanish, and Californian experiences demonstrate the need to have strong and timely policy and governance frameworks in place to maximize the technical advantages of conjunctive management.

Minimizing the impacts of groundwater extraction on streamflow typically requires conjunctive management of groundwater and surface water resources, and in particular, the capacity to regulate pumping volumes and timing. However, the introduction of conjunctive management faces technical challenges around aquifer characteristics. Long or poorly understood groundwater residence times mean that few management plans have trigger thresholds for regulating groundwater use based on the effects of its extraction on river flow. For example, in a review of 15 groundwater management plans across Australia, only two had explicit allocation restrictions based on streamflow conditions (White et al., 2016). The effectiveness of such management is dependent on the degree of hydraulic connectivity between the pumped aquifer and the river. In hydraulically well-connected systems, the residence times of alluvial groundwater and reaction times with streamflow events are quick and conjunctive management can more easily integrate groundwater and streamflow monitoring to protect periods of summer low flows from heavy groundwater extraction. In most other large groundwater systems interacting with rivers and streams, the groundwater residence time and fluxes to and from the river are not easily measured or are long relative to seasonal or annual time frames (Evans, 2007). As a result, the fine-tuning of groundwater allocations to minimize effects on low-flow requirements of rivers is difficult for managers and typically requires detailed modeling of the system and the collection of field data (i.e., tracers), in addition to streamflow and groundwater monitoring.

In the absence of detailed information on the aquifer hydrogeology and fluxes between the river and groundwater, exclusion zones (or with limited groundwater extraction) provide a blanket means of limiting the effects of groundwater pumping on streamflow. The greater the distance of a pumping bore to the river, the less effect a given pumped volume has on streamflow for a specified time period (Evans, 2007). This zonal approach can minimize short-term impacts on streamflow and environmental water regimes but not long-term impacts. The exclusion zones can be determined by travel times for different aquifer characteristics so that the method can be tailored for individual catchments or reaches.

Technical solutions

It is not only groundwater extractions that can drive hydrological alteration and the focus of environmental water regimes. In irrigation areas around the world, excess irrigation water percolating beyond the root zone can lead to groundwater mounding and soil salinization (Scanlon et al., 2007). These raised groundwater levels can lead to enhanced base flow in streams (Kendy and Bredehoeft, 2006). High levels of regional, saline groundwater can occur around lowland, semi-arid river systems, in part due to land clearing increasing diffuse recharge rates, and also to river regulation maintaining high river levels and preventing discharge of the saline groundwater into the river. As a result, salinization of floodplain soils can severely affect the health of riparian forests, which may already be under stress from lack of flooding due to river regulation. The poor health of the riparian forests has been a major driver of the release of environmental water and supporting infrastructure investment in the Murray River (Bond et al., 2014).

In cases of artificially enhanced groundwater levels, one of the main physical options is to pump out groundwater to maintain desired levels, particularly in areas where there are deleterious effects from saline shallow groundwater on vegetation health. Maintaining environmental health in riparian zones by pumping out saline groundwater has been investigated as a complementary method to supplying environmental water to floodplains and wetlands in poor health (Alaghmand et al., 2013; George et al., 2005). It can lead to enhanced lateral river recharge and increases in riparian tree health. However, the efficacy of this method depends on the transmissivity of the floodplain aquifer. Improvements in riparian tree health may also require a supply of fresh water through environmental flooding or an injection of fresh water into the aquifer (Berens et al., 2009). The groundwater lowering approach has drawbacks as it is spatially explicit, incurs high costs, and requires a sustainable disposal of the pumped saline groundwater.

A management option for dealing with rising groundwater levels is to limit the enhanced recharge by using reforestation to increase evapotranspiration from the soil profile. This approach can also be used to directly lower the groundwater table by evapotranspiration from phreatophytic trees upgradient of river systems (Khan et al., 2008; Peck and Hatton, 2003). Within irrigation districts, the lowering of groundwater tables can be largely accomplished by more efficient irrigation delivery systems (i.e., drip or spray irrigation compared with flood irrigation), although some percolation to the water table is required for irrigated crops to prevent solute buildup in the root zone (Oster and Wichelns, 2003). Strategic revegetation can also be used to limit recharge rates in both rain-fed and irrigated agricultural areas (Khan et al., 2008; Peck and Hatton, 2003).

20.5 MANAGING HYDROLOGICAL ALTERATION DUE TO URBANIZATION

Urbanization perturbs flow regimes primarily through the creation of impervious areas and hydraulically efficient drainage systems, which result in reduced infiltration, greatly increased runoff, and thus a significant increase in the temporal variability of *flashiness* of the stream hydrograph. Water quality is also degraded. Urbanization may also impact flow regimes through extractions for water use, or through discharge of wastewater to receiving waters, but the largest disturbance to flow regimes typically results from the problem of urban stormwater runoff (Walsh et al., 2012), and it is this impact that we primarily consider here. Further details on the hydrological impacts of urbanization have been provided in Chapter 3.

The problem is that urban stormwater runoff is normally catchment wide; impervious runoff flows to streams wherever these areas are drained by formal stormwater drains. Therefore, those on both private and public land where the stormwater is generated must be involved in finding the solution to its impact on flow regimes.

During recent decades, there has been a significant evolution in the way stormwater is managed, from an approach primarily focused on minimizing flooding (e.g., using retarding basins and pipe upgrades), to one where environmental concerns are increasingly considered, alongside landscape amenity, and the use of stormwater as a resource (Burns et al., 2012). This evolution has been taking place in many regions around the world, and while the name given to such approaches varies (e.g., *Low Impact Design* in North America, *Water Sensitive Urban Design* in Australia, and *Sustainable Urban Drainage Systems* in the United Kingdom; Fletcher et al., 2014a), the general principles tend to be similar:

1. Maintain the water balance and flow regime as close as possible to natural;
2. Maintain water quality as close as possible to natural;
3. Encourage water conservation and use of stormwater and wastewater as a resource;
4. Maximize the landscape amenity through integration of *stormwater control measures (SCMs)* into the urban landscape;
5. Add value to the urban landscape while minimizing drainage infrastructure costs;
6. Reduce runoff and peak flows and minimize or mitigate flooding.

Despite the seeming completeness of these principles, in practice a focus limited to reducing peak flows and reducing pollutant loads has been common (Burns et al., 2012), but more recently there has been increased attention paid to the need to design stormwater management systems to maintain or restore flow regimes as close as possible to natural (Burns et al., 2012; Fletcher et al., 2014b; Petrucci et al., 2013; Walsh et al., 2012). Achieving such objectives requires approaches that: (1) reduce the impervious area or (2) *break* the hydraulic connection between the impervious area and receiving waters, through the installation of technologies that retain, treat, use, evapotranspire, and infiltrate stormwater, rather than allowing it to discharge unmitigated to the receiving waters.

Numerous authors have pointed out the benefits of managing stormwater in a way that has much less impact on the flow and water quality regimes of receiving waters (Fletcher et al., 2014b; Grant et al., 2012; Vietz et al., 2015; Wong and Brown, 2009). First and foremost, the use of stormwater harvesting as a central approach to reducing the stormwater threat creates a simultaneous opportunity, by creating a *new* water resource with potential economic benefits. Retention of stormwater in the landscape can also enhance soil moisture and thus enhance vegetation health, potentially reducing the urban heat island effect (Endreny, 2008), while helping to reduce flooding (Burns et al., 2015b; Tourbier and White, 2007).

20.5.1 CHARACTERIZING STORMWATER IMPACTS

Studies demonstrating the impacts of urbanization on stormwater and thus on streamflow regimes go back several decades (Leopold, 1968), with the creation of impervious areas and hydraulically efficient drainage systems known to increase peak flows and total flow volume, while commonly reducing base flows. Urban stream hydrographs are known to be *flashy*, with highly variable flows.

The changes to flow regimes can be characterized using the flow metrics proposed by those such as Olden, Poff, and others (Konrad and Booth, 2005; Olden and Poff, 2003) to describe flow characteristics: magnitude, duration frequency, variability, and timing. Recently, for example, several authors have proposed frameworks to select appropriate ecohydrological metrics for analyzing urban impacts on flow regimes (Gao et al., 2009; Hamel et al., 2015).

However, perhaps the more important area of recent research is the development of *catchment metrics* that describe and predict the impacts of urban land use on flow regimes. Studies in this area first demonstrated that the *total imperviousness* (the proportion of a catchment made up of impervious surfaces) was a good predictor of changes in peak flows and flow volumes (Booth et al., 2002). More recently, a number of researchers have demonstrated that the nature of the hydraulic connection between the impervious surface and the receiving water is a primary driver of the hydrologic impact (Wong et al., 2000) and consequent impacts on channel morphology (Vietz et al., 2014) and ecosystem response (Burns et al., 2015a; Walsh and Kunapo, 2009). In other words, the extent to which the changes in site-scale hydrology of a given impervious surface propagate to the stream depend on whether there is a hydraulically efficient connection such as a pipe or constructed drain, or whether impervious runoff flows to adjacent pervious land, providing opportunities for infiltration and attenuation. Authors have thus developed variables such as *effective imperviousness* (Han and Burian, 2009), which is the proportion of the catchment made up of impervious areas draining directly to streams via pipes. Walsh and Kunapo (2009) went a step further, proposing *attenuated imperviousness*, which accounted for the degree of attenuation offered by overland flow path lengths.

There are also a number of commonly available tools for assessing urban stormwater impacts on stream hydrology, for example, Stormwater Management Model (US EPA, 2014) and the *Model for Urban Stormwater Improvement Conceptualization* (eWater CRC, 2014).

20.5.2 OPTIONS AVAILABLE TO ADDRESS HYDROLOGICAL IMPACTS FROM URBANIZATION

Regulation and licensing

First, there is a need for appropriate regulation that prevents unmitigated discharge of stormwater runoff to receiving waters, which is commonly the case for wastewater (Roy et al., 2008). Second, there is a need for economic models that encourage the use of stormwater as a resource (Walsh et al., 2012). Such use could either be within the urban landscape (e.g., for nonpotable household uses, industry, and irrigation of open space), or could be captured and transferred to nearby agricultural land to support water-demanding production activities such as cropping and horticulture. Indeed, the coupling of regulation to prevent stormwater runoff from being conveyed to receiving waters with economic benefit that could come from the harvesting and supply of stormwater as an alternative water resource creates an important synergy. Such models need to recognize the frequent mismatch between where stormwater is generated and managed, and the potential beneficiaries downstream of improved management. For example, creation of infiltration rain gardens to capture runoff close to the source might be an effective way to reduce degradation of downstream waterways (with benefits to those communities), but the costs of building and maintaining such systems may be inequitably borne by those in the upstream community.

Technical solutions

Reducing impervious areas can be achieved through careful site design to minimize the requirement for paved areas, creating smaller building footprints (e.g., promoting more compact double-story townhouse-type housing), or through specific techniques such as green (vegetated) roofs, porous pavements, and structural materials to allow porous areas to be trafficable.

Alternatively, specific systems can be constructed to intercept, retain, and treat stormwater, and then either used for various human needs, allowed to infiltrate into underlying soils, or evapotranspired by vegetation. We use the term *stormwater control measures* to describe such systems, although other terms such as *best management practices, alternative techniques,* or *sustainable drainage systems* are also commonly used (Fletcher et al., 2014a). Such techniques may be applied at a wide range of scales, from *at-source* to *end-of-pipe* (also called *end-of-catchment*), but in recent years, there has been increasing recognition of the need to emphasize intercepting and dealing with stormwater runoff as close as possible to its source, not only to improve performance (in terms of restoring a more natural flow and water quality regime), but also to increase the benefits to the community in terms of landscape amenity and urban microclimate (Fletcher et al., 2013).

Stormwater control measures include wetlands, ponds, and sediment basins, tanks for capturing stormwater runoff or runoff from roofs only (these are typically called *rainwater tanks*), swales, and infiltration systems (basins or trenches and rain gardens, also called bioretention of biofiltration systems). Each of these measures is briefly described in Table 20.1, along with their principal mechanisms. Systems based on detention (e.g., ponds) reduce the peak flow by providing storage to attenuate the discharge rate. Although detention can reduce flooding, in some cases it may increase damage to the stream channel by increasing the time above which the flow rate is sufficient to mobilize and transport sediments (Vietz et al., 2015). Water quality treatment is provided by sedimentation or through physical filtration and biochemical processes (i.e., conversion of nitrogen into nitrate and subsequently removal via denitrification).

Burns et al. (2012) and Walsh et al. (2012) recently demonstrated the need to design SCMs that are capable of returning the volume of runoff emanating from a given site toward the natural (predevelopment) level. Walsh et al. (2012) demonstrated that this was necessary in order to return ecologically important aspects of the flow regime (frequency of runoff, rate, and seasonality of base flows, variability of flows) to levels able to support a healthy stream ecosystem. A range of SCMs can be used to achieve this outcome, but given the necessity to reduce runoff volumes to natural systems, which promote stormwater harvesting (intercepting and storing stormwater for subsequent use in a range of nonpotable applications, or potentially for potable purposes, provided there is the necessary treatment processes and risk protections in place) is considered essential (Walsh et al., 2012). Evapotranspiration is the other primary mechanism by which stormwater runoff can be lost from the water balance; this may be by harvesting and then irrigation of vegetation, or might be specifically through the use of bioretention *rain gardens* and vegetated infiltration systems.

Stormwater infiltration, a widely used stormwater management technology, works by capturing stormwater runoff and allowing it (after treatment) to infiltrate into the ground, thus likely recharging groundwater and contributing to base flow (Hamel et al., 2013). Infiltration can thus help to reduce peak flows. However, infiltration needs to be combined with other techniques to reduce the overall volume (as described above); reliance on infiltration alone could otherwise potentially result in an increase in groundwater levels and stream base flows beyond their natural level (Bhaskar et al., 2016). Stormwater management strategies aimed at restoring more natural flow regimes will

Table 20.1 Commonly Used Stormwater Control Measures

Name(s)	Description and Principal Mechanisms
Rainwater/stormwater tank	Principal mechanisms: storage, detention, harvesting.
	Provide a source of water for a range of end users, thus reducing potable water consumption. Can be used to irrigate the urban landscape, thus contributing to urban amenity, microclimate amelioration, and restoration of soil moisture, groundwater, and base flow. Household-scale systems normally capture only roof runoff, whereas larger systems often harvest runoff from both roof and ground-level impervious areas.
Ponds and sediment basins	Principal mechanisms: detention and sedimentation, some biological treatment.
	Popular for their esthetic benefits, but are much less effective in terms of restoring more natural flow regimes and significantly improving water quality.
Wetland	Principal mechanisms: detention, sedimentation, physical and chemical filtration (through vegetation), some evapotranspiration.
	Can be a pretreatment for harvesting.
Rain garden/bioretention system/biofiltration system	Principal mechanisms: detention, physical and chemical filtration (high efficiency), biological treatment (typically high efficiency), infiltration, evapotranspiration.
	Bioretention systems are probably the most effective treatments currently available, and offer excellent flexibility in terms of configuration and layout. They can promote infiltration, or be lined where this is not appropriate. Modifications can be made to improve their water balance and maintain vegetation health.
Swale	Principal mechanisms: physical filtration, limited detention, may promote infiltration depending on underlying soils.
	Swales have the advantage of simultaneously providing some stormwater control while acting as a conveyance (can replace small stormwater pipes). They provide a useful at-source option for linear catchments such as roads, particularly when underlying soils permit effective infiltration.
Infiltration basins and trenches	Principal mechanisms: infiltration, evapotranspiration (if vegetated), physical filtration, detention (in ponding zone).
	Infiltration systems can be vegetated (as below) or unvegetated (e.g., using sand or gravel at the surface). Infiltration trenches are often *hidden*, being located underground and wrapped in geotextile.
Gross pollutant trap	Principal mechanisms: physical filtration.
	Gross pollutant traps are only suitable for pretreatment. They cannot provide a full range of flow regime and water quality treatment functions.
Sand/granular filter	Principal mechanisms: physical filtration (and chemical filtration in advanced granular filter media).
	Provide only water quality treatment, with no significant effect on flow regimes.
Other green infrastructure (e.g., green roofs and walls)	Principal mechanisms: physical and chemical filtration, evapotranspiration, some biological treatment.
	Green roofs and walls also have substantial microclimate and energy conservation benefits.

thus require an integrated suite of SCMs, designed to work together to restore ecologically important aspects of the regime (Fig. 20.4). It is therefore necessary to have appropriate design or performance targets to guide the design of the SCM strategy.

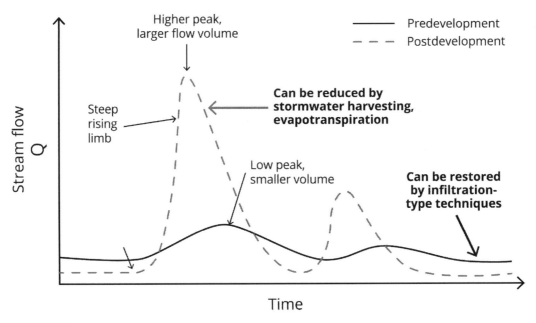

FIGURE 20.4

A combination of stormwater control measures to restore more natural flow regimes.

Source: Adapted from Marsalek et al. (2007)

Burns et al. (2013) present a framework for setting SCM performance objectives, based on the natural flow paradigm (Poff et al., 1997). Recognizing the need to compromise the complexity (and perhaps impossibility) of designing a system to match the exact natural flow regime (considering the large number of flow metrics that could be used), they developed an approach based on three overarching metrics (with the objective in each case being to mimic the predevelopment behavior):

1. Overall volume, expressed as the proportion of rainfall falling on an impervious area that should be lost via evapotranspiration or harvesting.
2. Contribution to base flow, expressed as the proportion of rainfall falling on an impervious surface, which is subsequently infiltrated.
3. A measure of the amount of rainfall required to generate runoff from a given area, expressed as the *initial equivalent loss (mm)*.

Burns et al. (2013) demonstrate how the target values of these three metrics can be calculated for a given site by drawing on local data on annual streamflow and evapotranspiration coefficients in forested and grassland catchments such as those presented in Zhang et al. (2001), as well as estimates of initial loss (rainfall necessary to generate surface runoff) in natural catchments such as those presented by Hill et al. (1998).

Collaborative management

Increasingly, urban stormwater is being managed within a more holistic framework known as integrated urban water management, where there is a strong focus on the interactions between the various components of the urban water cycle (Chocat et al., 2001). Integrated urban water management recognizes the ability to capture benefits such as water supply, urban amenity, and mitigation of the urban heat island (Fletcher et al., 2013), at the same time as delivering more traditional benefits of flood protection (Burns et al., 2010). This approach requires the involvement of disciplines such as urban planners, landscape architects, botanists, engineers, and urban designers. Most importantly, given the excess runoff volume generated by impervious surfaces, there is a need to encourage and facilitate (through regulation and financial incentives) collaboration between the *producers* of stormwater (those who create or manage a given impervious surface and potential consumers of the water resource that is created by it, e.g., a householder, a nearby industry, or agricultural producer).

20.6 MANAGING HYDROLOGICAL ALTERATION DUE TO REFORESTATION

Chapter 3 discussed the potential impacts of vegetation on streamflow. Forestry programs are commonly promoted for their environmental and social benefits (Calder, 2007), and there is a public perception that forests provide a positive impact across the board, including timber supply, recreation, carbon sequestration, and wildlife habitats. However, there is significant evidence that reforestation can have lasting impacts on water supply in the catchment through increased interception and evaporation (Albaugh et al., 2013; Calder, 2005; Robinson, 1998).

20.6.1 CHARACTERIZING REFORESTATION IMPACTS

There are a number of different methods that have been adopted to understand the hydrological impacts of vegetation change across a catchment. The appropriate method for a given catchment is guided by the resources and data availability. A few of the more common methods are briefly described below.

Paired catchment method

The paired catchment method involves using two neighboring catchments with similar characteristics, one used as a control catchment and the other as a treatment catchment. A calibration period tests that the two catchments respond in a hydrologically similar manner while catchment conditions remain the same. The relationship between streamflow in the two catchments is then used to establish a baseline for hydrological change in the treatment catchment once vegetation change occurs.

This method has been a common approach to understanding the impacts of vegetation change on streamflow, particularly in the foundation literature that conributed to the understanding of vegetation impacts on the water cycle (Zhang, 2015). Considerable resources and a long-term commitment are required to implement this approach (Zhang, 2015).

Time-trend analysis method

This method involves comparing prechange (calibration period) and postchange conditions within the same catchment. The relationship between streamflow and rainfall during the prechange period is compared with rainfall–streamflow outcomes in the postchange period.

The calibration period is usually based on a period of record prior to changes in vegetation occurring. In many catchments, the length of such a calibartion period may be too short or absent, making this method difficult to implement (Zhang, 2015).

Modeling

Vegetation impacts can be incorporated through a model that simulates the partition of rainfall between processes such as infiltration, evapotranspiration, reservoir storage, and surface water runoff, based on different land use and plant growth characteristics. Inputs to such models include climate information and catchment characteristics such as soil type (Albaugh et al., 2013).

20.6.2 OPTIONS AVAILABLE TO ADDRESS HYDROLOGICAL IMPACTS FROM REFORESTATION

Regulation and licensing

A number of countries have concurrent policies that promote reforestation, yet simultaneously recognize water shortages as a serious national issue (Calder, 2007). This may, in part, be due to a failure of scientists to adequately communicate forest hydrology knowledge to policy makers. There is also no current policy framework that assesses forestry schemes in a holistic manner, including consideration of water resource impacts (Calder, 2007). There are few water policies that incorporate vegetation water use within the allocation framework; however, some notable examples have considered reforestation. In South Africa, plantation forestry is considered a streamflow reduction activity under the *Water Act*, and therefore requires a license. The planning authorities predict the likely hydrological impact of new plantations and assess this against existing commitments and the reserve requirements for the catchment as part of the license determination (Albaugh et al., 2013). Similarly, in Australia, the National Water Initiative was a policy agreement had required jurisdictions to incorporate all significant water intercepting activities into the water planning framework. Australia saw a rapid increase in large-scale commercial plantation forests due to increasing domestic and export demands for sawlogs, paper woodchips, and paperboard products (Leys and Vanclay, 2010). This included investment from managed investment schemes (which grew through the 1990s) and superannuation funds (Timber Queensland, 2015). In South Australia, plantation forestry became a significant land use in some high-rainfall areas. This led to the implementation of a state-wide policy to incorporate plantations as a water user into the water planning framework. A similar approach has also been developed in western

Victoria. The management options identified included the following (Government of South Australia, 2009):

1. Implementing a forest water licensing scheme that requires plantation forest owners or managers to hold a tradeable water right.
2. Implementing a permit system that manages water-affecting activities, controlling their extent and nature.
3. Codes of practice and industry agreements.

These management options were implemented through the catchment water resource planning process, with the appropriate management option dependent on the state of water resource in the system. In highly impacted catchments, this effectively incorporates plantation water use into the formal water allocation framework and within any water-use cap.

20.7 **CONCLUSION**

Management of diffuse causes of hydrologic alteration has received little attention in environmental water management literature. The impacts on streamflow from groundwater, farm dams, urbanization, and forestation can all be significant. However, management options are complicated due to the spatial scale, diffuse nature of the impacts, and there are often multiple stakeholders involved.

In this chapter, examples of attempts to manage diffuse catchment changes using a variety of strategies have been provided, including direct regulation, economic instruments, education campaigns, and collaborative management. These approaches may be helpful to design policies addressing diffuse catchment changes in other situations. However, it is important that cultural, political, and biophysical contexts are considered when transferring a policy from one situation to another (Benson et al., 2012; Swainson and de Loe, 2011).

The management of diffuse catchment changes often requires a degree of acceptance and cooperation with water users. Consultation and education are helpful tools in addressing diffuse catchment changes, especially if combined with other policy instruments. It is important that the stakeholders involved know the consultation process is genuine rather than a token gesture. Furthermore, although agencies may wish to base management actions on *hard* data such as environmental monitoring, such data may be difficult and expensive to gather. Thus, agencies may need to obtain *soft* sources of information through consultation with landowners and field officers, and anecdotal evidence from water users.

Diffuse causes of hydrological alteration are often not included or do not fit neatly within historic licensing and regulation frameworks for water management. A key consideration for managing groundwater, farm dam, urbanization, and forestation impacts on instream flows is how these practices interact with the existing allocation mechanisms for environmental water in a given system. Importantly, diffuse causes of hydrologic alteration impact on river flow before it reaches a river. When defining and managing an existing cap with significant diffuse impacts, water uses, and impacts must be considered, particularly under the licensing and water-use arrangements that sit under the cap. If this does not occur, although use from the main stream is limited through a cap, the water use in the greater catchment may be able to continue to grow. Furthermore, if a cap

Table 20.2 Diffuse Causes of Hydrological Alteration and Options to Manage Impacts on Environmental Water

Cause	Management Options to Limit Use	Management Options for Excess Water/Changed Delivery
Farm dams	Licensing/integrate within cap 　• Regulation of new dams 　• Where required, voluntary decommissioning of dams 　• Encourage winter-fill dams Conditions on license holders (to protect instream low flows) 　• Low-flow bypass 　• Collaborative management rules	
Groundwater	Integrate within cap 　• Regulation of new licenses Condition on license holders 　• Limits on extraction rates and timing 　• Monitoring and compliance of consumption rates	Targeted extractions to manage mounding and salinity Conditions on license holders 　• Limit of recharge rates
Urbanization		Stormwater control measures 　• Rainwater/stormwater tanks 　• Ponds and sediment basins 　• Wetlands 　• Rain garden/bioretention and biofiltration systems 　• Swales 　• Infiltration basins
Reforestation	Integrate within cap 　• Require new reforestation areas to hold water-use licenses under the cap	

is in place to limit water use, it should be set well before the catchment enters a stressed or crisis point (Finlayson et al., 2008). Potentially, this could save public resources, provide increased certainty of supply for water users, and better protect the environment.

Where there are license conditions on individual water users that require a passing flow instream prior to pumping, reduced groundwater inflows or intercepted runoff may reduce reliability to instream water users and the environment. Table 20.2 summarizes the management options available, and where appropriate, how they link to the environmental allocation mechanisms described in Chapter 16.

REFERENCES

Alaghmand, S., Beecham, S., Hassanli, A., 2013. Impacts of groundwater extraction on salinization risk in a semi-arid floodplain. Nat. Hazard Earth Syst. 13, 3405–3418.

Albaugh, J.M., Dye, P.J., King, J.S., 2013. *Eucalyptus* and water use in South Africa. Int. J. Forest. Res. 2013, 11.

Arnold, J.G., Stockle, C.O., 1991. Simulation of supplemental irrigation from on-farm ponds. J. Irrig. Drain. Eng. 117, 408–424.

Ashraf, M., Kahlown, M.A., Ashfaq, A., 2007. Impact of small dams on agriculture and groundwater development: a case study from Pakistan. Agr. Water Manage. 92, 90–98.

Ayres, I., Braithwaite, J., 1992. Responsive Regulation: Transcending the Deregulation Debate. Oxford University Press, New York.

Beavis, S., Howden, M., 1996. Effects of Farm Dams on Water Resources. Bureau of Resource Sciences, Canberra, Australia.

Benson, D., Jordan, A., Huitema, D., 2012. Involving the public in catchment management: an analysis of the scope for learning lessons from abroad. Environ. Policy Governance 22, 42–54.

Berens, V., White, M.G., Souter, N.J., 2009. Injection of fresh river water into a saline floodplain aquifer in an attempt to improve the condition of river red gum (*Eucalyptus camaldulensis* Dehnh.). Hydrol. Process. 23, 3464–3473.

Bhaskar, A.S., Beesley, L., Burns, M.J., Fletcher, T.D., Hamel, P., Oldham, C.E., et al., 2016. Will it rise or will it fall? Managing the complex effects of urbanization on base flow. Freshw. Sci. 35 (1), 293–310.

Bond, N., Costelloe, J.F., King, A., Warfe, D., Reich, P., Balcombe, S., 2014. Ecological risks and opportunities from engineered artificial flooding as a means of achieving environmental flow objectives. Front. Ecol. Environ. 12, 386–394.

Booth, D.B., Hartley, D., Jackson, R., 2002. Forest cover, impervious-surface area, and the mitigation of stormwater impacts. J. Am. Water Resour. Assoc. 38, 835–845.

Braithwaite, J., 1985. To Punish or Persuade: Enforcement of Coal Mine Safety. State University of New York Press, Albany, NY.

Braithwaite, J., 2002. Restorative Justice and Responsive Regulation. Oxford University Press, Oxford.

Bredehoeft, J.D., Young, R.A., 1970. The temporal allocation of ground water – a simulation approach. Water Resour. Res. 6, 3–21.

Brunner, P., Simmons, C.T., Cook, P.G., Therrien, R., 2010. Modeling surface water– groundwater interaction with MODFLOW: some considerations. Ground Water 48, 174–180.

Burns, M., Fletcher, T.D., Hatt, B.E., Ladson, A., Walsh, C.J., 2010. In: Chocat, B., Bertrand-Krajewski, J.-L. (Eds.), Can allotment-scale rainwater harvesting manage urban flood risk and protect stream health? (La récuperation des eaux pluviales à l'echelle de la parcelle: peut-elle protéger contre les inondations et la dégradation des milieux aquatiques?). Novatech. GRAIE, Lyon, France.

Burns, M., Fletcher, T.D., Walsh, C.J., Ladson, A., Hatt, B.E., 2012. Hydrologic shortcomings of conventional urban stormwater management and opportunities for reform. Landsc. Urban Plan. 105, 230–240.

Burns, M.J., Fletcher, T.D., Walsh, C.J., Ladson, A.R., Hatt, B., 2013. In: Bertrand-Krajewski, J.-L., Fletcher, T. (Eds.), Setting objectives for hydrologic restoration: from site-scale to catchment-scale (Objectifs de restauration hydrologique: de l'échelle de la parcelle à celle du bassin versant). Novatech. GRAIE, Lyon, France.

Burns, M., Walsh, C., Fletcher, T.D., Ladson, A., Hatt, B.E., 2015a. A landscape measure of urban stormwater runoff effects is a better predictor of stream condition than a suite of hydrologic factors. Ecohydrology 8, 160–171.

Burns, M.J., Schubert, J.E., Fletcher, T.D., Sanders, B.F., 2015b. Testing the impact of at-source stormwater management on urban flooding through a coupling of network and overland flow models. WIREs Water 2, 291–300.

Calder, I.R., 2005. Blue Revolution – Integrated Land and Water Resources Management. Earthscan, London.

Calder, I.R., 2007. Forests and water—ensuring forest benefits outweigh water costs. Forest Ecol. Manage 251, 110–120.

Cartwright, I., Gilfedder, M., Hofmann, H., 2014. Contrasts between estimates of baseflow help discern multiple sources of water contributing to rivers. Hydrol. Earth Syst. Sci. 18, 15−30.

Cetin, L.T., Alcorn, M.R., Rahmanc, J., Savadamuthu, K., 2013. Exploring variability in environmental flow metrics for assessing options for farm dam low flow releases. Proceedings 20th International Congress on Modelling and Simulation. 1−6 December 2013, Adelaide, Australia, pp. 2430−2436.

Chin, A., Laurencio, L.R., Martinez, A.E., 2008. The hydrologic importance of small- and medium-sized dams: examples from Texas. Prof. Geogr. 60, 238−251.

Chocat, B., Krebs, P., Marsalek, J., Rauch, W., Schilling, W., 2001. Urban drainage redefined; from storm-water removal to integrated management. Water Sci. Tech. 43, 61−68.

Christian-Smith, J., Abhold, K., 2015. Measuring what matters: setting measurable objectives to achieve sustainable groundwater management in California. Union of Concerned Scientists.

Crase, L., O'Keefe, S., Dollery, B., 2012. Presumptions of linearity and faith in the power of centralised decision-making: two challenges to the efficient management of environmental water in Australia. Aust. J. Agric. Resour. Econ. 56, 426−437.

Cresswell, D., Piantadosi, J., Rosenberg, K., 2002. Watercress user manual. Available from: http://www.water-select.com.au.

DNRE, 2000. Sustainable Water Resources Management and Farm Dams: Discussion Paper. Department of Natural Resources and Environment, Melbourne, Australia.

DNRE, 2001. Irrigation farm dams; a sustainable future. Victorian Government response to Farm Dams (Irrigation) Review Committee Report. Department of Natural Resources and Environment, Melbourne, Australia.

Edeson, G., 2015. Updating integrated catchment management to improve the climate resilience of water-dependent communities. Water 42, 61−65.

Endreny, T., 2008. Naturalizing urban watershed hydrology to mitigate urban heat-island effects. Hydrol. Process. 22, 461−463.

Evans, R., 2007. The impact of groundwater use on Australia's rivers − technical report. Land and Water Australia, Canberra, Australia.

eWater CRC, 2014. Model for urban stormwater improvement conceptualisation (MUSIC). 6.0 ed. eWater Cooperative Research Centre, Canberra, Australia.

Famiglietti, J.S., 2014. The global groundwater crisis. Nat. Clim. Change 4, 945−948.

Finlayson, B.L., Nevill, C.J., Ladson, A.R., 2008. Cumulative impacts in water resource development. Water Down Under Conference. 14−17 April 2008, Adelaide, Australia.

Fletcher, T.D., Andrieu, H., Hamel, P., 2013. Understanding, management and modelling of urban hydrology and its consequences for receiving waters; a state of the art. Adv. Water Resour. 51, 261−279.

Fletcher, T.D., Shuster, W.D., Hunt, W.F., Ashley, R., Butler, D., Arthur, S., et al., 2014a. SUDS, LID, BMPs, WSUD and more − The evolution and application of terminology surrounding urban drainage. Urban Water J. 12, 525−542. Available from: http://dx.doi.org/10.1080/1573062X.2014.916314.

Fletcher, T.D., Vietz, G., Walsh, C.J., 2014b. Protection of stream ecosystems from urban stormwater runoff; the multiple benefits of an ecohydrological approach. Prog. Phys. Geogr. 38, 543−555.

Fowler, K., Morden, R., 2009. Investigation of strategies for targeting dams for low flow bypasses. Proceedings Hydrology and Water Resources Symposium. Newcastle, Australia. Engineers Australia, pp. 1185−1193.

Fowler, K., Donohue, R., Morden, R., Durrant, J., Hall, J., 2012a. Decision support and uncertainty in self-supply irrigation areas. Proceedings of the Australian Hydrology and Water Resources Symposium. Sydney, Australia.

Fowler, K., Donohue, R., Morden, R., Durrant, J., Hall, J., Narsey, S., 2012b. Decision support using Source for licensing and planning in self supply irrigation areas. Proceedings 15th International River Symposium. Melbourne, Australia.

Fowler, K., Morden, R., Lowe, L., Nathan, R.J., 2016. Advances in assessing the impact of hillside farm dams on streamflow. Aust. J. Water Resour. 19 (2), 96−108.

Gao, Y., Vogel, R.M., Kroll, C.N., Poff, N.L., Olden, J.D., 2009. Development of representative indicators of hydrologic alteration. J. Hydrol. 374, 136−147.

George, R., Dogramaci, S., Wyland, J., Lacey, P., 2005. Protecting stranded biodiversity using groundwater pumps and surface water engineering at Lake Toolibin, Western Australia. Aust. J. Water Resour. 9, 119−128.

Gonzales, A.L., Nonner, J., Heijkers, J., Uhlenbrook, S., 2009. Comparison of different base flow separation methods in a lowland catchment. Hydrol. Earth Syst. Sci. 13, 2055−2068.

Government of South Australia, 2009. Managing the water resource impacts of plantation forests − A state-wide policy framework. South Australia.

Grant, S., Saphores, J., Feldman, D., Hamilton, A., Fletcher, T.D., Cook, P., et al., 2012. Taking the "waste" out of "wastewater" for human water security and ecosystem sustainability. Science 337, 681−686.

Griffith, M., 2002. The European Water Framework Directive: an approach to integrated river basin management. Eur. Water Manage. 1−15.

Gunningham, N., Sinclair, D., 2005. Policy instrument choice and diffuse source pollution. J. Environ. Law 17, 51−81.

Habermehl, M.A., 1980. The Great Artesian Basin, Australia. BMR J. Aust. Geol. Geophys. 5, 9−38.

Habets, F., Philippe, E., Martin, E., David, C.H., Leseur, F., 2014. Small farm dams: impact on river flows and sustainability in a context of climate change. Hydrol. Earth Syst. Sci. 18, 4207−4222.

Hamel, P., Daly, E., Fletcher, T.D., 2013. Source-control stormwater management for mitigating the effects of urbanisation on baseflow. J. Hydrol. 485, 201−213.

Hamel, P., Daly, E., Fletcher, T.D., 2015. Which baseflow metrics should be used in assessing flow regimes of urban streams? Hydrol. Process. 29, 2367−2378.

Han, W.S., Burian, S.J., 2009. Determining effective impervious area for urban hydrologic modeling. J. Hydrol. Eng. 14, 111−120.

Hill, P., Mein, R., Siriwardena, L., 1998. How much rainfall becomes runoff? Loss modelling for flood estimation. Cooperative Research Centre for Catchment Hydrology (Report 98/5), Melbourne, Australia.

Hughes, D., Mantel, S., 2010. Estimating the uncertainty in simulating the impacts of small farm dams on streamflow regimes in South Africa. Hydrol. Sci. J. 55, 578−592.

ICOLD, 2007. World Register of Dams. International Commission on Large Dams, Paris.

Ignatius, A., Stallins, J.A., 2011. Assessing spatial hydrological data integration to characterize geographic trends in small reservoirs in the Apalachicola-Chattahoochee-Flint River Basin. Southeast. Geogr. 51, 371−393.

Jordan, P., Stephens, D., Morden, R., Sommerville, H., Fowler, K., 2011. Spatial Tool for Estimating Dam Impacts (STEDI) v.1.2. Sinclair Knight Merz, Melbourne, Australia.

Kahn, A.E., 1966. The tyranny of small decisions: market failures, imperfections, and the limits of economics. Kyklos 19, 23−47.

Kendy, E., Bredehoeft, J.D., 2006. Transient effects of groundwater pumping and surface-water-irrigation returns on streamflow. Water Resour. Res. 42, W08415.

Khan, S., Rana, T., Hanjra, M.A., 2008. A cross disciplinary framework for linking farms with regional groundwater and salinity management targets. Agr. Water Manage. 95, 35−47.

Konrad, C.P., Booth, D.B., 2005. Hydrologic changes in urban streams and their ecological significance. Amercian Fisheries Society Symposium 47. Amercian Fisheries Society.

Kumar, M.D., 2005. Impact of electricity prices and volumetric water allocation on energy and groundwater demand management: analysis from Western India. Energy Policy 33, 39−51.

Kustu, M.D., Fan, Y., Robock, A., 2010. Large-scale water cycle perturbation due to irrigation pumping in the US High Plains: a synthesis of observed streamflow changes. J. Hydrol. 390, 222−244.

Leblanc, M., Tweed, S., Van Dijk, A., Timbal, B., 2012. A review of historic and future hydrological changes in the Murray-Darling Basin. Glob. Planet. Change 80–81, 226–246.

Lee, S., 2003. Management of farm dams using low flow bypasses to improve stream health. Counterpoints 3, 72–79.

Lehner, B., Reidy Liermann, C., Revenga, C., Vörösmarty, C., Fekete, B., Crouzet, P., et al., 2011. High-resolution mapping of the world's reservoirs and dams for sustainable river-flow management. Front. Ecol. Environ. 9, 494–502.

Leopold, L.B., 1968. Hydrology for Urban Land Planning: a Guidebook on the Hydrological Effects of Urban Land Use. U.S. Geological Survey, Washington, DC.

Lett, R., Morden, R., McKay, C., Sheedy, T., Burns, M., Brown, D., 2009. Farm dam interception in the Campaspe Basin under climate change. 32nd Hydrology and Water Resources Symposium. Newcastle, Australia. Engineers Australia, p. 1194.

Leys, A., Vanclay, J., 2010. Land-use change conflict arising from plantation forestry expansion: views across Australian fence-line. Int. Forest. Rev. 12, 256–269.

Liebe, J., Van De Giesen, N., Andreini, M., Walter, M., Steenhuis, T., 2009. Determining watershed response in data poor environments with remotely sensed small reservoirs as runoff gauges. Water Resour. Res. 45, Available from: http://dx.doi.org/10.1029/2008WR007369.

Lowe, L., Nathan, R.J., 2008. Consideration of uncertainty in the estimation of farm dam impacts. Water Down Under Conference. 14–17 April 2008, Adelaide, Australia.

Mantel, S.K., Hughes, D.A., Muller, N.W., 2010a. Ecological impacts of small dams on South African rivers Part 1: Drivers of change – water quantity and quality. Water SA 36, 351–360.

Mantel, S.K., Muller, N.W., Hughes, D.A., 2010b. Ecological impacts of small dams on South African rivers Part 2: Biotic response – abundance and composition of macroinvertebrate communities. Water SA 36, 361–370.

Marsalek, J., Rousseau, D., Steen, P.V.D., Bourgues, S., Francey, M., 2007. Ecosensitive approach to managing urban aquatic habitats and their integration with urban infrastructure. In: Wagner, I., Marsalek, J., Breil, P. (Eds.), Aquatic Habitats in Sustainable Urban Water Management: Science, Policy and Practice. UNESCO/Taylor and Francis, Paris/London and New York.

Martinez-Santos, P., De Stefano, L., Llamas, M.R., Martinez-Alfaro, P.E., 2008. Wetland restoration in the Mancha Occidental Aquifer, Spain: a critical perspective on water, agricultural, and environmental policies. Restor. Ecol. 16, 511–521.

Maxted, J.R., McCready, C.H., Scarsbrook, M.R., 2005. Effects of small ponds on stream water quality and macroinvertebrate communities. New Zeal. J. Mar. Fresh. 39, 1069–1084.

McGlynn, B.L., McDonnell, J.J., 2003. Quantifying the relative contribution of riparian and hillslope zones to catchment runoff. Water Resour. Res. 39, 1310.

McMurray, D., 2004. Farm Dam Volume Estimations from Simple Geometric Relationships. Department of Water. Land and Biodiversity Conservation, Adelaide, South Australia.

McMurray, D., 2006. Impact of Farm Dams on Streamflow in the Tod River Catchment, Eyre Peninsula, South Australia. Department of Water, Land and Biodiversity Conservation, Adelaide, South Australia.

McNutt, M., 2014. The drought you can't see. Science 340, 1543.

MDBC, 2008. Mapping the Growth, Location, Surface Area and Age of Man Made Water Bodies, Including Farm Dams, in the Murray-Darling Basin. Murray-Darling Basin Commission, Canberra, Australia.

Melbourne Water, 2013. Annual report for Woori Yallock Creek Water Supply Protection Area Stream Flow Management Plan 2012: reporting period 28 March 2013 to 30 June 2013. Melbourne Water, Docklands, Australia.

Mulligan, K.B., Brown, C., Yang, Y.-C.E., Ahlfeld, D.P., 2014. Assessing groundwater policy with coupled economic-groundwater hydrologic modelling. Water Resour. Res. 50, 2257–2275.

Nathan, R., Jordan, P., Morden, R., 2005. Assessing the impact of farm dams on streamflows, Part I: Development of simulation tools. Aust. J. Water Resour. 9, 1–12.

Nathan, R.J., Lowe, L., 2012. Discussion paper: the hydrologic impacts of farm dams. Aust. J. Water Resour. 16, 75–83.

Neal, B., Nathan, R., Schreider, S., Jakeman, A., 2001. Identifying the separate impact of farm dams and land use changes on catchment yield. Aust. J. Water Resour. 5, 165–176.

Nelson, R.L., 2012. Assessing local planning to control groundwater depletion: California as a microcosm of global issues. Water Resour. Res. 48, W01502.

Nevill, C.J., 2009. Managing cumulative impacts: groundwater reform in the Murray-Darling Basin, Australia. Water Resour. Manage. 23, 2605–2631.

O'Connor, T.G., 2001. Effect of small catchment dams on downstream vegetation of a seasonal river in semi-arid African savanna. J. Appl. Ecol. 38, 1314–1325.

Odum, W.E., 1982. Environmental degradation and the tyranny of small decisions. BioScience 32, 728–729.

Olden, J.D., Poff, N.L., 2003. Redundancy and the choice of hydrologic indices for characterizing streamflow regimes. River Res. Appl. 19, 101–121.

Oster, J.D., Wichelns, D., 2003. Economic and agronomic strategies to achieve sustainable irrigation. Irrig. Sci. 22, 107–120.

Ostrom, E., 1990. Governing the Commons: the Evolution of Institutions for Collective Action. Cambridge University Press, Cambridge.

Ostrom, E., 1993. Design principles in long-enduring irrigation institutions. Water Resour. Res. 29, 1907–1912.

Ostrom, E., 2011. Background on the institutional analysis and development framework. Policy Studies J. 39, 7–27.

Peck, A.J., Hatton, T., 2003. Salinity and the discharge of salts from catchments in Australia. J. Hydrol. 272, 191–202.

Petrucci, G., Rioust, E., Deroubaix, J.-F.D.R., Tassin, B., 2013. Do stormwater source control policies deliver the right hydrologic outcomes? J. Hydrol. 485, 188–200.

Poff, N.L., Allan, J.D., Bain, M.B., Karr, J.R., Prestegaard, K.L., Richter, B.D., et al., 1997. The natural flow regime. BioScience 47, 769–784.

Pokherl, Y.N., Koirala, S., Yeh, P.J.F., Hanasaki, N., Longuevergne, L., Kanae, S., et al., 2015. Incorporation of groundwater pumping in a global land surface model with the representation of human impacts. Water Resour. Res. 51, 78–96.

Ring, I., Schröter-Schlaack, C., 2011. Instrument Mixes for Biodiversity Policies. Helmholtz Centre for Environmental Research, Leipzig, Germany.

Robinson, M., 1998. 30 years of forest hydrology changes at Coalburn: water balance and extreme flows. Hydrol. Earth Syst. Sci. 2, 233–238.

Rochford, F., 2004. 'Private' rights to water in Victoria: farm dams and the Murray Darling Basin Commission cap on diversions. Aust. J. Nat. Resour. Law Policy 9, 1–25.

Rodell, M., Velicogna, I., Famiglietti, J.S., 2009. Satellite-based estimates of groundwater depletion in India. Nature 460, 999–1002.

Ross, M., Martinez-Santos, P., 2010. The challenge of groundwater governance: case studies from Spain and Australia. Reg. Environ. Change 10, 299–320.

Roy, A.H., Wenger, S.J., Fletcher, T.D., Walsh, C.J., Ladson, A.R., Shuster, W.D., et al., 2008. Impediments and solutions to sustainable, watershed-scale urban stormwater management: lessons from Australia and the United States. Environ. Manage. 42, 344–359.

Scanlon, B.R., Jolly, I., Sophocleous, M., Zhang, L., 2007. Global impacts of conversions from natural to agricultural ecosystems on water resources: quantity versus quality. Water Resour. Res. 43.

Schreider, S.Y., Jakeman, A.J., Letcher, R.A., Nathan, R.J., Neal, B.P., Beavis, S.G., 2002. Detecting changes in streamflow response to changes in non-climatic catchment conditions: farm dam development in the Murray−Darling Basin, Australia. J. Hydrol. 262, 84−98.

Shapoori, V., Peterson, T.J., Western, A.W., Costelloe, J.F., 2015. Top-down groundwater hydrograph time-series modeling for climate-pumping decomposition. Hydrogeol. J. 23, 819−836.

Sinclair Knight Merz, 2012. Improving State Wide Estimates of Farm Dam Numbers and Impacts − Stage 3 − Improving Farm Dam Model Inputs. Sinclair Knight Merz, Melbourne, Australia.

Singh, A., 2014. Conjunctive use of water resources for sustainable irrigated agriculture. J. Hydrol. 519, 1688−1697.

Skurray, J.H., 2015. The scope for collective action in a large groundwater basin: an institutional analysis of aquifer governance in Western Australia. Ecol. Econ. 114, 128−140.

Souza Da Silva, A.C., Passerat De Silans, A.M., Souza Da Silva, G., Dos Santos, F.A., De Queiroz Porto, R., Almeida Neves, C., 2011. Small farm dams research project in the semi-arid northeastern region of Brazil. The International Union of Geodesy and Geophysics Symposium. July 2011, Melbourne, Australia, pp. 241−246.

State of Victoria, 2003. Irrigation and commercial farm dams compendium of ministerial guidelines and procedures. May 2003.

State of Victoria, 2008. Guidelines for meeting flow requirements for licensable farm dams. Public Report. May 2003, 34 pp.

Swainson, R., De Loe, R.C., 2011. The importance of context in relation to policy transfer: a case study of environmental water allocation in Australia. Environ. Policy Governance 21, 58−69.

Thompson, J.C., 2012. Impact and management of small farm dams in Hawke's Bay, New Zealand. PhD thesis. Victoria University of Wellington, New Zealand.

Timber Queensland, 2015. Timber Queensland − Plantation Ownership, Investment (accessed 20.04.16).

Tourbier, J., White, I., 2007. Sustainable measures for flood attenuation: sustainable drainage and conveyance systems SUDACS. In: Ashley, R., Garvin, S., Pasche, E., Vassilopoulis, A., Zevenbergen, C. (Eds.), Advances in Urban Flood Management. Taylor and Francis, London.

US EPA, 2014. Stormwater Management Model (SWMM). US Environmental Protection Agency. Available from: http://www2.epa.gov/water-research/storm-water-management-model-swmm (accessed 04.11.14).

Venot, J.-P., Molle, F., 2008. Groundwater depletion in the Jordan highlands: can pricing policies regulate irrigation water use? Water Resour. Manage. 22, 1925−1941.

Victorian Farm Dams (Irrigation) Review Committee, 2001. Farm dams (irrigation) review committee final report. Melbourne, Australia.

Vietz, G., Sammonds, M., Fletcher, T.D., Walsh, C.J., Rutherfurd, I., Stewards, M., 2014. Ecologically relevant geomorphic attributes of streams are impaired by even low levels of watershed effective imperviousness. Geomorphology 206, 67−78.

Vietz, G.J., Walsh, C.J., Fletcher, T.D., 2015. Urban hydrogeomorphology and the urban stream syndrome treating the symptoms and causes of geomorphic change. Prog. Phys. Geogr. Available from: http://dx.doi.org/10.1177/0309133315605048.

Voss, K.A., Famiglietti, J.S., Lo, M., De Linage, C., Rodell, M., Swenson, S.C., 2013. Groundwater depletion in the Middle East from GRACE with implications for transboundary water management in the Tigris-Euphrates-Western Iran region. Water Resour. Res. 49, 904−914.

Wada, Y., Bierkens, M.F.P., 2014. Sustainability of global water use: past reconstruction and future projections. Environ. Res. Letters 9.

Walsh, C.J., Kunapo, J., 2009. The importance of upland flow paths in determining urban effects on stream ecosystems. J. N. Am. Benthol. Soc. 28.

Walsh, C.J., Fletcher, T.D., Burns, M.J., 2012. Urban stormwater runoff: a new class of environmental flow problem. PLoS One 7, e45814.

Walter, R.C., Merritts, D.J., 2008. Natural streams and the legacy of water-powered mills. Science 319, 299–304.

Wee, H., 2015. California landowners resist efforts to monitor groundwater. CNBC. Available from: http://www.cnbc.com/2015/05/12/the-growing-tension-over-california-water-metering-.html.

White, E., Peterson, T.J., Western, A.W., Costelloe, J.F., Carrara, E., 2016. Can we manage groundwater? Water Resour. Res. 52, 4863–4882.

Wong, T.H.F., Brown, R., 2009. The water sensitive city; principles for practice. Water Sci. Tech. 60, 673–682.

Wong, T.H.F., Lloyd, S.D., Breen, P.F., 2000. Water Sensitive Road Design – Design Options for Improving Stormwater Quality of Road Runoff. Cooperative Research Centre for Catchment Hydrology, Melbourne, Australia.

Young, R.A., Bredehoeft, J.D., 1972. Digital computer simulation for solving management problems of conjunctive groundwater and surface water systems. Water Resour. Res. 8, 533–556.

Zhang, L., Dawes, W.R., Walker, G.R., 2001. Response of mean annual evapotranspiration to vegetation changes at catchment scale. Water Resour. Res. 37, 701–708.

Zhang, X., 2015. Conjunctive surface water and groundwater management under climate change. Front. Environ. Sci. 3, 1–10.

MANAGING INFRASTRUCTURE TO MAINTAIN NATURAL FUNCTIONS IN DEVELOPED RIVERS

21

Gregory A. Thomas

The Natural Heritage Institute, San Francisco, CA, United States

21.1 INTRODUCTION

On a planet where ecosystems are everywhere imperiled, none have suffered species decline as acutely or precipitously as the freshwater environments: rivers and their associated floodplains, wetlands, estuaries, and deltas. Freshwater species are facing extinctions at a rate that exceeds even those occurring in the oceans or tropical forests (Allen et al., 2012; Darwall et al., 2009; Dudgeon et al., 2006; Loh et al., 2005; Strayer and Dudgeon, 2010; Zoological Society of London and WWF, 2014). The stressors are many and various, including contaminants from industrial and municipal sources, toxic runoff from farms and cities, excessive harvesting of edible species and alien invasions. Yet, the predominant factor has been the alteration of the natural biophysical processes by hydraulic infrastructure (see, e.g., Chapter 1 of Allen et al., 2012; Zoological Society of London and WWF, 2014). Dams, diversion works, levees, locks, and the like are built and operate to regulate and *control* these intrinsically dynamic natural systems.

In the main, the river systems of the world have been subjugated in these ways for the economically important purposes of supplying water to cities and farms, generating clean and renewable power, preventing flood damage and navigation, but with scant consideration of their environmental consequences (see WCD, 2000). Yet, riverine ecosystems are shaped by, and dependent on, the natural flows of water, sediments, nutrients, and migratory fish. Excessive alteration of the magnitudes, duration, frequency, timing, and locus of these flows can substantially impair the biological productivity of these systems. To a greater or lesser extent, infrastructure reduces the dynamism and diversity characteristic of river systems in their natural state into habitats that are comparatively static, simplified, and uniform. Elimination of annual floods deprives riparian forest and wetlands of seasonal inundation, effectively disconnecting the river from its productive floodplains and disrupting the estuarine hydrodynamics that are the engine for the exceptional biological fertility at the freshwater–salt water interface. Dams and reservoirs also trap and thereby deplete the sediments that replenish the floodplains and maintain the channel and deltaic morphologies, and the associated nutrients that drive the riverine and marine food chains. When this happens, native

Water for the Environment. DOI: http://dx.doi.org/10.1016/B978-0-12-803907-6.00021-8

fisheries suffer, cultivation and pastoral use of the floodplain are eliminated, groundwater levels decline, and recreational and aesthetic values of a living river basin are compromised (see Chapters 3—5).

Sixteen years ago, the *World Commission on Dams (WCD)* chronicled both the economic benefits attributable to the 45,000 large dams already built in over 140 countries in the past half century, and the damage they have caused to water-dependent ecosystems and human food production systems (WCD, 2000). At that time, some 1800 additional dams were under construction. In the near future, this number is going to grow exponentially in some regions, notably Eastern and Southern Asia, Amazonia, and throughout the African continent, to meet the soaring demand for electricity and food production in these regions. At present, 1.1 billion people worldwide lack basic electricity service (The World Bank Group, 2015a), with access rates falling as low as 5% in some African countries, such as South Sudan and Chad (The World Bank Group, 2015b). The current capacity to store and manage water is also limited: storage averages less than 900 m^3/per capita across parts of Latin America, Asia, and Africa compared with 6000 m^3/per capita in North America and 4800 m^3/per capita in Australia (where $1\ m^3 = 0.001$ ML; World Bank 2015 in UNEP, 2008). An additional driver of accelerated river development is the growing global movement toward carbon-free sources of power. Electricity generated from falling water comprises 85% of renewable power today (International Energy Agency, 2016). China, which is home to almost half of the large dams, intends to double its hydropower capacity within the next 20 years, in keeping with its commitment to reduce greenhouse gas emissions from coal-based generators. India is on a similar trajectory. Another factor is the control of floods, which still cause more deaths and property damage than all other natural disasters combined (CRED, 2015). The classic way to tame a raging river is to confine it with dams and levees.

There is a modest counter trend in some advanced industrial nations, where a few dams are being dismantled (American Rivers and Trout Unlimited, 2002; Planet Experts, 2014). These are generally old, obsolete, and/or dangerous structures that obstruct the reproductive cycle of valued anadromous species such as salmon (American Rivers and Trout Unlimited, 2002; McCully, 2015). The most impressive of these initiatives in terms of scale is the settlement agreement to dismantle four large hydropower dams on the Klamath River at the California—Oregon border (see Box 21.1). These dam removals are remarkable in many ways, including their rarity compared with the massive inventory of dams in these countries and the accelerating new development that is occurring, largely in the developing nations.

BOX 21.1 KLAMATH HYDROELECTRIC SETTLEMENT AGREEMENT

For nearly 50 years, the hydropower dams on the Klamath River at the California—Oregon border in the United States have denied what was once the largest Chinook salmon migration in the continental United States access to the cold water habitats they need for spawning. Under the agreed settlement terms, the removal of four of the large dams on the Klamath was the solution preferred by the United States, the states of California and Oregon, the tribes, and even the power company, PacifiCorp, rather than the regulatory alternative, which would have required a substantial change in operations to control downstream temperatures and flow conditions. The primary terms of the settlement are: up to $200 million in increased power costs to customers, removal to be accomplished without cost to the power company, and immunity of the power company from liability for any damages caused by dam removal (see Klamath Hydroelectric Settlement Agreement, 2010, pp. 23—25). Indeed, the Public Utility Commissions of the two states determined that dam removal under these terms would be more economical than relicensing and continued operations (see Public Utility Commission of Oregon, 2010,

(Continued)

BOX 21.1 (CONTINUED)

pp. 8–11). Implementation of the settlement agreement depends on funding of $450 million (see Klamath Hydroelectric Settlement Agreement, 2010, pp. 23–30). Of that amount, $200 million (including interest) will be collected from PacifiCorp's power customers (Klamath Hydroelectric Settlement Agreement, 2010, pp. 23–24). The Oregon and California Public Utilities Commissions have approved that customer contribution, which will be fully collected by 2020 (see California Public Utilities Commission, 2012; Public Utility Commission of Oregon, 2010, pp. 8–11). The other contribution to the total budget is California's bond funds totaling $250 million (see Klamath Hydroelectric Settlement Agreement, 2010, pp. 24–25). In November 2014, California enacted Proposition 1, which provides those funds.

21.2 A BRIEF OVERVIEW OF RIVER BASIN INFRASTRUCTURE AND IMPLICATIONS FOR RIVER FUNCTIONS

This chapter will focus on the types of river basin infrastructure that are most consequential for the natural processes of rivers: dams, diversions, and levees.

21.2.1 DAMS

Large dams are built to control the flow of the river for three main purposes: irrigation, hydropower, and flood control, and about 20% serve multiple functions (WCD, 2000). The purpose determines the operating characteristics of the dam and the extent to which it stores and releases water on a schedule that distorts natural flows. The most important distinction is between storage dams, which are built to modify flow patterns, and run-of-river dams, which are not. There are many gradations between these modes of operation.

Dams that are operated for water supply or flood control must necessarily store water. They are operated to capture water from snowmelt and monsoon rains and hold it for use later during the hot, dry growing seasons, to buffer the seasonal or interannual variability in river flows. If the reservoir storage capacity is large relative to the inflow, the reservoir may also be operated to carry over stored water from years of abundance to years of scarcity. Half of the world's large dams are for irrigation, with the largest number in China, India, Pakistan, and the United States (WCD, 2000). Storage dams regulate the flow pattern to satisfy downstream requirements for irrigation, industry, or public supply, or to maintain sufficient depth for navigation. During the drier portions of the year, the flow is greater than it would be naturally. Too much flow at the wrong time can be just as ecologically damaging as too little flow.

Flood control reservoirs capture the peak flow events that would otherwise inundate downstream structures or land use—with potentially large economic consequences or risks to life—and release this water more slowly after the storm passes. These reservoirs are operated to reserve a large *conservation* storage capacity at the beginning of the flood season. They release water as necessary between storm events to maintain this conservation reserve in anticipation of the next flood event. Toward the end of the flood season, however, the dam will be operated to fill the conservation storage area as much as possible—without incurring a risk of overtopping in the event of a late season storm event—to maximize storage levels going into the dry season.

By shaving off the peak flow events, flood control dams disrupt the natural hydrograph. However, they can be operated to provide seasonal inundation of floodplains to keep them biologically active. Land uses in the floodplain, enabled by the flood control infrastructure, that may have to be modified to accommodate such controlled inundations will be discussed later in this chapter. Where the former floodplain has been *extensively* developed with houses and other high-value structures such as roads or bridges, or where permanent crops such as orchards have been planted, reoperating the dam to revert to historic inundation of the floodplain may simply be impractical.

Hydropower dams create hydraulic *head* for power production, either at the dam or through diversion of water to some point downstream. The head is the elevation difference between the reservoir level and the tailwater level (the water surface elevation in the river immediately downstream of the turbines). The higher the head, the greater the power production from a given amount of water discharged to the powerhouse.

These dams can be operated either as storage dams or as run-of-river dams, with gradations between these modes. When they can, most large hydropower reservoirs store and release water in a pattern to generate power at a consistent rate, more or less, throughout the year. They do this by storing water during the wet season and releasing it during the dry season, thereby counteracting the natural seasonal variation in flows. These dams can also be operated to store and release water on a daily cycle (carrying water over from times of low power demand to times of high power demand). This is referred to as *hydropeaking*. Electrical demands generally peak when air conditioners are turned on in the afternoon or lights are turned on in the evening, and during particular seasons such as summer for cooling or winter for heating. These plants create large daily fluctuations in flow as they are turned on and off. Hydropower is particularly well-suited for meeting peak demands in the grid because it can be switched on and off very quickly, unlike coal-fired or nuclear power stations. The reservoir inflows are generally out of phase with these releases at the dam, creating unnatural flow patterns that can be highly disruptive to downstream ecosystems and floodplain livelihoods. Peak power hydropower operations are generally the most impactful because of the rapidity and frequency of changes in flow patterns. Hydropower dams can also be operated to *backstop* other renewable sources of power that are intrinsically intermittent such as solar electric stations (which only generate power when the sun shines), or wind generators (which only generate power when the wind is blowing). Ironically, these otherwise environmentally preferred energy options can entail large environmental costs in the rivers that back them up.

In contrast to storage dams, run-of-river dams are operated in rhythm with the natural variations in river flow. By discharging water at roughly the same rate as the inflow to the reservoir, the dam essentially just passes the hydrograph through the reservoir, relatively undistorted. Ideally, the storage levels in the reservoir are held as high as possible in the run-of-river reservoirs to maximize the hydraulic head. Operating the dam so that the releases mimic the inflow to the reservoir does not require a complicated monitoring system. This can be done simply by maintaining a static storage level in the reservoir.

It is important to recall the definition of *environmental water regime* introduced in Chapter 1 as "the quantity, timing, and quality of water required to sustain freshwater and estuarine ecosystems and the human livelihoods and well-being that depend on these ecosystems." This flow regime can be delivered through *environmental water* (a volume of water legally allocated to the environment), or as this chapter will make the case, through the reoperation of a water resource system to provide dual benefits to the environment and other water users. The most efficient and effective apportionment of reservoir storage to the downstream environment would aim to achieve specified environmental water regime benefits (see, e.g., Acreman, 1996; Alfredsen et al., 2011; Cain and Monohan, 2008). For instance, the targets and schedule might include floodplain inundation to a specified areal extent each year for a period of 10 days in the July–August time frame (with specified maximum ramping rates, downstream transport of eggs and larvae during some period of the natural flow recession period for a specified length of time each year, large pulse releases for geomorphic benefits and sediment mobilization at any convenient time, but not more often than once every 5 years, etc.). To achieve the desired hydrograph, the sophisticated dam operator will release the water in combination with rainfall and runoff events as they occur in the downstream landscape and tributaries. Thus, a heavy storm may be supplemented with just enough additional reservoir release to create the desired pulse flows or geomorphic flows.

The picture would not be complete without also mentioning pumped storage facilities. One of the most ecologically destructive ways in which a dam can be operated is to toggle the releases on and off radically during the course of a day to meet peak power demands (or backstop an intermittent source of renewable power). This is attractive financially because the time value of power is much greater during these periods of high demand. Turning the water flows on and off daily will impair a riverine ecosystem and fishery because the fish cannot adapt to the radical and frequent changes in flows. For example, fish often get stranded on gravel bars when flows are suddenly reduced. However, there is a way in which a dam can be operated in this manner without greatly disrupting the downstream hydrograph; by operating it together with a pumped storage facility. The major dam would be operated as a run-of-river facility or, more likely, as a power dam that releases a uniform volume of water during the daily cycle (a base load generator). During the time of lowest demand in the power grid (e.g., 9 p.m.–7 a.m.) the power that is generated is used mostly to pump some of the reservoir water into another reservoir located at a higher elevation. Then, during the peak demand period (e.g., 3 p.m.–9 p.m.) water is released from the pumped storage reservoir through turbines and discharged back into the main reservoir. Properly operated, pumped storage can allow much better flow conditions to be maintained below the main dam. The disadvantage is that at least 20% of the power is lost to the pumping operations (MWH, 2009; NHA, 2012), so the efficiency of the dam is reduced significantly. As more energy is used to pump uphill than is regained, the overall process is a net loss of energy, but it provides storage that allows energy to be used at the most valued times.

21.2.2 DIVERSION STRUCTURES

As the name implies, diversion dams extract water out of the river. There are many kinds and sizes, some of which are more *fish friendly* than others. Both hydropower and water supply dams may have diversion structures, but flood control dams generally will not.

Hydropower diversion projects are an alternative to installing a powerhouse at the dam site. In these facilities, the dam may be relatively small and the consequent storage likewise. The dam is used primarily just to divert a large fraction of the inflow—sometimes all of it—into a tunnel or canal that conveys the water by gravity, but at essentially the same contour as the diversion site, to a location well downstream in the same watershed, or sometimes into an adjoining watershed, where the elevation differential between the canal and the river creates a large hydraulic head. The water is then dropped down through penstocks into the powerhouse, which may be located off the river. The discharge from the power plant flows back into the river. The advantage of these facilities is that larger amounts of power can be produced per unit of water due to the high head. The disadvantage is that the intervening reach of river, between the point of diversion and the point of return flow, may become largely dewatered (depending on the percentage of water diverted or released at the diversion structure). This technique, if operated only to maximize power generation, can be devastating to the intervening fishery.

Such facilities are common in the mountainous areas of the western United States, Yunnan Province in China, and in many other places in the world with steep topography. It is in these settings that the concept of *minimum instream flow* standards for environmental protection originated, and continues to be the staple of hydropower regulation institutions.[1] It is increasingly recognized, however, that this principle has become obsolete. The new environmental flow paradigm now aims not for instream standards, but for operations that allow the river to flow out of its banks seasonally in a natural flood regime, and aims for *variable* flows that emulate the natural hydrologic pattern instead of *minimum* flows (Thomas, 1996). This goal is extremely hard to attain in the diversion-type hydropower facilities described above, particularly at the smaller scale of many micro-hydropower schemes.[2]

Diversion structures are an essential feature of water supply projects. Some divert the water at the storage dam; some divert the water at a point downstream. In the latter case, the river channel is used as the water conveyance system. Both types of schemes are common, but their environmental consequences (and therefore, mitigation strategies) can be quite different. Both deplete the natural flows below the point of diversion into the irrigation canal. However, downstream diversion projects also result in flow *alterations*, which may be even more damaging than the flow depletions. Where water is diverted at the dam or reservoir, the volume of water that can then be dedicated to downstream flows will be diminished, but it can still be discharged in a pattern that maintains some degree of dynamism and floodplain interaction. Where the water is diverted at a downstream location, the river environment both above and below the point of diversion is likely to be severely impacted. Here the temporal mismatch between environmental and irrigation demands becomes important. Aquatic environments need pulse flows during the rainy season, and low flows during the dry season (refer to Chapter 4). Irrigation systems demand the opposite. During the rainy season, they aim to store water behind the dam rather than release it to the downstream river channel. During the dry season, the irrigation system aims to release an unnaturally high volume of water into the conveyance system, which will include the river channel in the case of the downstream diversion schemes. So, the downstream river is receiving either too little or too much water to maintain its natural functions. Reoperating the system to restore environmental water regimes requires these distortions to be counteracted. A description of how the desired environmental water regime can be restored will be discussed below.

---

[1] See *PUD No. 1 of Jefferson County v. Washington Department of Ecology*, 511 U.S. 700 (1994); *S.D. Warren Company v. Maine Department of Environmental Protection*, 547 U.S. 370 (2006).

[2] It is widely believed that these micro-hydro facilities are relatively benign from an environmental standpoint. In actual practice, they are often among the most impactful on a per kilowatt hour basis due not only to the dewatering of the river downstream of the diversion point, but also because of poor construction practices for the canal and accompanying roadway. Unless good practices are followed, these construction projects can result in massive erosion into the river channel below. Some of the worst examples can be found in Yunnan Province in China.

21.2.3 LEVEES

The third category of infrastructure with large implications for the environment is *levees*. These are artificial walls constructed to confine a river to its channel and prevent it from interacting with its historic floodplain. The reason for these walls is to confine the river so that the flat, flood-prone areas can be developed with structures and farms. Indeed, one of the most perverse trends in river basin development is the attraction of development into floodplains due to the construction of these levees. Floodplain levees that separate flood water storage areas from the river may also make flooding worse downstream (Acreman et al., 2003). Once this happens, it is generally impossible or very difficult and expensive to *undevelop* the floodplains so that the levees can be removed and the river can once again interact with its landscape. This dynamic interaction is the single most important factor in the biological productivity of a river. In many situations, the natural functions simply cannot be restored without removing or setting back these levees to release the river from its straightjacket. Indeed, it is fair to say that the single greatest barrier to river restoration all over the world is the lack of effective methods for removing floodplain structures to higher ground or making the farming operations compatible with seasonal flooding. This is a much neglected area of environmental research and innovation (Fig. 21.1).

Note also the poignant irony. One of the great Achilles heels of reservoir development is the challenge associated with relocating settlements from the reservoir area. Indeed, much of the

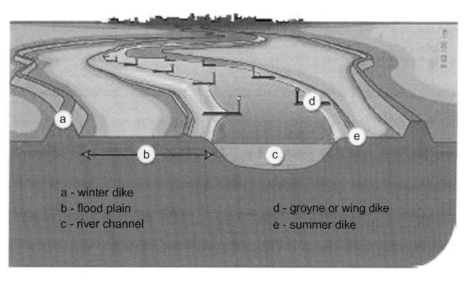

a - winter dike
b - flood plain
c - river channel
d - groyne or wing dike
e - summer dike

FIGURE 21.1

Sketch of a river, floodplain, and levees.

Source: Loucks et al. (2005)

antidam activism[3] is aimed precisely at the social costs of these relocation schemes, few of which provide opportunities for resettled households to improve their standards of living.[4] Yet, the key to reversing the environmental damage of dams—restoring the natural functions—is to reactivate the floodplains as an essential part of the river dynamic. As the flood constriction that resulted from the dam has acted as a magnet for floodplain settlement, it is usually the case that river restoration will necessitate relocating human settlements and land uses! Resistance to this may well hinder the restoration. In other words, the same issue that is used to try to prevent dam development may also be applied to prevent river restoration.

Yet, the cases are actually quite distinguishable. In the case of resettlement for reservoir development, the occupants are generally relocated some distance away from the river and in a manner that disengages them from their livelihoods and food production systems. In the case of floodplain relocation, the residents are asked (required?) to move to higher ground in the same vicinity. To the extent that they derive their livelihoods either directly or indirectly from the bounty of the river, their situation may actually be substantially improved. Arguably, the fishery will become more productive, as will the farmland as it benefits from the replenishment of soil and nutrients that the restored annual floods will provide.

21.3 MANAGING INFRASTRUCTURE TO PRESERVE OR RESTORE NATURAL FUNCTIONS IN DEVELOPED RIVER SYSTEMS

In considering how to site, design, and operate infrastructure to preserve or restore the health of rivers, this chapter will focus on those natural functions of rivers on which biological productivity depends: hydrology, sediment and nutrient processes, and access to essential habitats for aquatic animals. It will leave other aspects of water quality associated with man-made structures such as temperature effects, de-oxification, stagnation, and so on to other chapters of this book.

Aside from the decision whether to build a dam at all, or how many, the choices that have the largest consequences for the riverine environment are essentially of three types: where to site, how to design, and how to operate the dams. The options available depend crucially on when the decisions are made. Substantially different options emerge at the planning stage than those that are left after the infrastructure is built. For instance, the choice of location of the facility becomes irreversible once it is constructed. Concrete once poured cannot be unpoured. Design decisions are difficult, expensive, and often impractical to alter after the facility is built. This point is elaborated at length below in the section on dam design for sediment management. Operational policies, while alterable in theory, actually entail reoperation of the entire water management system, not just the physical infrastructure. This is illustrated below for all of the types of infrastructure that have been discussed in this chapter. For siting choices in all cases, design choices in most cases, and operational choices in some cases, the key lesson is that it is "critically important to get it right from the

---

[3]For example, International Rivers since its founding in 1985 has championed the cause of dam-affected people. See: http://www.internationalrivers.org/the-movement-for-rivers-and-rights.
[4]In a comparative study of dam-induced resettlement projects, Thayer Scudder (2005) found that three of 44 settlements analyzed had sufficient opportunities available for resettled households to raise their living standards.

start" (Ledec and Quintero, 2003; Opperman et al., 2015). Today's decisions may foreclose better options in the future. Mistakes are at best costly and often irreversible for many decades, after which the physical properties of the river may be altered forever. Therefore, this section of the chapter will end with advice on how to avoid precluding the more sustainable options.

The siting decision for river infrastructure is often driven by considerations of topography, hydrology, and geology. Put the facility where it will work best, is the usual engineering judgment. However, there is another pragmatic consideration, which is to put the facility where it will do the least harm. In the case of a relatively pristine river system with high natural values that are still largely intact, the best development strategy may be to deflect dams to adjacent, already impaired watersheds. This choice often arises in nations where river development is most pressing. Where some degree of development is necessary and inevitable, concentrating it in the areas that are already compromised is the pragmatic choice. The Mekong River system—the most productive freshwater fishery in the world—is an excellent example of this situation. Most of this system has already been highly compromised, but there are portions, particularly the Xe Kong tributary, that are still largely intact. Further development is inevitable. Concentrating it above existing dams is the smart approach.

For those river systems, or portions of river systems, that are already heavily engineered, the reality is quite different. Except in those limited situations where dam removal is feasible—generally dams that are obsolete and/or dangerous—the tactical option is not to try to *unengineer* the system, but to *reengineer* it (see Acreman et al., 2014). The toolkit for doing so, which will be described in this section, aims to achieve the goals enshrined in many policy guidelines such as the International Hydropower Sustainability Assessment Protocol of 2010 (see http://www.hydrosustainability.org/Protocol) and the World Bank's Social and Environmental Safeguards. In a field dominated by conflict, stalemate, and zero-sum thinking, these techniques offer creative solutions to bridge the professional gulf between engineers and ecologists, and advance the integration of biodiversity concerns into large hydraulic infrastructure projects.

Even the best siting, design, and operation of infrastructure will exact a toll on the natural functions of a river. As noted, these functions are highly complex and dynamic. The techniques discussed in this chapter cannot hope to preserve or restore all of the natural complexity and dynamism without compromising the economic value of the infrastructure. For example, by reoperating storage dams to function in a run-of-river mode, it might be possible to maintain or restore the floodplain inundation patterns that naturally occur every 10 years, but one generally cannot hope to preserve the larger and less frequent events that the dam is designed to contain. The extent of feasible reoperation may be limited by the encroachment of the floodplain by structures and land development. It may be limited by the capacity of the penstocks and power houses to discharge flood pulses without sacrificing power generation.

21.3.1 RULES OF THUMB FOR DAM SITING

In this section, some general principles for infrastructure siting will be considered, with a focus on dams. Considerations for siting diversion works and levees are found in the accompanying text box (Box 21.2).

As noted, the choice of a dam site will necessarily depend on such factors as topography, geology, and hydrology. For environmental performance, considerations should also include sediment management and fish passage. To factor in these considerations, a spatial scale much larger than the reservoir and its immediate environs should be adopted. When analyzing the effects of a proposed dam, the river basin should be analyzed for its sediment production, the effects of upstream dams, and the downstream impacts all the way to the coastal zone. Likewise, reservoir sustainability and downstream impacts should be analyzed over a sufficiently long temporal scale (300 years or more) to capture long-term consequences.

The article by Kondolf et al. (2014) (p. 276, 277) reports that:

> *Decisions about siting reservoirs largely determine future reservoir performance. Sediment problems can be minimized by giving preference to river channels with lower sediment loads (e.g. in less erodible areas, and perhaps higher in the catchment) and to sites where sediment passing is more feasible (e.g. steep gorges instead of low-gradient reaches). For a given level of hydroelectric generation within a river basin, it may be possible to minimize impacts by concentrating dams in a smaller number of rivers (preferably with naturally low sediment yields), allowing other rivers to flow freely, contributing sediment and supporting habitat.*
>
> *Dams in a series should be operated in concert to achieve management of sediment transport along the river system. Poor results and conflicts between upstream and downstream dams will result if dams are operated independently. Therefore, when a series of dams are developed along any river, particular attention should be given to establishing the appropriate coordination and data sharing among the parties, including both the historical and real-time monitoring data required to determine the efficiency of the operation and to identify means to improve the operation and pass sediment.*

Drawing upon those observations, one can posit certain principles for siting of dams to minimize their environmental impacts:

1. The headwater reaches of rivers are by definition in the more mountainous areas, where the river valleys are generally steeper, both in terms of the longitudinal (downstream) slope and in terms of the valley side slopes. Thus, reservoirs built in such reaches will tend to be longer, narrower, and deeper than comparably sized reservoirs in lowland areas, where the river slopes and topography are more gentle, and the reservoirs will tend to be shallower and wider. The steeper, narrower, deeper reservoirs of the headwaters are better suited to generating high velocities through the reservoir for flushing or sluicing the sediments and nutrients through sediment discharge gates installed for that purpose. Drawdown *flushing* focuses on scouring and resuspending deposited sediment and transporting it downstream. It involves the complete emptying of the reservoir through low-level gates that are large enough to freely pass the flushing discharge through the dam without upstream impounding, so that the free surface of the water is at or below the gate. *Sluicing* involves discharging high flows through the dam during periods of high inflows to the reservoir, with the objective of permitting sediment to be transported through the reservoir as rapidly as possible while minimizing sedimentation. Some previously deposited sediment may be scoured and transported, but the principal objective is to reduce trapping of incoming sediment rather than to remove previously deposited sediment. It is easier to drawdown smaller reservoirs because they depend less on year-to-year carryover storage. It is easier to maintain high volumes of flow through a smaller and steeper reservoir so that the sediments will be transported rather than falling out of suspension.

2. Reservoirs in the headwater terrain will tend to capture less sediment in the first place. As a general rule, the trap efficiency of a reservoir (for sediments and nutrients) will be a function of the ratio of the reservoir storage capacity to the average annual flow. This concept is quantified in the Brune curve, which is widely used to estimate trap efficiencies. Headwater reservoirs tend to be smaller and thus have lower trap efficiencies. In addition, the effects of both sediment/nutrient capture and flow alteration will be attenuated by any downstream tributaries.

3. Although there are notable exceptions, the lower reaches of rivers generally have higher fish biodiversity, biomass, and more migratory species, than the higher reaches. It is best to site the dam far upstream from the features that are most biologically productive such as floodplains, wetlands, estuaries, and deltas. Therefore, siting the dam higher in the catchment is generally better than siting it lower.

4. Smaller reservoirs are preferable to larger ones. The change from a flowing river to a still lake habitat presents major challenges for migratory fish, which are adapted to flowing water. Minimizing the length of passage through the still water to access to the habitats upstream will improve the success of fish migrations.

5. Cascades of dams compound the problem. Maintaining minimum lengths (e.g., 300 km) of flowing water between dams is essential to allow for migration upstream and larval drift downstream, as well as to complete other aspects of the life cycle. Notably, concentrating dams in already developed catchments to save virgin catchments from development has just this deleterious effect, which is why the former are sometimes referred to as *sacrifice basins*.

6. Perhaps the most important rule of thumb is to site dams above existing barriers to migrating fish, and avoid inundation of important spawning and rearing reaches. These barriers may be natural biogeographic features such as waterfalls, or may be artificial barriers such as existing dams for which fish passage facilities are not effective. This rule of thumb is not a universal solution, however. Some fish such as salmon can access steep rapids, and dams and reservoirs can certainly disrupt the habitat conditions for resident species that do not migrate to complete their life cycle. However, particularly in the exceptionally fecund tropical river systems, long-range migrations of a large variety of fishes are commonplace. The Mekong River system is a notable example, where 87% of the fish that have been studied are known to migrate often substantial distances to complete their life cycle (Baran and Borin, 2012).

BOX 21.2 SITING DIVERSION WORKS AND LEVEES

Diversions and levees are summarized here and will be discussed in greater detail later in the chapter.

Diversions: Operating diversion works to maintain environmentally beneficial flows downstream of the point of diversion is generally easier to achieve if the water is diverted into a conveyance channel at the storage reservoir rather than from a point on the river downstream of the reservoir. Using a river as a conveyance channel tends to cause entrainment of aquatic organisms and distortion of the natural flows both above and below the point of diversion.

Levees: Levees should be set back as far as practical from the river channel itself to allow the river to interact with a portion of its historic floodplain, where that would afford high habitat value and low risk of flood damage to structures. A classic example is the Lower Yellow River in China, where the gigantic levees below the Xiamenxia and Xiaolangdi Dams are sometimes as much as 22 kilometers apart, allowing an active river meander to take place between them. The author will elaborate this point in the discussion of reoperation strategies for flood control systems below.

7. Where a tributary or entire river is already partially compromised by major infrastructure, it may be better to concentrate future development in this same basin rather than to impair a pristine reach of the system. For a given amount of power, intensive development is less impactful than diffuse development. Cascades of dams may sacrifice one reach to leave other reaches with higher biodiversity value unimpaired.

21.3.2 DESIGN OF INFRASTRUCTURE FOR ENVIRONMENTAL COMPATIBILITY

Dams

For environmentally compatible dam design, we focus on the flows of sediments and their associated nutrients in recognition of their fundamental importance to the health and productivity of river systems (see Chapters 5 and 12). By trapping sediment in reservoirs, dams interrupt the continuity of sediment transport through rivers, resulting in downstream channel erosion, alteration of the morphological requirements for fish spawning and rearing, reduction in the nutrients that fuel the aquatic food chain, and ultimately, even loss of reservoir storage and hydropower production (Kondolf et al., 2014). In sum, the sediment processes in rivers maintain the morphology of the river channel and alluvial floodplains, and these provide the substrate for the diversity of habitats, which drive species abundance and diversity. The associated nutrients fuel the food web that nourishes this aquatic life (Kondolf et al., 2014; see Chapter 4).

In 2012, the *Natural Heritage Institute (NHI)* and China's *Yellow River Conservancy Commission (YRCC)* co-convened a workshop of Chinese and international experts to review the global experience and distill recommendations on siting, designing, and operating hydropower dams to facilitate the passage of sediments and nutrients. The consensus conclusions of these experts have been published in the American Geophysical Union *Earth's Future* journal (Kondolf et al., 2014). This chapter draws extensively on these findings to derive the guiding principles.

Dams are essentially immutable structures, and can be regarded as permanent features of the landscape within relevant planning horizons. This means that, unless the dam can be retrofitted, the river will be stuck with the original dam design indefinitely, even as river management values and needs evolve. A fundamental example of the imperative to get it right from the start is the incorporation of low-level gates to allow discharge of sediments. If such features are not built into the initial construction, retrofit may be very costly in the case of concrete dams, or impossible from an engineering and safety standpoint for rock-filled dams. Yet, there are notable examples of such retrofits. Xiamenxia Dam on the Yellow River has been revamped three times to install gates for sediment discharge to extend the useful life of the reservoir, which otherwise would have filled with sediment within a few decades. Where retrofit of low-level gates is not practical, it is sometimes possible to construct tunnels around the dam for sediment discharge. Another design consideration is installation of not just one level gate for sediment discharge, but multilevel gates so that the relatively clear water can be discharged at the higher level(s) to dilute the turbid water discharged at the lower level(s). This can substantially reduce the adverse effects on fish when the dam discharges sediment starved-water during routine operations, and then suddenly discharges unnaturally turbid water during flushing operations, without a period of transition to facilitate adaptation by the fish.

The penalty for failure to design a dam for sediment discharge is illustrated most graphically in some of the settings where dam construction occurred earliest and most intensely. For instance, in the United States, few if any of the hundreds of dams constructed by the US Army Corps of

Engineers or the US Bureau of Reclamation have been designed or are operated to discharge sediment. The consequence of this omission has been documented by the National Research Council (2011), which reports that sediment accumulation behind the six mainstem dams on the Missouri River in the United States amounts to about 4565 GL (or approximately 6 trillion tons). This is reported to have caused channel incision and bed degradation downstream of the dams, and the decline of three native fish and bird species (National Research Council, 2011),[5] and substantial loss of reservoir storage capacity. The storage capacity of Lewis and Clark Reservoir has been reduced more than 20% as Gavins Point Dam (on the Nebraska–South Dakota border) was closed in 1955 (National Research Council, 2011).

Regarding the design of gates for sediment discharge, the summary of workshop findings by Kondolf et al. (2014, pp. 276–277) counsels that:

> [A] long-term [sediment] equilibrium profile (...) should be calculated in advance for every project. Gates should be placed and sized with respect to the requirement to achieve the desired long-term profile. There is no standard location for the proper placement of gates, because this will depend on the situation at each dam, but in general gates should be set low enough and with sufficient hydraulic capacity to establish the desired equilibrium profile and support the type of sediment management operation identified for long-term use. For example, if flushing is to be performed during a low-flow period, the gates may be smaller and placed very low in the dam section, while gates for (...) sluicing will have much greater hydraulic capacity and will probably be set at a higher level. In many cases, an array of large radial gates at the bottom of the dam may be the best option. Their high initial capital costs are likely to be offset by the longer economic lifetime of the reservoir.

It is remarkable that even with the accelerated pace of dam development now taking place, the world is actually losing reservoir storage capacity year by year due to the accumulation of sediment (Morris and Fan, 1998). When this occurs, the finite inventory of suitable reservoir sites is diminished, making hydropower, in effect, a nonrenewable resource to that extent. However, with adequate maintenance and management of sedimentation, the usable life of a reservoir can be extended for a much longer period, even in perpetuity (Annandale, 2013; Palmieri et al., 2003).

It has been previously noted the adverse environmental consequences of sediment trapping to the downstream channel, floodplain, and deltaic morphology, and the consequent loss of habitat quality, complexity, and diversity. In addition to dam siting, design opportunities for enhancing sediment flows through reservoirs have been discussed.[6]

Fig. 21.2 shows techniques for reducing sediment capture, enhancing sediment flows, and removing sediment accumulation in reservoirs. These were explicated at length at the workshop

---

[5]The pallid sturgeon, the least tern, and the piping plover are listed as endangered or threatened under the federal Endangered Species Act and their decline has been "attributed to the river engineering projects (on the mainstem of the Missouri River) that created a colder, deeper, and less turbid river, and to the loss of large areas of sandbar habitat" (National Research Council, 2011, p. 15).

[6]For more detailed descriptions and explanations of various sediment management techniques to preserve reservoir capacity, see Morris and Fan (1998), Wang and Hu (2009), Annandale (2011), and Sumi et al. (2012). Sediment management classifications of Morris and Fan (1998) and Kantoush and Sumi (2010) both distinguish among three broad categories: (1) approaches to minimize the amount of sediment arriving to reservoirs from upstream; (2) methods to route sediment through or around the reservoir; and (3) methods to remove sediments accumulated in the reservoir to regain storage capacity.

sponsored by the NHI and the YRCC at the fifth International Yellow River Forum in September 2012, alluded to above. These principles are explained more definitively in the published proceedings paper by Kondolf et al. (2014) and are summarized in Fig. 21.2. This workshop brought together reservoir sediment management experts from China, Taiwan, Japan, Europe, Africa, and North and South America for two fruitful days of presentations and discussion to exchange collective experience in sustainably managing sediment in reservoirs and addressing problems of downstream sediment starvation.

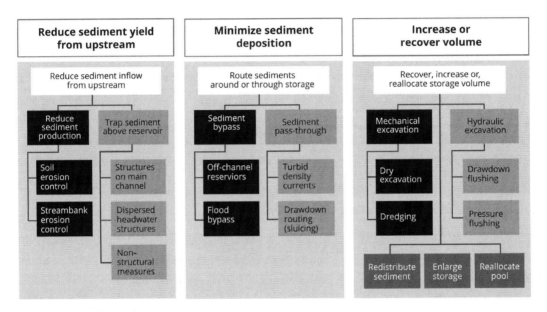

FIGURE 21.2

Classification of strategies for sediment management.

Source: Kondolf et al. (2014)

Fish passage facilities

As previously noted, dams block the migration of fish in and out of the habitats they need to spawn, feed, and seek refuge. Without access to these habitats, they cannot complete their life cycle and reproduce. Dams also fundamentally alter these critical habitats by creating a lake environment where there was previously a flowing river. The still water of the reservoir adversely impacts the survival of the larvae that need a velocity of flow to drift downstream to the reaches of the river where they can mature. Including in the design of the dam fish passage facilities that afford passage both upstream and downstream through both the dam and reservoir can—at least partially—mitigate the barrier to migration. Although the migration patterns of freshwater fish throughout the world are remarkably diverse, common approaches and principles can be identified in designing fish passage facilities that apply to all fish species and river systems. Mallen-Cooper in his work on the Mekong River with NHI has described these in a forthcoming publication, and the many factors that make fishway design a complicated bioengineering problem will be considered below (Figs. 21.3 and 21.4).

There are two components of effective fish passage:

Attraction, which involves designing fish passes to ensure that the hydraulic conditions (flow paths and turbulence) near the dam and powerhouse guide fish to the fishway entrance or entrances and

Passage, which involves the hydraulic and physical design of the fishway itself:

1. Fish passage must be designed to accommodate the range of needs and swimming abilities of the many species of concern. The smallest migrants usually have the weakest swimming ability and this determines the maximum water velocity, turbulence, and gradient of upstream fishways. The largest migrants and the volume of migratory biomass desired will determine the size, depth, space, and flow required in the fishway.
2. Headwater and tailwater levels determine the depths, operating range, length, and gradient of fishways.
3. Fishway entrances and designs may be substantially different for fish migrating at high flows during the wet seasons and fish migrating at low flows during the dry season.
4. For fish that only migrate during the daylight hours or the nighttime hours, the fishway may need to incorporate large resting pools.

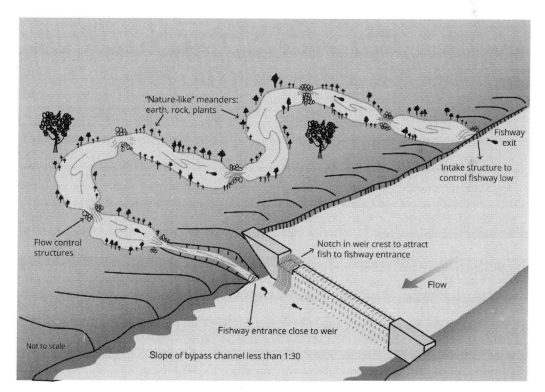

FIGURE 21.3

Conceptual layout of a bypass fishway.

Source: Thorncraft and Harris (2000)

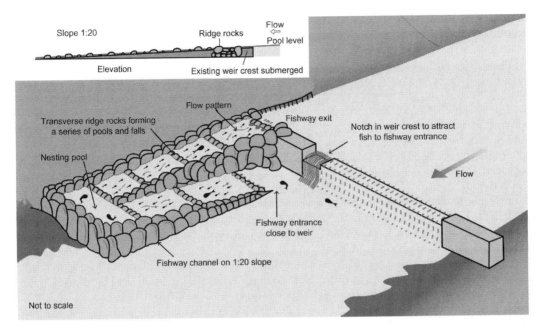

FIGURE 21.4

Conceptual layout of a partial-width rock-ramp fishway.

Source: Thorncraft and Harris (2000)

Fish migration is cyclic, involving upstream and downstream movements or return lateral movements on and off floodplains. These migrations are most commonly seasonal, but can occur once per generation for species that spawn only once and die. A common migration pattern in tropical and semi-tropical rivers such as the Mekong is for adult fish to migrate upstream to spawn, feed, and return downstream, alongside immature fish (subadults) that are dispersing often throughout the basin. Eggs, larvae, and juveniles drift downstream. Hence, to meet fish passage objectives, upstream passage needs to consider adult fish (of varying sizes) and downstream passage needs to consider returning adults, and eggs, larvae, and juveniles.

At large dams, there are two components of downstream passage to consider:

1. Passage through the impoundment or reservoir, which has changed from a flowing river that enabled larvae to drift passively downstream, to a lake where larvae stop drifting and can die from settling out on the bottom or lack of food;
2. Passage through the dam itself, which will have three paths: turbines, spillway, and sluice gates for sediment.

In sum, infrastructure—and here the focus is principally on dams and reservoirs—can be designed to enable the natural processes of rivers to be maintained to at least some degree. Rarely can these design features be incorporated after the dam is already built. The key imperative for those who construct and operate dams is to minimize both capital and operating costs. This creates perverse incentives that undermine *sustainable* infrastructure design, and these are in evidence broadly around the world. Designs that maintain natural functions exact increased capital costs, for installation of sediment discharge gates or fish bypasses, for instance—and in operating costs—in the form of loss of some stored water for sediment discharge or watering the fish passage facilities. It must be the function of those agencies that authorize, regulate, and finance dams to counteract these perverse incentives as, for instance, the Social and Environmental Safeguards policies of the World Bank Group seek to do (Hurwitz, 2014; Scheumann and Hensengerth, 2014).

21.4 OPERATING DAMS TO MAINTAIN NATURAL FUNCTIONS IN THE DOWNSTREAM ENVIRONMENT

Chapter 11 describes the methods for estimating environmental water regimes to preserve or restore components of the river hydrologic rhythms deemed important for desired ecosystem functions. Yet, without practical means to achieve them, these restoration targets will never result in healthy, functioning river systems. The burden of this chapter is to illustrate some techniques for operating (or reoperating) major dams to achieve those targets and thereby restore a substantial measure of the ecosystem functions and human livelihoods in the downstream river system that have been lost all over the world. Think of the flow targets as the environmental water demand side of the ledger, and the infrastructure reoperations as the environmental water supply side.

In this chapter, the toolkit of water management innovations will implement two of the key recommendations of the WCD: first, the objective of restoring, improving, and optimizing the benefits from existing large dams and identifying opportunities for mitigation, restoration, and enhancement and, second, the objective of sustaining the livelihoods of river-dependent communities by "releasing environmental flows to help maintain downstream ecosystem integrity and community livelihoods" (WCD, 2000).

The largest opportunities for reversing the damage of the past are not dam removals, which will remain a rare exception, but dam reoperation; and this will be most widely embraced where it can be accomplished without substantially diminishing the economic benefits for which the dams were constructed. Reoperation of these infrastructure systems to restore lost ecosystem functions can be called *reoptimization*.

About half of the major dams in the world were built primarily for agricultural irrigation, and the rest were built primarily for hydropower or flood control. Many are operated to provide multiple benefits. The challenge now is reoptimization of these dams by adding another benefit to the operational objectives, namely restoration of the downstream environment, ideally without significantly diminishing the economic benefits for which they were originally designed and operated.

As already shown, the function of dams is to store water, which inevitably distorts the natural hydrograph. Obviously, the degree that a dam alters the natural flow patterns is a key determinant of its adverse impact on the downstream ecosystems. Reoptimization of dam operations entails reversing this distortion: essentially restoring the key components of the natural hydrograph. This can be achieved by changing the storage regime. Essentially, the goal is to move the operations of

---

BOX 21.3 TYPES OF RESERVOIR OPERATION: STORAGE FACILITY VERSUS RUN-OF-RIVER

The extent to which a reservoir is operated as a storage facility or run-of-river facility depends on the percentage of the annual inflow that the reservoir is capable of storing. At the extreme end of the spectrum are reservoirs such as those on the Colorado River in the southwestern United States; a substantial river but not one of the largest. Lake Mead and Lake Powell can each store about 2 years of inflow (approximately 3520 GL and 3198 GL, respectively) (Allen, 2003). In contrast, the reservoirs on the largest rivers of the world may be comparably large, but still function as run-of-river hydropower facilities because the inflow is too large to store. On the mainstream of the Mekong River (the world's eighth largest river), for instance, the hydropower dams will be quite large (e.g., well over 1000 MW of capacity), yet the amount of storage will be relatively small. The Xayaburi Dam, the first to be constructed on the mainstream of the Lower Mekong, will create a reservoir 30 m in depth and 60—90 km long, and will have a total storage capacity of 703.2 GL and an active storage capacity of 225 GL (Baran et al., 2011), and generate some 1200 MW of power, but the storage levels will only vary about 4 m between the high- and low-flow periods. Large though they are, such dams will not actually alter the natural hydrograph of the river substantially. During the peak flow period, such dams will have to spill an appreciable fraction of the inflow around the powerhouse. As a consequence, they have a relatively low *capacity factor* (the power generation as a function of installed capacity of the turbines). Upstream in the headwaters, however, the gigantic power dams being built in China (e.g., Xiaowan at 4200 MW and Nuozhadu at 5850 MW) are already producing a noticeable distortion in both the high- and low-flow levels in the Mekong River (Räsänen et al., 2012). Indeed, it is the reduction in the annual variability of flows from these dams that makes the rest of the mainstream especially attractive for hydropower development.

This exemplifies a frequent configuration for a cascade of hydropower dams: the large storage dams are sited at the top of the cascade, to reduce the seasonal variability in flows through the rest of the system, and then the downstream dams are operated essentially as run-of-river facilities, *returbining* the same water over and over again. The problem this creates is that at the bottom of the cascade, the flow regime is radically altered from the natural pattern, to the detriment of the riverine ecosystems, perhaps all the way through the estuary or delta. This result can be countered, however, by operating the last reservoir in the cascade as a reregulation dam, essentially storing and releasing water in a pattern that restores some semblance of the natural variability. Notice, however, that this reregulation dam will then store the summer flows and release the winter flows through the generator, which is usually opposite of the power grid demand pattern.

---

the dam in the direction of releasing water at the same rate at which it flows into the reservoir. Ideally, the dam just passes the natural hydrograph through the reservoir and into the downstream channel with as little distortion as possible. As previously noted, this is commonly called run-of-river operations. Even run-of-river facilities are not wholly benign, however. They create a barrier for species migration, intercept sediment flows, which are essential to the downstream ecomorphology, and adversely affect temperatures and other water quality parameters (Box 21.3).

21.4.1 IMPORTANT FACTORS FOR DAM REOPERATION

Reoperation to restore lost ecological functions should be pursued tactically. It should focus on those downstream river reaches where flow restoration will produce the largest ecological and human livelihood benefits. River reaches that formerly supported the richest array of native species before development and suffered the greatest loses as a result of development are presumptively the ones that would benefit most from flow restoration. (Unfortunately, there is often a paucity of data on baseline conditions before dam construction that would permit comparisons with conditions after construction.) This principle does not deny that there may be factors other than flow that limit ecological function such as water quality.

In general, the river reaches that nourish (or could nourish) broad alluvial floodplains (e.g., those associated with savannahs in Africa and flood-prone semi-tropic basins in Asia), wetlands systems, deltas, and estuaries are the most important for biodiversity and food production (Ward, 1998; Welcomme, 1979). Based on research conducted mainly in the tropics, scientists have described a *flood pulse advantage* in which rivers that are connected to functioning floodplains produce a significantly greater biomass of fish per unit area than do rivers disconnected from floodplains or lentic water bodies such as reservoirs (Bayley, 1991). With notable exceptions such as the Niger inner delta in Mali and the Pantenal in Brazil, these floodplain and wetland features tend to be found in the lower reaches of the rivers; deltas and estuaries are always found there. Storage dams immediately upstream of such features are likely to be most damaging as their effects are not much attenuated, and dams immediately downstream of floodplains and wetlands are likely to create barriers to fish migration in both directions. This makes them particularly important targets for reoperation, fish passage retrofits, or decommissioning.

The reoperation potential of existing projects is complicated by two factors that merit emphasis in this chapter.

Reservoirs are just the storage features

Since the WCD report, much effort and impressive progress has been made in the applied science of defining environmental water requirements in various types of rivers (Davis and Hirji, 2003), whereas comparatively little progress has been made in the art and science of defining the operational toolkit for achieving these targets. Why is this? One reason is that reoperation to restore desired environmental water regimes entails reoperating not just the dams that have caused the flow alterations and depletions, but the entire water management system for which the reservoirs are just the storage element. Thus, to reoperate a hydropower dam, it is necessary to figure out how to reoperate the entire electrical grid system; if the goal is to reoperate an irrigation dam, then it is necessary to figure out how to reoperate the entire irrigation system; and to reoperate a flood control dam, it is necessary to figure out how to reoperate the entire land use regulation and flood management system. Reoperation of entire systems is technically challenging and therefore has been rarely undertaken. This chapter aims to change that by showing how it can, indeed, be done.

In hydropower systems, the dams and reservoirs store water for power generation later or, in run-of-river operations, to create the hydraulic head. To reoperate these dams to generate environmental water regimes, it may be necessary to also reoperate the entire array of generators powering the grid system, the transmission lines, the interconnections with neighboring grids, the power distribution system, and so on. This chapter will describe in greater detail just how that can be done and the limiting factors that must be taken into account. Likewise, flood control dams are only a part of the artificial and natural infrastructure that can be used to prevent flood damage. To reintroduce a flood regime that emulates the seasonal variability of rivers in their natural state, and allows the rivers to interact with their landscapes so as to create the conditions conducive to propagation of aquatic life, the entire flood control system will have to be reoperated. These other features include, for instance, levees to confine the flows away from structures and land uses that are important to protect. However, perhaps the greatest potential for restoring natural functions to developed rivers is to make use of the *natural infrastructure* that can also absorb floods in a way that provides auxiliary environmental benefits (see Opperman et al., 2013). This infrastructure consists of the floodplains that were active in their natural state (before the flood control infrastructure was built) and the associated groundwater system. A description of a technique of removing man-made infrastructure to allow periodic inundation of the floodplain and to infiltrate the recharge zones of aquifers (to the extent possible within the capacity constraints of the dam's release valves) will be discussed below.

The same is true of irrigation systems. The reservoir is merely the storage device. The rest of the system consists of the diversion works, the conveyance system, the distribution system, the on-farm application system and, very important, the groundwater system underlying the irrigated farmland. The detail of how all of these components can be reoperated together to augment storage capacity, capture a larger fraction of the runoff hydrograph, and add operational flexibility such that water releases can be increased or decreased to emulate a more natural and environmentally beneficial downstream flow pattern will be discussed below. This allows better control of the magnitude, duration, frequency, and timing of downstream flows than would otherwise be possible without adverse consequences to irrigation. In addition, it is important to add, reoperating all these components together can also substantially improve water supply reliability for the farms and stabilize and restore groundwater tables. Again, the point is that reoperation is not only about the dams and reservoirs.

Most large dams are multipurpose

A second complication in developing a reoperation strategy is that many large dams perform multiple functions. These dams may store water for consumptive use, regulate floods, and/or generate hydropower. The reoperation techniques for each of these infrastructure functions are more or less unique. In these cases, which are the appropriate reoperation techniques to apply?

That is not an intractable problem. With few exceptions, the multiple purposes are hierarchical. Flood control is usually the overriding mandate. Indeed, this is enshrined in law in the United States and many other countries. In the United States, all dams that perform a flood control function are subject to regulation by the US Army Corps of Engineers, a federal regulatory body that prescribes the operating rules (the *rule curve*) for these dams. That rule determines when water must be stored and when it must be released.[7] Making this rule curve the overriding operational mandate makes sense. It would be an imprudent dam operator that would risk damage to downstream properties or peril to lives to eke out yet another kilowatt of power or cubic meter of water supply. Therefore, it is safe to assume that, if one wishes to improve downstream environmental releases during flood control operations, one must apply one of the reoperation techniques appropriate to flood control operations, as described below (Fig. 21.5).

The second operational priority for a multipurpose dam will almost always be water supply. In a grid system, there may be many generators that can be called upon to meet power demand at any instant of the day. However, most water systems are rather severely storage limited. Given the unpredictability of rainfall and runoff conditions, maximizing water storage to abide both seasonal and interannual variability is generally indispensable. So, one can safely assume that at times when the dam is not being operated for flood control purposes (i.e., after the large runoff events), it is probably being operated to maximize storage for water supply. During these times, the techniques appropriate to augment storage flexibility, for example the conjunctive management of reservoirs with the groundwater system, will be the tools of choice.

---

[7]Flood Control Act of 1944, P.L. 78-534 (58 Stat. 887), 33 U.S.C. § 709; US Army Corps of Engineers, Engineer Regulations § 1110-2-240.

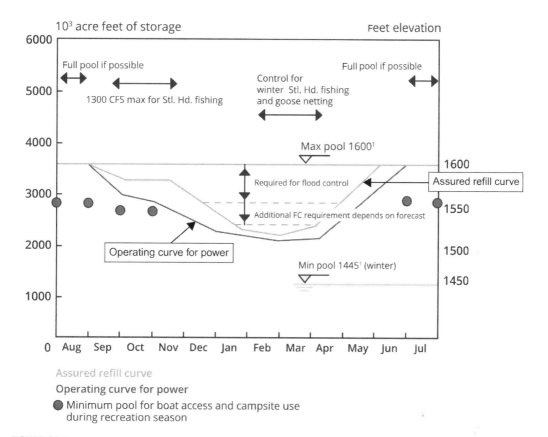

FIGURE 21.5

Example of a reservoir rule curve (where 1 acre foot = 1.233 ML and 1 foot = 0.305 m).

Source: Loucks et al. (2005)

This leaves hydropower operations as the lowest priority. Indeed, power generation is rather *opportunistic* in these multipurpose reservoirs. Power is generated as water is released for the higher priority purposes. This is somewhat ironic as it seems that most of the attention to dam reoperation has been paid to hydropower generation, whereas dam reoperations for the other purposes may actually hold the larger potential for improving environmental conditions below multipurpose dams.

To summarize, the operational purpose of the reservoir that is causing the flow distortions that need to be remedied will usually determine the proper tools in the toolkit to use. Box 21.4 itemizes the tools and techniques available in the reoperation toolkit according to that operational purpose.

21.4.2 OPERATION AND REOPERATION OF HYDROPOWER SYSTEMS

In this chapter, it has been observed several times that the simplest way to envision reoperation of hydropower dams to restore a semblance of the natural downstream hydrograph is to convert the operations from a storage mode to a run-of-river mode to the extent feasible. That will cause

---

BOX 21.4 EXAMPLES OF REOPERATION TECHNIQUES

Irrigation dams can be reoperated by:

1. Integrating groundwater and surface storage through conjunctive water management and groundwater banking (aquifer recharge and recovery);

   Irrigation projects that were built to counteract preexisting groundwater depletion give rise to the possibility of conjunctive water management and the integration of surface and groundwater management. This type of storage flexibility may enable a considerable degree of environmental water restoration by increasing the stored water yield and thus create new value in the system that can offset the costs of reoperation (Purkey et al., 1998). We shall consider this approach in more detail below.
2. Reducing physical losses or water from irrigation systems (e.g., reducing evaporative losses in water conveyance and applications, reducing deep percolation to salty aquifers, tail water recovery, etc.);
3. Relocating points of diversion and return flow;
4. Creating variability in release patterns;
5. Restricting rates of change in these release patterns;
6. Transferring entitlements to store, divert, or consume water;
7. Retiring waterlogged or salinized irrigation land.

Hydroelectric dams can be reoperated by:

1. Changing the role and function of the hydro dam in the mix of generation facilities for the grid;
2. Substituting thermal generators for daily peaking facilities;
3. Building reregulation reservoirs downstream of hydroelectric dams;
4. Building pumped storage facilities to reduce the need to operate dams to follow electrical load curves;
5. Improving coordination of cascades of dams to permit more flexible operation.

Flood control dams can be reoperated by:

1. Flood easements;
2. Flood routing and storage in retention basins;
3. Levee setbacks.

---

the hydropower dam to generate power in rhythm with the inflow of water into the reservoir. The natural hydrograph is just passed through the reservoir, relatively undistorted. In this mode, the function of the dam is primarily to create the hydraulic head needed for power generation; to the extent compatible with flood management—if that is a paramount purpose of the reservoir—the reservoir should be maintained at maximum storage levels to maximize the head.

However, to change operations in this manner, the role of the hydropower dam in meeting power demands in the electricity grid may have to be altered. During the rainy season, a dam that is reoperated to act more like a run-of-river dam will generate power at its maximum capacity, while substantially reducing power generation during the low-flow season. This adversely affects power reliability unless other generators in the grid adjust their operating schedule to compensate, or additional power plants are constructed to help meet the power demand during the dry season. Gas-fired power plants may be the best option for backstopping run-of-river hydropower plants as they have low capital costs relative to their operating (fuel) costs. Alternatively, the additional capacity may be obtained by interconnecting with neighboring electrical grids. Although this may seem an improbable scenario at first glance, augmenting the capacity of hydropower plants is necessary for system reliability whether or not the hydro dam is operated in a run-of-river mode. In fact, grid systems that are dominated by hydropower generators are inherently unreliable because of the large annual variability in rainfall and runoff (which is being exacerbated by global climate change). These power systems must add substantial *stand-by* capacity anyway to buffer the interannual variability in hydropower output. That being the case, the stand-by generators can also be operated to buffer the seasonal variability in power output when the hydropower generators are operated in a run-of-river mode.

To reoperate hydropower dams that are operated to follow the daily load curve (hydropeaking dams), the aim is to avoid or reduce the daily fluctuations in discharges into the downstream river, which is devastating to its ecosystem. Again the major challenge is to change the role that the hydropower dam plays within the mix of generators feeding the grid. Other generators must provide the peak power so that daily flow distortions can be avoided, or pumped storage facilities must be added to counteract the flow distortions, as described previously in this chapter.

A frequent occurrence around the world is that a cascade of hydropower dams is built on the same river and feeds into the same grid. Typically, one of those dams (usually a larger one near the top of the cascade) effectively controls flows through the rest of the cascade and into the downstream channel and floodplain. Under these circumstances, the final dam in the cascade may need to be reoperated as a *reregulation* dam to counteract the distortions in the hydrograph and release a more natural flow pattern into the downstream river system.

Whether it is feasible and ecologically beneficial to reoperate a hydropower dam depends on a host of practical considerations. There is a need for a rapid assessment tool that can be applied to any particular dam, or cascade of dams, to determine reoperation potential and the best approach to use. Here, a conceptual model is presented that can be applied to any river basin or geographic region to quickly and efficiently screen a large number of hydropower dams to identify those that are the most promising opportunities for reoperation to enhance ecosystem functions, human livelihoods, and traditional food production systems, without a significant reduction in hydropower benefits. This tool enables the user to determine where to invest limited financial resources to conduct technical feasibility studies leading to the development of an implementable reoptimization plan for hydropower dams.

This conceptual model is called the "Rapid Evaluation Tool for Screening the Potential for Reoptimizing Hydropower Systems" or "REOPS" (NHI, 2009a).[8] It assesses both the environmental desirability and technical feasibility of reoperation, leading to the development of an

---

[8]The schematic for the REOPs tool and explanatory text are included in the online resource accompanying this book.

- REOPS Explanation & Guidance Note for Hydropower Systems: http://n-h-i.org/wp-content/uploads/2017/02/REOPS-schematic-for-Hydropower-Systems.pdf
- REOPS schematic for Irrigation Systems: http://n-h-i.org/wp-content/uploads/2017/02/REOPS-schematic-for-Irrigation-Systems.pdf

implementable reoperation plan. Although this screening model was developed for initial application in Africa, it is designed to be used in any river basin throughout the world. REOPS serves as a reconnaissance level tool that can be used by individuals with varying levels of training and expertise, ranging from local community leaders and nongovernment organizations to project operators, development assistance officials, national planning agency officials, or other experts. It requires only information that is readily available from the open literature or from site inspection, and does not require detailed technical analysis.[9] As a rapid assessment tool, it intentionally oversimplifies the subtleties and complexities of the many physical requisites for successful reoperation. Thus, it necessarily (but not often) misses some opportunities (false negatives) and selects in favor of some dams that will prove to be infeasible on closer inspection (false positives). In sum, it sacrifices some precision for greater speed and efficiency.[10]

REOPS is a *dichotomous key*, in which one proceeds through a logic pathway by answering a series of queries in either the affirmative or negative. Depending on the answer, one will either default out of the pathway, with the conclusion that the dam is not a good candidate for reoptimization, or will be directed to the succeeding cell. The cells themselves are organized around four major considerations:

1. Whether the dam controls flows into downstream river features of exceptional biological productivity;
2. Whether the powerhouse itself is or can be made suitable for reoperation;
3. Whether the land uses in the river basin, downstream and upstream of the dam, are amenable to reoperation;
4. Whether techniques for accommodating changes in the schedule for power generation are feasible.

Thus, there are two converging lines of inquiry, one concerning the physical characteristics of the affected river basin, and one concerning the physical characteristics of the dam, reservoir, and powerhouse.

It is also important to note that REOPS only assesses the physical requisites for successful reoptimization of hydropower dams. Facilities that survive this technical analysis must then also be subjected to an economic feasibility analysis that will weigh the costs and benefits of reoptimization to see where the breakeven point may lie. That will often determine whether, and to what extent, reoperation is economically justified. For instance, in a typical reoptimization case, the array of costs and benefits might include:

Cost to:

1. Increase turbine capacity;
2. Upgrade turbine efficiency;
3. Increase transmission capacity;
4. Increase thermal generation;
5. Remove floodplain levees.

---

[9]Nonetheless, the authors have found that there are often cases where the requisite information is not readily available without interviewing dam operators.
[10]For instance, application of the tool in Africa revealed that there are six individual dams and three complexes of dams that appear to be worthy of further intensive investigation. There are also a number of cases (five altogether) where the tool did not yield a clear verdict. These will require some preliminary operational modeling to resolve.

Benefits:

1. Improved ecosystem function;
2. Improved livelihoods/food production;
3. Increased total power output;
4. Improved interannual reliability;
5. Reduced flood risk;
6. Climate change adaptation.

There may yet be a further sieve before the dam is finally selected for a detailed reoperation study related to the legal, political, and institutional feasibility of reoptimization. In NHI's experience, however, hydropower dams that appear to be good prospects for operational improvements on physical and economic grounds are unlikely to face insurmountable legal or political resistance.

21.5 REOPERATIONS OF IRRIGATION INFRASTRUCTURE FOR ENVIRONMENTAL RESTORATION

The purpose of an irrigation dam is to store water during the seasons of rainfall and snow melt so it can be conveyed to the fields during the hot and dry growing season, and also to make food production more reliable by storing water from wetter years for use during drier years. Irrigation projects may affect river flows downstream of the dam in two ways: by both distorting and depleting the natural flows.

1. Where the irrigation water is conveyed via the downstream river and then diverted to the fields, the flows to the point of diversion will be unnaturally high during the growing season, and unnaturally low during the reservoir refill season. The flows below the point of diversion will be depleted during both the growing season and the refill season.
2. Where the irrigation water is conveyed out of the reservoir by a canal, the flows in the downstream river will be depleted during both the growing season and the refill season. So, to restore more natural flows, it is necessary to counteract both the distortion and depletion effects. In a sense, restoring environmental water regimes amounts to an additional demand on the water supply. The challenge is to do this without diminishing the irrigation benefits. Where can this additional water for the environment be found without building more dams? And how can it be managed to restore a more natural downstream flow pattern?

21.5.1 CONJUNCTIVE WATER MANAGEMENT THROUGH GROUNDWATER BANKING

Stated simply, it is practical to augment water supply in irrigation dams for both irrigation purposes and for release as environmental water by operating the reservoir conjunctively (Box 21.5) with the aquifer underlying the irrigation lands, so that, in combination, they can capture and store that portion of the runoff hydrograph that is currently uncontrolled; that is, the water that is spilled from the reservoir

---

BOX 21.5 BENEFITS OF RESERVOIR REOPERATION IN CONJUNCTION WITH GROUNDWATER BANKING

The concept of reservoir reoperation in conjunction with groundwater banking has been shown to be feasible and effective for application in the Sacramento Valley in California in a joint study by the NHI and the Glenn-Colusa Irrigation District. It is documented in the report "Feasibility investigation of reoperation of Shasta and Oroville Reservoirs in conjunction with Sacrament Valley groundwater systems to augment water supply and environmental flows in the Sacramento and Feather rivers" (NHI and Glenn-Colusa Irrigation District, 2011).

Properly conducted, this type of reservoir reoperation in conjunction with groundwater banking will bring the groundwater system into balance, assure that the full irrigation demands will be met, and increase the amount of water in storage enough to permit environmental water release targets to be met in whole or in part. The environmental water objective is to restore at least some degree of the high flows that naturally occur during the spring rainfall and snowmelt period or during the usual summer monsoon season, and then also restore the low-flow regime during the dry season.

However, irrigation systems demand water in exactly the opposite pattern. They require that water be kept in storage when it is raining, and released during the dry periods. These mismatches can be reconciled by giving the farmers more flexibility in their source of water. An environmental water pulse can be released at just the right time during the wet season and in the right volume without compromising irrigation supplies by storing water that would have otherwise been spilled randomly for flood control purposes. In addition, the reservoir can avoid making large irrigation releases during key portions of the dry season by switching the farmers to the groundwater supplies at that time (and then switching back to surface water before the refill season to lower carryover storage levels). What have been lost are the periodic unmanaged flood flows. What have been gained are managed flows, tailored to achieve targeted ecosystem benefits.

---

during the rainy season (likely in a manner dictated by the flood control operations of the reservoir). More storage space can be created in the reservoir before the onset of the refill season by delivering additional water to farmers who would otherwise pump groundwater during this period. Ideally, the reservoir would be operated to reduce carryover storage to the dead storage level so that the capacity of the reservoir to capture flood flows is maximized. The additional water thus released from the reservoir is delivered to the farmers toward the end of the growing season to substitute for groundwater pumping. During these times, these farmers turn off their wells and take the surface water instead. This reduces the drawdown of the groundwater table. When the rains come, the reservoir will fill up again, but because there is now more storage space, a larger fraction of the flood flows can be captured and the water supply benefits are increased for all purposes, including restoration of downstream environmental water regimes. The rains will also infiltrate and replenish the groundwater table. The combined effect of less groundwater pumping and the annual replenishment will be improved groundwater tables.

Banking can be accomplished either directly, through percolation at spreading basins, or through the substitution of surface water for groundwater that would otherwise be pumped in areas that rely heavily on groundwater pumping, as described above. This is called in lieu groundwater recharge. For several practical reasons, the preferred method of banking groundwater for environmental water, as well as consumptive purposes, is through the in lieu recharge technique. To make this work, it may be necessary to first reengineer the irrigation system so that some or all of the farmers have access to both the surface water delivery system and the groundwater system. This can be done in two ways. If the irrigation system currently only relies on the surface water system, it will be necessary to drill production wells into the aquifer below these fields. If the irrigation system currently relies entirely on groundwater, it will be necessary to expand an adjacent surface water distribution system into the groundwater use area.

As noted, more water in storage due to this conjunctive management technique means that more water can be released from the reservoir for environmental purposes. During the rainy season, the reservoir can meet environmental water regime targets by releasing flows that, together with the uncontrolled runoff from downstream tributaries, will inundate the floodplain in a controlled manner.

Groundwater banking has at least two economic advantages compared to surface water storage: it reduces losses from evaporation, thus allowing for long-term storage, and it is generally less expensive than surface storage. As with all water storage systems, however, the main purpose of groundwater banking is to convert a fluctuating input of water from precipitation and snowmelt, into a steady supply stream that responds to a water demand pattern, different from the input stream. Also, in keeping with other forms of storage, groundwater banking occurs when water is plentiful, and produces stocks to tap when water is scarce.

The potential for groundwater banking is quite large for irrigation systems overlying aquifers with dewatered storage space due to past dependence on groundwater. In contrast, in areas where there is a high degree of interaction between the surface water in the rivers and the groundwater, the water table tends to recover from groundwater extractions quickly. Wet years compensate for depletion that may occur during dry years when heavy pumping occurs. The type of conjunctive management described above utilizes dewater aquifer space that persists from year to year and liberal interactions between rivers and aquifers may preclude using such aquifers as a storage. Additionally, where the geohydrology of the basin remains poorly documented, accounting for the stored water can be a significant problem. It may not be clear whether the groundwater pumper is taking water he/she previously *banked* or that of his/her neighbor.

Yet, this type of groundwater banking, which can help satisfy both consumptive and environmental water needs, has not yet been developed on a significant scale anywhere in the world. This may be because it entails more aggressive operations of the storage and release regime of the reservoirs with the attendant risk that next year's rains will not be sufficient to refill the reservoir. There must be some insurance to buffer that risk. That insurance mechanism is the groundwater bank itself. In years when refill is not sufficient, the groundwater bank pays the deficit by replacing some of the surface irrigation water supply. Ideally, enough irrigation demand can be met with the banked groundwater to allow the available surface water to be used to meet critical environmental water requirements in those dry conditions. As long as the groundwater system is allowed to replenish in wet years as much as it is drawn down in dry years, the groundwater levels will remain in balance over the long run. The limiting factor in the scale of such groundwater banking programs is the rate at which water can be extracted from the aquifer. Well-designed groundwater banking programs will be operated to assure that the potential reservoir storage deficit never exceeds the ability of the bank to repay it.[11]

21.5.2 REOPS TOOL FOR IRRIGATION SYSTEMS

The REOPS schematic can also be used as a tool to rapidly screen irrigation systems for reoperation potential (see NHI, 2009b). It is similar in design to the hydropower REOPS model and, indeed, the first several cells are identical because in both cases it is necessary to first ascertain whether reoperation of the dam will have a beneficial effect on the downstream ecosystems.

---

[11]It should also be noted that the dependence of anadromous fish on cold water releases from reservoirs has forestalled consideration of aggressive reservoir reoperation in areas such as California and the Pacific Northwest of the United States, where protection of these endangered stocks is an operational imperative.

Also like the hydropower model, the irrigation REOPS is designed for application to river basins all over the world by persons with varying levels of training and expertise; requires only information that is fairly readily available from the open literature or from site inspection; and does not require detailed technical analysis. The irrigation REOPS is also a dichotomous key, in which one proceeds through a logic pathway by answering a series of queries in either the affirmative or negative. When application of this screening tool indicates that an irrigation system is technically amenable to beneficial reoperation, an economic feasibility analysis must be applied to see whether the incremental costs are worth the incremental benefits. Some of the relevant factors are as follows:

Incremental costs:

1. Drilling groundwater wells;
2. Power costs for lifting groundwater;
3. Expanding irrigation water conveyance infrastructure;
4. Increased risks of irrigation supply shortfalls (depending on how the system is operated).

Incremental benefits:

1. Increased reliability of water supply for irrigation;
2. Increased water supply for environmental water releases;
3. Reduced distortion of natural hydrograph in the downstream river system;
4. Reduced flood hazards;
5. Resilience to increased incidence of floods and droughts due to climate change.

The schematic and explanatory text for the REOPs tool for irrigation systems is provided in the online resource accompanying this book.

21.6 REOPERATIONS OF FLOOD CONTROL INFRASTRUCTURE FOR ENVIRONMENTAL RESTORATION

Flood control infrastructure typically consists of reservoirs to capture and attenuate peak flow events and downstream levees to wall off the vulnerable floodplain structures and crops from the flood waters. Floods are ecologically important processes, but large floods often result in property damage and human casualties. Thus, reducing the economic and human toll wrought by flooding is a social and political imperative, but traditional flood management approaches are only partially effective and often conflict with ecological and water supply objectives. Flood control dams and levees disconnect rivers from their critical floodplain and the riparian and wetland habitats, which are ecologically vital parts of natural river systems for a host of fish species and birds and amphibians. In addition, to prevent erosion and overtopping of levees, flood management agencies suppress riparian vegetation and resist efforts to restore habitat along flood channels.

The solution to these environmental disadvantages is quite easy to state and exceedingly difficult to accomplish: give rivers more room to meander and convey water, keep development off the floodplain, and let the rivers flood. In sum, allow rivers to act like rivers again, which will be explored further below, but first, some background on flood control operations.

21.6.1 FLOOD CONTROL OPERATIONS

During the rainfall and snow melt seasons, prudent reservoir operators will always maintain sufficient storage space, called the flood reservation, to capture the floods as they come down the river. As it is not possible to forecast the timing or magnitude of big storms very far in advance, quite a lot of empty reservoir storage space must be reserved to reduce the risks of a catastrophic flood. At the same time, the dam operators want to keep the water levels as high as possible for irrigation supplies and power generation in the dry season. The maximum safe storage level will depend on the rate at which stored water can be released from the reservoir when a large flood is coming down the river. Water must be released from storage at rates that will avoid flood damage downstream. The levees wall off the surrounding lands so that larger volumes of the floodwater can be released in this manner. That release rate depends on the capacity of the outlets from the dam, and the magnitude of flood pulses that the downstream floodplain can accommodate without damage to lives or property. So operators can hold more water in storage—for water supply and hydropower—if they can release it faster, and they can release it faster if they can *flood proof* the downstream floodplain.

Although dams and levees make floods less frequent, they can actually make floods even more catastrophic when they do occur. That is because infrastructure for containing the *typical* flood event usually attracts additional settlement near the river bank and behind the levees. Yet, the inducement to build structures or plant permanent crops within the floodplain is based on a false promise that the river will not take them back. The great Mississippi floods of 1993 and 2008 and Hurricane Katrina in 2005 have sadly exposed the vulnerability of overreliance on levees. Rapid urbanization of floodplains combined with global warming promise to make those lessons even more graphic in the future.

Flood risk is calculated as the probability of a given flood magnitude multiplied by its economic and social consequences. Both the dam operations and the levee system are designed to control a specified size of flood risk. The existing flood management framework focuses on reducing the consequences of the high probability flood events. This may be the maximum size of floods that occurs every 50 or 100 years on average. Yet, there will eventually be a flood event that is larger than the system can control, maybe a hundred years from now, but maybe even next year. In 2016, the lower Mississippi basin experienced what is estimated to be a 500-year flood event. Flood control infrastructure cannot eliminate these low probability events with their much higher consequences, as illustrated by Hurricane Katrina. Moreover, building flood control levees in upstream areas to protect agricultural fields may just channel the floodwaters downstream into densely populated areas of the floodplain where the consequences of flooding are far greater. Finally, levees and the development they attract require flood control reservoirs to be operated at lower storage levels than would be necessary if the downstream floodplain were available to store and attenuate flood releases. This reduces potential water supply. These challenges are likely to become more acute as global climate change increases the frequency of extreme weather events and causes snowpack to melt more quickly.

Existing flood control systems are even more hazardous unless they receive constant and expensive maintenance. It only takes one weak spot in a levee to let in a flood that it was supposed to prevent. Engineers know this when they design these structures. It is the government officials who tend to forget; as they did before the Mississippi River flood of 1993 and again in 2008.

Devastating floods also occur in most of the densely populated floodplains of the world such as the Yangtze River in China and in the Limpopo in Mozambique. These catastrophes are a convergence of three factors: inducement of massive habitations below the high water mark in these floodplains, deferred maintenance of the levees, and a flood that exceeded the design specifications of the flood control infrastructure.

21.6.2 OPTIONS FOR REOPERATING FLOOD CONTROL SYSTEMS

To reoperate a flood control system to restore ecological functions in the river, it is usually necessary to first selectively liberate the river from the levee system so that it can interact with its floodplain. This is accomplished by moving the levees back from the river so that a portion of the floodplain now occupied by the farmlands can be hydrologically reconnected, while leaving the levees in place where necessary to protect structures or land uses that cannot tolerate periodic inundation and cannot be moved to higher ground. These measures make it possible to reoperate the reservoir to allow *controlled* flood events to be reintroduced into the downstream system.

Levee setbacks and flood bypasses will not only attenuate flood peaks, reduce velocities, and lower flood stage, they will also provide important habitat for fish, birds, and other wildlife. Use of flood bypasses is neither new nor untested, but simply neglected in the modern era.[12] The Yolo Bypass near Sacramento, California, is a classic example (see Opperman et al., 2009). The bypass was first created to divert floodwaters away from Sacramento, but now also provides tens of thousands of acres of habitat for both fish and birds, including critical rearing habitat for salmon and a suite of endangered fish species.

To reintroduce a *properly* controlled flood regime for ecological restoration, it is necessary to flood proof the downstream floodplain by restricting or relocating structures or permanent crops. That is exceedingly difficult for governments to accomplish, even when they are willing to try. The easiest situation is farmland that grows annual crops. These lands can tolerate and even benefit from floodwaters that replenish the soil moisture and nutrients. The right to temporarily flood private lands can often be purchased. These are called *flood easements*, and they compensate farmers for deferring planting for a brief period to allow the flood events to come in and subside. Synchronizing the growing season with the flood periods is a common practice in many floodplains around the world, including the Senegal River, the Mekong, the Irrawaddy, and parts of the Central Valley of California.

It is more difficult to reintroduce controlled floods into areas of the floodplain that have permanent crops such as orchards, or structures such as buildings and roads. Sometimes it is possible to reopen historic flood depressions to route floodwaters around cities or towns. Sometimes it is feasible to build ring levees around settlements and allow the rest of the landscape to flood. And, sometimes it is feasible to relocate structures or entire communities higher up in the floodplain. Where these techniques are not feasible, the existing land uses may preclude or severely limit the magnitude of restoration floods that can be reintroduced into the system.

It is important to emphasize that these flood-proofing techniques are designed to allow controlled floods to be reintroduced. They do not prevent the rare but catastrophic uncontrolled floods

---

[12]California's first state engineer, William Hammond Hall, advocated such an approach in his seminal report to the state legislature in 1882 as reported in Robert Kelly's book, "Battling the Inland Sea."

that overwhelm the reservoir storage capacity. Yet, by removing at least some of the floodplain development, these techniques do reduce the damage these uncontrolled floods will cause. A new focus on managing risk rather than controlling floods offers great promise to reduce these consequences of conventional flood control and to restore the ecological function of floodplains. Flood-based risk management includes a mix of structural, nonstructural, and operational strategies coordinated across an entire river basin. The goal is to protect densely settled areas while accommodating periodic flooding of undeveloped areas through a coordinated system of flood bypasses, overflow basins, levee setbacks, and reservoir reoperation strategies. These measures can better protect public safety, reduce long-term maintenance costs, and provide significant ecological and water supply benefits. Achieving these goals, however, requires integrated and basin-wide planning involving many agencies at various levels of government, which is often more difficult to implement than conventional approaches.

In sum, to foster a new era in flood management requires policies to:

1. Reduce rather than exacerbate the conflicts between private property and river dynamics and the resulting damages and risks to life;
2. Accommodate rather than control the smaller and more frequent flood events that can reconnect the floodplains to their rivers;
3. Invest in more sustainable flood management alternatives such as flood bypasses, flood easements, use of natural depressions for flood retention, levee setbacks, and reservoir reoperation to accommodate the largest flood events;
4. Limit future development in flood-prone areas.

Conventional flood control tries to keep the floods away from the people; the next era of flood management must try to keep the people away from the floods.

21.7 CONCLUSION

The pace of river development is moving faster than the tools at our disposal for ameliorating its risks to natural riverine processes. More disquieting, the pace is sure to accelerate with growing aspirations for carbon-free power and irrigated food production. Moreover, these challenges are occurring disproportionately in the countries least equipped to identify, assess, and choose the least impactful options.

At this juncture, the knowledge base for estimating environmental water requirements is reasonably mature, albeit dependent on field data that are often sparse and incomplete. These flow requirements can be regarded as the environmental water *demands*. In contrast, the work to generate the environmental water *supplies* is in its infancy. This is the work to advance understanding of how to site, design, and operate river basin infrastructure—or, where it is already in place, to reoperate it—to maintain natural riverine processes.

To move from theory and analysis to practice and achievement, the world is hungry for case examples. Nothing will speak as eloquently as reoperation innovations that have proven themselves on the ground, or more precisely, in the river. This is where the development assistance agencies—bilateral and intergovernmental—could have the greatest impact on the future course of river basin development: by supporting *learning laboratories* for river management innovations. Environmental and social safeguards policies emanating from these agencies are to be applauded, but are frequently irrelevant as private sources of investment have come to dominate the field. Sustainable development principles and protocols such as those developed by the International Hydropower Association, likewise are a positive influence (and grist for no end of international conference mills), but again, do not have strong leverage.

This chapter is intended as a checklist of promising techniques to move from theory to application. The clock that marks aquatic species extinctions is ticking loudly in the background.

ACKNOWLEDGMENTS

The concepts and assessments in this chapter owe much to the inspirational work of former colleagues at the NHI, including David Purkey and John Cain, and to the heroic work of current affiliates and colleagues in NHI's Global Dam Reoptimization Initiative projects, including Dr. George Annandale, Professor G. Mathias Kondolf, and Dr. Martin Mallen-Cooper. The chapter has also benefitted immeasurably from the insights provided by Dr. Jeff Opperman at The Nature Conservancy and the figures and illustrations provided by the NHI Director, Dr. Daniel Peter Loucks, Cornell University Professor Emeritus. In addition, the author wishes to acknowledge the indispensable contribution of Jessica Peyla Nagtalon at NHI to the formatting and editing of this chapter. Her deft and discerning touch permeates both the text and figures.

In the brave new paradigm of environmentally conscious river basin development described in this chapter, we are all learning together. Nothing will speak more eloquently than demonstrated successes in the application of these techniques on the ground, or more precisely, in the river. The challenge of converting theory into practice lies ahead. When this chapter is updated 10 years from now—as it must be—my hope is that it will abound with case examples that can serve as the guideposts leading to a world where rivers functions like rivers again, hard though they may work in the service of mankind.

REFERENCES

Acreman, M.C., 1996. Environmental effects of hydro-electric power generation in Africa and the potential for artificial floods. Water Environ. J. 10 (6), 429–434.

Acreman, M.C., Booker, D.J., Riddington, R., 2003. Hydrological impacts of floodplain restoration: a case study of the River Cherwell, UK. Hydrol. Earth Syst. Sci. 7 (1), 75–86.

Acreman, M., Arthington, A.H., Colloff, M.J., Couch, C., Crossman, N.D., Dyer, F., et al., 2014. Environmental flows for natural, hybrid, and novel riverine ecosystems in a changing world. Front. Ecol. Environ. 12, 466–473.

Alfredsen, K., Harby, A., Linnansaari, T., Ugedal, O., 2011. Development of an inflow-controlled environmental flow regime for a Norwegian River. River Res. Appl. 28, 731–739. Available from: https://www.researchgate.net/publication/230504492_Development_of_an_inflow-controlled_environmental_flow_regime_for_a_Norwegian_river (accessed 07.07.16).

Allen, D.J., Smith, K.G., Darwall, W.R.T., (Compilers), 2012. The Status and Distribution of Freshwater Biodiversity in Indo-Burma. International Union for Conservation of Nature, Cambridge and Gland, Switzerland. Available from: https://portals.iucn.org/library/sites/library/files/documents/RL-2012-001.pdf (accessed 16.09.15).

Allen, J., 2003. Drought Lowers Lake Mead. NASA Earth Observatory, 13 November 2013. Available from: http://earthobservatory.nasa.gov/Features/LakeMead/ (accessed 19.09.15).

American Rivers and Trout Unlimited, 2002. Exploring Dam Removal: A Decision Making Guide. Available from: http://www.americanrivers.org/assets/pdfs/dam-removal-docs/Exploring_Dam_Removal-A_Decision-Making_Guide6fdc.pdf?1ef746 (accessed 19.09.15).

Annandale, G.W., 2011. Going Full Circle. International Water Power and Dam Construction, April 2011 30–34.

Annandale, G.W., 2013. Quenching the Thirst: Sustainable Water Supply and Climate Change. CreateSpace, North Charleston, SC.

Baran, E., Larinier, M., Ziv, G., Marmulla, G., 2011. Review of the fish and fisheries aspects in the feasibility study and the environmental impact assessment of the proposed Xayaburi Dam on the Mekong mainstem. Report prepared for the WWF Greater Mekong. Gland, Switzerland: World Wildlife Fund.

Baran, E., Borin, U., 2012. The importance of the fish resource in the Mekong River and examples of best practice. In: Gough, P., Philipsen, P., Schollema, P.P., Wanningen, H. (Eds.), From Sea to Source; International Guidance for the Restoration of Fish Migration Highways, pp. 136–141.

Bayley, P.B., 1991. The flood pulse advantage and the restoration of river-floodplain systems. Regul. Rivers Res. Manage. 6, 75–86.

Cain, J., Monohan, C., 2008. Estimating Ecologically Based Flow Targets for the Sacramento and Feather Rivers. Natural Heritage Institute.

California Public Utilities Commission, 2012. Decision granting PacifiCorp's request to modify decision 11-05-002 in order to revise the Klamath Surcharge Rate and period over which such surcharge is collected, decision 12-10-028 (October 25, 2012). In the matter of the application of PacifiCorp (U901E), an Oregon Company, for an order authorizing a rate increase effective January 1, 2011 and granting conditional authorization to transfer assets, pursuant to the Klamath Hydroelectric Settlement Agreement, application 10-03-015 (filed 18 March 2010).

CRED, 2015. Centre for Research on the Epidemiology of Disasters. Available from: http://emdat.be/human_cost_natdis (accessed 19.09.15).

Darwall, R.T., Smith, K.G., Allen, J., Seddon, M.B., Reid, G.M., Clausnitzer, V., et al. (Eds.), 2009. Wildlife in a Changing World: an Analysis of the 2008 IUCN Red List of Threatened Species. International Union for Conservation of Nature, Gland, Switzerland, pp. 43–53.

Davis, R., Hirji, R., 2003. Environmental flows: flood flows. Water Resources and Environment Technical Note C.3. The World Bank, Washington, DC.

Dudgeon, D., Arthington, A.H., Gessner, M.O., Kawabata, Z., Knowler, D., Lévêque, C., et al., 2006. Freshwater biodiversity: importance, threats, status and conservation challenges. Biol. Rev. 81, 163–182.

Hurwitz, Z., 2014. Dam standards: a rights-based approach: a guidebook for civil society. International Rivers. Available from: http://www.internationalrivers.org/files/attached-files/intlrivers_dam_standards_final.pdf (accessed 03.13.16).

International Energy Agency, 2016. Renewables—hydropower webpage. Available from: https://www.iea.org/topics/renewables/subtopics/hydropower (accessed 08.25.16).

Klamath Hydroelectric Settlement Agreement, 2010. Available from: https://klamathrestoration.gov/sites/klamathrestoration.gov/files/Klamath-Agreements/Klamath-Hydroelectric-Settlement-Agreement-2-18-10signed.pdf (accessed 07.05.2016).

Kantoush, S.A., Sumi, T., 2010. River morphology and sediment management strategies for sustainable reservoir in Japan and European Alps, Annuals of Disaster Prevention Research Institute, Kyoto University, No. 53B.

Kondolf, G.M., Gao, Y., Annandale, G.W., Morris, G.L., Jiang, E., Zhang, J., et al., 2014. Sustainable sediment management in reservoirs and regulated rivers: experiences from five continents. Earth's Future 2 (5), http://dx.doi.org/10.1002/2013EF000184. Available from: http://onlinelibrary.wiley.com/doi/10.1002/2013EF000184/full (accessed 01.10.15).

Ledec, G., Quintero, J.D., 2003. Good dams and bad dams: environmental criteria for site selection of hydroelectric projects. Latin America and the Caribbean Region Sustainable Development Working Paper No. 16. The World Bank, Washington, DC. Available from: http://siteresources.worldbank.org/LACEXT/Resources/25853-1123250606139/Good_and_Bad_Dams_WP16.pdf (accessed 30.06.16).

Loh, J., Green, R.H., Ricketts, T., Lamoreux, J., Jenkins, M., Kapos, V., et al., 2005. The Living Planet Index: using species population time series to track trends in biodiversity. Philos. Trans. R. Soc. B 260, 289–295.

Loucks, D.P., van Beek, E., Stedinger, J.R., Dijkman, J., Villars, N., 2005. Water Resource Systems Planning and Management. UNESCO Press, Paris, France. Available from: https://ecommons.cornell.edu/handle/1813/2804 (accessed 23.08.16).

McCully, P., 2015. Dam Decommissioning. International Rivers. Available from: http://www.internationalrivers.org/dam-decommissioning (accessed 19.09.15).

MWH, 2009. Technical analysis of pumped storage integration with wind power in the Pacific Northwest. Final Report prepared for US Army Corps of Engineers NW Division, Hydroelectric Design Center. Available from: http://www.hydro.org/wp-content/uploads/2011/07/PS-Wind-Integration-Final-Report-without-Exhibits-MWH-3.pdf (accessed on 01.10.15).

Morris, G.L., Fan, J., 1998. Reservoir Sedimentation Handbook: Design and Management of Dams, Reservoirs and Watersheds for Sustainable Use. McGraw-Hill Book Co, New York.

National Research Council, 2011. Missouri River Planning: Recognizing and Incorporating Sediment Management. The National Academies Press, Washington, DC.

NHI, 2009a. Explanation and guidance in the use of the rapid evaluation tool for screening the potential for reoptimizing hydropower dams. Adapted from Thomas, G., DiFrancesco, K., 2009. Rapid evaluation of the potential for reoptimizing hydropower systems in Africa. Natural Heritage Institute. Final Report to The World Bank, Washington, DC. Available from: http://www.n-h-i.org/programs/redesign-and-reoperation-for-major-river-basin-infrastructure.html.

NHI, 2009b. Explanation and guidance in the use of the rapid evaluation tool for screening the potential for reoptimizing irrigation systems. Adapted from Thomas, G., DiFrancesco, K., 2009. Rapid evaluation of the potential for reoptimizing hydropower systems in Africa. Natural Heritage Institute. Final Report to The World Bank, Washington, DC. Available from: http://www.n-h-i.org/programs/redesign-and-reoperation-for-major-river-basin-infrastructure.html.

NHI and Glen-Colusa Irrigation District, 2011. Feasibility investigation of re-operation of Shasta and Oroville Reservoirs in conjunction with Sacramento Valley groundwater systems to augment water supply and environmental flows in the Sacramento and Feather rivers. Northern Sacramento Valley Conjunctive Water Management Investigation. Natural Heritage Institute. Funded by California Department of Water Resources and Bureau of Reclamation. Available from: http://www.n-h-i.org/uploads/tx_rtgfiles/NSVCWMP_Report_Final.pdf.

NHA Pumped Storage Development Council, 2012. Challenges and opportunities for new pumped storage development. White Paper. National Hydropower Association. Available from: http://www.hydro.org/wp-content/uploads/2012/07/NHA_PumpedStorage_071212b1.pdf (accessed 01.10.15).

Opperman, J.J., Galloway, G.E., Duvail, S., 2013. The multiple benefits of river-floodplain connectivity for people and biodiversity. In: second ed. Levin, S. (Ed.), Encyclopedia of Biodiversity, vol. 7. Academic Press, Waltham, MA, pp. 144–160.

Opperman, J.J., Galloway, G.E., Fargione, J., Mount, J.F., Richter, B.D., Secchi, S., 2009. Sustainable floodplains through large-scale reconnection to rivers. Science 326, 1487–1488.

Opperman, J.J., Grill, G., Hartmann, J., 2015. The power of rivers: finding balance between energy and conservation in hydropower development. The Nature Conservancy, Washington, DC. Available from: http://www.nature.org/media/freshwater/power-of-rivers-report.pdf (accessed on 30.06.16).

Oregon Public Utilities Commission, 2010. Application to implement the provisions of Senate Bill 76. Order No. 10-364. Entered 9.16.2010. Available from: http://apps.puc.state.or.us/orders/2010ords/10-364.pdf (accessed 07.05.16).

Palmieri, A., Shah, F., Annandale, G.W., Dinar, A., 2003. Reservoir Conservation: the RESCON Approach. World Bank, Washington, DC.

Planet Experts, 2014. Restoring River Ecosystems by Dismantling America's Dams. August 4, 2014. Available from: http://www.planetexperts.com/restoring-river-ecosystems-dismantling-americas-dams (accessed 19.09.15).

Purkey, T., Thomas, G., Fullerton, D., Moench, M., Axelrod, L., 1998. Feasibility Study of a Maximal Program of Groundwater Banking. Natural Heritage Institute, San Francisco, CA.

Räsänen, T.A., Koponen, J., Lauri, H., Kummu, M., 2012. Downstream hydrological impacts of hydropower development in the Upper Mekong Basin. Water Resour. Manage. 26, 3495—3515. In: International River, 2014. Environmental and social impacts of Lancang Dams. Research Brief. Available from: http://www.internationalrivers.org/files/attached-files/ir_lancang_dams_researchbrief_final.pdf (accessed 26.09.15).

REOPS Explanation & Guidance Note for Irrigation Systems, 2015. Available from: http://n-h-i.org/wp-content/uploads/2017/02/REOPS-Explanation-Guidance-Note-for-Irrigation-Systems_Sept.-2015.pdf.

REOPS schematic for Hydropower Systems, 2009. Available from: http://n-h-i.org/wp-content/uploads/2017/02/REOPS-schematic-for-Hydropower-Systems.pdf.

Scheumann, W., Hensengerth, O., 2014. Evolution of Dam Policies: Evidence from the Big Hydropower States. Springer Science & Business Media, Berlin, Heidelberg.

Scudder, T. (Ed.), 2005. The Future of Large Dams: Dealing with Social, Environmental, Institutional and Political Costs. Earthscan, London.

Strayer, D.L., Dudgeon, D., 2010. Freshwater biodiversity conservation: recent progress and future challenges. J. N. Am. Benthol. Soc 29, 344—358.

Sumi, T., Kantouch, S.A., Suzuki, S., 2012. Performance of Miwa Dam sediment bypass tunnel: evaluation of upstream and downstream state and bypassing efficiency. ICOLD, 24th Congress, Kyoto, Q92-R38, pp. 576—596.

The World Bank Group, 2015a. Energy access. Progress toward sustainable energy: Global Tracking Framework 2015. Available from: http://trackingenergy4all.worldbank.org/energy-access (accessed 09.19.15).

The World Bank Group, 2015b. Access to electricity (% of population). Sustainable energy for all (SE4ALL) database from World Bank, Global Electrification database. Available from: http://data.worldbank.org/indicator/EG.ELC.ACCS.ZS (accessed 09.19.15).

Thomas, G.A., 1996. Conserving aquatic biodiversity: a critical comparison of legal tools for augmenting streamflows in California. Stanford Environ. Law J. 15 (1).

Thorncraft, G., Harris, J.H., 2000. Fish Passage and Fishways in New South Wales: A Status Report. Technical Report 1/2000. Cooperative Research Centre for Freshway Ecology, Australia. Available from: http://citeseerx.ist.psu.edu/viewdoc/download?doi = 10.1.1.522.9318&rep = rep1&type = pdf (accessed 08.23.16).

UNEP, 2008. Vital Water Graphics—An Overview of the State of the World's Fresh and Marine Waters, second ed. United Nations Environment Programme, Nairobi, Kenya.

Wang, Z., Hu, C., 2009. Strategies for managing reservoir sedimentation. Int. J. Sediment Res. 24 (4), 369—384.

Ward, J.V., 1998. Riverine landscapes: biodiversity patterns, disturbance regimes, and aquatic conservation. Biol. Conserv. 83 (3), 269—278.

Welcomme, R.L., 1979. Fisheries Ecology of Floodplain Rivers. Longman Group Ltd, London, UK.

WCD, 2000. Dams and development: a new framework for decision-making. World Commission on Dams. Earthscan, London. Available from: http://www.unep.org/dams/WCD/report/WCD_DAMS%20report.pdf (accessed 04.03.16).

Zoological Society of London and WWF, 2014. Living planet report. Available from: http://www.livingplanetindex.org/projects (accessed 16.09.15).

ENVIRONMENTAL WATER AND INTEGRATED CATCHMENT MANAGEMENT

22

Michael J. Stewardson[1], Wenxiu Shang[2], Giri R. Kattel[1,3], and J. Angus Webb[1]

[1]The University of Melbourne, Parkville, VIC, Australia [2]Tsinghua University, Beijing, China
[3]Nanjing Institute of Geography and Limnology Chinese Academy of Sciences (NIGLAS), Nanjing, China

22.1 INTRODUCTION

Integrated catchment management (ICM) is an approach to sustainable land and water management, recognizing flow-mediated connections through catchments and the need for interdisciplinary and community-based collaboration (Commonwealth of Australia, 2000; Falkenmark, 2004; Jakeman and Letcher, 2003; Kattel et al., 2016). Originating in Australia, South Africa, and England over the last 30 years, ICM is an alternative to the traditional program-based approach. The key features of the ICM approach are that it: (1) seeks to find the proper balance between humans and the impacts that their activities cause to ecosystems; (2) does not consider land, water, and biodiversity management as separate activities; (3) facilitates dialogue between scientists, stakeholders, and policy makers; (4) supports coordinated actions across levels of government and nongovernment organizations; and (5) regards the catchment as the critical landscape unit across which coordination is required.

Although this book is concerned primarily with environmental water management, it is important to understand that hydrological alteration is not the only stress on freshwater ecosystems. Other catchment disturbances, including the expansion and intensification of agriculture, urbanization, and climate change, have added additional stress to freshwater ecosystems via altered sediment supply, degraded water quality, channelization of rivers, removal of riparian vegetation, and the introduction of exotic species (Nilsson and Renofalt, 2008), and the infrastructures constructed in river systems to impound water for human use often cause nonflow-related problems, including acting as a barrier to fish movement and transport of materials, and fragmenting river networks (Nilsson and Renofalt, 2008). In this chapter, we argue that environmental water management will be more successful if planned within a broader program of ICM. In particular, we emphasize the need to consider nonflow stressors and fragmentation of the freshwater ecosystems that can act to undermine the benefits of environmental water delivery.

The ICM approach is specifically intended to overcome the limitations of narrowly focused river management programs. It acknowledges the complex basin-wide interactions between water resource and other stressors on river ecosystems as well as social, economic, and political concerns (Jakeman and Letcher, 2003). ICM policies are widely reported as sustainable, with increased levels of integration between groups of natural and social scientists, land and water users, land and water managers,

Water for the Environment. DOI: http://dx.doi.org/10.1016/B978-0-12-803907-6.00022-X

and planners and policy makers across spatial scales (Macleod et al., 2007). Although ICM policies often produce highly positive outcomes in management, the success of ICM in practice is still questioned (Jakeman and Letcher, 2003) and results can be disappointing if, for example, there is poor integration between conflicting and competing policy measures (Macleod et al., 2007).

Despite the wide promotion of an ICM approach, it is still common for river basin management programs to focus on single stressors or interventions. For example, Australia's Murray–Darling Basin Plan is primarily a water resource plan, although it does provide water quality targets (Hart, 2015). Developed by the *Murray–Darling Basin Authority (MDBA)*, an agency within Australia's federal government, the Basin Plan specifies the Sustainable Diversion Limits for the basin's water resources and thereby protects environmental water requirements. However, the MDBA does not have authority over other important stressors affecting the Murray–Darling Basin's river ecosystems because these are mostly state government responsibilities. This separation of land, biodiversity, and water management functions across different government agencies (and levels of government in some cases) is widespread. Without catchment-level coordination across these agencies, there is little capacity for environmental water planning and management agencies to consider complementary actions such as riparian restoration, management of pest species, or removal of barriers to fish movement. Weak integration of environmental water management with management programs to address other nonflow stressors is a serious impediment to successful restoration of river basins worldwide.

This chapter makes the case for considering environmental water management within a broader ICM approach, providing the necessary starting point for any environmental water manager interested in implementing ICM. It focuses on two ubiquitous features of river ecosystems in disturbed catchments that are best addressed with an ICM approach: the co-occurrence of multiple flow and nonflow stressors; and strong flow-mediated environmental interactions throughout a river network. The chapter begins by demonstrating the many interactions between multiple stressors affecting river ecosystems, including hydrological alteration (Section 22.2). This includes a review of studies evaluating the ecological outcomes of environmental water delivery. In particular, it examines the influence of nonflow stressors on the responses to environmental water. Despite the relatively small sample of studies published in the literature, there is remarkably consistent evidence that nonflow stressors can seriously undermine the success of environmental water delivery (Section 22.2.1). In Section 22.2.2, we discuss methods for diagnosing the important stressors acting on a freshwater ecosystem. This is a possible starting point for evaluating the need for integrated planning of environmental water with other catchment interventions to alleviate all stressors. Section 22.3 describes the many flow-mediated connections throughout a river network that can only be considered through planning at a catchment scale. Finally, Section 22.4 provides a case study and some principles concerning the implementation of ICM.

Clearly, implementing ICM is not easy because, among other challenges, it requires the linking of traditionally isolated areas of *Natural Resources Management (NRM)* including land, water, biodiversity, water quality, environment, agriculture, and forestry. It must also include consideration of governance, institutions, modeling tools, stakeholder engagement, monitoring, science, adaptive management, and many of the other elements discussed in this book, but applied beyond the specific context of environmental water management. This chapter does not attempt to provide a guide for how to do ICM because this is a big topic covered comprehensively elsewhere (e.g., Smith et al., 2015). Instead, we provide the case to encourage management of environmental water as one element of a broader NRM challenge within an ICM framework.

22.2 MULTIPLE STRESSORS

22.2.1 IMPACT OF NONFLOW STRESSORS ON THE EFFECTIVENESS OF ENVIRONMENTAL WATER DELIVERY

Hydrological alteration rarely occurs in isolation from other stressors for two important reasons. First, the human activities that create a demand for water supply (e.g., irrigation), hydropower (e.g., human settlement), or flood mitigation (e.g., floodplain agriculture), all of which are the primary drivers of hydrological alteration (see Chapter 3), will also inevitably introduce nonflow stressors into aquatic environments (e.g., waterborne contaminants, removal of native riparian vegetation and changes to channel morphology). Urbanization is an extreme example producing severe hydrological alteration in combination with disturbances to almost every other aspect of a river catchment and the riparian corridor (Walsh et al., 2005).

Second, there are complex sequences of disturbances where water resource development leads to changes in river connectivity, water quality, sediment regime, riparian vegetation, channel morphology, and the spread of exotic species. All of these act as additional stressors on aquatic ecosystems, interacting with the direct effects of hydrological alteration. Table 22.1 illustrates some of the many causal interactions that can occur between these stressors, where a response in one attribute becomes a stressor on another aspect of the aquatic ecosystem. This table demonstrates the many ways in which individual catchment disturbances can lead to a sequence of other catchment disturbances, which all act as stressors on freshwater ecosystems.

Understanding and addressing the widespread co-occurrence of multiple stressors on river ecosystems is critical to successful environmental water management. Importantly, there is a growing literature reporting on outcomes from environmental water delivery and nonflow stressors are often implicated as the cause of environmental water projects falling short of expected restoration targets (discussed below). For example, experimental environmental water projects have failed as a result of poor longitudinal hydrological connectivity inhibiting dispersal for recolonization by fish (Bradford et al, 2011; Decker et al., 2008; Rolls and Wilson, 2010; Rolls et al., 2012) and invertebrates (Brooks et al., 2011; Decker et al., 2008; Mackie et al., 2013). Lateral connectivity can also be a critical limiting factor for native fish populations in the case of wetland flooding events (Vilizzi et al., 2013). The presence of exotic fish species is a common problem that compromises the effectiveness of environmental water pulses designed to restore native fish communities (e.g., Bice and Zampatti, 2011). Even if nonnative species are initially reduced, reinvasion of exotic species can occur if there is incomplete elimination of the species (Cross et al., 2011; Korman et al., 2011; Valdez et al., 2001). Environmental water released from water at depth in a reservoir can have low dissolved oxygen and water temperatures, compromising any intended benefits from the release (Bednarek and Hart, 2005; Bradford et al., 2011; Rolls and Wilson, 2010). Other sources of poor water quality can also compromise environmental watering objectives such as the effect of elevated soil salinities on germination success of riparian vegetation (Raulings et al., 2011). An impoverished soil seedbank can limit the possibilities for germination of riverbank vegetation when water regimes are restored both for aquatic species following multiple years of dry conditions (Siebentritt et al., 2004), and for flood-intolerant species subject to extended periods of inundation prior to a drawdown event (Raulings et al., 2011).

Table 22.1 This Table Gives a Few Representative Examples of the Many Interactions between Stressors in Rivers. The Existence of Interacting Stressors Is Likely a Common Feature of All Modified Rivers, but That These Specific Interactions and Their Details Are Specific to the Rivers Being Studied

Stressor	Connectivity	Water Regime	Water Quality	Sediment Regime	Riparian Vegetation	Channel Morphology	Exotic Species
			Response				
Dams (papers show that dams' influence on vegetation and morphology is mainly by changing flow and sediment)	Dams hinder the migration of biota (Gehrke and Harris, 2000)	Operation of a dam reduced flow and eliminated flow pulses and floods (Ortlepp and Mürle, 2003)	Trapping of silicate in reservoir leads to a reduction in downstream concentrations (Humborg et al., 1997)	A sediment trapping efficiency of 99% is estimated for a dam (Yang et al., 2006)			
Hydrological alteration	Loss of floods reduces exchange of carbon between the river channel and floodplain (Sam et al., 2000)		High salinity levels occurred in years when dam releases were low (Lind et al., 2007)	An artificial flood increased gravel transportation (Ortlepp and Mürle, 2003)	Reduced cottonwood abundance and an absence of young trees occurred in the riparian ecosystem was caused by river damming and water diversion (Rood et al., 2003)	An artificial flood changed the morphology of the river bed (Ortlepp and Mürle, 2003)	Summer-spawning nonnative fishes in the San Juan River, US, responded positively to extended low summer flows (Propst and Gido, 2004)
Degraded water quality				Clay and silt particles that are efficiently transported in suspension in river flow may be deposited	Salinity increased suddenly and considerably during wetland water level drawdowns limited the success		Exotic fishes almost disappeared in observation sites when cold

Altered sediment regime	Velocity profile of flow is influenced by suspended sediment concentration (Coleman, 1981)	Flushing of fine sediment might account for observed decrease in particulate organic matter and Particulate Phosphorous after the artificial floods (Robinson and Uehlinger, 2008)	when chemical environment changes (Einstein and Krone, 1962)	in restoring riparian vegetation (Raulings et al., 2011)	Increased sediment inputs and decreased sediment transport are thought to be the reason for the reduction of bedrock habitat (McLoughlin et al., 2011)	hypolimnetic high-flow pulse was released (King et al., 1998)
Degraded riparian vegetation	Runoff is released faster to the channel where riparian vegetation is absent (Schlosser and Karr, 1981)	Lots of research has shown riparian vegetation could remove chemical load from water (Daniels and Gilliam, 1996)	Riparian vegetation could remove sediment from water effectively (Daniels and Gilliam, 1996)	Riparian vegetation is affected by the deposition of alluvial soil (Richardson et al., 2007)	Newly formed native riparian vegetation produced changes to the channel and floodplain morphology (Rood et al., 2003)	Plant invasion may manifest as a symptom of the degradation of local riparian vegetation (Richardson et al., 2007)
Channel engineering (indirect influences on most factors. Channelization	The central canal excavated in the Kissimmee River cut	Water velocities over a range of discharge conditions were lower in	Channelization is believed to have facilitated nutrient transport in the Kissimmee	Channelization in the Kissimmee River caused absence of flow in	Channelization in the Kissimmee River resulted in drier conditions over much of the floodplain, making	

(Continued)

Table 22.1 This Table Gives a Few Representative Examples of the Many Interactions between Stressors in Rivers. The Existence of Interacting Stressors Is Likely a Common Feature of All Modified Rivers, but That These Specific Interactions and Their Details Are Specific to the Rivers Being Studied *Continued*

Stressor	Response						
	Connectivity	Water Regime	Water Quality	Sediment Regime	Riparian Vegetation	Channel Morphology	Exotic Species
usually changed other factors by changing flow	through the naturally meandering channel, leaving the remnant channels without flow and floodplain without inundation (South Florida Water Management District, 2006)	the naturalized segment than in the channelized segment (Bukaveckas, 2007)	River (South Florida Water Management District, 2006)	remnant river channels, making organic matter accumulate on the channel bottom and ending active point bar formation (South Florida Water Management District, 2006)	former wetland plant communities convert to upland communities (South Florida Water Management District, 2006)		
Exotic species		Increased biomass of exotic species in the riparian zone resulted in increased water use, which in turn altered the hydrology (Richardson et al., 2007)		In South Africa, thickets of the alien shrub *Sesbania punicea* trapped sediments, increasing habitats for exotic species (Richardson et al., 2007)	The invasion of an aggressive exotic plant caused degradation of the riparian zone vegetation (Rood et al., 2003)	Alien *Tamarix* trees reduced the width of river channels in large areas of the arid western US (Richardson et al., 2007)	

These recent experimental environmental water studies demonstrate the need to consider environmental water delivery within an ICM approach in order to coordinate management of nonflow and flow stressors. Experience suggests that failure to consider nonflow stressors can seriously undermine the effectiveness of environmental watering. Indeed, these stressors may be the critical limiting factor and should be tackled prior to or concurrent with environmental water delivery. Importantly, nonflow stressors (including cases where these are secondary stressors produced by water resource development) are often not mitigated by the provision of environmental water alone.

Failure to tackle important nonflow stressors will not only lead to poorly performing environmental water programs; there is also a risk that poor outcomes will reduce societal support for environmental watering programs, including in other rivers and possibly even other countries. This is a serious concern as we enter a phase where there is a growing number of experimental environmental water projects worldwide reporting on project outcomes.

22.2.2 DIAGNOSING CRITICAL STRESSORS TO TARGET IN CATCHMENT MANAGEMENT

Successful ICM requires an ability to determine the primary stressors affecting environmental function within a catchment. However, diagnosing the important stressors in any natural or impacted environment is difficult. For the same reasons that make the detection of benefits of environmental water releases difficult (see Chapter 25), it is often not possible to identify stressors of primary importance in catchments based purely upon empirical data collection in the field. Large-scale data syntheses in the form of mensurative experiments may be able to detect large-scale patterns (e.g., Konrad et al., 2011), but it is often not possible to draw a coherent signal from a collection of studies undertaken by different researchers (Poff and Zimmerman, 2010).

Environmental scientists have drawn from principles of epidemiology to test hypotheses and diagnose causes and effects in uncontrolled environmental systems (Adams, 2003; Cormier et al., 2010). Epidemiology faces many of the same challenges as environmental science in identifying primary stressors and causal relationships, and has developed several ways of approaching these challenges. Of primary interest for ICM is the concept of the *causal pie* (Rothman, 2002), where each slice of the causal pie is a candidate cause for an observed effect (e.g., altered flow, reduced water quality, barriers to dispersal are all *sufficient* causes of reduced fish abundance). This model also acknowledges that potential causes do not operate in isolation, but focuses attention on identifying (or diagnosing) the stressor/s responsible for most of the observed effect.

If sufficient data are available, computationally intensive statistical methods can be used to identify the most important stressor in a system. For example, hierarchical partitioning (MacNally, 2000) is a multiple model-based approach that tests many combinations of potential causal variables to identify those variables that explain the most variation in the endpoint (e.g., fish abundance) both individually (the main effect of the stressor) and interactively (two or more stressors acting in concert). Similarly, other data-mining-based approaches such as Boosted Regression Trees (Elith et al., 2008) can also examine patterns in large data sets to identify the most important stressors in a system from a list of candidate causes.

However, these data-mining methods require extensive data sets, which are often not available. Other methods have been developed to take advantage of different sources of information. The US

Environmental Protection Agency's CADDIS method (Norton et al., 2008) is designed specifically to diagnose the most likely cause of environmental impairment in aquatic ecosystems. It uses a combination of local field data, data from laboratory experiments, and evidence from the literature to reach conclusions. The method is most often employed in a regulatory context for identifying infringements of environmental protection laws, but it could also be employed for catchment management planning.

There are also purely literature-based methods that could be turned to the issue of diagnosing the most likely cause of environmental impairments. A systematic literature review (Khan et al., 2003) is an analytical approach to testing hypotheses across sets of studies (i.e., papers) that is prevalent in several disciplines, most notably the health sciences. This approach is a fundamental component of *evidence-based medicine* that has seen an *effectiveness revolution* in patient management since the early 1970s (Stevens and Milne, 1997). A systematic review uses objective, transparent, and repeatable search methods to locate evidence regarding one or more specific hypotheses to be tested; then employs statistical or qualitative methods to combine this evidence to reach a conclusion on the hypothesis/es. Qualities of objectivity, transparency, and repeatability, along with the use of an explicit method for combining evidence, distinguish systematic reviews from the *narrative* reviews that dominate environmental science and management (Khan et al., 2003).

Methods of systematic literature review have been adapted from the medical model for use in environmental sciences (CEBC, 2010), and an increasing number of systematic reviews are available through the website of the Collaboration for Environmental Evidence (http://www.environmentalevidence.org). However, the time and expense involved in undertaking a full systematic review has meant that these methods have not been widely adopted in environmental management. As a result of this, recent years have seen an increasing interest in *rapid review* methods that have the same philosophy as systematic reviews, but are less costly and can provide an answer in the time frames usually required by environmental management (i.e., weeks rather than months or years). The Collaboration for Environmental Evidence has recently established a rapid reviews working group, and we can expect to see more of these approaches appearing in the literature.

One such method is Eco Evidence (Norris et al., 2012; Webb et al., 2015). This approach uses an eight-step method that is consistent with the aims and philosophy of systematic review for hypothesis setting, evidence gathering and synthesis, and reporting on conclusions. Several Eco Evidence reviews have tested the effects of changes in flow regimes on different ecological endpoints (Greet et al., 2011; Grove et al., 2012; Miller et al., 2013; Webb et al., 2012a, 2012b). In particular, one example (Webb et al., 2013) was able to draw much stronger conclusions from an extensive set of flow-response studies than the original review that used ad hoc methods for evidence synthesis (Poff and Zimmerman, 2010). Unlike CADDIS, Eco Evidence and *traditional* systematic review methods are designed to test specific hypotheses, rather than to identify a primary stressor per se. However, a literature review could be applied to a series of hypotheses, one for each of the candidate causes, and thereby identify the most likely cause.

Individually or collectively, the methods outlined above provide a toolkit of approaches that could be used to identify dominant causes of environmental degradation in river ecosystems as part of an ICM-based approach. Identification of the stressor/s of importance provides an immediate pathway toward intervention through management practices and complementary measures, allowing managers to focus limited resources where they will be of most use. Such an approach could see complementary works (e.g., habitat restoration, water quality improvement) implemented alongside environmental water deliveries, with the outcome being a greater environmental improvement than would be achieved by considering issues separately.

22.3 ENVIRONMENTAL INTERACTIONS ACROSS CATCHMENT

A major advantage of ICM is that it can account for critical hydrologically induced spatial connections within river networks. These connections act at multiple scales mediating the movement of water, sediment, energy, oxygen, nutrients, contaminants, and biota along stream channels, laterally with floodplains and vertically with the hyporheic zone, influencing all physical and biological aspects of the river ecosystem (Brierley et al., 2010; McCluney et al., 2014). Ecosystem processes also act at a hierarchy of spatial scales including individual gravel pieces, bed form features, reaches, river segments, and the entire network (Frissell et al., 1986; McCluney et al., 2014; Zavadil and Stewardson 2013). These connections are key to understanding trajectories of river ecosystems both as they respond to human disturbances, but also in response to river restoration activities including environmental water.

The primary spatial connection through river networks is the downstream drainage of water accumulating as tributary streams combine in larger rivers. Virtually all human activities that produce hydrological alteration (see Chapter 3) cast a downstream hydrological shadow that can persist all the way to the basin outlet in the case of major disturbances, for example, by large dams. The most obvious result of this downstream effect is in the alteration of the hydrological water regime, but other changes occur such as hydrological connectivity between the river channel, floodplain, and groundwater. Also, modification to tributary inflows can alter the dominant source of water and hence its biochemical properties, including the quality and concentration of nitrogen, phosphorus, and carbon. Similarly, environmental water provisions will produce changes in hydrological regime, connectivity, and water source that can extend over hundreds of kilometers of river downstream of the point at which environmental water is directly manipulated. In large river basins the environmental water needs of tributary and mainstem rivers need to be considered together. Environmental water manipulations in multiple tributary streams will have a cumulative impact on the mainstem river. If the accumulated water regime does not meet the mainstem river environmental water needs, decisions need to be made regarding the tributaries from which additional environmental water should be sourced.

The significance of these spatial connections for environmental water management is clearly demonstrated if we consider the impact of dams on downstream sediment regimes and river channel morphology (discussed in Chapter 5). Although a site-based environmental water plan might focus on delivering an environmental water regime to provide for critical instream physical habitats, these habitats can be dramatically altered as a result of channel adjustments downstream of the dam. Petts and Gurnel (2005) provide a synthesis of the complex spatial sequence and temporal evolution in downstream channels that can extend for hundreds of kilometers in response to sediment trapping within a dam and altered sediment transport capacity of the river produced by hydrological changes. Such channel adjustments will affect physical habitats that are targeted by environmental water regimes and it may be necessary to consider water requirements to maintain channel morphological conditions including bed forms and river channel geometry (discussed in Chapter 12). These complex interactions between water and sediment regimes downstream of dams demonstrate the need to consider spatial connections within networks when planning environmental water management.

An understanding of flow-mediated transport of water contaminants, dissolved oxygen, and energy through river networks is central to addressing water quality impairment associated with hydrological alterations (discussed in Chapter 6). The delivery of material through river networks including nutrients and carbon is also essential for aquatic life. Rivers not only transport materials but they can also hold them in transient or permanent storages, and many contaminants can be transformed into different forms while in transit through a catchment or entirely removed, for example, in the case of nitrogen fixation. These connections are best addressed within an ICM approach to evaluating impacts of hydrological alterations, identifying nonflow stressors (i.e., poor water quality), and the design and implementation of environmental water regimes and complementary land management actions to address water quality concerns.

Mobile organisms such as fishes and invertebrates move along rivers either by swimming for motile forms or drift (Peterson et al., 2013 and discussed in Chapter 4). Such movements are often important in providing access to aquatic habitats for particular life stages or supporting species dispersal across the freshwater landscape. Barriers such as weirs and dams fragment the network (Ward, 1998), and prevent organisms from accessing critical habitats (Schlosser and Angermeier, 1995). Poor outcomes for experimental environmental water projects (discussed earlier) are commonly attributed to barriers for movement of biota through river networks. In particular, barriers to longitudinal connectivity through the river network (e.g., dams and weirs) can inhibit the dispersal of fish and invertebrates, limiting recolonization in river reaches restored with environmental water regimes (Bradford et al, 2011; Brooks et al., 2011; Decker et al., 2008; Mackie et al., 2013; Rolls and Wilson, 2010; Rolls et al., 2012).

Unfortunately, environmental water planning is often undertaken at a site scale using readily available stream flow gauging information and site-based surveys of physical habitats and biological communities. Such studies have little consideration of the important catchment-scale processes that both affect and are affected by conditions at the site, including those described in the preceding paragraphs. A site-based approach presents three major risks for the effectiveness of environmental flow planning. First, critical stressors controlling conditions at this site, associated with land and water management activities elsewhere in the catchments, may not be addressed. Second, adverse effects of environmental water management plans on other parts of the catchment may be overlooked. Third, opportunities to coordinate environmental water delivery from multiple sources and other land, biodiversity, and river management activities to meet needs in multiple locations through a catchment will be unachievable. As a result, expectations of what can be achieved from such site-based environmental water programs are likely to be unrealistic. In situations where environmental water delivery comes at a high cost, disappointing outcomes can undermine community support to expand, or even continue, investment in environmental water delivery elsewhere.

Although an ICM approach cannot completely eliminate these risks, it is a sound response to ensure important interactions between water, land, and biodiversity management, and the response of freshwater ecosystems across catchments are considered in planning environmental water releases. However, moving to a catchment-scale approach considerably increases the complexity of the planning task including the social challenges of coordinating activities across multiple agencies and the technical challenge of assessing the consequences of these environmental connections.

Confronting these challenges will be necessary if the full benefits of environmental water delivery are to be achieved at the catchment scale.

22.4 IMPLEMENTING INTEGRATED CATCHMENT MANAGEMENT

Management of environmental water within an ICM approach has been attempted in various parts of the world. For instance, the Chinese government is trying to implement ecologically sustainable management by setting ecological red-lines that define acceptable changes of stressors such as red-lines for water quantity and quality along with riparian zone protection (Central Committee of the Communist Party of China and Chinese State Council, 2010). Australian state governments have been practicing ICM for at least two decades with a variety of institutional approaches (HRSCEH, 2000). In Victoria, Australia, *catchment management authorities (CMAs)* have been established to coordinate land, water, and biodiversity management in an integrated way at the catchment scale. This Victorian example provides a useful case study of the ICM approach (Box 22.1).

BOX 22.1 ENVIRONMENTAL WATER MANAGEMENT WITHIN AN INTEGRATED CATCHMENT MANAGEMENT FRAMEWORK IN VICTORIA, AUSTRALIA

Since 1994, in Victoria, Australia, 10 CMAs (originally named Catchment Protection Boards) have applied an integrated catchment management approach to protecting Victoria's land, water, and biodiversity resources (*Catchment and Land Protection Act 1994* [Vic]). Their statutory authority includes responsibility for management of regional waterways, floodplains, drainage, and environmental water (*Water Act 1989* [Vic]). The relatively long history of experience in ICM including responsibility for managing significant volumes of environmental water make these Victorian CMAs a useful case study for this chapter.

Every 6 years, the CMAs are required to develop a regional catchment strategy (Victorian Auditor-General, 2014) that addresses a diverse range of environmental management issues including:

1. Biodiversity and native vegetation;
2. Soil health and salinity;
3. Threatened plant and animal species;
4. Waterway health including environmental water management;
5. Fire recovery and flood response and recovery (Victorian Auditor-General, 2014).

These strategies are required to:

1. Assess the condition and use of regional land and water resources and identify areas for priority attention;
2. Identify objectives for the quality of regional land and water resources, along with measures for achieving these;
3. Identify priority partners and their roles in strategy implementation;
4. Identify arrangements for reviewing the strategy and monitoring implementation;
5. Link with relevant federal, state, and regional legislation, policies, strategies, and plans;
6. Summarize the key findings from the review of the previous regional catchment strategy.

The *Water Act 1989* requires that the CMA's prepare regional waterway management strategies. These strategies consider the range of stressors affecting rivers. They include prioritization of environmental water management actions to address the impacts of hydrological alterations. However, these recommendations are aligned with actions to tackle

(Continued)

BOX 22.1 (CONTINUED)

other nonflow stressors including works to improve river channel and riparian condition, flood and floodplain management, drainage works, response to natural disasters and extreme events where they affect waterways, and management of water quality.

As a regional agency, the CMAs are required to plan and implement these strategies in close consultation and partnership with regional communities. This is guided by a Community Engagement and Partnerships Framework (VGVCMA, 2012), which includes principles and measures of success. A recent review concluded that the CMAs had different approaches to community engagement, but the most comprehensive approaches include an annual engagement strategy that set engagement goals and supporting actions to achieve these, along with processes for regularly reviewing the strategy and the effectiveness of engagement activities.

Low levels of funding and its uncertainty from year-to-year are among the main constraints limiting the effectiveness of the Victorian CMAs (Victorian Auditor-General, 2014). Across the state with a total area of close to 240,000 km^2, the 10 CMAs receive only AU$200 million each year, mostly as grants from state and federal government, to develop and implement these strategies (Victorian Auditor-General, 2014). The CMAs are unable to access stable funding through a levy on ratepayers, despite the headline outcome of national inquiry on coordination of catchment management recommending this approach (HRSCEH, 2000).

Successful ICM requires multiple elements in addition to the identification of critical stressors and understanding of environmental interactions throughout catchments (discussed in the previous sections). These include: authoritative objectives; policy settings and institutional arrangements to allow coordination of land and water management; effective stakeholder engagement in planning; and sustained investment in the capacity (skills, tools, and knowledge) to plan and implement ICM. With increasing awareness of ecological security and better understanding of ecological science, approaches to dealing with multiple stressors and catchment-scale interactions continue to be developed. There is no simple recipe for ICM that will work across all river catchment settings. However, 12 principles are provided as part of the *Ecosystem Approach* that is endorsed for implementation by the International Convention on Biodiversity (Secretariat of the Convention on Biological Diversity, 2004). These principles provide the basis for effective integrated management of land, water, and living resources (Box 22.2).

BOX 22.2 TWELVE PRINCIPLES OF THE ECOSYSTEM APPROACH

Principle 1: The objectives of management of land, water, and living resources are a matter of societal choices.
Different sectors of society view ecosystems in terms of their own economic, cultural, and society needs. Indigenous peoples and other local communities living on the land are important stakeholders and their rights and interests should be recognized. Both cultural and biological diversity are central components of the ecosystem approach, and management should take this into account. Societal choices should be expressed as clearly as possible. Ecosystems should be managed for their intrinsic values and for the tangible or intangible benefits to humans, in a fair and equitable way (see also Chapter 10).

Principle 2: Management should be decentralized to the lowest appropriate level. Decentralized systems may lead to greater efficiency, effectiveness, and equity. Management should involve all stakeholders and balance local interests with the wider public interest. The closer the management is to the ecosystem, the greater the responsibility, ownership, accountability, participation, and use of local knowledge.

(Continued)

BOX 22.2 (CONTINUED)

Principle 3: Ecosystem managers should consider the effects (actual or potential) of their activities on adjacent and other ecosystems. Management interventions in ecosystems often have unknown or unpredictable effects on other ecosystems; therefore, possible impacts need careful consideration and analysis. This may require new arrangements or ways of organization for institutions involved in decision making to make, if necessary, appropriate compromises.

Principle 4: Recognizing potential gains from management, there is usually a need to understand and manage the ecosystem in an economic context. Any such ecosystem management program should: reduce those market distortions that adversely affect biological diversity; align incentives to promote biodiversity conservation and sustainable use; and internalize costs and benefits in the given ecosystem to the extent feasible. The greatest threat to biological diversity lies in its replacement by alternative systems of land use. This often arises through market distortions, which undervalue natural systems and populations and provide perverse incentives and subsidies to favor the conversion of land to less diverse systems. Often those who benefit from conservation do not pay the costs associated with conservation and, similarly, those who generate environmental costs (e.g., pollution) escape responsibility. Alignment of incentives allows those who control the resource to benefit and ensures that those who generate environmental costs will pay.

Principle 5: Conservation of ecosystem structure and functioning, in order to maintain ecosystem services, should be a priority target of the ecosystem approach. Ecosystem functioning and resilience depends on a dynamic relationship within species, among species, and between species and their abiotic environment, as well as the physical and chemical interactions within the environment. The conservation and, where appropriate, restoration of these interactions and processes is of greater significance for the long-term maintenance of biological diversity than simply protection of species.

Principle 6: Ecosystem must be managed within the limits of their functioning. In considering the likelihood or ease of attaining the management objectives, attention should be given to the environmental conditions that limit natural productivity, ecosystem structure, functioning, and diversity. The limits to ecosystem functioning may be affected to different degrees by temporary, unpredictable, or artificially maintained conditions and, accordingly, management should be appropriately cautious.

Principle 7: The ecosystem approach should be undertaken at the appropriate spatial and temporal scales. The approach should be bounded by spatial and temporal scales that are appropriate to the objectives. Boundaries for management will be defined operationally by users, managers, scientists, and indigenous and local peoples. Connectivity between areas should be promoted where necessary. The ecosystem approach is based upon the hierarchical nature of biological diversity characterized by the interaction and integration of genes, species, and ecosystems (see also Chapter 10).

Principle 8: Recognizing the varying temporal scales and lag-effects that characterize ecosystem processes, objectives for ecosystem management should be set for the long term. Ecosystem processes are characterized by varying temporal scales and lag-effects. This inherently conflicts with the tendency of humans to favor short-term gains and immediate benefits over future ones.

Principle 9: Management must recognize that change is inevitable. Ecosystems change, including species composition and population abundance. Hence, management should adapt to the changes. Apart from their inherent dynamics of change, ecosystems are beset by a complex of uncertainties and potential *surprises* in the human, biological, and environmental realms. Traditional disturbance regimes may be important for ecosystem structure and functioning, and may need to be maintained or restored. The ecosystem approach must utilize adaptive management (see Chapter 25) in order to anticipate and cater for such changes and events, and should be cautious in making any decision that may foreclose options, but at the same time, consider mitigating actions to cope with long-term changes such as climate change.

Principle 10: The ecosystem approach should seek the appropriate balance between, and integration of, conservation and use of biological diversity. Biological diversity is critical both for its intrinsic value and because of the key role it plays in providing the ecosystem and other services upon which we all ultimately depend. There has been a tendency in the past to manage components of biological diversity either as protected or nonprotected. There is a

(Continued)

BOX 22.2 (CONTINUED)

need for a shift to more flexible situations, where conservation and use are seen in a broader context and the full range of measures is applied in a continuum from strictly protected to human-made ecosystems.

Principle 11: The ecosystem approach should consider all forms of relevant information, including scientific and indigenous and local knowledge, innovations, and practices. Information from all sources is critical to arriving at effective ecosystem management strategies. A much better knowledge of ecosystem functions and the impact of human use is desirable. All relevant information from any concerned area should be shared with all stakeholders and actors, taking into account, inter alia, any decision to be taken under Article 8(j) of the Convention on Biological Diversity. Assumptions behind proposed management decisions should be made explicit and checked against available knowledge and views of stakeholders.

Principle 12: The ecosystem approach should involve all relevant sectors of society and scientific disciplines. Most problems of biological diversity management are complex, with many interactions, side effects, and implications, and therefore should involve the necessary expertise and stakeholders at the local, national, regional, and international level, as appropriate.

Source: Secretariat of the Convention on Biological Diversity (2004)

22.5 CONCLUSIONS

In this chapter, we argue that environmental water management should be integrated within a broader ICM approach. The presence of multiple flow and nonflow stressors in river systems is common because the human developments that produce flow alteration usually also alter other aspects of river catchments. Even in the case of a catchment where flow alteration is achieved without any other direct human impact, flow alteration can lead to a sequence of other changes in the riverine environment that produce stressors that will not respond to environmental water alone. Experience shows that environmental water programs fail when managers fail to address nonflow stressors such as barriers to movement of aquatic fauna, poor water quality, and exotic species. Effective ICM requires consideration of these multiple stressors and attribution of their role in catchment impairment.

Implementing an integrated approach requires a significant policy and institutional infrastructure, which makes use of the available science and engages widely drawing on the knowledge of all stakeholders. There are international examples of integrated approaches to catchment management that include environmental water delivery. However, local environmental and social settings have a big influence on the feasibility of these approaches. Although its importance is widely recognized, the promise of ICM is yet to be fully realized. Significant commitment is required by catchment management agencies, governments, and the societies they serve, to build and sustain this integrated approach over multiple decades, and ultimately, to restore degraded catchments.

REFERENCES

Adams, S.M., 2003. Establishing causality between environmental stressors and effects on aquatic ecosystems. Hum. Ecol. Risk Assess. 9, 17–35.

Bednarek, A.T., Hart, D.D., 2005. Modifying dam operations to restore rivers: ecological responses to Tennessee River dam mitigation. Ecol. Appl. 15 (3), 997−1008.

Bice, C.M., Zampatti, B.P., 2011. Engineered water level management facilitates recruitment of non-native common carp, *Cyprinus carpio*, in a regulated lowland river. Ecol. Eng. 37 (11), 1901−1904.

Bradford, M.J., Higgins, P.S., Korman, J., Sneep, J., 2011. Test of an environmental flow release in a British Columbia river: does more water mean more fish. Freshw. Biol. 56, 2119−2134.

Brierley, G., Reid, H., Fryirs, K., Trahan, N., 2010. What are we monitoring and why? Using geomorphic principles to frame eco-hydrological assessments of river condition. Sci. Total Environ. 408 (9), 2025−2033.

Brooks, A.J., Russell, M., Bevitt, R., Dasey, M., 2011. Constraints on the recovery of invertebrate assemblages in a regulated snowmelt river during a tributary-sourced environmental flow regime. Mar. Freshw. Res. 62 (12), 1407−1420.

Bukaveckas, P.A., 2007. Effects of channel restoration on water velocity, transient storage, and nutrient uptake in a channelized stream. Environ. Sci. Technol. 41 (5), 1570−1576.

CEBC, 2013. Guidelines for systematic review in environmental management. Version 4.2. Centre for Evidence-Based Conservation & Collaboration for Environmental Evidence, Bangor, Wales. Available from: www.environmentalevidence.org/Authors.htm

Central Committee of the Communist Party of China and Chinese State Council, 2010. No.1 Central Document for 2011: decision on accelerating the development of water reform. Available from: http://www.gov.cn/gongbao/content/2011/content_1803158.htm (in Chinese).

Coleman, N.L., 1981. Velocity profiles with suspended sediment. J. Hydraul. Res. 19 (3), 211−229.

Commonwealth of Australia, 2000. Co-ordinating catchment management: report of the inquiry into catchment management. House of Representatives Standing Committee on Environment and Heritage, Canberra. Available from: http://www.aphref.aph.gov.au-house-committee-environ-cminq-cmirpt-report-fullrpt.pdf.

Cormier, S.M., Suter, G.W., Norton, S.B., 2010. Causal characteristics for ecoepidemiology. Hum. Ecol. Risk Assess. 16, 53−73.

Cross, W.F., Baxter, C.V., Donner, K.C., Rosi-Marshall, E.J., Kennedy, T.A., Hall Jr., R.O., et al., 2011. Ecosystem ecology meets adaptive management: food web response to a controlled flood on the Colorado River, Glen Canyon. Ecol. Appl. 21 (6), 2016−2033.

Daniels, R.B., Gilliam, J.W., 1996. Sediment and chemical load reduction by grass and riparian filters. Soil Sci. Soc. Am. J. 60 (1), 246−251.

Decker, A.S., Bradford, M.J., Higgins, P.S., 2008. Rate of biotic colonisation following flow restoration below a diversion dam in the Bridge River, British Columbia. River Res. Appl. 24, 876−883.

Einstein, H.A., Krone, R.B., 1962. Experiments to determine modes of cohesive sediment transport in salt water. J. Geophys. Res. 67 (4), 1451−1461.

Elith, J., Leathwick, J.R., Hastie, T., 2008. A working guide to boosted regression trees. J. Anim. Ecol. 77, 802−813.

Falkenmark, M., 2004. Towards integrated catchment management: opening the paradigm locks between hydrology, ecology and policymaking. Int. J. Water Resour. Dev. 20, 275−281.

Frissell, C.A., Liss, W., Warren, C.E., Hurley, M.D., 1986. A hierarchical framework for stream habitat classification: viewing streams in a watershed context. Environ. Manage. 10 (2), 199−214.

Gehrke, P.C., Harris, J.H., 2000. Large-scale patterns in species richness and composition of temperate riverine fish communities, south-eastern Australia. Mar. Freshw. Res. 51, 165−182.

Greet, J., Webb, J.A., Cousens, R.D., 2011. The importance of seasonal flow timing for riparian vegetation dynamics: a systematic review using causal criteria analysis. Freshw. Biol. 56, 1231−1247.

Grove, J.R., Webb, J.A., Marren, P.M., Stewardson, M.J., Wealands, S.R., 2012. High and dry: an investigation using the causal criteria methodology to investigate the effects of regulation, and subsequent environmental flows, on floodplain geomorphology. Wetlands 32, 215−224.

Hart, B.T., 2015. The Australian Murray−Darling Basin Plan: challenges in its implementation. Int. J. Water Resour. Dev. 20015, 1−16.

HRSCEH, 2000. Co-ordinating Catchment Management: Report of the Inquiry Into catchment Management by the House of Representatives Standing Committee on Environment and Heritage. Commonwealth of Australia, Canberra Australia, p. 182.

Humborg, C., Ittekkot, V., Cociasu, A., Bodungenet, B.V., 1997. Effect of Danube River dam on Black Sea biogeochemistry and ecosystem structure. Nature 386 (6623), 385−388.

Jakeman, A.J., Letcher, R.A., 2003. Integrated assessment and modelling: features, principles and examples for catchment management. Environ. Model. Softw. 18, 491−501.

Kattel, G.R., Dong, X., Yang, X., 2016. A century-scale, human-induced ecohydrological evolution of wetlands of two large river basins in Australia (Murray) and China (Yangtze). Hydrol. Earth Syst. Sci. 20. pp. 2151−2168.

Khan, K.S., Kunz, R., Kleijnen, J., Antes, G., 2003. Five steps to conducting a systematic review. J. R. Soc. Med. 96, 118−121.

King, J., Cambray, J.A., Impson, N.D., 1998. Linked effects of dam-released floods and water temperature on spawning of the Clanwilliam yellowfish *Barbus capensis*. Hydrobiologia 384, 245−265.

Konrad, C.P., Olden, J.D., Lytle, D.A., Melis, T.S., Schmidt, J.C., Bray, E.N., et al., 2011. Large-scale flow experiments for managing river systems. BioScience 61, 948−959.

Korman, J., Kaplinski, M., Melis, T.S., 2011. Effects of fluctuating flows and a controlled flood on incubation success and early survival rates and growth of age-0 rainbow trout in a large regulated river. Trans. Am. Fisheries Soc. 140, 487−505.

Lind, P.R., Robson, B.J., Mitchell, B.D., 2007. Multiple lines of evidence for the beneficial effects of environmental flows in two lowland rivers in Victoria, Australia. River Res. Appl. 23 (9), 933−946.

MacNally, R., 2000. Regression and model-building in conservation biology, biogeography and ecology: the distinction between − and reconciliation of − "predictive" and "explanatory" models. Biodivers. Conserv. 9, 655−671.

Macleod, C.J.A., Scholefield, D., Haygarth, P.M., 2007. Integration for sustainable catchment management. Sci. Total Environ. 373, 591−602.

Mackie, J.K., Chester, E.T., Matthews, T.G., Robson, B.J., 2013. Macroinvertebrate response to environmental flows in headwater streams in western Victoria, Australia. Ecol. Eng. 53, 100−105.

McCluney, K.E., Poff, N.L., Palmer, M.A., Thorp, J.H., Poole, G.C., Williams, B.S., et al., 2014. Riverine macrosystems ecology: sensitivity, resistance, and resilience of whole river basins with human alterations. Front. Ecol. Environ. 12 (1), 48−58.

McLoughlin, C.A., Deacon, A., Sithole, H., Gyedu-Ababio, T., 2011. History, rationale, and lessons learned: thresholds of potential concern in Kruger National Park river adaptive management. Koedoe 53 (2), 69−95.

Miller, K.A., Webb, J.A., de Little, S.C., Stewardson, M.J., 2013. Environmental flows can reduce the encroachment of terrestrial vegetation into river channels: a systematic literature review. Environ. Manage. 52, 1201−1212.

Nilsson, C., Malm Renöfält, B., 2008. Linking flow regime and water quality in rivers: a challenge to adaptive catchment management. Ecol. Soc. 13 (2), 18. Available from: http://www.ecologyandsociety.org/vol13/iss2/art18/.

Norris, R.H., Webb, J.A., Nichols, S.J., Stewardson, M.J., Harrison, E.T., 2012. Analyzing cause and effect in environmental assessments: using weighted evidence from the literature. Freshw. Sci. 31, 5−21.

Norton, S.B., Cormier, S.M., Suter II, G.W., Schofield, K., Yuan, L., Shaw-Allen, P., et al., 2008. CADDIS: the causal analysis/diagnosis decision information system. In: Marcomini, A., Suter II, G.W., Critto, A. (Eds.), Decision Support Systems for Risk-Based Management of Contaminated Sites. Springer, New York, pp. 351−374.

Ortlepp, J., Mürle, U., 2003. Effects of experimental flooding on brown trout (*Salmo trutta fario* L.): The River Spöl, Swiss National Park. Aquat. Sci. Res. Across Bound. 65 (3), 232−238.

Peterson, E.E., Ver Hoef, J.M., Isaak, D.J., Falke, J.A., Fortin, M.J., Jordan, C.E., et al., 2013. Modelling dendritic ecological networks in space: an integrated network perspective. Ecol. Lett. 16 (5), 707−719.

Petts, G.E., Gurnel, A.M., 2005. Dams and geomorphology: research progress and future directions. Geomorphology 71, 27−47.

Poff, N.L., Zimmerman, J.K.H., 2010. Ecological responses to altered flow regimes: a literature review to inform the science and management of regulated rivers. Freshw. Biol. 55, 194−205.

Propst, D.L., Gido, K.B., 2004. Responses of native and nonnative fishes to natural flow regime mimicry in the San Juan River. Trans. Am. Fisheries Soc. 133, 922−931.

Raulings, E.J., Raulings, E.J., Morris, K., Roache, M.C., Boon, P.I., 2011. Is hydrological manipulation an effective management tool for rehabilitating chronically flooded, brackish-water wetlands? Freshw. Biol. 56 (11), 2347−2369.

Richardson, D.M., Holmes, P.M., Esler, K.J., Galatowitsch, S.M., Stromberg, J.C., Kirkman, S.P., et al., 2007. Riparian vegetation: degradation, alien plant invasions, and restoration prospects. Divers. Distrib. 13 (1), 126−139.

Robinson, C.T., Uehlinger, U., 2008. Experimental floods cause ecosystem regime shift in a regulated river. Ecol. Appl. 18 (2), 511−526.

Rolls, R.J., Leigh, C., Sheldon, F., 2012. Mechanistic effects of low-flow hydrology on riverine ecosystems: ecological principles and consequences of alteration. Freshw. Sci 31, 1163−1186.

Rolls, R.J., Wilson, G.G., 2010. Spatial and temporal patterns in fish assemblages following an artificially extended floodplain inundation event, Northern Murray-Darling Basin, Australia. Environ. Manage. 45 (4), 822−833.

Rood, S.B., Gourley, C.R., Ammon, E.M., Heki, L.G., Klotz, J.R., Morrison, M.L., et al., 2003. Flows for floodplain forests: a successful riparian restoration. BioScience 53 (7), 647−656.

Rothman, K.J., 2002. Epidemiology: an Introduction. Oxford University Press, New York.

Schlosser, I.J., Angermeier, P.L., 1995. Spatial variation in demographic processes in lotic fishes: conceptual models, empirical evidence, and implications for conservation. Am. Fisheries Soc. Symp. 17, 360−370.

Schlosser, I.J., Karr, J.R., 1981. Riparian vegetation and channel morphology impact on spatial patterns of water quality in agricultural watersheds. Environ. Manage. 5 (3), 233−243.

Secretariat of the Convention on Biological Diversity, 2004. The Ecosystem Approach, (CBD Guidelines). Secretariat of the Convention on Biological Diversity, Montreal, Canada.

Siebentritt, M.A., Ganf, G.G., Walker, K.F., 2004. Effects of an enhanced flood on riparian plants of the River Murray, South Australia. River Res. Appl. 20 (7), 765−774.

Smith, L., Porter, K., Hiscock, K., Porter, M.J., Benson, D. (Eds.), 2015. Catchment and River Basin Management: Integrating Science and Governance. Routledge, Oxford.

South Florida Water Management District, 2006. Kissimmee River restoration studies: Evaluation Program Executive Summary. Available from: http://my.sfwmd.gov/portal/page/portal/xrepository/sfwmd_repository_pdf/krr_exec_summary.pdf.

Stevens, A., Milne, R., 1997. The effectiveness revolution and public health. In: Scally, G. (Ed.), Progress in Public Health. Royal Society of Medicine Press, London, pp. 197−225.

Valdez, R.A., Hoffnagle, T.L., McIvor, C.C., McKinney, T., Leibfried, W.C., 2001. Effects of a test flood on fishes of the Colorado River in Grand Canyon, Arizona. Ecol. Appl. 11 (3), 686−700.

VGVCMA, 2012. Community Engagement and Partnerships Framework for Victoria's Catchment Management Authorities. Victorian Government and Victorian Catchment Management Authorities, Victoria, Australia.

Victorian Auditor-General, 2014. Effectiveness of Catchment Management Authorities. Victorian Auditor General's Report PP No 364, Session 2010−14. Victorian Government Printer, Melbourne, Australia.

Vilizzi, L., McCarthy, B.J., Scholz, O., Sharpe, C.P., Wood, D.B., 2013. Managed and natural inundation: benefits for conservation of native fish in a semi-arid wetland system. Aquat. Conserv. Mar. Freshw. Ecosyst. 23 (1), 37–50.

Walsh, C.J., Roy, A.H., Feminella, J.W., Cottingham, P.D., Groffman, P.M., Morgan, R.P., 2005. The urban stream syndrome: current knowledge and the search for a cure. J. N. Am. Benthol. Soc. 24, 706–723.

Ward, J.V., 1998. Riverine landscapes: biodiversity patterns, disturbance regimes, and aquatic conservation. Biol. Conserv. 83, 269–278.

Webb, J.A., Wallis, E.M., Stewardson, M.J., 2012a. A systematic review of published evidence linking wetland plants to water regime components. Aquat. Bot. 103, 1–14.

Webb, J.A., Nichols, S.J., Norris, R.H., Stewardson, M.J., Wealands, S.R., Lea, P., 2012b. Ecological responses to flow alteration: assessing causal relationships with Eco Evidence. Wetlands 32, 203–213.

Webb, J.A., Miller, K.A., King, E.L., de Little, S.C., Stewardson, M.J., Zimmerman, J.K.H., et al., 2013. Squeezing the most out of existing literature: a systematic re-analysis of published evidence on ecological responses to altered flows. Freshw. Biol. 58, 2439–2451.

Webb, J.A., Miller, K.A., de Little, S.C., Stewardson, M.J., Nichols, S.J., Wealands, S.R., 2015. An online database and desktop assessment software to simplify systematic reviews in environmental science. Environ. Model. Softw. 64, 72–79.

Yang, Z., Wang, H., Saito, Y., Milliman, J.D., Xu, K., Qiao, S., et al., 2006. Dam impacts on the Changjiang (Yangtze) River sediment discharge to the sea: the past 55 years and after the Three Gorges Dam. Water Resour. Res. 42, 4.

Zavadil, E., Stewardson, M., 2013. The role of geomorphology and hydrology in determining spatial-scale units for ecohydraulics. In: Maddock, I., Harby, A., Kemp, P., Wood, P.J. (Eds.), Ecohydraulics: an Integrated Approach. John Wiley & Sons, Chichester, pp. 125–142.

ACTIVE MANAGEMENT OF ENVIRONMENTAL WATER

PLANNING FOR THE ACTIVE MANAGEMENT OF ENVIRONMENTAL WATER

23

Jane M. Doolan[1], Beth Ashworth[2], and Jody Swirepik[3]

[1]University of Canberra, Canberra, ACT, Australia [2]Office of the Victorian Environmental Water Holder, East Melbourne, VIC, Australia [3]Clean Energy Regulator Australia, Canberra, ACT, Australia

23.1 INTRODUCTION

Establishing environmental water allocations is clearly a difficult issue, requiring considerable technical, policy, and institutional advances and decisions, as previous chapters have outlined. However, this is not the end of the story. Depending on how those allocations have been made, there is still considerable work required to ensure that these allocations are managed and used effectively and to demonstrate to governments and the community that the allocation of water to the environment has been worthwhile and is achieving its ecological objectives.

The level of actual management required depends on how the environmental water allocations have been established. Chapter 17 identified five mechanisms for allocating environmental water:

1. License conditions on consumptive users;
2. Conditions on storage operators or water resource managers;
3. An ecological reserve;
4. A cap on consumptive use;
5. Environmental water rights.

Mechanisms 1, 2, and 4 involve placing constraints or obligations on consumptive users, which effectively provide a residual environmental water regime for the river or wetland system. As outlined in Chapter 19, environmental water allocations provided in this *rules-based* way do not require any real-time water management decisions. The key issue for environmental water managers is ensuring that the conditions are complied with by consumptive users and that the flow is actually left in the river system. This has been described as *passive management* in Chapter 19.

By contrast, the provision of an ecological reserve held in storage or the provision of environmental water rights where the environment is issued with water rights of the same structure and legal properties as those of consumptive water users, requires robust and rigorous decision making on the use of that water in real time. This has been described in Chapter 19 as

Water for the Environment. DOI: http://dx.doi.org/10.1016/B978-0-12-803907-6.00023-1

active environmental water management. In these cases, environmental water managers can have considerable discretion in how, where, and when they use the environmental water.

Active environmental water management is only possible in regulated (i.e., dam-controlled) water systems where construction of dams and regulators enables the storage of water and the ultimate movement and use of that water across interconnected water systems. These regulated systems can range from low levels of regulation (e.g., a small dam on a single river) to very large, highly regulated systems with a series of dams and weirs along the mainstream and major tributaries coupled with significant water delivery infrastructure linking irrigation areas and towns within a highly complex interconnected network. Well-known examples of the latter include the *Murray−Darling Basin (MDB)* system in Australia, the Colorado River, and the Sacramento-San Joaquin Delta in the United States.

These large, highly regulated water systems are the water systems where historic systems of water rights have developed, where highly developed water markets have been established, where there is significant competition for water among a range of uses, and where water has a known market value. These are often the systems that are fully allocated or overallocated, that is they are already either at or over sustainable levels of resource development, as described in Chapter 17. It is in some of these systems where there is ongoing controversy and debate about the need for improved environmental water regimes and where there has been significant effort in recovering water for the environment.

Chapters 17 and 18 outline the ways in which environmental water can be increased in a variety of ways including changing conditions on license holders and storage operators, lowering a cap, increasing the ecological reserve, investing in water savings or acquiring water rights from existing users. In general, the model used in these large, highly regulated water systems is to work within the existing water rights framework either by investing in water efficiencies to create environmental water rights or by buying consumptive water rights on the water market and providing these to the environment. This has been the route taken in the MDB, and, to some extent, in the Colorado and the Sacramento-San Joaquin situations. This strategy, although expensive and often controversial, does not undermine the security of existing consumptive users or their business enterprises and is politically more feasible than taking water away from existing users.

Providing the environment with environmental water rights in these large, highly regulated systems means that environmental water managers can hold very significant water assets with known market values. In the case of the MDB, where there has been over AU$13 billion allocated to a rebalancing and reset of the entire MDB, the *Commonwealth Environmental Water Holder (CEWH)* currently holds over 2400 GL of water rights of varying reliabilities (at February 29, 2016) that are worth billions of AU$ (DEE, 2016). In addition, because of the interconnected nature of these large, regulated systems, environmental water managers can have a lot of discretion in how, where, and when they use their water. They can make choices between which parts of the flow regime to focus on (e.g., releasing water to provide summer baseflow(s) or augment spring flushes), which river reaches or wetlands to water, and in highly connected water systems, even between river systems. They can make choices between how much water to use in any one year and how much could be carried over in storages for the succeeding year. In water systems where water markets operate, environmental water managers can even decide on whether some portion of water allocation should be traded (i.e., sold) to consumptive users.

Given the value of the environmental water rights and the wide range of options for its use, environmental water managers holding water rights will be subject to high levels of public scrutiny and their decisions are required to be transparent as part of normal public accountability requirements. This raises a number of issues and challenges associated with the active management of environmental water including:

1. Understanding institutional imperatives to implement active environmental water management;
2. Ensuring that environmental water is actively managed as part of local integrated river management programs to achieve agreed river management objectives;
3. Developing policy and planning frameworks governing the use of environmental water;
4. Operationalizing environmental water delivery from environmental water rights;
5. Monitoring the effects of using environmental water to support adaptive environmental water management and demonstrate the overall benefits of providing water for the environment.

This chapter looks briefly at institutional arrangements required to support active environmental water management and concentrates on describing the principles and key elements required in the policy and planning frameworks to govern active environmental water management and ensure that it is undertaken as part of a local integrated river management program. Chapters 24 and 25 describe the issues associated with operationalizing environmental water delivery and monitoring effects.

In this chapter, the case examples used to illustrate the issues and possible solutions are taken from the MDB. As mentioned previously, it is a highly regulated, and in its southern part, highly interconnected river system with clear water entitlements and an active and mature water market. In recent years, it has been subject to a major environmental water recovery program to recover ultimately around 2800 GL (billion liters) of water entitlements from the consumptive pool (mostly irrigation) to be held by the Australian CEWH and used to improve the environmental condition of the rivers and wetlands within the Basin (DEE, 2016). State governments in the southern MDB also have environmental water holders (or equivalent) holding their own environmental entitlements and together with the Commonwealth, they jointly hold 500 GL of environmental entitlements through a special initiative where deployment must be agreed through consensus. Although each entity has its own statutory obligations, they have to act collaboratively to plan and deploy their environmental water across the rivers and wetlands in their state and across the MDB.

These environmental water management arrangements in the MDB represent the most complex arrangement for ongoing, active environmental water management in the world today. The principles and policies developed to deal with ecological, accountability, and community issues in this situation are relevant and able to be applied to the full range of regulated systems from the simple one-storage one-river system to the well-known, large, highly regulated systems.

23.2 INSTITUTIONAL IMPERATIVES FOR ACTIVE ENVIRONMENTAL MANAGEMENT

Chapter 19 has outlined the potential range of institutions that can be set up to manage environmental water, and in doing so, established the principle that there must be a match between the organizations, their powers and capacities, scale of operation, and the type of environmental water for which they are responsible. Where significant volumes of environmental water rights have been provided, the institutional arrangements need to be robust and commensurate with both the value of the water assets being managed and the discretion in use that can be exercised. For example, good governance practice suggests that it would not be appropriate to have public assets worth millions of dollars being managed by small local groups with little formal training and no accountability requirements (VPSC, 2016). However, this may be an appropriate arrangement for a small instream license aimed at the management of a specific local wetland where the value of the asset and the level of discretion on its use are relatively low.

Chapter 19 showed that environmental water organizations that undertake active management typically possess some form of legal personhood (to enable the holding of water rights) and have legal standing. At the extreme end of the institutional range, the arrangements in Australia have resulted in the creation, under Commonwealth legislation, of the CEWH, and in the state of Victoria, the creation of a state equivalent, the *Victorian Environmental Water Holder (VEWH)* under state legislation. Both these entities have a range of statutory powers and obligations that are proportionate to the value of the water assets they hold. They are provided with broad statutory guidelines covering the way in which they must undertake their functions, but within that they have a lot of freedom in their decision making on the use of environmental water, including the decision to trade, without reference to the government of the day. They are also subject to the normal public accountability and audit provisions of any government authority. This means they are required to demonstrate that they are making the best use of water toward achieving the environmental outcomes for which their environmental water rights have been provided.

Environmental water managers holding water rights need to demonstrate efficient and effective use for two critical reasons. First, as custodians of a public resource, they have a clear duty to do so. Second, the manner in which they undertake their responsibilities will determine their ongoing *social license* to operate. Community support (or at least lack of opposition) is critical to the capacity of environmental water managers to implement long-term active environmental watering programs. The way they go about their job and the decisions they make have to make sense to the community. This is particularly important in active environmental water management rather than in passive situations. In passive situations, from a community perspective, once the environmental water regime has been agreed, it simply happens. Compliance is the key issue in this case. However, in active management, decisions are being made continually on where and when the environmental water is used. Those decisions can be challenged by the community if not well understood. The active environmental water manager, as the decision maker, is both a presence and a public face in the local community and will be the subject of strong views at the local level. Given this, the development of the sensible and effective policy and planning frameworks for the active management of environmental water rights and the manner in which decisions are communicated to the community are critical issues in both achieving the desired environmental outcomes and in generating community support and acceptance of the need to provide water for the environment.

23.3 ACTIVE ENVIRONMENTAL WATER MANAGEMENT IN A LOCAL INTEGRATED WATERWAY MANAGEMENT CONTEXT

A key principle in planning and managing environmental water is that it needs to occur within a broader context of *integrated waterway management (IWM)* aimed at delivering a set of agreed river/wetland management objectives at the local level. These objectives provide the basis not only for determining how much water may be required for the environment, guiding environmental water management and evaluating performance, but also for a suite of complementary water quality and habitat management programs.

It is important to recognize that just providing environmental water may not be sufficient to achieve environmental objectives in many river/wetland systems because they are generally impacted by a range of degrading factors, not just a changed flow regime. These factors can include instream and riparian habitat loss, changes to longitudinal and floodplain connectivity, water quality problems, and issues associated with catchment land use.

This means that the planning and management of environmental water must occur within the broader context of IWM undertaken at the local scale. In this way, agreed river/wetland management objectives guide environmental water management as a key component of a broader program including complementary on ground works to achieve those objectives. In a true IWM framework, this would involve genuine community consultation to establish objectives and develop the management/restoration plan to achieve them. Undertaking this objective setting step within an IWM context ensures that:

1. There is engagement with the community on their aspirations for the river/wetland and they understand the role of flow in that process.
2. Delivery of environmental water and achievement of the environmental objectives is feasible on the ground and within the constraints of related catchment and land management.
3. The environmental water will be managed within a broader river management program and therefore achievement of the environmental water objectives are less likely to be limited by other degrading factors, including poor habitat and/or water quality. This provides confidence that, where environmental water is provided, it will achieve the community's objective for the river or wetland. This is particularly important where environmental water has been sourced from consumptive uses in controversial circumstances.

Previous chapters have discussed ways in which this river/wetland management objective setting can be undertaken, the technical methodologies for assessing environmental water requirements and water allocation processes, which provide the final agreed environmental water arrangements, including the combination of passing flow arrangements and the allocation of environmental water rights.

Victoria provides an illustrative case study demonstrating how environmental water management can be undertaken within a broader IWM context. Here, the state-wide policy framework for the management of rivers and wetlands states in its guiding principles that environmental water management will be comprehensively integrated with complementary works programs and that priorities for environmental water management will be identified through regional waterway planning processes (DEPI, 2013).

In Victoria, to implement the state-wide policy, local *Catchment Management Authorities (CMAs)* develop Regional Waterway Strategies in consultation with their local communities. These strategies identify in a consistent way, the environmental, economic, cultural, and social values associated with all the major river reaches and wetlands in their region and the threats to these values. On this basis, they identify priority reaches and wetlands for protection and restoration for the following planning period—in this case, 8 years. This includes setting long-term resource condition targets and management objectives, and establishing work programs to meet them. Work programs include the provision and management of environmental water and complementary on ground works such as habitat restoration, water quality improvement, floodplain/associated land management, and infrastructure protection as required (DEPI, 2013).

Scientific environmental flows assessment studies recommend an environmental water regime to maintain the community values at a relatively low level of risk and can be used to identify where additional environmental water may be required. These studies provide the basis to predict likely environmental outcomes under a range of water sharing and climate change scenarios. These inform Regional Sustainable Water Strategies, undertaken every 10 years or so using a highly consultative process, to consider the balance between consumptive and environmental water and set passing flow requirements, establish environmental entitlements in regulated systems and where there is agreement on the need for additional environmental water, also commit to volumes and methods of further environmental water recovery (DSE, 2009, 2011a, 2011b).

Environmental entitlements (i.e., water rights) provided as a result of these planning processes are held and actively managed by the VEWH who can use them to provide water across the entire Victorian regulated water grid for a large range of river reaches and wetlands across the state. To make decisions on environmental water use, the VEWH works with the local CMAs who develop annual proposals for the use of environmental water in rivers and wetlands in their region to meet the river objectives identified in their Regional Waterway Strategies. Their annual proposals are informed by their scientific environmental flows assessment studies and by discussions with river operators who need to consider how the environmental water can be delivered within a set of broader water management objectives and rules. The VEWH will then make decisions on where and how water from the environmental entitlements will be used across the rivers and wetlands proposed by the CMAs. Once these decisions are made, the CMAs then manage the local delivery of the environmental water (in collaboration with their river operator), undertake complementary on ground works programs, manage the ongoing community engagement process, and monitor the environmental outcomes. This planning hierarchy is illustrated in Fig. 23.1.

Managing environmental water as part of a local IWM program ensures that environmental outcomes for which the water has been provided are known, are supported by the community, and can be delivered on the ground.

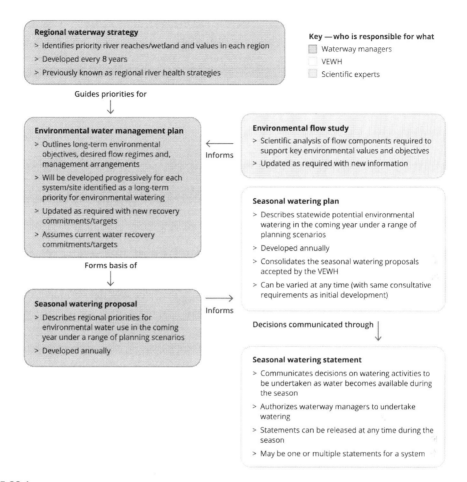

FIGURE 23.1

Planning arrangements showing how environmental water is managed in Victoria, Australia, as part of a broader integrated water management program aimed at achieving agreed river and wetland objectives at the local scale.

Source: VEWH (2015b)

23.4 POLICY FRAMEWORK FOR ACTIVE ENVIRONMENTAL WATER MANAGEMENT

The previous section outlines how environmental water management needs to be delivered within a local IWM program aimed at achieving agreed river and wetland objectives. This is an important policy principle, but in itself, is insufficient to govern active decision making in environmental water management. Environmental water managers holding environmental water rights in large, highly regulated systems will often have the capacity to make decisions between river reaches and wetlands. They will need to make choices between how much water to use in any one year, and in some systems, how much should be held (or carried over) in storages for the succeeding year/s and how much, if any, should be traded to consumptive users. It is critical that they have a sensible, transparent policy framework for making these decisions that provides a clear rationale for the decision that can stand up to scrutiny.

23.4.1 **OVERALL POLICY OBJECTIVE**

Assuming that environmental water rights have been provided either by government or under legislation, with the stated purpose to improve or secure environmental outcomes, then the key objective for such a policy framework is that it should deliver the *greatest environmental benefit for the volume of water applied.*

Criteria can then be developed to compare and evaluate potential, competing uses of environmental water across the relevant water system. In Australia, in the early days of environmental water management, that is the mid-2000s, a set of criteria were developed, which were aimed at identifying "the highest value environmental use" for a volume of available environmental water. They were based on a common sense approach, were easily explained to the community and used to guide decision making at the highest levels. They include:

1. Extent and significance of the environmental benefit expected from the watering action, for example the area watered, the size of the breeding event to be triggered, and/or the conservation status of the species that would benefit.
2. Certainty of achieving the environmental benefit from the watering action and ability to manage other threats, for example there is a high level of certainty about the environmental outcome and relevant complementary measures are being undertaken at the site.
3. Ability to provide ongoing benefits at the site at which the watering action is to take place, for example there is known capacity to provide for watering over the long term.
4. Implications of not undertaking the potential watering action at the site, for example the potential for critical or irreversible loss of important environmental values and the conservation significance of that loss.
5. Feasibility of the watering action, for example the dependence on infrastructure capacity and/or actions of other users, flexibility of timing of the delivery, operational requirements, and constraints.
6. Overall cost-effectiveness of the watering action, for example the likely benefit to be achieved considered against the cost of the watering action (including the volume of water to be used and any delivery and risk management costs).
7. Risks associated with watering (e.g., blackwater events, mobilization of salinity).

These criteria were used extensively over the past decade to evaluate a range of environmental water demands competing for a limited volume of available environmental water. They are now included in all major policy frameworks for the management of environmental water in Australia (CEWO, 2013b; MDBA, 2014a; VEWH, 2015b) and have been captured into the Basin-wide Environmental Watering Strategy for the MDB (MDBA, 2014b). They are used at the highest levels to justify choices made by environmental water managers.

23.4.2 POLICY GUIDANCE FOR ACTIVE ENVIRONMENTAL WATER MANAGEMENT IN A VARIABLE CLIMATE

The objective and criteria outlined above provide the high-level policy principles for making decisions on the allocation of environmental water between competing environmental demands.

Application of these policy principles requires the prediction of the environmental benefits occurring as a result of an environmental watering action. In making these predictions, environmental water managers use the scientific environmental flows assessment studies that were the rationale for providing the environmental water rights in the first place. For the most part, they use best available science to describe a preferred water regime to maintain or improve the key environmental values of a river or wetland. However, experience in Australia during the Millennium Drought (1997–2009) showed some significant limitations in these methodologies in that, in general, they provided preferred flow regimes based on average flow conditions. In the midst of the most severe drought on record, this was shown to be inadequate for a number of key reasons including:

1. There was insufficient environmental water available to provide any significant part of the preferred flow regimes in many systems;
2. They did not provide any clear direction on what the environmental priorities should be in extreme water scarcity;
3. There was significant community backlash to the use of environmental water in times of extreme water scarcity and any attempt to apply water had to be based on impeccable logic and clear necessity.

In severe drought, the interpretation of the principle of choosing environmental watering actions that *provided the greatest environmental benefit* actually became choosing those that *avoided the greatest environmental loss*. However, environmental water managers realized that in making these decisions, they had no underlying ecological model of how these systems worked built into the policy framework. Consequently, it was not clear in making decisions to avoid loss, which decision would be the most important for long-term environmental sustainability. As a result, environmental water managers further developed the policy framework to deal with climate variability, particularly water scarcity and drought, clearly and unambiguously.

They went back to how natural freshwater systems in Australia coped with variability, explicitly recognizing the role of drought refuges in periods of water scarcity, and the importance of resilience to enable recovery in more favorable periods. This translated into a new broad policy goal, which aimed to ensure that priority environmental assets were able to survive water scarcity and that river and wetlands systems were resilient enough to recover in wetter years. This did not change the long-term river and wetland health objectives referred in the previous section, but did provide considerable guidance in annual planning for the use of environmental water and in the annual river restoration activities under different water resource scenarios. It became a *seasonally adaptive* approach where in extremely dry years, priority would be given to the protection of drought refuges, avoiding irreversible loss and catastrophic events. In dry years, priority would be given to maintaining ecological functioning and then in average and wetter years, priority given to recruitment and recovery. This is summarized in Fig. 23.2. More detailed descriptions of the actions required under the seasonally adaptive approach and its implications for broader IWM are outlined in the *Northern Regional Sustainable Water Strategy* (DSE, 2009).

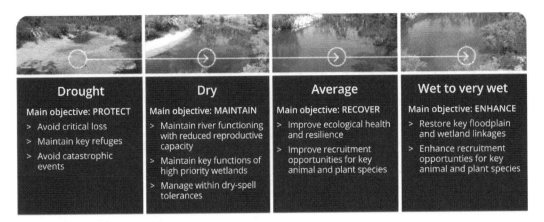

FIGURE 23.2

Seasonally adaptive approach showing objectives for annual environmental water use under different climatic scenarios.

Source: VEWH (2015b)

This approach has been adopted universally in Australia and governs the use of both state and Commonwealth environmental water (CEWO, 2013b; MDBA, 2014a; VEWH, 2015b).

The adoption of this framework provided a number of clear benefits and improved the social license for environmental water managers to operate. It made sense to the community, everybody was making hard decisions in extreme drought, including the environment. It was based on clear knowledge of how Australian systems operated and it paralleled similar decision frameworks that irrigators were using in their businesses. It also provided the basis to begin to consider how these systems could be managed under climate change.

The adoption of this framework also meant that there was a flow-on requirement for refinements to river health planning and environmental flows assessment methodologies and a number of new knowledge gaps. It was necessary to map important drought refuges, particularly at a system scale, and to know the minimum survival regimes for these systems and, as seasonal conditions improved, which elements of the flow regime should be reinstated first.

The policy objective and criteria, coupled with the seasonally adaptive approach outlined above, meant that environmental water managers had a robust policy framework for making decisions on the use of environmental water under all climatic conditions. Using this framework in the middle of the Millennium Drought, environmental water managers made a number of difficult environmental trade-off decisions for the use of a limited volume of environmental water. Decisions were made to save a small fish species (the Murray Hardyhead) from extinction, mitigate a blackwater (i.e., low oxygen event) in the Campaspe River, and maintain a number of drought refuges (DSE, 2008; MDBA, 2012).

The framework was also used in one of the most challenging decisions made during the Millennium Drought. This involved making decisions in 2007 and 2008 on the use of a small volume of available environmental water between watering a number of Ramsar-listed wetlands to provide drought refuges along the River Murray or providing water into the system of

Ramsar-listed large lakes at the lower end of the River Murray (Lower Lakes), which were at high risk of widespread acidification. The Lower Lakes were below sea level (at that time, at -0.5 m) due to very low inflows and the need to provide upstream communities with water for critical human needs. Acidification was predicted to occur when lake levels reached -1.75 m, which was anticipated at the end of summer if the very dry conditions prevailed.

The environmental water managers involved in this decision had 16.5 GL in 2007/2008 and 6 GL in 2008/2009 of water allocated against their environmental water rights. The critical time for both options was to provide the water in spring. For the drought refuges, it would enable breeding of some endangered and vulnerable species of birds. For the Lower Lakes, it may have reduced the likelihood of acidification over the summer, depending on the very high level of evaporation and the actual inflows, which were clearly unknown at the time. There was considerable stakeholder interest in the outcomes of the decision; some strongly supporting providing water to the Lower Lakes (including the South Australian Government who are the custodians of the lakes), and others calling for the environmental water to be handed to irrigators.

The environmental water managers centered their decision on determining what volume of water would make a *material* difference to the levels of the Lower Lakes. Data analysis and modeling established that under average climate conditions, 30 GL would normally increase the height of the Lower Lakes by about 3.8 cm and would take about 2 weeks to evaporate. However, under the intense drought conditions being experienced, 30 GL was more likely to add 4.5 cm to the lakes but it would evaporate in about 5 days, providing little advantage to the Lower Lakes. As a result, given the water available throughout those 2 years, environmental water managers decided that *the best environmental use* was to water the series of wetlands to provide drought refuges, although all watering decisions were subject to the test of whether the water available (at that time or in the foreseeable future) could make a material difference to the Lower Lakes.

In general, these decisions stood up to the strong level of stakeholder scrutiny and resulted in bird breeding in the drought refuges, which provided a small amount of environmental good news in an otherwise bleak period. The situation finally improved in 2009, when floods in the northern part of the MDB provided some water to the parched system. The available environmental allocations in that year were 48 GL and this, for the first time, met the test of making a material difference to the Lower Lakes. In that year, the environmental water was provided to the lakes and that decision coupled with higher river flows meant that there was confidence that the lakes would be able to last out the rest of the summer without widespread acidification.

After the drought broke in early 2010, the availability of environmental water increased and relieved environmental water managers were able to move their objectives for environmental water management from the avoidance of loss to promoting recovery and enhancing environmental conditions.

23.4.3 CONCLUSION

The examples discussed above show that environmental water managers actively managing environmental water rights can be required to make some hard decisions, trading off between a number of potential uses for their available environmental water, particularly in times of water scarcity. It is critical that these decisions are guided by a clear policy framework. Although the policy framework described above was developed in Australia to suit Australian conditions and the issues arising at the time, it does provide some important pointers for others developing their own policy frameworks. The key elements are applicable in any situation where there is ongoing, active management of environmental water, whether this is in a simple environmental reserve situation or the management of environmental water rights in a complex water system. These include:

1. Decision making should be aimed at achieving the "greatest environmental benefit for the volume of environmental water available" and clear criteria need to be developed to enable comparisons between competing environmental options.
2. The decision-making framework needs to be
    a. transparent, logical, and easily understood by the community;
    b. able to be applied in all climatic conditions;
    c. able to provide guidance on environmental trade-off in all situations including extreme water scarcity.
3. Decisions made under the framework need to be based on good scientific evidence.

23.5 ANNUAL PLANNING FOR THE USE OF ENVIRONMENTAL WATER

Although having a clear and transparent policy framework for the allocation of available environmental water is a critical component of active environmental water management, an annual planning and prioritization process is also critical to identify the suite of options for the use of environmental water (to which environmental water managers then apply the policy framework). This provides transparency around environmental water use and provides confidence that the choice of sites is planned, and their needs are evidenced, so that the public can understand the unfolding watering activities.

Ideally, annual planning is built from the *bottom up* with the local river and wetland managers identifying the needs of their wetlands and river reaches based on antecedent conditions and their technical environmental flows assessment studies. It needs to incorporate all the river reaches and wetlands within the geographic area and water systems across which the environmental water rights can potentially be used. This provides the portfolio of environmental options that environmental water managers can assess against the available environmental water to identify those that represent the greatest environmental benefits. In large systems with interests of many jurisdictions or levels of government this will require a coordinated effort to ensure local information is fed into the broader plan.

However, at the time of planning, the actual volume of environmental water provided against their environmental water rights will not be known, given that it is dependent on system inflows. Therefore, the environmental water managers will need to assess the portfolio of environmental options under a range of climatic scenarios, aligned to the seasonally adaptive approach outlined in Fig. 23.2. They will need to identify a suite of priorities under drought, dry, average, or wet scenarios. This is necessary because just allocating the water is not the end of the story—it also needs to be delivered on the ground. This may require community notifications, agreement with other environmental water managers, negotiation with storage operators, coordinating with other water deliveries (e.g., for town water supply or irrigation), risk mitigation strategies to be in place on site (e.g., to avoid salinity peaks or low oxygen events), and the use of local infrastructure.

Providing a plan outlining priorities under different climatic scenarios enables all the players (including the local river/wetland managers, the storage operators, local infrastructure operators) to undertake the preparatory work to enable the actual delivery of environmental water to the agreed sites. Chapter 24 describes in more detail some of the issues in actual operational delivery of environmental water.

In the MDB, annual planning for the use of environmental water follows this broad outline.

Each year, environmental water holders look at a range of possible water resource scenarios that are developed from climate predictions coupled with hydrological modeling that predicts water availability for all water entitlement holders. They put together their watering plans for that year using proposals from local river and wetland managers outlining the specific water requirements for individual rivers or river reaches and wetlands in line with local objectives and environmental flows assessment studies. These proposals outline the water required under the potential range of water availability scenarios: drought, dry, average, and wet. They are developed in consultation with the community either through the formal establishment of environmental watering advisory groups or through discussion with existing local community groups.

The portfolio of local proposals is then considered by environmental water holders against the policy framework to identify best environmental use under the range of water availability scenarios. Decisions to commit water for use at particular sites in the current year must be weighed against decisions to hold (or carry over) water for potentially higher priorities in the following year or trade water to consumptive users. In the MDB, where there are multiple environmental water holders with differing geographic scope for their water use, they work in consultation to determine their individual and collective priorities and, in doing so, ensure that they can meet their own statutory obligations while also achieving the optimal suite of environmental outcomes across the landscape, regardless of jurisdictional boundaries.

The outcomes of these decision-making processes are then published by each environmental water holder in a seasonal or annual watering plan together with the objectives and rationale for decision making. These plans outline in detail the river reaches or wetlands that will be watered under each water availability scenario, the volume that will be provided and the expected environmental outcomes of the watering. Figs. 23.3 and 23.4 show the outcomes of this planning for the Loddon River, Victoria, taken from the VEWH Seasonal Watering Plan 2012−13, which also indicates opportunities for the Commonwealth to partner to provide the agreed flow regime (VEWH, 2012).

The seasonal watering plans of the VEWH and the annual watering use options of the CEWH provide this level of information on all planned watering activities. In line with their statutory and accountability requirements, these plans are publicly available on their websites. The work shows the clear input of local river managers, the level of local consultation undertaken, the significant amount of background technical information, and provides a high level of transparency around decision making. This is backed by a recent audit by the Australian National Audit Office (ANAO, 2013) of Commonwealth environmental watering activities, which concluded that the CEWO's water-use planning and decision-making approach is:

1. sound and appropriately underpinned by an assessment framework that is mostly applied as intended; and
2. has been progressively strengthened and enhanced over time.

Planning scenarios			
Drought	**Dry**	**Average**	**Wet**
Expected availability of Water Holdings	9461 mL Victorian Water Holdings 820 ML Commonwealth Environmental Water Holdings	9991 mL Victorian Water Holdings 1619 ML Commonwealth Environmental Water Holdings	10,349 mL Victorian Water Holdings 2159 ML Commonwealth Environmental Water Holdings	10,349 mL Victorian Water Holdings 2159 ML Commonwealth Environmental Water Holdings
Environmental objectives	Maintain channel form Maintain in-stream and riparian vegetation Reduce encroachment of terrestial vegetation Maintain water quality	Maintain channel form Maintain in-stream and riparian vegetation Reduce encroachment of terrestial vegetation Maintain water quality	Maintain channel form Maintain in-stream and riparian vegetation Reduce encroachment of terrestial vegetation Maintain water quality Sediment flushing in pool to allow for fish habitat	Maintain channel form Maintain in-stream and riparian vegetation Reduce encroachment of terrestial vegetation Maintain water quality Sediment flushing in pool to allow for fish habitat
Priority watering actions	Autumn/winter/ spring low flows Spring fresh Winter low flows (2013–14)	Autumn/winter/ spring low flows Spring fresh Winter low flows (2013–14)	Autumn/winter/ spring low flows Spring fresh Winter/spring low flows (2013–14) Spring fresh (2013–14) Summer freshes	Autumn/winter/ spring low flows Spring fresh Winter/spring low flows (2013–14) Spring fresh (2013–14) Summer freshes
Possible volume required from the Water Holdings	10,418 ML	10,418 ML	11,498 ML	11,498 ML
Possible carryover into 2013/14	392 ML	1721 ML	1539 ML	1539 ML

FIGURE 23.3

Planned watering scenarios and environmental objectives for the Loddon River, Victoria, Australia, in 2012—13.

Source: VEWH (2012)

It should also be noted that, although considerable effort goes into planning, decisions may change in response to the unfolding season. For example, a planned environmental watering may no longer be necessary if enough rainfall occurs in the local catchment at the right time or a flood event may suddenly provide opportunities that were not possible at the start of the season. Environmental water holders meet regularly throughout the year to consider the climatic conditions and review whether there are new and better opportunities for watering. Where variations are made to their seasonal watering plans, environmental water holders are required to inform the public on their websites. Actual environmental water deliveries are reported in the subsequent year together with known environmental outcomes (CEWO, 2013a, 2015b; MDBA, 2012, 2014a; VEWH, 2013, 2014, 2015a).

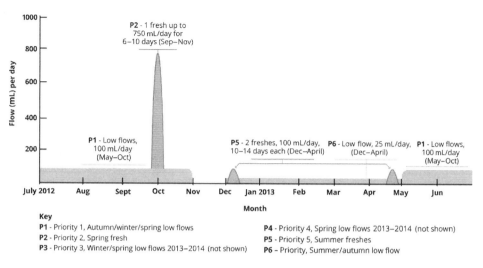

FIGURE 23.4

Planned water scenarios and prioritized flow components for the Loddon River, Victoria, Australia, in 2012–13.

Source: VEWH (2012)

23.5.1 PLANNING FOR THE NEXT FEW YEARS: DECISIONS ON CARRY OVER AND TRADE

Planning must also ensure that the future requirements of priority river and wetland assets are recognized and factored into annual planning processes. This is generally dealt with by allocating environmental water to be carried over into the subsequent year (i.e., water allocated in one year can be kept for use in the following year, subject to certain conditions). This can only occur where the environmental water rights allow for it and is generally only a feature in water systems where there is significant storage available.

Carry over provides flexibility and enables environmental water to be delivered at a time that is of the greatest value to the environment. For example, carry over can help ensure that environmental water holders can meet high winter and spring demands when there is a risk that there will be little water available under entitlements at the beginning of the water year. Each year, as river and wetland managers put together their proposals for their local sites, they try to outline the likely water requirements in subsequent years. Together with the long-term water availability outlook, this helps environmental water holders make decisions on how much water should be carried over in each water system at the end of each season and for what purpose (Fig. 23.3).

Environmental water holders in large, regulated systems with an active water market can also use water trading to move water to the rivers and wetlands where it is most needed, and to smooth out some of the variability in water availability across systems and across years.

Environmental water holders can use water trade in two key ways:

1. Enabling environmental water to be moved across river systems and/or between environmental water holders for environmental purposes (i.e., there is no financial consideration associated

with the trade). These are known as administrative water transfers and are required to operationalize many environmental water decisions outlined in seasonal watering plans. These administrative transfers can actually constitute a significant portion of the allocation trade occurring in any one year within the MDB.

2. Selling environmental water allocations to consumptive users or buying water on the temporary water market where it is in line with their statutory objectives (i.e., if it benefits the environment).

In the MDB, both the VEWH and the CEWH have sold water on the temporary market in situations where all foreseeable environmental demands in that water system were able to be met in that year, including sufficient water to carry over. The VEWH has bought or sold a small amount of water allocation each year since it was established in 2011. The CEWH announced its first sale in 2014. The VEWH has also, on very rare occasions, purchased water to enhance environmental outcomes in systems where insufficient environmental water was available. Revenue raised by selling water allocation can be used to purchase water to meet any environmental water shortfalls, or if legislation allows, to invest in measures (i.e., monitoring, technical or small structural works, or other improvements to the performance of the environmental watering program).

Decisions to sell can be very sensitive, particularly when significant public investment has been used to procure the entitlements in the first place. In times of drought, there can be significant pressure from irrigators to put environmental water on the market. This means that any decisions to sell or buy need a clear policy framework and great transparency. The VEWH has developed a water allocation trading strategy (VEWH, 2015c). Fig. 23.5 shows the key considerations guiding carry over and trade decisions (VEWH, 2015b). The CEWH developed a Commonwealth Environmental Water Trading Framework following the release of a discussion paper for public consultation (CEWH, 2014).

A key principle is that if environmental water managers have the opportunity to use the water market, they need a robust policy framework that is transparent and well understood to support their decisions and demonstrate that in making these decisions, they are still achieving the greatest environmental benefits for the water involved in line with the overall policy objective.

23.5.2 PLANNING FOR COMPLEX MULTISITE WATERING AT A SYSTEM SCALE IN HIGHLY INTERCONNECTED SYSTEMS

Finally, the level of interconnectedness in some of the highly regulated water systems can provide opportunities for planning highly complex, multijurisdictional, multisite watering events where environmental water releases are coordinated from different tributaries to provide significant benefits at multiple sites along a large river section or managed to meet a number of environmental targets in different locations as the water moves downstream.

These exercises require a huge amount of planning and commitment by all players in the system, including the various environmental water managers, river operators, local river and wetland managers, and the support of a range of stakeholders. However, if successful, they can achieve major environmental outcomes in highly efficient ways and can also garner significant public interest and support.

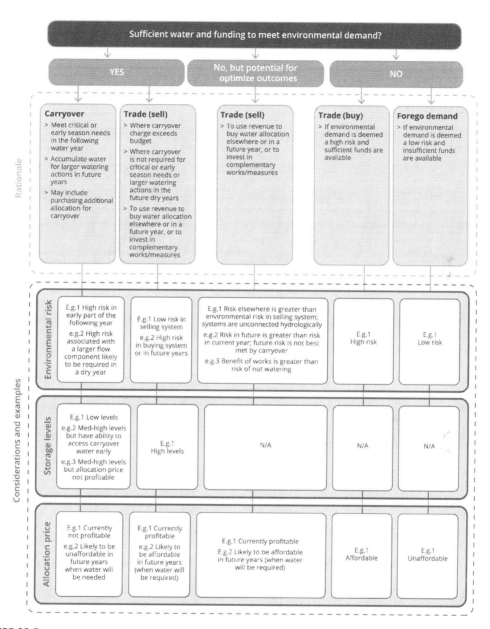

FIGURE 23.5

Victorian decision process for making decisions on carry over or trade of environmental water.

Source: VEWH (2015b)

For example, in the MDB in 2014–15, waterway managers, environmental water holders, and storage managers in northern Victoria worked together to deliver a multisite watering that achieved effective and efficient environmental outcomes in three key ways:

1. Reusing environmental water after it had returned from wetlands to the river;
2. Maximizing environmental benefits from consumptive water on its way to users;
3. Combining environmental and consumptive water releases to achieve greater results (i.e., environmental water was *piggybacked* on other water deliveries to achieve larger flows required to trigger fish breeding).

Fig. 23.6 shows how this was undertaken and the planned environmental outcomes it achieved.

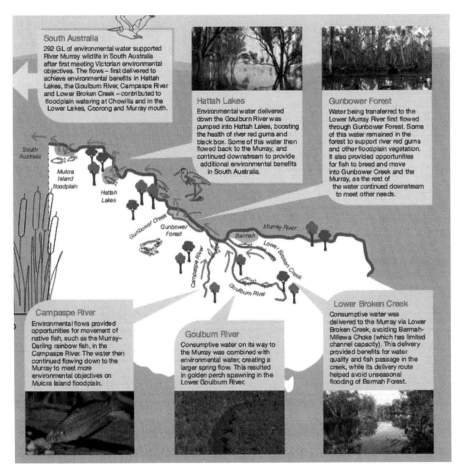

FIGURE 23.6

Coordinated multisite watering event conducted in northern Victoria, Australia, in 2015 and the ecological outcomes and benefits it generated.

Source: VEWH (2015a)

Similar large, high-flow pulse releases have been coordinated and managed in the Colorado River in the United States, with the most recent 2014 release delivering 132 GL all the way through to the river delta in Mexico. The immense effort involved in this is described in more detail in Chapter 22.

In large-scale, highly regulated systems and interconnected systems, the possibilities for environmental water managers are almost endless. In 2014, the VEWH utilized the Victorian water grid and negotiated a substitution of its water held in a southern catchment for water held by urban retailers in the north to underpin a floodplain watering event requiring large volumes of water. In the end, it was not used because natural rainfall occurred. However, the proposal showed that the potential benefits of this type of approach will only increase as environmental water managers learn more about their systems, storage operators learn how to integrate a new suite of environmental demands into their operations, and other water right holders recognize the opportunities that can be provided.

23.5.3 CONCLUSION

Active environmental water management occurs whenever environmental water managers have some discretion or choice about the use of environmental water. This can occur in a simple one-river, one-dam system with an environmental contingency reserve requiring only decisions on release triggers or in highly complex water systems where environmental water managers hold environmental water rights and have considerable discretion about how, when, and where they can use them. In either case, but particularly in the latter, where they hold valuable water assets on behalf of the public, they need to have robust institutions to enable them to undertake their role and clear and transparent policy, and planning frameworks to guide their decision making. These policy and planning frameworks need to ensure that:

1. Environmental water use at local sites is aimed at achieving clear agreed river and wetland management objectives and is actively managed locally as part of a river or wetland restoration program including complementary water quality and habitat improvement actions.
2. Decision making on the use of environmental water is aimed at achieving the "greatest environmental benefit for the volume of environmental water available."
3. There is a transparent, logical decision-making process that is easily understood by the community and based on the best available scientific information.
4. There are robust planning arrangements that ensure local river and wetland requirements are fully understood and considered in decision making on the use of environmental water across river systems.
5. Opportunities for collaboration between environmental water holders, storage operators, and other water right holders are encouraged and innovative solutions can be tested.

The water systems in which active environmental water management can occur are often the most highly modified and least natural systems. Although this can be a challenge, it can provide many opportunities for active environmental water managers. However, it does require them to think differently, to understand how the water system operates, what other water right holders are doing, or needing to develop innovative ways to utilize their water and get the best environmental outcome possible.

23.6 OPPORTUNITIES AND CHALLENGES

Active environmental water management is effectively a new management discipline and people involved in it now are setting the policy and management frameworks that will continue to be refined over future years. A body of practice is being developed, which is summarized in this and other chapters. Although huge advances have been made in the planning and use of environmental water over the last decade, there are still significant challenges and opportunities facing environmental water managers, which need to be taken into account as they develop their policy and planning frameworks. These include:

1. The need to maintain their social license to operate;
2. Understanding what is meant by greatest environmental benefit;
3. Increasing community demands on environmental water use;
4. Emerging knowledge gaps.

23.6.1 SOCIAL LICENSE TO OPERATE

As they are actively making decisions on the use of environmental water, which can affect local communities, environmental water managers in these situations need to realize that they have a public face and local profile. How they coordinate with local river and wetland managers and how they consult and inform the public will all affect the level of community support for their operations and ultimately determine their social license to operate. This reinforces the need to have logical and transparent decision-making frameworks, which take account of local needs and where decisions can be explained and understood by local communities.

Environmental water managers need to be efficient and effective, transparent and accountable in the planning and management, and sensitive to local issues. They also need to monitor community attitudes. Experience in Victoria during the Millennium Drought showed that communities supportive of environmental water releases in periods of good water availability later called for environmental water to be returned to towns and irrigators when scarcity hit.

This means that environmental water holders need to be canny in their operations, alert to community attitudes, and able to recognize emerging issues before they become full-on community controversies. Social license to operate is a concept that is difficult to define well. It becomes more clearly recognized after it is lost. However, experience shows that it is relatively easy to lose and very difficult to regain it once lost. This is discussed further in Chapter 27.

23.6.2 UNDERSTANDING GREATEST ENVIRONMENTAL BENEFIT

A key principle in active environmental management is the need to achieve the greatest environmental benefit. This requires predicting environmental outcomes of competing watering options and being able to compare them. The criteria and approaches described in this chapter are first-cut attempts at doing this. However, challenges still remain, particularly in the comparison of potential watering options. These include:

1. Predicting and valuing environmental benefits in current years compared with future years.
2. Understanding the relative importance of key ecosystem elements, for example making decisions to create the flow conditions for bird breeding events versus fish migration.
3. Understanding the changing benefit values as water availability increases. In a drought, all options are generally valuable and the choice is made on which is most valuable. In wetter conditions, the environmental values and benefits become more marginal.
4. Predicting and valuing reduction in ecological risk, for example providing an increase in baseflow(s) in a stream may not trigger any ecological event with obvious benefits. It may simply reduce the general risk of ecological decline. This may be important over the longer term but will generally rate lower than more obvious benefits.

A further consideration is the longevity of benefits, given climate change and the question as to whether we should be adjusting our environmental watering programs now to more fully prepare for a climate change future. This would require consideration and decisions on some very difficult scientific and societal questions. Will it be possible to maintain current management objectives under climate change? If so, then how much additional water will be required? Should there be further water recovery for the environment and at what financial, social, and economic cost? When do we change our management objectives and to what? All these questions are legitimate management questions but will require a mature community debate with the scientific community making an informed and well-considered input.

Similarly, there will be a constant pressure to demonstrate to the community and governments that the operations are efficient and effective and that ecological outcomes are being achieved. Chapter 25 deals with the scientific issues around establishing monitoring programs to show success. However, although scientific validity of monitoring programs and results are obviously important, they can take considerable time to implement and can be relatively obtuse to the community. Environmental water managers need scientific support but also need to show local and immediate effects, use local community groups as volunteer monitors, collect citizen data and local anecdotes. They need to show and celebrate success and involve their local communities even where it may not be a full scientifically valid experiment. Environmental water managers in Australia provide annual watering reports showing ecological outcomes from their watering (CEWO, 2015a; MDBA, 2014a; VEWH, 2015a).

23.6.3 INCREASING COMMUNITY DEMANDS ON ENVIRONMENTAL WATER USE

An interesting dynamic is emerging in Australia where some communities are happy to support environmental water management where it provides regional benefits including recreation and tourism. Through the Millennium Drought in Australia, as local lakes dried up often for the first time, communities became fully aware of how important those environments were for amenity, regional tourism and recreation, and how much influence they had on local economies. These communities are now advocating for social and cultural benefits to be taken into account when setting priorities for the use of environmental water. Previous chapters recognized that providing good environmental conditions can also provide some social benefits. However, as active environmental water management decisions are being taken in some areas of Victoria, communities are actively trying to influence them to ensure that their particular lakes and wetlands are selected over others. This reinforces the need for clear and transparent policy frameworks for decision making. However, it also raises the issue of how these secondary benefits should be taken into account in the development of these frameworks. For example, in Victoria, the opportunity to provide social and cultural benefits is assessed but only after consideration of all the other environmental criteria (DEPI, 2013).

In many cases, the intrinsic community benefits provided by environmental water thereby secure community support. Environmental water holders need to be aware of these benefits as well as benefits provided to indigenous communities and factor it into their decision making wherever they can, while still meeting the primary environmental purpose of their entitlements. Where it is not possible, it needs to be very carefully communicated to a community in order not to create more long-term opposition to their operations and/or lose their social license.

23.6.4 NEW KNOWLEDGE

Environmental water management is a new and growing discipline. Currently, it is mining the available science generated through environmental flows assessment studies aimed at securing environmental water. Although this is clearly very important, efficient management under a range of climate conditions will require new and more applied science. There is clearly a role for a partnership with scientists, but to be effective, it will need to be more agile and responsive to emerging practice and more integrated with river management and operations than has been traditionally the case.

Finally, in the past decade, there has been a rapid evolution in the policy and planning frameworks for the management of environmental water. This has occurred in response to a rapid increase in the volume of water entitlements made available to the environment and the corresponding pressure to demonstrate good governance and efficient and effective use. This pressure will continue to mount in the future as the environmental water managers become more sophisticated in their operations, and, in particular, as they start to use water markets as part of their portfolio for achieving their environmental objectives. It will also increase as the community becomes more comfortable with their operations and starts to take more interest in their decisions and demand more from them in terms of social benefits. Environmental water management is an area where technical advances will always have to be coupled with corresponding advances in governance and accountability measures, and where high-quality communication with local communities will be one of the hallmarks of success.

REFERENCES

ANAO, 2013. Commonwealth Environmental Watering Activities Canberra. Australian National Audit Office, Australia.

CEWH, 2014. Commonwealth Environmental Water Trading Framework. Commonwealth Environmental Water Holder, Canberra, Australia.

CEWO, 2013a. Environmental Outcomes Report 2012—13. Commonwealth Environmental Water Office for the Australian Government, Canberra, Australia.

CEWO, 2013b. Framework for Determining Commonwealth Environmental Water Use. Commonwealth Environmental Water Office for the Australian Government, Canbera, Australia.

CEWO, 2015a. Improved outcomes for native fish, birds, frogs and habitat from environmental watering 2014−15 Outcomes Snapshot. Commonwealth Environmental Water Office for the Australian Government, Canberra, Australia.

CEWO, 2015b. Monitored Outcomes 2013−14. Commonwealth Environmental Water Office for the Australian Government, Canberra, Australia.

DEE, 2016. *Environmental water holdings* [Online]. Australian Government Department of the Environment and Energy. Available from: http://www.environment.gov.au/water/cewo/about/water-holdings (accessed 01.04.16).

DEPI, 2013. Victorian Waterway Management Strategy. Victorian Department of Environment and Primary Industries, Melbourne, Australia.

DSE, 2008. Environmental Watering In Victoria 2007/08. Victorian Department of Sustainability and Environment, Melbourne, Australia.

DSE, 2009. Northern Region Sustainable Water Strategy. Victorian Department of Sustainability and Environment, Melbourne, Victoria.

DSE, 2011a. Gippsland Region Sustainable Water Strategy. Victorian Department of Sustainability and Environment, Melbourne, Victoria.

DSE, 2011b. Western Region Sustainable Water Strategy. Victorian Department of Sustainability and Environment, Melbourne, Victoria.

MDBA, 2012. The Living Murray environmental watering in 2010−11. Murray-Darling Basin Authority, Canberra, Australia.

MDBA, 2014a. 2014−15 The Living Murray Annual Environmental Watering Plan. Murray-Darling Basin Authority, Canberra, Australia.

MDBA, 2014b. Basin-Wide Environmental Watering Strategy. Murray-Darling Basin Authority, Canberra, Australia.

VEWH, 2012. Seasonal Watering Plan 2012−13. Victorian Environmental Water Holder, Melbourne, Australia.

VEWH, 2013. Reflections—Environmental Watering in Victoria 2012−13. Victorian Environmental Water Holder, Melbourne, Australia.

VEWH, 2014. Reflections—Environmental Watering in Victoria 2013−14. Victorian Environmental Water Holder, Melbourne, Australia.

VEWH, 2015a. Reflections—Environmental Watering in Victoria 2014−15. Victorian Environmental Water Holder, Melbourne, Australia.

VEWH, 2015b. Seasonal Watering Plan 2015−16. Victorian Environmental Water Holder, Melbourne, Australia.

VEWH, 2015c. VEWH Water Allocation Trading Strategy 2015−16. Victorian Environmental Water Holder, Melbourne, Australia.

VPSC, 2016. Public entity types, features and functions [Online]. Victorian Public Sector Commission. Available from: http://vpsc.vic.gov.au/governance/public-entity-types-features-and-functions/ (accessed 01.04.16).

ENVIRONMENTAL WATER DELIVERY: MAXIMIZING ECOLOGICAL OUTCOMES IN A CONSTRAINED OPERATING ENVIRONMENT

24

Benjamin B. Docker and Hilary L. Johnson

Commonwealth Environmental Water Office, Canberra, ACT, Australia

24.1 INTRODUCTION

Freshwater biodiversity is some of the most threatened globally, in large part due to catchment and water resource development aimed at increasing economic possibilities for communities (Vörösmaty et al., 2010). Overexploitation has often resulted from this development, and in many instances led to detrimental impacts on human well-being, expected to be exacerbated by climate change (Daily, 1996; Poff et al., 2010; Postel and Richter, 2003; Vörösmaty et al., 2000, 2010). Mitigating these impacts and restoring ecosystems to health has become a priority for governments around the world and the provision of water for the environment is a mechanism sometimes used to help achieve this. The principle of environmental water is simple; more water for instream, flood-plain, and wetland uses than is otherwise provided for, due to development of the resource for other purposes.

Delivering this water through environmental water releases is often based on the premise of the natural flow regime (Poff et al., 1997) and the reinstatement of ecologically important components of it (Bunn and Arthington, 2002). In highly modified rivers a *designer* approach that targets specific ecological and ecosystem service outcomes may be better suited to achieving ecological conservation where limiting change from natural conditions is no longer feasible (Acreman et al., 2014). Considerable work has gone into understanding how much water should be allocated and the way in which it should be provided to the environment, with many different methodologies and tools developed for this purpose (Tharme, 2003). However, once the theoretical flow requirements have been determined (Chapter 11) and plans developed (Chapter 23), much less focus has been applied to the task of how to actually implement environmental water regimes. This is particularly important where the environmental water allocation mechanism requires ongoing active management (i.e., environmental water rights), and this forms the focus of this chapter.

Water for the Environment. DOI: http://dx.doi.org/10.1016/B978-0-12-803907-6.00024-3

563

Implementing environmental water rights poses challenges when regulatory environments and water resource governance structures were designed for different purposes and objectives. These systems create a number of challenges for the environmental water manager, which would not exist if a regulatory and governance model was designed with all resource uses in mind, with system sustainability as a fundamental tenet. In historical terms, environmental water users are newcomers and so often find themselves operating in a system not designed to meet their particular requirements. Existing water users' rights need to be respected, but that means the overall design of institutional arrangements and the management of the resource (e.g., how water orders are made and how volumes deemed to be used are accounted for in different parts of the system) is likely to be less than optimal as adjustments and temporary arrangements with high transaction costs (i.e., new methods to account for instream use rather than water extracted) are made to accommodate the new user. In this context, the environmental water manager needs to be flexible and adaptive, and at times creative, to get the best outcome while working collaboratively with other users and river operators to ensure that the operating regime evolves over time to a more efficient and productive paradigm.

Experience over a number of years in different river basins around the world has highlighted the challenges and the innovative and ad hoc approaches that have been put in place in order to achieve environmental outcomes from environmental water delivered in systems not specifically designed to support them. Understanding these challenges is important for informing how new environmental water programs might be developed and implemented in river basins that would benefit from them, particularly in places where new storages are being built and there is still an opportunity to design an operating regime for meeting environmental as well as social and economic needs. According to the World Watch Institute, dam building around the world is increasing rapidly. Between 2003 and 2010 the total installed hydropower capacity increased by 3.5% annually (WWI, 2012). Designing operating and regulatory arrangements that encompass all uses, including environmental uses, from the outset is likely to significantly reduce the transaction and operational costs from retrofitting systems that have been operating for some time, leading to better triple bottom line outcomes.

Studies elsewhere have examined the implementation challenges associated with acquiring water for the environment, notably in the United States, for example, Horne et al. (2008) and Hardner and Gullison (2007) in the Columbia Basin, Neuman and Chapman (1999), Neuman (2004), Neuman et al. (2006), and King (2004) in Oregon and Washington, and Ferguson et al. (2006) in Montana. Other than a brief overview of the Murray–Darling Basin, Australia, in Banks and Docker (2014), very little has been documented about the use of environmental water that gives effect to those acquisitions through operational management of flows. This is a pertinent area of concern for governments and citizens if they are to fully realize the benefits of often significant investments in acquiring rights to water for the environment.

This chapter seeks to outline the key implementation challenges for environmental water rights, with a view to considering opportunities to address these challenges where they are already present and to avoid them in places where they are not yet manifest. It does so by drawing on examples from case studies of the Murray–Darling Basin and two basins in western North America. These examples are presented to illustrate the operational arrangements generally required for implementing environmental water releases through the provision of environmental water rights (Section 24.2), to outline the delivery challenges both in terms of maximizing environmental outcomes and managing within physical, legislative, and societal constraints (Section 24.3), and to document continuous improvement within this context, including through adaptive management in practice (Section 24.4).

24.1.1 THE MURRAY−DARLING BASIN

The development of the water resources of the Murray−Darling Basin occurred primarily for navigation and irrigation purposes from the early days of the colonies. The implementation of environmental water releases has taken place much more recently, but still over a period of more than 30 years. Environmental water delivery occurred first at a catchment scale with flows focused on specific environmental assets (e.g., floodplain wetlands), then through state-wide programs aimed at improving river and wetland health more broadly and ultimately through cross-border initiatives at a subbasin and then whole-of-basin scale (Docker and Robinson, 2014). This progression has seen greater volumes of water recovered from consumptive users for the environment as governments have sought to address the recognized imbalance in water use (MDBA, 2010a, 2010b).

Initially, environmental water regimes were conceived as operating rules in statutory water-sharing plans, but with the separation of property rights in water from land and the development of the water market, the environment has increasingly come to rely on the use of legal water entitlements as the means to accessing and using water both instream and on land. Although conditions on storage operators and private water abstractors remain the primary mechanisms to provide environmental water (MDBA, 2010a, 2010b), environmental water rights are seen as increasingly important due to the capacity for active management and therefore more efficient use (Docker and Robinson, 2014). Within the Murray−Darling Basin, entitlements for the environment have been principally recovered through investments in irrigation efficiency and through direct market purchases, a process that is planned to continue until 2024. Entitlement-based environmental water is now being held and delivered each year, subject to water availability, across 16 of the 22 river catchments within the Basin guided by overall objectives to protect and restore water-dependent ecosystems and ecosystem functions, and ensure their resilience to climate change (Commonwealth of Australia, 2012). Expected basin-wide environmental outcomes have been described for fish, water birds, vegetation, and hydrological connectivity (MDBA, 2014a), although they are only indicative of broader ecosystem health and support to the productive base of the water resources.

24.1.2 THE COLUMBIA AND COLORADO BASINS

Environmental water regimes are implemented across several river basins in western North America, including through the operational management of reservoirs to support fish habitat and passage and through the acquisition of water extraction rights from farmers to allow water to remain instream at critical fish spawning locations. This chapter draws on examples of implementation challenges from the Columbia and Colorado River basins.

The Columbia River Basin was developed for both irrigation and hydropower from the early days of the twentieth century (BPA et al., 2001). It is governed by both Canada and the United States, with an agreement between the two countries on how water releases will occur between upstream and downstream storages, principally for flood control and power production objectives (BPA et al., 2001). Environmental water releases have been implemented primarily due to concerns for fish populations and the capacity for migration and spawning due to altered habitat resulting from the large infrastructure works and their operations. Environmental water is provided through operating regimes applied to major storages, particularly on the main river channels, and through market mechanisms largely in upstream tributaries. Unlike the government-led approach to the use of market mechanisms implemented in Australia, efforts in the western United States since 2002 have focused on providing a framework to support local organizations to acquire water for instream use at a river reach scale. Environmental water in this context is that which has been left instream as a result of the acquisition from irrigators of the rights to extract water. In this regard, they are similar to the acquisition of *unregulated* water rights (a right to extract passing flow) in the Murray–Darling Basin where there are limited operational decisions to be made, and implementation following acquisition is concerned primarily with decisions about when to *activate* or *call* the right that has been acquired and to ensure its instream protection from other users.

The Colorado River Basin has been described as one of the most regulated rivers in the world (Blinn and Poff, 2005). Storages in the Basin can hold over four times the average annual flow of the river, the vast majority of which is held in Lake Mead and Lake Powell, the two largest reservoirs in the United States (Adler, 2007). Over 30 million people rely on the Basin's rivers for drinking water, which also support irrigated agriculture and hydroelectric power generation (Adler, 2007; Jacobs, 2011). Environmental water releases have been implemented through operating regimes applied to major storages such as the Glen Canyon Dam on Lake Powell. Located upstream of the Grand Canyon, trial releases (termed *high-flow experiments [HFEs]*) from Glen Canyon Dam have been undertaken since 1996. The releases are primarily focused on reinstating "controlled floods" for the purpose of restoring sandbars and beaches along the Colorado River corridor in the Grand Canyon National Park and Glen Canyon National Recreational Area. These sand features and associated backwater habitats can provide key wildlife habitat, potentially reduce erosion of archeological sites, enhance riparian vegetation, maintain or increase camping opportunities, and improve the wilderness experience (Reclamation, 2011). Although the HFEs change the timing and rate of releases from the dam, they do not change the annual volume of water released and made available to downstream users. However, to maximize the rate of release from the dam (and hence the magnitude of the flow peak), water can be delivered through penstocks that lead to hydroelectric generators and through bypass tubes that are not equipped with power-generating capability, with the latter leading to reduced power production (Reclamation, 2011).

In most years, 90% of the Colorado River's flow has been consumed by the time it reaches the US–Mexico border; less than 1% of historical flows now make it to the river's lower delta and estuary (Buono and Eckstein, 2014). A 2012 amendment to the 1944 treaty between the United States and Mexico governing the Colorado River, Minute 319, provided a one-off, high-volume *pulse flow* to support the Colorado River Delta (IBWC, 2012). The water was provided by the United States and Mexican governments, as a result of water savings generated from repairs to infrastructure. The pulse flow delivered 132 GL to the Colorado River Delta in 2014 and has been complemented by base flows provided by the Colorado River Delta Water Trust, a binational coalition of nongovernment organizations that have been purchasing water rights since 2008 (Buono and Eckstein, 2014).

24.2 OPERATIONAL FLOW MANAGEMENT FOR THE ENVIRONMENT
24.2.1 GOVERNANCE AND IMPLEMENTATION ARRANGEMENTS

When considering governance arrangements for environmental water management, it is necessary to first consider the broader legal and regulatory environment. This legal and regulatory environment has a critical impact on the capacity to effectively deliver environmental water and the need for constant innovation. Issues include how water is made available to different uses, when and where it can be taken, stored, and traded, and the approvals required and protections in place for different types of users. Managing these issues is invariably different in different jurisdictions, but generally there will be a regulatory architecture, backed by legislative or common law practice that characterize the property rights to water, whether for public or private use.

The Murray–Darling Basin

In the Murray–Darling Basin, there are a number of entities that have responsibilities for actively managing environmental water (Banks and Docker, 2014). This includes the *Commonwealth Environmental Water Holder (CEWH)* (a federal entity established under the *Water Act 2007* (Commonwealth of Australia, 2007) to manage the Commonwealth's environmental water holdings), the Victorian Environmental Water Holder and state agencies in New South Wales and South Australia (which manage state-owned water entitlements and/or allocations), and The Living Murray initiative (MDBA, 2011), a joint Commonwealth–state program pre-dating the *Water Act 2007*, that provides environmental water to six icon sites along the Murray River. Each water holder manages a portfolio of water entitlements of varying characteristics that must be used to give effect to the Basin Plan 2012 (Commonwealth of Australia, 2012). This can involve making decisions on the delivery, carryover, and trade of the resource, while working cooperatively with other environmental water holders.

The Basin Plan 2012 includes an Environmental Watering Plan, which requires the *Murray–Darling Basin Authority (MDBA)*, a federal entity with responsibility to develop and facilitate the implementation of the Basin Plan, to provide overall guidance on environmental water use through a basin-wide environmental watering strategy, and requires states to provide guidance at a valley scale through long-term environmental watering plans for each catchment. Annual priorities for environmental water use are also prepared at a basin scale by the MDBA and at a valley scale by the states. These collective priorities represent the current understanding of environmental demands for water and therefore inform the decisions of environmental water holders and managers on the most effective use of their water portfolios for the coming year and over the longer term. Government bodies are advised by environmental water advisory groups, which differ in form and function in different valleys and states. These are quasi-representative/technical committees often made up of government officials, scientists, landholders, and indigenous members who draw upon local and on-ground knowledge to advise governments on the most appropriate environmental water use.

The delivery of environmental water generally requires an environmental water manager to make a decision on the use of water for a particular outcome and a river operator to implement the decision as requested through the normal processes by which entitlement holders order water. Once a decision to make water available for use has been made, then an operational advisory group may be formed to support the implementation of a watering action. These are event-specific stake-holder groups, some more formal than others, that ensure open communication between relevant parties so that the event can be effectively implemented and actively managed in response to changing circumstances. They generally include river operators and environmental water managers and other operational and technical agencies, as required. In 2014—15, for the Murray River alone, the MDBA convened 230 Operational Advisory Group meetings (IRORG, 2015).

In the Murray—Darling Basin, the delivery of environmental water is a government-managed process. Although community input and advice are critical (particularly through local advisory groups, natural resource management bodies, and in some cases nongovernment organizations), decision making is undertaken by government agencies acting in the broader public interest. State governments are concerned with valley-scale and connected valley outcomes within their jurisdiction, whereas the CEWH takes a basin-wide perspective with regard to coordinating environmental water releases through multiple states, balancing upstream and downstream environmental objectives, and making trade-offs in local objectives, where necessary, for the benefit of the wider basin environment (Connell, 2011; Docker and Robinson, 2014).

Western United States

Within the Columbia River Basin, environmental water exists within a system governed by the overarching Columbia River Treaty, which specifies arrangements for operating large dams on both the Canadian and US sides of the border in a way that minimizes flood risk and maximizes power production, which is then shared between the parties. A series of federal structures in the United States operate under this agreement, although for the most part this arrangement is about regulating instream flows rather than managing consumptive use, which remains the responsibility of the states (primarily Idaho, Oregon, Washington, and the Canadian province of British Columbia). The Columbia River Treaty only mentions irrigation once and most diversions for irrigation occur in the tributaries, not the mainstream. Indeed, although Washington has one large irrigation project, the Columbia Basin Project, diverting 74% of all irrigation water from the Basin (NRC, 2004), the state of Oregon effectively has a ban on diverting from the mainstream of the Columbia River (MacDougal and Kearns, 2014).

For the federal dams there are two agencies responsible for operations, the Federal Bureau of Reclamation and the US Army Corp of Engineers. Along with the *Bonneville Power Administration (BPA)*, which markets the power produced, these are known as the Action Agencies. Collectively, they seek the efficient operation of the system, and since 1995 this has involved operations to support the recovery of fish listed under the *Endangered Species Act 1973*. In providing flows for fish they are guided by Biological Opinions provided by National Oceanic and Atmospheric Administration Fisheries and the National Fish and Wildlife Service. These Opinions include documented recommendations for actions that the agencies can take to improve outcomes for different species of fish. A *Technical Management Team* with interagency representation advises the Action Agencies on operations that optimize passage conditions for anadromous and resident fish. It meets year round to facilitate the different seasonal requirements for fish, although at different times of the year different requirements (i.e., fish flows, flood control, power generation) will take priority.

The environmental water regime in the mainstem is provided by conditions on storage operators rather than through actively managed environmental water rights. In the Columbia Basin, at the start of the year, *rule curves* are established to guide operations to manage flood control, power generation, and water levels for fish, both migratory and stationary. An annual operating plan is produced, which outlines the general approach to meeting multiple objectives throughout the year.

Given the large number of structures and operating entities, both federal and local, within the Columbia Basin, the parties have entered into what is known as the Pacific Northwest Coordination Agreement. This is an agreement between different operators to enable synchronized operations that maximize power generation while meeting other nonpower needs, including environmental water releases for fish and water supply for irrigation as efficiently as possible. Although most irrigation supply is undertaken by local operators rather than through the federal system and individually the impacts these operations have on the functioning of the whole system is not considered significant, collectively, it can have a large (particularly seasonal) impact as irrigation supply requirements do not necessarily align with other system uses in terms of the timing of filling and releasing water from storage (NRC, 2004). On smaller tributaries, the conflict between demand for irrigation withdrawals and releases for fish passage can be particularly acute, hence the establishment of the *Columbia Basin Water Transaction Program (CBWTP)*.

The CBWTP, a partnership between the BPA and *National Fish and Wildlife Foundation (NFWF)*, is managed by the NFWF working through *qualified local entities (QLE)*. Potential QLEs include state agencies, tribes, water trusts, water districts, watershed councils, irrigation districts, and other interested parties. QLEs are responsible for determining which flow components they wish to protect and the water rights that will enable that protection. They put forward applications for funding to be used to enter into a transaction with irrigation districts, landowners, and others to acquire the instream water rights. Funding is provided by the BPA and donors.

As with the Murray–Darling Basin, water rights acquired from irrigators for environmental use in the Columbia Basin are governed by a regulatory regime largely designed for extracting and diverting water rather than leaving it instream. The regulatory arrangements include limits on where, when, and how much water can be taken by a license holder, based on a system of priority rights. The licenses that are acquired are therefore limited by the conditions in place as designed for another purpose and can offer little protection to water left in the river. The Columbia is also complicated by the lack of a basin-wide framework for water allocations across two countries, seven states, and 13 federally recognized native Indian tribes (BPA et al., 2001).

The Colorado River is managed and operated under numerous compacts, federal laws, court decisions, and decrees, contracts, and regulatory guidelines collectively known as the *Law of the River*. This collection of documents, which includes the Colorado River Compact of 1922 and the Mexican Water Treaty of 1944, apportions the water and regulates the use and management among the seven basin states (Arizona, California, Colorado, Nevada, New Mexico, Utah, and Wyoming) and Mexico (Reclamation, 2008).

With respect to environmental water releases from the Glen Canyon Dam, the *Grand Canyon Protection Act 1992* mandates that the dam be operated in a manner that protects and mitigates adverse impacts to, and improves the values for which Grand Canyon National Park and Glen Canyon National Recreation Area were established. This obligation must be implemented consistent with and subject to the Law of the River. A total of five US Department of the Interior agencies and one US Department of Energy agency have responsibilities under the *Grand Canyon Protection Act 1992* (Interior, 2014) to implement this obligation.

The Protocol for HFE Releases from Glen Canyon Dam through 2020 was approved in 2012 and provides the framework for conducting and evaluating HFEs, so as to comply with the *Grand Canyon Protection Act 1992*. The Protocol was developed by the Department of Interior following analysis of HFEs conducted in 1996, 2004, and 2008. It sets the parameters of individual HFEs, including the maximum release rate and the maximum duration of the release (Reclamation, 2011, 2012). These rates reflect and are constrained by both the outlet capacity of the dam, as well as the need to minimize the impact on hydroelectric generation.

To support the coordinated implementation of the Protocol, the Secretary of the Interior issued a Directive in 2012 to establish the Glen Canyon Leadership Team. The six government agencies all have membership on the team. The Leadership Team is directed to work together to ensure that appropriate coordination is undertaken to implement the commitments set out in a protocol and to ensure appropriate external coordination (Interior, 2014). External coordination includes consultation with affected tribes.

The Bureau of Reclamation found that adverse effects to sacred sites could result from the HFE Protocol, primarily from limitation of access of tribes to sacred sites during the period of HFE releases. Reclamation completed the HFE Protocol Memorandum of Agreement (MOA; Reclamation, 2012) with affected tribes and other parties to address these effects. The MOA requires that a meeting be conducted with the parties after each HFE event, to review the effects of the HFE, and use the results of the meeting to inform monitoring for future HFEs, and to design and implement any measures necessary to prevent or control adverse effects of future HFEs. It also requires all the parties are notified at least 30 days in advance of any planned HFEs, and consult with tribes to resolve any conflicts with tribal access to or uses of the Colorado River (Reclamation, 2012).

24.2.2 ADMINISTRATIVE ARRANGEMENTS

To deliver environmental water it is often not simply enough to acquire a water license. Additional approvals and regulatory requirements need to be met. These can include water-use permits that are often tied to a particular piece of land and authorize the extraction at that location, and works licenses, which authorize the use of particular infrastructure for manipulating water flows, be it a pump or some other diversionary structure. These are often the recognized point where water is deemed to have been used and accounted for and where the river operator is seeking to ensure that there is sufficient water instream to meet the order. For an environmental order this point may be a piece of in-river infrastructure (e.g., a diversionary weir) rather than a pump location as it often is for irrigators. As essential regulatory requirements, it is important that these approvals are in place well before an environmental water release is due to occur.

Often the point at which an environmental water manager is seeking an outcome is not the point at which the water right was previously used. The capacity to use the water in a different location is therefore required, and where it exists, the mechanism of trade can be used to transfer the water from one license to another, either between accounts within the environmental manager's own portfolio or between accounts held by different environmental water managers. Within the Murray—Darling Basin, the administrative arrangements required to do this are the same as those used by irrigators when they sell water to another entitlement holder and all transfers are accounted for on state government water registries; a process that can take a day or two within a state jurisdiction and up to 2 weeks for an interstate transaction (NWC, 2011). Licenses to use water also need to identify the specific infrastructure that may be used to extract or divert the water, which is a process that can take weeks, if not months, depending on the jurisdiction (e.g., NSW, 2015a).

River operators are also subject to statutory requirements for managing releases from headwater storages. These requirements often specify the conditions on water releases including maximum flow rates and seasonal requirements as specified in water resource plans. They may include operating targets for water levels both within the reservoir and in the river downstream at different times of year to support the provision of human uses and meet environmental objectives. For instance, the Murray—Darling Basin Plan 2012 specifies that river operators must have regard to water quality targets for dissolved oxygen, salinity, and algae at different locations in the river while also meeting other operational requirements (Commonwealth of Australia, 2012).

24.2.3 RISK MANAGEMENT

In planning environmental water releases, comprehensive risk management is essential and a key feature at multiple governance levels (ANAO, 2013). The most significant risks are operational ones, in particular the risk of unintended flooding causing economic loss, environmental harm through water quality deterioration associated with floodplain inundation, and amenity and social impacts such as disruption of recreational fishing and camping activities due to fluctuating water levels.

Risks should be assessed and discussed with relevant parties during the planning process and prior to decisions to either make water available or deliver it. Mitigating measures can then be agreed between the parties involved and this can involve conditions being placed on approvals to make water available prior to an environmental water release. Such conditions may involve important design considerations including the use of buffers between flow heights sought and those where impacts are possible; changes to the timing of flows to avoid impacts on other water users, for example recreational fishing events; changes to release patterns, for example rates of rise and fall to minimize bank erosion and notching; and implementing complementary measures such as communications to riparian landholders or hikers and campers in National Parks and other recreation areas. Some of these are discussed more in Section 24.3.

24.2.4 OPERATIONAL MONITORING

Monitoring the operational aspects of flow events both during and immediately after they have occurred is also a key risk management measure. Ensuring that the water goes to where it is intended is the first step in any monitoring program and essential both for achieving the desired environmental objectives and for ensuring that unintended consequences do not arise. In practice, it often involves the use of gauge readings, generally available online, and observations from on ground personnel combined with modeled understanding of what the hydrology would have been in the absence of the environmental water release. This takes on particular significance in relation to the protection of environmental water left instream. To the extent that this water is extracted by a downstream rights holder, either legally or illegally, then it is unlikely to achieve the maximum benefit as intended. This issue is discussed further in Section 24.3.

24.3 DELIVERY CHALLENGES IN IMPLEMENTING ENVIRONMENTAL WATER REGIMES

There are often a range of delivery methods available to an environmental water manager (Table 24.1). The methods for delivering water depend on the infrastructure available and the legal and regulatory framework that governs access to water. In some cases multiple methods are used. For example, environmental water in the Goulburn River in the Murray–Darling Basin has been released from an upstream storage to achieve certain vegetation and fish outcomes in the Goulburn River itself. The water that continues to flow downstream to the Murray River is then pumped out for use at Hattah Lakes, a Ramsar-listed wetland on the mid-Murray floodplain several hundred kilometers downstream. Of course, implementing such measures requires supportive accounting and administrative arrangements, and the design of an event needs to take these arrangements into consideration, including, for example, with respect to the accounting for losses along the way and the transaction times and procedures required to ensure that the correct volumes are debited from the appropriate accounts at the correct locations (DSE, 2009).

Environmental water releases do not occur in a vacuum. Any flow that is provided specifically for the environment is done so amidst the background of natural stream flows caused by rainfall, and other operational deliveries made for a variety of purposes including hydropower, urban, industrial, and agricultural use. In this context, each of these delivery methods has implementation challenges associated with its use. Environmental water managers employ a range of strategies to address these challenges and maximize environmental outcomes, including:

1. Delivering water in response to natural cues;
2. Delivering water to target multiple sites and outcomes;
3. Coordinating environmental water delivery with other water sources.

Table 24.1 Main Types of Environmental Water Use in the Murray–Darling, Australia, and Columbia, Colorado, and Walker Basins, United States

	Murray–Darling Basin	Columbia Basin	Colorado Basin
Releasing stored water either standalone or in addition to tributary flows	Y	Y	Y
Using infrastructure to divert water onto floodplain wetlands and creeks	Y	—	—
Preventing diversions through market transactions and leaving water in-river	Y	Y	Y
Raising and lowering weirpools and other reservoirs	Y	Y	—

The ability to maximize outcomes can, however, be limited by physical, legislative, and societal constraints that need to be factored into environmental watering decisions and implementation. Factors that can limit the delivery of environmental water include:

1. Avoiding unintentional flooding;
2. Infrastructure limitations (e.g., dam outlets) and maintenance;
3. Competition for available channel capacity;
4. Minimizing social and economic impacts.

24.3.1 DELIVERING WATER TO MAXIMIZE ENVIRONMENTAL OUTCOMES

Delivering environmental water in response to natural cues

Many ecological responses such as fish spawning or migration and colonial water bird nesting are triggered by hydrological and biological cues. This could include changes in water levels, river flows, water temperature, or carbon and nutrient inputs following rainfall and HFEs (MDBA, 2014a). Some of these cues are complex and not always well known, however, typically they align with the natural hydrology. To maximize the likelihood of success, managers aim to time the delivery of environmental water with natural inflows (see Box 24.1) or inputs (see Box 24.2) into the river system, either observed or modeled.

BOX 24.1 DESIGNING EVENTS IN RESPONSE TO NATURAL CUES IN THE MURRAY–DARLING BASIN, AUSTRALIA

The natural seasonal flow regime in the southern Murray–Darling Basin is typically characterized by high flows in winter and early spring. These flows are particularly important to native fish—they improve the habitat and water quality of deep pools, provide food resources to support the survival of juvenile fish and enable adult fish to grow and enhance their body condition in the lead up to spawning. This ultimately contributes to successful native fish breeding, nesting, and juvenile recruitment. Regulation of the rivers and creeks of the southern Murray–Darling Basin has dampened or even reversed natural flow seasonality, with water now stored or extracted in the winter months and released in the summer for irrigation supply (MDBA, 2014c).

In winter 2015, a novel approach to using environmental water to respond to natural hydrological cues was trialed for the first time in the Murray River. Environmental water releases from one of the main storages on the river, Hume Dam, were guided by inflows to mimic a proportion of the modeled natural flow downstream (i.e., a modeled estimate of the seasonal flows downstream of Hume Dam, if the dam was not present). Releases were initially managed to below channel capacity to minimize the risk of bank slumping and avoid affecting infrastructure works and public access for firewood collection programs. Releases were increased during spring, but were still managed to levels that minimized the risk of third-party impacts. These limits meant that releases followed the pattern but not the magnitude of the modeled natural flow (Fig. 24.1). Although the spawning of golden perch and silver perch was detected during the event, the full ecological outcomes from this approach are not yet known. Nevertheless, some benefits identified by agency staff to-date include:

1. Better alignment in meeting environmental demands, particularly in restoring flow seasonality;
2. Mimicking of natural hydrological patterns of variability and unpredictability;
3. Reduced administrative burden as some operational decisions are automated;
4. Greater certainty for other water users, river operators, and water holders on how environmental water will be used and the basis for decisions.

(Continued)

BOX 24.1 (CONTINUED)

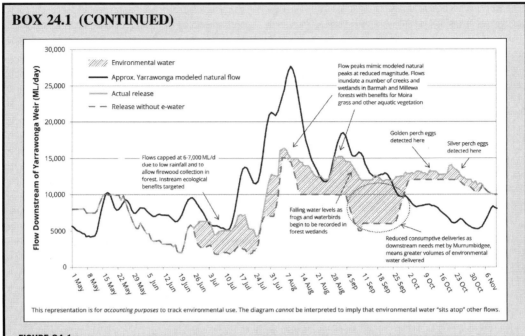

FIGURE 24.1

Hydrograph of actual flows relative to modeled natural and predicted releases in the absence of environmental water in the Murray River, Australia, downstream of Yarrawonga weir between June and November 2015.

Source: Data from Murray–Darling Basin Authority

BOX 24.2 RESPONDING TO NATURAL CUES AS INFORMED BY MONITORING AND MODELING IN THE COLORADO RIVER, UNITED STATES

The principal driving variables for *high-flow experiments (HFEs)* in the Colorado River are sediment and flow. Nearly all the natural sediment load once transported by floods is now trapped behind the Glen Canyon Dam. The Paria and Little Colorado rivers, located downstream from the dam, are now the primary sources of sediment to the Colorado River in the Grand Canyon (Blinn and Poff, 2005). The aim of the HFEs is to restore managed floods that are appropriately timed to build sandbars by mobilizing recently accumulated sand and to avoid eroding older sand deposits (Interior, 2014).

To time the HFE releases for maximum sandbar-building effect, scientists undertake both monitoring and modeling to identify when, how much, and for how long water must be released in a controlled flood to export approximately the same amount of sand as was supplied by tributaries during the input season. This includes continuous monitoring of sand flux at the downstream end of Marble Canyon (located between Glen Canyon Dam

(Continued)

BOX 24.2 (CONTINUED)

and the confluence with the Little Colorado River); modeled estimates of sand inputs from Paria River floods within 24 h of a flood event based on observed correlations among discharge, sand concentration, and sand grain size (with estimates further refined by field measurements of sand concentrations and particle sizes in water samples); and use of a sand routing model to predict sand export from Marble Canyon for different possible controlled flood scenarios (Grams et al., 2015).

The absence of a natural cue does not preclude the delivery of environmental water. The development and modification of river systems may have removed or significantly reduced natural cues and environmental water delivery could be necessary to arrest environmental decline. However, the choice of water delivery method should still be influenced by the nature of the season and therefore the environmental demand. For instance, in circumstances where there is highly variable year-on-year hydrology, environmental water regimes should generally be designed to mimic the extant conditions. This means that in dry times environmental water is mostly used for maintaining low flows and providing refuge habitat. In moderate conditions, environmental water is used to restore variability to in-channel river flows and provide connectivity along rivers and creeks and to low-lying wetlands. During wet conditions, environmental water provides the capacity to contribute to higher flows targeting higher floodplain wetlands, mitigating the water quality impacts often associated with floods, and managing the recession of flows to allow the completion of ecological processes such as water bird breeding and fish movement back to the river (CEWO, 2013a).

Designing a flow to achieve a prescribed environmental outcome necessitates flexibility. Often there is a limited time window in which the event can occur, when the river conditions are expected to be most suitable and when demand from the environment is most pressing. Within this window, releases may be designed to occur in response to hydrological, biological, or geomorphological triggers, and to achieve multiple objectives, recognizing that this may involve trade-offs between competing environmental outcomes (Box 24.3).

BOX 24.3 RESPONDING TO NATURAL CUES WHEN CONSIDERING MULTIPLE OBJECTIVES

In the Macquarie River, northern Murray—Darling Basin, Australia, in 2014, in a break from past practice where environmental water has traditionally been released with the sole objective of inundating the Macquarie Marshes, a Ramsar-listed wetland, a multiobjective flow scenario was devised. The approach sought to balance outcomes for the vegetation communities and habitat for water birds in the Marshes with beneficial outcomes for fish populations in the river channel. The focus was on providing flows at a time when water temperatures would give native fish species an advantage over alien species (European carp).

Overall, approximately 30,000 ML of environmental water was delivered (some from the New South Wales government and some from the federal government's environmental water accounts). The delivery was to be released in response to a volumetric flow trigger if that occurred before October 5, 2014. The trigger was designed to ensure appropriate temperature and chemical cues for native fish. In the absence of the trigger by the due date, flows were to be released regardless. An additional fish contingency of almost 10,000 ML was also available if the trigger event occurred by October 5 and after that date if a smaller trigger subsequently occurred (Fig. 24.2).

(Continued)

BOX 24.3 (CONTINUED)

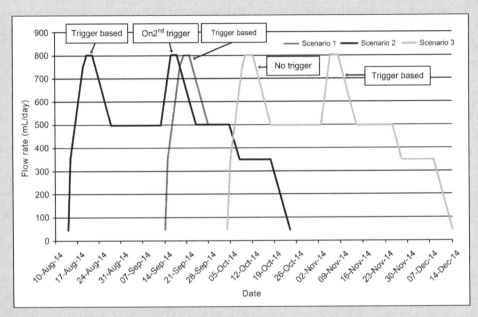

FIGURE 24.2

Planning scenarios for the release of environmental water in the Macquarie River, Australia, in spring 2014, illustrating the need to plan for multiple contingencies, particularly when outcomes for multiple ecosystem components with different water requirements are sought. *Scenario 1*: Release if trigger occurs before September 15 and second trigger occurs during event; *Scenario 2*: Release if trigger occurs after September 15; and *Scenario 3*: Release if no trigger occurs by 5 October and second trigger occurs during event.

Ultimately, the flow trigger did not occur and only the initial 30,000 ML was released as planned from October 5, inundating more than 7000 ha of the south and north marshes (NSW, 2015b). If the objective had only been for the marshes then the event would most likely have been released earlier in the season when conditions were cooler. Although the target inundation extent was largely reached, the desired duration was not, due to hot and dry weather leading up to the event. However, a compromise on timing had to be made in the hope of achieving beneficial outcomes for fish, in addition to vegetation and bird habitat. Although the triggers for commencing delivery of environmental water or for using the additional native fish contingency water did not eventuate in 2014, environmental water was still delivered at a time when water temperatures were more suitable for the promotion of movement and breeding of native fish. This outcome was facilitated by the use of a new cold water pollution mitigation curtain (a piece of infrastructure that allows surface rather than deep water to be released) installed at Burrendong Dam, the major upstream storage.

Multiple use of water—recrediting and reuse

To maximize the efficient use of any given volume of environmental water, the reuse of that water for environmental benefit as it flows downstream is often sought. In the Goulburn River example referred to earlier, this was possible because the State of Victoria has enacted *return-flow* provisions that enable environmental water use at multiple locations along a river, accounted for using the existing trading framework. With this approach, the volume that was estimated to have reached the confluence of the Goulburn and Murray rivers was accounted for at a license tagged to that location and then immediately transferred to a license downstream through a zero-dollar trade, to enable extraction of water at the downstream location. This can occur several times along the length of a river as water is used multiple times for outcomes instream and in wetlands throughout the system. Other states within the Murray–Darling Basin do not yet have similar arrangements in place, although there is a work program underway, coordinated by the MDBA, to address the lack of these arrangements across the Basin (MDBA, 2014b). Where these arrangements do not exist, the water released is no longer available for the environment beyond the first point where it is deemed by the river operator or state framework to have been used, and as a result is at risk of reregulation or extraction for consumptive use downstream, thereby limiting the overall environmental outcomes achieved. As illustrated in both the western United States and Murray–Darling Basin (Box 24.4), often the protection of environmental water from reregulation or extraction by other users needs to be negotiated on a case-by-case basis.

BOX 24.4 NEGOTIATING THE PROTECTION OF ENVIRONMENTAL WATER RELEASES

The Australian government identified the *shepherding* or protection of instream flows of environmental water as a key administrative challenge associated with the recovery of water for the environment in *unregulated* river systems. Shepherding was defined by the New South Wales (NSW) and federal government as "the delivery of a volume of water from a nominated license location to a downstream delivery location where it can be made available for use by the environment" (NSW, 2012a). That is, ensuring that environmental water left instream is not extracted by rights holders downstream and is accounted for, minus any losses, at the end of the system. To support the work on identifying an appropriate methodology, several shepherding trials were conducted between 2009 and 2011 (NSW, 2012b):

1. In March 2009, 11,400 ML was shepherded from Toorale Station at the junction of the Warrego and Darling Rivers downstream to the Menindee Lakes, resulting in 5976 ML being available for environmental use in the Murray River.
2. In the summer of 2009–10, 38,000 ML was shepherded from Toorale Station, resulting in 30,400 ML being delivered to the Menindee Lakes.
3. In September 2010, 7672 ML was shepherded from Toorale Station, resulting in 6580 ML being released to the Great Darling Anabranch downstream of the Menindee Lakes.

The different proportions of the total volume that was accounted for downstream generally reflect different antecedent conditions within the river at different times—less water makes it downstream in drier conditions. These trials were only possible due to the operation of Menindee Lakes (a large, shallow water storage on the Darling River in western NSW) being in control of the NSW government, rather than the *Murray–Darling Basin Authority (MDBA)* working on behalf of all Basin governments, which would normally be the case in wetter

(Continued)

BOX 24.4 (CONTINUED)

conditions. If the Lakes had been under the control of the MDBA, the rules in place under the Murray–Darling Basin Agreement (Commonwealth of Australia, 1993) would have resulted in this water being automatically shared with the other state governments for subsequent allocation to consumptive users. The Murray–Darling Basin Agreement describes the circumstances under which the MDBA or the NSW government has control over the Menindee Lakes resource. The Darling River was identified as an initial location to develop and trial methodologies for addressing the shepherding problem, but is by no means unique. Any river within the Murray–Darling Basin that allows the taking of water subject to river flow conditions, and to some extent they all do, has this issue.

Compliance and enforcement of instream flows to ensure that the water is not extracted by other users has also been identified as a key potential issue in the Columbia Basin (Hardner and Gullison, 2007). Although not necessarily a significant problem at the time of their review, Hardner and Gullison (2007) posited that this may be simply because many transactions are specifically designed to avoid the risk of take from downstream users. Local implementing entities reported in response to questioning that good flow data also meant that they would be in a strong position to take action against downstream rights holders, if necessary. Nevertheless, in response to the Hardner and Gullison review and several other programmatic requirements, the Columbia Basin Water Transactions Program developed and released a new accounting framework aimed at strengthening monitoring and compliance of transactions. The framework (McCoy and Holmes, 2015) is structured around four tiers of compliance: (1) contractual compliance to ensure the terms of any agreement are adhered to; (2) flow accounting, to ensure that flow targets in the target reach are met; (3) aquatic habitat response, to understand the extent to which aquatic habitat responded to changes in flow; and (4) ecological function, which integrates transaction and flow-specific data with broader monitoring efforts to understand improvements in habitat condition for target species.

Coordinating environmental water delivery

The volume of environmental water that a manager has available to actively manage is typically a small percentage of the total water in the system. To maximize the benefit and efficiency of environmental water use, managers look to coordinate its delivery with other sources, e.g., natural flows, operational releases, or other sources of environmental water.

Adding water to high natural flows

In order to achieve large flow events, managers often require the release of environmental water from storage into a river that is already high as a result of tributary inflows. In addition to reducing the volume of environmental water that needs to be released to achieve a flow event, it also has the advantage of aligning with natural cues.

When coordinating the delivery of environmental water with natural flows, timing of water releases becomes critical. Water must be released from storage so that it combines with the tributary flow at just the right time to increase the height of the river and/or to extend the duration at the location required to achieve the environmental objective in river habitats or ecological functions, or inundation of riparian areas and connected floodplain wetlands. Often there is a window of opportunity when a high stream flow occurs at an optimal time for the environmental features being targeted (Box 24.5).

BOX 24.5 PIGGYBACKING ENVIRONMENTAL WATER ON OTHER RIVER FLOWS

In the southern Murray—Darling Basin, Australia, the window of opportunity for adding environmental water to high natural flows is often in late winter or spring, which is early in the water year (July—June). As water is allocated to each license holder only as it becomes available through the year, there is a risk that early in the water year there will be insufficient water allocated to environmental accounts to enable sufficiently large environmental water releases to occur. In the preceding year, it is therefore necessary for environmental water holders to carryover enough water in storage (i.e., bank it from the previous year) to be able to provide for these early season flows, but within the account limits prescribed in water-sharing plans. For the *Commonwealth Environmental Water Holder (CEWH)*, this can involve transferring water between accounts with different carryover limits to ensure that water is in the valley that it is most likely to be required in the upcoming year. For example, in the Murrumbidgee valley in New South Wales, where there is a carryover limit of 30% of an entitlement volume, the CEWH carried over 56 GL in 2013—14 (CEWO, 2014a), just below its 62 GL limit, with a view to making this water available in conjunction with early season allocation and state government environmental water to support a *piggyback* event targeting the mid-Murrumbidgee wetlands in spring of the following year.

For 3 years in a row, since the commencement of the Murray—Darling Basin Plan, the Murray—Darling Basin Authority has identified the mid-Murrumbidgee wetlands as a high priority environmental asset requiring environmental water. Owing to infrastructure constraints (i.e., a low-lying road crossing downstream of the major storages) delivering water to these wetlands requires adding water to tributary inflows that connect with the main river channel downstream of both the dam and of the crossing. In spring 2013, 150,000 ML of Commonwealth environmental water was approved for use. The release of this water was subject to the occurrence of a suitably sized natural flow event occurring between 15 August and 15 October. A target volume of around 15,000 ML/day at Wagga Wagga was considered a suitable trigger flow for the event. The regulated environmental water releases from Burrinjuck and Blowering dams would have involved an increase in flows over 2—3 days to reach a peak discharge of 28,200 ML/day (or 5 m gauge height) at Wagga Wagga for 3—5 days, followed by a recession that mimicked natural rates of fall (nominally, a rate of 10—15% decrease in flow per day). The total duration of the event was anticipated to be approximately 20 days.

It was determined that a peak flow event on 20 September 2013 at Wagga Wagga of 13,437 ML/day would have been a sufficient trigger for the environmental water release to proceed. However, the event was the subject of stakeholder concerns about potential flooding of private property and an approval to exceed current flow constraints at Yanco Creek downstream had not been provided by the relevant State Minister. The higher flows would have inundated some low-lying farmland, cutting off access to areas of river flat for a few days, causing inconvenience and the potential for economic loss. As a result, a decision was made on September 25, 2013 that the piggy bank event would be canceled pending further consultation with riparian landholders, and subsequently, alternative water use actions were pursued.

A similar event was again planned in spring 2014. On this occasion, landholder consent was obtained for a trial flow event, although it was obtained only after a trigger event had already passed. As there was no further trigger event during the planned delivery window, again the event did not proceed. As a result, some water was allocated to pumping to several small wetland areas to maintain habitat, but this was much smaller in magnitude and overall environmental outcome. In spring 2015, stakeholder concerns meant the event could again not proceed even though a trigger event occurred. For 3 years in a row, operational constraints on delivery prevented the planned outcomes being achieved in any substantial way.

The higher the tributary inflow the less water needs to be released from storage for a piggyback event and vice versa. It is therefore not possible to know with certainty how much water will be required until, and often after the event has concluded. Ensuring sufficient environmental water reserves prior to an event is therefore important. Good rainfall—runoff models and prediction services are also useful in this regard. The Australian Bureau of Meteorology provides 7-day stream flow forecasts for a number of rivers around Australia. These forecasts, in addition to local rainfall predictions, are an important aid to decision making.

An additional consideration in delivering in conjunction with natural flow events is the accounting treatment of the water delivered (Box 24.6).

BOX 24.6 ACCOUNTING FOR ENVIRONMENTAL WATER USE IN THE MURRAY–DARLING BASIN

Generally speaking, within the Murray–Darling Basin entitlement holders do not have a right to order water from a particular storage to increase instream flows (MDBA, 2013). If a water user requires water, the river operator will first seek to ensure that demand is met by flows already in the river. This means that natural stream flows downstream of the dam will be the first source of water available for use and if the river operator does not need to release additional water, they will not do so. The primary outcome sought for the operation of the Murray River, as with other rivers in the Basin, is the "conservation of water and minimization of losses" (MDBA, 2015). However, for an environmental water manager who wishes to add water to that which is already instream, additional releases from storage often need to be made. Doing this regularly has the potential to place a greater burden on storage releases than historically occurred when those licenses were previously used for irrigation. Over time this may have an impact on the overall resource availability for all users. To protect the rights of other users, the river operator may instead insist that the environmental manager is charged the full (or a greater) volume of the flow from its account, even if some of that flow would have occurred naturally regardless (Fig. 24.3). This clearly impacts negatively on water availability for the environment by drawing excessively on the environmental account and so is a deterrent to efficient water use for the environment.

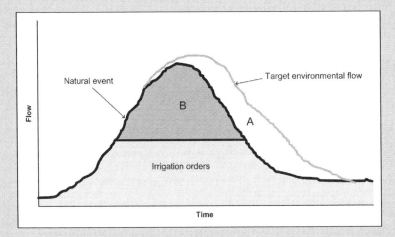

FIGURE 24.3

Schematic representation of accounting for environmental water where additional water is instream from a natural rainfall event. An environmental water manager seeks the target flow event as indicated, building on the natural flow. Instead of being debited volume A from the environmental account, the water authority debits volume A + B because in normal circumstances any water order above existing orders would simply be met from the water already in the river (volume B). If additional water (volume A) must be released from storage, this could over time have an impact on overall resource reliability, impacting other users. To compensate for this the environmental account may be charged extra.

(Continued)

BOX 24.6 (CONTINUED)

Within the Murray River, case-by-case decisions have been made to allow environmental water managers to release water on already high stream flows when conditions are likely to have minimal effect on other users. However, this is not the norm and a work program is underway through the implementation of the Basin Plan to examine how and under what conditions this can be made standard operating practice so that all users' rights are protected (MDBA, 2014b).

Coordinating environmental flows with other storage releases

Achieving environmental outcomes in a working river requires consideration of other operational flows that may be occurring around the time of the desired event. This includes consideration of what those other releases might contribute to achieving the environmental objectives sought, as well as any potential dis-benefits that might arise through the interaction of releases being made for different purposes (Box 24.7).

BOX 24.7 DELIVERING ENVIRONMENTAL FLOWS IN CONJUNCTION WITH IRRIGATION ORDERS

In the Gwydir Valley, New South Wales, Australia, in spring 2014, up to 20,000 ML of environmental water was made available for in-river environmental flows targeting outcomes for fish. Owing to dry conditions in the catchment the river operator had advised all users that water would only be available for extraction during specified *block releases* (i.e., short time windows when releases would be made from storage) unless a user was willing to pay the full conveyance losses from their own account for the water to be delivered. Rather than seek releases outside of the block release and therefore pay the additional losses, the Commonwealth Environmental Water Holder made a decision to add environmental water both to the top of the block release and immediately after, in order to achieve a hydrograph shape considered most beneficial for fish. This involved a short sharp peak and a gradual recession (Fig. 24.4A). Delivery for irrigation orders in these circumstances would normally have involved a sharp increase followed by an extended flow at a stable height and then a sharp decline in water level as the flow is passed through. Such a flow is not ideal for fish because it can leave nesting sites on in-channel bars, benches, and snags stranded as the water recedes too quickly. A steadier drawdown period is preferred (Southwell et al., 2015). The actual release hydrograph from spring 2014 is depicted in Fig. 24.4B where it can be seen that the flow attenuates downstream, but largely holds its shape to the end of the system.

This contrasts with the previous year when environmental water was released in advance of irrigation orders but was subsequently swamped by irrigation releases from November onward, effectively removing the gradual flow recession that was desired, at least in the upper reaches of the river (Fig. 24.4C). This is likely to have reduced the benefits to fish as the response from Bony Bream and Spangled Perch was clearer in downstream sections where the actual hydrograph was closer to the target shape (Southwell et al., 2015).

(Continued)

BOX 24.7 (CONTINUED)

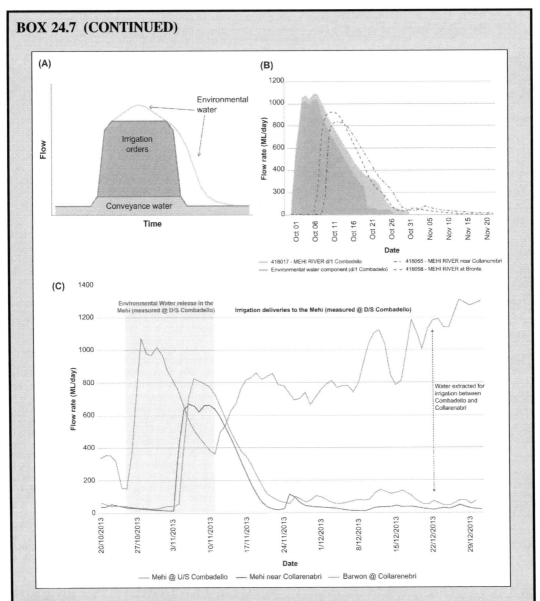

FIGURE 24.4

(A) Desired shape of hydrograph for flows in the Mehi River and Carole Creek, Australia, in spring 2014; (B) hydrograph along the Mehi River from Combadello Weir to the confluence with the Darling River, spring 2014, with irrigation volume in purple and environmental volume in green provided for illustrative purposes only; and (C) hydrograph along the Mehi River, spring 2013, showing environmental water attenuating downstream and extraction of consumptive water mid-river.

Source: Data from NSW Department of Primary Industries, Office of Water: http://realtimedata.water.nsw.gov.au/water.stm

Water delivered downstream for consumptive purposes, when at the right time and targeting the right hydrograph, can reduce the overall volume of environmental water required to achieve an environmental outcome. However, as these examples in the Gwydir Valley show (Box 24.7), consumptive flows can also interact adversely with environmental water, reducing the potential benefit that might otherwise have been achieved. Having strong communication mechanisms between environmental water managers, irrigators, and river operators is therefore important in preventing negative, while maximizing positive, interactions. As identified above, this can be through formal committees and structures such as those used in the Columbia Basin (see Box 24.8).

BOX 24.8 COORDINATING ENVIRONMENTAL WATER RELEASES AND OPERATIONS IN THE COLUMBIA RIVER, WESTERN UNITED STATES

Within the Columbia Basin, the interaction of flow releases for fish habitat with other releases for hydropower, flood mitigation, and irrigation is considered by the Technical Management Team, consisting of experts from all relevant federal and state agencies. Systems Operations Requests that outline proposed operational guidelines to give effect to Biological Opinions are put to the Committee, often by the National Fish and Wildlife Service or a state wildlife agency in relation to fish, for review and decision. The Action Agencies explain to members how they would give effect to the guidelines if approved and a poll is then taken either approving or denying the request. Systems Operations Requests can be quite specific, detailing flow rates, timing, and duration required, given the prevailing and expected hydrological conditions.

Coordination is then necessary between the operating of structures in the Federal Columbia River Power System, as there are short distances between the dams, most of them are run-of-the-river, with little or no storage, and the outflow from one dam is usually the start of the pool behind the next downstream dam (NRC, 2004). To facilitate coordination each year, the Action Agencies prepare an annual water management plan consistent with relevant Biological Opinions and with a Fish Passage Plan prepared by the US Corp of Engineers. The water management plan specifies the objectives and strategies that each dam will seek to implement throughout the year and how conflicts between competing priorities will be dealt with. For example, the draft 2016 plan identifies that where there is a conflict between refilling Libby Dam and providing spring sturgeon spawning flows, water release for spawning flows will take priority over refilling by early July (BPA et al., 2015). As inflow conditions cannot be known with much certainty in advance, the Technical Management Team provides advice to the Action Agencies as conditions unfold during the year.

Coordinating multiple sources of water for a greater outcome

In some circumstances there can be multiple sources of water available to achieve an environmental outcome. This is the case in the southern Murray–Darling Basin where there are multiple environmental water managers at state and federal levels, each with different portfolios of water entitlements. Coordination is important to ensure that each *bucket* of environmental water is being used in a complementary and not counterproductive way. Key elements to support coordination between environmental water managers include having common objectives and targets, agreed roles and responsibilities, and communication channels, either through formal committees or informal mechanisms (see Box 24.9).

BOX 24.9 COORDINATING MULTIPLE SOURCES OF ENVIRONMENTAL WATER

Within the Murray–Darling Basin, Australia, coordination among environmental water managers occurs through three primary mechanisms:

1. The Basin Plan environmental management framework, which ensures alignment of overarching objectives, targets, and expected outcomes. This framework ensures that all managers are working toward the same ends.
2. The administrative mechanisms for making water available. Use of the federal government's environmental water is mostly undertaken by transferring it to a state government account under the management of the state agency. This ensures that use of the federal government's water is consistent and complementary to state government plans and uses for state environmental water and can be combined with it for greater effect.
3. Ongoing conversation between parties relevant to a flow event through phone, email, and meeting discussions. These conversations are supported by various committees and forums in different catchments. For example, the Southern-connected Basin Environmental Water Committee is a forum for information exchange among environmental water holders and relevant river operators in the Murray system and its major tributaries, the Goulburn and Murrumbidgee.

The Commonwealth Environmental Water Holder has partnership agreements in place with the Victorian and New South Wales Government Environmental Water Holders to support these arrangements through complementary planning, decision making, and delivery arrangements. Fig. 24.5 illustrates an outcome from the coordinated use of

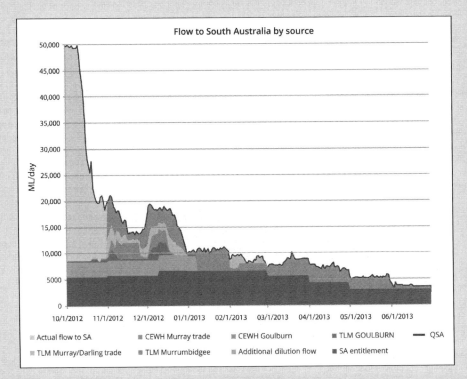

FIGURE 24.5

Flows at the South Australian (SA) border from October 2012 to June 2013 illustrating the contribution of water from different environmental water portfolios and different rivers to the overall hydrograph. SA entitlement is the monthly volume that must be delivered to SA under the Murray–Darling Basin Agreement; additional dilution flow is planned environmental water that must be delivered under the Murray–Darling Basin Agreement in particular circumstances. The name of the tributary river source is as indicated.

Source: Adapted from MDBA (2013)

(Continued)

BOX 24.9 (CONTINUED)

water from multiple environmental water portfolios and multiple river catchments in the southern Murray—Darling Basin. It shows flows on the Murray River at the South Australian border with the different sources of water each making up a portion of the overall hydrograph. The objective in this case was to ensure a more gradual recession from the natural flow event that occurred in October 2012, while maintaining higher but fluctuating flows for longer. Without environmental water the natural flow event in early October would have dropped back rapidly, first to additional dilution flow volumes that are provided as planned environmental water (under the Murray—Darling Basin Agreement) to improve in river salinities, and then to the South Australian government's basic entitlement flow. Contributions of water from the Murray, Murrumbidgee, and Goulburn Rivers, however, ensured a more gradual flow recession and some more natural variability in flow heights for the remainder of the water year, which contributed to golden perch breeding and Murray cod recruitment (Ye et al., 2015).

24.3.2 LIMITATIONS ON ENVIRONMENTAL WATER RELEASES

Avoiding unintended flooding

The example in Box 24.5 illustrates the sociological dimension to environmental water delivery, an important and indeed often limiting factor in implementation. The need for environmental water programs is generally highest in highly developed systems where the flow regime has been most disturbed and where people have moved onto and developed the floodplain for commercial enterprise. The presence of people and businesses on the floodplain means that the capacity to deliver flows that are beneficial to the environment is necessarily constrained. River operators are governed by operating rules that generally limit how much they can release so as not to have detrimental impacts on landholders. Environmental water managers may seek to have flows delivered close to these limits to maximize the environmental outcome. However, a conservative approach is often warranted in recognition of the often uncertain legal or political environment for intentional inundation of private land.

Environmental water managers should seek to ensure that there is sufficient buffer between the operational flow limit and the targeted flow height to guard against the possibility of unexpected rainfall or river behavior causing unintentional flooding. Any inundation of private land should be agreed through negotiated consents with potentially affected landholders. This can be a time-consuming process, particularly if there are many landholders in the relevant river reach. One holdout can disrupt the implementation of flows and associated public benefits throughout the whole reach. Nevertheless, achieving community and landholder support is important for the sustainability of environmental water programs in the longer term (Box 24.10).

BOX 24.10 "GOOD NEIGHBOUR POLICY": BUILDING A SOCIAL LICENSE TO OPERATE IN THE MURRAY—DARLING BASIN, AUSTRALIA

As a new member of the water management community, and one that can sometimes be considered as being in conflict with the needs of other water users and landholders, the Commonwealth Environmental Water Holder (CEWH) recognizes the importance of building a social license to operate. This means the focus of the CEWH is not just

(Continued)

BOX 24.10 (CONTINUED)

maximizing environmental outcomes from its available water resources, but doing so in a way that has the trust and confidence of communities. Being viewed as a legitimate, diligent, and valued member of the community will support the ongoing existence of the CEWH and provide it with greater standing when looking to negotiate changes to current limitations to environmental water delivery.

To build this social license, the CEWH has developed and committed to a good neighbor policy (CEWO, 2016). The policy is a set of practices that guide the management of Commonwealth environmental water. It aims to promote mutually respectful and harmonious relationships with other water users and landholders and is built around key concepts such as minimizing third-party impacts, maximizing mutually beneficial outcomes, negotiating consent, flexible water management, local engagement and collaboration, and transparency in decisions and operations.

In addressing these issues, a comprehensive risk assessment will need to be undertaken at multiple levels (local, state, federal, as appropriate) to understand the potential likelihood and consequence of unintended flooding and the management strategies necessary to minimize the risk. These strategies include changing the event design to target a lower flow rate, ensuring a buffer between the maximum desired flow rate and the rate at which impacts could occur, ensuring adequate monitoring of rainfall and river conditions throughout the event, and consultation with landholders to deliver flows at times when their business is not likely to be affected. For example, in the Gwydir Valley in October 2011, an environmental water release targeting the Gwydir wetlands was suspended due to a local rainfall event that threatened to increase the height of the Gwydir River above bank-full level, risking inundation of adjacent winter cereal crops. In the Murray Valley, the maximum operational flow rate has been reduced in recent years from 18,000 ML/day to 15,000 ML/day downstream of Yarrawonga weir due to concerns from local landholders about reduced access to parts of their property as environmental water holders seek to inundate parts of the Barmah and Millewa (Central Murray) Forests Ramsar sites.

Infrastructure limitations and maintenance

As noted above, environmental water delivery can be limited by infrastructure such as the size of dam outlets. These can present hard and long-term limits on flow rates. However, there can also be short-term limits to flow rates due to factors such as infrastructure maintenance (see Box 24.11).

BOX 24.11 WORKING AROUND INFRASTRUCTURE MAINTENANCE IN THE COLORADO RIVER, UNITED STATES, AND MURRAY–DARLING BASIN, AUSTRALIA

The release rate from Glen Canyon Dam can be limited due to maintenance activities. The eight generators or *units* are subject to annual and ongoing maintenance, which can mean that they are unavailable for extended periods of time (ranging from weeks to months to more than a year). Given the age of the power plant (nearly 50 years) and scheduled and unplanned maintenance, it is reasonable to expect that in the 10-year period the *High-Flow Experiment (HFE)* protocol is in place, at least one unit would be unavailable for HFE releases at any given time, reducing the release

(Continued)

BOX 24.11 (CONTINUED)

capacity from 110,000 ML/day to below 104,000 ML/day (Interior, 2014). In addition to maintenance, the rates of flow releases are also influenced by changes in the elevation of the reservoir. The elevation affects the units' efficiencies, with raised tail water elevation during an HFE event decreasing unit efficiency and hence flow rates (Interior, 2014).

Infrastructure maintenance has also influenced environmental flows in the Murray–Darling Basin. For example, planned winter flows in the Goulburn River in 2015 had to be reduced in duration due to unforeseen maintenance of a bridge undertaken during school holidays to minimize disruption to road users as the bridge is a critical route to school for local residents. Flows were originally targeted at above 6600 ML/day for 14 days, however, this was reduced to just 4 days. Although monitoring results from the event have yet to be assessed, it is expected that the reduced duration of the flow will have reduced its effectiveness in achieving one of its aims, which was the drowning out of terrestrial vegetation that had encroached into the riverbanks during the drought from 2000 to 2010.

Again, this emphasizes the need for environmental water managers to maintain open lines of communication with relevant infrastructure operators, so that they are aware of and can potentially influence the timing of scheduled maintenance that could affect water deliveries.

Channel capacity constraints

Delivering environmental water in conjunction with other operational releases can raise the potential for conflict between the needs of competing users, particularly when channel capacity is limited and when users have time-critical needs (Box 24.12). Often environmental managers have a larger window of opportunity than commercial users and so can be more flexible about when flows are released. However, this is not always the case. For example, if a flow is required to sustain a bird-breeding event on a floodplain wetland, then the provision of water can be very time critical to prevent water recession that might cause birds to abandon their nests prior to fledglings becoming self-sufficient.

BOX 24.12 COMPETITION FOR CHANNEL CAPACITY IN THE MURRAY–DARLING BASIN, AUSTRALIA

In 2014–15, flows to South Australia from upstream were inhibited by channel capacity constraints as a result of hot and dry conditions generating increased irrigation demand. More environmental water could have been delivered over summer to benefit the Coorong, Lakes Alexandrina and Albert Ramsar site at the end of the system, if there had been less demand from irrigators over that period. In the upstream reaches irrigation deliveries can serve an in-channel environmental need as the water flows through the system, but as it is often diverted for irrigation mid-river, the benefits are limited spatially and as noted above can be detrimental to the environment in some cases. For instance, although there was environmental demand for flows to reduce salinity in the Coorong lagoon at the mouth of the Murray River, higher flows in the Goulburn River to meet that demand would not have been beneficial for the environment over the summer period. This spatial limitation on the role of irrigation deliveries is particularly evident in the above Gwydir Valley example (Box 24.7), where the vast majority of the flow during spring and early summer along the Mehi River in 2013–14 was removed from the system by irrigators between the flow gauges at Combadello and Collerenabri (Fig. 24.4C).

Of course, environmental water acquired from irrigators is not additional water in the system. When the entitlements were previously held by irrigators, the delivery of water associated with those entitlements would in any case have been competing for channel capacity at the same time as other commercial users. Indeed, its subsequent use for environmental demands, which are often at times other than during peak irrigation season, is likely to have lessened competition for channel capacity rather than increased it. This is a direct commercial benefit to many irrigators that is rarely accounted for in discussion of net impacts on water users associated with the acquisition and use of water for environmental purposes.

Minimizing social and economic impacts

The framework for environmental water releases typically places limits on its delivery, which are often in response to minimizing negative social and economic impacts. In most cases, this is reflected in the volume of water set aside for the environment. However, even within the broader operating framework, decisions on individual environmental water releases are made with the aim of avoiding or minimizing negative social and economic impacts. For example, the 2015 winter flows in the Murray River were reduced in magnitude to avoid affecting firewood collection in the Barmah—Millewa Forest (see Fig. 24.1). Spring freshes in the Goulburn River in 2013 and 2014 were timed to avoid the opening of the Murray cod fishing season, when recreational fishers are seeking stable river conditions. Similarly, releases from Glen Canyon Dam are optimized to minimize negative impacts on electricity generation (see Box 24.13).

BOX 24.13 MINIMIZING SOCIAL AND ECONOMIC IMPACTS FROM ENVIRONMENTAL WATER DELIVERY

In the case of the Colorado River, the *High-Flow Experiment (HFE)* protocol caps the duration for flow releases at 96 h. This is to limit the loss of potential future power generation associated with bypassing water around the turbines.

However, even within the broader operating framework, decisions on environmental water regimes are made with the aim of avoiding or minimizing negative social and economic impacts. This often requires negotiation with the affected parties. In developing the 2014 HFE, a peak magnitude of 78,000 ML/day was initially considered due to expected maintenance at Glen Canyon Dam and other limitations due to power regulation and reserves. This was later increased to 91,000 ML/day as a result of the Bureau of Reclamation shifting scheduled maintenance and the Western Area Power Administration slightly shifting power requirements to increase the Glen Canyon release capacity and thus the peak magnitude of a potential HFE. The start date of the HFE was specifically timed with better conditions for the Western Area Power Administration to market hydropower generated by the event—by scheduling the commencement on a Monday resulted in an anticipated cost savings of US$200,000 (Reclamation, 2014).

Although such examples may not present legal or physical barriers to environmental water deliveries, failure to consider and accommodate them can lead to the loss of local community support. For what is a relatively new natural resource management activity, community support can be critical for the enduring nature of reforms and could underpin negotiations on the removal of legislative or physical barriers to environmental water releases over the longer term.

24.4 CONTINUOUS IMPROVEMENT

24.4.1 FROM FLOW EVENTS TO FLOW REGIMES

In highly developed river basins, the decline in the environmental condition has often occurred over many decades. Redressing this decline is a long-term proposition. The MDBA's Basin-wide Environmental Watering Strategy identifies that widespread restoration of ecosystem responses and populations of key species through implementation of the Murray–Darling Basin Plan will likely take 20 or more years (MDBA, 2014a). As is evident in the building of sandbars in the Colorado Basin, some ecological responses can be achieved with a single flow event. However, the long-term maintenance of sandbars is reliant on flows across multiple years. This emphasizes the importance of providing flow regimes across seasons and multiple years to achieve long-term ecological responses. At a basin scale, this requires consideration of seasonally appropriate flow regimes across multiple catchments, many of which are hydrologically connected.

This concept underpins the Commonwealth Environmental Water Outcomes Framework (CEWO, 2013b). The framework identifies a hierarchy of generic outcomes that can be achieved from environmental water over different temporal scales. At the lowest levels, these are outcomes that can be achieved in less than 1 year (e.g., fish movement and spawning) and between 1 year and 5 years (e.g., larval and juvenile fish recruitment). The cumulative implementation of these shorter-term outcomes over successive years and across multiple catchments is expected to contribute to the long-term, Basin-wide objectives identified in the Murray–Darling Basin Plan. The framework both guides the planning on environmental water and informs the monitoring and evaluation of outcomes.

To deliver a flow regime that supports the full life cycles and ecological processes of the river systems requires environmental water managers to take a multiyear approach to planning (CEWO, 2015). Consideration needs to be given to the flows delivered and outcomes achieved over successive years, and then planning for the range of environmental demands under different climatic scenarios and potential flows into future years. Critically, it emphasizes that environmental water releases cannot be managed in isolation, as one-off events, and need to be supported by a robust adaptive management process.

24.4.2 ADAPTIVE MANAGEMENT IN PRACTICE

As discussed in Chapter 25 adaptive management is important in reducing uncertainty and supporting the continual improvement of environmental water management. There are challenges for environmental water managers applying adaptive management—watering events often lack adequate replication or controls, and monitoring data can be limited or noisy. Despite these limitations, monitoring results can be used to provide a weight-of-evidence approach that improves understanding for some aspects of ecosystem function, helps identify the most likely causal factors, strengthens overall inferences, and identifies or adapts future watering actions (Melis et al., 2015; see Box 24.14). This is relevant not only in managing for environmental outcomes but also in responding to operational issues and community concerns (Box 24.15).

BOX 24.14 ADAPTIVE MANAGEMENT OF GLEN CANYON DAM, UNITED STATES

The Glen Canyon Dam Adaptive Management Program was established in 1997. The program provides for monitoring of the results in interventions undertaken at Glen Canyon Dam and includes an Adaptive Management Work Group, a federal advisory committee, which has representation from each of the cooperating federal agencies, Colorado River Basin states, environmental groups, recreation interests, and contractors for federal power from Glen Canyon Dam. The group works collectively to identify and recommend appropriate management strategies, monitoring and research programs, and changes to operating criteria and plans (Interior, 2015).

Despite various criticisms of the program (Camacho et al. 2010; Susskind et al., 2012), it has provided key inputs into future management decisions and has been critical in identifying outcomes that differed significantly from what was expected.

The first *high-flow experiment (HFE)* in 1996 demonstrated that high-flow releases could build sandbars. However, sand delivered by tributaries did not accumulate on the riverbed over multiple years, as had been assumed. Rather, the sandbars were built from sand from existing sandbars rather than from sand stored on the riverbed. On the basis of these findings, the HFEs in 2004 and 2008 were timed to follow tributary floods that provided "new sand to the system, which allowed sandbars to be built before the sand is carried downstream" (Melis, 2011; Melis et al., 2011). The increase in sandbar volumes following these two events indicated that releases timed to follow sand inputs are an effective sandbar-building strategy (Grams et al., 2015). The monitoring also showed that sandbars are built relatively quickly (hours to a few days) under these sand-enriched conditions, but they also tend to erode quickly (days to several months) following an HFE (Melis, 2011; Melis et al., 2011). This suggested the need for more frequent HFEs.

These findings all informed the development of the Protocol for HFE Releases from Glen Canyon Dam through 2020, which was approved in 2012. In addition to timing releases to follow sand inputs from tributary flows, the Protocol also allows more frequent HFEs, that is in 2012, 2013, and 2014. Monitoring has found that many sandbars have increased in size following each controlled flood, and the cumulative results of the three consecutive releases suggest that sandbar declines may be reversed if controlled floods can be implemented frequently enough when sediment conditions are favorable (Grams et al., 2015).

BOX 24.15 ADAPTIVE MANAGEMENT IN THE GOULBURN RIVER, MURRAY–DARLING BASIN, AUSTRALIA

The design and delivery of environmental water in the Goulburn River in the Murray–Darling Basin has evolved significantly since 2012–13. A large number of lessons have been learned in terms of the timing, height, and duration of flow events to achieve desired outcomes for native fish, bank vegetation, and geomorphology. A number of practices have also been put in place to address community concerns that included the risk of bank slumping and notching, the disruption of local angling events, and irrigators' access to pumps (Fig. 24.6). The adapting of practice required the collective efforts of the local catchment management authority (Goulburn–Broken Catchment Management Authority), the three environmental water managers operating in the catchment (the Commonwealth Environmental Water Holder, the Victorian Environmental Water Holder, and The Living Murray program), and the river operator, Goulburn–Murray Water, to review the environmental monitoring results of the previous year as well as community issues raised, and redesign the delivery to respond to the findings. The involvement of the river operators is particularly important—the operator is responsible for bulk water releases to meet needs in downstream catchments (referred to as *intervalley transfers*), and these have been managed to complement environmental water releases and support the achievement of environmental water requirements. The river operator also plays an important communications role as it notifies river diverters of upcoming releases via letters, emails, phone calls, and/or sms alerts. This communication has proved important in providing sufficient warning time for irrigators to plan their watering needs around environmental water releases during which their pumps are not accessible. The progressive changes made to environmental water delivery in the Goulburn River from 2012 to 2015 have resulted in both an improvement in the achievement of environmental outcomes and a decrease in concerns from irrigators.

(Continued)

BOX 24.15 (CONTINUED)

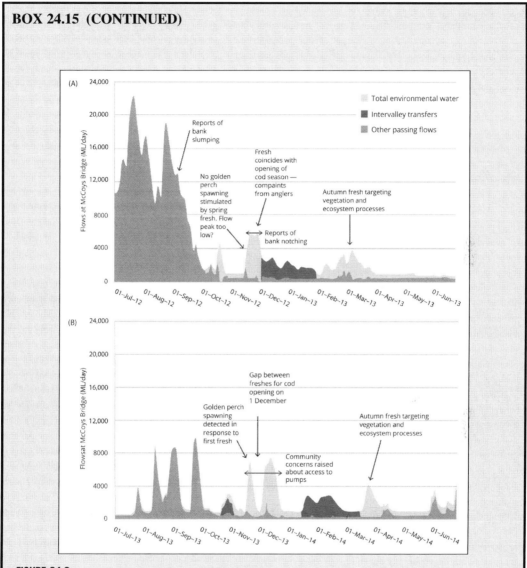

FIGURE 24.6

Hydrograph of environmental water releases in the Goulburn River, Australia, over 3 successive years: (A) 2012–13; (B) 2013–14; and (C) 2014–15, illustrating adaptive management of the approach in response to scientific advice and community concerns.

Source: Using hydrological data provided by Goulburn–Murray Water

(Continued)

BOX 24.15 (CONTINUED)

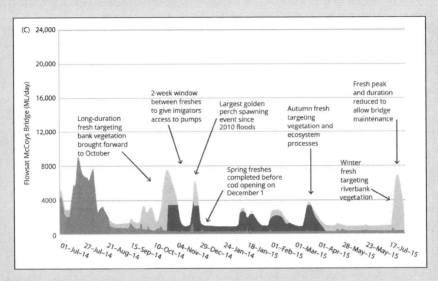

FIGURE 24.6

(Continued).

As illustrated in Fig. 24.6, environmental water releases in the Goulburn River are coordinated with releases from storages to meet consumptive demands in downstream valleys (intervalley transfers). In 2012−13 (Fig. 24.6A), environmental water regimes were primarily focused on spring freshes to stimulate the spawning of golden perch and support the recovery of in-channel vegetation, which had suffered declines during the millennium drought and 2010 flood. No golden perch spawning was detected (Stewardson et al., 2014), possibly due to the peak height of the flow being too low (noting that conditions were very dry during spring and summer 2012−13; Fig. 24.6A). At the same time, landholders reported bank slumping following high flows early in the water year, and bank notching during the environmental water releases. Anglers also complained about environmental water releases disrupting the opening of the Murray cod fishing season at the start of December. Environmental water managers commissioned an investigation into the cause of bank slumping and notching. The slumping was caused in part by the lack of semi-aquatic vegetation on the riverbanks. The notching was believed to be a result of maintaining the flow at a constant height for extended periods.

In response to lessons learned in 2012−13, a spring fresh targeting golden perch spawning in 2013−14 (Fig. 24.6B) was delivered with a higher peak than the previous year. A second fresh was provided to encourage semi-aquatic vegetation to recolonize the banks. To avoid notching, water levels were not held at the same height for greater than 5 days and peak height was varied between the two freshes. The freshes were timed either side of the opening of the Murray cod fishing season, to provide stable river conditions for anglers. The first freshes in November successfully triggered long-distance migration of native fish and a golden perch spawning event. New growth of semi-aquatic native vegetation was observed in the weeks after the two environmental water releases (Webb et al., 2015). However, community concerns continued to be raised around bank slumping as well as irrigators having limited access to pumps

(Continued)

BOX 24.15 (CONTINUED)

due to high water levels for extended periods of time. In planning for the 2014—15 water year (Fig. 24.6C), advice was sought from scientists involved in monitoring in the Goulburn River on improving the ecological responses to the flows. They suggested that a sequence of *events* or flows would increase the likelihood of a positive fish response. A longer duration fresh with a gradual recession targeting semi-aquatic vegetation outcomes was delivered earlier in the year to support vegetation growth and condition before the onset of warmer weather in summer. This also aimed to be a *priming flow* to support fish condition prior to the subsequent fresh targeting native fish spawning in November. The earlier timing of the first spring fresh also minimized the delivery of environmental water during the warmer months when irrigation demand can be higher. The second fresh commenced 2 weeks later, which provided an opportunity for irrigation before river levels rose again. This relatively short and sharp fresh targeted at fish spawning was completed prior to the opening of the cod fishing season. In response to these changes, the largest golden perch spawning event since floods in 2010 was observed following the second fresh. No significant community concerns were raised in relation to environmental flows and anglers reported that the fishing was "the best in years."

24.4.3 ACHIEVING RESULTS

Although the administrative, legal, and physical limitations on water management present a challenge in optimizing environmental water releases, the demonstrated outcomes to date show the benefits that can be achieved even within these constraints (e.g., Stewardson et al., 2014; Wassens et al., 2014; Ye et al., 2015). Since 2009, over 5000 GL of Commonwealth environmental water has been delivered to the rivers, wetlands, and floodplains of the Murray—Darling Basin. This has been complemented by environmental water provided under state governments and The Living Murray programs. During this time, a range of beneficial outcomes have been achieved for native fish, water birds, vegetation, and water quality (CEWO, 2013c, 2014b), including:

1. Supporting native fish spawning and movement, and providing habitat diversity, including refuge habitat from poor water quality and drought;
2. Providing foraging and breeding habitat for water birds, including colonial nesting species;
3. Improving the condition of aquatic and riparian vegetation;
4. Flushing of salt from the basin and supporting appropriate salinity gradients in the Coorong (the Murray River's estuary);
5. Improving dissolved oxygen levels;
6. Providing whole of system longitudinal connectivity;
7. Supporting habitat and ecosystem processes by mimicking natural flow patterns.

In the Colorado River Basin, the adoption and implementation of the 2012 HFE Protocol on the Glen Canyon Dam is showing positive results in increasing the size of downstream sandbars (see Box 24.14). There have also been some encouraging results in response to the Colorado River pulse flow (under Minute 319 of the US—Mexico Water Treaty of 1944). Preliminary findings in response to the 8-week event that commenced in March 2014 include the river connecting to the sea, the recharge of groundwater, the germination of new vegetation (both native and nonnative) and an increase in vegetation greenness (IBWC, 2014).

24.5 **CONCLUSION**

This chapter has sought to outline the key implementation challenges for environmental water rights, with a view to considering opportunities to address these challenges where they are already present and to avoid them in places where they are not yet manifest. Determining environmental water requirements and targets through scientific analysis and modeling is an important first step in environmental water management. However, specificity is less important than broad principles and rules of thumb when planning for environmental water use. This is because environmental water delivery is more about active management in response to changing circumstances and the need to be able to respond appropriately to conditions on the ground, both environmental and sociological, than following a predesigned script or modeled output. It also involves operating within a complex set of rules and procedures that were not designed to facilitate the outcomes being sought. As a result, compromises, trials, and constant innovation and ad hoc solutions are a feature of present-day environmental water operations. Consequently, the management of the system is not as efficient and effective as it could be and will require adjustments over time as environmental water releases are integrated into normal system operations.

To better meet the needs of all users adjustments that are required in the operating environment include:

1. Clarifying that water rights are not simply a right to extract water but a right to access and use water whether that use is instream or on land;
2. Ensuring that all rights holders have their use of water protected from appropriation by other users, whether that use is instream or on land;
3. More completely integrating river operations and environmental water planning so that the desired outcome can be achieved through efficient use of multiple sources of water rather than only relying on specified environmental water rights;
4. More clearly defining triggers, rules, or *standing orders* for the release of water that contributes to the long-term flow regime;
5. A framework for negotiating the flooding of riparian lands with landholders in a way that facilitates mutually beneficial outcomes.

Within the current operating environment, environmental water managers and river operators need to be highly cognizant of flow conditions and social and economic uses of the river at the time of a planned event if the environmental objectives of a particular flow are to be realized. Other flows will impact on the volume of water required, and whether any water provided and therefore debited from an environmental account is additional to what would have occurred otherwise. The flow situation will also have an impact on the level of risk of unintended flooding and the potential impacts on other water and riparian users within the system. The use of the resource by other parties may impact the timing, magnitude, duration, and extent of a flow that is possible and this situation is highly dynamic. Taking account of this sociological dimension is just as important to decisions on environmental water use as the science that informs an understanding of flow requirements, as ultimately the *social license* for an environmental water manager to operate is dependent on community support and the cooperation of other water users.

Despite the constraints, good environmental outcomes are nevertheless demonstrably being achieved from the implementation of environmental water regimes. Monitoring reports identify increased fish spawning, improved vegetation condition, successful bird breeding and improved water quality. However, restoration of degraded river systems will take time to achieve and lasting outcomes need to be considered at 10-, 20-, and 30-year time horizons. In lesser time frames, operating rules and administrative arrangements need to evolve to accommodate environmental water users more effectively if outcomes are to be fully realized.

REFERENCES

Acreman, M., Arthington, A., Collof, M.J., Couch, C., Crossman, N.D., Dyer, F., et al., 2014. Environmental flows for natural, hybrid, and novel riverine ecosystems in a changing world. Front. Ecol. Environ. 12 (8), 466–476.

Adler, R.W., 2007. Restoring Colorado River Ecosystems: A Troubled Sense of Immensity. Island Press, Washington, DC.

ANAO, 2013. Commonwealth Environmental Watering Activities. Performance Audit Report No. 36, 2012–2013. Australian National Audit Office, Canberra, Australia.

Banks, S.A., Docker, B.B., 2014. Delivering environmental flows in the Murray-Darling Basin (Australia)—legal and governance aspects. Hydrol. Sci. J. 59 (3–4), 688–699.

Blinn, D.W., Poff, N.L., 2005. Colorado River Basin. Rivers of North America. Elsevier, Canada.

BPA, USACE, and USBR, 2001. The Columbia River System Inside Story, second ed. Bonneville Power Administration, US Army Corps of Engineers and US Bureau of Reclamation.

BPA, USACE, and USBR, 2015. 2015 Water management plan. Bonneville Power Administration, US Army Corps of Engineers and US Bureau of Reclamation.

Bunn, S.E., Arthington, A.H., 2002. Basic principles and ecological consequences of altered flow regimes for aquatic biodiversity. Environ. Manage. 30, 492–507.

Buono, R.M., Eckstein, G., 2014. Minute 319: a cooperative approach to Mexico–US hydro-relations on the Colorado River. Water Int. 39 (3), 263–276.

Camacho, A.E., Susskind, L.E., Schenk, T., 2010. Collaborative planning and adaptive management in Glen Canyon: a cautionary tale. UC Irvine School of Law Research Paper No. 2010–6. Colum. J. Environ. Law 35 (1), Available from: http://papers.ssrn.com/sol3/papers.cfm?abstract_id = 1572720 (accessed April 2016).

CEWO, 2013a. A Framework for Determining Commonwealth Environmental Water Use. Commonwealth Environmental Water Office, Canberra, Australia.

CEWO, 2013b. The Commonwealth Environmental Water Outcomes Framework. Commonwealth Environmental Water Office, Canberra, Australia.

CEWO, 2013c. Commonwealth Environmental Water Office 2012–13 Outcomes Report. Commonwealth Environmental Water Office, Canberra.

CEWO, 2014a. Commonwealth Environmental Water Holder annual Report 2013–14. Commonwealth Environmental Water Office, Canberra, Australia.

CEWO, 2014b. Commonwealth Environmental Water Office 2013–14 Outcomes Report. Commonwealth Environmental Water Office, Canberra, Australia.

CEWO, 2015. Integrated Planning for the Use, Carryover and Trade of Commonwealth Environmental Water: Planning Approach 2015–16. Commonwealth Environmental Water Office, Canberra, Australia.

CEWO, 2016. Portfolio Management Planning: Approach to Planning for the Use, Carryover and Trade of Commonwealth Environmental Water 2016–17. Commonwealth Environmental Water Office, Canberra, Australia.

Commonwealth of Australia, 1993. Murray-Darling Basin Agreement. Parliament of Australia, Canberra. Available from: https://www.legislation.gov.au/Details/C2004A04593 (accessed October 2015).

Commonwealth of Australia, 2007. Water Act. Parliament of Australia, Canberra. Available from: https://www.legislation.gov.au/Details/C2016C00469 (accessed October 2015).

Commonwealth of Australia, 2012. Basin Plan. Parliament of Australia, Canberra. Available from: https://www.comlaw.gov.au/Details/F2012L02240 (accessed October 2015).

Connell, D., 2011. The role of the commonwealth environmental water holder. In: Connell, D., Grafton, R.Q. (Eds.), Basin Futures: Water Reform in the Murray-Darling Basin. ANU Press, Canberra, Australia.

Daily, G.C., 1996. Nature's Services: Societal Dependence on Natural Ecosystems. Island Press, Washington, DC.

Docker, B., Robinson, I., 2014. Environmental water management in Australia: experience from the Murray-Darling Basin. Int. J. Water Resour. D 30 (1), 164–177.

DSE, 2009. Northern Region Sustainable Water Strategy. Department of Sustainability and Environment, Government of Victoria, Melbourne, Australia.

Ferguson, J.J., Chillcott Hall, B., Randall, B., 2006. Keeping fish wet in Montana: private water leasing: working within the prior appropriation system to restore streamflows. Pub. Land & Resources L. Rev. 27, 1–13.

Grams, P.E., Schmidt, J.C., Wright, S.A., Topping, D.J., Melis, T.S., Rubin, D.M., 2015. Building sandbars in the Grand Canyon. Eos 96, http://dx.doi.org/10.1029/2015EO030349. Available from: https://eos.org/features/building-sandbars-in-the-grand-canyon (accessed March 2016).

Hardner, J., Gullison, T., 2007. Independent External Evaluation of the Columbia Basin Water Transactions Program (2003–2006). Hardner and Gullison Consulting, LLC, Amherst, NH.

Horne, A., Purkey, A., McMahon, T.A., 2008. Purchasing water for the environment in unregulated systems: what can we learn from the Columbia Basin. Aust. J. Water Res. 1, 61–70.

IBWC, 2012. Minute No. 319 – Interim international cooperative measures in the Colorado River Basin through 2017 and extension of Minute 318 cooperative measures to address the continued effects of the April 2010 earthquake in the Mexicali Valley, Baja California. International Boundary and Water Commission United States and Mexico, Coronado, CA.

IBWC, 2014. Minute 319 Colorado River Delta environmental flows monitoring – initial progress report. International Boundary and Water Commission United States and Mexico. Available from: http://www.ibwc.gov/EMD/Min319Monitoring.pdf (accessed November 2015).

Interior, 2014. Memorandum approval of recommendation for experimental high-flow release from Glen Canyon Dam, November 2014. US Department of the Interior. Available from: http://www.usbr.gov/uc/rm/amp/amwg/mtgs/14aug27/HFE_Memo_14oct24_PPT.pdf (accessed November 2015).

Interior, 2015. Glen Canyon Dam adaptive management work group charter. US Department of the Interior. Available from: http://www.usbr.gov/uc/rm/amp/amwg/pdfs/amwg_charter.pdf (accessed April 2016).

IRORG, 2015. Review of River Murray operations 2014–15. Report of the Independent River Operations Review Group, Unpublished Report.

Jacobs, J., 2011. The sustainability of water resources in the Colorado River Basin. Bridge 41 (4), 6–12.

King, M.A., 2004. Getting our feet wet: an introduction to water trusts. Harv. Environ. Law Rev. 28, 495–534.

MacDougal, D., Kearns, Z., 2014. The Columbia River Treaty review: will the water users' voices be heard? Available from: http://www.martenlaw.com/newsletter/20141117-columbia-river-treaty-review (accessed November 2015).

McCoy, A., Holmes, S.R., 2015. Columbia Basin Water Transactions Program Flow Restoration Accounting Framework. National Fish and Wildlife Foundation, Portland, OR.

MDBA, 2010a. Guide to the Proposed Basin Plan—Volume 1. Publication No. 60/10. Murray-Darling Basin Authority, Canberra, Australia.

MDBA, 2010b. Guide to the Proposed Basin Plan—Volume 2. Publication No. 61/10. Murray-Darling Basin Authority, Canberra, Australia.

MDBA, 2011. The Living Murray Story—One of Australia's Largest River Restoration Projects. Murray-Darling Basin Authority, Canberra, Australia.

MDBA, 2013. Constraints Management Strategy 2013–2024. Murray-Darling Basin Authority, Canberra, Australia.

MDBA, 2014a. Basin-wide Environmental Watering Strategy. Murray-Darling Basin Authority, Canberra, Australia.

MDBA, 2014b. Constraints Management Strategy Annual Progress Report 2013–14. Murray-Darling Basin Authority, Canberra, Australia.

MDBA, 2014c. 2014–15 Basin Annual Environmental Watering Priorities: Overview and Technical Summaries. Murray-Darling Basin Authority, Canberra, Australia.

MDBA, 2015. Objectives and Outcomes for River Operations in the River Murray System. Murray-Darling Basin Authority, Canberra, Australia.

Melis, T.S. (Ed.), 2011. Effects of Three High-Flow Experiments on the Colorado River Ecosystem Downstream from Glen Canyon Dam. U.S. Geological Survey Circular 1366, Arizona.

Melis, T.S., Grams, P.E., Kennedy, T.A., Ralston, B.E., Robinson, C.T., Schmidt, J.C., et al., 2011. Three experimental high-flow releases from Glen Canyon Dam, Arizona — effects on the downstream Colorado River ecosystem. US Geological Survey Fact Sheet 2011–3012 4.

Melis, T.S., Walters, C.J., Korman, J., 2015. Surprise and opportunity for learning in Grand Canyon: the Glen Canyon Dam adaptive management program. Ecol. Soc. 20 (3), 22.

NRC, 2004. Managing the Columbia River: Instream Flows, Water Withdrawals and Salmon Survival. National Research Council. National Academies Press, Washington, DC.

Neuman, J.C., 2004. The good, the bad, the ugly: the first ten years of the Oregon Water Trust. Neb. Law Rev. 83, 432–484.

Neuman, J.C., Chapman, C., 1999. Wading into the water market: the first five years of the Oregon Water Trust. J. Environ. Law Litig. 14, 146–148.

Neuman, J.C., Squier, A., Achterman, G., 2006. Sometimes a great notion: Oregon's instream flow experiments. Environ. Law 36, 1125–1155.

NSW, 2012a. Water shepherding option and issues analysis report. Water shepherding in NSW — advice to the water shepherding taskforce. New South Wales Government. Available from: http://www.water.nsw.gov.au/__data/assets/pdf_file/0009/547596/recovery_water_shepherding_options_issues_analysis_report.pdf (accessed November 2015).

NSW, 2012b. Proposed arrangements for shepherding environmental water in New South Wales: draft for consultation. New South Wales Government. Available from: http://www.water.nsw.gov.au/__data/assets/pdf_file/0004/555745/recovery_water_shepherding_proposed_arrangements_shepherding_nsw_draft_for_-consultation.pdf (accessed November 2015).

NSW, 2015a. Water customer service charter report. Department of Primary Industries, New South Wales Government. Available from: http://www.water.nsw.gov.au/about-us/customer-service (accessed November 2015).

NSW, 2015b. Environmental water use in New South Wales: outcomes 2014–15. Unpublished Draft Report. New South Wales Office of Environment and Heritage, Sydney, Australia.

NWC, 2011. National Water Markets Report 2010–11. National Water Commission, Canberra, Australia.

Poff, N.L., Allan, J.D., Bain, M.B., Karr, J.R., Prestegaard, K.L., Richter, B., et al., 1997. The natural flow regime: a paradigm for river conservation and restoration. BioScience 47, 764–784.

Poff, N.L., Richter, B., Arthington, A., Bunn, S., Naiman, R., Kendy, E., et al., 2010. The ecological limits of hydrologic alteration (eloha): a new framework for developing regional environmental flow standards. Freshw. Biol. 55, 147–170.

Postel, S., Richter, B., 2003. Rivers for Life: Managing Water for People and Nature. Island Press, Washington, DC.

Reclamation, 2008. The Law of the River. US Department of the Interior, Bureau of Reclamation Lower Colorado Region. Available from: http://www.usbr.gov/lc/region/g1000/lawofrvr.html (accessed November 2015).

Reclamation, 2011. Environmental Assessment for the Development and Implementation of a Protocol for High-Flow Experimental Releases from Glen Canyon Dam, Arizona 2011 through 2020, 2011a. US Department of the Interior, Bureau of Reclamation, Salt Lake City, UT.

Reclamation, 2012. Finding of No Significant Impact for the Environmental Assessment for the Development and Implementation of a Protocol for High-Flow Experimental Releases from Glen Canyon Dam, Arizona Through 2020, 2012b. US Department of the Interior, Bureau of Reclamation, Salt Lake City, UT.

Reclamation, 2014. Memorandum: Approval of Recommendation for Experimental High-Flow release from Glen Canyon Dam, November 2014. Available from: http://gcdamp.com/images_gcdamp_com/c/cc/2014_Recommendation_HFE_Memo_%26_PPT_14oct24_PPT.pdf (accessed January 2017).

Southwell, M., Wilson, G., Ryder, D., Sparks, P., Thoms, M., 2015. Monitoring the Ecological Response of Commonwealth Environmental Water Delivered in 2013−14 in the Gwydir River system: A Report to the Department of Environment. Commonwealth of Australia, Canberra.

Stewardson, M.J., Jones, M., Koster, W.M., Rees, G.N., Skinner, D.S., Thompson, R.M., et al., 2014. Monitoring of Ecosystem Responses to the Delivery of Environmental Water in the Lower Goulburn River and Broken Creek in 2012−13. The University of Melbourne for the Commonwealth Environmental Water Office. Commonwealth of Australia, Canberra, Australia.

Susskind, L., Camacho, A.E., Schenk, T., 2012. A critical assessment of collaborative adaptive management in practice. J. Appl. Ecol. 49 (1), 47−51.

Tharme, R.E., 2003. A global perspective on environmental flow assessment: emerging trends in the development and application of environmental flow methodologies for rivers. River Res. Appl. 19, 397−441.

Wassens, S., Jenkins, K., Spencer, J., Thiem, J., Wolfenden, B., Bino, G., et al., 2014. Monitoring the Ecological Response of Commonwealth Environmental Water Delivered in 2013−14 to the Murrumbidgee River system. Draft final report. September 2014. Commonwealth of Australia, Canberra, Australia.

Webb, A., Vietz, G., Windecker, S., Hladyz, S., Thompson, R., Koster, W., et al., 2015. Monitoring and Reporting on the Ecological Outcomes of Commonwealth Environmental Water Delivered in the Lower Goulburn River and Broken Creek in 2013−14. The University of Melbourne for the Commonwealth Environmental Water Office. Commonwealth of Australia, Canberra, Australia.

WWI, 2012. Use and Capacity of Global Hydropower Increases. Vital Signs Online. World Watch Institute, Washington, DC. Available from: http://vitalsigns.worldwatch.org/vs-trend/global-hydropower-installed-capacity-and-use-increase (accessed July 2015).

Vörösmaty, C.J., Green, P., Salisbury, J., Lammers, R.B., 2000. Global water resources: vulnerability from climate change and population growth. Science 289 (5477), 284−288.

Vörösmaty, C.J., McIntyre, P.B., Gessner, M.O., Dudgeon, D., Prusevich, A., Green, P., et al., 2010. Global threats to human water security and river biodiversity. Nature 457, 555−561.

Ye, Q., Livore, J.P., Aldrige, K., Bradford, T., Busch, B., Earl, J., et al., 2015. Monitoring the Ecological Responses to Commonwealth Environmental Water Delivered to the Lower Murray River in 2012−13. Report 3 prepared for Commonwealth Environmental Water Office by South Australian Research and Development Institute. Commonwealth of Australia, Canberra, Australia.

PRINCIPLES FOR MONITORING, EVALUATION, AND ADAPTIVE MANAGEMENT OF ENVIRONMENTAL WATER REGIMES

25

J. Angus Webb[1], Robyn J. Watts[2], Catherine Allan[2], and Andrew T. Warner[3]

[1]The University of Melbourne, Parkville, VIC, Australia [2]Charles Sturt University, Albury, NSW, Australia [3]United States Army Corps of Engineers, Alexandria, VA, United States

25.1 INTRODUCTION

The importance of allocating water to improve the ecological health and ecosystem services of river systems is recognized among most freshwater scientists and water managers. Within the context of other social goals, environmental water can represent a huge investment of public funds for the environment. For example, as part of the Murray—Darling Basin Plan in Australia, the *Restoring the Balance Program* is expending AU$3 billion purchasing water entitlements from irrigators, with the water delivered as an environmental water regime (Skinner and Langford, 2013). Similarly, the implementation of environmental water regimes in the Upper Yuba River, CA, United States, would reduce revenue for hydropower generators by up to 45% (Rheinheimer et al., 2013). Although the extent of the outcomes—in the form of ecosystem health and economically and socially important services—from these investments is uncertain, management decisions must still be made regarding the short-term operation of these programs and the long-term planning cycles. Monitoring the environmental outcomes and related socio-economic benefits from environmental water regimes supports at least two objectives important to management.

First, the information gained from monitoring can be used to demonstrate the *return-on-investment* of environmental water in terms of environmental, economic, and social outcomes. In highly allocated river systems such as those in the Murray—Darling Basin and in the western United States, any water delivered to the environment must be recovered from consumptive uses (often irrigated agriculture). The economic impact of foregone productive capacity can be calculated. To determine if these impacts and the public investment in environmental water regimes are justified, it is important to evaluate the extent to which reallocating water to environmental purposes leads to tangible ecological benefits, the restoration of ecosystem services, and socially desired outcomes.

Second, monitoring outcomes will add to our growing knowledge regarding the effects of environmental water regimes. Although significant advances have been made in the understanding of flow—ecology relationships, especially during the past two decades, significant knowledge gaps

Water for the Environment. DOI: http://dx.doi.org/10.1016/B978-0-12-803907-6.00025-5

599

remain that continue to challenge management (see Chapters 14 and 15). Moreover, our understanding of outcomes in socio-ecological systems is poor in general, not just when considering the effects of environmental water. In order to make best use of the available environmental water, we need to improve this knowledge through monitoring and reflection. Reflection by individuals or groups is an active process of considering information such as that gained from monitoring, and building it into knowledge, understanding, and practice (Schön, 2008). The word *reflection* is purposefully used here to highlight a critical stage in managing environmental water that involves evaluating monitoring results and changing management decisions based on those results. The role of *reflector* is also highlighted below (Section 25.8.2) not as the person who carries out these tasks, but instead as the person explicitly responsible for ensuring that the tasks are completed in a timely manner and with appropriate rigor. Synthesis of monitoring data provides new knowledge and improved understanding of the observed responses to different flow events and under different circumstances. This improves our ability to predict responses to environmental water releases, contributing to both short-term environmental water delivery decisions, and also to long-term planning cycles. This in turn will increase the likelihood of achieving good ecological and social outcomes and improve the return-on-investment of environmental water for all stakeholders.

This cycle of monitoring, evaluation, learning, adjustment, and improvement in decisions and outcomes is best conceived through the lens of adaptive management. In this chapter, we provide an introduction to adaptive management, outline the importance of effective monitoring and evaluation as an input to the adaptive management cycle, and note challenges for learning that must be overcome for adaptive management of environmental water regimes to be successful. We also provide a set of general principles for monitoring and adaptive management of environmental water.

25.2 BACKGROUND: HISTORY OF ADAPTIVE MANAGEMENT

Adaptive management is an approach to natural resource management for people who must act despite uncertainty about what they are managing and the impacts of their actions (Allen and Garmestani, 2015). Adaptive management is an alternative to conventional, reductionist management that should enable effective action within complex socio-ecological systems (Pahl-Wostl, 2007). Adaptive management is not blind or aimless trial and error, but is purposeful and deliberate (Allan and Stankey, 2009) and structured (Pahl-Wostl et al., 2010). Whether learning is the result of planned experimental learning (i.e., *active* adaptive management) or from careful reflection on management actions (i.e., *passive* adaptive management), the central aim of adaptive management is to consider the consequences of actions undertaken for management, and to improve future management actions through that consideration.

Since the late 1970s, the concept of adaptive management has evolved in numerous directions, each emphasizing different aspects of assessment and action, but all centered on iterative learning about a system and making management decisions based on that learning (Williams and Brown, 2014). The adaptive process is often represented as a cycle of *plan*, *do*, *monitor*, and *learn* (Fig. 25.1). The simplicity of this cycle conceals the complexity of learning from practice in complex socio-ecological systems. Large programs and/or complex situations often contain multiple *mini-loops* (see examples in Fig. 25.1) operating at different scales, but nested within the larger adaptive framework (Bormann and Stankey, 2009). Stakeholders from different backgrounds, or with different goals or constrained areas of influence, are likely to prioritize some mini-loops over others or one aspect of the adaptive management cycle over another. These mini-loops enable the process to effectively go backward as well as forward, as learning occurs within the management cycle.

Raadgever et al. (2008) identify two separate interpretations of adaptive management in natural resource management discourse: one with a focus on science and the other focused more on social learning and co-management. The first highlights technical or scientific matters such as testing scenario modeling of systems (Rivers-Moore and Jewitt, 2007; Williams, 2011) and field-scale experimentation (Pollard et al., 2011). The second works with theories and practice of participatory learning and decision making (Stringer et al., 2006), social learning (Blackmore and Ison, 2012), evaluation (Bryan et al., 2009), and governance (Ison et al., 2013). Both interpretations are valid and useful, and in practice, adaptive management is effective when it acts as a framework within which these interpretations can be integrated.

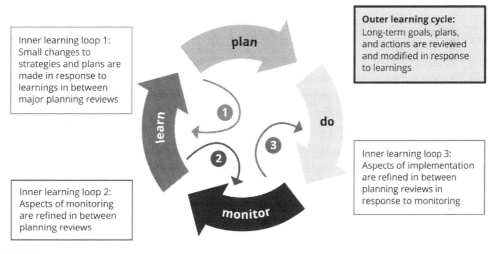

FIGURE 25.1

The adaptive management cycle showing the outer learning cycle where lessons inform the next formal phase of planning and implementation. The inner learning loops are small changes that are made based on learnings that occur between major planning reviews. The inner loops effectively allow progress in the outer loop in both directions.

25.3 MEASURING THE SUCCESS OF ADAPTIVE MANAGEMENT

For almost as long as adaptive management has been described as such, it has been assessed as failing to achieve its promise as a revolutionary approach to managing (Allan and Curtis, 2005; Walters, 2007; Westgate et al., 2013). This provides an impression of wholesale failure that is unwarranted, with examples of effective adaptive management reported in natural resource management broadly (Williams and Brown, 2014) and in the management of water (Failing et al., 2013; Smith, 2011).

Although acknowledging that adaptive management can be effective, published success stories are still exceptions rather than the norm, partly because of problems with defining success meaningfully. We have indicated that the adaptive management approach covers a wide variety of iterative long-term practices within complex socio-ecological systems, suggesting that assessing the *success* of adaptive management must, therefore, be complex as well. Some reasons for apparent or actual underperformance of adaptive management may include:

1. Adaptive management is happening in name only. The meta-analysis of McFadden et al. (2011) suggests that adaptive management as a concept for practice has moved from mere theory to being accepted as a practical approach, but that this has not yet led to full implementation or uptake. As it is such an appealing concept, the term *adaptive management* has been embraced across the numerous fields of natural resource management as something of a panacea for all complex issues (Pahl-Wostl et al., 2012). However, many policies and management plans that specify an adaptive management approach do not provide mechanisms for it to occur. Simply specifying an adaptive management approach in strategic documents does not guarantee that any learning or practice change will take place.

2. False expectations of what adaptive management can achieve and how this can be measured. For example, Gunderson (2015) suggests that the Grand Canyon Adaptive Management Program has been a success because, "The development of the flow treatments have informed managers about key resource dynamics and allowed them to learn that further releases are unlikely to help resolve endangered species management as there is a long-term sediment issue in the Grand Canyon." However, he notes that some critics of the program argue that adaptive management should be abandoned, as the large expenditure on it has failed to fix the critical resource issues. Effective adaptive management starts with framing good questions (Allan and Stankey, 2009), and assessments of the success of adaptive management need to reference the original questions and the capacity to act.

3. Adaptive management is anticipated to be large and expensive. High-profile and highly resourced projects such as the Grand Canyon Adaptive Management Program mentioned above raise expectations of large-scale activities and outcomes. However, the story we tell of the Dartmouth Dam adaptive management (Box 25.1) shows that adaptive management can operate at local scales and with modest investment, but useful outcomes.

4. Poor or fragmented assessment, documentation, and reporting of adaptive management obscure successes. Institutional arrangements, including *projectification* (sensu Allan, 2012)—for instance management of landscapes over time through discrete projects—may divide the system issues and management into operational units that can easily miss the learning and adaptation that is occurring within it. We were able to tell the story of the Dartmouth Dam example (Box 25.1) because social science researchers were included in the team, and both documented and reported evidence of adaptive management in the system over an extended period of time. Similarly, Allan and Watts (2016) note that the successes of adaptive management are often implicit and not reported because they do not necessarily occur within a named adaptive management structure (sensu Fig. 25.1). For example, managers of environmental water in the Edward−Wakool river system, New South Wales, Australia, can claim success with adaptive management of environmental water regimes. However, this success became apparent to

BOX 25.1 CASE STUDY: DARTMOUTH DAM, AUSTRALIA, VARIABLE FLOW TRIALS, AN EXAMPLE OF EMERGENT ADAPTIVE MANAGEMENT

Dartmouth Dam variable flow trials in south eastern Australia are an example of adaptive flow management, where managers used flexibility within existing river operating rules to learn about and alter the management of the dam for improved environmental outcomes, while simultaneously fulfilling social and economic objectives and requirements. Dartmouth and Hume dams are operated by agencies as part of the Murray River System under the direction of the Murray River Division of the Murray–Darling Basin Authority. The Mitta Mitta River is used to transfer water from Dartmouth to Hume Dam (Fig. 25.2), primarily to support irrigated agriculture. The volume, variability, and seasonality of flows in the Mitta Mitta have been heavily modified, with minimum flows traditionally being released from the dam at low constant flow rates at times when bulk water transfers are not required (Watts et al., 2010).

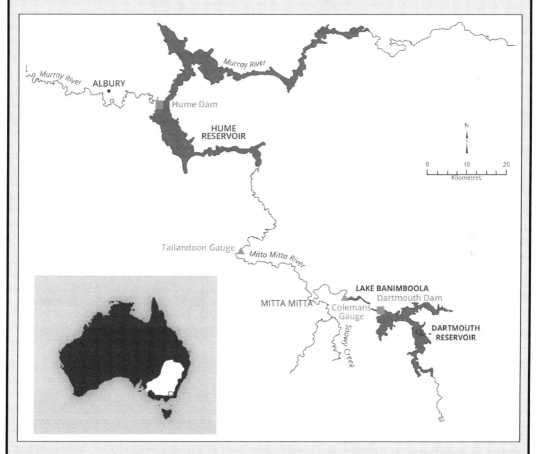

FIGURE 25.2

Dartmouth and Hume dams in the upper Murray River system, south eastern Australia.

Source: Watts et al. (2010)

(Continued)

BOX 25.1 (CONTINUED)

River operators working with scientists experimented with transferring consumptive water between Dartmouth and Hume dams using a variable hydrograph to reduce environmental impacts (Watts et al., 2010). Through a series of trials, much was learned about biophysical aspects of the Mitta Mitta River, including that managed variable flows are ecologically beneficial when compared to sustained periods of relatively constant flows during both the filling and releasing modes of operation of Dartmouth Dam (Watts et al., 2009). Spatial and temporal scale, and antecedent conditions, contributed to the strength and nature of these benefits. As the Dartmouth variable flow trials progressed, it became clear to the river operators and researchers that they were part of an active adaptive management program, even though the variable flow trials project had not been envisaged as such at its outset (Allan and Stankey, 2009). There was a cycle of learning from a series of monitored and evaluated variable releases. Monitoring directly informed the planning and implementation of subsequent variable flow trials and the development of new operational guidelines that continue to guide the operation of Dartmouth Dam. Although the project was tightly bounded to a relatively short stretch of river within existing dam operating rules, lessons from the Dartmouth variable flow trials were incorporated into the broader system operation and water reform framework, including the system-wide review of Murray River Systems Operations that commenced in 2007/08.

Even this relatively simple example of adaptive management is more complex than the concept presented in the outer learning loop of Fig. 25.1. This in part reflects the multiple scales in both space and time at which planning and action occur. Significant institutional review and planning tend to operate over medium to long time cycles, whereas during the life of a management plan, considerable shorter-term learning occurs through implementation of management decisions. In this case, institutional conditions encouraged a shift to adaptive management over a timescale that helped to achieve environmental, social, and economic objectives downstream of the dam. A key lesson from this case study is that adaptive management does not have to be specified a priori, but can emerge if there is a trusting relationship between stakeholders and they are willing and able to change their operational paradigm in the face of new knowledge (Watts et al., 2010). Including a social scientist in the project team to facilitate reflection on the processes of adaptive flow management, as well as on biophysical lessons, ensured that deliberate investigation and evaluation occurred.

observers outside the system only when evidence of learning from a range of sources was brought together and reported in a single document (Allan and Watts, 2016).

25.4 IMPROVING UPON PREVIOUS MONITORING AND ADAPTIVE MANAGEMENT

An essential element of adaptive management is monitoring, but it is widely acknowledged that monitoring and evaluation of ecological restoration projects has been deficient in the past. Bernhardt et al. (2005) reviewed river restoration projects in the United States, and found that only approximately 10% had any post-project monitoring of any sort, and that most of this monitoring was restricted to output monitoring, for example for a riparian restoration project, assessing whether the planned number of trees was planted. Failures of monitoring for ecological restoration can be traced to inadequate investment or appreciation of the need for monitoring, the opportunistic nature of many projects, an inability to collect data before restoration takes place, the fact that many of these projects are undertaken by community members and not-for-profit groups who are not trained in principles of effective monitoring, and the high cost of undertaking high-quality monitoring and assessment (Brooks and Lake, 2007). There have been gradual signs of improvement in this, with an Australian study reporting that 14% of projects included monitoring (Brooks and Lake, 2007), including a greater proportion in the later years of that sample. However, for both these examples, there is little sign of monitoring results being evaluated in an adaptive management sense to improve decision making and environmental outcomes in the future.

Flow restoration monitoring has been no exception to this underuse of evaluation results, and unfortunately has been found to be consistently lacking in its ability to advance science and/or support management (Souchon et al., 2008). For example, a global review of 113 large-scale flow experiments implemented at dams across 22 countries found considerable opportunity to improve the efficiency and value of these experiments for both scientific learning and improved management (Olden et al., 2014).

Many of the proposed reasons for the apparent failure of adaptive management mentioned in Section 25.3 also apply to monitoring ecological restoration. It can also be argued that a failure to monitor can also be a completely rational response for managers attempting to survive within the world of natural resource management (Rutherfurd et al., 2004). For example, funds are competitively allocated within management agencies. If a weak monitoring program fails to detect an environmental improvement, this may be used (incorrectly) as evidence that the management program is not working, and the funds may then be directed elsewhere.

Whatever the reasons for the past failure to monitor and adaptively manage environmental water (and other restoration) programs, this is not an option for the future. The highly contested nature of environmental water (Poff and Matthews, 2013) means that if environmental water programs are to survive, well-targeted monitoring must be undertaken to justify the public investment and to argue for the value of environmental water in terms of ecological outcomes, ecosystem services provided, and socio-ecological benefits against alternative consumptive uses.

Beyond monitoring and evaluation, however, we need to improve the mechanisms of uptake of new knowledge generated by monitoring. This would allow results to be used within adaptive management frameworks in order to improve decision making by managers, and therefore maximize the benefits that can be achieved from the limited volumes of environmental water available.

25.5 THE LONG-TERM AIM OF MONITORING AND ADAPTIVE MANAGEMENT

The description of the poor history of restoration monitoring outlined above (Section 25.4) establishes the expectation that all restoration projects should be monitored. Indeed, it has been suggested that a restoration project cannot be considered successful unless it is accompanied by strong monitoring (Palmer et al., 2005). This, however, is an unreasonable expectation. The cost of monitoring, the opportunistic nature of many restoration projects, and the sheer number of these projects taking place mean that a sizeable fraction of restoration projects are unlikely to be monitored. How can we use monitoring and adaptive management to improve outcomes for those projects?

High-quality monitoring, conducted in concert with targeted research, can be used to build general understanding of ecological responses to flow alterations. For rivers that are monitored, this will allow us to predict responses to future flow management, improving management of that river. More importantly, such understanding will allow us to predict responses in rivers that are not monitored. This type of approach is being implemented in the *Long-Term Intervention Monitoring (LTIM)* Project in the Murray–Darling Basin (Gawne et al., 2013; Box 25.2), and is at least proposed in the Ohio River Basin within the *Sustainable Rivers Project (SRP)* (Box 25.3). The logical consequence of this is that monitoring needs to be conducted, or at least the results synthesized, at large scales (Section 25.8.5). Thus, we should not aim to monitor everywhere, but to conduct high-quality monitoring at a smaller number of sites that are a sample of the range of sites to which we wish to extrapolate results. This does not imply that the sampled sites are identical to those to which results will be extrapolated, but are drawn from the distribution of conditions of those sites. Thus, results should be able to be extrapolated to rivers of similar hydrologic regime and geomorphic conditions, in similar climatic zones, and with similar biological communities (Poff et al., 2010).

If adaptive management is implemented well, then knowledge and wisdom will build over time and monitoring and evaluation programs will change. The aim is to move from a state of high uncertainty regarding environmental and related socio-economic outcomes, to improved knowledge where we can make predictions with higher confidence. Once this is achieved, it can be argued that the purpose of rigorous monitoring has been achieved. Monitoring efforts could be reduced when we are confident of what the outcomes of environmental water will be. The resources saved by reduced monitoring could then be directed toward monitoring of other outcomes or river systems

---

BOX 25.2 MONITORING RESPONSES TO ENVIRONMENTAL WATER REGIMES IN THE MURRAY–DARLING BASIN, AUSTRALIA

The Murray–Darling Basin Plan in Australia includes possibly the largest (in terms of volume relative to system flows) environmental water program implemented anywhere in the world. In the long term, approximately 2750 GL (approximately 8% of average annual inflows) will be allocated to the environment on top of existing entitlements. This water comes from a combination of direct purchases of existing irrigation licenses, and through the installation of infrastructure to improve efficiency of the use of both environmental and consumptive water (Skinner and Langford, 2013).

Demonstrating the value of this major investment of public funds in the environment is of paramount importance, and the Commonwealth Environmental Water Office has funded an initial 5-year monitoring program to assess ecosystem responses to environmental water releases—the LTIM Project (Gawne et al., 2013). Evaluation of socio-environmental effects of the Basin Plan is being done through other programs (MDBA, 2014).

Recognizing that monitoring cannot be conducted at all rivers receiving Basin Plan water allocations, the LTIM Project is undertaking monitoring at seven *selected areas* throughout the Murray–Darling Basin (see Map 1). Monitoring is coordinated through a central *Monitoring Advisor* group, with *Monitoring Providers* at each of the selected areas developing programs that both reflect local priorities and also inform basin-scale evaluation. Conceptual models were developed for all endpoints being monitored to identify the major flow and nonflow drivers of ecological condition (MDFRC, 2013). Endpoints that are being analyzed at the basin scale (fish, vegetation, birds, and ecosystem metabolism) are monitored using standardized methods to ensure compatibility of the data being collected in the different selected areas (Gawne et al., 2013). These standardized methods will ensure that monitoring being undertaken in the selected areas can:

1. Inform basin-scale analyses that will have much greater power to detect effects than can analyses for individual selected areas.
2. Inform basin-scale analyses for species that reproduce and recruit over much larger scales than the selected areas.
3. Be used to infer effects for rivers and sites that have not been monitored.

not so well-understood. Thus, the monitoring within an adaptive management program is also adaptive to learning over time.

This type of *lifecycle* approach to monitoring and adaptive management is seldom mentioned in the literature, possibly because there are so few published examples of adaptive management proceeding to the point where knowledge has improved sufficiently that we could consider monitoring less.

BOX 25.3 USING ELOHA CLASSIFICATIONS TO INFORM MONITORING INVESTMENT IN THE SRP

The SRP is a national effort of the *United States Army Corps of Engineers (USACE)*, in partnership with *The Nature Conservancy* and dozens of other organizations (Warner et al., 2014). Its principal aim is to define and implement environmental water regimes through adaptive reservoir management. The project initially focused on eight demonstration basins containing 36 USACE dams, with the aim of using this experience to help guide operational changes for as many as 600 dams impacting an estimated 80,000 km of rivers.

Across the 490,000 km² Ohio River Basin, United States, there are 84 USACE dams, the operations of which may be changed to improve ecological function downstream as part of the SRP. It has been realized that developing individual environmental water regimes for each dam would be prohibitively expensive and time-consuming, as would be monitoring outcomes of environmental water regime implementation.

In a nascent effort, SRP scientists have used the stream classification portion of the *Ecological Limits of Hydrologic Alteration (ELOHA)* framework (Poff et al., 2010; see also Chapters 11, 13, and 27) to classify rivers in the basin and characterize their unaltered flow patterns and to subsequently define environmental water requirements based on the classification. This classification—applied to date for only a portion of the basin (Dephilip, 2013)—can be expanded and used to support the definition of environmental water requirements for rivers across the basin. The classification might also be used to help target limited monitoring resources. This could allow targeted monitoring of selected rivers within a given class type, and extrapolation of results to other rivers within the same class. Another benefit of using a river classification system to help guide monitoring is that it can help direct resources toward river types that are unique, excessively threatened, or poorly understood.

Although this approach is conceptually sound, it is recognized that issues may arise. For example, the approach can unravel if different components of environmental water requirements are implemented on different rivers of the same class, but monitoring is only conducted on one of them. Different stresses across rivers of the same class type that are beyond flow alteration—such as water quality degradation or direct habitat destruction—could also undermine this approach. At the time of writing, ELOHA classifications had only been completed for 20 of the 84 USACE dams, but it is hoped that the classification can eventually support monitoring and adaptive management objectives for the entire Ohio River Basin, where resources do not allow for monitoring on all rivers.

25.6 CHALLENGES FOR MONITORING AND EVALUATION OF ECOSYSTEM RESPONSES TO ENVIRONMENTAL WATER RELEASES

The ecosystem responses to environmental water regimes are inherently difficult to monitor, both from technical and institutional points of view (Webb et al., 2010b). First, it needs to be noted that active *manipulative* experiments, with random allocations of treatments, are impossible. Environmental water is delivered to rivers determined by infrastructure availability, sufficiency of water, and management and stakeholder engagement. These rivers are very much not a random sample. *Mensurative* experiments (those where variables are measured under different conditions over a period of time; Konrad et al., 2011) are the best we can hope for. Here, contrasts are made across environmental factors or gradients to try to infer the effects of these *treatments*. The inferences may not be as strong as those gained from manipulative experiments, but are still very useful for progressing understanding of ecosystem response to flow alteration.

Monitoring of ecological impacts, including positive impacts of restoration programs, is most effectively done using factorial experimental designs, preferably based upon the *before–after control-impact* (*BACI*; Downes et al., 2002) group of experimental designs (Fig. 25.3A). This type of design has the best prospect for isolating the effects of the treatment from other variations caused by uncontrolled environmental factors, and therefore avoiding false conclusions. They do this by assessing change over time at one or more *impact* sites affected by the treatment, and comparing this change to changes experienced at one or more control sites. Unfortunately, in environmental water management this type of experimental design is very much the exception rather than the rule.

For large-scale environmental water programs, it is difficult to identify before–after boundaries. For example, the institutional process of water reallocation (e.g., buybacks of irrigation licenses, water savings from infrastructure upgrades to improve efficiency of water use) under the Murray–Darling Basin Plan is expected to take up to 7 years (Skinner and Langford, 2013), with environmental allocations gradually increasing over that time. Even when water recovery is complete, environmental allocations will vary from year to year and place to place (as do those for consumptive uses), dependent upon the amounts of water in storage, and forecast and real inflows to those storages.

Partly for these reasons, the clear delineation of impact and control rivers is also difficult. For example, can a river be considered to be experiencing an environmental water treatment if environmental water is delivered only occasionally? Beyond this, control systems are very hard to identify because of the scale of environmental water programs. Rivers are large and distinct entities in their lowland reaches where environmental water regimes are commonly delivered, and therefore it is extremely difficult to find a control system that is similar in all respects but for the absence of the experimental treatment. For the same reasons of scale, it is difficult to conceive of programs with reasonable replication of control and impact systems.

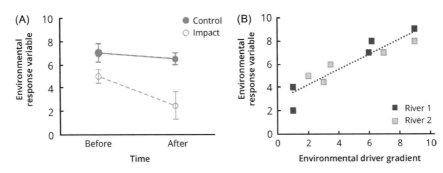

FIGURE 25.3

Comparison of monitoring designs. (A) *BACI* designs (and their derivatives) are recognized as being the strongest design for identifying the effects of environmental impacts (including positive impacts of restoration).
(B) *Gradient* designs rely on building continuous relationships between environmental drivers and responses.
They are not as inferentially strong as a BACI design but may often be the only applicable design for evaluating the effect of environmental water.

Thus, learning about the effects of environmental water delivery is most likely to occur through the application of *gradient* designs (Fig. 25.3B), where each site or river or sample through time forms a data point in a continuous statistical model (Webb et al., 2010a). Such designs provide replication and ensure that a wide range of conditions is included in the evaluation.

Although some of the replication described above can be gained by combining data across large scales, here we run into a jurisdictional problem. Many rivers are managed by local authorities, and responsibility for monitoring often rests with those authorities. This naturally leads to fragmentation of monitoring efforts, with different practitioners using different methods to attempt to monitor essentially the same outcomes (Webb et al., 2010a). This difference in approach makes it difficult to combine results from different programs to reach any sort of general understanding (Poff and Zimmerman, 2010). Stronger experimental designs are possible around individual case studies (e.g., revised operations of a single dam). At this smaller scale, there is a greater amount of control over data collection. However, it is difficult to extrapolate such individual results to other similar cases.

The rate of learning will be fastest in an active adaptive management program, which would ideally include comparison of outcomes from environmental water releases that are both optimal and suboptimal (in terms of volume, timing, or duration of flow events), therefore providing better knowledge of what works and what does not. These programs will be easier to monitor and evaluate because they more closely resemble familiar experimental designs, and also because we expect a difference in the treatment effect of the different water releases. However, water managers are expected to achieve the best environmental outcomes from available environmental water, and to deliberately do otherwise could lead to accusations of poor management. One may think that this reduces the opportunity to compare outcomes of optimal and suboptimal environmental water releases. However, given that broad societal considerations constrain options for environmental water delivery, our experience is that in many cases environmentally suboptimal water release decisions are actively being made as part of a negotiated process. Thus, there are opportunities to compare the outcomes of suboptimal and optimal water releases with outcomes of more natural flood events. This knowledge can then be used through active adaptive management to increase the rate of learning and inform better decision making in future years.

The impacts of environmental water releases are wider than simply the ecological responses. The water delivery occurs within complex socio-ecological systems, and there will be impacts on the wider ecosystem, and humans within the systems. A major challenge is to ensure that evaluation of the social and economic impacts of environmental water regimes is undertaken at the same time as biophysical monitoring and the results are integrated and synthesized to assist water managers make good decisions. However, the current focus is on monitoring biophysical endpoints, and this type of integration is not occurring (CEWO, 2015).

25.7 **CHALLENGES FOR LEARNING**

Learning (for simplicity here used to mean creating new knowledge and understanding) from monitoring, and acting on that learning, is just as challenging as undertaking monitoring itself. The first challenge is that we currently often conflate evaluation (reflecting on monitoring with an aim to learn) and auditing (Allan and Curtis, 2005). Auditing checks to see if what was supposed to happen did happen, and rests upon an implicit value judgment that compliance is good and noncompliance is not. Auditing is necessary and useful, but should be purely procedural to ensure that project resources are managed well. At least two challenges for learning arise when an auditing rather than evaluation framing is assumed. The first is that, within a project, adapting to new knowledge learned from implementation may be constrained by initial promises and predetermined targets and milestones. Second, the results from monitoring may lead to judgment rather than learning. For example, in an auditing framework, the nonoccurrence of a predicted positive outcome of an environmental water release is judged as a failure, whereas in an evaluation framing it is a lesson that can inform future practice. Conflation of auditing with evaluation can lead to perverse outcomes such as that described above when managers actively choose not to monitor to protect their projects from adverse judgment (Rutherfurd et al., 2004).

Auditing is an attractive framing because it is easy to see its value, and who benefits from it. Learning through evaluation is less straightforward, as the lessons from monitored actions may not be immediate, and there may be a wide range of potential beneficiaries, beyond temporal and physical boundaries. Thus, it is easier to measure clear indicators of milestones and targets than to consider if the trajectory of system response is the best or even an appropriate one (Madema et al., 2014). The complexities of learning from monitoring are such that it appears to be easier to obtain funds to monitor than it is to put resources into reflecting on what is being learned from that monitoring.

Evaluating may assist with learning, but there remain challenges with sharing and using the knowledge generated. In large projects and systems, the people with immediate experience of the lessons may not be those with the influence to act upon this new knowledge. Mediating processes such as modeling or narrative creation may be required to enable lessons to be understood or used. Even without such mediation, documenting and disseminating lessons in some form will be necessary, and this requires planning and resources. Ensuring sufficient resources for all stages of *learning* may require champions (Schultz and Fazey, 2009; Webb et al., 2010a), as this is not automatically funded. More problematic is how to combine lessons from monitoring with other forms of learning, especially within the framings of *scientific validity* and political necessity.

25.8 GENERAL PRINCIPLES FOR MONITORING AND ADAPTIVE MANAGEMENT OF ENVIRONMENTAL WATER REGIMES

Although the challenges laid out above are considerable, adaptive management that is based on learning from monitoring is possible. Ultimately, effective environmental monitoring and adaptive management rely upon strong and enduring partnerships between managers and researchers, and between the different disciplines necessary to understand the physical, environmental, and social impacts of environmental water management. A strong partnership will have the diversity of skills, motivations, and resources necessary to *stay the course* and see the completion (hopefully many times) of the adaptive management cycle. Some of the key factors that influence the adaptive management process include:

1. Recognition of the time necessary to build and maintain partnerships: the role of trust;
2. Appreciation of the importance of individuals, roles, skills, and experience;

3. Looking for understanding from other projects, but realizing that there is no single recipe that can be applied to other systems;
4. Recognizing that adaptive management may emerge spontaneously;
5. Coordination of monitoring and evaluation across large spatial and temporal scales;
6. Creation of programs of requisite simplicity;
7. Innovative approaches to analysis and evaluation to make the most of the available data;
8. Maintaining the adaptive aspect of adaptive management;
9. Ensuring regular reporting of observations, learnings, and processes in addition to the annual report.

These factors are outlined further below.

25.8.1 EXPEND THE TIME NECESSARY TO BUILD AND MAINTAIN TRUSTING PARTNERSHIPS

Ideally, partnerships should be established at the start of the adaptive management program. Direct contact between the individuals responsible in each organization and each disciplinary area is necessary, as it is individual interactions that determine the success or failure of collaborations (Cullen, 1990).

The first step in establishing an effective partnership is an open conversation regarding the expectations and ambitions of each member. Different team members will be hoping to achieve different things from the partnership: senior managers may be seeking successful achievement of policy goals; local managers will be most interested in outcomes for their local environment and stakeholders; researchers will be interested in achieving *impact*, both from the point of view of affecting environmental management practice and also in terms of high-quality publications in international journals (Webb et al., 2010a). Different disciplinary backgrounds of the team members will also introduce challenges for building the partnership because of differences in language, research traditions, and expectations. Although all members may want to *learn*, they may want to learn different things about the system, and have different understandings of what learning is.

At the intersection of these individual aims lies the shared vision that has motivated the individuals to become involved. For adaptive management of environmental water, that shared vision will probably revolve around improving management decisions for an environmental water regime; to maximize outcomes in terms of ecosystem responses, ecosystem services, and social outcomes. Although environmental water programs will often require trade-offs among different outcomes, sometimes opportunities exist for shared benefits that may not have been realized without an effective partnership of individuals with different goals (Box 25.4). It is important to articulate and continually rearticulate this vision over the course of the partnership. Otherwise, changing expectations of one partner might not be matched by others, leading to a mismatch between actions and emphases, and ultimately a less effective outcome.

However, a strong focus on meeting targets (which may be one of the *expectations* mentioned above) can reduce the chance of learning about the system, especially from learning through surprise. Knowledge emerges within complex socio-ecological systems, so the sharing of purpose and vision must be accompanied by discussions on responding to emergent knowledge, surprises, and *failures*.

By establishing partnerships at the start of any project, all team members have greater owner-ship of the decisions taken and actions implemented. Early establishment of the partnership also provides the time necessary to build trust among team members. The importance of trust cannot be overstated. Trust facilitates knowledge exchange among team members, allowing learning to be applied within the adaptive management loop. Trust is also created more easily when team mem-bers have something in common. There is a history of science—management partnerships that have not achieved their aims (Benda et al., 2002). However, we wonder whether in some regions the increasing prevalence of research higher degree qualified (i.e., Masters and PhD) staff within gov-ernment agencies may be contributing to a strengthening of multidisciplinary science—management partnerships.

BOX 25.4 SOCIO-ECONOMIC BENEFITS OF ENVIRONMENTAL WATER REGIME IMPLEMENTATION ON THE GREEN RIVER, KENTUCKY, UNITED STATES

As described in the introduction to this chapter, implementation of environmental water regimes is often assumed to involve trade-offs with other social objectives such as forgone water supply or reduced hydropower generation. However, there are conditions under which environmental water releases can align with other objectives such that flow restoration maintains or even enhances other reservoir purposes. The Green River, Kentucky, is one such example.

The Green River is a tributary of the Ohio River, draining a 23,400 km^2 watershed in the central eastern portion of the United States. The Green River Dam was built in 1969 by the USACE on the upper mainstem, primarily for flood risk management, recreation, and water supply. As the first SRP site, the principles of the natural flow paradigm (Bunn and Arthington, 2002; Poff et al., 1997; Postel and Richter, 2003; Richter et al., 1996) were applied on the Green River to define environmental water requirements based on the natural (predam) characteristics of seasonal flow patterns of the river, and a limited set of general life history traits for selected species. Particular attention was paid to spring and autumn when the reservoir pool is filled and evacuated, respectively, and historic patterns of flow are most altered. An iterative approach was used to model various modified operational scenarios for the reservoir, assessing both implications for other purposes and relative strength in achieving more natural patterns of flow and temperature (USACE, 2002).

The selected scenario of reservoir storage reallocation and modified operations involved: reducing the flood pool level by approximately 1.3 m (5% of original reservoir storage), delaying the autumn reservoir drawdown, and extending the spring filling schedule (Warner et al., 2014). These changes in operations were made for a 3-year trial period beginning in December 2002, and adopted in a revised water control plan in 2006.

Of ecological importance, the modified operations involve delaying a majority of the autumn drawdown until early November in order to restore both natural seasonal low-flow conditions through October and a more natural temperature regime as a result of the reservoir undergoing temperature destratification prior to the later releases. Additional changes in operations for ecological benefit include modifying the timing and pattern of spring reservoir fill and sculpting the shape of individual release events to mimic storms.

Beyond ecological benefits, the modified operations have enhanced both flood protection for large-magnitude events, by enabling higher minimum and maximum releases during the noncrop season (USACE, 2002); and water supply, by maintaining high pool levels throughout a majority of the year. Moreover, these changes have benefitted in-reservoir recreation, which is particularly important to the local economy. Not only has monitoring shown the revised operations have had no adverse impact on in-reservoir recreational fish populations, a socio-economic evaluation suggests the revised operations have increased visitor days, jobs, and directly related economic activity across a 48-km radius of the reservoir (Fig. 25.4). These recreational and related economic benefits are attributed in part to a 40% increase in the time that reservoir levels are maintained in the ideal recreation zone, especially through October as compared to original operations.

(Continued)

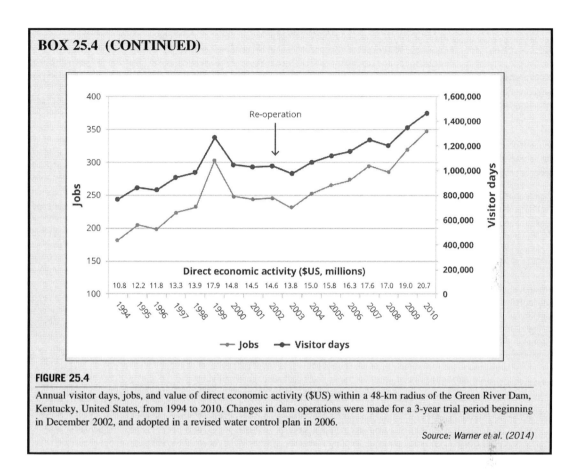

BOX 25.4 (CONTINUED)

FIGURE 25.4

Annual visitor days, jobs, and value of direct economic activity ($US) within a 48-km radius of the Green River Dam, Kentucky, United States, from 1994 to 2010. Changes in dam operations were made for a 3-year trial period beginning in December 2002, and adopted in a revised water control plan in 2006.

Source: Warner et al. (2014)

25.8.2 THE IMPORTANCE OF INDIVIDUALS, ROLES, SKILLS, AND EXPERIENCE

Holling and Chambers (1973) recognize the importance of individuals with particular skill sets within natural resource management partnerships, going so far as to provide humorous titles and cartoon sketches for each role. A *peerless leader* is required to create long-term surety of funding and protect programs from the cuts and restructures that are a common feature of most environmental management agencies. Similarly, the *compleat amanuensis* will go above and beyond the call of duty to complete unforeseen tasks, smooth over cracks in the program, and make sure that momentum is maintained. *Utopians* are blessed with optimism and ambition, and can push the partnership toward novel and unexpected outcomes, providing great opportunities to learn. However, this must be tempered by the pragmatic and budget-centric (i.e., *projectified*) outlook of the *blunt Scott*, who ensures that the partners continue to work toward the common vision, and achieve predetermined milestones.

There is no fundamental reason why any of these roles should be filled by either a manager or researchers. However, generally speaking we expect to see the peerless leaders being relatively senior environmental managers, who have the authority and funding to keep monitoring and adaptive management programs going—one of the often-cited reasons for failure (Schreiber et al., 2004). For the most part, utopians will be drawn from the ranks of research scientists, who, with a research-focused outlook, will be looking to push back the bounds of what is possible. The blunt Scott may be drawn from the ranks of environmental consultants; more so than any other group, consultants are used to working within tight time and budget constraints, and are heavily focused on project delivery. The compleat amanuensis could come from any of these groups; the main criterion for this individual is a deep passion for the long-term well-being of the system being managed, which makes it possible for this person to endure the frustrations that inevitably come with being the project gofer. Good project leaders may encompass many of these characteristics.

To these roles, should be added the role of *reflector*. This is the individual who, while part of the team, spends considerable time in reflection to ensure that adaptive management is taking place, and also what is being learned about adaptive management in the process. These are people who have the ability and interest to bring all the learnings together; and our experience suggests that this role is well-filled by social scientists. Indeed, at least part of the general failure to report adaptive management in the literature may stem from a lack of inclusion of reflectors in natural resource management teams, and a consequent failure to appreciate the adaptive nature of many programs (Allan et al., 2008). Therefore, a primary responsibility of the reflector ought to be ensuring documentation and reporting of adaptive management learning both within and beyond the project team.

Separate to these project roles will be early decisions about what disciplines are required within the multidisciplinary partnership. For environmental water management, scientific knowledge generally needs to cover hydrology and hydraulics, and various specialties (e.g., water chemistry, vegetation, macroinvertebrates, waterbirds, and fish). Webb et al. (Chapter 14) suggest that ecological modeling expertise should also be added to teams to take advantage of rapid advances within this field. The need for specific disciplines may vary among projects to some extent, but we reiterate that a socially focused reflector would improve adaptive management outcomes and learning for most projects.

Ideally, these various roles and disciplinary specialties would be filled by the same individuals for extended periods of time. Rapid staff turnover in natural resource management agencies is a problem for long-term programs (Webb et al., 2014). The loss of institutional knowledge that occurs when individuals move on can hamper projects, creating a change in emphasis and direction, or at the very least causing a delay while the new team member becomes familiar with the project.

Beyond these roles, individuals bring a mix of skills to an adaptive management team. Senior managers have a policy focus, and often are responsible for delivering on legislative requirements. Local managers often have in-depth knowledge of the systems being managed; so it is important to take advantage of this knowledge in developing monitoring and adaptive management programs (Fisher and Ball, 2003). Ultimately, managers are the decision makers within any adaptive management program, and so they are integral to any successful program. Research scientists bring technical skills and an innovative attitude that can prevent the ossification of *standard methods*—for instance those that have been used before—within monitoring and evaluation programs.

One of these skills is experience with adaptive management itself. If possible, every adaptive management team should contain at least one member with experience in a successful adaptive management program; in this case having experience in the field of adaptive management of an environmental water program. This person may well be the person who fulfills the role of reflector, as he or she will know what may work, what does not, and have a better chance of guiding the team toward reflection, learning, and practice change.

25.8.3 LEARNING FROM OTHER PROJECTS

The role of reflector outlined above highlights the value of being able to learn from the experiences of other projects, both successful and unsuccessful. Too often, adaptive management proceeds in isolation of the learning from other projects. As described above, many adaptive management projects are not sufficiently documented and reported, which prevents effective communication to other practitioners. Thus many projects, teams, and individuals make the same errors, and the rate of progress in adaptive management is slow. One or more team members fulfilling the role of reflector will be able to improve the use of previous knowledge and learning (where this can be found) to improve the chances of effective adaptive management. Just as important, they will be able to capture the learnings from their own project through documentation, and disseminate them through reporting to the wider adaptive management community.

Although such communication and learning is valuable for improving the effectiveness of adaptive management, it should not be an expectation that effective adaptive management practices in one project will translate immediately to another context. Although the idealized concept of adaptive management portrayed in Fig. 25.1 makes the process appear standardized, every case will be to some extent unique. Learning from other projects needs to occur within an understanding that all projects are different.

25.8.4 ADAPTIVE MANAGEMENT MAY EMERGE SPONTANEOUSLY

As outlined in Section 25.3, there is a widespread perception that adaptive management has often not been successful. However, Box 25.1 demonstrates that adaptive management sometimes occurs in the absence of a preconceived adaptive management program. We contend that adaptive management is more common than often perceived, and that it is occurring spontaneously in settings not expressly set up according to the principles of Fig. 25.1. Management programs established with a trusting set of partners, and a willingness to learn and be flexible, will be adaptive almost as a matter of course. These programs are less likely to follow the regimented *forward* circular pathway illustrated in Fig. 25.1, and instead have more reversals via the inner loops at the various stages of implementation. More importantly, they are unlikely to be documented and reported as successful examples of adaptive management, as they were not set up with this in mind.

Capturing the learning from such examples of *informal adaptive management* would be valuable for the progression of adaptive management as a whole, but is difficult because of this very informality. If all environmental managers were able to take on something of the reflector role, they would recognize informal adaptive management when it occurs, providing an opportunity to capture learning and disseminate this information to the wider adaptive management community. This would help to inform new perspectives on the practice of adaptive management as well as reduce the perception that the process is often unsuccessful.

25.8.5 COORDINATION OF MONITORING AND EVALUATION ACROSS LARGE SCALES

As described above, it is not possible or desirable to monitor everywhere. Nevertheless, managers must be able to infer outcomes for areas that have not been monitored. The best approach to achieve this, despite the limitations on monitoring, is to coordinate monitoring across large scales. Using compatible methods among multiple *sampling units* (e.g., rivers), and then synthesizing the results in large-scale data analysis, improves the statistical power to detect effects (through replication of data), and our chances of identifying the individual effect of different drivers of ecological response (Section 25.6). This approach is explored in detail in Webb et al. (2010a, 2014).

This approach also provides a robust method of extrapolating results from areas that have been monitored to the many areas that cannot be monitored, because we have a sample of monitoring locations allowing us to generalize to the population of potential locations. Moreover, such coordination may improve the efficiency of monitoring in general. Currently, with multiple monitoring programs run by multiple agencies at different levels of government and natural resource management agencies, there exists considerable potential for inefficiency. Each agency designs its own program, potentially going through the same processes of conceptual modeling, review of monitoring methods, selection of sites, etc.; project management is again replicated across programs; and finally different programs may end up monitoring the same endpoints in the same river, but in different ways. High-level coordination can prevent this inefficiency.

25.8.6 CREATE A PROGRAM OF REQUISITE SIMPLICITY

The different skills and motivations of the team members can be used to create the most appropriate level of sophistication in the monitoring program. Left to their own devices, research scientists are likely to design a monitoring program of great technical sophistication, which includes innovative approaches and the necessary robustness to pass peer review in the best scientific journals. Such a program is often too expensive and complex to implement over large scales. Conversely, time-poor managers faced with budgetary constraints may be likely to fall back upon previously used methods for monitoring, without assessing their utility for the new program. Somewhere between these two extremes, and probably not achievable without the input of both groups into the monitoring design process, lies a program of *requisite simplicity* (Rogers, 2007), a program with sufficient technical aspects to answer the questions at hand, but one that is still practical to implement and affordable (Fig. 25.5).

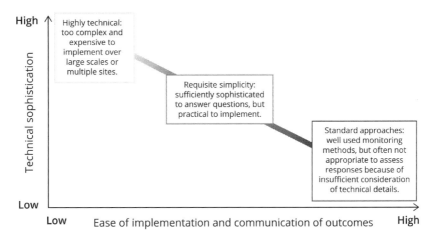

FIGURE 25.5

Finding the balance of technical sophistication and ease of implementation necessary to design a monitoring program capable of informing adaptive management of environmental water regimes.

Source: Webb et al. (2010a) with permission from CSIRO Publishing

25.8.7 INNOVATIVE APPROACHES TO ANALYSIS AND EVALUATION: MAKE THE MOST OF YOUR DATA

The limitations to the experimental design of environmental water programs outlined above (Section 25.6) means that we will often be unable to analyze data from monitoring programs using well-known statistical approaches. Here, the adaptive management partnership can again be useful, with research scientists better able to employ advanced data analysis methods to extract the best value from the monitoring data collected. Chapter 14 provides a brief overview of some of the more advanced statistical methods available for informing environmental flows assessments. Those methods (e.g., machine learning approaches, Bayesian methods) are equally able to be used for analyzing monitoring data within an adaptive management framework. Some of these methods require specialist training, experience and software, and qualities that are most likely to be found within the research-based members of the partnership.

25.8.8 MAINTAIN THE ADAPTIVE ASPECT OF ADAPTIVE MANAGEMENT

As described above (Section 25.5), monitoring programs should not be static, but should incorporate the learnings from the adaptive management loop. Over time, an adaptive management program would move from a state of low confidence in the likely outcome of a management action, through to higher confidence, and ultimately to the point where monitoring for that endpoint could be greatly reduced. This would allow the scarce monitoring resources to be directed to another endpoint of high uncertainty, or potentially to a new monitoring location for which there was previously insufficient funds to monitor. Such decisions should not be taken hastily. It will probably require several circuits of the outer adaptive management loop (Fig. 25.1) before we could be sufficiently confident that knowledge had advanced to the point where monitoring programs could be changed substantially. Even in circumstances when monitoring can be reduced, it is imperative to retain the adaptive aspect of learning from the simplified monitoring and evaluation program.

25.8.9 REGULAR REPORTING OF OBSERVATIONS, LEARNINGS, AND PROCESSES

The essential element for sharing learning with others is through reporting. Regular reporting of environmental water delivery plans, water release decisions, processes for the delivery of environmental water and field observations during environmental water releases will enable others to better understand the learning process. These different aspects of environmental water management are often reported in separate documents. A synthesis that brings together all of these elements in a single document is most useful for others, and can assist the process of inducting new team members to a project.

We have observed several times above that one of the reasons for the perception of failure of adaptive management has been the lack of documentation and reporting of adaptive management, whether formal or informal. The reflector role nominated above (Section 25.8.2) should also take responsibility for ensuring a regular two-way transfer of knowledge and learning within and beyond the project team. Such knowledge transfer does not need to be constrained to the annual reporting cycle, as adaptive management can be highly effective when done outside of a traditional reporting cycle (Box 25.1). This also short circuits the considerable time that can elapse between the writing and approval of an annual report, increasing the rate of learning.

25.9 CONCLUSIONS: ADAPTIVE ENVIRONMENTAL WATER REGIMES

In this chapter, we have argued that monitoring and adaptive management of environmental water regimes cannot be considered as optional. The importance of adaptive management of environmental water has also been recognized elsewhere (Summers et al., 2015). Without high-quality monitoring, evaluation, and adaptive management, there is a real risk that the benefits and outcomes of environmental water programs may not be understood, and these programs may be under threat. Equally important to gaining this understanding, is communicating it to the decision makers who must decide whether to continue to fund these programs.

However, the scale of the challenge should not be underestimated. Achieving a change in practice through adaptive management will not be easy (Section V of this book). If it were, then there would not be so much literature on the failure of adaptive management. All stakeholders will need to be committed to the process in terms of engagement and funding. They will also need a large amount of patience. An inner loop of the adaptive management cycle for environmental water will take a minimum of 1 year, with the outer loop taking at least 5–10 years. Inevitably, the rate of learning may be slow, and all stakeholders must be prepared for this.

A commitment to adaptive management will provide real opportunities to learn about environmental water regimes. As stated in the introduction to this chapter (Section 25.1), there is much to learn. Adaptive management also provides an opportunity to move away from the current focus on biophysical outcomes of environmental water regimes, to more fully encompass the complex socio-ecological systems. In Australia, the early stages of development of the Basin Plan (notably the "Guide to the Basin Plan") met with great resistance at least partly because of inadequate involvement of affected communities in the processes (Evans and Pratchett, 2013; Hart, 2016). The consequent emphasis on *localism* and key local individuals (sensu Evans et al., 2013) as a way of overcoming this will sit comfortably within an adaptive management framework.

As part of this learning, we must learn about adaptive management itself to help improve upon its rate of success. In Section 25.8.2, we argued for the inclusion of reflectors in adaptive management teams. The inclusion of such individuals would greatly increase the rate at which we learn about adaptive management, and would reduce the instances of adaptive management occurring without being reported (Allan and Watts, 2016). Arguing for funding to include such individuals in adaptive management teams will be challenging, and will require a greater appreciation at senior management levels (i.e., those that control the funding) of the importance of learning not just about the biophysical system, but about the socio-ecological system, and about adaptive management itself. Here, networks of practitioners have a role to play.

Networks of monitoring, evaluation, and adaptive management teams provide an effective means for rapidly disseminating information among multiple teams and increasing the rate of learning beyond that possible with normal documentation and reporting. It must be remembered that monitoring and evaluation of environmental water regimes is a young field, and adaptive management of environmental water programs even younger. Any opportunity to learn from the successes (and failures) of groups around the world should be taken. Practitioner networks have the potential to increase the appreciation of how adaptive management can be *done well*, increasing support at all levels.

The challenges for monitoring, evaluation, and adaptive management of environmental water regimes are many, but we should also be mindful of the tremendous opportunities presented. Adaptive management cannot proceed without a commitment to manage; the global investment in environmental water is huge as legislators increasingly comprehend the scale of the challenge facing the world's freshwater systems. Management decisions will continue to be made, and there is great potential to *learn by doing* within this setting. To maximize the rate of learning, and therefore the rate of improvement of management decisions, it is incumbent upon practitioners and researchers to grasp this opportunity.

25.10 SUMMARY

Environmental water programs are a major investment in the environment that must be delivered amid uncertainty as to the ecological, ecosystem service and social outcomes that they will provide. Adaptive management is seen as a means of reducing these uncertainties while delivering the programs. However, there are many examples in the natural resource management literature of adaptive management failing, or being perceived to have failed, to live up to its promise. We cannot afford for this pattern to be repeated with environmental water.

In this chapter, we introduced adaptive management and explored reasons why it has often not succeeded in the past. Furthermore, we outlined the challenges for monitoring and evaluation—an

---

BOX 25.5 IN A NUTSHELL

1. Environmental water represents a major investment of public funds in the environment, with flows being delivered amid uncertainty over the ecological, ecosystem service and social benefits that will occur. Adaptive management is seen as a means of delivering flows despite this uncertainty.
2. There are many examples of natural resource management projects where adaptive management has failed, or has been perceived to have failed, to deliver on expectations. We need to ensure that this is not repeated with environmental water.
3. Monitoring and evaluation of the benefits of environmental water is also challenging, with both technical and logistic issues making it difficult to draw strong conclusions regarding the biophysical benefits of environmental water regimes.
4. There has been too much of a focus on evaluating purely biophysical outcomes of environmental water programs. For adaptive management to succeed greater emphasis needs to be placed upon the socio-ecological system.
5. We have outlined general principles for successful monitoring, evaluation, and adaptive management of environmental water programs. With large investments in environmental water regimes across the world, there is great potential to use adaptive management to improve outcomes from these programs. However, this process takes considerable time, and will require commitment and patience from all stakeholders.

---

essential component of any adaptive management program—of environmental water regimes. We have outlined general principles that will, if followed, improve monitoring, evaluation, and adaptive management of environmental water regimes. One important consideration is moving away from the current focus on biophysical outcomes, to considering the entire socio-ecological system.

The major investments in environmental water programs will continue for some time, and present a great opportunity to use adaptive management to improve decision making in environmental water management, but also learn about adaptive management itself (Box 25.5).

REFERENCES

Allan, C., 2012. Rethinking the 'project': bridging the polarized discourses in IWRM. J. Environ. Policy Plan. 14, 231−241.

Allan, C., Curtis, A., 2005. Nipped in the bud: why regional scale adaptive management is not blooming. Environ. Manage. 36, 414−425.

Allan, C., Curtis, A., Stankey, G.H., Shindler, B., 2008. Adaptive management and watersheds: a social science perspective. J. Am. Water Resour. Assoc. 44, 166−174.

Allan, C., Stankey, G.H., 2009. Synthesis of lessons. In: Allan, C., Stankey, G. (Eds.), Adaptive Environmental Management: A Practitioners Guide. Springer, Dordrecht, pp. 341−346.

Allan, C., Watts, R.J., 2016. Seeking and communicating adaptive management; two cases from environmental flows. In: Webb, J.A., Costelloe, J.F., Casas-Mulet, R., Lyon, J.P., Stewardson, M.J. (Eds.), Proceedings of the 11th International Symposium on Ecohydraulics. The University of Melbourne, Melbourne, Australia, paper 26631.

Allen, C.R., Garmestani, A.S., 2015. Adaptive management. In: Allen, C.R., Garmestani, A.S. (Eds.), Adaptive Management of Social-Ecological Systems. Springer, Dordrecht, pp. 1−10.

Benda, L.E., Poff, N.L., Tague, C., Palmer, M.A., Pizzuto, J., Cooper, S., et al., 2002. How to avoid train wrecks when using science in environmental problem solving. BioScience 52, 1127−1136.

Bernhardt, E.S., Palmer, M.A., Allan, J.D., Alexander, G., Barnas, K., Brooks, S., et al., 2005. Synthesizing US river restoration efforts. Science 308, 636−637.

Blackmore, C., Ison, R., 2012. Designing and developing learning systems for managing systemic change in a climate change world. In: Wals, A., Corcoran, P.B. (Eds.), Learning for Sustainability in Times of Accelerating Change. Wageningen Academic Publishers, Wageningen, pp. 347−361.

Bormann, B.T., Stankey, G., 2009. Crisis as a positive role in implementing adaptive management after the Biscuit fire, pacific Northwest, U.S.A. In: Allan, C., Stankey, G. (Eds.), Adaptive Environmental Management: A Practitioner's Guide. Springer, Dordrecht, pp. 143−167.

Brooks, S.S., Lake, P.S., 2007. River restoration in Victoria, Australia: change is in the wind, and none too soon. Restor. Ecol. 15, 584−591.

Bryan, B.A., Kandulu, J., Deere, D.A., White, M., Frizenschaf, J., Crossman, N.D., 2009. Adaptive management for mitigating *Cryptosporidium* risk in source water: a case study in an agricultural catchment in South Australia. J. Environ. Manage. 90, 3122−3134.

Bunn, S.E., Arthington, A.H., 2002. Basic principles and ecological consequences of altered flow regimes for aquatic biodiversity. Environ. Manage. 30, 492−507.

Cullen, P., 1990. The turbulent boundary between water science and water management. Freshw. Biol. 24, 201−209.

CEWO, 2015. Monitored outcomes. Commonwealth Environmental Water Office, Canberra. Available from: http://www.environment.gov.au/system/files/resources/c01b7e11-f61c-407b-9a9d-de4835351b0e/files/monitored-outcomes.pdf.

DePhilip, M.A. Moberg, T., 2013. Ecosystem flow recommendations for the Upper Ohio River basin in western Pennsylvania. The Nature Conservancy, Harrisburg, PA.

Downes, B.J., Barmuta, L.A., Fairweather, P.G., Faith, D.P., Keough, M.J., Lake, P.S., et al., 2002. Monitoring Ecological Impacts: Concepts and Practice in Flowing Waters. Cambridge University Press, Cambridge, UK.

Evans, M., Marsh, D., Stoker, G., 2013. Understanding localism. Policy Stud. 34, 401−407.

Evans, M., Pratchett, L., 2013. The localism gap − the CLEAR failings of official consultation in the Murray Darling Basin. Policy Stud. 34, 541−558.

Failing, L., Gregory, R., Higgins, P., 2013. Science, uncertainty, and values in ecological restoration: a case study in structured decision-making and adaptive management. Restor. Ecol. 21, 422−430.

Fisher, P.A., Ball, T.J., 2003. Tribal participatory research: mechanisms of a collaborative model. Am. J. Comm. Psychol. 32, 207−216.

Gawne, B., Brooks, S., Butcher, R., Cottingham, P., Everingham, P., Hale, J., et al., 2013. Long term intervention monitoring project: logic and rationale document version 1.0. Report prepared for the Commonwealth Environmental Water Office. Murray-Darling Freshwater Research Centre 109. Available from: https://www.environment.gov.au/water/cewo/publications/long-term-intervention-monitoring-project-logic-and-rationale-document.

Gunderson, L., 2015. Lessons from adaptive management; obstacles and outcomes. In: Allen, C.R., Garmestani, A.S. (Eds.), Adaptive Management of Social-Ecological Systems. Springer, Dordrecht.

Hart, B.T., 2016. The Australian Murray-Darling Basin Plan: factors leading to its successful development. Ecohydrol. Hydrobiol. 16, 229−241.

Holling, C.S., Chambers, A.D., 1973. Resource science − the nurture of an infant. BioScience 23, 13−20.

Ison, R., Blackmore, C., Iaquinto, B.L., 2013. Towards systemic and adaptive governance: exploring the revealing and concealing aspects of contemporary social-learning metaphors. Ecol. Econ. 87, 34−42.

Konrad, C.P., Olden, J.D., Lytle, D.A., Melis, T.S., Schmidt, J.C., Bray, E.N., et al., 2011. Large-scale flow experiments for managing river systems. BioScience 61, 948−959.

Madema, W., Light, S., Adamowski, J., 2014. Integrating adaptive learning into adaptive water resources management. Environ. Eng. Manage. J. 13, 1801−1816.

McFadden, J.E., Hiller, T.L., Tyre, A.J., 2011. Evaluating the efficacy of adaptive management approaches: is there a formula for success? J. Environ. Manage. 92, 1354−1359.

MDBA, 2014. Murray-Darling Basin water reforms: framework for evaluating progress. Murray-Darling Basin Authority, Canberra.

MDFRC, 2013. Long term intervention monitoring project: generic cause and effect diagrams Version 1.0. Report prepared for the Commonwealth Environmental Water Office. Murray-Darling Freshwater Research Centre, 163 pp. Available from: https://www.environment.gov.au/water/cewo/publications/ltim-cause-effect-diagrams.

Olden, J.D., Konrad, C.P., Melis, T.S., Kennard, M.J., Freeman, M.C., Mims, M.C., et al., 2014. Are large-scale flow experiments informing the science and management of freshwater ecosystems? Front Ecol. Environ. 12, 176−185.

Pahl-Wostl, C., 2007. Transitions towards adaptive management of water facing climate and global change. Water Resour. Manage. 21, 49−62.

Pahl-Wostl, C., Kabat, P., Möltgen, J. (Eds.), 2010. Adaptive and Integrated Water Management: Coping With Complexity and Uncertainty. Springer, Berlin.

Pahl-Wostl, C., Lebel, L., Knieper, C., Nikitina, E., 2012. From applying panaceas to mastering complexity: toward adaptive water governance in river basins. Environ. Sci. Policy 23, 24−34.

Palmer, M.A., Bernhardt, E.S., Allan, J.D., Lake, P.S., Alexander, G., Brooks, S., et al., 2005. Standards for ecologically successful river restoration. J. Appl. Ecol. 42, 208−217.

Poff, N.L., Allan, J.D., Bain, M.B., Karr, J.R., Prestegaard, K.L., Richter, B.D., et al., 1997. The natural flow regime. BioScience 47, 769−784.

Poff, N.L., Matthews, J.H., 2013. Environmental flows in the Anthropocence: past progress and future pro-spects. Curr. Opin. Environ. Sustain. 5, 667−675.

Poff, N.L., Richter, B.D., Arthington, A.H., Bunn, S.E., Naiman, R.J., Kendy, E., et al., 2010. The ecological limits of hydrologic alteration (ELOHA): a new framework for developing regional environmental flow standards. Freshw. Biol. 55, 147−170.

Poff, N.L., Zimmerman, J.K.H., 2010. Ecological responses to altered flow regimes: a literature review to inform the science and management of regulated rivers. Freshw. Biol. 55, 194−205.

Pollard, S.R., du Toit, D., Biggs, H.C., 2011. River management under transformation: the emergence of strate-gic adaptive management of river systems in the Kruger National Park. Koedoe 53, 1−14.

Postel, S., Richter, B., 2003. Rivers for Life: Managing Water for People and Nature. Island Press, Washington, DC.

Raadgever, G.T., Mostert, E., Kranz, N., Interwies, E., Timmerman, J.G., 2008. Assessing management regimes in transboundary river basins: do they support adaptive management? Ecol. Soc. 13, 1−21.

Rheinheimer, D.E., Yarnell, S.M., Viers, J.H., 2013. Hydropower costs of environmental flows and climate warming in California's Upper Yuba River watershed. River Res. Appl. 29, 1291−1305.

Richter, B.D., Baumgartner, J.V., Powell, J., Braun, D.P., 1996. A method for assessing hydrologic alteration within ecosystems. Conserv. Biol. 10, 1163−1174.

Rivers-Moore, N.A., Jewitt, G.P.W., 2007. Adaptive management and water temperature variability within a South African river system: what are the management options? J. Environ. Manage. 82, 39−50.

Rogers, K.H., 2007. Complexity and simplicity: complementary requisites for policy implementation. In: 10th International Riversymposium and Environmental Flows Conference. The Nature Conservancy, Brisbane, Australia. Available from: http://archive.riversymposium.com/2007_Presentations/C1_Rogers.pdf.

Rutherfurd, I.D., Ladson, A.R., Stewardson, M.J., 2004. Evaluating stream rehabilitation projects: reasons not to, and approaches if you have to. Aust. J. Water Resour. 8, 57−68.

Schön, D.A., 2008. The Reflective Practitioner: How Professionals Think in Action. Basic Books, New York.

Schreiber, E.S.G., Bearlin, A.R., Nicol, S.J., Todd, C.R., 2004. Adaptive management: a synthesis of current understanding and effective application. Ecol. Manage. Restor. 5, 177−182.

Schultz, L., Fazey, I., 2009. Effective leadership for adaptive management. In: Allan, C., Stankey, G.H. (Eds.), Adaptive Environmental Management: A Practitioners Guide. Springer, Dordrecht, pp. 295−303.

Skinner, D., Langford, J., 2013. Legislating for sustainable basin management: the story of Australia's Water Act (2007). Water Policy 15, 871−894.

Smith, C.B., 2011. Adaptive management on the central Platte River — Science, engineering, and decision analysis to assist in the recovery of four species. J. Environ. Manage. 92, 1414—1419.

Souchon, Y., Sabaton, C., Deibel, R., Reiser, D., Kershner, J., Gard, M., et al., 2008. Detecting biological responses to flow management: missed opportunities; future directions. River Res. Appl. 24, 506—518.

Stringer, L.C., Dougill, A.J., Fraser, E., Hubacek, K., Prell, C., Reed, M.S., 2006. Unpacking "participation" in the adaptive management of social-ecological systems: a critical review. Ecol. Soc. 11, 719—740.

Summers, M., Holman, I., Grabowski, R., 2015. Adaptive management of river flows in Europe: a transferable framework for implementation. J. Hydrol. 531, 696—705.

USACE, 2002. Environmental assessment and finding of no signficant impact, reregulation. Green River Lake, Kentucky. US Army Corps of Engineers, Louisville, KY.

USACE, 2011. Economic impact analysis, reoperation of Green River Lake. Kentucky, Pilot Project for the Sustainable Rivers Project. US Army Corps of Engineers, Louisville, KY.

Walters, C.J., 2007. Is adaptive management helping to solve fisheries problems? Ambio 36, 304—307.

Warner, A.T., Bach, L.B., Hickey, J.T., 2014. Restoring environmental flows through adaptive reservoir management: planning, science, and implementation through the Sustainable Rivers Project. Hydrol. Sci. J. 59, 770—785.

Watts, R.J., Ryder, D.S., Allan, C., 2009. Environmental monitoring of variable flow trials conducted at Dartmouth Dam, 2001/02—07/08 Synthesis of key findings and operational recommendations. Institute for Land Water and Society Report No. 50. Charles Sturt University, Albury, NSW.

Watts, R.J., Ryder, D.S., Allan, C., Commens, S., 2010. Using river-scale experiments to inform the adaptive management process for variable flow releases from large dams. Mar. Freshw. Res. 61, 786—797.

Webb, J.A., Miller, K.A., de Little, S.C., Stewardson, M.J., 2014. Overcoming the challenges of monitoring and evaluating environmental flows through science-management partnerships. Int. J. River Basin Manage. 12, 111—121.

Webb, J.A., Stewardson, M.J., Chee, Y.E., Schreiber, E.S.G., Sharpe, A.K., Jensz, M.C., 2010a. Negotiating the turbulent boundary: the challenges of building a science-management collaboration for landscape-scale monitoring of environmental flows. Mar. Freshw. Res. 61, 798—807.

Webb, J.A., Stewardson, M.J., Koster, W.M., 2010b. Detecting ecological responses to flow variation using Bayesian hierarchical models. Freshw. Biol. 55, 108—126.

Westgate, M.J., Likens, G.E., Lindenmayer, D.B., 2013. Adaptive management of biological systems: a review. Biol. Conserv. 158, 128—139.

Williams, B.K., 2011. Adaptive management of natural resources—framework and issues. J. Environ. Manage. 92, 1346—1353.

Williams, B.K., Brown, E.D., 2014. Adaptive management: from more talk to real action. Environ. Manage. 53, 465—479.

DEFINING SUCCESS: A MULTICRITERIA APPROACH TO GUIDE EVALUATION AND INVESTMENT

26

Erin L. O'Donnell[1] and Dustin E. Garrick[2]

[1]*The University of Melbourne, Parkville, VIC, Australia* [2]*University of Oxford, Oxford, United Kingdom*

26.1 INTRODUCTION

Environmental water programs encompass the suite of actions from setting a vision for healthy rivers and communities, through to protection, restoration, and management of environmental water (Chapter 1). This book showcases the tremendous diversity of environmental water programs in operation around the world, and the importance of a range of public and private organizational actors in different institutional settings working together to maintain and enhance the health of aquatic ecosystems (see Chapter 19). As environmental water management has matured as a discipline, more and more programs are attempting to demonstrate the benefits they have achieved. As Chapter 25 explored, environmental water policy makers and practitioners are making increasing use of adaptive management principles to continually refine and improve environmental water management.

Despite the rapid growth and diversification in environmental water programs since the 1990s, there remains a real gap in defining the success of an environmental water program. There is still a pervasive desire for a one-off, *set and forget* policy solution to solve the environmental water problem; yet all the evidence to date suggests that sustained effort is required. The case studies explored throughout this book demonstrate the sheer complexity of environmental water programs, which require implementation over decadal timescales and a wide range of spatial scales, from the very local through to multijurisdictional and multinational transboundary aquatic systems. Successful environmental water programs deliver not only improved environmental water regimes in the short term, but a sustained improvement in aquatic health over time. Doing so requires going beyond the ecological and hydrological measures of efficacy, and depends on well-defined legal frameworks and organizational capacity matched to the socio-political and ecological context.

Demonstrating success of environmental watering is critically important, and it must be assessed in the specific context in which the environmental water program is taking place, including the broad aims and trade-offs of each specific program. Environmental water protection and recovery

Water for the Environment. DOI: http://dx.doi.org/10.1016/B978-0-12-803907-6.00026-7
625

can be achieved in a way that improves overall water resource management for all users, but changes to water resource management are likely to create winners and losers, and any definition of success needs to be explicit about this (and ideally, offset these losses).

This chapter focuses on the legal, institutional, and economic factors and influences on performance, and builds directly from the discussion in Chapters 17 and 19, which place the mechanisms of protecting and managing environmental water in an institutional and organizational context. Chapter 19 identified two distinct policy strategies for the management of environmental water:

1. Protection and maintenance, where the existing levels of environmental water are protected and maintained against future degradation;
2. Recovery and management, where additional environmental water is recovered and managed to improve environmental water regimes into the future.

These two policy strategies can be achieved using a mix of environmental water allocation mechanisms (Chapter 17), depending on the specific opportunities and ecological needs of the local context. Chapter 19 emphasized the role of environmental water organizations tasked with the responsibility to implement the policy strategies. Importantly, when the combination of policy and allocation mechanism requires ongoing decision making to deliver environmental water in the most effective way, the environmental water organization will need the capacity to actively manage its water (see Fig. 19.1). This flexibility brings its own challenges, as each decision to allocate and use environmental water comes under scrutiny.

As Chapter 25 argues, all environmental water is managed with significant uncertainty in terms of the benefits that it will ultimately deliver, to both communities and the aquatic ecosystems. Environmental water managers, policy makers, and practitioners around the world are asking: what defines success for environmental water? How do we know what success looks like? And how do we maintain successful environmental water programs over time?

The *Columbia Basin Water Transactions Program (CBWTP)* is a prime example of the growing need for comprehensive and integrated approaches to define and measure the success of environmental water programs. In response to accountability measures and reporting requirements of its primary funders (the Bonneville Power Administration and the National Fish and Wildlife Foundation), the CBWTP has developed increasingly rigorous and sophisticated frameworks for monitoring and evaluation. Initial monitoring focused on ensuring compliance with water right transfers to demonstrate the potential for water transactions to enhance streamflows for fish habitat (Garrick et al., 2009). The CBWTP has since expanded the commitment to track over 60 attributes of water transactions, while prioritizing a staged approach to monitor compliance, biological, and ecosystem outcomes within a comprehensive accounting framework (McCoy and Holmes, 2015). The CBWTP also shows the importance of tracking organizational capacity and strategic development, highlighted by the requests from the Independent Scientific Review Board of the *Northwest Power and Conservation Council (NPCC)* to monitor cost-effectiveness and transaction costs.

Defining and measuring success is not easy. This chapter aims to provide a structured approach to guide future developments in this space in a coherent manner, and enable practical lessons that have been learned to reach a broader audience. To do so, this chapter establishes a multicriteria framework to define and evaluate the success of environmental water programs over time, in terms of the often unique goals and trade-offs underpinning each program.

26.2 DEFINING SUCCESS: SIX CRITERIA

Historically, environmental water programs have been evaluated using fairly static criteria of efficiency and effectiveness. These criteria focus attention on the most easily measured outputs of environmental water programs: the volume of water recovered or protected at a particular point in time, and the financial cost of doing so. Yet, we know that this is not enough to determine whether an environmental water program has been successful, or whether it will *continue* to be successful.

Chapter 25 shows the critical importance of monitoring and evaluation to provide the necessary feedback into the planning and prioritization of environmental water management. However, as Chapters 17 and 19 demonstrate, environmental water management does not happen in a vacuum. It is dependent on a range of institutions, laws, and organizations, which in turn hinge on the capacity of the individuals engaged in the environmental water program.

This chapter shifts the focus from the standard performance criteria—efficiency and effectiveness—by introducing a dynamic perspective. Setting up efficient and effective programs and sustaining success over the long term may involve short-term efficiency losses, up-front investments in capacity building, and/or periodic organizational adaptations as part of an adaptive management process (Garrick and O'Donnell, 2016). The distributional issues and trade-offs associated with environmental water underscore the need for additional principles and criteria for defining success, including equity, legitimacy, and accountability as cornerstones of performance.

This broader context for evaluating water governance is also receiving increased international attention. *The Organization for Economic Co-operation and Development (OECD)* Water Governance initiative has recently developed specific governance principles and indicators to address a number of governance deficits and gaps for complex water planning, allocation, and management tasks (OECD, 2015). The initiative stipulates three primary criteria for water governance: effectiveness, efficiency, and trust and engagement (the latter being roughly analogous to legitimacy) (OECD, 2015). This combination of efficacy and legitimacy is essential to defining and maintaining success for environmental water programs.

In this chapter, we build on the general framework developed by the OECD Water Governance Initiative to identify six criteria that can be used to define success in environmental water programs:

1. Effectiveness;
2. Efficiency;
3. Legitimacy;
4. Legal and administrative frameworks;
5. Organizational capacity;
6. Partnerships.

The first three criteria reflect those specified by the OECD: efficiency, effectiveness, and legitimacy. We consider these criteria to be the essential *policy* elements of a successful environmental water program, shaping the strategic direction of the program at the highest levels.

The second three criteria reflect the challenge of building and maintaining capacity to implement the environmental water program: legal and administrative frameworks, organizational capacity, and partnerships. We describe these criteria as essential *practical* elements. Legal frameworks, organizational capacity, and partnerships are critical during the setup period for an environmental water program, and are therefore considered enabling conditions (Garrick et al., 2009). However, we know that environmental water management is not a one-off task, and is likely to require an ongoing adaptive response to changing conditions over time. Environmental water programs must be equipped with the capacity to adapt and to continue in their work as conditions change. Institutional capacity has been traditionally difficult to define and measure, and so has often been overlooked and underfunded despite the complexity associated with this work. By paying attention to institutional capacity at the outset, and continuing to invest in maintaining it over time, an environmental water program will have a much more robust framework for its long-term success.

Each of the six criteria is examined in more detail below.

26.2.1 EFFECTIVENESS

Effectiveness refers to the "contribution of governance to define clear sustainable water policy goals and targets at all levels of government, to implement those policy goals, and to meet expected targets" (OECD, 2015). In the context of environmental watering, effectiveness refers to the overall outputs (provision of an environmental water regime) and outcomes (ecosystem health and related social and economic benefits) supported by environmental watering. As Chapter 25 outlines, environmental watering assessments have traditionally focused on outputs (water protected or provided), with less detail on the more difficult to measure and often more uncertain long-term outcomes. Given the long-term nature and complex, multidimensional goals of environmental watering programs, effectiveness is best assessed using clear preintervention baselines, and by assessing long-term trends.

The precise outputs and outcomes comprising effectiveness will depend on the particular goal (protection versus recovery) and the tools used to achieve it. For example, demonstrating effectiveness in setting a cap entails the establishment and enforcement of a cap according to prescribed standards, and relies on appropriate water accounting technologies (for more detail on caps, see Chapter 17). Effective environmental water programs require both water accounting technologies and transparent reporting on allocation, reliability standards, and how the water was used.

In both cases, the monitoring of outcomes should be targeted at particular ecosystem elements that may be modified/improved by the environmental water allocation. This is likely to be easier to identify in the case where water is diverted out-of-stream for ecosystem benefits such as directing environmental water into an offstream wetland as is widely practiced in Australia. Broad improvements to freshwater or estuarine ecosystems achieved through instatement of a cap on water extractions can affect the aquatic ecosystems in a number of ways, and it will be important to identify these potential improvements specifically at the commencement of the policy change. Monitoring activities should be targeted toward the range of spatial and temporal scales over which an ecological outcome is likely to be observed.

26.2.2 **EFFICIENCY**

Efficiency "relates to the contribution of governance to maximize the benefits of sustainable water management and welfare at the least cost to society" (OECD, 2015). A focus on efficiency is paramount in an era of shrinking public budgets and complex public policy goals. In the context of environmental water, the public and private programs are held to a standard of value for money, which involves the twin objectives of achieving the least-cost path to a specified output or outcome, on the one hand, while maximizing the net benefits of a given increment in investment in environmental watering activity, on the other.

The earliest environmental water protection and recovery programs were often starting from the position of extremely low levels of environmental water (or even none at all), so that any additional environmental water was seen as good and justifiable. However, as environmental water management has matured, environmental water organizations, like consumptive water users, need to demonstrate that the environmental water is being used to achieve the maximum benefits achievable for the available water (Pittock and Lankford, 2010). It is worth noting that the metrics for efficiency are often very different for consumptive and environmental water users, as consumptive users can generally rely on whether they have been able to use their water to generate an income (e.g., as an irrigator), whereas environmental water programs will be assessed against generating value for money in terms of environmental improvements for public funds. This can play out very differently for public and private organizations, depending on the level of public scrutiny applied to the funding of the environmental water program. Often, privately funded environmental water programs may be more willing to engage in experimental activities than publicly funded programs. For example, the Australian Conservation Foundation ran one of the first privately funded attempts to use small-scale donations to purchase water for a wetland in the Murray–Darling Basin in Australia. This program demonstrated the success of using water markets to buy temporary water for the environment in Australia, but also highlighted the need for investment by government to generate the significant water volumes required to achieve lasting environmental improvements (Siebentritt, 2012).

Demonstrating efficiency is closely tied to generating effectiveness, and will also need to be targeted to the specific environmental water policy and allocation mechanism. Where the environmental water policy is based on protection, it will be essential to link the level of protection to a meaningful outcome (for communities and the ecosystems), even where the level is less than desired. Where environmental water is recovered, it is important to consider the total cost, including transaction costs, rather than merely the cost of the water alone, as water acquisitions involve high information gathering, negotiation, and monitoring and enforcement costs (Garrick and Aylward, 2012; Garrick and O'Donnell, 2016). When the environmental water can be actively managed, demonstrating efficiency means also showing that the environmental water was used to achieve the maximum benefit at the time. Horne (2009) showed that the ecological response varies depending on how much water is applied, when the water is applied, and how long it has been since the last watering event, and that this response is nonlinear, so some water is not always enough, and more water is not always better. Further, there are often trade-offs between short- and long-term efficiency. Economizing on program setup and engagement can render short-term efficiency gains vulnerable to retrenchment and political backlash over the long term.

26.2.3 LEGITIMACY

The third principle identified by the OECD Water Governance Initiative is trust and engagement, which are critical for "public confidence and ensuring inclusiveness of stakeholders through democratic legitimacy and fairness for society at large" (OECD, 2015). Legitimacy is rarely a key objective for environmental water programs, and it is often assumed that the legitimacy of the environmental water program itself will emerge organically from the broader political context. As Chapter 7 highlights, this is not necessarily the case. Hogl et al. (2012a; p. 280) argue that traditionally, environmental policies "usually fall short in one of two ways: effectiveness (i.e., the inability to achieve preset policy targets) and legitimacy (i.e., a lack of trust in and identification with governing procedures and their outcomes)." The two previous criteria have focused on efficiency and effectiveness, and legitimacy is the next most important criterion.

Environmental water programs require ongoing maintenance and investment over the long term. Chapter 25 shows that ecological systems and human values change over time, as does what we know about them, so environmental water management must be able to adapt to these new values. However, building this capacity for longevity requires explicit investment in the ability to maintain institutions, organizations, and environmental water allocation mechanisms over time. One of the keys to this longevity is legitimacy. When environmental water management is seen as a legitimate activity by regional and national communities (and sometimes international), it will be easier to build and retain community support.

Legitimacy can be functionally expressed as the combination of input legitimacy and output legitimacy (Scharpf, 1999). Input legitimacy focuses on the *process*, and whether it was acceptable to the affected people. Input legitimacy requires explicit consideration of access, equal representation, transparency, accountability, consultation and cooperation, independence, and credibility (Hogl et al., 2012b). Chapter 19 demonstrated the importance of these factors for environmental water organizations, both active and passive.

Output legitimacy focuses on the *solution*, and whether the intervention actually solved the problem, or otherwise achieved its goal. In addition to demonstrating effectiveness (discussed above), output legitimacy emphasizes awareness, acceptance, mutual respect, active support, robustness, and common approaches to shared problem solving (Hogl et al., 2012b). It is not enough to simply recover the required volume of environmental water, or alter the operation of the dam to provide the required water regime: the environmental water organizations need to build a bridge between the different community groups affected by water resource use. For example, the Deschutes River Conservancy is a nonprofit organization in central Oregon dedicated to the improvement of streamflow and water quality in the upper Deschutes River. To build a broad base of community support, it established a membership policy for its board of directors to ensure representation of key stakeholders from both the public and private sectors.

These three criteria represent significant policy goals for environmental water programs: they must be able to demonstrate that they are effective, efficient, and legitimate. The next three criteria are focused on practical elements of implementation, and comprise the necessary conditions and institutional capacities to enable, sustain, and scale up environmental water programs.

26.2.4 **LEGAL AND ADMINISTRATIVE FRAMEWORK**

The OECD considers that "water crises are often primarily 'governance' crises" (OECD, 2011, 2015), and the same can be said for environmental water programs. Building an environmental water program requires investment in an appropriate legal and administrative framework, both at the outset (to support the new environmental water program), but also over time. Traditionally, the legal and administrative framework has been considered as only an enabling condition for the successful implementation of environmental water programs, but continuing to invest in and support this legal framework is critical to ensure that the environmental water program can continue to deliver environmental outcomes.

Garrick et al. (2009) examined the case studies of the Columbia Basin and the Murray–Darling Basin to identify enabling conditions for environmental water recovery. Although there are specific conditions necessary to support using water markets, the success of the environmental water program was also constrained by the "physical, social and economic factors driving demand for environmental water allocation; administrative procedures, organizational development and institutional capacity to effect transfers; and adaptive mechanisms to overcome legal, cultural, economic, and environmental barriers" (Garrick et al., 2009; p. 366). The breadth of these enabling conditions means that environmental water programs need to be tailored to meet the needs of "the diversity of legal, administrative and organizational systems within and across countries" (OECD, 2015).

It may well be beyond the scope of environmental water programs to significantly alter the legal and administrative frameworks of their particular contexts. However, an environmental water program can target specific legal and administrative barriers, and require ongoing investment in the maintenance of the legal framework.

Fundamental to good governance is the rule of law (OECD, 2015), which establishes that everyone is subject to the law, which is clearly expressed and publicly available, and which limits the power of governments or other organizations to act arbitrarily in a manner that affects the rights of others (World Justice Project, 2016). Ensuring that governments and citizens can have faith in and respect for their political institutions is fundamental to the implementation of environmental water programs. The rule of law will always be a work in progress, but deficiencies in this most fundamental condition will need to be explicitly addressed by any environmental water program. For example, environmental water programs may need to build and maintain specific legal frameworks to support the monitoring of water use and the enforcement of water rights. Implementing a protection policy, for example, requires a legal commitment to water accounting (so that it is clear who is using what water across the system), and establishing an administrative forum for potential complaints to be heard and addressed.

Implementing a water recovery policy may require more significant legal interventions. In Montana, in the western United States, environmental water organizations were instrumental in achieving the legal changes necessary to enable water to be transferred from existing consumptive users to instream use for the environment. Building the support among water users for this new legislation, and maintaining these legal powers and accompanying administrative framework once it was established, was a critical success factor for the environmental water program in Montana (Malloch, 2005). In general, when environmental water programs focus on recovery via market transactions, Garrick et al. (2009) determined that success will depend on "(1) establishment of rights to and limits on freshwater extraction and alteration; (2) recognition of the environment as a legitimate water use; and (3) authority to transfer existing water rights to an environmental purpose."

26.2.5 ORGANIZATIONAL CAPACITY

Organizational capacity refers to the physical, institutional, human, and financial resources required to deliver specified objectives and outcomes at social desired levels and over the long term. Adequacy of organizational capacity means the ability to cover the transaction costs of delivering program outcomes (effectiveness) at least cost (efficiency) within a framework of trust and engagement (legitimacy). In short, every environmental water organization requires a minimum investment in capacity building to undertake its activities.

As discussed in Chapter 19, organizational capacity needs to be targeted to the specific activities of the environmental water organization(s), which in turn will be influenced by the environmental water policy and allocation mechanisms. Where the policy focuses on protection, the organization will need the capacity to influence water resource policy and engage with water planning debates. The organization will need sufficient technical skills (or access to technical studies) to demonstrate that the protection proposals are feasible and meaningful. Although public and private organizations may play different roles as part of an environmental water program, each organization must invest appropriately in building the organizational capacity to carry out its specific functions.

More specifically, implementation of protection and maintenance policies will depend on having the capacity to enforce environmental water mechanisms such as a cap or conditions on storage operators or licensed water users. When water is recovered for the environment, it will be even more critical for the environmental water organization to be able to ascertain that its water rights are protected. In the Columbia Basin, enforcement is often complaints based, so environmental water organizations need the capacity to check whether their water flow is present, and the capacity to complain and be heard if it is not. When the recovery policy is combined with the capacity to acquire and hold environmental water rights, the organization will also need the capacity to make decisions on the future management of environmental water. Active environmental water management will depend on building sufficient technical skills to plan and manage environmental water, as well as ongoing community engagement to demonstrate the outcomes and build support for the environmental watering program. Almost all environmental water programs involve public–private partnerships and hence require explicit consideration of the optimal distribution and coordination of that capacity, ideally based on the comparative advantages of each entity involved (Loehman and Charney, 2011). For example, local knowledge and social networks may position community organizations and nonprofits better to engage community and conduct supplementary site-level monitoring.

26.2.6 PARTNERSHIP

Chapter 19 demonstrated the critical importance of partnerships for the successful operation of environmental water organizations. Although it is possible for environmental water programs to be undertaken in isolation, this almost never happens in practice. Indeed, the very nature of environmental water programs requires them to operate in a nested context of overlapping responsibilities and functions. Partnerships also bring together complementary skill sets, which is especially important for the multidisciplinary field of environmental water. For example, environmental water organizations typically rely on other organizations to undertake water accounting and enforcement activities (to varying degrees), and to operate storages where environmental water can be held.

Furthermore, when decisions are made based on a number of different inputs, retaining the independence of the partner organizations means that the inputs will not be influenced by the overall decision maker. This approach can be observed in Victoria, Australia, where the Victorian Environmental Water Holder makes decisions on how to use environmental water each year, based on inputs from the catchment management authorities (on environmental watering priorities) and the water corporations (on the availability of the water). Refer to Fig. 19.2 in Chapter 19 for more detail on this arrangement. Partnerships can also help tackle the thorny problem of legitimacy. Government agencies can partner with nongovernment organizations to help advocate for environmental water policy, recover water from existing users, and manage water once recovered. Government organizations can provide the necessary connection back to the democratic principles and the rule of law, but nongovernment organizations often have the capacity to engage more effectively with individuals and communities. Partnerships between government and nongovernment environmental water organizations are a highly successful approach to the implementation of environmental water programs (Garrick and O'Donnell, 2016). For example, the state of Colorado has a government agency with sole responsibility for protecting and recovering water for the environment, the *Colorado Water Conservation Board (CWCB)*. However, the CWCB relies on a nongovernment agency, the Colorado Water Trust, to facilitate water transactions from existing users that will improve environmental water levels. Individual irrigators are much more comfortable with the idea of discussing a potential transaction with the Colorado Water Trust to determine whether it will go ahead, before bringing in the formal arrangement with the government agency.

Finally, partnerships enable the implementation of the twin principles of subsidiarity and complementarity. Subsidiarity refers to the principles of assigning governance tasks to the lowest level possible, whereas complementarity refers to the higher-level coordination of local capacities. For instance, local small-scale organizations may be best placed to provide the necessary input to decisions that may need to be made at a larger spatial scale to be most effective. Partnerships between local, regional, national, and, occasionally, multinational organizations support effective engagement with local knowledge, while at the same time, enabling higher-level agreements to be reached when necessary. As discussed in Chapter 19, the Murray—Darling and Colorado River basins are excellent examples of the possibilities of these partnerships.

Partnerships help to bring diverse roles and responsibilities together to deliver the environmental water program, without compromising the independence or diluting the focus of any particular organization. Building and maintaining the necessary partnerships to maintain the programs is an important element of success in its own right.

These six criteria, across the two broad dimensions of policy and implementation, offer a new perspective on performance, providing a basis for evaluating the success of environmental water programs, and linking success with different patterns of institutional and organization development to reform legal frameworks, build capacity, and strengthen partnerships. Fig. 26.1 demonstrates these two dimensions, and the way the investment in policy criteria and implementation capacity interacts to generate different end states for an environmental watering program.

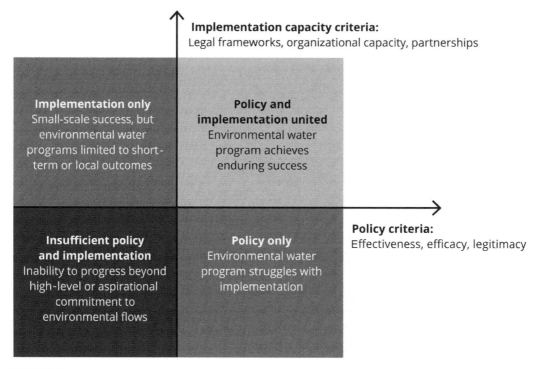

FIGURE 26.1

Policy criteria (horizontal axis) and implementation capacity criteria (vertical axis) as two dimensions of success for environmental watering programs.

Insufficient investment in any of these criteria will undermine the success of the environmental watering program. Many programs around the world struggle to implement a high-level policy commitment to environmental water management, leaving the environmental watering program as little more than an aspirational goal (Le Quesne et al., 2010). The challenge of implementing the centrally mandated policies and laws stipulating environmental water regimes without the necessary investment in both policy and implementation criteria can be seen in the Yellow River in China, and the ongoing challenges of implementing South Africa's ecological reserve (for more on these examples, refer to Chapter 17). Investment in policy criteria without a corresponding investment in the implementation elements can likewise undermine implementation, as it is difficult to achieve the broad goals of efficacy, efficiency, and legitimacy without an investment in the legal frameworks or organizational capacity necessary to achieve them. Early efforts in the Murray—Darling Basin prioritized specific volumetric recovery of environmental water, but only recently has this been coupled with recognition of the importance of building robust environmental organizations that work in partnership with local communities and other water organizations. However, investment in the implementation criteria without the commitment to the broader policy criteria is likely to see short-term, small-scale success that is difficult to build on. Early attempts to recover water for the environment in the western United States experienced successes with pilot programs and proof-of-concept for a diverse range of environmental water transactions; some programs struggled to scale up their achievements by focusing on implementation deals without parallel investments in policy criteria and strengthening and adapting enabling conditions (Garrick, 2015a). The enduring success of an environmental water program critically depends on investing in both the policy and implementation criteria over the life of the program.

The next section of this chapter illustrates these potential states by drawing on the examples of the Columbia River and the Murray–Darling Basin. Although neither case represents a perfect environmental watering program, each environmental water program has explicitly engaged with a range of these criteria.

26.3 ILLUSTRATING THE CRITERIA IN THE FIELD: THE COLUMBIA AND MURRAY–DARLING BASINS

The Columbia River Basin and the Murray–Darling Basin illustrate the conditions and criteria for success in mature environmental water programs, providing insight about success across multiple stages of design, implementation, and adaptation (for more detail on these case studies, please refer to the discussion in Chapters 17 and 19). Using case studies helps to draw lessons about the capacity and effectiveness of different approaches to environmental watering, undertaken in different contexts. These lessons focus on the application of the six criteria in the policies of protection and maintenance, and recovery and management.

The Columbia Basin has undertaken almost 50 years of legal and administrative reforms to protect environmental water for depleted salmon habitat, followed by almost two decades of active efforts to recover and manage environmental water rights (Neuman et al., 2006). The Murray–Darling Basin features a more accelerated set of legal reforms since the mid-1990s to recognize and establish the environment as a legitimate use, followed by the development of an interim cap on diversions, which has been updated based on newly calculated sustainable diversion limits. These system-wide reforms have been paralleled by efforts to recover and manage environmental water. These long-established initiatives illustrate the need to define success, establish a baseline and indicators to measure progress, and build and sustain institutional capacities to scale up and adapt. Before delving into the case studies and the lessons they demonstrate, a short primer on applying concepts of performance management to environmental water programs is provided (Box 26.1).

BOX 26.1 APPLYING CONCEPTS OF PERFORMANCE MANAGEMENT TO EVALUATE ENVIRONMENTAL WATER PROGRAMS

The expenditure of public funding to manage environmental water creates the need for accountability and reporting. The push for accountability underpins efforts to evaluate the performance of environmental water programs. Program evaluation assesses whether a program achieves its intended outcomes. The outcomes of environmental water programs can take diverse forms, ranging from a *healthy working river* to the survival of an endangered species, to recreational uses; and typically include multiple goals. Defining criteria for success is a prerequisite to measuring them, and program evaluation requires explicit statement of all intended outcomes.

Applying the concepts of performance management to evaluate environmental water programs implies a series of steps or considerations. First, performance management and program evaluation are aimed at informing decision makers about

(Continued)

BOX 26.1 (CONTINUED)

program outcomes, organization development, capacity building, and feeding into strategic planning. Therefore, performance management is considered an iterative process that links evaluation with decisions, as illustrated by the periodic use of reports prepared by the Productivity Commission and Australian National Audit Office of the Market Buyback and Environmental Watering Arrangements in the Murray–Darling Basin to guide improvements (e.g., ANAO, 2011).

Second, multiple, complementary forms of evidence are needed, often spanning Western and traditional knowledge, as well as local and system-wide monitoring capacities. For example, in Australia the Commonwealth Environmental Water Office's framework for monitoring and evaluation explicitly acknowledges the roles of diverse partners, including local community groups, in monitoring the outcomes of environmental watering activities (CEWO, 2012).

Third, program evaluation aims to assess effectiveness by establishing a causal relationship between the design of the program and observed outcomes, determining whether the change in the outcomes (e.g., a healthy working river, or survival of endangered fish) is explained by the activities of the environmental watering program (e.g., delivery of water to a network of ecologically important sites). If, for example, the change in endangered fish populations is attributed to ocean temperatures rather than streamflow restoration, the program may not have been necessary to achieve the intended outcome. Efforts to assess the causal link between program activities and outcomes have led to the identification of a *program process* involving inputs (e.g., resources), program activities, outputs, and outcomes over an iterative evaluation cycle.

Finally, program evaluation implies a diagnostic process to guide interventions through a sequence of increasingly specialized questions designed to link the need for the program with the criteria for assessing whether and how those needs are met in different contexts.

When designing a program evaluation, it is necessary to consider each of these elements. Understanding the program logic and testing its validity through the evaluation process requires an explicit identification of each element of the program, starting with the need to develop and test performance measures that represent the intended outcome. Fig. 26.2 uses an example of the intended outcome of a healthy fish population, and identifies the series of needs, objectives (linking the outcome to the environmental water regime), inputs (funding, technical studies), activities (water transactions), and outputs (water recovered).

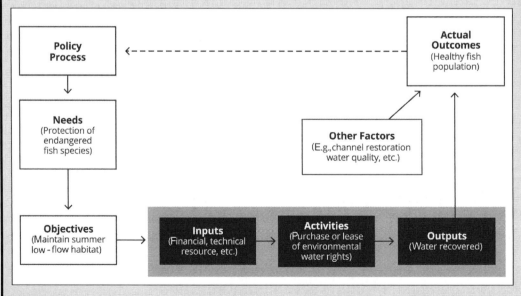

FIGURE 26.2

Understanding the elements of an environmental water program.

Source: Adapted from McDavid et al. (2012)

26.3.1 **COLUMBIA BASIN**

In its 2012 annual report, the CBWTP reported on its first 10 years, highlighting a decade of outcomes since its creation in 2003 (Columbia Basin Water Transactions Program, 2012). Water recovery, measured in terms of cumulative volume protected over the lifespan of the transactions implemented since 2003, offered the overarching measure of effectiveness. Over 7400 GL of water were protected, described as enough water to fill Safeco Field (a baseball stadium) almost 216 km high. Water recovery was not the only measure of effectiveness; the development of a diverse portfolio of transactional tools was cited as critical for restoring instream flows, ranging from leases of instream flows, to water stored in reservoirs for environmental water releases. Finally, the program's effectiveness was described in terms of its growing extent (136 streams) and activity (344 transactions) across four states and 19 (of 62) sub-basins in the Columbia. Qualitative measures of program effectiveness included the stories and case studies of individual transactions featuring before-and-after images and descriptions of the biological changes experienced in the sites targeted for streamflow restoration. Together these quantitative and qualitative measures were linked as measures of a decade of improvement. Efficiency and cost-effectiveness, the final of the overarching criteria, were less explicit. Over US$35 million were spent on environmental water transactions (including purchases and leases of water rights), and transaction costs ranged from 13% to 70% of total costs (Garrick et al., 2013).

Underpinning these indicators of program effectiveness were commitments to trust, legitimacy, and building capacity. The director of the Freshwater Trust, Joe Whitworth, stated that the "equation is simple: no trust, no deal." Capacity building was touted as being as important as efficiency, with Stan Bradshaw of Trout Unlimited Montana noting that "operational support allows us to go out and build relationships ...[with] a multiplier effect in communities as neighbors talk to neighbors." The CBWTP also emphasized the need for strategic partnerships as a key ingredient for lasting effectiveness, noting over 1400 partnership activities in the program's first decade. The Walla Walla Watershed Management Partnership in southwestern Washington is a prime example of an innovative local partnership involving a new statutory authority to promote flexible approaches to water allocation. Together, these multiple dimensions of effectiveness, efficiency, and legitimacy, enabled and sustained by legal frameworks that enabled the environmental water transactions, organizational capacity, and partnerships across different regions and sectors of the community, contributed to the enduring success of the CBWTP. Building on this success has required the development of new accounting procedures for tracking the outcomes as well as the outputs of flow transactions. The shift from tracking only changes in environmental water to accounting for impacts on fish passage and habitat change, outcomes that are only partially influenced by changes in flow levels, required much more sophisticated and sustained approaches to benchmarking and performance.

Since its 10-year anniversary in 2012, the Columbia Basin has developed a comprehensive framework to "create and implement an accounting methodology that uses well-defined measures of progress to track the effectiveness of flow restoration as a tool to improve aquatic habitat conditions for targeted fish populations" (McCoy and Holmes, 2015). This framework uses a logic path approach to account for performance across four sequential stages of water recovery: contractual compliance (legal accountability), flow accounting, aquatic habitat response, and ecological function. Paralleling these efforts, the CBWTP has actively monitored price and transaction costs data to track economic trends, capacity building, and strategic development.

The basin-wide performance trends for the CBWTP can obscure the substantial variability across geographic priority regions, and over time. This variation across sub-basins includes examples of basins in all four categories depicted in Fig. 26.1. The Deschutes River is noted for its investment in both policy and implementation criteria, balancing transaction development with partnership, planning, and policy reform. However, sub-basins in western Montana have until recently prioritized implementation criteria, and experienced initial successes before encountering barriers due to persistent deficits in the policy criteria and the legal and administrative framework. The multidimensional picture of performance that emerges from the Columbia reinforces the need for assessing effectiveness, efficiency, and legitimacy, while also tracking the development and strengthening of essential implementation capacity—legal frameworks, organizational capacity, and partnerships—over time.

26.3.2 MURRAY–DARLING BASIN

The scale of environmental water recovery in the Murray–Darling is unprecedented internationally, with 2750 GL to be recovered for the environment under the Murray–Darling Basin Plan (MDBA, 2012). In the most recent water year (2014–15) on record, the Murray–Darling Basin Authority reported the use of over 2000 GL for basin environmental watering activities. As a comparison, total surface water diversions in the Murray–Darling Basin for 2014–15 were 7415 GL (excluding interception activities and farm dams) (MDBA, 2015). Monitoring and evaluation efforts have tracked the effectiveness of environmental water use by assessing site interventions and system-wide impacts (CEWO, 2012). The emphasis on ecological and hydrological outcomes has been matched by increasing concern for governance arrangements (PC, 2010).

Effectiveness, efficiency, and legitimacy criteria need to be considered across all phases of environmental watering, including acquisition, delivery, and trading decisions. Effectiveness has required institutional and organizational capacity guided by frameworks for decision making, as well as public engagement (CEWO, 2013).

Efficiency and cost-effectiveness criteria have emphasized value for money. In the acquisition phase, efficiency criteria were used to compare water entitlement buybacks and irrigation efficiency projects as means of securing water for the environment. Irrigation efficiency projects were estimated at AU\$5600/ML, while water buybacks were just over a third of this price (Crase et al., 2012; Hart, 2015). Irrigation efficiency projects nevertheless became an attractive means of securing legitimacy amidst political backlash from irrigators concerned about the loss of water and livelihoods as part of the recovery efforts (Marshall, 2013).

Effectiveness, efficiency, and legitimacy have proven as important for assessing performance in the management and trading of the environmental water reserves. The creation of two entities—the Commonwealth Environmental Water Holder and Victoria Environmental Water Holder—represents an important institutional innovation to coordinate the delivery of environmental water. Both were created with the intent of improving the efficiency and effectiveness of environmental water management, by finding ways to coordinate and streamline environmental water delivery (O'Donnell, 2012, 2013). However, both organizations initially focused on efficiency and effectiveness as the sole indicators of legitimacy, which was not enough to build trust and engagement with local communities. Addressing this legitimacy deficit has taken time and effort, and is really only just beginning. One of the pathways toward legitimacy taken by the Commonwealth Environmental Water Holder has been to build stronger partnerships, and embed local arrangements as part of the broader strategy for environmental water management.

Although there are basin-wide environmental watering priorities set by the Murray–Darling Basin Authority, there remains some discretion in how the Commonwealth Environmental Water Holder uses its water. In 2012, the Commonwealth Environmental Water Holder formalized an agreement with a local nongovernment organization in South Australia, Water For Nature, where 10 GL of water would be made available each year for use in local South Australian wetlands (CEWO and Nature Foundation SA, 2012). Water For Nature then became responsible for managing the delivery of this water, involving the local community in deciding where, when, and how to undertake environmental watering, and reporting back to the Commonwealth Environmental Water Holder. The Commonwealth Environmental Water Holder has continued to seek out partnerships with specific organizations, and one of the most recent involves a partnership with the Ngarrindjeri Regional Authority (an organization representing indigenous Australians), to jointly manage environmental water to meet cultural flow needs (for more on cultural flows, see Chapter 9). This partnership is based on managing water in the Coorong and Lower Lakes at the mouth of the Murray–Darling Basin, and is set to last for 3 years (CEWO and Ngarrindjeri Regional Authority, 2016).

As with the Columbia Basin, implementation success in the Murray–Darling hinges upon the development, strengthening, and periodic adaptation of enabling conditions and capacities. The 1994 *Council of Australian Governments (COAG)* reforms authorized the environment as a legitimate water use, and were followed by successive local, state, and federal (Commonwealth) environmental water acquisition and delivery programs. The legal authority and organizational capacity established through these reforms depended in turn on effective partnerships, foremost between the states and the Australian government, and illustrated by the intergovernmental agreements creating the National Water Initiative (COAG, 2004) and the 500 GL *First Step* for The Living Murray program (MDBC, 2007). Over time, partnerships with regional communities via catchment management authorities, irrigators, and not-for-profit environmental groups have proven critical, particularly since the passage of the *Water Act 2007* and Basin Plan 2012, which relied on new federal authority and encountered resistance and retrenchment by states and stakeholders. The importance of building community support for environmental water activity was recently highlighted once again, with the passage of legislation in 2015 that capped the volume of water that could be recovered by purchase from existing users at 1500 GL. This attempt to limit the extent of environmental water acquisition in the Murray–Darling Basin underscores the importance of building legitimacy over the long term, and continuing to maintain the legal frameworks that enable environmental water recovery.

The next section of the chapter examines the way that these six criteria play out in the field, and draws three important lessons about the application of these criteria.

26.4 WHAT GETS MEASURED, GETS MANAGED: MOVING FROM CRITERIA TO EVALUATION

The foregoing criteria and illustrations in the field point to the need for rigorous assessment of program effectiveness over time (including efficiency and legitimacy); with attention to the institutional capacity needed to scale up activity and sustain progress. Measuring the criteria for success with a high spatial and temporal resolution is needed, as are efforts to identify the institutional arrangements and capacities to explain performance trends. A multicriteria approach to assess program effectiveness has several implications for benchmarking and evaluation.

26.4.1 ESTABLISH A BASELINE

First, effectiveness depends on a clear baseline and targets for assessing trends (see Box 26.2). Several other chapters have discussed the importance of establishing a baseline in terms of the science of adaptive management, by establishing a set of data against which future interventions can be assessed. However, baselines are also important in terms of building consensus on the existing state of nature, and reaching agreement on the next steps. Furthermore, a baseline needs to include not only the ecological condition, but also an accurate assessment of and accounting for water extraction within the basin as well.

> ### BOX 26.2 THE IMPORTANCE OF ESTABLISHING A BASELINE
>
> The Columbia Basin, United States, has a patchwork of administrative rules and instream flow plans to define environmental water requirements at the watershed (catchment) level. The physical dewatering of tributaries has left some stream reaches completely dry, providing a clear hydrological baseline due to chronic over allocation. However, inconsistent approaches to the assessment of environmental needs have hindered the establishment and harmonization of ecological and hydrological baseline at the basin scale. As a consequence, clear targets and baseline conditions have been established only for selected tributaries. For example, the Lemhi River of the Salmon sub-basin in the Columbia has a statutory target of 35 cubic feet per second (approximately 1 cubic meter per second) against which annual and long-term instream flow agreements can be assessed. There is also significant legal uncertainty about the nature of water rights, stemming from the *use it or lose it* requirements, and a lack of administrative capacity to measure, monitor, and enforce them where they do exist. Progress toward targets remains hard to gauge due to the lack of standardized targets and inadequate capacity.
>
> In the Murray–Darling Basin, Australia, there has been a struggle to define an effective baseline, as the impetus for environmental water policies was the emerging reliability and water quality problems of the mid-1990s. Although ecological and hydrological modeling has been used to highlight the impact of water extractions on the ecological health of the river, debate has tended to focus on the difference between the pristine form of the aquatic ecosystems of the river, and the way they are now. There has been an attempt to define the river as a *working river* (Hillman, 2008), and the Murray–Darling Basin Plan established June 2009 as the baseline against which implementation of the plan would be assessed, but establishing a comprehensive baseline of environmental condition that reflects the variability of dry and wet years and defining the goal of environmental water protection and recovery policies remains challenging. This inability to clearly state the intent and likely outcomes of the policies has led to lower targets for water recovery; for example, the Murray–Darling Basin Authority indicated that a minimum recovery of 3000–4000 GL was required to improve the health of the Murray–Darling Basin, and the ultimate recovery target established by the Murray–Darling Basin Plan was only 2750 GL.

26.4.2 MEASURE SUCCESS ACROSS MULTIPLE DIMENSIONS AND TIME FRAMES

Second, implementing environmental water management requires broad measures of success across multiple dimensions to translate incremental improvements and outputs into long-term trends and outcomes (Box 26.3). Success is more than the volume of environmental water protected or recovered, or the cost of doing so. Narrow visions of success can be easily undermined by changing community values, but broad definitions of success that actively engage with legitimacy, governance capacity, and partnerships are more likely to be enduring. Success depends on all of the criteria and capacities identified above.

BOX 26.3 MULTIPLE DIMENSIONS AND METRICS OF SUCCESS

The Columbia Basin Water Transactions Program, United States, transactions dataset stores over 60 variables related to hydrological, biological, legal, and economic attributes of water recovery projects. A comprehensive accountability framework adopted in 2015 prioritizes data and monitoring to assess legal compliance, biological impacts, and ecological function. The high costs of monitoring and enforcement functions (described as *stewardship* activities by practitioners in the Program) have posed a prime capacity challenge, requiring new modes of governance to coordinate private and public monitoring activities. The Independent Scientific Review Panel of the NPCC provided guidance to the Program urging cost monitoring to better gauge value for money and to compare alternative transactional tools used for water recovery (Northwest Power and Conservation Council, 2010). This has caused the Program to develop new approaches for defining and measuring *program effectiveness* to supplement its legal, ecological, and economic assessments of water recovery. While effectiveness and efficiency are amenable to quantifiable indicators and metrics, legitimacy and partnership strength are less tractable, posing important questions about the indicators and tools for assessing program effectiveness across contrasting geographies, political economic conditions, and organization structures.

The Murray—Darling Basin, Australia, has seen incremental improvements in environmental water allocation, and an even slower transition beyond the volumetric emphasis of water recovery programs. In 1995, water extraction was capped in the Murray—Darling Basin, but, although states were required to report on the cap, there was no associated reporting of environmental outcomes. In 2004, the continued decline of the health of the Basin again demanded action, this time in the form of the First Step Living Murray water recovery, which aimed to provide environmental water to six icon sites (MDBC, 2007). Again, reporting focused on the progress to achieve the water recovery targets rather than the use of the water to achieve ecological goals. Finally, in 2007, the degraded state of the Basin required urgent action, and a new and faster program for water recovery was undertaken. It was not until 2012, however, that the Commonwealth Environmental Water Holder finally developed a broad framework for monitoring and evaluation of the actual outcomes of the water recovery (CEWO, 2012). An example of how environmental water reporting has transitioned from a volumetric focus to a broader record of effectiveness and legitimacy is the Victorian Environmental Water Holder annual report *Reflections*. The Victorian Department of Sustainability and Environment released the first environmental watering booklet in 2009, showing the volumes of environmental water available and the locations of use in the water year 2007—08 (State of Victoria, 2009). The booklet now reports on the effectiveness of environmental watering (both in the short term and as part of longer-term monitoring programs), the efficiency goal of environmental watering (making "every drop count"), and the ongoing legitimacy of environmental watering, by explicitly identifying the shared benefits of environmental watering for the broader community (VEWH, 2015).

26.4.3 LONG-TERM SUCCESS DEPENDS ON LEGITIMACY

Third, the effectiveness of environmental water programs cannot be assessed in the short-run (Box 26.4). Instead, it requires attention to legitimacy, creation, and maintenance of legal and administrative frameworks, and organization capacity. Efforts to economize on legitimacy have often proven short-sighted and threatened the lasting progress.

BOX 26.4 FOCUS ON THE LONG-TERM

The Columbia Basin, United States, has pioneered the use of market-based water reallocation for environmental water restoration, dating to the passage of the Oregon *Instream Water Rights Act* in 1987, which authorized private acquisition of high reliability water rights for instream flows. Initial experiments were premised on the ideas of *free market environmentalism*—the need for clear, private, tradable, and secure property rights to incentivize water allocation and create an opening for private groups to enter the marketplace on behalf of the environment. Reflecting on the initial expectations and the gap between the logic and limits of free market environmentalism, Neuman (2004) observed that it has turned out to be much harder than expected to spend the available public and private funding to acquire environmental water rights due to cultural, legal, and economic barriers (Neuman, 2004). The contrasting experiences of western Montana (Bitterroot) and central Oregon (Deschutes) illustrate the importance of investments in capacity building despite the potential for inefficiencies or higher up-front costs. In the 5-year period from 2003 to 2007, the Bitterroot and Deschutes sub-basins of the Columbia recovered roughly the same amount of water, and were among the top performing sub-basins in terms of water recovery across the Columbia (approximately 46 and 36 cubic meters per second, respectively, first and second highest of 13 sub-basins) (Garrick and Aylward, 2012). In this case, the flow (cubic meters per second) refers to a *net present value* of the flow rate protected over the term of the contract (Brewer et al., 2008). This metric of water recovery enables comparison of temporary and permanent acquisitions on an apples-to-apples basis by treating water transactions like a financial annuity, by calculating the net present value of the flow rate protected over the full term of the contract.

However, transaction costs and program budgets were much higher per unit of water in the Deschutes than the Bitterroot, in part due to up-front investments in studies, water planning, and water banking institutions. Unpacking the 5-year period into annual trends revealed that the Deschutes was able to lower transaction costs and steadily increase water recovery over the period, whereas the Bitterroot recovered the vast majority of its total in the first 2 years before hitting the wall. In a follow-up study, the divergence widened between the Deschutes, which reached flow targets and scaled up its activities, and the Bitterroot, which encountered stubborn institutional barriers. The higher transaction costs in the Deschutes during the early period represented an investment in the institutional frameworks and collaborative governance needed for legitimacy and adaptive efficiency (Garrick, 2015b).

The Murray–Darling Basin, Australia, has had environmental water policies in place since 1994, with incremental improvements in water for the environment over that period. Since 2007, water recovery for the environment has proceeded fairly rapidly in the Murray–Darling Basin, thanks largely to the National Water Initiative and the creation of transferable water rights and active water markets. The Australian government initially set aside AU$3.2 billion dollars for purchasing water from existing users, and AU$5.8 billion to invest in water efficiency savings (Australian Government, 2010). By November 23, 2015, the Commonwealth Environmental Water Holder had over 1600 GL (long-term average yield) (Commonwealth of Australia, 2015), most of which has been recovered using a range of tender programs, where a decision to purchase was made based on the price of water in the market. However, as discussed above, the speed of water recovery has not been accompanied by similar investment in maintaining the legitimacy of the environmental water program. This came to a head in 2014, when the Australian government limited the total volume of water recovered via purchase to 1500 GL (Commonwealth of Australia, 2014).

26.5 CONCLUSION

Performance management of environmental water programs is still an emerging field. The intent of this chapter is to provide a structure for evaluating the success of environmental water programs, to support the evolution of this discipline in a coordinated and strategic manner.

Environmental water programs operate to achieve multiple goals over the long term. Success depends on demonstrating not only efficacy and efficiency, but also legitimacy to sustain the activity of the program over long time periods. To manage this complexity, this chapter proposes a multicriteria approach to defining success and evaluating environmental watering programs.

This chapter identifies six criteria that enable environmental water practitioners to build a robust framework for the long-term success of their environmental water programs:

1. Effectiveness;
2. Efficiency;
3. Legitimacy;
4. Legal and administrative frameworks;
5. Organizational capacity;
6. Partnerships.

These criteria emphasize the essential elements of the policy and governance indicators of success (effectiveness, efficiency, and legitimacy), as well as criteria that affect implementation capacity (legal frameworks, organizational capacity, and partnerships). The application of these criteria is illustrated using case studies of the Columbia River and Murray–Darling basins. These case studies demonstrate the importance of engaging with all the six criteria in a range of water resource management contexts.

REFERENCES

ANAO, 2011. Restoring the Balance in the Murray-Darling Basin. Australian National Audit Office, Canberra, ACT.

Australian Government, 2010. Water For the Future. Canberra: Department of Sustainability, Environment, Water, Population and Communities.

Brewer, J., Kerr, A., Glennon, R., Libecap, G.D., 2008. Water Markets in the West: Prices, Trading, and Contractual Forms. Econ. Inquiry 46, 91–112.

COAG, 2004. Intergovernmental agreement on a national water initiative, Council of Australian Governments.

Columbia Basin Water Transactions Program, 2012. Annual report 2012: A decade of outcomes 2003–2012. National Fish and Wildlife Foundation, Portland, Oregon. Available from: http://cbwtp.org/jsp/cbwtp/library/documents/NLB_CBWTP_Annual12_R6.pdf.

CEWO, 2012. Commonwealth Environmental Water – Monitoring, Evaluation, Reporting and Improvement Framework May 2012 V1.0. Commonwealth Environmental Water Holder for the Australian Government, Canberra.

CEWO, 2013. Framework for Determining Commonwealth Environmental Water Use. Commonwealth of Australia, Canberra, Australia.

CEWO & Nature Foundation Sa, 2012. Press release: Commonwealth works with Nature Foundation South Australia to deliver water (24 October). Commonwealth Environmental Water Office and Nature Foundation SA, Adelaide.

CEWO & Ngarrindjeri Regional Authority 19 April 2016. Joint Media Release: Achieving environmental and cultural water benefits in the lower River Murray region. Available from: http://www.environment.gov.au/water/cewo/media-release/achieving-environmental-and-cultural-water-benefits-lower-river-murray-region/?utm_source = CRMSCEWODatabase&utm_medium = email&utm_content = PartnershipArticle2Link&utm_campaign = 2016AutumnNewsletter.

Commonwealth Of Australia, 2014. Water Recovery Strategy for the Murray-Darling Basin. Department of Environment, Canberra, Australia.

Commonwealth Of Australia. 2015. Environmental Water Holdings [Online]. Available from: https://www.environment.gov.au/water/cewo/about/water-holdings: Commonwealth of Australia (accessed 23.11.15).

Crase, L., O'Keefe, S., Kinoshita, Y., 2012. Enhancing agrienvironmental outcomes: Market-based approaches to water in Australia's Murray-Darling Basin. Water Resour. Res. 48, W09536. Available from: http://dx.doi.org/10.1029/2012WR012140.

Garrick, D., 2015a. Water Allocation in Rivers Under Pressure: Water trading, transaction costs, and transboundary governance in the Western USA and Australia. Edward Elgar, Cheltenham, UK.

Garrick, D.E., 2015b. Water Allocation in Rivers under Pressure: Water Trading, Transaction Costs and Transboundary Governance in the Western US and Australia. Edward Elgar Publishing, Cheltenham, UK.

Garrick, D., Aylward, B., 2012. Transaction costs and institutional performance in market-based environmental water allocation. Land Econ. 88, 536−560.

Garrick, D., O'Donnell, E., 2016. Exploring private roles in environmental watering in Australia and the US. In: Bennett, J. (Ed.), Protecting the Environment, Privately. World Scientific Publishing.

Garrick, D., Siebentritt, M.A., Aylward, B., Bauer, C.J., Purkey, A., 2009. Water markets and freshwater ecosystem services: Policy reform and implementation in the Columbia and Murray-Darling Basins. Ecol. Econ. 69, 366−379.

Garrick, D., Whitten, S.M., Coggan, A., 2013. Understanding the evolution and performance of water markets and allocation policy: A transaction costs analysis framework. Ecol. Econ. 88, 195−205.

Hart, B., 2015. The Australian Murray−Darling Basin Plan: challenges in its implementation (part 1). Int. J. Water Resour. Dev. Available from: http://dx.doi.org/10.1080/07900627.2015.1083847.

Hillman, T., 2008. Ecological requirements: creating a working river in the Murray-Darling Basin. In: Crase, L. (Ed.), Water Policy in Australia: The Impact of Change and Uncertainty. Resources For the Future, Washington DC.

Hogl, K., Kvarda, E., Nordbeck, R., Pregernig, M., 2012a. Effectiveness and legitimacy of environmental governance − synopsis of key insights. In: Hogl, K., Kvarda, E., Nordbeck, R., Pregernig, M. (Eds.), Environmental Governance: The Challenge of Legitimacy and Effectiveness. Edward Elgar, Cheltenham, UK, online ed.

Hogl, K., Kvarda, E., Nordbeck, R., Pregernig, M., 2012b. Legitimacy and effectiveness of environmental governance − concepts and perspectives. In: Hogl, K., Kvarda, E., Nordbeck, R., Pregernig, M. (Eds.), Environmental Governance: The Challenge of Legitimacy and Effectiveness. Edward Elgar, Cheltenham, UK, online ed..

Horne, A. 2009. An approach to efficiently managing environmental water allocations. PhD, University of Melbourne.

Le Quesne, T., Kendy, E., Weston, D., 2010. The Implementation Challenge: Taking stock of government policies to protect and restore environmental flows. World Wildlife Fund UK, Godalming, Surrey.

Loehman, E.T., Charney, S., 2011. Further down the road to sustainable environmental flows: funding, management activities and governance for six western US states. Water Int. 36, 873−893.

Malloch, S., 2005. Liquid Assets: Protecting and Restoring the West's Rivers and Wetlands through Environmental Water Transactions. Trout Unlimited, Arlington, VA.

Marshall, G.R., 2013. Transaction costs, collective action and adaptation in managing complex social−ecological systems. Ecol. Econ. 88, 185−194.

McCoy, A., Holmes, S.R., 2015. Columbia Basin Water Transactions Program Flow Restoration Accounting Framework. Oregon National Fish and Wildlife Foundation, Portland. Available from: http://cbwtp.org/jsp/cbwtp/library/documents/Flow%20Restoration%20Accounting%20Framework_version2_0_August2015.pdf

McDavid, J.C., Huse, I., Hawthorn, L.R.L., 2012. Program Evaluation and Performance Measurement: An Introduction to Practice. Sage Publications, UK.

MDBA, 2012. Murray-Darling Basin Plan. Murray-Darling Basin Authority, Commonwealth of Australia, Canberra, Australia.

MDBA, 2015. Towards a healthy, working Murray-Darling Basin: Basin Plan Annual Report 2014-15. Murray-Darling Basin Authority, Canberra, ACT.

MDBC, 2007. The Living Murray Business Plan: 25 May 2007. Murray-Darling Basin Commission, Canberra.

Neuman, J.C., 2004. The good, the Bad, and the Ugly: The First Ten Years of the Oregon Water Trust, The. Neb. L. Rev. 83, 432.

Neuman, J.C., Squier, A., Achterman, G., 2006. Sometimes a Great Notion: Oregon's Instream Flow Experiments. Environ. Law 36, 1125—1155.

Northwest Power And Conservation Council, 2010. Final RME and Artificial Production Categorical Review Report. In: PANEL, I. S. R. (ed.).

O'Donnell, E., 2012. Institutional reform in environmental water management: the new Victorian Environmental Water Holder. J. Water Law 22, 73—84.

O'Donnell, E., 2013. Australia's environmental water holders: who is managing our environmental water? Aust. Environ. Rev. 28, 508—513.

OECD, 2011. Water Governance in OECD Countries A Multi-level Approach. In: PUBLISHING, O. (ed.).

OECD, 2015. OECD Principles on Water Governance: welcomed by Ministers at the OECD Ministerial Council Meeting on 4 June 2015. Online: Directorate for Public Governance and Territorial Development.

Pittock, J., Lankford, B.A., 2010. Environmental water requirements: demand management in an era of water scarcity. J. Integr. Environ. Sci. 7, 75—93.

PC, 2010. Market Mechanisms for Recovering Water in the Murray-Darling Basin (Final Report, March). Productivity Commission, Canberra.

Scharpf, F.W., 1999. Governing in Europe: Effective and Democratic? Oxford University Press, Oxford and NewYork.

Siebentritt, M.A. (Ed.), 2012. Water trusts: what role can they play in the future of environmental water management in Australia? Proceedings of a workshop held on 1 December 2011. Water Trust Alliance and Australian River Restoration Centre, Canberra.

State of Victoria, 2009. Environmental watering in Victoria 2007/08. Department of Sustainability and Environment, Melbourne.

VEWH, 2015. Reflections: Environmental Watering in Victoria 2014—15. Victorian Environmental Water Holder, Melbourne, Australia.

World Justice Project, 2016. What is the rule of law? [Online]. Available from: http://worldjusticeproject.org/what-rule-law (accessed 2.03.16).

REMAINING CHALLENGES AND WAY FORWARD

MOVING FORWARD: THE IMPLEMENTATION CHALLENGE FOR ENVIRONMENTAL WATER MANAGEMENT

27

Avril C. Horne[1], Erin L. O'Donnell[1], Mike Acreman[2], Michael E. McClain[3], N. LeRoy Poff[4,5], J. Angus Webb[1], Michael J. Stewardson[1], Nick R. Bond[6], Brian Richter[7], Angela H. Arthington[8], Rebecca E. Tharme[9], Dustin E. Garrick[10], Katherine A. Daniell[11], John C. Conallin[3], Gregory A. Thomas[12], and Barry T. Hart[13]

[1]The University of Melbourne, Parkville, VIC, Australia [2]Centre for Ecology and Hydrology, Wallingford, United Kingdom [3]IHE Delft, Delft, Netherlands [4]Colorado State University, Fort Collins, CO, United States [5]University of Canberra, Canberra, ACT, Australia [6]La Trobe University, Wodonga, VIC, Australia [7]Sustainable Waters, Crozet, VA, United States [8]Griffith University, Nathan, QLD, Australia [9]Riverfutures, Derbyshire, United Kingdom [10]University of Oxford, Oxford, United Kingdom [11]The Australian National University, Canberra, ACT, Australia [12]The Natural Heritage Institute, San Francisco, CA, United States [13]Water Science, Echuca, VIC, Australia

27.1 INTRODUCTION

Like most environmentally centered disciplines, the field of environmental water management is relatively young. The origins of modern environmental water praxis stem from efforts in the 1970s, mostly to mitigate the impacts of onstream dams by releasing minimum flows to maintain habitat for individual species (Acreman and Dunbar, 2004; Tennant, 1976; Tharme, 2003). Much of the early work was led by water engineers who recognized that hydraulic properties of the river, particularly depth and velocity, were important in defining physical habitat availability for sport fish such as salmon and trout (Bovee and Milhous, 1978; Milhous et al., 1989). This work evolved into the fields of ecohydraulics and ecohydrology. Ecohydraulics extended into microscale processes such as flow turbulence (Wilkes et al., 2013). In a somewhat parallel research stream, ecohydrology saw the growing collaboration between hydrologists and ecologists focused on broader relationships between river flow and ecosystem condition, moving beyond the hydraulic habitat requirements of individual organisms (Dunbar and Creman, 2001; Nestler et al., in review; Chapter 11). The importance of the environmental water regime in sustaining overall ecosystem structure and process became widely accepted in the 1990s (Richter et al., 1997), underscored by the seminal paper of Poff et al. (1997) on the natural flow paradigm, and followed by publication of flow—ecology principles (Bunn and Arthington, 2002; Nilsson and Svedmark, 2002).

Water for the Environment. DOI: http://dx.doi.org/10.1016/B978-0-12-803907-6.00027-9

649

The concept of environmental water regimes as essential elements of sustainable water resource management in river basins was consolidated in the 2000s, with such regimes increasingly recognized for both their ecological and societal values (King et al., 2003; Poff and Matthews, 2013). In systems that remain relatively natural, environmental water regimes can be established to protect key river objectives. However, there has been increasing recognition that few rivers retain their natural flows or ecological integrity because of anthropogenic alteration of the environment. In many instances, therefore, conserving or restoring the range of flow variability experienced by a river under natural conditions is simply not realistic or societally desirable. Instead, there is a need to place more attention on the flow management of emerging and future hybrid and novel ecosystems (Acreman et al., 2014). Here, as in the case of heavily modified rivers, the objective may be to conserve targeted aspects of a river ecosystem for specific, predefined purposes such as recovering endangered species or maximizing benefits to people in line with the *Sustainable Development Goals (SDGs)*, for instance, rather than attempting to restore the river's condition prior to major development.

In 2007, the Brisbane Declaration established an international consensus on the definition and principles of environmental flows (or water levels) and defined environmental flow management as providing:

> ... *the water flows needed to sustain freshwater and estuarine ecosystems in coexistence with agriculture, industry, and cities. The goal of environmental flow management is to restore and maintain the socially valued benefits of healthy, resilient freshwater ecosystems through participatory decision making informed by sound science.*

To achieve this, the Brisbane Declaration (2007) called for commitment to a number of key actions for maintaining and restoring environmental water regimes, many of which were aimed at expanding the number of river systems where environmental water regimes are being implemented, broadening stakeholder engagement and participatory decision making, and building the networks and capacity required to initiate, maintain, and enforce environmental water implementation.

In the decade since the Brisbane Declaration was drafted, how far have we come in meeting this commitment? This is a difficult question to answer, as there is no clear, globally consistent approach to measuring how successfully environmental water practice has responded, and no widely accepted global database for sharing information on where and how environmental water is provided. The lack of international benchmarks and databases is evidence of a still-fragmented landscape for environmental water management, as well as of insufficient prioritization of investment in the assessment of its global impact. However, the wide range of experts, approaches, and case studies across different subdisciplines of environmental water brought together in this book have demonstrated that there has been significant progress on several fronts.

First, there has been a clear broadening in the concept of *environmental water management* to reflect the rationale of the Brisbane Declaration. Methodologies for environmental flows assessment now incorporate stakeholder values, participation, and codesign, and recognize the dual role of environmental water in supporting ecological and societal values and benefits, especially for those people who directly rely on river ecosystems for their livelihoods (King et al., 2003; Poff and Matthews, 2013; Ziv et al., 2012). Second, there has been significant progress in the development of the science that underpins the assessment of environmental water requirements (Arthington, 2015; Stewardson and Webb, 2010). Third, we have seen environmental water discussed in high-level forums and incorporated into national and international water policy and legislation across the globe (Hirji and Davis, 2009; Le Quesne et al., 2010; O'Donnell, 2013). This is, for example, reflected in the rapid growth in the number of government agencies and nongovernment entities funding environmental water projects (Garrick and O'Donnell, 2016; Garrick et al., 2011; Pahl-Wostl et al., 2013; Richter and Thomas, 2007; see Chapters 17–19).

Despite this progress, there remain significant challenges in the delivery of environmental water regimes on the ground. This persistent, overarching implementation challenge is revealed in each section of this book. This final chapter focuses on directions to achieve successful implementation of environmental water policies into the future, and is organized around six key questions:

1. How much water do rivers need?
2. How do we increase the number of rivers where environmental water is provided?
3. How can we embed environmental water management as a core element of water resource planning?
4. How can knowledge and experience be transferred and scaled more easily?
5. How can we enhance the legitimacy of environmental water programs?
6. How can we support the inclusion of adaptive management as standard practice?

Addressing any one of these questions requires the experience and knowledge held in shared skills across multiple disciplines. This is a recurring theme throughout this book: environmental water management is more successful when it actively seeks to incorporate a diverse range of expertise and interests.

27.2 HOW MUCH WATER DO RIVERS NEED?

Twenty years ago, Richter et al. (1997) asked the question "How much water does a river need?" In Chapter 9, Jackson recast this question as "How much water does a culture need?" to highlight the social and cultural complexity of societal relationships with water, and the need for greater integration of social dimensions in environmental water management. The challenge of defining an environmental water regime is thus twofold; understanding and incorporating the shared objectives and values of the river, along with understanding and incorporating the scientific links between flow and ecosystem condition (Finn and Jackson, 2011). Although there have been considerable advances in both these aspects of environmental water management, there remain areas for advancement.

A central element of water management is the establishment of a shared vision for the river system that acknowledges the diverse uses of the resource and ways in which cultures value and benefit from the natural environment. The amount of water a river needs is inherently linked to the sort of river the society wants, while also recognizing the intrinsic rights of nature. There are many challenges when defining a vision for a river and objectives for environmental water. First, our values change over time both due to our priorities and changes in the way we interact with nature. For example, it is well documented that willingness-to-pay for environmental outcomes increases with wealth and economic security (Whittington, 2010). There are more options available in a river system that is not heavily developed or relied upon for human uses. In heavily developed systems, while we increasingly recognize the value of a healthy environment to support our lives, we know that restoring systems to a natural state is unrealistic if they are to simultaneously meet modern human water needs. Increasing population growth and climate change are also driving systems to evolve in novel ways, making the relevance of the term *natural* for understanding river system objectives uncertain (Chapter 11). It is also possible that there will be substantial shifts in the way communities interact with nature in the not-so-distant future. For example, robotics is increasingly being used to support agriculture, changing the way that farms and rural communities may be structured. Artificial environments are being created to meet the needs of residents in urban centers, either through created parklands, or in the future, through documentaries and augmented or virtual reality. Simultaneously, transport options may make many remote river basins more readily accessible for recreation. We do not claim to know how these aspects will shape the way river basins are used and valued, but rather highlight that planning and policy must recognize that there will be shifting values and objectives, possibly including those pertinent to environmental water management, as technologies, economies, and societies change.

Second, much of the early work in environmental water science and the development of environmental flows assessment methods took place in developed countries in temperate climatic regions; other methods have been developed in semi-arid and tropical regions and/or applied in less developed countries (Tharme, 2003). It is important to consider that people in different countries often think about and experience the environment and river systems differently, because of differing cultural values, beliefs, and practices, along with differing needs for river natural resources (Daniell, 2015; Enserink et al., 2007). Varying climatic regimes also lead to a diversity of river socio-ecological systems and flow regimes, and different relationships between them. Knowledge of the environment in different settings is shaped through personal experience, and then often transferred through generations. Values and objectives developed and transmitted through customary approaches may be difficult for scientists and managers to capture or fully comprehend using the current common tools for environmental flows assessments (Christie et al., 2012). Finally, and perhaps in a similar vein, there has been limited involvement of indigenous communities in defining water needs and values within the framework of environmental water assessments. This is despite river systems being central to many indigenous cultures, livelihoods, and values (Finn and Jackson, 2011). The utilitarian values that underpin many of our existing water resource management systems are oftentimes at odds with the relational values of many indigenous communities with river systems, where they support relationships between each other and nonhuman nature (Chapter 9). Reconciling these distinct approaches to valuing a river system and creating a shared vision and objectives for a river is an essential element of progressive environmental water management.

Many of the world's river systems are managed through water resources infrastructure such as dams, with thousands of new dams in planning and construction (Winemiller et al., 2016; Zarfl et al., 2015). In these systems, over time the environment both downstream and upstream of these dams becomes increasingly modified from its original state. Where water infrastructure is still in the planning stage, greater opportunity exists to avoid and/or minimize the potential socio-ecological effects of flow alteration at a system scale, particularly through appropriate dam placement (Opperman et al., 2015) and the introduction of dam design features and operational rules that enable environment water delivery (Richter and Thomas, 2007; Chapter 21). Once the construction stage is reached, opportunities to reverse ecological degradation and shift systems in the direction of their predam state are often severely constrained by the design features or operational requirements of the dam to meet its economic purpose. In this situation, the relevance of the natural flow regime as a reference or target may not be appropriate for dams that severely modify downstream biophysical processes. It may be more appropriate to *design* an environmental water regime that meets the multiple objectives (consumptive and environmental) of the system (Acreman et al., 2014). "The idea of being able to define and quantify the components of the flow hydrograph and assemble them into an environmental flow regime that meets a particular set of ecological and social objectives can be thought of as a 'designer' approach, producing environmental flows that support desired ecosystem states or provide desired ecosystem services" (Acreman et al., 2014). Although this concept is appealing and implicitly underpins the development of many environmental water regimes in practice (e.g., GBCMA, 2014), there remains a significant challenge as to how to design and manage a flow regime to ensure that the complex needs of the environment are supported in the longer term (Acreman et al., 2014; Arthington, 2015; Arthington et al., 2006; Harman and Stewardson, 2005). This will usually require trade-offs between different river-level objectives (e.g., agriculture, hydropower, urban, and environmental), and indeed between different elements of the river ecosystem (e.g., fish and vegetation). A manager must decide how to operate the water resource system and its storages to achieve a socially desirable outcome that meets both environmental and societal needs (Acreman et al., 2014; Poff et al., 2016). This highlights the need for decision frameworks that facilitate these trade-off decisions and account for the various societal objectives that intersect with water resource management.

Although our understanding of flow—ecology relationships has significantly improved over the past 20 years (Arthington, 2015), there are still gaps in our knowledge of the ecological effects of flow alterations (Poff and Zimmerman, 2010; Webb et al., 2013). A major challenge is to undertake controlled water management experiments at the catchment scale, with much research still reliant on the occurrence of changing flow conditions during floods or droughts for advancing the knowledge base (Konrad et al., 2011; Olden et al., 2014). More collaboration between dam owners/operators, landholders, and scientists is needed to codevelop hypotheses and provide robust tests of these via flow manipulation experiments (Poff et al., 2003). In addition, new environmental and management challenges in the future will require different scientific information to support management decisions. The research agenda has to anticipate these challenges in order to support managers as the implications of new challenges are realized. Here, we focus on just two of these future challenges: (1) active environmental water management and (2) climate change.

Chapter 19 describes the emergence of active environmental water management. Recall that *active* environmental water management refers to those allocation mechanisms that require ongoing decision making concerning when and how to use environmental water to achieve the desired outcomes. As more river systems are altered by water resource development and/or reach their limits of sustainability, it is likely that there will be an increase in jurisdictions adopting allocation mechanisms for environmental water that require active management. Chapter 11 discusses some of the specific planning and operational challenges of actively managing environmental water. Importantly, active management highlights two distinct, but nested cycles of adaptive management (Chapter 25). The first is the planning cycle, for instance setting the longer-term objectives and priorities for the system (Chapter 23). The second is the implementation cycle, for instance the ongoing decisions around how to operationalize and use the environmental water from day to day, seasonally, and from year to year (Chapter 24).

Both the planning and implementation cycles of environmental water management can be improved through the use of conceptual models that (1) link the decisions available (e.g., to release environmental water at different spatial and temporal scales) to the objectives being managed for and (2) provide quantitative information to show the benefits of one flow decision over another. However, the resolution or granularity of information required differs between the planning and implementation cycles, with finer-scale information needed for implementation (Horne et al., in press). The implementation cycle, in contrast to long-term planning, has the advantage of being able to adjust the environmental water regime in a dynamic way to account for feedbacks and state transition information for environmental endpoints (Overton et al., 2014; Shenton et al., 2012). It also allows the flexibility to adjust the details of flow events (e.g., the peak magnitude of a flow event and its duration) to meet real-time assessments of adequacy. However, managers require information to inform the precise timing of when flow is required, and consideration of releases for other users and unregulated inflows in a particular season. Transparent and detailed information on the marginal return of a decision (e.g., whether delivery of half the water would provide half the benefit) thus becomes important for implementation, with these calculations needing to take place at a within-year timescale. As the value of water increases, so too does the importance of these decisions. As it will usually not be possible to provide the complete desired environmental water regime in all years, making the best use of the water available will require an understanding of the benefits or risks of providing one component of the flow regime without (or instead of) another, or providing one flow component, but at less than the recommended magnitude (Gippel et al., 2009; Richter et al., 2006). Multiyear sequencing of environmental water releases also becomes relevant; in how many years out of five does a particular flow event need to be provided? Future environmental flows assessment methods will need to respond to these information needs to allow adaptive management at the implementation scale. Moreover, there will be a substantial role for improved decision support tools to assist environmental managers to assimilate and assess the implications of this fine-grained information in order to implement active management (Horne et al., 2016). In situations where the available environmental water is constrained, the question changes from how much water is needed to how do we best manage the water available for the environment.

Climate change is also creating new challenges for environmental water science (see Chapter 11 and 14) and management. Over the history of environmental water management, the assumption of a stationary climate has prevailed, and this has allowed statistical analysis of historical flow regimes to be used in planning future flow allocations. Contemporary environmental flows assessment approaches such as *Ecological Limits of Hydrologic Alteration* (*ELOHA*; Poff et al., 2010) and sustainability boundaries (Richter et al., 2012) still rely on this simplifying assumption. It is now recognized that the climate is changing and future hydrologic regimes are likely to deviate substantially from historical *reference* conditions in many regions (e.g., Reidy Liermann et al., 2012). Thus, from a water management point of view, uncertainty about future flows complicates the planning and implementation cycles necessary to achieve and sustain effective environmental water regimes. Uncertainty over available future water will likely intensify debates around environmental water allocations, and around water allocation more broadly.

Beyond the water management challenge, however, other challenges are emerging for the practice of environmental water management as the climate changes. Warming temperatures are modifying species performance and fitness, and changing interactions among species in ways that are already causing shifts in species distributions and ecosystem processes. Thus, the notion of reference ecological conditions is also shifting and creating a need to move away from a simple statistical association between average hydrologic regime components and time-averaged ecological states toward a more *process-based* understanding of the linkages between hydrologic regime dynamics (extreme events, antecedent flows) and ecological processes (see Chapter 11).

The shifting landscape of hydrologic and ecological baselines (Kopf et al., 2015) creates new impetus for environmental water scientists and managers to engage stakeholders in defining socially desirable ecological targets that can be achieved under future water regimes. To date there are few examples where these issues have begun to be considered, and large uncertainties are liable to remain a feature of forecasting efforts for some time. New decision support tools will be required, to aid in the identification of realizable objectives and assess the likelihood of achieving them through well-defined management interventions (Poff et al., 2016).

Finding the *answer* to the question of how much water a river needs and for what purposes is thus a complex undertaking with each river being different, and also having different needs over time. Both environmental and societal drivers will define the appropriate environmental water regime, but the challenge remains to develop and use the tools necessary to answer this question for any single application. Rather than providing a *definitive answer*, refining the approach to objective setting, and strengthening our scientific knowledge on what different benefits flow regimes will be likely to deliver, will improve the robustness of our approach to providing environmental water under changing futures (Davies et al., 2014).

27.3 HOW DO WE INCREASE THE NUMBER OF RIVERS WHERE ENVIRONMENTAL WATER IS PROVIDED?

The Brisbane Declaration argued for considerable and rapid expansion of the number of locations around the world where environmental water regimes are in place. How do we best support the implementation of environmental water into new locations? Moore (2004) surveyed 272 individuals involved in environmental water management, across a range of organizations, and asked respondents how the concept of environmental water was initially established in their various river basins and countries. The results show that public awareness and recognition of the importance of river flows to local livelihoods were both important factors. Interestingly, the introduction of environmental flows assessment projects and expertise was also seen as a major driver. Environmental flows assessments, particularly when there is stakeholder engagement, can lead to significant, positive changes in community attitudes concerning water resource management (King et al., 1999; Moore, 2004).

Moore (2004) also asked respondents to identify the major difficulties and obstacles to implementing environmental water regimes in their region (Fig. 27.1). The respondents indicated that increased public awareness would help build political will and effective stakeholder engagement (both identified as obstacles in Fig. 27.1). Similarly, recognition of the importance of local livelihoods (a success factor identified in Moore, 2004) is linked to having an understanding of the socio-economic costs and benefits associated with environmental water (an identified obstacle; Fig. 27.1). Institutional arrangements, including inadequate or inappropriate funding of environmental water projects, along with insufficient technical capacity, are also identified as common obstacles to the implementation of environmental water regimes.

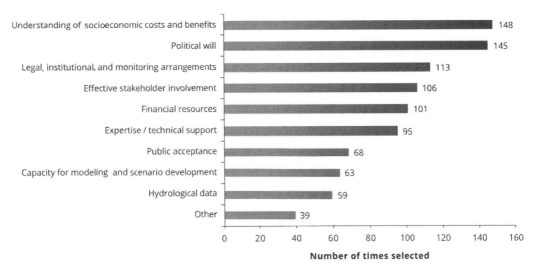

FIGURE 27.1

Survey results showing the major difficulties and obstacles to implementation of environmental water programs within the respondents' areas.

Source: Moore (2004)

Governments need sufficient evidence to make the case for reform (OECD, 2012). Much of the world's species richness is located in developing countries, and is thus under pressure from efforts to stimulate economic growth and alleviate poverty (Fazey et al., 2005; Winemiller et al., 2016). In systems where there is high demand for water, allocation of water to the environment will likely require a trade-off with consumptive water uses (see Chapter 16). Ecosystem services—the contributions of biodiversity, ecosystem structure, and function to human well-being (see Chapter 8)—provides one potential mechanism to assess the benefits of environmental water regimes using a value and measurement system that aligns more closely with traditional economic outcomes from water resource use. Many developing countries are grappling with the challenges of poverty alleviation, human well-being, and economic development. There may thus be a heavier reliance on water resources to meet immediate human needs, making it difficult to consider longer-term ecosystem needs and ecosystem services for future generations (Christie et al., 2012). This has led some experts to conclude that the goals of poverty reduction and environmental management are incompatible (Dalal-Clayton and Bass, 2009). In these cases, identifying ecosystem services provided by environmental water may improve community understanding and support, as well as provide the evidence to support reform. People often unknowingly rely on ecosystem services. There remain significant challenges for identifying and communicating these ecosystem services. Moreover, there are further challenges for economic valuation techniques (as discussed in Chapter 8), and these are exacerbated in developing countries where limited valuation of environmental goods and services has occurred (Christie et al., 2012; Kenter et al., 2011) and other or complementary means of valuing river resources and ecosystem integrity are recognized. These limitations can, in part, be addressed through local engagement and capacity building in the identification and valuation of ecosystem services (Christie et al., 2012), and through the use of nonvaluation techniques (see Chapter 8).

Although ecosystem services are likely to be an important tool for expanding the implementation of environmental water, it is important to recognize the long-running and unresolved debate about how to value nature. There are those who suggest nature is important for the benefits it provides humans, and others who see a moral responsibility to protect nature for its own sake, for instance its intrinsic value (Davidson, 2013). There is debate around how well ecosystem services concepts capture intrinsic biodiversity and environment values (Davidson, 2013; Dudgeon, 2014). As an extreme position, Krieger (1973) suggested that if the *value* of trees is purely around services to humans, it may be that a plastic replica of a tree could provide some of these same services. We suspect that most people would have some underlying discomfort with this position, as they perceive some value from knowing that nature exists and knowing it will be there for future generations. This remains a core challenge for river basin communities; how much importance should they place on the *existence* or *intrinsic* value of nature compared to the immediate human needs from the river? In New Zealand, the Whanganui River has now been recognized as a person when it comes to the law, giving it rights and interests for its own sake. This recognizes the river as a living force that should be respected and protected in its own right, not just for the benefit of development and human well-being (Calderwood, 2016).

Financing of water reform and implementation remains an ongoing struggle in many countries (OECD, 2012), and a recognized barrier to the implementation of environmental water management (Moore, 2004). Where major infrastructure projects are funded by donor agencies, the implementation of an environmental water program could be linked to the funding criteria and included within the project costings (Hirji and Davis, 2009). An alternative approach is to purchase water for the environment from existing uses, managed through government or private actors such as *nongovernment organizations* (*NGOs*; Chapter 18). This is readily achieved in systems that are well-licensed with existing water markets (e.g., the Murray—Darling Basin; Hart, 2015), but can also be achieved through strong stakeholder engagement in systems where there is no existing water market (e.g., NGOs operating in parts of the western United States; Garrick et al., 2009; Horne et al., 2008). Such purchasing may currently be occurring in only a handful of locations (most notably the western United States and Australia), but is expected to become more important in the future. Where funds or other resources are limited, there may be a role for the rapid rollout of precautionary environmental flows assessments that are inexpensive to implement and can be done with limited empirical information (Richter et al., 2012; Tharme, 2003). These would then be followed later by more robust environmental flows assessments (a triage type approach). However, it is important to recognize that where there are significant economic trade-offs with providing environmental water, it is likely that the precautionary approach will fall short of providing adequate information to justify limiting economic growth. Even with a precautionary environmental water regime determined, there would be an ongoing financial cost associated with implementation and enforcement.

A key opportunity for achieving implementation of environmental water regimes is to build on global commitments that require environmental water. The SDGs (officially known as Transforming Our World: the 2030 Agenda for Sustainable Development) provide a set of 169 targets developed by an intergovernmental process under United Nations Resolution A/RES/70/1 of September 25, 2015 (http://www.un.org/sustainabledevelopment/). Healthy terrestrial and aquatic ecosystems are explicit in Goal 15 and implicit within Goal 3 (Good Health and Well-being) and Goal 6 (Clean Water and Sanitation). Goal 15 specifies targets such as by 2020 "ensure the conservation, restoration and sustainable use of terrestrial and inland freshwater ecosystems and their services," "reduce the degradation of natural habitats, halt the loss of biodiversity," "introduce measures to prevent the introduction and significantly reduce the impact of invasive alien species on land and water ecosystems," and "integrate ecosystem and biodiversity values into national and local planning, development processes, poverty reduction strategies and accounts." A key challenge is to embed the importance of environmental water into the language of the SDGs through the use of environmental water indicators for measuring performance of the SDGs to demonstrate how healthy river ecosystems can deliver human health and well-being. One example of success with this type of approach is the recognition of the importance of environmental water for achieving the goals of the European Water Framework Directive, namely that all rivers in Europe achieve *good* ecological status by 2020. The European Commission is engaging with the environmental water science community to determine just how to achieve this; a guidance document has recently been published by a pan-European ecological flows group (European Commission, 2015). A further challenge lies in providing accessible and transferrable tools and techniques to achieve these goals (discussed further in Section 27.5).

Until recently, environmental water expertise and experience was held by a few groups distributed around the world, principally in North America, Europe, South Africa, Australia, and New Zealand. Capacity building has been increasingly incorporated in the implementation of environmental water projects (e.g., Acreman et al., 2006). In parallel, environmental water issues have been incorporated into academic teaching and research programs. In addition, programs such as the International Union for Conservation of Nature's Water and Nature Initiative have provided workshops and involvement in environmental water projects in 12 river basins covering 30 countries worldwide (http://www.waterandnature.org). These forums play a role in building awareness of the issue of environmental water and its importance at a global scale. Similarly, aid organizations, universities, and NGOs have built capacity in China and India (Gippel and Speed, 2010; Gopal, 2013). Specific training has been complemented by broader activities such as the establishment of environmental water networks. However, short-term funding has limited ongoing viability of such networks.

27.4 **HOW CAN WE EMBED ENVIRONMENTAL WATER MANAGEMENT AS A CORE ELEMENT OF WATER RESOURCE PLANNING?**

Successful implementation of environmental water management requires institutional structures that facilitate the allocation of environmental water, but importantly, integrate this with the broader area of water resource management and catchment/land-use planning. In its most basic form, this requires institutions to account for and monitor what water is available and where that water is going (Chapter 16). In many countries, water policy is developing faster than the data infrastructure to support policies, with the result that policy implementation often occurs in a data vacuum (Chapter 16). Designing and funding these information systems will be an essential element for increasing the implementation of environmental water.

To implement effective and sustainable reforms to provide environmental water, environmental mainstreaming is required; "the informed inclusion of relevant environmental concerns into the decisions of institutions that drive national and sectoral development policy, rules, plans, investment and action" (Dalal-Clayton and Bass, 2009). As a result of the initial advocacy role environmental organizations played in establishing the importance of instream river health, environmental water management is often siloed within water resource organizations. This limits the ability for novel integrated solutions and effective policy debate (Dalal-Clayton and Bass, 2009). Ideally, the governance structure for managing water resources would support concepts of conjunctive water uses (maximizing environmental outcomes from the delivery of water to consumptive uses as discussed in Chapter 21), rather than the usual competitive paradigm that is reinforced through institutional boundaries delineating environmental versus productive uses (Richter, 2010, 2014). Integrating environmental water within broader water resource management would also allow more nimble and informed responses to change (Dalal-Clayton and Bass, 2009).

As climate change impacts continue to create great uncertainty about (and change in) water availability, active discussions will be required around how these changes to supply are distributed among water users, and what changes are required to deliver water for both societal and environmental purposes (Poff et al., 2016). The institutional arrangements and environmental water allocation mechanisms should reflect conscious societally informed decisions around how water resources will be managed adaptively over time (see Chapters 17 and 19).

There are also challenges and opportunities for developing synergies between urban and rural water management, and development regimes that address keeping water in the landscape to reflect natural patterns of hydrological water balance (Grafton et al., 2015). Specifically, impervious urban areas typically have issues with excess flow production and insufficient capacity to retain and infiltrate water in the environment during storm events, whereas many rural areas have attenuated flows due to greater (constructed) water storage capacity and irrigation demand.

Importantly, water resource policy extends well beyond the government sectors that manage water resources, with many other sectors relying on water or influencing water resource outcomes. For example, energy, food, and water policies are all inherently linked. All efforts need to be made to ensure that they are coordinated with a shared vision so that policies from one sector do not adversely impact on other sectors and rather attempt to find mutual benefits (Hussey and Pittock, 2012; Pittock et al., 2013). The processes established to do this must be nimble enough not to hinder water resource policy with layers of bureaucracy.

27.5 HOW CAN KNOWLEDGE AND EXPERIENCE BE TRANSFERRED AND SCALED?

The challenge and urgency of protecting environmental water regimes is global, but significant advances in environmental water science and practice have been unevenly distributed among countries and global biophysical, social, cultural, and political settings (McClain and Anderson, 2015). A present-day cartogram of scientific knowledge regarding ecological responses to flow alteration and efforts to implement best environmental flows assessment practices would be heavily skewed toward North America, Europe, and Australia (Konrad et al., 2011; Poff and Zimmerman, 2010). In Africa, the Republic of South Africa passed ground-breaking legislation explicitly protecting river flows for basic human needs and ecosystems (RSA, 1998), and similar legislation has spread to neighboring countries in southern and eastern Africa (GoZ, 2002, 2011). South Africa has also contributed significantly to the development of holistic methodologies for environmental flows assessment (King et al., 2003, 2008), although implementation has been limited in that country. In Asia, China is emerging as a national leader in both environmental flows assessment research and cooperation between researchers and resource managers to improve practice in water resources management (Hou et al., 2007; Opperman and Guo, 2014; Sun et al., 2008). Research in the Mekong has significantly advanced knowledge of the coupled effects of basin-scale flow alteration on ecosystem function and human livelihoods (Molle et al., 2012; Ziv et al., 2012). In Latin America, Mexico has recently launched a scientifically rigorous and politically innovative effort to protect environmental water regimes nationwide, but research and practice across the rest of Latin America has been limited and irregular. Little or no information is available for other parts of the world.

Certainly, some prerequisites for achieving universal environmental water implementation are more ubiquitous. For example, United Nations processes emphasizing sustainable development over nearly 30 years (Drexhage and Murphy, 2010) have resulted in supportive international conventions and nearly universal reform of national laws and policies to require ecosystem protection. Dating from even earlier efforts, policy and practice in most countries acknowledge that a minimum level of flow should be maintained in rivers regardless of mounting pressures from other water uses. This perspective derives mainly from 19th- and 20th-century efforts to protect fishery resources and water quality in the interest of livelihoods and public health (Chandler, 1873; State of California, 1914). Although the specified minimum flows (e.g., Q95, Q10, or 10% of mean annual runoff) are generally insufficient to achieve modern social–ecological objectives, they do ensure that some amount of flow remains in perennial rivers everywhere. This international policy reform and acknowledgment of minimum flow requirements forms the foundation on which greater environmental water regime protection can be built.

Knowledge and experience gradually spread globally via numerous formal and informal mechanisms embedded in academic, professional, and governmental processes. However, the rapid pace of water resource development worldwide, and corresponding decline in freshwater ecosystems, require new strategic action on the part of the environmental water community to accelerate the pace of knowledge transfer, applied research, and implementation on the ground. Two interrelated elements are proposed as the pillars of an international strategy, with peripheral actions to extend the impact more broadly. The first element is to mainstream environmental water science and practice into major international science and development initiatives, thereby raising the profile of the topic and engaging the highest echelon of scientists and highest levels of practitioners and decision makers. Second, is to establish a global network of living laboratories (and related community of practice) at locations where advanced science, tools, and practice are being codesigned and codeveloped. These living laboratories could also include demonstration sites of best practice scaled to local conditions.

As mentioned earlier, the global community has agreed to the United Nations Sustainable Development Agenda 2030, which includes specific targets for freshwater ecosystem protection under Goal 6 and coastal resources under Goal 14 (http://www.un.org/sustainabledevelopment/). Environmental water requirements appear explicitly in indicators 6.4.2 (Level of water stress: Freshwater withdrawal as a percentage of available freshwater resources) and 6.6.1 (Change in the extent of water-related ecosystems over time; UN Water, 2016). These goals and targets align with the long-standing efforts of the Convention on Biological Diversity, Ramsar Convention, DIVERSITAS, and the more recent Intergovernmental Science-Policy Platform on Biodiversity and Ecosystem Services. Together these initiatives represent a strong international commitment to protect and sustainably develop freshwater and coastal resources; they are also an invitation to the science community to engage in research supporting the achievement of stated environmental goals. In response, the international science community has launched multiple coordinated research initiatives. Examples include the eighth phase (2014–21) of UNESCO's International Hydrological Programme; the decade *Change in Hydrology and Society* of the International Association of Hydrological Sciences (2013–22); and Future Earth, the new program of the International Council for Science. Elements of water research appear in each of these initiatives, and their continued growth in conjunction with policy initiatives holds promise for extending environmental water science and practice more widely. Other efforts and initiatives appear at national and regional scales (Arthington et al., 2010).

Efforts to transfer and expand knowledge and experience in environmental flows assessment and water management should be further enhanced by establishing a network of living laboratories and demonstration sites focused on the science and practice of environmental water management. The term *living labs* refers to structured collaborations between researchers and the users of research outputs to codesign and codevelop research activities (Van der Walt et al., 2009). This approach helps to embed environmental water research in ongoing water resource management and policy making so that research products are better tailored to the needs of end users and can be implemented more effectively and efficiently. An early example of this approach is the Sustainable Rivers Project, which partnered scientists from The Nature Conservancy with water managers of the US Army Corps of Engineers to investigate and demonstrate how dams could be reoperated to meet improved ecological objectives among multiple options (Warner et al., 2014; see also Chapter 25). Today there are examples of such partnerships around the world. Linking these efforts in a coordinated network would provide the research infrastructure to apply commonly adopted approaches to hypothesis setting and align monitoring programs regionally and globally, converting quality data into usable tools.

Increased attention must also be given to better understand the social dimensions of environmental water management. Wescoat (2009) argued that the 21st century would be a period of increasing circulation of expertise about water challenges and governance responses. Environmental water management is a prime example, where researchers and practitioners have exchanged lessons about the policy and governance approaches to establish and manage environmental water across diverse geographical and governance settings, and with different degrees of water scarcity over time. Successful environmental water management requires effective frameworks and methods for policy transfer, or preferably translation (Mukhtarov and Daniell, 2017), that allow for adaptations of the policies to fit the different contexts, nature of the problems, and the range of relevant policy settings and instruments (Swainson and de Loe, 2011). For this reason, Australia and the United States have been extensively compared, illustrating the contrasting roles of public and private organizations in the acquisition of delivery of environmental water in river systems, coupled with the common need for resources, accountability, and legitimacy (Garrick et al., 2009; Horne et al., 2008). A network of living labs offers a further opportunity to forge communities of practice needed to translate lessons and learn from similarities and differences in both rural and urban settings. The frameworks, approaches, and case studies collected by this book demonstrate this potential.

The peer-reviewed scientific literature is an essential dissemination channel for credible research results, but it is insufficient to reach the full spectrum of international scientists, practitioners, and decision makers. Living labs or other mechanisms to share successes, challenges, and tools for implementation allow outcomes to be tailored to those implementing environmental water regimes. Importantly, they also allow discussion of the institutional arrangements needed to support environmental water implementation.

27.6 HOW CAN WE ENHANCE THE LEGITIMACY OF ENVIRONMENTAL WATER PROGRAMS?

The primary normative guideline for governance. . . is legitimacy.

Wolf (2002, p. 40)

Measuring success for environmental water programs has typically focused on two metrics: effectiveness and efficiency. Did the environmental water provided perform the ecological function that was required, for instance effectiveness (McCoy, 2015)? And did it do so efficiently, for instance at a cost that was acceptable, and minimized where practicable (e.g., Aylward et al., 2016)? In debates about the need for environmental water, these metrics are essential to the decision to provide water to the environment, especially when doing so means that it is not provided to other consumptive uses. Indeed, one of the signs of increasing maturity in environmental water management is that it now actively considers whether the environmental water is being provided in a way that maximizes environmental benefit and minimizes costs (Garrick, 2015; Horne et al., 2016; Pittock and Lankford, 2010). However, to date there has been limited reporting on the broader social benefits of healthy rivers in the landscape as a result of environmental watering activities.

We are starting to understand that legitimacy is crucial to the long-term success of environmental water programs. In 2015, the *Organization for Economic Cooperation and Development (OECD)* identified trust and engagement, alongside efficiency and effectiveness, as the core elements of good water governance (OECD, 2015). Chapter 26 defines legitimacy as both *input* and *output* legitimacy

(Scharpf, 1999). Output legitimacy includes measures of efficacy, and is often relied on to assert the legitimacy of environmental water programs: they are considered to be legitimate because they worked. However, output legitimacy also includes awareness, acceptance, and a common approach to shared problem solving (Hogl et al., 2012), all of which depend on the *process* used to implement the environmental water program. The process is a function of input legitimacy, and requires explicit consideration of access, equal representation, transparency, accountability, consultation and cooperation, independence, and credibility (Hogl et al., 2012).

Chapter 26 places legitimacy alongside effectiveness and efficiency as one of the core measures of success for environmental water programs. There are two important practical lessons that emerge from this inclusion of legitimacy. First, it is essential to invest in building legitimacy at every stage in an environmental water program from environmental water agenda setting and the environmental flows assessment, through to the mechanisms to provide them, to the ongoing implementation and management of environmental water. In particular, this requires acknowledgment that the first step for all environmental water programs is the development of a shared awareness and acceptance by stakeholders that the environment itself needs water in particular quantities, timing, and qualities to meet specific objectives, including those of ecosystem maintenance and functionality. Failing to take the time to ensure that such legitimacy for the program has been established at this point of agenda setting may undermine all future phases of policy making and implementation of the proposed program. Indeed, it may leave it vulnerable to defeat in a political battle where a coalition of stakeholders opposing the environmental water has its agenda represented in the policy decisions (Daniell et al., 2014). It may also lead to feelings of injustice if some stakeholders consider that they have been treated unfairly by the development of the environmental water program, having not been adequately involved in its development (Lukasiewicz and Dare, 2016). We now understand that legitimacy is much more than a function of the outcomes of an environmental water program or the quality of the scientific evidence underpinning it, and is unlikely to be built without significant investment in stakeholder engagement and communication (see Chapter 7).

However, building and maintaining legitimacy may well increase the upfront cost of an environmental water program. Garrick and O'Donnell (2016) showed that investing in legitimacy may increase the transaction costs of water recovery for the environment, and may extend the duration of water recovery programs, so that it takes more time and more money to reach the intended goal. However, failing to invest in legitimacy, specifically in the implementation of a range of stakeholder engagement methods to build and maintain it (Daniell, 2011), may undermine an otherwise successful environmental water recovery program at a later time, deferring costs until the future. For example, the environmental water recovery program in the Murray—Darling Basin in Australia used water trading to rapidly acquire large volumes of water for the environment. However, this water recovery came to be viewed as deeply problematic by irrigators, who successfully lobbied the government to impose a new, lower limit on the volume of water that could be recovered for the environment by purchasing it from other users (The Hon. Greg Hunt MP and The Hon. Bob Baldwin MP, 14 September 2015). Remaining environmental water to be recovered under the Basin Plan must be secured through infrastructure-based efficiency measures, which are more expensive than direct purchases, and may not deliver the desired ecological benefits (Bond et al., 2014). Thus, the failure to secure the support of other water users will now be a substantial cost to the Basin Plan water recovery program. In the Columbia River Basin in North America, initial implementation of environmental water regimes involved acquisition of water rights for instream uses under the principle of *buy-and-dry*, triggering political and legal resistance from communities affected by the reduction of irrigation and associated industries, and threatening to reverse the regulatory reforms enabling environmental water transactions (Pilz, 2006). In the decade since, there has been considerable work to rebuild the legitimacy and trust that had been eroded, including an explicit rejection of the buy-and-dry philosophy by the National Fish and Wildlife Foundation and the Columbia Water Transactions Program.

Developing effective stakeholder engagement processes can thus be one important means of enhancing the legitimacy of an environmental water program. However, building a real partnership between stakeholders, so that they are all committed to achieving a successful environmental water outcome, takes time, effort, trust, and humility (see Chapter 7). It requires experts to admit that they may not have all the answers, and policy makers and practitioners to listen to the diversity of views that can inform environmental water management. One often missing element is a specific investment in building partnerships with indigenous peoples, particularly in the context of historical colonization and disenfranchisement (Robinson et al., 2015). The legitimacy of an environmental water program will likely depend on giving all voices equal opportunity to be heard and to influence the design and implementation of the program, accepting that there are many different forms of knowledge. The acceptance by practitioners and scientists that environmental water management is primarily a social process within a wider complex social–ecological system as opposed to a scientific management process has been slow to resonate within the field (Arthington, 2015). Participatory decision making thus remains poorly integrated into many existing environmental water policies and programs. However, there are increasing numbers of instances where stakeholder engagement has been adequately funded and integrated into such programs. These include instances where stakeholder engagement is a compulsory part of legal frameworks such as in the European Union Water Framework Directive (EU, 2000, 2002; see also von Korff et al., 2012), and where conflict and transaction costs have been reduced and legitimacy enhanced through stakeholder engagement investments in environmental water (see Chapter 7).

Chapter 19 introduced the concept of active management of environmental water, which requires environmental water organizations to choose how to use environmental water to achieve the best outcomes each year. This flexibility is extremely important for effectiveness and efficiency, but it also requires the decision maker to actively engage with stakeholders and local communities both before and after decisions are made and implemented. Environmental water organizations are getting better at making these decisions transparent, but real legitimacy requires expanding the sphere of influence, so that local communities are invested in making the best decisions for environmental water in their local context. By focusing on both input (the process) and output (the outcome) legitimacy, environmental water policy makers and practitioners can embed legitimacy throughout the environmental water management process.

To facilitate the adoption of legitimacy as a measure of success, there are two foci for potential future activity. First, we should build a database of demonstration catchments that show how legitimacy has (or has not) been embedded and resourced throughout environmental water programs. Despite the continuing challenges of implementation, environmental water programs are now sufficiently widespread and diverse that the capacity for collective learning is enormous (see Chapter 25). Second, where environmental water programs are already underway, policy makers and practitioners need to actively include legitimacy as a measure of success, and report on both input and output legitimacy. There are many ways to monitor and report on legitimacy. A starting point might be using indicators such as media articles (positive and negative), and short surveys to report stakeholder numbers, diversity of stakeholder groups, and to gauge the opinions of stakeholders involved in the ongoing program.

27.7 CAN ADAPTIVE MANAGEMENT BECOME STANDARD PRACTICE?

Adaptive management centers on the concept of iterative learning resulting in improvements in management (Allan and Stankey, 2009; Pahl-Wostl et al., 2007). It is essential in situations where management decisions must be made under uncertainty. Under the umbrella of adaptive management, there is the simple approach of *learning by doing* through to more complex and rigorous processes involving management experiments and sophisticated ecological response models (Allan and Stankey, 2009). Adaptive management is particularly well suited to problems such as environmental water management where the outcomes are responsive to management decisions, but there is uncertainty about the outcomes of alternative decisions (Williams and Brown, 2014). There are multiple sources of uncertainty affecting environmental water management, including climatic uncertainty affecting future water availability and demands for consumptive use, and scientific uncertainty concerning ecological responses to changing patterns of flow variability (see Chapter 15). Despite the well-recognized potential benefits of adaptive management, few successful examples have been reported (see Chapter 25), although the extent to which this reflects limited successes or simply limited formal assessment and reporting of successes is unclear (Allan and Watts, in review). There are a number of particular challenges that need to be considered to support adaptive management becoming standard practice within environmental water management.

First, and perhaps the most significant challenge for adaptive management, is establishing the legitimacy of the environmental water program, including both social and institutional settings that allow for both success and failures, include a focus on learning, and provide the necessary funds to support the effort (Poff et al., 2003). Providing such institutional arrangements will be essential to the success of adaptive management (Ladson, 2009). One of the key benefits of adaptive management is its potential to facilitate learning through a structured dialogue between scientists and managers (Ladson, 2009; Pahl-Wostl et al., 2007). There has been significant progress in the uptake of knowledge from scientific literature into environmental flows assessment methods (see Chapter 11–14); however, there remains some distance between the growing body of scientific research and translation of this knowledge to address the information needs of managers (Acreman, 2005; Williams and Brown, 2014, 2016). Achieving the full benefit of adaptive management will require novel collaborative arrangements between scientists and managers, as well as with other stakeholders, including local communities, to better incorporate a range of knowledge bodies into management processes (Vietz et al., in review).

Second, documentation of hypotheses, decisions, and outcomes of adaptive management must be improved to facilitate learning and knowledge transfer across catchments. Chapter 25 recommends the inclusion of *reflectors* in adaptive management teams to document and disseminate learnings, and Allan and Stankey (2009) suggest "careful documentation processes" is one of the core elements of successful adaptive management. This requires a documented hypothesis or predictive model that links alternative management actions to management objectives, which in the case of environmental water requirements, links flow decisions to environmental objectives (Allan and Stankey, 2009; Williams and Brown, 2014). The model need not be numeric; it could even take the form of a list of predictions under different management options (i.e., suggested as an option in ELOHA; Poff et al., 2010). The documentation of the predictions, not so much how they are arrived at, is the important aspect. This documented model, complete with its inherent uncertainties, plays an essential role in presenting researchers' and managers' understanding of how a system behaves, and in building consensus and shared understanding among those involved in the management process (Beven and Alcock, 2012; Liebman, 1976). There has been considerable growth in the number of scientific publications examining the environmental effects of flow alteration (Poff and Zimmerman, 2010; Stewardson and Webb, 2010; Webb et al., 2013); however, the challenges of linking this scientific knowledge to management decisions are considerable (Acreman, 2005). In the case of environmental flows assessments, while there are methods that support transparent empirically based frameworks (i.e., ELOHA and DRIFT), the majority of environmental water regime recommendations and many management decisions are currently based on expert judgments that draw from the experts' cumulative experience and understanding of current literature (Stewardson and Webb, 2010). Often there is an implicit conceptual model that captures the causal pathways that underlie these expert judgments, but this is rarely documented as part of the decision-making process. The explicit representation of such models is a crucial element of adaptive management of environmental water, and one that is particularly important for active environmental water management.

Finally, monitoring and evaluation is an essential element of adaptive management, but traditional agency-led programs are time-consuming and expensive (Williams and Brown, 2014). Without monitoring, there can be no adaptive learning, no way to complete the adaptive management cycle, and no way to update future management in the light of new knowledge. One reason identified for the failure of adaptive management is the unfortunately common lack of commitment to monitoring by management agencies (Schreiber et al., 2004). These monitoring programs need to support both the inner and outer adaptive management loops (see Chapter 25) of short-term implementation and long-term planning. Effective monitoring can help to demonstrate the benefits of providing environmental water (enhancing legitimacy), and can also add to the growing knowledge around ecological responses with flow (see Chapter 25). The design, funding, and administration of such monitoring programs needs to be identified as early as possible, and a commitment made to long-term engagement (Davies et al., 2014). Citizen science efforts at developing models and monitoring their outcomes could further enhance the potential of intelligence gathering, dissemination, and learning to underpin effective adaptive management (e.g., Liu et al., 2014). Exploring options to enhance the resourcing, local support, and implementation of monitoring has the potential to allow adaptive management to occur in places where the monitoring cost may otherwise be prohibitive.

27.8 CONCLUSION

Clearly, on-the-ground implementation of policy aspirations is the foremost global challenge to achieving environmental flows.

Le Quesne et al. (2010)

There has been significant progress in environmental water management across the globe as evidenced by the increasing scientific knowledge and the number of countries that now recognize environmental water in their policies and legislation. Despite this progress, there remains a persisting implementation challenge, even as the need for political will and renewed policy commitments and resources becomes increasingly urgent. It is difficult to assess the progress and remaining challenges, as there is no central repository of information on the level of environmental water regime implementation across the globe. Moving forward, a benchmarking study would allow greater assessment of progress and improve opportunities to share and coordinate across regions.

There is, however, a range of examples of successful environmental water management and ongoing research into the challenges of environmental water management as highlighted throughout this book. There are great opportunities to build on this experience through better mechanisms to translate this knowledge across geographies and policy settings. One possible approach is to establish living labs as a means to communicate science, management tools, and social engagement strategies that have been successful (or unsuccessful!), and the journey of implementation over time. An important aspect of making the knowledge and tools gained at living labs transferrable is to develop a consistent framework and language.

There is an inherent trade-off in the implementation of environmental water management between pragmatism and efficiency on the one hand, and aiming for the *best case solution*. Indeed, in writing this chapter it was apparent that different authors have different opinions on the correct balance. In reality, this is a location-specific question. It is important to recognize that in those countries where environmental water regimes have been implemented, it has often been a lengthy and iterative process. Whatever the first step is (be it a precautionary and readily implemented option, or a well-researched and consultative option), the process should allow for learning and changes over time. These changes may be in response to changing social values, changing climate, or new knowledge. A key aspect of adaptive environmental water management will be ensuring that suitable institutional and governance arrangements are in place, and that the program maintains (or establishes) legitimacy. This is an aspect of environmental water management that still requires research and demonstration in the field. Importantly, getting the institutional and governance arrangements right will ensure ongoing funding and community support for an environmental water program (Pahl-Wostl et al., 2013).

This book has provided an overview of the complete environmental water management process—*from policy and science to implementation and management*. In this final chapter, we have highlighted some of the reoccurring themes throughout the book and discussed these in terms of the ongoing implementation challenge. Many of these challenges are not technical in nature, but rather related to concepts of engagement, partnership, legitimacy, sharing of knowledge, and enabling institutional structures. This highlights the importance of Ostrom's (1990) concepts of engagement and social agreements (as introduced in Chapter 1), and provides a positive lens to work through toward the sustainable management of our water resources.

The most powerful force ever known on this planet is human cooperation—a force for construction and destruction.

Jonathan Haidt (2012)

REFERENCES

Acreman, M., 2005. Linking science and decision-making: features and experience from environmental river flow setting. Environ. Model. Softw. 20, 99–109.

Acreman, M., Dunbar, M.J., 2004. Defining environmental river flow requirements—a review. Hydrol. Earth Syst. Sci. 8, 861–876.

Acreman, M., Arthington, A.H., Colloff, M.J., Couch, C., Crossman, N.D., Dyer, F., et al., 2014. Environmental flows for natural, hybrid, and novel riverine ecosystems in a changing world. Front. Ecol. Environ. 12, 466–473.

Acreman, M.C., King, J., Hirji, R., Sarunday, W., Mutayoba, W., 2006. Capacity building to undertake environmental flow assessments in Tanzania. *Proceedings of the International Conference on River Basin Management, Morogorro, Tanzania, March 2005.* Sokoine University, Morogorro. Available from: https://cgspace.cgiar.org/handle/10568/36183.

Allan, C., Stankey, G., 2009. Synthesis of lessons. In: Allan, C., Stankey, G. (Eds.), Adaptive Envionmental Management: A Practitioner's Guide. Springer Science.

Allan, C.A., Watts, R.J., in review. Revealing adaptive management of environmental flows. Environ. Manage.

Arthington, A., 2015. Environmental Flows: Saving Rivers in the Third Millennium. University of California Press, Berkley, California.

Arthington, A.H., Bunn, S.E., Poff, N.L., Naiman, R.J., 2006. The challenge of providing environmental flow rules to sustain river ecosystems. Ecol. Appl. 16, 1311–1318.

Arthington, A.H., Naiman, R.J., McClain, M.E., Nilsson, C., 2010. Preserving the biodiversity and ecological services of rivers: new challenges and research opportunities. Freshw. Biol. 55, 1–16.

Aylward, B., Pilz, D., Kruse, S., McCoy, A.L., 2016. Measuring cost-effectiveness of environmental water transactions. Ecosystem Economics LLC. Report prepared for California Coastkeeper Alliance and Lkamath Riverkeeper.

Beven, K.J., Alcock, R.E., 2012. Modelling everything everywhere: a new approach to decision-making for water management under uncertainty. Freshw. Biol. 57, 124–132.

Bond, N., Costelloe, J., King, A., Warfe, D., Reich, P., Balcombe, S., 2014. Ecological risks and opportunities from engineered artificial flooding as a means of achieving environmental flow objectives. Front. Ecol. Environ. 12, 386–394.

Bovee, K., Milhous, R., 1978. Hydraulic Simulation in Instream Flow Studies: Theory and Techniques. US Fish and Wildlife Service, Fort Collins, CO.

Bunn, E.S., Arthington, A., 2002. Basic principles and ecological consequences of altered flow regimes for aquatic biodiversity. Environ. Manage. 30, 492–507.

Calderwood, K., 2016. Why New Zealand is granting a river the same rights as a citizen. Radio National, Sunday Extra, Tuesday 6 September.

Chandler, C.F., 1873. Report Upon the Sanitary Chemistry of Waters, and Suggestions with Regard to the Selection of the Water Supply of Towns and Cities. American Public Health Association.

Christie, M., Cooper, R., Hyde, T., Fazey, I., 2012. An evaluation of economic and non-economic techniques for assessing the importance of biodiversity and ecosystem services to people in developing countries. Ecol. Econ. 83, 67–78.

Dalal-Clayton, B., Bass, S., 2009. The Challenges of Environmental Mainstreaming – Experience of Integrating Environment into Development Institutions and Decisions. International Institute for Environment and Development, London.

Daniell, K.A., 2011. Enhancing collaborative management in the Basin. In: Connell, D., Grafton, R.Q. (Eds.), Basin Futures: Water Reform in the Murray-Darling Basin. ANU E-Press, Canberra, Australia.

Daniell, K.A., 2015. Designing stakeholder engagement processes for river basin management: using culture as an analytical tool. Proceedings of the 36th Hydrology and Water Resources Symposium "The Art and Science of Water". Engineers Australia, Hobart, Australia.

Daniell, K.A., Coombes, P.J., White, I., 2014. Politics of innovation in multi-level water governance systems. J. Hydrol. 519, 2415−2435.

Davidson, M.D., 2013. On the relation between ecosystem services, intrinsic value, existence value and economic valuation. Ecol. Econ. 95, 171−177.

Davies, P.M., Naiman, R.J., Warfe, D.M., Pettit, N.E., Arthington, A.H., Bunn, S.E., 2014. Flow-ecology relationships: closing the loop on effective environmental flows. Mar. Freshw. Res. 65, 133−141.

Drexhage, J., Murphy, D., 2010. Sustainable development: from Brundtland to Rio 2012. Background paper prepared for consideration by the High Level Panel on Global Sustainability at its first meeting, 19 September 2010.

Dudgeon, D., 2014. Accept no substitute: biodiversity matters. Aquat. Conserv. Mar. Freshw. Ecosyst. 24, 435−440.

Dunbar, M.J., Acreman, M.C., 2001. Applied hydro-ecological science for the 21st Century. In: Acreman, M.C. (Ed.), Hydro-ecology: Linking Hydrology and Aquatic Ecology, 1–18. IAHS Publ. 266, IAHS Press, Wallingford, UK.

Enserink, B., Patel, M., Kranz, N., Maestu, J., 2007. Cultural factors as co-determinants of participation in river basin management. Ecol. Soc. 12.

EU, 2000. Directive 2000/60/EC of the European Parliament and of the Council, of 23 October 2000: establishing a framework for Community action in the field of water policy, L 327, 22.12.2000. OJEC, 1−72.

EU, 2002. Guidance on public participation in relation to the Water Framework Directive: active involvement, consultation, and public access to information. Final version after the Water Directors' meeting, December 2002.

European Commission, 2015. Ecological Flows in the Implementation of the WFD. European Commission, Brussels, Belgium.

Fazey, I., Fischer, J., Lindenmayer, D.B., 2005. Who does all the research in conservation biology? Biodivers. Conserv. 14, 917−934.

Finn, M., Jackson, S., 2011. Protecting Indigenous values in water management: a challenge to conventional environmental flow assessments. Ecosystems 14, 1232−1248.

Garrick, D., 2015. Water Allocation in Rivers Under Pressure: Water Trading, Transaction Costs, and Transboundary Governance in the Western USA and Australia. Edward Elgar, Cheltenham, UK.

Garrick, D., O'Donnell, E., 2016. Exploring private roles in environmental watering in Australia and the US. In: Bennett, J. (Ed.), Protecting the Environment, Privately. World Scientific Publishing.

Garrick, D., Siebentritt, M.A., Aylward, B., Bauer, C.J., Purkey, A., 2009. Water markets and freshwater ecosystem services: policy reform and implementation in the Columbia and Murray-Darling Basins. Ecol. Econ. 69, 366−379.

Garrick, D., Lane-Miller, C., McCoy, A.L., 2011. Institutional innovations to govern environmental water in the western United States: lessons for Australia's Murray−Darling Basin. Econ. Pap. 30, 167−184.

GBCMA, 2014. Goulburn River: Seasonal Watering Proposal 2014−15. Goulburn-Broken Catchment Management Authority, Shepparton, Australia.

Gilvear, D.J., Webb, J.A., Smith, D.L., Nestler, J.M., Stewardson, M.J., 2016. Ecohydraulics exemplifies the emerging "paradigm of the interdisciplines". J. Ecohydraul. 1, 1−2.

Gippel, C.J., Speed, R., 2010. Environmental flow assessment framework and methods, including environmental asset identification and water re-allocation. ACEDP (Australia-China Environment Development Partnership), River Health and Environmental Flow in China, International Water Centre, Brisbane, Australia.

Gippel, C.J., Cosier, M., Markar, S., Liu, C., 2009. Balancing environmental flows needs and water supply reliability. Int. J. Water Resour. Dev. 25, 331−353.

GoK, 2002. The Water Act. Government Printer, Government of Kenya, Nairobi, Kenya.

Gopal, B. (Ed.), 2013. Environmental Flows: an Introduction for Water Resources Managers. National Institute of Ecology, New Delhi, India.

GoZ, 2011. The Water Resources Management Act. Government Printer, Government of Zambia, Lusaka, Zambia.

Grafton, R.Q., Daniell, K.A., Nauges, C., Rinaudo, J.-D., Chan, N.W.W. (Eds.), 2015. Understanding and Managing Urban Water in Transition. Springer, Dordrecht, The Netherlands.

Harman, C., Stewardson, M., 2005. Optimizing dam release rules to meet environmental flow targets. River Res. Appl. 21, 113−129.

Hart, B.T., 2015. The Australian Murray−Darling Basin Plan: challenges in its implementation. Int. J. Water Resour. Dev. 32, 819−834.

Hirji, R., Davis, R., 2009. Environmental Flows in Water Resources Policies, Plans and Projects: Findings and Recommendations. The International Bank for Reconstruction and Development/World Bank, Washington, DC.

Hogl, K., Kvarda, E., Nordbeck, R., Pregernig, M. (Eds.), 2012. Environmental Governance: the Challenge of Legitimacy and Effectiveness. Edward Elgar, Cheltenham, UK.

Horne, A., Purkey, A., Mcmahon, T.A., 2008. Purchasing water for the environment in unregulated systems— what can we learn from the Columbia Basin? Aust. J. Water Resour. 12, 61−70.

Horne, A., Szemis, J.M., Kaur, S., Webb, J.A., Stewardson, M.J., Costa, A., et al., 2016. Optimization tools for environmental water decisions: a review of strengths, weaknesses, and opportunities to improve adoption. Environ. Model. Softw. 84, 326−338.

Horne, A., Szemis, J., Webb, J.A., Kaur, S., Stewardson, M., Bond, N., et al. (in press). Informing environmental water management decisions: using conditional probability networks to address the information needs of planning and implementation cycles. Environ. Manage.

Hou, P., Beeton, R.J.S., Carter, R.W., Dong, X.G., Li, X., 2007. Response to environmental flows in the lower Tarim River, Xinjiang, China: ground water. J. Environ. Manage. 83, 371−382.

Hussey, K., Pittock, J., 2012. The energy−water nexus: managing the links between energy and water for a sustainable future. Ecol. Soc. 17, 31.

Kenter, J.O., Hyde, T., Christie, M., Fazey, I., 2011. The importance of deliberation in valuing ecosystem services in developing countries—Evidence from the Solomon Islands. Glob. Environ. Change 21, 505−521.

King, J., Tharme, R., Brown, C., 1999. Definition and implementation of instream flows. Contributing paper: dams, ecosytem functions and environmental restoration. World Commission on Dams.

King, J., Brown, C., Sabet, H., 2003. A scenario-based holistic approach to environmental flow assessments for rivers. River Res. Appl. 19, 619−639.

King, J.M., Tharme, R.E., De Villiers, M.S., 2008. Environmental flow assessments for rivers: manual for the Building Block Methodology. Water Research Commission, Pretoria, Republic of South Africa.

Konrad, C.P., Olden, J.D., Lytle, D.A., Melis, T.S., Schmidt, J.C., Bray, E.N., et al., 2011. Large-scale flow experiments for managing river systems. BioScience 61, 948−959.

Kopf, R.K., Finlayson, C.M., Humphries, P., Sims, N.C., Hladyz, S., 2015. Anthropocene baselines: assessing change and managing biodiversity in human-dominated aquatic ecosystems. BioScience 65, 798−811.

Krieger, M.H., 1973. What's wrong with plastic trees? Rationales for preserving rare natural environments involve economic, societal, and political factors. Science 179, 446−455.

Ladson, T., 2009. Adaptive management of environmental flows − 10 years on. In: Allan, C., Stankey, G. (Eds.), Adaptive Environmental Management: a Practitioner's Guide. Springer Science.

Le Quesne, T., Kendy, E., Weston, D., 2010. The Implementation Challenge − Taking Stock of Government Policies to Protect and Restore Environmental Flows. The Nature Conservancy and WWF.

Liebman, J.C., 1976. Some simple-minded observations on the role of optimization in public systems decision-making. Interfaces 6, 102−108.

Liu, H.-Y., Kobernus, M., Broday, D., Bartonova, A., 2014. A conceptual approach to a citizens' observatory — supporting community-based environmental governance. Environ. Health 13, 1—13.

Lukasiewicz, A., Dare, M., 2016. When private water rights become a public asset: stakeholder perspectives on the fairness of environmental water management. J. Hydrol. 536, 183—191.

McClain, M.E., Anderson, E.P., 2015. The gap between best practice and actual practice in the allocation of environmental flows in integrated water resources management. In: Setegn, S.G., Donoso, M.C. (Eds.), Sustainability of Integrated Water Resources Management. Springer International Publishing.

McCoy, A.H.S.R., 2015. Columbia Basin Water Transactions Program Flow Restoration Accounting Framework. National Fish and Wildlife Foundation, Portland, Oregon.

Milhous, R., Updike, M., Schneider, D., 1989. Physical Habitat Simulation System Reference Manual—Version II. US Fish and Wildlife Service, Fort Collins, Colorado.

Molle, F., Foran, T., Kakonen, M., 2012. Contested Waterscapes in the Mekong Region: Hydropower, Livelihoods and Governance. Earthscan, Abington, UK, New York.

Moore, M., 2004. Perceptions and Interpretations of Environmental Flows and Implications for Future Water Resource Management — A Survey Study. Linkoping University, Linkoping, Sweden.

Mukhtarov, F., Daniell, K.A., 2017. Diffusion, adaptation and translation of water policy models. In: Conca, K., Weinthal, E. (Eds.), The Oxford Handbook of Water Politics and Policy. Oxford University Press, Oxford, UK.

Nilsson, C., Svedmark, M., 2002. Basic principles and ecological consequences of changing water regimes: riparian plant communities. Environ. Manage. 30, 468—480.

O'Donnell, E., 2013. Common legal and policy factors in the emergence of environmental water managers. In: Brebbia, C.A. (Ed.), Water and Society II. WIT Press, Southampton, UK.

OECD, 2012. Meeting the Water Reform Challenge.

OECD, 2015. OECD Principles on water governance: welcomed by Ministers at the OECD Ministerial Council Meeting on 4 June 2015. Online: Directorate for Public Governance and Territorial Development.

Olden, J.D., Konrad, C.P., Melis, T.S., Kennard, M.J., Freeman, M.C., Mims, M.C., et al., 2014. Are large-scale flow experiments informing the science and management of freshwater ecosystems? Front. Ecol. Environ. 12, 176—185.

Opperman, J., Guo, Q., 2014. Carp and collaboration. Stockholm Water Front 3.

Opperman, J.J., Grill, G., Hartmann, J., 2015. The power of rivers: finding balance between energy and conservation in hydropower development. The Nature Conservancy, Washington, DC. Available from: http://www.nature.org/media/freshwater/power-of-rivers-report.pdf (accessed 30.06.16).

Overton, I., Pollino, C., Roberts, J., Reid, J., Bond, N., McGinness, H., et al., 2014. Development of the Murray-Darling Basin Plan SDL adjustment ecological elements method. Report prepared by CSIRO for the Murray-Darling Basin Authority. CSIRO.

Pahl-Wostl, C., Sendzimir, J., Jeffrey, P., Aerts, J., Berkamp, G., Cross, K., 2007. Managing change toward adaptive water management through social learning. Ecol. Soc. 12 (2), 30.

Pahl-Wostl, C., Arthington, A., Bogardi, J., Bunn, S.E., Hoff, H., Lebel, L., et al., 2013. Environmental flows and water governance: managing sustainable water uses. Curr. Opin. Environ. Sustain. 5, 341—351.

Pilz, R.D., 2006. At the confluence: Oregon's instream water rights in theory and practice. Environ. Law. J. 36, 1383—1420.

Pittock, J., Lankford, B.A., 2010. Environmental water requirements: demand management in an era of water scarcity. J. Integrat. Environ. Sci. 7, 75—93.

Pittock, J., Hussey, K., McGlennon, S., 2013. Australian climate, energy and water policies: conflicts and synergies. Aust. Geogr. 44, 3—22.

Poff, N.L., Matthews, J., 2013. Environmental flows in the Anthropocene: past progress and future prospects. Curr. Opin. Environ. Sustain. 5, 667—675.

Poff, N.L., Zimmerman, J.K.H., 2010. Ecological responses to altered flow regimes: a literature review to inform the science and management of environmental flows. Freshw. Biol. 55, 194–205.

Poff, N.L., Allan, J.D., Bain, M.B., Karr, J.R., Prestegaard, K.L., Richter, B.D., et al., 1997. The natural flow regime. BioScience 47, 769–784.

Poff, N.L., Allan, J.D., Palmer, M.A., Hart, D.D., Richter, B.D., Arthington, A.H., et al., 2003. River flows and water wars: emerging science for environmental decision making. Front. Ecol. Environ. 1, 298–306.

Poff, N.L., Richter, B.D., Arthington, A.H., Bunn, S.E., Naiman, R.J., Kendy, E., et al., 2010. The ecological limits of hydrologic alteration (ELOHA): a new framework for developing regional environmental flow standards. Freshw. Biol. 55, 147–170.

Poff, N.L., Brown, C.M., Grantham, T.E., Matthews, J.H., Palmer, M.A., Spence, C.M., et al., 2016. Sustainable water management under future uncertainty with eco-engineering decision scaling. Nat. Clim. Change 6, 25–34.

Reidy Liermann, C.A., Olden, J.D., Beechie, T.J., Kennard, M.J., Skidmore, P.B., Konrad, C.P., et al., 2012. Hydrogeomorphic classification of Washington state rivers to support emerging environmental flow management strategies. River Res. Appl. 28, 1340–1358.

Richter, B.D., 2010. Re-thinking environmental flows: from allocations and reserves to sustainability boundaries. River Res. Appl. 26, 1052–1063.

Richter, B.D., 2014. Chasing Water: A Guide for Moving from Scarcity to Sustainability. Island Press, Washington, DC.

Richter, B.D., Thomas, G.A., 2007. Restoring environmental flows by modifying dam operations. Ecol. Soc. 12. Available from: http://www.ecologyandsociety.org/vol12/iss1/art12/.

Richter, B.D., Baumgartner, J.V., Wigington, R., Braun, D.P., 1997. How much water does a river need? Freshw. Biol. 37.

Richter, B.D., Warner, A.T., Meyer, J.L., Lutz, K., 2006. A collaborative and adaptive process for developing environmental flow recommendations. River Res. Appl. 22, 297–318.

Richter, B.D., Davis, M.M., Apse, C., Konrad, C., 2012. A presumptive standard for environmental flow protection. River Res. Appl. 28, 1312–1321.

Robinson, C.J., Bark, R.H., Garrick, D., Pollino, C.A., 2015. Sustaining local values through river basin governance: community-based initiatives in Australia's Murray–Darling basin. J. Environ. Plan. Manage. 58, 2212–2227.

RSA, 1998. National Water Act. Government Printer, Pretoria, Republic of South Africa.

Scharpf, F.W., 1999. Governing in Europe: Effective and Democratic? Oxford University Press, Oxford and New York.

Schreiber, E.S.G., Bearlin, A.R., Nicol, S.J., Todd, C.R., 2004. Adaptive management: a synthesis of current understanding and effective application. Ecol. Manage. Restor. 5, 177–182.

Shenton, W., Bond, N.R., Yen, J.D.L., MacNally, R., 2012. Putting the "ecology" into environmental flows: ecological dynamics and demographic modelling. Environ. Manage. 50, 1–10.

State of California, 1914. Fish and Game Commission twenty-third biennial report. California State Printing Office, California.

Stewardson, M., Webb, J., 2010. Modelling ecological responses to flow alteration: making the most of existing data and knowledge. In: Saintilan, N., Overton, I. (Eds.), Ecosystem Response Modelling in the Murray-Darling Basin. CSIRO Publishing, Melbourne, Australia.

Sun, T., Yang, Z.F., Cui, B., 2008. Critical environmental flows to support integrated ecological objectives for the Yellow River Estuary, China. Water Resour. Manage. 22, 973–989.

Swainson, R., De Loe, R.C., 2011. The importance of context in relation to policy transfer: a case study of environmental water allocation in Australia. Environ. Policy Govern. 21, 58–69.

Tennant, D.L., 1976. Instream flow regmines for fish, wildlife, recreastion and related environmental resources. Fisheries 1, 6–10.

Tharme, R.E., 2003. A global perspective on environmental flow assessment: emerging trends in the development and application of environmental flow methodologies for rivers. River Res. Appl. 19, 397−441.

The Hon. Greg Hunt MP, The Hon. Bob Baldwin MP, 2015. Joint media release: coalition delivers election commitment with 1500GL water buyback cap. Commonwealth of Australia, Canberra, Australia.

UN Water, 2016. Integrated monitoring guide for SDG 6 targets and global indicators. Available from: http://www.unwater.org/publications/publications-detail/en/c/424975/.

Van Der Walt, J.S., Buitendag, A.A., Zaaiman, J.J., Van Vuuren, J.J., 2009. Community living lab as a collaborative innovation environment. Issues Inform. Sci. Inform. Technol 6, 421−436.

Vietz, G.J., Lintern, A., Webb, J.A., Straccione, D., in review. River bank erosion and the influence of environmental flow management. Environ. Manage.

Von Korff, Y., Daniell, K.A., Moellenkamp, S., Bots, P., Bijlsma, R.M., 2012. Implementing participatory water management: recent advances in theory, practice, and evaluation. Ecol. Soc. 17 (1), 30.

Warner, A.T., Bach, L.B., Hickey, J.T., 2014. Restoring environmental flows through adaptive reservoir management: planning, science, and implementation through the Sustainable Rivers Project. Hydrol. Sci. J. 59, 770−785.

Webb, J.A., Miller, K.A., King, E.L., De Little, S.C., Stewardson, M.J., Zimmerman, J.K.H., et al., 2013. Squeezing the most out of existing literature: a systematic re-analysis of published evidence on ecological responses to altered flows. Freshw. Biol. 58, 2439−2451.

Wescoat, J.L., 2009. Comparative international water research. J. Contemp. Water Res. Educ. 142, 61−66.

Whittington, D., 2010. What have we learned from 20 years of stated preference research in less-developed countries? Ann. Rev. Resour. Econ. 2, 209−236.

Wilkes, M.A., Maddock, I., Visser, F., Acreman, M.C., 2013. Incorporating hydrodynamics into ecohydraulics: the role of turbulence in the swimming and habitat selection of river-dwelling salmonids. In: Kemp, P., Harby, A., Maddock, I., Wood, P.J. (Eds.), Ecohydraulics: an Integrated Approach. Wiley.

Williams, B.K., Brown, E.D., 2014. Adaptive management: from more talk to real action. Environ. Manage. 53, 465−479.

Williams, B.K., Brown, E.D., 2016. Technical challenges in the application of adaptive management. Biol. Conserv. 195, 255−263.

Winemiller, K., McIntyre, P., Castello, L., Fluet-Chouinard, E., Giarrizzo, T., Nam, S., et al., 2016. Balancing hydropower and biodiversity in the Amazon, Congo and Mekong. Science 351, 128−129.

Wolf, K.D., 2002. Contextualizing normative standards for legitimate governance beyond the state. In: Jürgen, R.G., Gbikpi, B. (Eds.), Participatory Governance: Political and Societal Implications. Leske and Budrich, Opladen, Germany.

Zarfl, C., Lumsdon, A.E., Berlekamp, J., Tydecks, L., Tockner, K., 2015. A global boom in hydropower dam construction. Aquat. Sci. 77, 161−170.

Ziv, G., Baran, E., Nam, S., Rodríguez-Iturbe, I., Levin, S.A., 2012. Trading-off fish biodiversity, food security, and hydropower in the Mekong River Basin. Proc. Natl. Acad. Sci. 109, 5609−5614.

MAPS

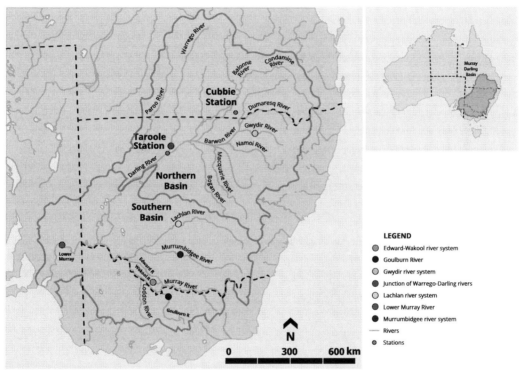

MAP 1:

The Murray–Darling Basin, Australia. The map shows major tributaries and locations of the seven "Selected Areas" for the Commonwealth Environmental Water Office long-term intervention monitoring (LTIM) project. Data from these locations are being used to evaluate the effectiveness of government investment in environmental water being delivered under the Murray–Darling Basin Plan. See Chapter 25, and in particular Box 25.2, for more details. The map also shows Taroole and Cubbie stations (see Chapter 17).

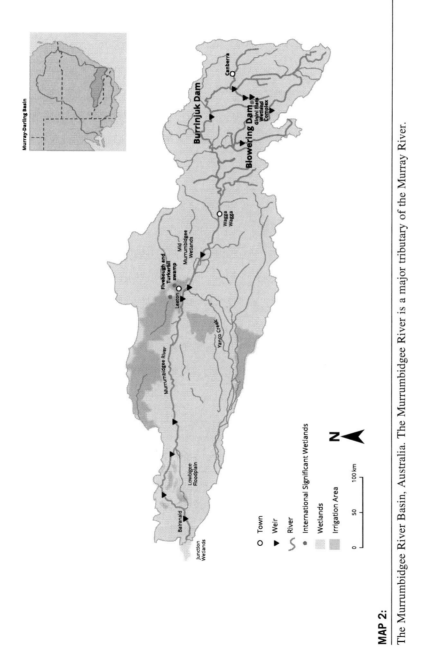

MAP 2:

The Murrumbidgee River Basin, Australia. The Murrumbidgee River is a major tributary of the Murray River.

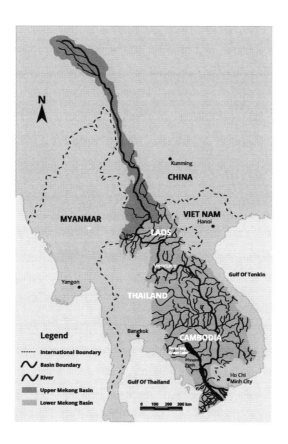

MAP 3:

The Mekong River Basin, Southeast Asia.

MAP 4:

The Yellow River Basin, China.

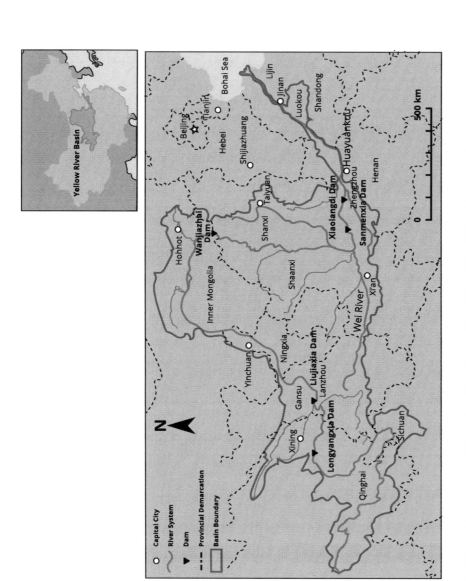

MAP 5:

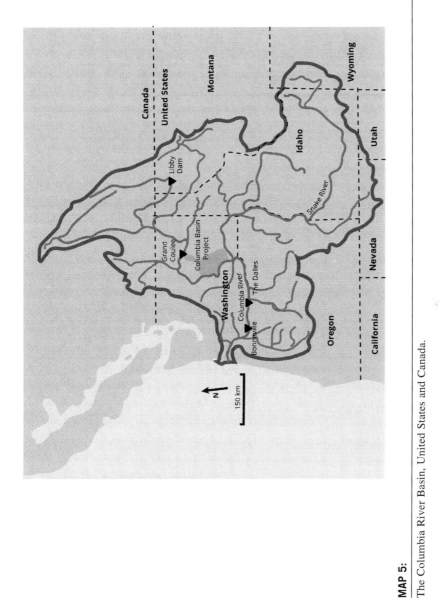

The Columbia River Basin, United States and Canada.

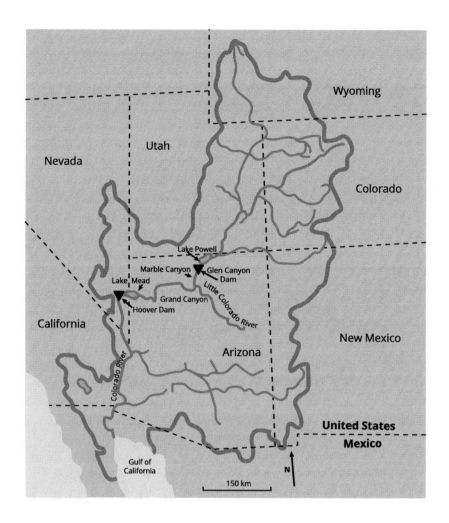

MAP 6:

The Colorado River Basin, United States and Mexico.

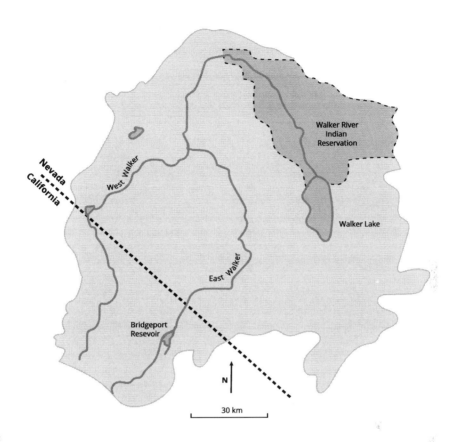

MAP 7:

The Walker River Basin, United States.

Abbreviations

AIC	Aikake's Information Criterion
ANAO	Australian National Audit Office
ANN	Artificial Neural Network
BACI	Before–After Control-Impact
BAGS	Bedload Assessment in Gravel-bedded Stream
BBM	Building Block Methodology
BIC	Bayesian Information Criterion
BN	Bayesian Networks
BPA	Bonneville Power Administration
CAMS	Catchment Abstraction Management Strategies
CBWTP	Columbia Basin Water Transactions Program
CEM	Conceptual Ecological Models
CEWH	Commonwealth Environmental Water Holder
CEWO	Commonwealth Environmental Water Office
CMA	Catchment Management Authority
COAG	Council of Australian Governments
CPT	Conditional Probability Tables
CWCB	Colorado Water Conservation Board
DGA	Dirección General de Aguas
DIC	Deviance Information Criterion
DRIFT	Downstream Response to Imposed Flow Transformation
DWA	Department of Water Affairs
EI	Effective Imperviousness
ELOHA	Ecological Limits of Hydrologic Alteration
ENEE	Empresa Nacional de Energía Eléctrica
ES	Ecosystem Service
ET	Evapotranspiration
EWO	Environmental Water Organization
FLM	Functional Linear Model
FSR	Flow Stress Ranking
GEP	Good Ecological Potential
GES	Good Ecological Status
GLM	General Linear Model
GMP	Groundwater Mitigation Program
GMS	Greater Mekong Subregion
HFE	High-Flow Experiment
HFSR	Habitat Flow-Stressor Response
HM	Hierarchical Model
HOF	Hands-Off Flow
IBFA	Integrated Basin Flow Assessment
ICM	Integrated Catchment Management
ICMA	Intentionally Created Mexican Allocation
IFC	International Finance Corporation

IFIM	Instream Flow Incremental Methodology
IHA	Indicators of Hydrologic Alteration
IPBES	Intergovernmental Science-Policy Platform on Biodiversity and Ecosystem Services
IUCN	International Union for the Conservation of Nature
IWRB	Idaho Water Resource Board
IWM	Integrated Waterway Management
IWRM	Integrated Water Resource Management
LHWP	Lesotho Highlands Water Project
LiDAR	Light Detection and Ranging
LTIM	Long-Term Intervention Monitoring
MDB	Murray–Darling Basin
MDBA	Murray–Darling Basin Authority
MDBC	Murray–Darling Basin Commission
ML	Machine Learning
MOA	Memorandum of Agreement
MWR	Ministry of Water Resources
NFWF	National Fish and Wildlife Foundation
NFWS	National Fish and Wildlife Service
NGO	Non-Governmental Organizations
NHI	Natural Heritage Institute
NPCC	Northwest Power and Conservation Council
NRM	Natural Resources Management
NWA	National Water Act
NWI	National Water Initiative
OECD	Organization for Economic Cooperation and Development
PDF	Probability Density Function
PHABSIM	Physical Habitat Simulation
QLE	Qualified Local Entities
RAM	Resource Assessment and Management
RIF	Regional Investment Framework
RQA	Resource Quality Objectives
RVA	Range of Variability Approach
SAM	Strategic Adaptive Management
SCM	Stormwater Control Measures
SDG	Sustainable Development Goals
SDL	Sustainable Diversion Limits
SEWDS	Seasonal Environmental Water Decision Support
SMART	Specific, Measurable, Achievable, Realistic, and Time-bound
sMDB	Southern Murray–Darling Basin
SRA	Sustainable Rivers Audit
SRP	Sustainable Rivers Project
SWMM	Stormwater Management Model
TI	Total Imperviousness
TIR	Thermal Infrared
TMT	Technical Management Team
TNC	The Nature Conservancy
TPC	Thresholds of Potential Concern
USACE	United States Army Corps of Engineers

VEWH	Victorian Environmental Water Holder
VSTEEP	Values, Social, Technical, Environmental, Economic, Political
WCC	Water Conservancy Commission
WCD	World Commission on Dams
WEPS	Water Evaluation and Planning System
WFD	Water Framework Directive
WMA	Water Management Area
WSUD	Water Sensitive Urban Design
WWF	World Wildlife Fund
WWI	World Watch Institute
YRCC	Yellow River Conservancy Commission

Glossary

Below, we provide definitions for commonly used terms in this book. The definitions are not universal, and we expect that some readers will disagree with some of them. The definitions are also made within the context of the subject of this book—environmental water management; many of the terms would have definitions that apply equally in other fields.

Active environmental water management   The decisions made by the environmental water organization on how to use its water to achieve maximum environmental benefits are active and ongoing (usually by deciding to reacquire a temporary water right, to release water from storage for use instream or extraction into a wetland, or to trade its water rights to other users).

Adaptive management   A systematic process for continually improving management policies and practices by learning from the outcomes of previously employed policies and practices (http://www.millenniumassessment. org/documents/document.776.aspx.pdf). Adaptive management is often represented graphically through an adaptive management cycle.

Afforestation   Planting of forests on land that has historically not contained forests (www.globalwaterforum.org/ resources/glossary/).

Ambiguity   Uncertainty related to multiple interpretations of a situation.

Bank storage   The saturated zone that forms close to stream channels due to infiltration of water from the channel and floodplain. The groundwater level in the bank storage zone will fluctuate with stream stage.

Bayesian networks   Graphical node-link models that represent probabilistic causal relationships among variables. Probabilities may be estimated subjectively or informed by data, and they can be updated using Bayes' theorem when new information becomes available.

Bayesian statistics   "Provides a mechanism for combining knowledge from subjective sources with current information to produce a revised estimate of a parameter" (Burgman, 2005). Bayesian models are generally more flexible and powerful than other more commonly used statistical approaches, but are consequently more difficult to construct.

Confidence interval   "In the long run, someone who computes 95% confidence intervals will find that the true value of parameters lie within the computed intervals 95% of the time." (Burgman, 2005).

Consumptive use   Water withdrawn from surface or groundwater resources (such as the water diverted from rivers, which is used by crops and pasture, urban centers, etc.).

Continuous variable   A variable that can take any value within a given interval.

Credible interval   The Bayesian equivalent of a confidence interval. It is the interval that encompasses a certain proportion of the probability distribution (e.g., 95%) of a random variable estimated by a Bayesian model.

Determinism   Situation where everything is known precisely.

Discrete variable   A variable with a finite number of values.

Ecological response model   A relationship, either qualitative or quantitative, between an ecological endpoint (e.g., spawning of a fish species) and one or more environmental drivers, including aspects of the water regime. The relationship may be informed by data, expert opinion, or a combination of both.

Ecosystem services   The benefits people obtain from ecosystems. These include provisioning services such as food and water; regulating services such as flood and disease control; cultural services such as spiritual, recreational, and cultural benefits; and supporting services such as nutrient cycling that maintain the conditions for life on Earth (www.globalwaterforum.org/resources/glossary/).

Effectiveness   The overall outputs (provision of an environmental water regime) and outcomes (ecosystem health and related social and economic benefits) supported by environmental watering.

Efficiency   Achieving the *least-cost path* to a specified output or outcome, while maximizing the net benefits of a given increment in investment in environmental watering activity.

Environmental allocation mechanism   The policy tool/s available to provide environmental water including: caps on consumptive users, conditions on storage operators, conditions on private license holders, reserves, and environmental water rights.

Environmental flows assessment   The process used to determine the environmental water requirement (see below) for an ecological endpoint. This may be based upon a combination of hydrological, hydraulic, and ecological knowledge; possibly in combination with the use of expert knowledge and opinion. By convention, we retain the use of "flow" although we define this term to include water volume requirements for wetlands and other ponded water bodies.

Environmental water   Encompasses all water legally available to the environment through the array of possible allocation mechanisms. Each year, the precise volume of environmental water actually allocated under these legal mechanisms may vary depending on overall water availability.

Environmental water management   The process of determining, allocating, implementing, and managing environmental water. This management activity is located on a spectrum from passive to active. *Passive* management is associated with establishing the long-term plans and rules that do not require further action to provide environmental water. *Active* environmental water management refers to those allocation mechanisms that require ongoing decision making concerning when and how to use environmental water to achieve the desired outcomes.

Environmental water organizations   Identifiable organizations that can be held accountable for specific tasks in implementing environmental water policies.

Environmental water regime   The quantity, timing, and quality of water required to sustain freshwater and estuarine ecosystems and the human livelihoods and well-being that depend on these ecosystems (Brisbane Declaration 2007). The key aspect is that the word "regime" captures the time-varying nature of flow. This builds on environmental water requirements to include consideration of combined ecosystem objectives for the river. Broadly, we favor the term *environmental water* (a water volume) instead of environmental flow (a discharge) as this concept is applicable across both ponded and flowing water bodies like wetlands.

Environmental water release   A release from storage made specifically for the purposes of meeting a downstream environmental outcome. The environmental water regime can be delivered through a combination of environmental flow releases and *exogenous flows* (including unregulated inflows and releases for other water uses, such as agriculture or hydropower).

Environmental water requirement   The water regime required to sustain a given ecological endpoint within the freshwater and estuarine ecosystem (based on an environmental flows assessment).

Epilimnion   The upper layer of water in a thermally stratified lake or reservoir.

Epistemic uncertainty   Uncertainty that can be reduced with improved knowledge.

Farm dam   Small earth structures designed to impound water on private land for livestock, domestic purposes, aquaculture, irrigation, erosion control, aesthetic appeal, and other purposes. They are sometimes given other names depending on their location and how they are used, including small dams, small reservoirs, private dams, catchment dams, hillside dams, small ponds, and runoff dams.

Flow stressors   Features of water regimes in freshwater ecosystems, usually alterations as a result of human disturbances that contribute to impaired ecological condition.

Fuzzy bounds   Range between 0 and 1, and bounds the values in the fuzzy set, which are used to describe the concepts for a given fuzzy variable.

Hierarchical model   A *statistical model* (in this book we consider hierarchical Bayesian models primarily) that considers multiple *sampling units* (e.g., rivers) simultaneously. Sampling units are considered to be partially dependent, and so the information from one unit helps to strengthen estimates of model parameters for other units. Hierarchical models have the ability to fit relationships and test hypotheses both within and between sampling units.

Hypoliminion   The lower layer of water in a thermally stratified lake or reservoir, typically cooler than the water above.

Hydraulic conductivity   The ease with which water can move through pore spaces or fractures.

Hyporheic zone   The saturated region immediately beneath the sediment−water interface in which streamborne water and groundwater mix.

Input legitimacy   Explicit consideration of access, equal representation, transparency, accountability, consultation and cooperation, independence, and credibility.

Institution   A set of constraints and incentives structuring human interactions; rules encoded in policies and regulations.

Integrated catchment management   An approach to tackling sustainable environmental management problems by coordinating land, water, and biodiversity management and recognizing flow-mediated connections through catchments and the need for interdisciplinary and community-based collaboration.

Learning   In the context of *adaptive management* learning is the improvement in knowledge of the technical and nontechnical aspects of the *socio-ecological system* being assessed that come about through the iterations of the adaptive management cycle.

Living labs   Structured collaborations between researchers and the users of research outputs to codesign and code-velop research activities.

Measurement uncertainty   Error caused by imprecise and inaccurate instruments and operators (Burgman, 2005).

Metalimnion   A distinct middle layer in a thermally stratified lake or reservoir in which there is a relatively rapid decrease in temperature with depth compared to the above or below layers. Also called thermocline.

Minimally informative prior   A prior probability distribution that contains very little information regarding the value of the parameter (it is impossible for it to contain no information). This is typically achieved by specifying very high uncertainty in the distribution (e.g., specifying a prior as normal with mean 0 and variance 10,000).

Monitoring and evaluation   The process of collecting data on ecological or social responses to environmental water management (monitoring) and the synthesis of those data (evaluation) to improve knowledge of the system. Monitoring and evaluation is an essential component of *adaptive management*.

Monte Carlo analysis   Uses probability distributions to represent different kinds of uncertainty, combining them to generate estimates of a risk (Burgman, 2005).

Natural resource management   The management of any pool of natural resources, but in the context of this book, generally referring to the management of catchments and the contents thereof. *Environmental water management* is a special case of natural resource management.

Natural variability   The inherent randomness and unpredictability of natural systems. It can be characterized with greater accuracy, but not reduced.

Nonflow stressors   Environmental characteristics of freshwater ecosystems, other than those related to water regimes that contribute to impaired ecological condition.

Organization   An identifiable public or private entity with its own objectives.

Organizational capacity   The physical, institutional, human, and financial resources required to deliver specified objectives and outcomes at social desired levels and over the long term.

Output legitimacy   Effectiveness, awareness, acceptance, mutual respect, active support, robustness, and common approaches to shared problem solving.

Passive environmental water management   Activities related to maintaining the legal and policy framework in which environmental water is acquired and managed (and in some instances, may be the legal *owner* of environmental water).

Phreatic zone   The saturated zone below the water table.

Posterior probability distribution   In Bayesian modeling, the prior probability distribution is combined with data via the *likelihood function* to produce the posterior probability distribution. This is the probability distribution that represents an updated estimate (from the prior) for a model parameter.

Precautionary principle The management concept stating that in cases "where there are threats of serious or irreversible damage, lack of full scientific certainty shall not be used as a reason for postponing cost-effective measures to prevent environmental degradation," as defined in the Rio Declaration (www.globalwaterforum.org/resources/glossary/).

Prior probability distribution In Bayesian modeling, the prior is a probability distribution specifying our initial belief concerning the distribution of a model parameter before including the data collected. A prior may be informed by previous research and/or expert opinion.

Probability density function A function that describes the relative likelihood for a random variable to take on a given value.

Probability distribution For a random variable, a probability distribution describes the range of values, and their relative likelihoods, that may be observed. It assigns probabilities to the values of a variable (for a discrete variable) or to a range of values of a variable (for a continuous variable). Probability distributions are often displayed graphically, but can also be represented mathematically as the *probability density function*.

Quadrat A series of squares (or other shapes) of a uniform size placed across a habitat of interest and used to identify the distribution of species within a larger area.

Rating curve Relationship between the water level and discharge rate in a river.

Reflection Related to *monitoring and evaluation*, reflection is the process of considering information gained (e.g., through monitoring), and building it into knowledge, understanding, and practice. It goes beyond the simple assessment of whether or not ecological goals of environmental water programs have been achieved.

Reforestation Planting of forests on lands that have previously contained forest but have since been converted to some other use (compare Afforestation) (www.globalwaterforum.org/resources/glossary/).

Reservoir reoperation The redesign of reservoir operation rules to achieve a balance between the original purposes of a reservoir and emerging concerns such as the restoration of natural flow regimes (www.globalwaterforum.org/resources/glossary/).

Rule of law Everyone is subject to the law, which is clearly expressed and publicly available, and limits the power of governments or other organizations to act arbitrarily in a manner that affects the rights of others.

Science–management partnership A collaboration between natural resource managers and researchers aimed at conducting research to improve knowledge and simultaneously improving management outcomes.

Socio-ecological (or Social-ecological) An ecosystem, the management of this ecosystem by actors and organizations, and the rules, social norms, and conventions underlying this management (http://www.millenniumassessment.org/documents/document.776.aspx.pdf). Note that the social and ecological systems are equally important but interrelated systems.

Statistical model A relationship based on empirical data that relates changes in one variable (e.g., an ecological endpoint) to changes in one or more other variables (e.g., descriptors of the water regime). These are generally created using specialist software. Statistical models are the primary means we describe for creating *ecological response models*.

Stormwater control measures Systems constructed to intercept, retain, and treat stormwater, which is then either used for various human needs, allowed to infiltrate into underlying soils, or evapotranspired by vegetation.

Thermal destratification The process of mixing of thermally stratified lakes or reservoirs.

Thermal stratification Change in water temperatures at different depths in a lake or reservoir due to the change in water's density with temperature, with denser cooler water in the lower layers and warmer water in the upper layers.

Thermocline See Metaliminon.

Third-party impacts (or externalities) The wider impacts imposed on others from private or individual actions that are not transmitted through market prices (www.globalwaterforum.org/resources/glossary/).

Uncertainty "Any departure from the unachievable ideal of complete determinism" (Walker et al., 2003).

Vadose zone The unsaturated zone that lies between the land surface and the groundwater table (or phreatic zone).

Water regime Refers to characteristics of the temporal and spatial variability in water volume and its movement within an aquatic habitat, including wetlands, aquifers, and estuaries; *stream-flow regime* is more specific to the water regime of river channels.

Index

Note: Page numbers followed by "*f*," "*t*," and "*b*" refer to figures, tables, and boxes, respectively.

dominant, 92
effective, 92–93
environmental flows assessments and, 39
environmental water management altering, 57t
low baseflows and reduced, 71–72
Stream morphology, 84
Stream restoration, for sediment management, 256
Stream salinity. *See* Salinity
Stream-flow gauging, uncertainty sources in, 323b, 323f
Stream-flow regimes, 690
 causes of, 39
 components of, 67–68
 definition of, 39
 urbanization changes to, 43–45, 44f
Streamflow series
 hydrological alteration and, 40b
 rainfall runoff model for, 40
 reversing water use and regulation for modeling, 41
 streamflow transposition for modeling, 41
Streamflow transposition, for streamflow series, 41
Structured expert elicitation, 325b
Subjective judgment, for uncertainty quantification, 333–334
Submerged weirs, 109–110
Success
 adaptive management measuring, 601–604
 Columbia River Basin and criteria for, 635–639
 defining
  criteria for, 627–635, 634f
  effectiveness for, 628
  efficiency for, 629
  legal and administrative framework for, 631–632
  legitimacy for, 630, 663
  organizational capacity for, 632
  overview of, 625–626
  partnership, 632–635
 of environmental water management programs, 662
 evaluation of, 639–641
  baseline for, 640, 640b
  legitimacy for long-term, 641, 642b
  across multiple dimensions and time frames, 640–641, 641b
 of ICM, 519–520
 investment in criteria for, 634–635
 MDB and criteria for, 635–639
Sulm River, Austria, 277–278, 278f
Supporting services, 153
 indicators of, 155t
Surface pumps, 109–110
Surface water rights, 400
Suspended sediment concentrations (SSC), 244
Suspended-load sediment transport, 239, 244
Sustainability boundaries, 214
Sustainable Development Goals (SDGs), 20, 357, 650, 683
 environmental water regime targets of, 658

global community agreeing to, 661
Sustainable diversion limits (SDLs), 355, 364, 379–380, 410, 683
Sustainable Groundwater Management Act, US, 463
Sustainable Rivers Audit (SRA), 270–273, 272f, 683
Sustainable Rivers Project (SRP), US, 438, 607b, 683
Sustainable Urban Drainage Systems, UK, 466
SVP. *See* Shared vision planning
Swatchh Bharat (Clean India Mission), 26
SWMM. *See* Stormwater Management Model
Systemic review, 526

T

Tamarix spp., 71–72
Tanks. *See* Farm dams
Tanzania, 21
Targets, for environmental water management, 189–190
 for Cedar River, Washington, 196b
 hierarchical approach linking objectives to, 193, 194f
 for MDB, Australia, 195b
 setting, 192–193
 SMART, 192
 terminology of, 190f
Tasmania, Mersey River in, 30
Tax deductible water products, 407, 415
Technical Management Team (TMT), 568, 683
Technical solutions
 for diffuse catchment changes management, 455
 for farm dam hydrological alteration using low-flow bypasses, 459–460, 460f, 461f
 for groundwater diversion hydrological alteration management, 465
 for urbanization hydrological alteration management, 468–471
Temperature. *See* Water temperature
Teri Dam, India, 26
Thailand, Nam Songkhram River Basin in, 29
Thermal destratification, 109–110, 690
Thermal infrared (TIR) imagery, 110, 683
Thermal stratification, 116, 690
Thermocline, 690
Third-party impacts (externalities), 337b, 401–403, 573b, 690
Thompson River, Australia, 294, 294f
Three Gorges Hydroelectric Plant, on Yangtze River, China, 47–48, 91, 94
Threshold bed shear stress, 240, 240f
Threshold stream power, 239–240
Threshold velocity, 239–240, 240f
Thresholds of potential concern (TPC), 157, 683
TI. *See* Total imperviousness
Time-trend analysis method, reforestation hydrological alteration characterized by, 472
TIR imagery. *See* Thermal infrared imagery

67304921R00418

Made in the USA
Lexington, KY
07 September 2017